OUTLINE OF INDUSTRIAL ORGANIC CHEMISTRY

# OUTLINE

## OF

# INDUSTRIAL ORGANIC CHEMISTRY

Alfred Rieche

Director of the Institute of Organic Chemistry of the German Academy of Science in Berlin, and of the Institute of Chemical Engineering of the Friedrich Schiller University at Jena

Third improved and enlarged edition

Translated by

F. Sterne, F.I.L.

Technical Editor for English Edition

K. Littlewood, B.Sc., Ph.D., A.R.I.C., A.M.Inst.F.

Dept. of Fuel Technology and Chemical Engineering
University of Sheffield

CHEMICAL PUBLISHING CO. INC.
200 PARK AVENUE SOUTH
NEW YORK, N.Y.

ENGLAND: BUTTERWORTH & CO. (PUBLISHERS) LTD.
LONDON: 88 Kingsway, W.C.2

AUSTRALIA: BUTTERWORTH & CO. (AUSTRALIA) LTD.
SYDNEY: 20 Loftus Street
MELBOURNE: 343 Little Collins Street
BRISBANE: 240 Queen Street

CANADA: BUTTERWORTH & CO. (CANADA) LTD.
TORONTO: 14 Curity Avenue, 16

NEW ZEALAND: BUTTERWORTH & CO. (NEW ZEALAND) LTD.
WELLINGTON: 49/51 Ballance Street
AUCKLAND: 35 High Street

SOUTH AFRICA: BUTTERWORTH & CO. (SOUTH AFRICA) LTD.
DURBAN: 33/35 Beach Grove



Suggested U.D.C. numbers
661·7
662·6/·7
664·1/·2
665
677/678

First American Edition, 1970
Chemical Publishing Co. Inc.




Printed in the German Democratic Republic by
Leipziger Druckhaus

## Preface to the Third Edition

Since the 1961 Edition was almost sold out within three years, the publication of a Third Edition has proved necessary. In its essential aspects it corresponds to the Second Edition but takes into account modern developments.

Despite the suggestions made on various sides for omission of the technological parts the author could not yet see his way, at least in the present Edition, to omit outdated processes and to forgo the fairly detailed description of processes. The object of this outline and textbook is to give information on the origin and development of branches of industry up to the present, and not to describe the physico-chemical basis and the basic equipment and methods of industrial organic chemistry. A number of good specialized books have been published covering these aspects.

A knowledge of the development of industry and its processes should not only be part of the general education of the chemist and chemical engineer but it is a condition for the proper assessment of problems of equipment technology, of effecting reactions, and of the profitability of the processes.

In the present book the reader will find many hints of this type which are derived from practical experience. Almost imperceptibly the student, too, is brought to consider these problems. Even the experienced industrial chemist can quickly obtain a general picture of the development and present state of those branches of industry, or processes, which are of interest to him.

Justification for the author's views in this respect is provided by the agreement expressed by many colleagues in all parts of the world, and also by the fact that the book is now being published in the most important languages.

Both publishers and author hope, therefore, that the present Edition too, like its predecessors, will receive a warm welcome.

A large number of friends and colleagues have again made useful suggestions from which the Third Edition has benefited. Some of these were gentlemen mentioned gratefully in the Preface to the Second Edition, and I wish to express to them once more my thanks.

My special thanks are due to Prof. Dr. O. Neunhoeffer, Saarbrücken, Prof. K. Thinius, Leipzig, Prof. Dr. E. Baumann, Dresden, and Dipl.-Ing. Sobol, Bitterfeld, who all offered many very important suggestions.

Above all I also have to thank my colleagues at the Institute for Organic Chemistry in the German Academy of Science, Berlin-Adlershof, Miss Dr. A. Martini, Dr. E. Höft, and Dipl.-Chem. H. Schultze, who revised the new Edition with me and very carefully carried out the proof-reading. I thank Dr. H. Seeboth for his assistance and Miss Dipl.-Chem. I. Eichhorn for many bibliographical references.

I am grateful to my publishers, S. Hirzel, who recently celebrated their 110 th anniversary, for their ever agreeable, expert, and generous co-operation.

It is inevitable that this Edition, too, will contain some defects and mistakes, although great care was taken to avoid them. Publishers and author will be grateful for any suggestions to eliminate these.

Berlin and Jena, July, 1964 A. Rieche

# Preface to the Second Edition

The first edition of this book was received with almost unanimous approval by my colleagues and above all by the students so that it soon was out of print.

Most reviewers liked the selection and arrangement of the material and the method of presentation. I, therefore, retain these in principle in this second edition, in spite of various criticisms. For didactic reasons, the sections on equipment and unit operations are again inserted where the use of certain equipment is particularly typical for a process. More detailed information on these questions can be obtained, if required, from the specialized books available. This also applies to further information on the physico-chemical basis of the processes.

Many colleagues expressed the wish to see the book enlarged so as to include all sections of industrial organic chemistry, and to have a larger part devoted to the chemistry of natural gas, petroleum and the simple aliphatic hydrocarbons, which is of special interest at the present time. This has been done in the new edition by substantially enlarging these sections and by including modern international developments. Entirely new chapters have been added on organic dyes, dyeing, and textile printing, tanning and the manufacture of leather, pest control materials, and gunpowder and explosives. Thus the book now covers the whole field of organo-chemical technology. Only perfumes and essential oils have been omitted as I have not yet found a way to treat this very complex and extensive field in a compressed form.

Although some less important chapters were shortened the present volume is larger than the first edition because of the above-mentioned extensions and the inclusion of the new chapters. I hope, nevertheless, that the readability of the book has not thereby been impaired and that students can still work through the essential parts.

A number of colleagues have kindly reviewed the first edition and pointed out various deficiencies and mistakes. Their valuable suggestions have been incorporated in the present second edition.

During the preparation of the second edition I have again benefited by the support of friends and colleagues who gave advice, reviewed critically the chapters relating to their respective fields, and assisted me in drafting parts of the manuscript.

I should like to express my sincere gratitude to: Prof. Dr. Asinger, Aachen, Prof. Dr. Bergander, Berlin, Dr. Grimm, Schkopau, Dipl.-Chem. Grosse, Leipzig, Dr.-Ing. Hammer, Berlin-Charlottenburg, Prof. Dr. Heimann, Karlsruhe, Dr. Hennig, formerly Bitterfeld, Prof. Dr. Hilgetag, Berlin, Prof. Dr. v. Hornuff, Dresden, Prof. Dr. Klare, Teltow-Seehof, Prof. Dr. Kölbel, Berlin-Charlottenburg, (Miss) Dr. Martini, Berlin, Dr. Naumann, Jena, Prof. Dr. Neunhoeffer, Saarbrücken, Dr. Püschel, Berlin, Dr.

H. Schmitt, Böhlen, Dr. M. Schulz, Berlin, Dipl.-Ing. Sobol, Bitterfeld, Prof. Dr. Stather, Freiberg (Sa.), Dr. Thinius, Leipzig, Prof. Dr. Voss, Dessau.

Ing. W. Künzel, Wolfen, again executed with much understanding and care the additional 25 line drawings which represent an important improvement in the new edition.

My colleagues, Dr. E. Naumann and Dipl.-Chem. M. Landbeck, Jena, and Dr. E. Höft, Berlin, assisted me energetically during the preparation of the book for the printers and by reading the proofs, for which I thank them sincerely. I also owe thanks to friends and colleagues, not named here, who made many suggestions and co-operated in procuring material and adding final touches to the manuscript.

Finally, recognition and thanks are also due to the publishers for their generous and understanding co-operation. It is above all due to them that the much enlarged book can be supplied in similar form and at a similar price as the first edition, a fact that will be welcomed mainly by the students. Although I hope that all important defects of the first edition have been eliminated in the second, I am certain there is still room for improvement. I shall always be grateful for suggestions.

Berlin and Jena, January, 1961 A. Rieche

# Preface to the First Edition

Excellent textbooks and manuals on industrial organic chemistry have been published in recent years. However, students of chemistry lack so far a short textbook that is wholly within their capacities. For this reason I am publishing in bookform my lectures, given at Jena University, on industrial organic chemistry. The present 'Outline of Industrial Organic Chemistry' does not claim to be a complete textbook. Rather is it a collection of lectures summarizing the development of the most important techniques used in the organic chemical industry; the grouping by branches of industry follows the arrangement of Winnacker-Weingaertner's 'Chemische Technologie' which has proved so useful. Basic industries are given special prominence. Some branches of industry have been treated in greater detail than their general technological importance merits. This applies, for instance, to the sections on chemical processing of wood, in particular the utilization of sulphite waste liquor and the microbiological protein synthesis. The reason is that this was the author's special field during his more than twenty years' work in industry and it seems to him that these processes have been treated too briefly in other books. Moreover, microbiological techniques have assumed so great and general importance in other countries that it appeared appropriate to deal in greater detail with the microbiological production of protein yeast, the first method of this type to be used.

Within the framework of a textbook comprising a few hundred pages it is impossible to give an account of every aspect of industrial organic chemistry. Several special branches were, therefore, omitted deliberately, in particular those which, because of their diversity, did not allow of short treatment, such as for instance organic dyes. The techniques used in the production of pharmaceutical products, pest control materials, perfumes, and explosives are in many respects similar to those used in the intermediates chemical industry and the same equipment is employed. They have, therefore, been treated only briefly or not at all. Whilst chemical methods are used in dyeing, bleaching, textile printing, the manufacture of cosmetics and of leather, these industries do not form part of chemical industry proper and will not be dealt with here.

In this book, unit operations and equipment have not been treated in isolation from special organic technology because I believe that it is more instructive, and also more interesting for the reader, when the sections in which chemical unit operations are described, are distributed throughout the book. They will be found where the use of certain methods and equipment is particularly typical for a process. The scope of the book is limited to chemical unit operations since it is written mainly for chemists.

The main emphasis of the book then is on special chemical technology, particularly extensive treatment being given to materials. This still seems to me to be the best method of imparting a genuine chemical-technical general education to the student. The book is intended to guide him towards such a way of thinking, without burdening him with too many facts. A special effort has, therefore, been made to present it in an easily intelligible form. Over a hundred simple line drawings of typical equipment and of the course of important processes which might be called 'picture reportage of a chemical process' give graphic support to the text.

During the writing of this book I was assisted by suggestions and advice from friends and colleagues in the chemical industry who critically reviewed the chapters referring to their respective fields. I should like to express my gratitude for their valuable co-operation and support to Dr. Bemmann, Leuna, Dr. Bittner, Dessau, Ober-Ing. Ehrlich, Wolfen, Ober-Ing. Gillitzer, Wolfen, Dr. Grimm, Schkopau, Dr. Grunewald, Böhlen, Dr. Hilgetag, Berlin, Dr. Pattok, Wolfen, Dr. Prinzler, Böhlen, Dr. H. Schmitt, Böhlen, Dr. Thinius, Leipzig, and Prof. Dr. Voss, Dessau.

The line drawings were prepared with great understanding and much care by Ing. Künzel, Wolfen. My assistant, Dipl.-Chem. Naumann, Jena, gave me most effective support in preparing the book for the printers and in reading the proofs. Other friends and collaborators, not named here, also helped me in the final stages of the manuscript. To all these are due my sincere thanks. Finally, I must express my special gratitude to the publishers for the generous way in which they complied with all my wishes and for their ever understanding and harmonious collaboration.

The book of which the first edition is now ready will be subjected to many a criticism. I shall always be grateful for any suggestions that might lead to its improvement.

Berlin and Jena, December, 1955 A. Rieche

# Contents

# Introduction

Chemical engineering is the practical application of chemical knowledge. Its function is the creation of consumer goods by means of chemical methods. In mechanical engineering commodities are produced by transformation without changing their composition; in chemical engineering, however, the materials are converted. The ultimate purpose of all technology must be to satisfy the continuously increasing requirements of man, to render his life easier, richer, healthier, and safer from natural misfortunes. Its results should be only of benefit to mankind, unfortunately they are often used for destruction. The main aim of chemical engineering is the upgrading of raw materials. In this, use is also made of many mechanical-physical methods. The conversion of materials in chemical engineering then is brought about by physical effects and chemical reactions. This gives it its extraordinary variety.

All technology aims at achieving the greatest possible economy. Maximum quantity and optimum quality is to be produced with the minimum expenditure of power and raw materials. This has been man's guiding principle since he started producing staple products. Thus the basic economic problems are always:

(1) the quantity of raw material used
(2) the quantity of energy (steam, electricity, coal, gas, and water) used
(3) the requirements in buildings and equipment (investment, interest)
(4) wear and tear (the cost of repair and replacement)
(5) the requirements in human labour (salaries and wages)
(6) the use of transport, problems of storage and packing
(7) the yield of by-products, removal of solid, gaseous and liquid wastes.

The total expenditure represents the cost price of a product. The economy of a process is best ascertained by preparing an 'estimate'.

### Model of an estimate

Basic material (cost of materials)
Consumption per 100 kg finished product of: material A
B
C
D

(1) Total basic material = price of materials for 100 kg finished product
Basic costs (calculated per 100 kg finished product)

Power, e.g. water
gas
steam
electricity
compressed air
ice

Fuel

Wages, including bonuses (for dirt, danger, heat)

Cost of repairs (wages and materials)

Cost of transport (transport of raw materials and transport within the works)

Other costs (lubricants and cleaning materials, packing materials, laboratory materials, filter materials, protective clothing, cleaning of tank wagons, laboratory costs, expenditure on dyeing, etc.)

Depreciation, interest, taxes

(2) Total basic costs for 100 kg finished product

Overhead costs

Departmental overhead costs

Works overhead costs (management, administration, research, expenditure for hygienic, social, and cultural purposes, fire protection, works police)

(3) Total overheads

Total of:

(1) Basic materials per 100 kg
(2) Basic costs per 100 kg
(3) Overhead costs per 100 kg

---

Costs of production per 100 kg
+ costs of marketing
tax on turnover

---

Factory cost per 100 kg finished product.

The technological application of a process is always preceded by a scientific effort. The process must be devised or developed. Even the greatest inventor and innovator must build on the foundations laid by his predecessors. This is the reason for the particularly close connection between technology and science in chemical engineering. Intimate co-operation between chemical science and technology was first developed in Germany. Our great chemical undertakings owe their success to having realized the importance of this principle. Chemical research at German universities has been particularly highly developed for many decades as a result of J. von Liebig's teaching. The universities supplied both the chemical basis for technological development, and intellectually and scientifically well-trained men capable of independent research work. The chemical firms themselves have maintained very large and expensive research laboratories and technological departments employing excellent specialists. Favourable conditions for this development were created by the healthy economical basis of the German chemical industry. The structure of German research and technology served as an example which only much later was followed by the rest of the world.

One has to distinguish between 'inventions' and 'discoveries'. If 'discovery' is defined as the demonstration of a law of Nature, the concept of an 'invention' can be defined as the practical exploitation of laws of Nature for industrial purposes.

Only inventions can be patented; according to the German Patent Law, dated 7. 4. 1891, para. 1:

'Patents are granted for new inventions the commercial exploitation of which is possible'.

The long view taken by German patent law and the exemplary testing for originality by experienced testers has greatly contributed to the favourable development of German industry.

Although patent legislation had been demanded by German industry, in particular the dyestuffs industry, for a very long time, it was not introduced in Germany till 1877 whilst in England patents had been granted long before that time. It was an important feature of German patents that it was not the material that was protected but the method of its production.

Thus progress and competitions were ensured and no inventor or industrialist could 'rest on his laurels'. Nobody could secure for himself a monopoly for all time. There was always the necessity to find a new process or to improve an existing one.

A further advantage which German patent legislation had over the older English one was very careful examination for originality, and the obligation to publish, together with the possibility of raising objections within a limited period, before any patent was granted. After it had been granted a patent was, and still is, largely secure. It can only be challenged by an invalidity suit.

The first German patent for dyestuffs, granted in 1877, the year of the foundation of the German Patent Office, was for the production of methylene blue by BASF, No. 1886.

In the German Democratic Republic (GDR), the Patent Law of September 6, 1950, provides for two types of patents: 'Patents are granted either as commercial patents or as exclusive patents, for new inventions that are capable of being exploited industrially'.

Nationalized undertakings, State research institutes and other public bodies, or inventors subsidized from public funds, have to take out commercial patents in respect of their inventions. Exclusive patents can be claimed only by the owner of the patent rights, e.g. private individuals or private firms.

Commercial patents may be exploited by the owner of the patent rights and his legal successors, and by people or bodies authorized by the Patent Office.

The exploitation of such a patent by the owner of the patent rights as defined above requires authorization by the Patent Office. The person or body authorized to exploit a commercial patent has to pay a fee to the owner of the patent rights. The exploitation of exclusive patents is the subject of free negotiation.

An amendment of the Patent Law, introduced in the GDR in 1963, provides for the granting of a patent in the first instance without examination; only when the invention is to be exploited a thorough investigation of the patent claim is carried out.

A textbook of technology should also include a history of inventions. It should describe how a process originated, how it changed, how and why it attained its present form. Progress in the development of a process may be due to new scientific discoveries, possibilities opened up by new equipment, or changes in the supply of power or raw materials. Modification of a process may also be brought about by problems connected with by-products and waste products, questions of space, and dangerous effects.

*Development of a technological process*

The task of developing a chemico-technical process from observations made in a chemical laboratory presents many problems and is often very difficult. When the basic reaction has been determined, equipment that will make production possible in optimum conditions must be designed, and an optimum mode of operation must be found. This design must allow for later changes, to reduce production costs and to make adjustments necessitated by changes in the economic and raw material situation.

The centre of gravity of chemical engineering is shifting more and more from processes based on materials to processes the nature of which is determined by the equipment used, in particular the continuous catalytic processes. These do not permit, as was possible in the traditional technology of aromatic intermediates, dyestuffs and pharmaceuticals, the use of existing plant and equipment for many purposes. Special equipment must be developed. The useful life of a process is being constantly reduced by rapid technological progress. (In 1964, 40 per cent of the turnover of the large chemical works of the German Federal Republic consisted of products developed and marketed within ten years.) This means that processes must be developed as fast as possible. The optimum conditions for a process must, therefore, be recognized ever more rapidly as chemical engineering develops. A large part of the chemical engineering processes in use today is the result of laborious, expensive, and protracted experiments.

The classical working method based purely on experiment, the success of which is due mainly to experience and intuition, is no longer adequate. Even if occasionally some problems may still be solved by the lucky touch of a chemist of genius, it is nevertheless necessary to make full use of all the experiences made in chemical reaction techniques, operational techniques, organization and economy.

The concept 'chemical reaction technology' comprises the investigation of the conditions for a chemical-technological process and the search for means to use it in such a way as to derive maximum economical benefit from it. Its basis is physico-chemical:

| | |
|---|---|
| thermodynamics | reaction kinetics |
| hydrodynamics | mass transfer (e.g. diffusion) |
| heat transfer | |

The great importance of thermodynamics and reaction kinetics rests in their ability to indicate whether a reaction can take place at all under given conditions and if so, to what extent. The actual course of the reaction is determined by mass and heat transfer and flow. From all these data, duration and conditions of the reaction, the shape and size of the reaction vessels can be calculated in advance. Moreover, separation methods must be established and material and power balance sheets drawn up.

Finally, questions of material and corrosion connected with the equipment must be settled by experiment and by drawing on accumulated experience. As a rule, the following problems must be solved:

Development of suitable equipment and its gradual enlargement,
Reaction, heating, refrigeration,
Choice of equipment and raw materials, possibility of corrosion
Comminution, stirring and mixing
Isolation, separation (e.g. filtration, washing, distillation)
Purification of the product (e.g. dissolving and reprecipitating, recrystallizing)
Recovery and regeneration of raw materials and solvents; losses involved
Requirements of raw and auxiliary materials, their storage, mechanical preparation, dosage
Control of the process, determination of operational samples and intermediate analyses
Problems of quality, purity, specifications for the initial, intermediate, and finished products
Problems of safety, protection at work, dangers
Consumption of power (electricity, steam, gas, water, cooling brine)
Duration of the process, time schedule, utilization of equipment, efficient use of space and time, batch or continuous mode of operation
Removal of waste products, utilization of by-products.

The designing of a process and its application in large-scale operation usually proceeds in these steps:

(1) Determination of the scientific basis
(2) Process on a laboratory scale

(3) Semi-pilot plant scale
(4) Pilot plant (experimental works)
(5) Large-scale plant

On the basis of existing knowledge of operational techniques it is in many cases today possible to omit some development stages and to calculate and plan parts of the large-scale process in advance. The basic data for a process are nowadays not always obtained by the firm's own research and development department. More and more frequently the results from laboratory or pilot plant, even the whole process, the 'known how', are acquired with the complete project since the customary procedure is much too slow in view of the rapid development of technology.

It is often necessary to repeat these preliminary stages several times until the smooth running of the process on a large scale is assured. Each enlargement of the charge and of the plant brings new difficulties and problems.

The theoretical examination of a process is composed of the following stages:

Interpolation or extrapolation of results already known,
application of the theory of similarity,
rough calculation using simplified assumptions,
detailed mathematical treatment.

These calculations often require a large number of individual numerical operations, and computers are, therefore, used also in chemical engineering.

In whatever manner the experiences and the knowledge required for a process are gained, they are finally reduced in each case to instructions for operation, a process flow sheet and an equipment flow sheet. To complement these, material and energy

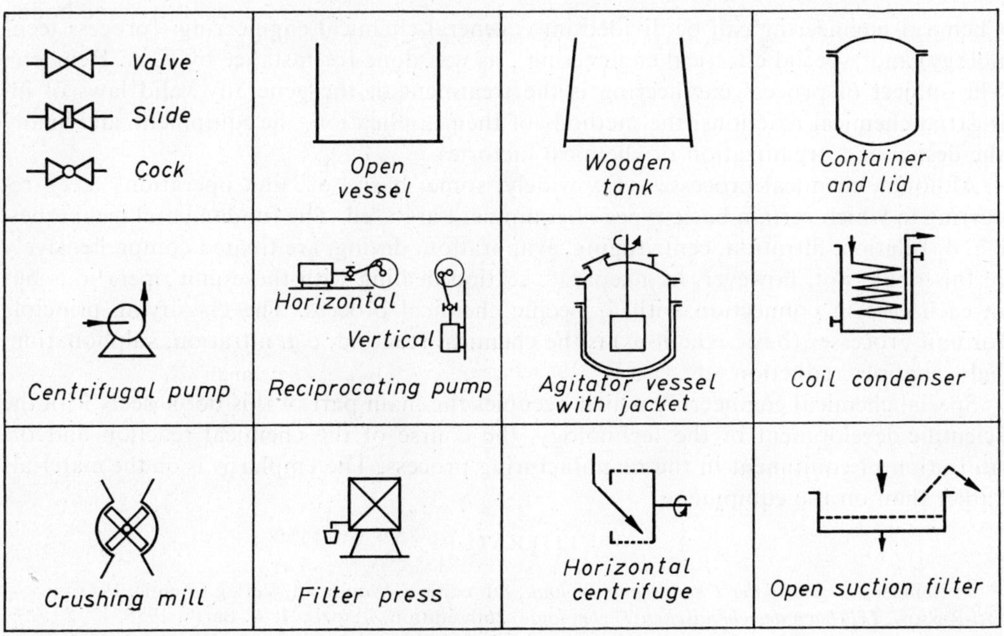


Symbols for equipment

flow sheets are also usually prepared. The process flow sheet shows essentially the individual operations, the equipment flow sheet contains information on the type of equipment, its disposition, the movement of materials, cables, control and shut-off devices, measurement points, means of transport, etc. These features are represented by very simple symbols which every chemist must learn. Some of these symbols are shown in the following table.

The illustration following shows the process flow sheet of a simple process, side by side with the equipment flow sheet. It describes the production of the sodium salt of $\beta$-naphthalene sulphonic acid ($\beta$-salt) from naphthalene by sulphonation. As described in more detail on pages 266–268, the $\beta$-salt is converted by alkaline fusion into the important intermediate $\beta$-naphthol.

The process flow sheet shows that the process consists of the following seven operations: melting of naphthalene (1), sulphonation with 96 per cent sulphuric acid (2), some naphthalene being separated at this stage. The dissociation back into naphthalene of the $\alpha$-acid formed (3) and the expulsion of the excess naphthalene by steam (4) are performed in one operation. In the same vessel the $\beta$-salt is then precipitated with a salt solution or $\beta$-naphthol mother liquor (5) and in a separate operation (6) the $\beta$-salt is cooled and crystallized. Finally, the $\beta$-salt is separated in a centrifuge (7).

The process flow sheet shows the course of the process, the principles underlying it, and the individual operations. How the equipment is used for the performance of these operations can be seen from the equipment flow sheet. On the basis of the data available a preliminary plan is then drawn up and a design for a plant is worked out.

## Classification of chemical engineering

Chemical engineering can be divided into 'general chemical engineering' (process technology) and 'special chemical engineering', as was done for instance by F. A. Henglein. The subject of process engineering is the treatment of the generally valid laws of industrial chemical reactions, the methods of their application, the equipment, and finally the design and organization of chemical factories.

Although chemical processes vary widely, some 'basic' or 'unit operations' keep recurring in which certain basic types of equipment are used. The fundamental techniques, e.g. distillation, filtration, centrifuging, evaporation, drying, are treated comprehensively in this book, not, however, in a separate section dealing with these unit operations but in each case in connection with a specific chemical process. The classifying principle for unit processes (basic reactions) is the chemical reaction, e.g. nitration, sulphonation, halogenation, reduction, etc.

Special chemical engineering which occupies the main part of this book deals with the scientific development of the technology, the course of the chemical reaction and the utilization of equipment in the manufacturing process. The emphasis is on the materials rather than on the equipment.

### LITERATURE

F. A. Henglein, *Grundriß der Chemischen Technik*, 8th edition, Weinheim, Verlag Chemie, 1958.

Ost-Rassow, *Lehrbuch der chemischen Technologie*, 26th edition, Leipzig, J. A. Barth, 1955.

W. Langenbeck, *Einführung in die organisch-technische Chemie*, 3rd edition, Dresden and Leipzig, Th. Steinkopff, 1954.

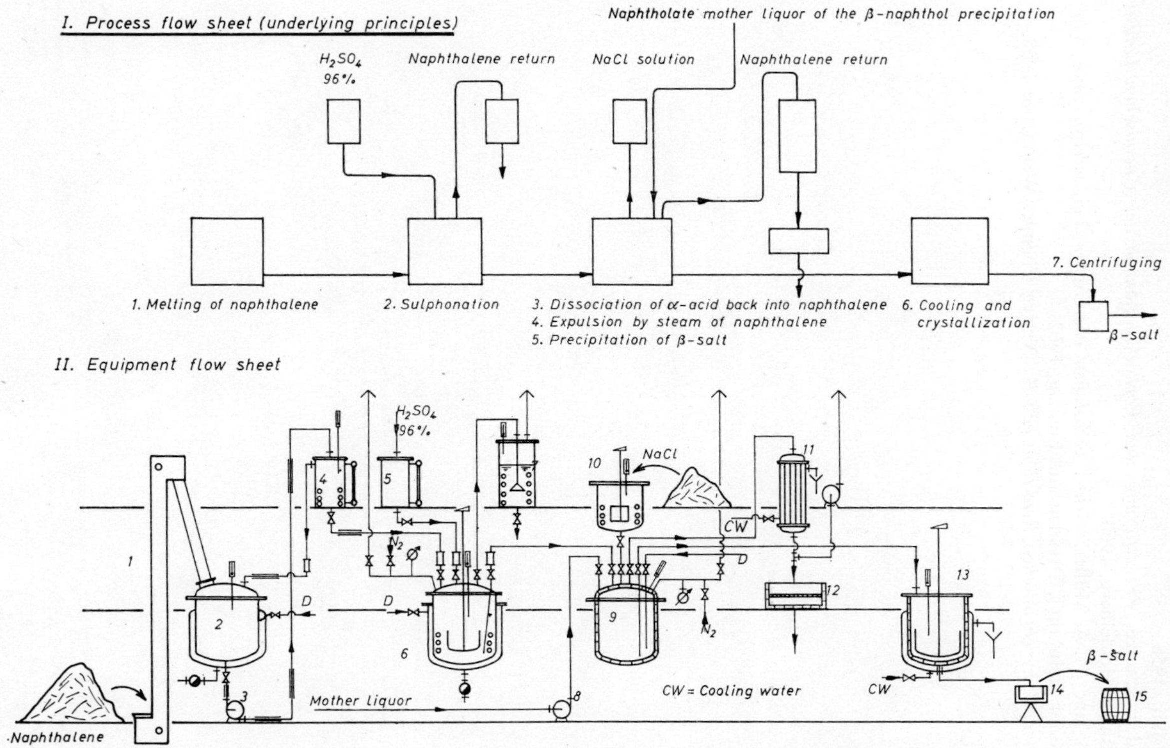


Schematic representation of the process for the manufacture of β-naphthalene sulphonic acid sodium (β-salt) (simplified)

*1.* Elevator
*2.* Jacket-heated fusion kettle, steam trap and thermometer
*3.* Hot-stage centrifugal pump
*4.* Hot-stage measuring vessel for naphthalene and thermometer
*5.* Measuring vessel for $H_2SO_4$
*6.* Cast iron sulphonation tank, jacket and coil heated, agitator, steam trap and thermometer
*7.* Absorption vessel filled with water, can be heated and cooled
*8.* Centrifugal pump
*9.* Brick-lined expulsion tank with direct steam feed and thermometer
*10.* Salt solution vessel with agitator and heating coil
*11.* Refrigerator with turbo-ventilation
*12.* Brick-lined suction filter
*13.* Brick-lined crystallization tank, agitator and cooling jacket
*14.* Centrifuge
*15.* Decanter

Winnacker-Weingaertner, *Chemische Technologie*, Volume 1, p. 1–260, Volumes 3 and 4, München, C. Hanser, 1952; Winnacker-Küchler, 2nd edition, Volume 1, 1959.

Ullmann, *Encyklopädie der technischen Chemie*, 3rd edition, München, Urban & Schwarzenberg; since 1951; in particular Volume 1, Chemischer Apparatebau und Verfahrenstechnik.

A. G. Kassatkin, *Chemische Verfahrenstechnik* I and II, 5th edition, Deutscher Verlag für Grundstoffindustrie, Leipzig, 1963.

Brown *et al.*, *Unit Operations.* Wiley & Sons, New York, 1951.

P. H. Groggins, *Unit Processes in Organic Synthesis.* 5th edition, McGraw-Hill Book Co., New York, 1958.

O. Fuchs, *Physikalische Chemie als Einführung in die Chemische Technik.* Sauerlander & Co., Aarau and Frankfurt/Main, 1958.

H. Kölbel and J. Schulze, *Projektierung und Vorkalkulation in der Chemischen Industrie.* Springer, Berlin/Göttingen/Heidelberg, 1960.

W. R. A. Vauk and H. A. Müller, *Grundoperationen chemischer Verfahrenstechnik, eine Einführung.* Verlag Theodor Steinkopff, Dresden and Leipzig, 1962.

W. Brötz, *Grundriß der chemischen Reaktionstechnik.* Verlag Chemie, Weinheim, 1958.

# *A. The chemistry of coal and crude oil and the production of chemical raw materials and intermediates*

# I. Technology and upgrading of coal

## FORMATION, MINING, AND PROPERTIES OF COAL

Coal was formed during geological periods by the transformation of the remains of higher plants, e.g. wood. According to age and geological effects almost all transition stages are found, from lignin-type coals, very rich in oxygen and hydrogen, to lignite, bituminous coal, anthracite, and finally to graphite which is pure carbon. Accumulations of carbon are found even in the oldest geological era, the Azoic = Archean, in which no organic life existed. It is still possible to show the vegetable origin of all these forms except graphite. The relationship of the solid fuels to the geological periods is shown in the following table:

| Period | Subdivision | Duration | Fuel |
|---|---|---|---|
| (1) Paleozoic period | | | |
| | Carboniferous | (about 70 million years) | Bituminous coal (estimated duration of formation of 1 m seam about 10,000 years) |
| | Permian | total 140 million years | |
| (2) Mesozoic period | | | |
| | Triassic<br>Jurassic<br>Cretaceous | total 160–180 million years | |
| (3) Neozoic period (Cenozoic) | | | |
| Tertiary | Paleocene | Old tertiary | Brown coals (estimated duration of formation of 1 m seam about 1,000 to 2,500 years) |
| | Eocene | | |
| | Oligocene | 60 million years | |
| | Miocene | | |
| | Pliocene | New tertiary | |
| Quaternary | Diluvium<br>Alluvium | 1 million years | Peat, wood |

Coal was formed under the influence of biological factors, hydrogenesis, molecular rearrangements, rock pressure, temperature, and time. It is generally assumed that the formation of bituminous coal is not just a question of time, but mainly a consequence of the effects of rock pressure and temperature. Moscow coal originated during the Carboniferous period, i.e. it is very old, yet, lying at the surface, it has remained lignite. Recently, however, doubts have been expressed regarding this explanation. On the other hand there are young (tertiary) coals that have been transformed into bituminous coals by geothermal effects, e.g. volcanic intrusions, such as Upper Bavarian pitch coal and Kassel coal. As a rule, coal is the more mature the deeper it lies.

Whilst coal was formed by the transformation of organic matter in littoral regions and contains, therefore, only such mineral constituents as were present in the cell (intrinsic ash), augmented by minerals which penetrated into the cell (extrinsic ash), bituminous rock originated in deep sea regions by the penetration of organic micro-organisms, e.g. plankton, of prehistoric seas, into porous mineral sediments where they were decomposed, polymerized, and resinified.

Bituminous coal seams were formed several hundred million years ago from enormous swampy forests that luxuriated in a humid warm climate, probably also favoured by the higher $CO_2$ content of the atmosphere which is assumed to have been present during the Carboniferous period. The vegetation was submerged by settlement of the swamp floor and with reduced access of air decay began: rotting, formation of peat, ageing under the influence of micro organisms, called humification. This is the first stage of the process of coalification. (See H. Bode, *Braunkohle*, 1930, **29**, 982.)

It is likely that decomposition by bacterial effect took place mainly in the cellulose whilst a residue of lignin, waxes, and resins remained and formed the main base of coal; this theory is, however, still not generally accepted. A large part of the coal substance was formed from pollen and seeds which are particularly resistant and contributed largely to the formation of the tar content.

As long as the vegetation remained above the water level the layer of the forest swamp that was being converted into peat could continue to grow. When the swamp floor settled further, the vegetation was submerged and decayed.

Such settlement often occurred as a result of rhythmical folds of the surface, giving rise to long trench-like depressions marginal to mountain systems. The formation of the bituminous coal deposits at the edge of the Rhenish Slate Mountains (Rheinisches Schiefergebirge) can, for instance, be explained in this way. The formation of 1 m of a bituminous coal seam is assumed to have taken 10,000 years. After settlement the seam was covered by silt from rivers or by inundation by the sea, until the velocity of settlement decreased and new forest swamps began to grow. The periodic alternation of swamp formation and inundation is probably the cause of the multiplicity of seams, often interspersed by thick layers of rock in the bituminous coal fields. The total duration of the formation of bituminous coal in the Carboniferous period is estimated by geologists to be about 70 million years. Subsequently the coal deposits were subjected to displacements and folds. After humification, a second important stage of coalification, petrification, takes place under the influence of rock pressure and increased temperature. The degree of transformation of the coal is determined by the degree and extent of petrification. A critical threshold value is the temperature between 300 and 325°C when 'metamorphism' takes place and which is supposed to be the dividing line between the formation of bituminous coal and that of brown coal. Some doubt the decisive effect of pressure and temperature. Since coalification is an endothermic reaction they assume that the life processes in the anaerobic medium themselves are the source of energy for the removal of oxygen and the carbonization. In some cases the geologists have found that the process of coal formation took place very rapidly and it is considered possible that for the conversion of peat into bituminous coal even the period between the formation of the peat swamp and its flooding and covering by sand and rock was adequate.

It must be said, however, that the plants concerned in the formation of brown coal were different from those concerned in the formation of peat (which can be observed even today) or of bituminous coal. In the xylites (lignites) frequently found in brown coal deposits the structure and chemical constituents of wood are still recognizable. The main period of formation of brown coal was the Tertiary when gigantic swamp cypresses, palm trees, mammoth trees, and the first pine trees flourished whilst bituminous coal was formed in the Carboniferous, Jurassic, and Cretaceous periods. The original constituents of bituminous coal are nearly completely destroyed and the vegetable origin can be confirmed only under the microscope.

It is estimated that the 100 m thick seam of Geisel Valley coal was formed about 50 million years ago over a period of 125,000 to 250,000 years.

Anthracite represents the last stage of coalification; the vegetable origin can be demonstrated only under the microscope in polarized light. Graphite is pure re-crystallized carbon and is therefore completely ash-free.

'Ageing' and 'maturing' of coal means, therefore, that its carbon content increases and its hydrogen and oxygen contents decrease. During coalification, separation occurs first of water, then of $CO_2$ and finally of hydrocarbons, particularly of methane ($CH_4$), known in the mine as firedamp. The calorific value of the coal increases correspondingly.

The largest deposits of bituminous coal are in the Rhineland-Westphalia region (reserves 73·2 milliard tons), in the former Upper Silesia (72·7) and in England (190). There are also very large reserves in the U.S.S.R., mainly in the Donetz and Kusnetz regions in the Altai. The total reserves of the various States of the U.S.A. amount to 970 milliard tons. Large deposits are also found in China, smaller ones in Japan, India, and Australia. World reserves are estimated at 7,413 milliard tons.

Composition and calorific value of solid fuels
(from Gums, *Kurzes Handbuch der Brennstoff- und Feuerungstechnik*, Springer-Verlag, 1942)

| Fuel | As used basis | | | | Volatile matter | Pure-coal basis | | | | | Calorific value kcal/kg | |
|---|---|---|---|---|---|---|---|---|---|---|---|---|
| | Ash % mean | Water % mean | Calorific value kcal/kg gross | net | % | C % | $H_2$ % | $O_2$ % | $N_2$ % | S % | gross | net |
| Air-dried wood | 0·4 | 15 | 4,026 | 3,657 | >70 | 50·43 | 6·17 | 43·28 | 0·12 | 0 | 4,760 | 4,430 |
| Peat | 4·5 | 25 | 3,930 | 3,566 | 70 | 58·8 | 5·7 | 33·4 | 1·7 | 0·4 | 5,574 | 5,270 |
| German raw brown coal | 4 | 51·5 | 3,098 | 2,652 | 56–58 | 68·78 | 5·82 | 21·71 | 1·01 | 2·68 | 6,410 | 6,130 |
| Brown coal briquettes from Central Germany | 9·5 | 14 | 5,340 | 5,020 | 56–58 | 68·78 | 5·82 | 21·71 | 1·01 | 2·68 | 6,410 | 6,130 |
| Gas coal and free burning coal[1] | 6 | 4 | 7,550 | 7,270 | 35 | 84·66 | 5·41 | 7·26 | 1·60 | 1·07 | 8,390 | 8,100 |
| Fat coal | 6 | 4 | 7,800 | 7,530 | 25 | 88·55 | 5·00 | 3·83 | 1·57 | 1·05 | 8,665 | 8,395 |
| Forge coal[1] | 6 | 4 | 7,820 | 7,580 | 16 | 90·26 | 4·38 | 2·95 | 1·4 | 1·01 | 8,685 | 8,450 |
| Lean coal[1] | 6 | 4 | 7,785 | 7,570 | 11 | 91·18 | 4·00 | 2·50 | 1·41 | 0·91 | 8,650 | 8,435 |
| Anthracite[1] | 6 | 1·3 | 7,981 | 7,776 | 9·5 | 90·62 | 3·97 | 3·06 | 1·27 | 1·08 | 8,609 | 8,397 |
| Graphite | | | | | | ~100 | | | | | 8,100 | 8,100 |

Brown coal deposits are found in almost every part of the world. The largest deposits in Germany are in the Central German and Nordrhein coalfields; the Central German deposit continues into Czechoslovakia. With the exception of the Scandinavian countries most European countries have brown coal deposits although some of these are not yet being exploited. Both the U.S.A. and the U.S.S.R. are very rich in brown coal. In the Soviet Union exploitation on a large scale did not start until after 1945. So far most of the brown coal produced has come from Europe although only about 12·5 per cent of

[1] Nut coal from the Ruhr-Aachen coalfield

the world's reserves are situated there. In America, probably the country richest in brown coal, production has been insignificant since the exploitation of petroleum and bituminous coal is evidently more economical.

*Mining of bituminous coal*

Bituminous coal is extracted almost exclusively by deep mining down to depths of 1,300 m, with an average of 600 to 800 m. In the United States some bituminous coal is found at a depth of only 100 m below the surface so that opencast mining is possible there. The thickness of seam most economically worked, as for instance in the Ruhr coalfied, is between 1·5 and 3 m. Seams of 4 to 10 m thickness, such as are found in the former Upper Silesia and in the United States, require special methods of working. A similarly favourable situation as regards depth and seam thickness as in the United States exists in some bituminous coalfields in the Soviet Union.

Before the decision is taken to sink a mine, the terrain is explored by sample borings (for more details about drilling technique see p. 80). For this, rotary drills are used consisting of screwed pipes with a crown either of hard metal or fitted with diamonds. Water is pumped down through the pipe which flushes the slurry upwards. A clay or barium sulphate suspension is used to increase the buoyancy of the water. From time to time a solid drill core is brought to the surface to obtain a picture of the composition of the rock. Once the extent, position, direction, thickness, and number of the seams are known a start is made with the development of the mine. Each mine has two shafts, an intake shaft for the supply of ventilation air and a return shaft from which used air is exhausted. While both shafts can be used for the hoisting of men and materials, coal is usually wound in the intake shaft. Air passes through every working place, following one of several routes between the intake and return shafts. The exploration and preparatory work for a shaft depends on the types of the seams which may be embedded in the rock in various thicknesses and often contain faults.

For the exploration from a shaft which requires a large investment of capital, usually two levels are developed, the lower one connected to the winding shaft, the upper one to the ventilation shaft. Informative sketches of a bituminous coal mine shaft will be found in Winnacker-Weingaertner, *Chemische Technologie*, vol. III, and in Gumz and Regul. On each level a tunnel is driven at right angles to the line of dip. These are the main cross cuts which should cut across all the seams. Main headings are driven parallel to the direction of the seam and from these further cross cuts. All these roadways are lined for protection against caving-in, and provided with rail tracks. For the preparatory work the seams are opened up from the lower to the upper level. The tunnels run, therefore, through the dead rock. Only with smaller thickness of seam is it possible to work the longwall method which aims at complete extraction, leaving no pillars of coal standing since these not only involve losses but also present a hazard. The working (advance) of the seam proceeds in three shifts over a width of 150–300 m. A strip of 1–2 m width is extracted from the face and secured against collapse by props which used to be made of timber and are now made of steel. Then the conveyor is brought up to the new face and the extracted cavity (goaf) is either caved in by withdrawing the supports (timber drawing), or stowed with stone, rubble, etc. ('gobbing'). This is done either manually or, if fine material is used, by pneumatic stowing with compressed air.

The extraction of the coal was carried out until 10 to 15 years ago mainly by men using pneumatic picks. Since then, full mechanization has made great progress. The coal is undercut by the coalcutting machine and extracted by the coal plough. In some cases, it is necessary to loosen the seam first by shotfiring using special explosives. In spite of steadily increasing mechanization much still depends on the skill, experience, and physical effort of the miner. Until fairly recently, the main cost factor in underground mining used to be wages (about 65 per cent). Increased mechanization, which made most progress at first in the United States, has succeeded in greatly reducing the proportion of wages in the total costs of bituminous coalmining.

The extracted coal is brought by belt conveyor, shaking conveyor, chain haulage, etc., to the tubs which usually have a capacity of 1–6 tons. These are then hauled by locomotives to the pit bottom and raised to the surface.

In thicker seams, for instance in the former Upper Silesia, a different method of working is used. Working proceeds directly into the coal. Several points, the 'pillar sections', are worked simultaneously by breaking through from one heading to the height of the roof over a width of about 5 m to the next heading. A pillar of several metres thickness is left between the adjacent workings; this leads to great losses of coal but has certain advantages in ground control. Here, too, props are used as supports to prevent roof falls. When a sector has been extracted, the pillar is cut down as far as possible, the supports are removed, and the roof is caved in. In view of the size of the collapsing chambers, damage on the surface is unavoidable. The extracted cavities are, therefore, in many cases stowed hydraulically with materials of small particle size, usually sand.

At the second stage of this method of working the advance proceeds from the extremity being extracted back towards the haulage road; this is called the bord and pillar method.

Bituminous coal production in millions tons

| Country | 1913 | 1938 | 1950 | 1955 | 1960 | 1964 |
|---|---|---|---|---|---|---|
| Germany | 190 | 162 | — | — | — | — |
| German Democratic Republic | — | 35[1] | 2·8[1] | 2·8 | 2·7 | 2·3 |
| German Federal Republic | — | 137 | 111 | 131 | 142·3[4] | 142·2 |
| European Coal and Steel Community | — | — | — | 246 | 234 | — |
| U.S.S.R. | 30 | 114 | 220[2] | 300 | 375 | 408·9 |
| U.S.A. | 516 | 355 | 500 | 441 | 391 | 454·7 |
| Gr. Britain | 292 | 231 | 220 | 225 | 196·8 | 196·7 |
| France | 40 | 46 | 51 | 55 | 56 | 53·0 |
| Poland | 9 | 38 | 78[2] | 94 | 104·4 | 117·3 |
| China | — | — | — | 94 | 420 | — |
| Europe | 606 | 683 | 773 | 908[3] | — | |
| Asia | 56 | 124 | 95 | 164 | — | |
| Africa | 8 | 18 | 29 | 37 | — | |
| World | — | — | — | 1,591 | 1,970 | |

The West German bituminous coalmining industry sustained great losses in the war and has been able to recover only very slowly. In many mines it was impossible to lift the mine waters and the ventilation failed. By order of the Occupying Powers, exploration and preparatory work had to be greatly restricted so that the mines had to be worked without regard to economic considerations, to provide the compulsory exports. By the end of the war, 76 per cent of all miners' houses had been destroyed, a fact which had a very detrimental effect on the miners' morale and performance.

As a result of the competition by oil, the bituminous coalmining industry of the Western countries finds itself in a critical situation. This is a consequence of technical progress, for oil firing in power stations

[1] Former Upper and Lower Silesia and Saxony, 1950 Saxony
[2] Estimated
[3] With U.S.S.R.
[4] German Federal Republic and Saar

and heating plants has great advantages compared with coal burning. Part of the coal produced became unsaleable, leading to increases in stocks and finally to the closure of pits. Coal can be stored in dumps only for limited periods since it deteriorates in storage, and because spontaneous ignition may occur.

*Mining of brown coals*

Formerly, brown coal was less economical as a fuel than bituminous coal because of its smaller calorific value (see Table on p. 13) and its greater content of non-heat-producing constituents, water and ash. Its production became much cheaper than that of deep-mined bituminous coal only at the beginning of the century, as a result of the introduction of fully mechanized opencast working; brown coal then began to be used increasingly for the generation of electricity and steam which are today produced much more cheaply from brown coal than from bituminous coal. This has been achieved by complete mechanization of opencast working. Brown coal, too, was at first produced by deep mining. In 1890, 75 per cent of the brown coal produced in Germany still came from deep mine, 25 per cent from opencast mine. In those days, opencast mining was used only where the coal was lying almost directly below the surface. Twentyfive years later the proportions were reversed and today deep-mined brown coal represents only a small fraction of the total.

The development of an opencast mine requires very careful preliminary geological investigation which should furnish an accurate picture of the size and thickness of the seam, nature and stratification of the overburden, and the situation concerning mine water. The costs of construction, purchase of the large earth-moving equipment, tubs and rails, and the removal of the overburden are enormous, tying down large amounts of capital for long periods before a return on them can be expected. Profitability is critically dependent on the ratio of the quantity of overburden to coal. The ratio of overburden to coal (O:C ratio), is used to indicate how many metres of overburden have to be removed for one metre of coal. This ratio has tended to increase with increasing mechanization of opencast mining. In working with conveyor bridges the ratio can be even greater than with traction haulage without impairing the profitability. Thus for instance a ratio of 4·5:1 is nowadays considered quite favourable, 9:1 as still just profitable. The deposits of 100 m thickness in the Geisel Valley and near Cologne are examples of exceptionally favourable conditions. Profitability is, of course, a function not only of the O:C ratio but also of the depth at which the coal lies and of the thickness of the seam. A total depth of 140 m with a seam thickness of 20 m is still quite acceptable.

According to the type of seam, different working methods are used, varying in the method of earth removal, transport, and disposal of the overburden, whilst the actual extraction of the coal is more or less uniform. When moving the overburden one always aims at filling the extracted part of the mine again with overburden, to level it, and to cover it with topsoil so that it can once more be used for agricultural purposes. Formerly, this was not always done and one can still see in many places the dead, barren, gigantic dumps and overburden tips by the side of extracted sites which are devoid of vegetation due to the formation of acid from iron pyrites.

The three most common methods of removal of overburden are:

(1) with excavator, traction haulage, and dumping by tippers,
(2) overburden conveyor bridge,
(3) cable dredger.

Methods (2) and (3) eliminate, therefore, haulage. Various types of excavators are used for the removal of the overburden and extraction of the coal. According to the working method adopted, the excavators used are (a) upward-cutting excavators that lift off the coal or overburden above, (b) downward-cutting excavators that remove the material below, and (c) rotating and compound excavators. The excavators are equipped with bucket chains, shovels, often of very large capacity, and bucket wheels. In Germany, the most common type are bucket chain excavators. The excavators run either on rails which are moved forward, together with the ties, by rail-shifting machines as the workings advance, or on caterpillar tracks. Bucket chain excavators are so large that they must always be moved on rails. In Germany, excavators are built nowadays with a power of 4,500 kW and a performance of 100,000 m$^3$ per day.

In the removal of the overburden by haulage the overburden is carried by trains consisting of special tipping tubs, often of very large size (up to 60 tons payload), to the extracted side of the site. There the overburden is dumped and thrown over the side of the mine by dumping excavators, rotating excavators on rails, or recently also by excavators on caterpillar tracks. Conveyor dumpers with built-in conveyor belts and jibs of 90 m length are also used. With these, dumping of the overburden at a great height and over a large area is possible.

A bridge conveyor spans the entire site, moves about on its rails, and carries the overburden, removed by the excavators, by means of a conveyor belt across the site to the extracted side. These conveyor bridges are impressive, bold constructions which may have a total length of up to 600 m (e.g. at Espenhain), but usually 200–300 m (see Figure 1). At the end of the war in 1945, 21 conveyor bridges were in use in Germany.

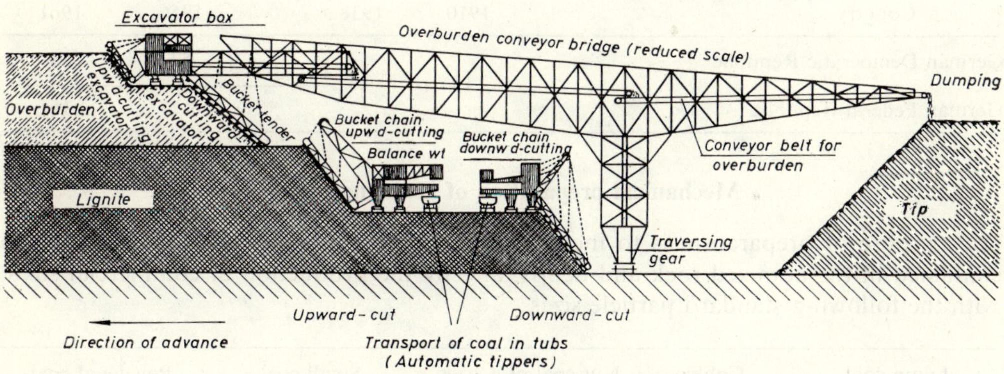


Figure 1. Diagram of a brown coal opencast mine and conveyor bridge (width much reduced)

Cable excavators perform a dual function, mining and removal of the overburden. The opencast site is not spanned by a rigid bridge but by wire ropes extending between two towers which are placed on large rail- or full-track vehicles and travel backwards and forwards on both sides of the site. A crane carriage is suspended from supporting ropes and operated by a traction rope; to it is attached an excavating bucket with which overburden or coal can be dug, lifted from the mine and hauled to the edge of the site. Cable excavators are very versatile machines, particularly useful when the seam is irregular, but their performance is not very good (up to 5,000 m$^3$ of overburden per day).

The same types of excavators are used for the mining of coal, the most commonly used being the bucket chain excavator, employed for either upward- or downward-cutting. Bucket wheel excavators, running on caterpillar tracks and equipped with a belt conveyor, are becoming increasingly popular. The coal is often cut on several levels, starting usually from the level of a rail track, with one upward-cutting and one downward-cutting excavator ('stepped working'). For the subsequent preparation of the coal it is important to carry out a preliminary separation at the mining stage according to the ultimate use of the coal. This applies particularly to coal destined for low-temperature carbonization plants since they work on the principle: 'Low-temperature carbonization of the coal begins in the mine', i.e. by the correct separation into high-bituminous lignite and flame coal.

### *Economic aspects*

The brown coal reserves of the world are estimated provisionally at 1,200 milliard tons, but are certainly much higher. Of these, the largest deposits are in the United States (646 milliard tons), the Asiatic part of the Soviet Union is next with 288 milliard tons, then Germany. Before the first World War, German reserves were estimated at 112 milliard tons. The largest increase in brown coal mining after the war occurred in the Soviet Union. Before the war, the pithead price of 1 ton rough coal was about RM 2–3, briquettes cost about RM 12.

Brown coal production (in millions of tons)

<table>
<tr><th>Country</th><th>1913</th><th>1943</th><th>1946</th><th>1952</th><th>1964</th></tr>
<tr><td>German Democratic Republic</td><td rowspan="2">86</td><td rowspan="2">252</td><td>110</td><td>150</td><td>256·9</td></tr>
<tr><td>German Federal Republic</td><td>52</td><td>83</td><td>110·9</td></tr>
</table>

Brown coal briquettes (in millions of tons)

<table>
<tr><th>Country</th><th>1910</th><th>1938</th><th>1956</th><th>1961</th></tr>
<tr><td>German Democratic Republic</td><td rowspan="2">15</td><td rowspan="2">44</td><td>52</td><td>58</td></tr>
<tr><td>German Federal Republic</td><td>17</td><td>16</td></tr>
</table>

## Mechanical preparation of bituminous coal

The purpose of preparation is to increase the value of the coal by extracting the ash contained in the admixed rock and debris, and by the production of commercial sizes, with the following standard particle sizes:

| Lump coal | Cobbles | Nut coal of 4 sizes | Small coal | Powdered coal |
|---|---|---|---|---|
| larger than 120–80 mm | 120–80 mm | down to 10 mm | up to a size of 10 mm | less than 5 mm |

The methods used in mechanical preparation include comminution in crushers, sieving in screening machines, concentration in settling tanks, sluices, or separators for the extraction of ash-carrying materials, and grinding, mixing, and de-dusting. Several methods are in general use. Dry preparation by grading and mechanical preparation using compressed air is expensive due to the high price of compressed air and because separation is difficult. Wet preparation is preferable. In this method the separation, based on the different specific gravities of the solid constituents, is carried out in a liquid in motion either in settling tanks, by washing in troughs, or by flotation. Nature provides an example, in the sedimentation processes going on in rivers and the sea, of how wet separation of solid matter can be carried out.

Figure 2 shows the principle of elutriation of a mixture in a horizontal stream of water. This method is suitable for all solid materials, not only for coal. The mixture is passed through several settling cones increasing in size and connected in series. The flow velocity *V* in each cone is smaller than the settling velocity of the particular grain size. If the buoyancy of the water is inadequate, which is the case with

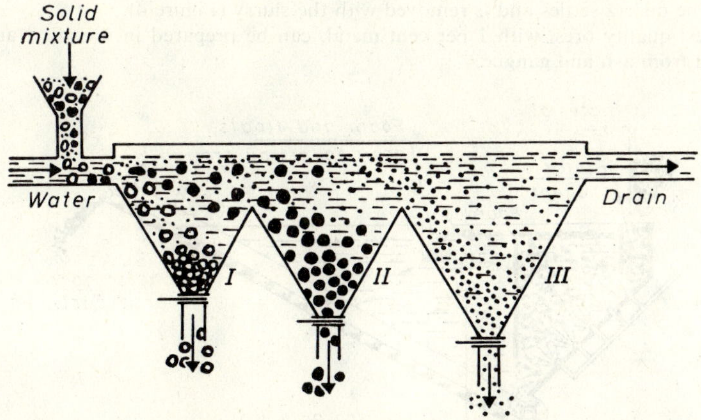


Figure 2. Elutriation in a horizontal stream of water. The specific gravity of water can be increased by suspensions of chalk or clay: 'viscous-liquid preparation'. Reduction of flow velocity from I to III is done by increasing the volume of the conical vessels

large and heavy material, the density of the water can be raised by suspensions of chalk or clay. This is called viscous-liquid preparation. The same principle is used in rising-stream classifiers, illustrated in Figure 3. Several vessels of increasing diameter are connected in series, thus reducing the velocity of the stream and causing the settling of increasingly fine particles. In gaseous media the same principle is applied in air separators. The third method is the flotation process. It is mainly used for the concen-

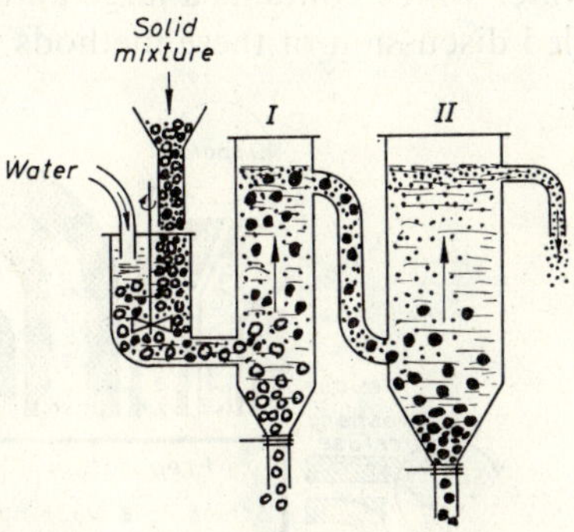


Figure 3.
Elutriation in a rising stream of water. The water velocity is smaller than the settling velocity. The velocity of water in II is less than in I

tration of minerals but a brief description will be given here in the context of the methods of preparation of solids. Flotation utilizes the difference in wettability by water of the components of a mixture to bring about their separation.

Minerals have first to be comminuted down to 0·3 mm; for coal >1 mm is sufficient. The material is mixed in a chamber fitted with a fan, with water which contains 'collectors'. These (e.g. xanthates) adsorb by their polar group to the surface of the particle and render it hydrophobic by their non-polar group. The hydrophobic particles which are now less wettable rise with the air bubbles and concentrate in a layer of foam. In order to improve the dispersion of the air into as small bubbles as possible 'foaming agents' are added. These agents, e.g. Nekal BX or Nekalit, designed to reduce the surface tension of the water against air, promote the formation of tiny air bubbles but should not encourage the formation of large ones. The addition of certain oils renders the foam more buoyant.

In a mixture of, for example, copper pyrites and quartz the copper pyrite rises and is collected with the foam layer, the quartz settles and is removed with the slurry (Figure 4).

Even the lowest-quality ores, with 1 per cent metal, can be prepared in this way, and coal can be cleanly separated from ash and gangue.

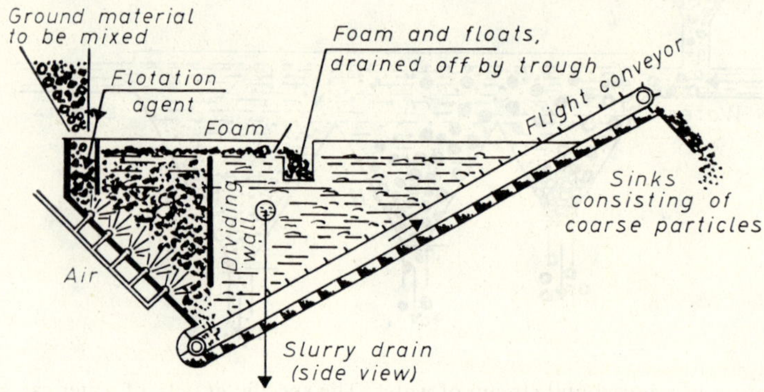


Figure 4. Diagram of a flotation classifier

All concentrates produced in wet preparation are drawn off by filters (e.g. plate filters, revolving filters, and rotating disc filters). The residual moisture of the coal will be 20–25 per cent, ash can be removed down to 0·3–0·5 per cent. Further drying of the prepared coal takes place, for instance, in draining and during storage. A modern coal preparation and washing plant for coking coals is very extensive because the clarification of the wash water which contains a large amount of fine material requires a large plant. A more detailed discussion of these methods would be out of place here.

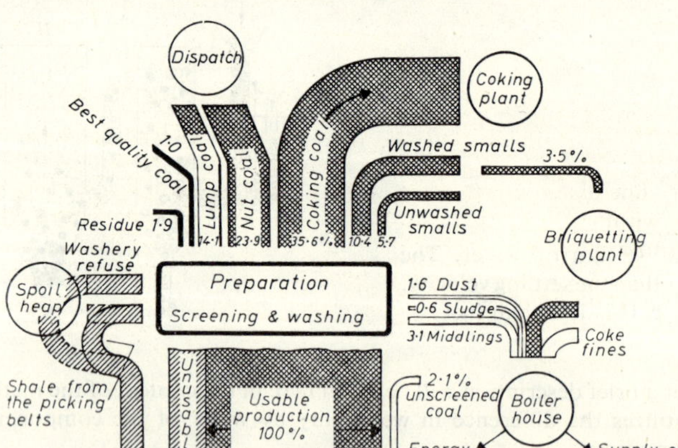


Figure 5. Quantitative flow diagram for the preparation of bituminous coal

The 'quantitative flow diagram' (Figure 5) shows the quantities of different types of coal and of waste materials produced in the preparation of bituminous coal (P. Damm, in Winnacker-Weingaertner, *Chem. Technologie*).

The most important use of prepared bituminous coal is for coking plants (about 35 per cent). Of the small coal (25 to 50 per cent), gas coal is used as fuel by collieries and power stations, caking coal for coking. At times it is difficult to sell lean coal and steam coal with volatile contents of less than 20 per cent, or gas coal which does not yield hard coke. Thus, for instance, before the war the heaps of fine coal in Upper Silesia were enormous. An outlet for surplus fine coal is offered by briquetting. Briquettes open up new possibilities for chemical utilization by coking and low-temperature carbonization.

## Mechanical preparation of brown coal

In contrast to the preparation of bituminous coal, that of brown coal is not nearly so extensive and is limited to classification by particle size, separation of coal and gangue, and division according to structural constituents, e.g. dull coal (durain), vitrain, and ligneous components. Comminution is carried out in crushers and grinders. Mechanical separation according to particle size, applied in general only for the separation of wood or ligneous constituents, is by shaking screens and riddlers, vibration shifters or pressure sieves. Pyrites can be removed by magnetic separation. An important part in the preparation of brown coal, because of its high water content (often more than 50 per cent) is played by drying. The heat of evaporation of the water contained in the coal absorbs a considerable part of the energy used in drying, i.e. part of the coal must be sacrificed for it. Drying is necessary whether the product is to be used for briquetting or put through another process. For briquetting, 13–15 per cent water is left in the coal to act as binder, but chemical preparation methods require more intensive drying. First, attempts are made in the mine to achieve as efficient as possible drying by mechanical means, e.g. by drainage trenches, but this method cannot be used during rainy periods. Thermal drying is always necessary and is carried out in special steam driers.

## Drying of coal

In general, the water contained in the coal is evaporated by passing the coal over steam-heated surfaces. Hot gases can also be used for drying.

A problem yet to be solved is the utilization of the heat content of the waste steam. The recovery of the heat contained in the evaporated moisture of the coal is very difficult because the vapours carry large quantities of very fine dust, despite the use of dust removal installations. Heat exchangers would be rendered ineffective within a short time by dust deposits. For reasons of thermal economy, therefore, the steam is saturated to a maximum degree and the process is so arranged that a maximum quantity of residual water is evaporated as the coal cools. The use of the highly compressed steam for driving motors is particularly advantageous for the same reason. The driers are then coupled in series using the decompressed steam.

It is possible to run brown coal briquetting plants more economically by decompressing the high-pressure steam in backpressure turbines to 3–4 atmospheres, thus driving the briquetting presses by electricity and supplying at the same time surplus current to the grid. In this way economical use is also made of the coal and dust unsuitable for briquetting.

During drying the volume of brown coal decreases by 30 per cent or more, because of its colloidal structure. If drying is carried out rapidly, the coal dries more on the outside than on the inside, leading to shrinkage, cracking, and powdery decomposition of the coal granules: this phenomenon is used when pulverization is required, for instance in pulverizing driers. Non-uniformly dried coal granules do not yield stable briquettes. Low-carbonization plants require very stable briquettes because the throughput of the oven depends on this property. In briquetting, uniform moisture content of the granules is, therefore, the aim and one tries to achieve it by:

(1) leaving the coarse granules in the drier longer than the fine granules, by sieving off the fine granules which dry more quickly during their passage through the drier,
(2) crushing the partially-dried coarse granules during the drying process.

### *Types of driers*

Two types of driers are used preferentially in briquetting plants: disc driers and multiple-tube driers. Figure 6 shows the schedule of operations for a disc drier. It consists of about 40 heated slotted discs,

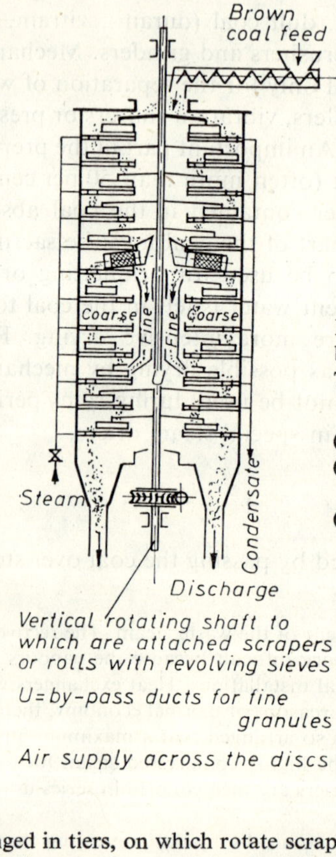


Figure 6. Flow sheet of a steam disc drier for briquetting plants. 40 heated discs in 3–4 groups. To obtain a uniform moisture content of both coarse and fine granules:

(1) Longer residence of coarse granules in drier
(2) Sieving off of the more readily dried fine granules by revolving sieves
(3) Crushing of the partially dried coarse granules during the drying process

arranged in tiers, on which rotate scrapers attached to arms. First, the coal drops over 24 discs during which time the fine particles dry. Those smaller than 2 mm are sieved off by revolving intermediate sieves and drop through a duct by-passing the next six discs. The main mass of the coal reaches another disc with revolving sieves another ten discs down. There, particles of less than 4 mm are removed. The remainder passes over another six discs by which time it is dry. Adjustable flaps in the outer jacket of the drier provide for the supply of air. The diameter of the driers is about 5 m, and their height varies from 5 to 11 m; 5–20 tons of raw brown coal are dried per hour on a heating surface of 400 to 950 m$^2$, the heat consumption amounting to 800 kcal per kg of evaporated water.

### *Multiple-tube driers*

Multiple-tube driers are cylindrical revolving steam vessels with an axis inclined $<10°$. Pipes are rolled into the bases of the casing through which coal and dry air pass in the same direction. The drum runs on rollers (see Figure 7). The hot steam is fed through the hollow shaft on the coal feed side and condenses on the outer pipe walls. The condensate is discharged on the other side. The moist brown coal slides into the drying pipes through a ground feed-hopper. From time to time it is necessary to remove blockages by pushing a rod through the pipes. The continuously revolving coal passes down through the pipes and drops through a sluice at the bottom whilst the waste steam from which the dust is still to be removed escapes upwards.

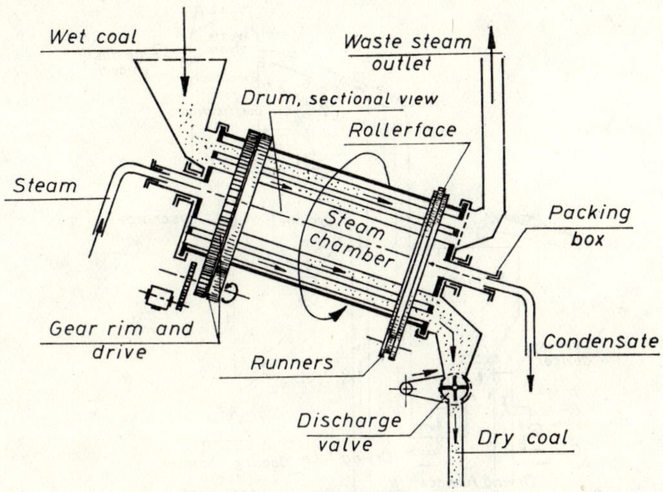


Figure 7. Multiple-tube steam drier (scheme for brown coal)

### *Drum driers*

Combustion gases can be used to remove water from coal in drum driers (see Figure 102). A drum drier consists usually of a very long revolving pipe which is slightly inclined downwards and contains sloping foils. These raise the material to the highest point; it then drops and moves slowly towards the discharge point. Vibrating hammers or chains are often fixed inside the drum to prevent clogging. Hot gases from a combustion chamber or air heated by steam over a calorifier are fed into the inlet. The dry material is discharged through the bottom of the drum. The dust must be removed from the gases by a cyclone.

The use of coal dust in chemical preparation necessitates rapid drying; on the other hand, the use of finely divided dry coal entails a great risk of ignition and explosion. Only waste gases poor in oxygen are, therefore, suitable as carriers of the drying heat. The combustion gases used in the drum drier are sufficiently poor in oxygen. Although the gas feed temperature is 900°C, very little heating (80°C) takes place in the drum drier with coal containing 60 per cent moisture. 90 per cent saturation by steam of the dry gases is achieved.

### *Pulverizing-drying mill (e.g. the Lurgi process)*

In one operation the coal is dried by evaporating the moisture, crushed, dried again by evaporating the residual moisture, and fine dust for briquetting and dust for coal-dust furnaces or coal hydrogenation is obtained. This is achieved by combining the drying by hot gas with grinding and screening in a crusher blower. A kind of pneumatic separation causes the coarse particles to drop back repeatedly into the crusher blower (see Figure 8). Some of the waste gases are fed back into the combustion chamber of the drier which they leave almost free of oxygen. Despite an initial temperature of 900 °C there is hardly any carbonization and the loss of tar produced is less than 2 per cent. The coarser the coal in the vertical drying tube attached to the crusher blower the slower it moves; this satisfies the main requirement for uniform drying. Measured quantities of raw brown coal are delivered from the bunker into the coal drying furnace. Hot gases from the gas burner flow around the pipe leading from it. These gases and the

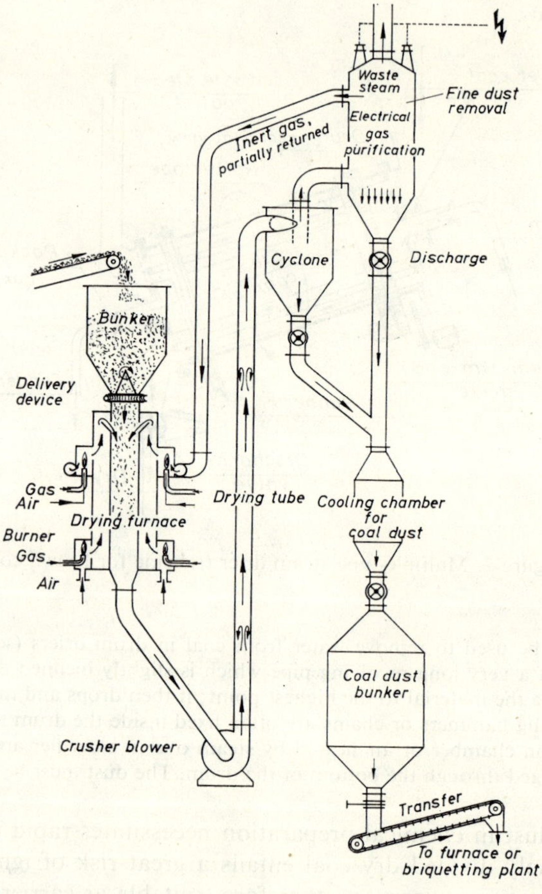


Figure 8. Operation of the Lurgi crushing and drying process for coal

coal move in the same direction through the drying tube to the fan which drives the dust towards the separator. The finest dust is separated in an electrical precipitator, coarser dust is delivered straight into the cooling chamber. Part of the gas stream is deflected back into the drying furnace. The dried dust is collected in a bunker filled with inert gas and is transported from there to the presses or other points of consumption. In a similar manner the raw brown coal for coal-dust-fired boilers is dried by hot air or by flue gases returned from the combustion chamber by suction, and simultaneously ground into combustible dust. One of the most difficult problems of brown coal technology is dust control and the prevention of explosions.

## Briquetting

Sufficient moisture is contained in brown coal to act as a binder for briquetting; on the other hand, the production of briquettes from bituminous coal requires an additional binding agent. It is usual to use for this purpose ground bituminous coal tar pitch in quantities of 5–8 per cent; however, the sulphite cellulose spent liquor from the sulphite digestion of wood can also be used in combination with clay. In certain circumstances, briquetting is possible without a binder by using very high pressures. Unwashed crushed bituminous coal can be pressed into briquettes directly, washed bituminous coal must first be dried in a drum drier and crushed to a particle size of 0–10 mm. Pitch is added to the coal in a steam kneading mill at about 90°C; at this temperature the mixture is plastic. Ovoid briquettes are produced on roll presses, quadrangular briquettes, weighing from 0·5 kg up to as much as 10 kg, on stamping presses.

No binder is required for sufficiently stable briquettes to be produced from brown coal; for this its own moisture content of about 15 per cent is adequate. To obtain optimum values for strength of the briquettes, pressure and moisture content must be harmonized when a certain type of press is used. For both brown coal and brown coal low temperature coke, tar or concentrated sulphite waste liquor can be used as binder; in this case, however, it is advisable to press at an elevated temperature but no high pressure is required.

Two types of presses are in general use: stamping presses and roll presses. The stamping press (Humboldt) is based on the invention of the Bavarian railway engineer Carl Exeter who designed the first stamping press for pressing peat on the principle of a plodder. The first press was installed in 1856 in the Haspelmoor (near Munich) peat plant, the first brown coal briquettes were pressed in 1858 at the 'Theodor' Colliery, now Brown Coal Colliery Ammendorf. In briquetting plants, toggle-lever presses with two dies are most popular. They are designed on the principle of the plunger pump without valves or slides and are used mainly for larger furnace briquettes. The feed device delivers at each stroke the quantity of coal necessary for one briquette into the horizontal open moulding channel into which it is pushed by the die. The required pressure is obtained by an adjustable constriction in the mould and a hump in the channel. The briquettes push each other out of the open moulding channel without fritting together. The moulds are subject to considerable wear.

In roll presses the coal is pressed between a ring actuated by rollers and a wheel which is itself forced against the ring by strong springs; segmented moulds produce a notched strip which is cut up into pieces. This method is used mainly for the manufacture of briquettes from brown coal suitable for carbonization to be used in the production of lump coke.

A briquetting plant consists basically of the crushing plant, a number of steam disc driers or multiple tube driers, and the stamping or roll presses: from the driers the coal is transported by worm-conveyor to the cooling plant where the residual moisture evaporates from the hot coal during its passage through air-cooled louvred coolers. Trough flight conveyors then transfer the coal from the cooling chamber to the presses. Dust is removed by suction from all points where it deposits, e.g. from the dies and the opening of the presses. The warm air from the engine, press, and drier houses is utilized for drying.

Brown coal briquette production in (all) Germany
(see also p. 18)

| Year | 1885 | 1910 | 1935 | 1957 | 1961 |
|---|---|---|---|---|---|
| mill. tons | 0·75 | 15 | 33 | 69 | 74 |

## Coal as a fuel

The generation of steam and electricity by the combustion of coal still forms the basis for most chemical works. Fuel technology and the development of steam generation will, therefore, be discussed here quite briefly.

In the generation of electrical energy one has to distinguish between condensation power stations in which the waste steam of the turbines is condensed back to boiler feed water in cooling towers, and power stations with back-pressure type steam turbines which also produce heating steam (see p. 27). For a long time, brown coal was a more expensive fuel than bituminous coal. Not until brown coal was mined on a very large scale in opencast mines did its 'heat price' drop to such a level that steam and electricity produced in large brown coal-fired generating stations cost less than half when bituminous coal was used as fuel. This is a development of recent years. During this period the large stations in the brown coal regions of Central Germany and of the Rhineland were built. Electricity produced from brown coal is often nearly competitive with electricity produced by water power so that those branches of industry that use large amounts of electricity, such as the production of aluminium and magnesium, carbide and calcium cyanamide, and synthetic ammonia, and the electrolysis of alkali-metal chlorides, have developed rapidly in these regions during the last few decades. The effect of this was felt also in the organic industry. The cheap chlorine produced by electrolysis formed the basis for the production of chlorinated solvents and the intermediates containing chlorine.

However, in the long run there is no justification for the mere burning of coal suitable for carbonization, without extracting refined products. Arrangements will have to be made to utilize coal for this purpose and to combine coal upgrading and power generation, as is done already for instance in the brown coal Combine Böhlen. A particularly difficult problem is the combustion of the little utilized saliferous coals containing the sodium salt found in Central Germany. They will certainly have to be used soon as fuel since in some mining areas (e.g. in the Geisel Valley) high-grade coal will be exhausted in the foreseeable future. This will necessitate new boiler plants.

The invention of briquetting created the conditions for the universal use of brown coal as a domestic fuel, for boiler plants, destructive distillation, and gasification. It became possible to use raw brown coal in large boiler plants only when special furnaces had been built designed to take account of the characteristics of brown coal, its small energy to mass ratio, its high moisture content, the varying particle size and the high ash content.

### *Types of furnaces*

The burning material rests on the grate, the purpose of which is to carry it until it is burnt up, to remove the ash or to permit it to drop through, and to allow access of air (forced draught). Grates are placed usually not under the boiler but in front of it. Heat transfer is by convection and radiation. A short description and some simplified sketches will explain the various types of grate firing.

High performance of a grate requires transport of the coal across it adjusted to the combustion, e.g. by gravity or by mechanical movement. There are, accordingly, inclined grates, travelling grates, and thrust grates, and finally pulverized coal firing which dispenses with grates.

The high moisture content of many coals necessitates preliminary drying which is carried out by the radiation from the burning material and the hot walls, or better still by hot combustion gases, waste gases, or pre-heated combustion air. According to the state of the combustion, air is supplied through the grate as updraught and above the burning layer as 'secondary air'.

Straight-grate firing (1) is suitable only for small furnaces. The fuel is fed on top of the burning material. The thermal output is limited by the performance of the stoker and reaches at best $4 \times 10^6$ kcal/h.

The inclined grate (2) has a slope of 30–45°. The fuel is fed through a funnel-shaped slide at the top and slides slowly downwards; at the bottom, complete combustion of the coal and removal of the ashes takes place on a horizontal part of the grate. A special form of the inclined grate is the step grate (3), the grates being arranged in terraces at gradually diminishing angles. With this type of grate the area of combustion is exceptionally large and step grates are, therefore, suitable for fuels of low calorific value (raw brown coal). The size of the grate surface depends essentially on the calorific value of the fuel. However, lengths over 7 m are uncommon. The thermal output of step grates may reach $20 \times 10^6$ kcal/h. The trough grate (5) is a combination of a straight grate in the centre and two inclined grates at the ends; in a firing arrangement several of these are placed side by side. Finally, there are the travelling grates, such as the chain grate (4), in which small straight grates form an endless chain. These are, however, suitable only for medium and high grade fuel of a calorific value greater than 4,000 kcal/kg. The coal drops from a funnel-shaped hopper on to the charging end and moves through a combustion

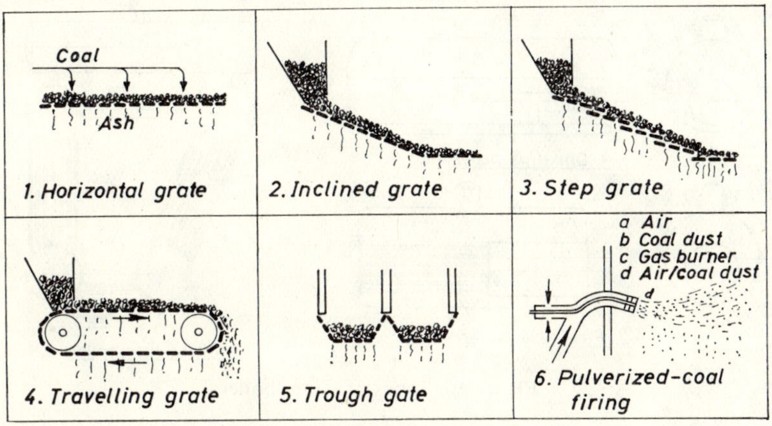


Figure 9. Different types of furnace

chamber with adjustable velocity. Ash and slag drop off at the end. Such grates are made up to 6 m long and wide, and under favourable conditions their thermal output may reach up to $54 \times 10^6$ kcal/h.

The coal can also be moved along by push feeders, either over or under the burning fuel (overfeed or underfeed stoking). In powdered coal firing (6) ground coal is blown into the combustion chamber. The residence time of the coal particles is only about one second. With outputs of up to $400 \times 10^6$ kcal/h this is a very efficient and versatile form of firing for large boiler plants. A special form of this type of firing is pulverized fuel firing, a combination of coal pulverization with drying by recirculated combustion gases or air pre-heated by the boiler waste gases, and screening of the dust by gravity, the coarse particles being returned each time to the pulverizer for further comminution (see Figure 8).

### *Steam boilers*

The development of fuel engineering proceeded simultaneously with that of steam generators. The trend was towards ever-increasing sizes of grate and heating surfaces, greater pressures and more steam produced per m$^2$ of heating area. The production costs of steam were reduced proportionately. Large electricity and steam generating plants are now built with boilers designed for maximum pressures of 150 atmospheres or more. In plants where electricity and large quantities of steam are required for internal use, the pressure difference between for instance 100:20 atmospheres is utilized for driving back-pressure turbines. Medium and low pressure steam is fed into the internal supply line. This makes possible a very economic supply of power. Only about 1,200 kcal are required for 1 kW as against several times this amount in an efficient condenser power plant.

The essential components of a steam boiler plant are firing, boiler, superheater in which the saturated steam is freed of entrained water droplets by superheating, and pre-heater. Feed water and combustion air are pre-heated by waste gases. A boiler plant also comprises a flue dust separator which may be an electrostatic precipitator, a smoke stack for the discharge of the waste gases, and a water softening plant. For the efficient running of high pressure and maximum pressure boiler plants water softening as well as the removal of silica must be carried out very carefully. Chalk and soda, phosphates, Permutit filters, and, best of all, synthetic ion exchangers (e.g. Wofatite) are used for water softening.

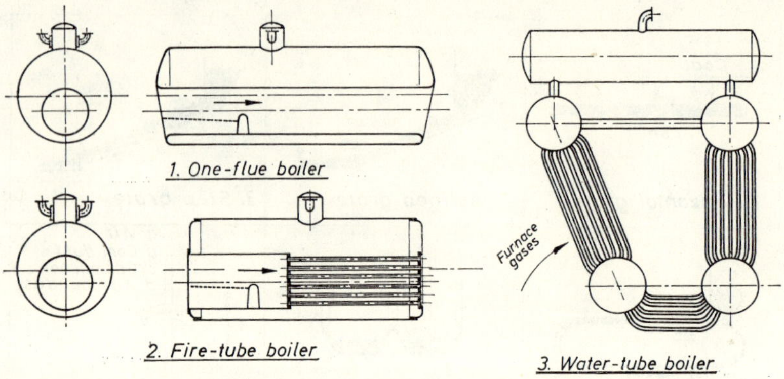


Figure 10. Types of steam boilers

The simplest type of steam boiler is the one-flue boiler (Figure 10). By increasing the number of flues the heating surface was enlarged (multiple-flue boiler). The continued increase in the number of flues resulted in the development of fire-tube boilers in which the gases pass through many water-cooled tubes. The volume of water in these boilers is relatively large (large water-space boilers). Development in another direction led to the reduction of the water space by splitting it up into many tubes outside the boiler proper in the combustion chamber, and thus increasing the heating surface. The fire tubes are combined into drums both at the top and at the bottom (water-tube boiler 3).

In these, the water passes through the tubes around which the combustion gases flow. The output of the boilers, calculated in tons of steam per hour, depends on the heating surface in m$^2$. The largest boiler plants in Germany have a heating surface of about 3,200 m$^2$ per boiler and under full load can produce 1,000,000 lb. of steam per hour.

## LITERATURE

Paul Kuckuk, *Geologie, Mineralogie und Lagerstättenlehre*. Springer, Berlin, 1951.

Lehmann, *Leitfaden der Kohlengeologie*. Knapp, Halle, 1953.

K. Kegel, *Lehrbuch des Braunkohlentagebaues*. Knapp, Halle, 1953.

Grumbrecht, *Leitfaden des Braunkohlentagebaues*. Knapp, Halle, 1949.

Gumz, *Kurzes Handbuch der Brennstoff- und Feuerungstechnik*. Springer, Berlin 1942.

Winnacker-Weingaertner, *Chem. Technologie 3*, Org. Technologie I. Karl Hanser, München, 1952; 2nd ed., Winnacker-Küchler, vol. **3**, Org. Techn. I, 1, 1959.

Ullmann, *Encyklopädie der techn. Chemie*, 3rd ed., vol. 4, Urban & Schwarzenberg, München–Berlin, 1953.

Krist, *Die festen Brennstoffe*. Fachbuchverlag Leipzig, 1951.

O. Neunhoeffer, *Grundlagen der Schwimmaufbereitung*, Verlag Th. Steinkopff, Dresden, 1948.

W. Gumz und R. Regul, *Die Kohle, Entstehung, Eigenschaften, Gewinnung und Verwendung*. Verlag Glückauf G. m. b. H. Essen, 1954.

# 1. CARBONIZATION OF COAL

Coal and other solid fuels can be converted into fuel gases or gases for use in synthesis processes by carbonization or gasification.

One of the oldest forms of chemical processing of the solid fuels, wood, peat, brown coal, and bituminous coal is carbonization. It consists in subjecting them to thermal treatment in a closed space without access of air, to prevent the occurrence of combustion processes. Carbonization of a solid fuel leaves a non-volatile residue which contains, in addition to carbon, all the ash constituents, charcoal in the case of wood, coke in the case of peat and coal.

The first patent for the carbonization of bituminous coal to produce tar was taken out in 1681 in England by the German J. J. Becher but the time was not ripe for the utilization of the gas. Long before its use for lighting purposes was thought of and its production for this purpose begun, it was available in England as coke-oven gas.

## Coal gas, town gas

The first man to produce bituminous coal gas in continuous operation for lighting purposes was probably the Dutchman J. P. Minckelers. The Englishman Murdoch who had used gas for lighting his house and workshop even earlier, installed the first gas lighting plant in Birmingham in 1798. The first street lighting followed in London in 1813, in Berlin and Hanover in 1826, and in Leipzig in 1838.

Formerly, illuminating gas was produced by distillation from bituminous coal, except for small quantities from oils and fats. Nowadays, a mixed gas is used consisting of bituminous coal gas and water gas or producer gas (see below). Carbonization gas is often added where available. Thus, for instance, carbonization gas from Edderitz is fed into the long-distance gas supply grid of the Magdeburg Gas Co., connected with the former Continental Gas Co., Dessau. Jena town gas is mixed with gas from the Böhlen combine. In bituminous coal areas scrubbed coke oven gas often serves as town gas. Brown coal briquettes are frequently used for gasification. Town gas is also produced by the Lurgi pressure gasification method from brown coal by gasification at 22 atmospheres with oxygen and steam, e.g. at Böhlen for the supply of the Leipzig district (see also under Gasification).

In the German Federal Republic the supply of gas for domestic, commercial and industrial purposes from colliery and blast-furnace coking plants is closely integrated with that from gas works and gas-producing oil refineries. A very extensive and ramified long-distance gas grid is available. Over long distances, compressed gas is supplied through seamless steel pipes.

Approximate composition of town gas in vol.-per cent

| Type of gas | $H_2$ | CO | $CH_4$ | Higher hydrocarbons | $CO_2$ | $N_2$ |
|---|---|---|---|---|---|---|
| Town gas from bituminous coal mixed with producer gas | 41 | 13 | 23 | 2 | 3 | 18 |
| Town gas from bituminous coal mixed with water gas | 51 | 21 | 18 | 2 | 3 | 5 |

In the United States, 'carburetted water gas' has been used for decades as town gas. Thermocarburettion is destructive distillation of oils, e.g. tar oil or petroleum; this produces a mixture of water gas and oil gas. In 'cold carburettion' the gas is passed through hot benzene of which it absorbs about 4 g per m$^3$.

Oil gas itself has been produced since 1815 (Taylor) by heating vegetable and animal fats and oils to destruction; it was subsequently adopted for lighting many towns. This method of producing gas soon declined, however, as a result of the increase in prices of fats and oils. It was given a new lease of life by the use of mineral oils, but only for lighting railway carriages, light houses, and beacons (J. Pintsch, VEB Gaselan). More recently, the production of town gas from oil and hydrocarbons has been steadily gaining ground (*Brennstoffchemie*, 1960, **41**, 101).

As long as illuminating gas was burnt in swallow-tail burners, the presence of materials which lent luminosity to the flame by forming glowing carbon particles, was of great importance. This aspect ceased to matter with the introduction of the Welsbach incandescent light and the increasing use of town gas for cooking and heating purposes in private houses and commercial premises. The first consideration was the calorific value. With town gas, this should be above 3,000 kcal/Nm$^3$, normally 4,200 to 4,400 kcal/Nm$^3$.

*Retorts and ovens*

At first, Murdoch used iron crucibles, later horizontal iron retorts. Today, iron has been replaced everywhere by fireproof materials, fireclay, or silica bricks. For many decades, horizontal fireclay cylinders, at first with circular, later with oval cross-section, were in common use as retorts. Nine retorts, each holding 180 kg of coal, were bricked in a furnace chamber. The waste gases of the furnace firing were passed through regenerators, a reversing process being thus used to heat the combustion air.

Further progress was brought about by the introduction of vertical retorts. They have a height of about 5 m, the cross-section is oval, and they hold about 500 kg of coal. An oven usually contains 18 retorts. Vertical ovens yield more gas and tar than horizontal ones. The retorts are charged from above and after carbonization the coke is discharged at the bottom into a conveyor chute. The gas escapes into a foul-gas main, where the major part of the tar is collected, and is then purified in the usual way. Few vertical retorts are still in use.

Since 1910, large gas works employ by-product ovens, with large, narrow chambers and indirect heating. They were developed initially from the inclined 'Munich chambers', then followed the vertical chambers, and eventually the horizontal chambers which resemble in very large plants the coke ovens of coking plants and hold up to 13 tons of coal. Medium-sized plants use smaller vertical-chamber ovens.

Five ovens with six chambers each, holding 1·5 tons, i.e. with a total charge of 45 tons, are usual and adequate for the gas supply of a medium-sized town. They are heated by producer gas, and the carbonization period is about 12 hours. Super-heated steam is then injected generating water gas and reducing the temperature of the coke by 100°C. After discharge the coke is quenched with water.

An oven suitable for medium-sized and small gas works is the Koppers continuous oven with vertical chamber, built of silica bricks and with a through-put of about 10 tons of coal per oven per day. It is likely to become increasingly popular.

This oven (see Figure 11) consists of a chamber that is heated indirectly by producer gas. Every four hours a batch of coal is fed from a hopper through an airtight charging valve into the oven. At the same intervals of time coke is discharged from the bottom of the chamber into a truck. Steam, fed into the bottom part of the chamber containing the devolatilized incandescent coke, is converted into water gas in the top part. The water gas fraction can be controlled by adjusting the steam supply. From the upper part of the oven the tar is extracted into the tar collecting tank; the gas escapes at the top.

## *Purification of gas*

Figure 11 shows a simplified diagram of gas purification. Crude gas contains dust, $NH_3$, $H_2S$, $CS_2$, HCN, $SO_2$, as well as benzene and its homologues, and naphthalene. These impurities can be removed by a relatively simple method.

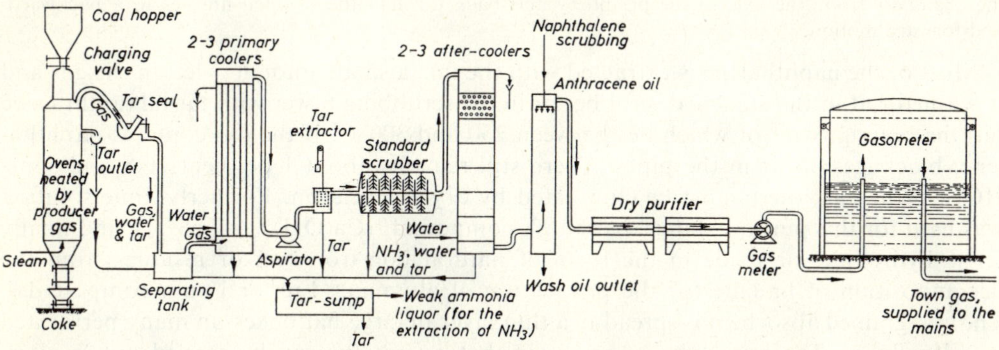


Figure 11. Flow diagram of a gas works

After cooling to about 80°C by injection of water or weak ammonia liquor the gas passes via a water seal (tar seal), where most of the tar and liquor is separated out, into the primary cooler (e.g. a cooler with air-cooled annulus, or with water jet scrubbing, or a tubular cooler). From the tar seal the greater part of the tar passes through a decanter into the tar sump. In the primary coolers cooling down to about 30°C takes place. A large part of the water condenses here and passes partly into the decanter with the tar and partly into the tar sump directly. From the primary coolers the gas passes either into further coolers where it is cooled down to 15–20°C (in gas works practice Reutter coolers are used) or first into tar fog extractors and then into standard ammonia scrubbers (usually brush scrubbers).

A Reutter cooler consists of iron boxes placed one on top of the other, into the front walls of which were rolled numerous horizontal cooling tubes. These cooling tubes through which the water flows are so arranged that the gas is in contact with as many of them as possible. They are particularly suitable for the cooling of gases. The numerous tubes act as baffle plates causing an efficient separation of entrained liquids. The gas enters at the top and escapes from the cooler at the bottom; the cooling water, on the other hand, rises from the bottom to the top through elbow pipes from one box to the next. The outer walls of the cooling tubes are, moreover, sprayed in the gas chamber with weak ammonia liquor which is discharged from the cooler at the bottom, together with the tar. The same system is used in many producer gas and water gas plants.

Between the primary and the after cooler there is often a fan (exhauster) and the tar extractor. In Pelouze and Audouin's tar extractor, the gas in which the tar is suspended in the form of fine droplets (aerosol) issues from numerous small perforations of a metal sheet. Opposite each perforation is a baffle plate.

The perforated sheets and baffle plates are suspended in the shape of a bell from a rope with counterweight and partly immersed in a circular tank filled with water. The pressure drop which determines the degree of purification, can be adjusted by the counter-weights, i.e. by raising or lowering the extractor bell.

The final operation is scrubbing which removes mainly $NH_3$, HCN, naphthalene and other impurities. This is done in the standard scrubber (Bamag-Meguin) in which the gas either passes through a liquid curtain or is led past very large liquid surfaces which are being constantly renewed, as is also done in the Feld and Ströder scrubbers.

The standard scrubber consists of a horizontal cylinder divided into separate chambers. The partitions are recessed in the centre through which passes a rapidly rotating shaft with, in each chamber, a pair of metal discs, surrounded by wooden hurdles. The scrubber is half-filled with the washing fluid which is sprayed by the rotating discs on to the wooden hurdles. The gas passes successively through all the chambers from the axis to the periphery and back through the wooden hurdles in a backward-and-forward motion.

Most of the naphthalene is extracted with the tar; a small amount is left in the gas and is washed out in the standard scrubber or in the scrubbing tower with tar oil (anthracene oil) the boiling point of which lies between 250 and 300°C. If the gas contains naphthalene, blockages occur in the pipes. There still remains about 1 per cent $H_2S$ and some HCN in the gas; these must be eliminated by dry purification. Formerly, lime solution was used for this purpose, later 'Lanning's compound' ($Ca(OH)_2 + FeSO_4$), and finally ferric hydroxide, or oxide in the form of natural bog iron ore, or residues from the decomposition of bauxite for the production of alumina, 'Lux or Lauta compounds'. The fine-grained absorbent is spread in a thin layer in large flat boxes on many perforated wooden floors. The boxes are so arranged that operations can be carried out in alternate layers. The chemical process is as follows:

$$2\,Fe(OH)_3 + 3\,H_2S = Fe_2S_3 + 6\,H_2O$$

The absorbent is regenerated with limited quantities of air in which the ferric oxide itself acts as a catalyst and converts the $Fe_2S_3$ back into $Fe_2O_3$, with simultaneous formation of sulphur:

$$Fe_2S_3 + 3\,O = Fe_2O_3 + 3\,S$$

Another method is to supply continuously 1–2 per cent of air with the gas into the boxes; the reaction then is:

$$H_2S + O = H_2O + S$$

The gas then contains 8 per cent of $N_2$. When 40–55 per cent of sulphur have accumulated in the absorbent it is spent and must be replaced. It is used for the production of sulphur or sulphuric acid.

The gas still contains most of the benzene. After the dry purification the benzene can be washed out in a scrubbing tower with wash oil which can be used again after the benzene has been driven out.

The purified gas passes through a gas-meter (see Figure 52) into a gasholder and from there through a pressure-regulating valve into the grid. Two types of gasholders are commonly used: wet bell-shaped gasholders (see Figure 11) and dry disc-type gasholders (M.A.N.). In the latter type a horizontal disc which is sealed with tar against the wall of the gasometer moves like a piston in the cylindrical gas chamber. The largest gasholder of this type built so far holds 600,000 $m^3$.

By-products of town gas manufacture are ammonia, tar, gas, coke, spent gas purification material, and benzene which is sometimes produced separately. Ammonia is usually combined with sulphuric acid and made into ammonium sulphate. This ammonium sulphate produced in gas works was the first 'synthetic nitrogen fertilizer'. Tar is supplied to tar distillation plants. It is similar to coke-oven tar and is used for the manufacture of the usual tar products (see under Carbonization of bituminous coal).

*Scrubbing and purification of gases*

Gas is scrubbed by passing it over large liquid surfaces. Either liquid films are produced by spraying the liquid countercurrent to the gas flow in a scrubbing tower filled with packing material (packed tower), or the gas is passed through bubble tray columns. These scrubbing devices offer the gas a certain resistance. The gas can also be passed through towers in which the scrubbing liquid is atomized.

Of great importance for the scrubbing of gases and even more for the precipitation of droplets (mist) and solid particles (dusts) is a continuous change of direction of the gas flow by letting it impinge on deflecting surfaces. Tubular coolers with horizontal tubes normal to the gas flow are, therefore, more efficient than the more common tubular coolers with vertical tubes (see pp. 52 and 66, e.g. Reutter coolers, Bamag-Meguin).

Considerably more efficient are special scrubbers in which the liquid is spread over very large surfaces, such as the standard scrubber, already discussed on p. 32, or those in which the gas passes through a liquid curtain. The Ströder scrubber (the 'crossed-curtain scrubber', made by the Bayer Co., Leverkusen) contains in one housing two shafts rotating in opposite directions, to which are attached spray discs; these eject the scrubbing liquid tangentially so that two crossing mist curtains are formed. The discs are staggered and dip into the liquid. There is no pressure drop as in scrubbing towers. A great number of different scrubbing liquids can be used, and also drying agents, e.g. concentrated sulphuric acid for drying chlorine. The largest Ströder scrubbers have a gas output of 20,000 $m^3/h$.

Other scrubbers with rotating discs or plates producing liquid mist curtains are Theisen and Feld scrubbers, employed mainly in carbonization plants and gas works. They are used for instance for the precipitation of atomized tar and work even with as little as 50 $mg/m^3$ of gas.

Gas scrubbers can be used for the precipitation of dusts, the condensation of mists, and the washing out of soluble gases (absorbents), e.g. hydrochloric acid from the waste gases produced in chlorination processes.

Dusts are also eliminated from gases by mechanical aerosol precipitation (L. Dietrich, *Mechanische Industrie-Entstauber*, *Chem. Ing. Tech.* 1953, **52**, 185):

(1) gravity
(2) shock and impact
(3) centrifugal force
(4) filtration
(5) electric charges

The choice of method depends mainly on the particle size and the velocity of the gas flow. A dust is characterized by the particle size distribution, by measuring of which a characteristic curve is obtained. Precipitation by gravity is carried out in stilling basins (dust chambers). This method can be used only for dusts with diameters of more than 200 $\mu$. By placing several chambers in tandem other separation similar to wet separation (stream classification, Figures 2, 3, 4) can be carried out.

Dusts can also be separated by deflecting and impacting the gas flow on surfaces. Numerous such designs exist, all based on splitting up the gas flow into many narrow streams which are suddenly deflected by baffle plates on which dust particles up to 50 $\mu$ are precipitated.

The most commonly used separator, the cyclone, is based on precipitation by centrifugal force.

The gas flow enters the upper part of a cylindrical vessel with conical bottom tangentially. The particles are driven outwards by centrifugal force whilst the gas flows in a spiral downwards along the walls. The particles are precipitated on the wall and slide downwards into the conical discharge tube. The revolving gas flow is then reversed, rises through the centre of the cyclone and escapes by the upper discharge tube.

In addition to the large standard cyclone many special cyclones using very ingenious gas distribution methods have been developed. Particularly efficient are multiclones (Lurgi), multiple-cell separators consisting of many small cyclones. The gas is violently rotated by revolving guide vanes and passes into many small cyclones where the dust is precipitated and collected by collecting pipes. This is, therefore, a centrifuging method.

Very fine dusts are best filtered in fabric filters wich can be either pocket filters, frame filters, or tube cloth filters. The dust adhering to the filters is dislodged by shaking and tapping devices.

*Electrical dust collection*

In almost all cases where large quantities of gas containing very fine dusts or mists are to be cleaned, the scrubbers are combined with electro-filters (EGP filters). On the basis of Hohlfeld's experiments, Cottrell (United States) and Möller (Germany) developed industrial equipment. This became possible when Siemens produced machines for the generation of high-voltage direct current, with voltages of 30,000 to 80,000 V. Alternating

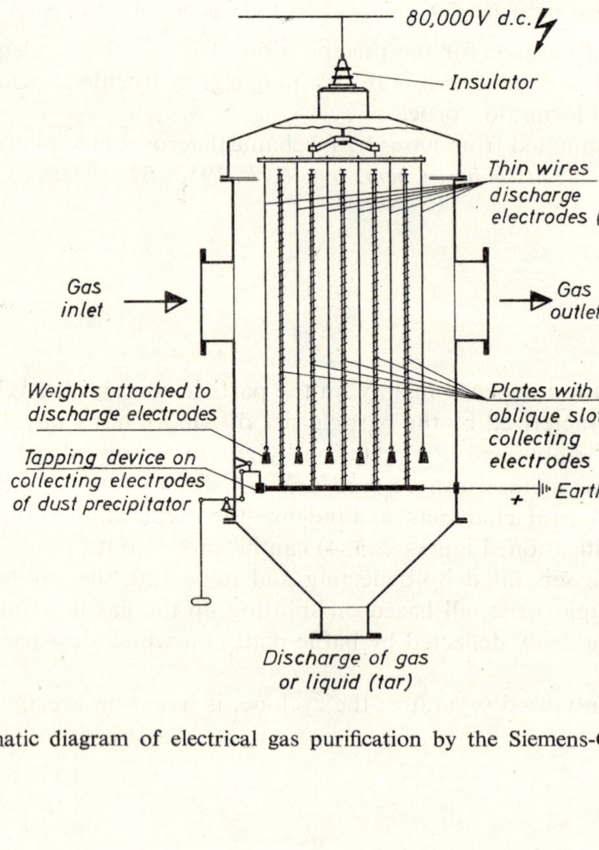


Figure 12. Schematic diagram of electrical gas purification by the Siemens-Cottrell-Möller method

current synchronous rectifiers are most commonly employed. Electrical dust collection by the Siemens-Cottrell-Möller method has been in general use since 1906 (see Figure 12) (*Erdöl und Kohle*, 1955, **8**, 802).

There are tube separators and chamber separators. In the latter type, chrome-nickel wires, several metres long, strained by weights and electrically insulated, are suspended in chambers. These are the discharge electrodes. Next to them are metal sheets or wire nets, the collecting electrodes. The discharge electrodes are connected to the negative pole of the d.c. generator, the collecting electrodes to the positive pole; the latter are earthed.

A corona glow discharge is produced around the discharge electrode and the gases are ionized. The dust or mist particles acquire a negative charge and migrate to the collecting electrode where they are discharged and precipitated. Gravity or vibration cause them to drop into a collecting hopper from where they are discharged.

LITERATURE

Ost-Rassow, *Lehrbuch der Chemischen Technologie*, 25th ed., J. A. Barth, Leipzig, 1953.

Schuster, *Energetische Grundlagen der Gastechnik*. Knapp, Halle, 1950.

Krist, *Die festen Brennstoffe*. Fachbuchverlag, Leipzig, 1951.

Thau, *Die Stadtgasindustrie*. VDI-Verlag, Berlin, 1935.

Ullmann, *Encyklopädie der Techn. Chemie*, 3rd ed., vol. 4, p. 663, Urban & Schwarzenberg, München–Berlin, 1953.

## Coking, low temperature carbonization

Two methods can be used for the treatment of coal: carbonization, i.e. coking and low temperature carbonization, and direct hydrogenation, or the hydrogenation of the products of coal distillation. To these are added gasification for the production of fuel gases, of gases for chemical synthesis, and of hydrogen. Carbonization can be considered only partly as a refining process since the coal constituents are completely decomposed and tar, petrol, etc., are obtained only as by-products. On these fundamental industrial processes of coal carbonization, hydrogenation, and gasification is based the greater part of the industrial organic chemistry of aromatics and to a smaller extent that of aliphatic materials. Figure 13 shows the industrial processes used for the production of basic aliphatic materials from the raw materials coal and petroleum.

The aromatic basic materials obtained by the carbonization of bituminous coal and also, although to a lesser extent, by the low temperature carbonization and hydrogenation of brown coals, particularly by the aromatization processes described later, are not included in the diagram; they will be treated in the appropriate chapters. On the other hand, the diagram shows the production of aliphatic materials from wood and carbohydrates which, together with coal, crude oil and natural gas, form the third raw material basis of industrial organic chemistry.

## Chemical processing of bituminous coals

Initially, bituminous coal was carbonized only for the production of metallurgical coke to be used in the smelting of ore. This is the oldest form of treatment of coal and its origins go back to the 16th century. Tar and gas were at first by-products and only later became the basis for the whole of aromatic chemistry. The low temperature carbonization of brown coal on the other hand was in the beginning used mainly for the manufacture of hard paraffin as a substitute for stearin in the production of candles, and coke

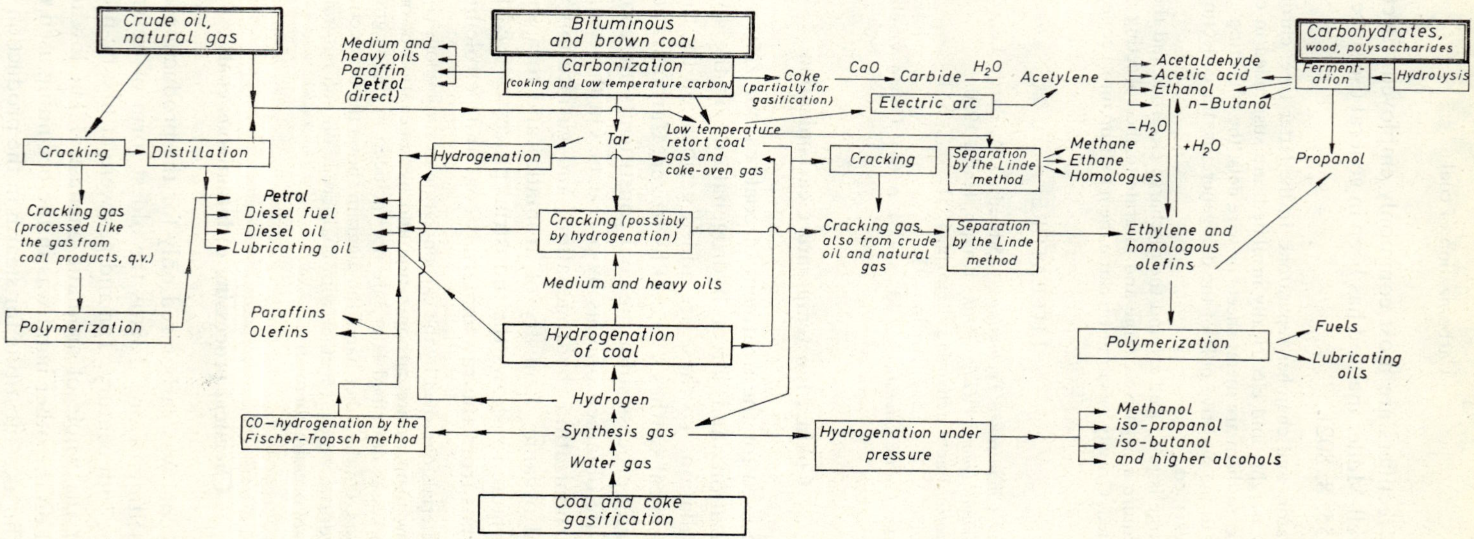


Figure 13. The sources of natural raw materials for the basic aliphatic materials and diagram showing their production

was a by-product used later for the manufacture of fuel for ships. The by-products, bituminous coal tar and coke-oven gas, brown coal for the low temperature retort process and brown coal tar oil, later became the basis for the development of the aliphatic and aromatic materials chemical industry in Germany. In countries where crude oil was produced it became the basis for the aliphatic, and partly also for the aromatic, chemical industry. The following diagram shows the possibilities for chemical processing of bituminous coal (*Erdöl und Kohle*, 1955, **8**, 382).

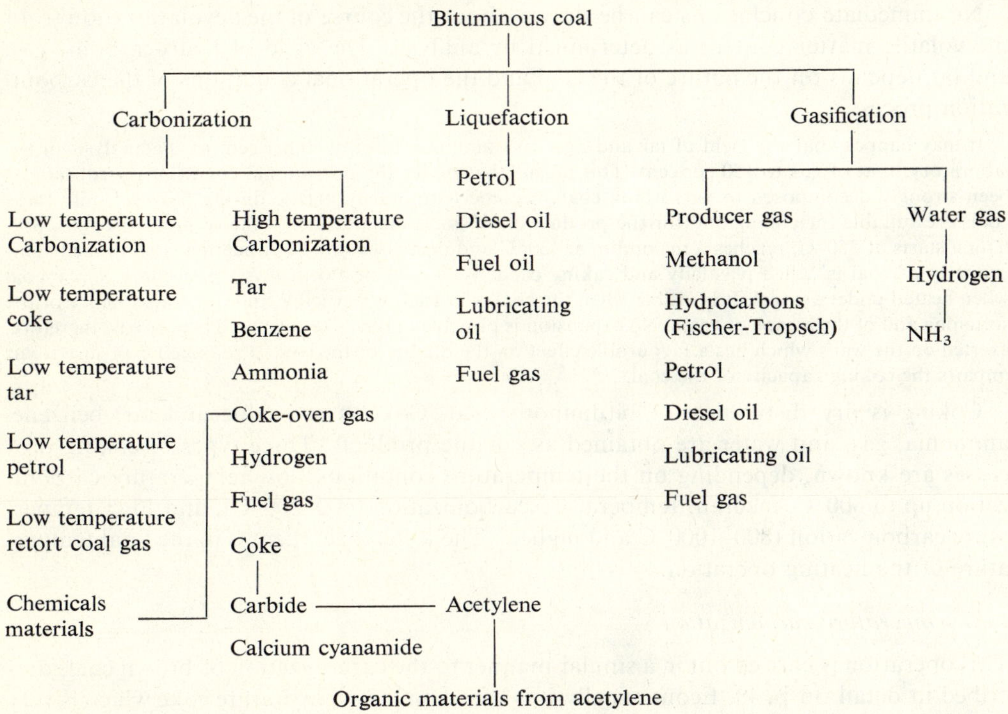


The composition, the properties, and the behaviour during heating are the determining factors for the treatment of bituminous coal. Many methods have been developed to characterize the behaviour of the coal for various purposes. The only factors of interest for coal as a fuel are its content of water, ash, and volatile matter, and its calorific value. For coking, low temperature carbonization, and gasification, however, other factors are of importance, e.g. the course of devolatilization, the yield of hydrocarbons, the bitumen content, softening capacity, and the pressure exerted during heating and sintering.

Bituminous coal, unlike brown coal, contains little water (3–6 per cent). The ash (3–8 per cent) consists partially of the mineral constituents of the original vegetation, and this part cannot be removed. Most of the ash, however, originates from the overburden and the accompanying rock, and can often be removed by the processes described above. The type and quantity of ash determine the slag formation. Ash consists of: $SiO_2$, $Al_2O_3$, $Fe_2O_3$, CaO, MgO, $K_2O + Na_2O$ in the form of sulphates, and traces of many other elements. A comparison of the composition and the calorific values of brown and bituminous coals has already been made in the table on p. 13.

The volatile content depends mainly on the age of the coal and the course of the metamorphosis of the coal. Brown coals, being the youngest coals, least changed by pressure and high temperature, have the highest volatile matter content.

The decisive factor for coking of bituminous coals is their yield of tar, and they are, therefore, roughly classified by their volatile matter content:

| | |
|---|---|
| Free-burning coal | >35 per cent |
| Gas coal | 28–35 per cent |
| Fat or coking coal | 19–28 per cent |
| Steam coal | 12–19 per cent |
| Lean coal | 10–12 per cent |
| Anthracite | <10 per cent |

No immediate conclusions can be drawn about the course of the devolatilization from the volatile matter content as determined by analysis. The yield of hydrocarbons, gas, and tar depends on the nature of the coal and the operational conditions of the carbonization process.

It may happen that the yield of tar and light oils amounts to only 50 per cent of the analysis in the laboratory, that of gas to 150 per cent. This means that under the operational conditions used, tar has been strongly decomposed to gas. Many coals become temporarily plastic during heating; only these coals are suitable for coking and for the production of good metallurgical coke of great strength. Softening starts at 350°C, reaches a maximum at 420°C and decreases rapidly if heating is continued. This property of coal is called plasticity and caking capacity. Connected with it is the swelling of the coal when heated under air seal. It is visible when the coal is heated in a crucible and is a consequence of the softening and of the escape of gases. No expansion is possible in the coke-oven and pressure is, therefore, exerted on the walls which has a favourable effect on the quality of the coke. Prolonged exposure to air impairs the coking capacity of the coal.

Coking is dry distillation of bituminous coal. Coke is a residue and tar, benzene, ammonia, gas, and water are obtained as volatile products. Three types of coking processes are known, depending on the temperature conditions: low temperature carbonization up to 600°C, medium temperature carbonization (600–800°C), and high temperature carbonization (800–1000°C and higher). The difference, then, is in the final temperature of the heating operation.

*Low temperature carbonization*

This operation is carried out in a similar manner to the carbonization of brown coal, described in detail on p. 48. Economically it is based on low-temperature coke which is very reactive and has many specialized uses. The yield of volatile components is smaller than in the low temperature carbonization of brown coal in which the volatile components carry the main part of the production costs. The processes occurring during low temperature carbonization are the same as during high temperature carbonization in the initial stage up to 600°C:

Stages of low temperature carbonization:

| | |
|---|---|
| 100°C | water is eliminated |
| 200°C | gases contained in the coal begin to escape |
| 300°C | onset of decomposition |
| 375–450°C | beginning of tar formation proper |
| 550–600°C | tar formation complete, water of decomposition and gas formation |
| 600°C upwards | $NH_3$ in larger quantities |

Of great importance for the quantity and nature of the volatile components is their secondary decomposition in the ducts after they have left the low temperature carbonizer.

Two groups of low temperature carbonization methods are distinguished by the type of heating used:

(1) Carbonization through heated surfaces, with indirect heating, i.e. heat transfer from the fuel gases across the walls of the oven, and
(2) Carbonization with circulating gases in which heat is transferred from the fuel gases direct to the coal.

Up to the present, the equipment used for the carbonization of bituminous coal through heated surfaces was different from that used for brown coal.

The carbonization cells consist of narrow chambers of 2–4 m length, 2 m height, and 30–120 mm width. The walls are made of cast steel or firebricks, with flues heated by gas at 600–650°C. Charging is done automatically by hopper trucks; the coke is discharged into the coke truck by opening a bottom plate of the chamber. Carbonization takes 2–6 hours. An oven with six cells produces about 12 tons in 24 hours. A low-temperature carbonization chamber oven is, therefore, very similar in design to the coke oven of coking plants, the latter is, however, considerably larger. The volatile components escape through the top of the oven and are utilized as described in the section on the low temperature carbonization of brown coal.

Carbonization with circulation gases and the design of the Lurgi oven is also described in detail, together with the production of tar and low-temperature petrol, in the section on the low temperature carbonization of brown coal.

The following table shows the yields from low temperature carbonization and coking of bituminous coal:

| Products obtained | Low temperature carbonization (per cent) | Coking per cent |
|---|---|---|
| Coke | 75–85 | 73·0 |
| Tar | 6–12 | 3·5 |
| Light oils | 1·0 | 1·2 |
| Gas | 6–15 | 19·5 |

Carbon contents (per cent)
Bituminous coal, 85; low temperature coke, 90; high temperature coke, 97.

The table shows clearly that with low temperature carbonization an intermediate stage is reached which, depending on the method of operation used, is subject to wide variations, whilst high temperature coking leads to a terminal condition, with a much more uniform composition of the distillation products. Moreover, there are great differences in the nature of the tar produced by carbonization with indirect heating and of that produced by carbonization with circulation gases. These are characterized mainly by the fact that in carbonization with indirect heating cracking causes secondary reactions at the hot oven walls. Bituminous low temperature tars and oils contain only very small amounts of aromatic hydrocarbons which during high temperature carbonization form subsequently in larger quantities at the hot coke oven walls and in the ducts. Low temperature petrol contains many unsaturated compounds and tends to resinify.

The formation of resinous products in petrol is called 'gumming'. It is the result of the polymerization of olefin groups, introduced by peroxides which form by autoxidation, and of Diels Alder reaction of dienes. Gumming can be prevented by the introduction of antioxidants which stop the polymerizing chain reaction introduced by peroxides. An efficient antioxidant for motor fuel is for instance p-benzyl-amidophenol.

As with brown coal products, high quality diesel oils and petrols are produced only by additional hydrogenation, i.e. by hydrogenative refining.

Economically, there is no great difference between low temperature carbonization and coking of bituminous coal. Both methods can supplement each other since in low temperature carbonization use can be made of coals which are less suitable for coking. Low temperature carbonization is, however, used only to a small extent.

Medium temperature carbonization has not so far found industrial application.

*High temperature carbonization*

Until the middle of last century iron was produced with charcoal and a large part of the forests of most industrial countries has fallen victim to pile charring. As early as 1850, 80,000 tons of coke were produced from bituminous coal in Germany, rising to 12 million in 1900, 35 million in 1913, and 50 million in 1940 (in each of the years 1953 and 1954 the German Federal Republic produced about 40 million tons in coking plants and gas works). Technical progress in the coking industry consisted mainly in the development of oven units of ever increasing size, the acceleration of the process, the production of better quality coke, and the improvement in the yield of liquid by-products and gas. Whilst the first coke oven had a capacity of 4–6 tons of coal, with a coking time of 48–72 hours, modern ovens hold 16–30 tons and a coking period of 14–24 hours is sufficient.

To be certain of the high quality of the coke to be produced it is necessary to use well-cleaned coal of uniform quality. Such coal is made available by preparation. The only suitable coals are those which soften at a certain temperature, i.e. have caking capacity, such as the fat coals.

In the Zwickau coalfield (German Democratic Republic) only coal slacks are produced; these cannot be carbonized directly. A method for the coking of coal slacks has been developed there which consists in tamping the coal firmly with a special machine into the hopper truck before the latter is pushed into the oven chamber.

The coke oven is a horizontal chamber oven of 12 m length, with a height of 4–6 m, and 45 cm width. 20–30 such ovens form a battery, and batteries form groups, large plants consisting thus of upwards of 500 ovens. The average yield of an oven of this size is 20 tons of coke in 24 hours, i.e. with 500 ovens the output of the whole plant is 10,000 tons of coke per day. The lateral walls of the coking chamber are built of refractory silica bricks and contain vertical heating flues in which burn gas burners. The gases have a temperature of 1,200–1,400°C.

A large number of coke ovens has been developed, differentiated mainly by the method of heating, or by the arrangement of the flues for the heating gases and the combustion air. They all have in common that the waste heat of the gases from the combustion chambers is conducted over brick chequer regenerators and is utilized for the pre-heating of the air and, in combination ovens, of the lean gases (producer gas, waste gas). Figure 14 shows the design and method of operation of a regenerative coke oven.

The combustion flues flanking the oven are divided longitudinally into two halves. One half of these, together with the corresponding halves of all the ovens in the battery, is heated by gas and pre-heated air, and the waste gases pass through the second half of the flues into the stack, having heated the corresponding regenerator (heat storage) group. The process is reversed after 1/2–1 hour, and the second half of the flues is heated whilst the waste gases pass through the first half (see Figure 14).

The regenerative coke oven is heated by the gas of the coke oven block. The gas is burnt in the burner chambers which stand in a row between the coking chambers and heat their walls. Combustion air which has been pre-heated to about 1,000°C in the recuperators (regenerators) underneath is fed into the burner chambers. The hot waste gases are collected at the top in a horizontal duct, pass through the heating flues of the second half of the ovens from top to bottom, pre-heat the regenerators and are collected in a lower horizontal outlet duct (first period). After reversal of all inlets and outlets the

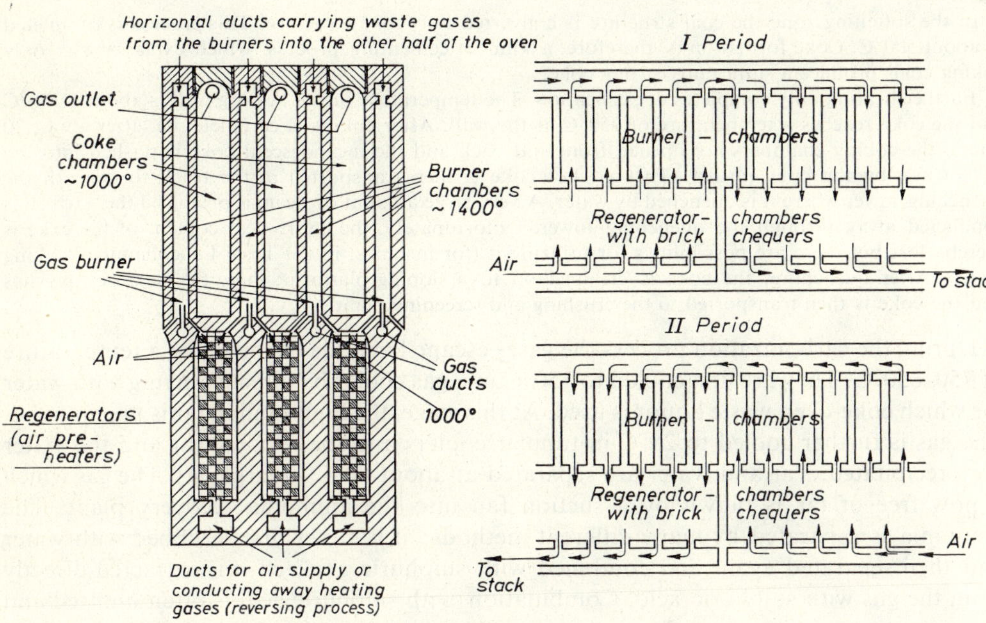


Figure 14.
Left: Cross section of regenerative coke oven.
Right: Path of heating gases in a regenerative coke oven during the two periods of heating

gases flow in the opposite direction (second period). The flue and the access of air for each heating wall must be very carefully adjusted by means of dampers or, for the gas feed, valves. Combination ovens are designed in a very similar way but they can be heated by their own gases of about 5,000 kcal/Nm$^3$ or by lean gases of about 900–1,000 kcal/Nm$^3$. This releases part of the valuable coke-oven gas for other purposes.

Coke oven technology has been particularly highly developed in Germany by famous firms, and from this progress have benefited coke oven plants all over the world. The numerous designs of coke ovens differ mainly in the type of heating used, i.e. the arrangement of the ducts for the combustion gases, and of the combustion chambers and firings, the development of the flames, multi-stage combustion of the gases to achieve as far as possible uniform heating of the chamber walls, care of furnace linings, saving of heating gases, and utilization of gases of the lowest possible quality.

After lifting off the lid, the oven is charged from a hopper truck which is supplied by a coal bunker. Then the oven is lit, or in continuous operation the heating is continued, and carbonization takes place slowly from the heated walls towards the interior. A softened zone, parallel to the heated walls, proceeds gradually from the walls between the finished coke and the uncoked coal towards the centre in a horizontal direction.

There are, therefore, two softening zones which eventually meet in the centre, so that all stages in the transformation of the coal up to the finished coke co-exist almost simultaneously in the chamber. Similarly, the channels in the charge through which the gaseous products escape proceed through the mass.

In the softening zone the coal structure is converted into coke structure. This process is completed at about 500°C. Coke formation is, therefore, a kind of cementing process, which explains why only coking coals produce a sufficiently strong coke.

Further heating only completes degasification. The temperature of the heating wall is about 1,350°C and the coke reaches a temperature of 950°C at the wall. After coking is complete, i.e. after about 20 hours, the coking chambers are opened front and back and the incandescent coke is pushed into the coke truck by the coke pusher. The red hot coke is then transported in the truck to beneath the quenching tower where it is quenched by water. A very large amount of steam is produced there which is conducted away through the quenching tower. Unfortunately, the entire heat content of the coke is thereby lost but there are possibilities for utilizing it (for instance, in the large Lauchhammer coking plant). After quenching, the coke is dropped on to a sloping platform; there the steam evaporates and the coke is then transported to the crushing and screening plant.

During the carbonization process the gases escape from the chambers at a temperature of 750–850°C. They are first cooled to 150°C in a gas cooler by direct spraying with water for which coke-oven waste liquor is used. At this stage, the high-boiling tar is precipitated. The gas is further cooled to 25°C in tubular coolers and the residual tar and the water are precipitated. Tar and water are separated at about 80°C in large pits. The gas which is now free of tar is blown by a suction fan into the ammonia recovery plant. The ammonia is separated by many different methods; it is either first scrubbed with water and then separated again and combined with sulphuric acid, or it is extracted directly from the gas with sulphuric acid. Combination with sulphuric acid is often omitted and aqueous ammonia is given off.

Benzene and its homologues are contained in the gas in the vapour state, about 25–40 g/m$^3$. Its recovery takes place in scrubbing towers of 20–30 m height, lined with wooden hurdles (spray towers); in these the gas is scrubbed with middle oil on the counter-current principle. Special high-performance scrubbers are also used.

The benzene is driven out of the wash oil with steam in stills. The crude benzene is then distilled once more and worked up to pure benzene, pure toluene, and xylene. 40 per cent of the residual gas which has a high calorific value is used for underjet firing of the coke-ovens. The remainder is supplied to industry and town grid. It is a valuable source of aliphatic compounds; methane, ethane, ethylene, and the higher homologues are produced from it for aliphatic syntheses. Since coke-oven gas is actually too valuable for firing, efforts are being made to replace it in the coking plants to a large extent by producer gas so as to be able to offer more coke-oven gas to industry.

The increase in the yield of crude tar and crude benzene illustrates best the technical progress in coking plants during the last few decades. The average yield in Germany was respectively: in 1925—3·5 and 0·875 per cent, in 1938—4 and 1·24 per cent. These figures may appear insignificant; they represent, however, a considerable improvement in the percentage yield. This improvement was achieved by modernizing the coke-oven plants, perfecting the condensing plant, and above all by constantly improving the chemical control of the coking process. The following table shows the coal output and coal production in million of tons, and the production of tar, crude benzene and nitrogen in the form of ammonia in Germany.

| Country | Bitum. coal in mill tons | Coke in mill tons | Tar[1] in 1,000 tons | Crude benzene in 1,000 tons | Nitrogen as ammonium sulphate |
|---|---|---|---|---|---|
| 1936: Germany (undivided) | 158·3 | 35·8 | 1,427 | 421 | 478 |
| 1947: German Federal Republic | 71·1 | 14 | 506 | 140 | — |
| 1951: German Federal Republic | 119 | 31·7 | 1,400 | 389 | 396 |
| 1954: German Federal Republic | 128 | 38·9 | 1,607 | 435 | 438 |
| 1961: German Federal Republic | 143 | 44 | 1,784 | 591[2] | — |

The production of gas in coking plants and gas works in the German Federal Republic in 1959 amounted to 20·5 billion $m^3$, GDR 3·17 billion $m^3$.

Output of bituminous coal, production of coke, and percentage of coal carbonized in the most important countries in 1950 and 1960

| Country | Coal in million tons | | Coke in million tons | | Coal carbonized (in %[3]) |
|---|---|---|---|---|---|
| | 1950 | 1960 | 1950 | 1960 | 1950 |
| BRD | 110·7 | 142·0 | 27·3 | 44·7 | 32·8 |
| U.S.S.R. | 220·0 | 375·0 | 30·0 | 56·2 | 18·1 |
| U.S.A. | 499·5 | 391·0 | 70·0 | 55·4 | 18·6 |
| Great Britain | 219·7 | 196·8 | 16·7 | 51·9 | 10·1 |

The first table shows clearly the decline in production in Germany due to the war and its subsequent growth. The second table shows that the proportion of bituminous coal that is carbonized is higher in Germany than in any of the other countries. The output of coke in the German Federal Republic had by 1952 again reached a daily rate of 100,000 tons which corresponds approximately to the level of production in 1936.

About 50 per cent of the coke produced is used in steelworks, mainly for the reduction of the iron ore in blast furnaces. During this reduction the carbon is not completely oxidized to $CO_2$; the waste gases contain a large amount of CO. These waste gases are very cheap although of low quality, and are used for underjet firing of coke ovens and as fuels for gas engines. Like iron ore, other ores are also reduced by coke to metal, e.g. zinc, tin, and copper.

*Constituents of bituminous coal tar and their extraction*

The bituminous coal tar from coke ovens and gas works is worked up centrally in a few processing plants (see p. 47). It contains a large number of chemical compounds, in particular aromatics, of which several hundred are very well known and have been thoroughly investigated. In his *Lehrbuch der Organischen Chemie*, 13th ed., Karrer lists 282 known compounds.

[1] Crude tar production in the German Federal Republic in 1958 was 1,893,000 tons
[2] From coke ovens and gas works
[3] In 1955, 114 million tons of bituminous coal were produced in the German Federal Republic (value 5·6 billion DM) of which 56 million tons were carbonized.

The distillation of crude tar yields up to 60 per cent residue (called pitch) and 40 per cent distillates (tar oils). The main constituent of refined products is naphthalene with nearly 6 per cent, then phenanthrene with 3·8 per cent, acenaphthene and fluorene with 1·5 per cent each, anthracene 1·1 per cent, cresol 1·2 per cent, carbazol about 1 per cent, benzene and homologues only 0·68 per cent. Most of the benzene and the benzene derivatives are contained in the gas (about 30 times the amount contained in the tar). All other compounds are present only in fractions of one per cent, but because of the enormous yield of tar the quantities are, nevertheless, considerable. Figure 15 shows some of the aromatics and other materials that can be produced from bituminous coal tar.

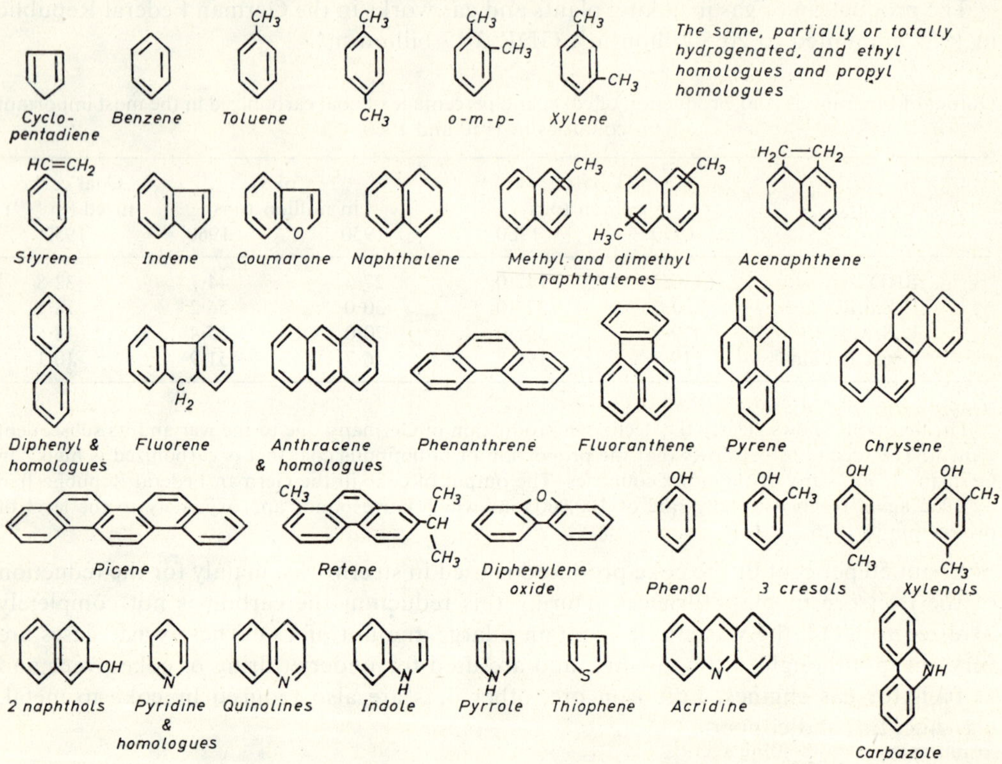


Figure 15. Some carbocyclic and heterocyclic hydrocarbons in bituminous coal tar

The tar industry (e.g. A.G. für Teerverwertung and the Rütgers Works) isolated these materials. Before the 1939–1945 war a separate chemistry of the so-called 'heavy hydrocarbons' had begun to develop. Many more uses were found by the intermediates and dyestuff industries and in the field of plant protection. The subsequent chemical treatment of the aromatic hydrocarbons is dealt with generally in a separate section on aromatic organic intermediates.

Gas works tars have a similar composition to tars from coke ovens but are much less uniform.

*Processing of tar*

After removal of 5 per cent of water the tar is distilled under a vacuum in large stills of 50–70 tons capacity. Figure 16 shows the process diagram of bituminous coal tar and the tar products obtained. The distillation residue is pitch for which there are many applications. The distillate is divided into four main fractions: up to 180°C light oils, 180–230°C medium oils, 230–270°C heavy oils, above 270°C crude anthracene oil.

The percentage composition of light oil, boiling up to 180°C, is as follows:

| | |
|---|---|
| Benzene | 45–55 |
| Toluene | 6–12 |
| Xylene | 2–5 |
| Naphthalene | 12–15 |
| Phenols | 8–12 |
| Pyridine bases | 1–3 |

Most of the benzene is extracted by washing the gas with medium oil. In 1938, Germany led world benzene production with 520,000 tons, U.S.A., 1938: 400,000 tons. Federal German Republic, 1954: 435,000 tons, 1958: 547,000 tons (U.S.A., 1958: 1,100,000 tons, of which about 650,000 tons from coking plants, the remainder from crude oil and gases by aromatization.) The production of aromatics in petroleum refineries continues to increase. In addition to the aromatic hydrocarbons, the crude benzene contains also partially hydrogenated benzenes, pyridine bases, and phenols. Formerly, concentrated sulphuric acid was used for the purification; this caused a loss of about 20 per cent when resinifying compounds were removed by polymerization and unsaturated compounds by sulphonation. For motor benzene the content of unsaturated compounds is, however, a favourable factor since they improve its anti-knock properties. For this reason, milder washing methods are used nowadays which result in a loss of only about 10 per cent. The distillate is purified up to 180°C in many small- and medium-sized plants by various methods which take into account the composition of the light oil. They all have in common washing by sulphuric acid and neutralization before fractionation, or first a preliminary fractionation followed by washing and neutralization of the fractions. Washing with diluted acid is used for the extraction of the pyridine bases and with diluted caustic soda solution for that of the phenols (tar acids).

The distillation residue of light oil is coumarone resin. It is produced by the sulphuric acid treatment (condensation) from coumarone and indene. It is an important cheap resin used for paints, lacquers, printing inks, and binding agents, and every effort is made to obtain it in as light-coloured, high-melting, and transparent a form as possible. Due to autoxidation it has a tendency to turn yellow, reacting with the oils used. It can be refined by hydrogenation, and it is also condensed with phenol, fatty acid chlorides, etc.

In the first light oil under 80°C it is possible to isolate cyclopentadiene and $CS_2$. As with all benzenes and oils attempts are being made to purify the crude benzene without chemical treatment. The most efficient of these methods is hydrogenating purification because no hydrocarbons are lost and only the olefins and hydro-aromatics are transformed into saturated hydrocarbons and sulphur compounds are desulphurized.

The yield of crude benzene depends on the carbonization temperature. When it is relatively high (e.g. 1,100°C), thermal dehydrogenation causes the formation from the hydro-aromatics, olefins, etc. of more benzene and, due to demethylation, fewer homologues. The toluene content varies, therefore, according to the carbonization conditions between 5 and 30 per cent. The approximate composition of a coke-oven benzene is: benzene 80 per cent, toluene 15 per cent, xylenes 3 per cent, higher solvent naphthas 2 per cent.

**Processing of bituminous coal tar**

(After P. Damm in Winnacker-Weingaertner)

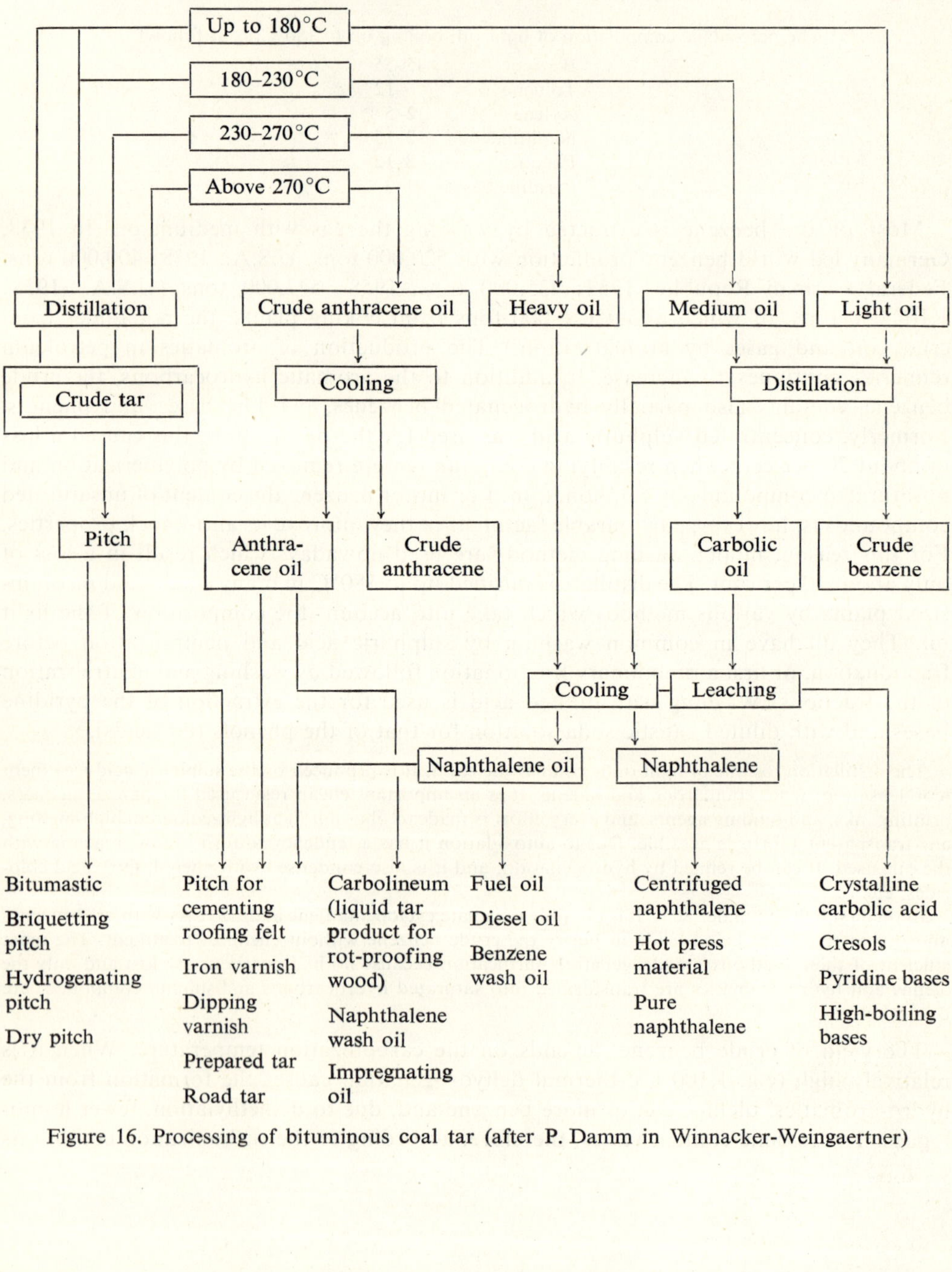


Figure 16. Processing of bituminous coal tar (after P. Damm in Winnacker-Weingaertner)

Benzene is used mainly in fuel mixtures as motor benzene when it is possible to use all the homologues in the same proportions as they are produced. Other products are pure distillates of benzene, toluene, xylene for chemical syntheses (about a quarter of the total production in 1938). In general, efforts are made to obtain as much toluene as possible. During periods of increased trinitrotoluene production (explosive) there was always a shortage of toluene. Syntheses have even been devised for the conversion of benzene into toluene. Frequently, however, alkyl benzenes produced petrochemically are now de-alkylated.

Medium oil which boils between 180 and 230°C, contains 30–40 per cent naphthalene, 15–25 per cent phenols, 5 per cent bases. It is usually treated by cooling in pans; naphthalene is crystallized out and is separated by centrifuging and pressing. This process produces 'hot pressed material'. Naphthalene is the cheapest aromatic raw material.

Prices for 100 kg, 1939: Naphthalene, crude, 6·50–7·50 RM, hot press material (ex Ruhr) 11–12·50 RM, pure, flakes, 19·00 RM, 1954 (German Federal Republic): naphthalene hot pressed material 32·50 to 36·00 DM, pure, flakes, 44·00–52·00 DM. For comparison, 100 kg '90' benzene costs (without duty) 1939: 24·25 RM, 1954: 36·00 DM; toluene pure (duty free), 1939: 44·00–45·00 RM, 1954: 153·00 to 163·00 DM.

The effluent from the hot pressed material (boiling range 160–1,210°C) is fractionated. A diluted solution of caustic soda is used to extract from it 30–40 per cent 'tar acids', i.e. phenols which are then precipitated with $CO_2$ from the alkaline solution.

Although the boiling point of pure pyridine bases is considerably lower, they are a constituent of medium oil since they have a loose bond with phenol. The dephenolized oil is washed with diluted sulphuric acid and yields bases, which are then liberated with ammonia, and fractionated. The second fraction of the medium oil again contains naphthalene and phenols which are separated with alkali.

The first fractions of heavy oil (230–270°C) are added to the medium oil. The second fraction, naphthalene oil 2, boils up to 280°C. Heavy oil contains the most important highly condensed aromatic hydrocarbons, alkylated naphthalenes, diphenyl, diphenylene oxide, acenaphthene, dinaphthyl, fluorene, naphthol, quinoline, indole, scatol, etc.

Anthracene oil (270–360°C), is cooled and 40–45 per cent crude anthracene is separated out, containing mainly carbazole (20 per cent), in addition to phenanthrene and other highly condensed aromatic hydrocarbons. Crude anthracene is an important raw material for the dyestuffs industry. For this purpose it is separated from carbazole and other hydrocarbons, either by purification with pyridine or by melting with caustic soda; in the latter case carbazole forms a sodium salt which is insoluble in certain organic solvents.

Pure anthracene can be oxidized to anthraquinone, the intermediate for many vat dyes, using either a vanadium pentoxide catalyst or chromic acid. At first, the carbazole-containing anthracene residues posed great problems, until methods were developed to produce from them dyes and plant protection agents, with the result that there is now at times even a shortage of carbazole.

Efforts are still being made to find uses for numerous other hydrocarbons. Anthracene oil and naphthalene oil 2 are used in the production of carbon black.

The largest German factories for the working up of bituminous coal tar are: VEB Teerdestillation and Chemische Fabrik Erkner and at Niederau, Dresden, Rütgers at Pasing, Munich, Rauxel in Westphalia, the Gesellschaft für Teerverwertung mbH, Duisburg-Meiderich (largest plant producing about 1 million tons per year), and two works belonging to Thyssen, at Meiderich and Rauxel.

Coke oven gas is becoming increasingly important in the fuel economy and as a raw material for syntheses: 40 per cent is used in the under-jet firing of coke ovens. By using it in combination with producer gas and blast furnace gas, consumption can be reduced considerably, resulting in an increased supply to the national grid. Towards the end of the 1939–1945 war 5,100 km of long-distance gas mains for the supply of industry and commerce had already been built in Germany.

Coke oven gas is fractionated by the Linde method for the production of hydrocarbons, either directly or after cracking for the extraction of olefins, in particular ethylene and propylene. It is also used for the production of synthesis gas, gas for the synthesis of ammonia, for the production of hydrogen for hydrogenation and for the production of synthesis gas for the Fischer-Tropsch synthesis. Finally, it is also used for the production of acetylene in the arc furnace. In the DBR, coke oven gas forms a valuable supplement to natural gas. Like brown coal gases, it is an important source of sulphur. In 1936, coke oven plants supplied 58 per cent of the German sulphur requirements (26,000 tons). The production of sulphur from coal amounted in 1960 to 280,000 tons in the DBR and to 112,000 tons in the GDR.

*Carbonization of brown coal, low temperature carbonization*

Low temperature carbonization of coal (see p. 38) is devolatilization for the production of coke (low temperature coke) and volatile products, both liquid (tar), and gaseous (low temperature retort coal gas) at an upper temperature limit of 700°C. At higher temperatures the tar decomposes excessively and the coke is not sufficiently reactive for further use. The volatile products are collected, and separated by distillation, producing low temperature tar, light oil, benzene, low temperature gas, and foul water. The residue is low temperature coke. During the 1939–1945 war about 20 million tons of pre-dried brown coal were used annually in Germany for low temperature carbonization, yielding 2·3 million tons of tar.

Formerly, the economic object of low temperature carbonization was the production of kerosene and hard paraffin. During the 1914–1918 war, low temperature carbonization was carried out in the German provinces of Sachsen-Anhalt and Thuringia for the production of low temperature tar as a raw material for marine fuel oil, diesel oil, and lubricant because of a shortage of petroleum products due to the blockade. A period of stagnation lasting twenty years followed since the cost of production could not be recovered from the liquid low temperature carbonization products. Development did not start again until demand for low temperature tar as a raw material for hydrogenation increased strongly and brown coal lump coke produced an adequate return as a smokeless stable fuel. Due to its high ash content brown coal coke was formerly not used for metallurgical purposes. However, at present a low temperature carbonization plant is in operation in the German Democratic Republic, at Lauchhammer, where metallurgical coke is produced from brown coal briquettes. This coke can be used in 'low shaft' furnaces for the smelting of iron ores (e.g. in the Iron Works Calbe a. d. Saale). ('Low shaft' furnaces are low wide furnaces with a short shaft. They have been developed since 1949 in the German Democratic Republic for the utilization of coke made from brown coal briquettes which is not as stable as the coke normally used for blast furnaces.)

Outside Germany, low temperature carbonization is hardly practised, only at Most (Czechoslovakia) is there a very large low temperature carbonization plant.

The low temperature carbonization process consists of the following stages:

(1) Residual drying of coal or briquettes
(2) Separation, by heating of tar, light oil, gas, water of constitution and reaction; the heat can be supplied from an external source
(3) Discharge and cooling of the coke, and rendering it inert

It proved impossible to supply the heat across the iron walls of the low temperature carbonization retort because of the strong corrosive action of tar and gas on iron. For this reason, low temperature carbonization was carried out for decades in the cylindrical retort designed by Rolle. This retort is lined with fireclay and heat is supplied externally. The yield of tar is poor (60 per cent) and output is low.

The calculation of the tar yield in low temperature carbonization retorts is based on a comparison with the yield obtained by standard distillation in an aluminium retort in the laboratory. A 100 per cent yield means therefore that it is the same as in the laboratory.

Great progress was achieved by the introduction of the coal treatment retort designed by Geissen-Kosag. This retort consists of a rotary heating cylinder, made of heat-resisting cast iron, heated directly by gas; the coal trickles downwards along the wall, being continuously turned over and kept in motion. Such indirect heating retorts are being built for daily throughputs of 150 tons of raw brown coal. Tar yield is 85–95 per cent.

A marked improvement was brought about by the Borsig-Geissen retort (Rheinmetall-Borsig). This retort consists of a vertical, rotating Sichromal steel pipe with a diameter of 40 cm and a height of 10 m, heated directly by a gas burner. The pipe is surrounded by a bell-type distributor which permits the coal to slide rapidly past the heated tube in a layer of only 1–2 cm thickness (Figure 17). Carbonization is complete within a few minutes. Tar yield is 110 per cent, throughput, however, is only 35 tons per day. The high tar output and relatively low gas formation are due to the fact that the volatile products are very quickly removed from the influence of the decomposing action of hot brick or metal walls. The amount of gas produced in the new Borsig-Geissen retort is, therefore, only about half that obtained in the old Geissen-Kosag retort.

The larger yield of high-quality tar is not the only advantage of indirect heating low temperature carbonization. The gas obtained is also very concentrated and rich in energy. By scrubbing it with water under pressure its calorific value can be increased from 4,000 to 7,800 kcal/m$^3$. This makes it possible to use it with particular advantage as town gas or for the production of saturated and unsaturated hydrocarbons. Thus, for instance, low temperature carbonization gas from Edderitz is supplied about 30 km to Dessau and fed into the central supply grid of Magdeburg. It is diluted with water gas to town gas standards.

Sulphur can be removed from the low temperature carbonization gases by dry purification with 'Lauta-compound' (ferric hydroxide from the decomposition of bauxite for the production of aluminium) which after exhaustion is extracted with $CS_2$. The gases can also be purified by wet scrubbing, e.g. by the 'alkazid' process (potassium salts of amino acids) or by the Koppers method with $K_2CO_3$ liquor.

This method of indirect heating low temperature carbonization is used only in a very few older-type plants where a larger number of retorts is combined into batteries sharing common feed and discharge installations. No new plants have been built for this process in recent years. Not the least reason for this is the fact that the heat-resistant special steel tubes can at present be made only in relatively small sizes, with the result that the throughput of a Geissen retort is small compared with the performance of a circulating gas retort (the Borsig-Geissen retort has a performance of about one tenth that of a Lurgi circulating gas furnace). If new special steels should make possible the construction of retorts with a much higher performance, the indirect heating low temperature carbonization process should again have a future since it offers a number of advantages. Alternatively, the carbonization of brown coal could be carried out in the units employed in the low temperature carbonization of bituminous coals; these retorts are built in batteries similar to those of coke ovens.

*Circulating gas low temperature carbonization*

In large low temperature carbonization plants the Lurgi recirculation gas process has been used exclusively for the past fifteen years. Its advantage lies in the fact that residual drying and low temperature carbonization are separated so that relatively small quantities of foul water are produced since it is possible to discharge the vapours from the drying zone and the foul water separately. Owing to the short residence period the tar is well preserved. Lump coke is obtained by briquetting of the coal. The performance of

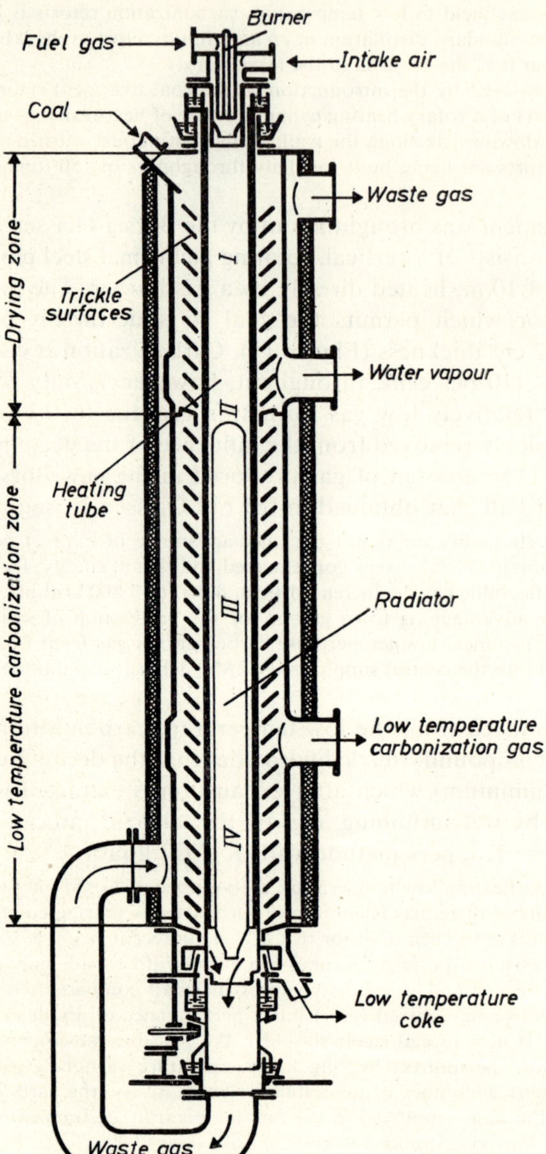

Figure 17

recirculation gas low temperature carbonization ovens, initially 300 tons per day, has been continuously increased in the course of the last few years and amounts at the present time to about 500 tons per day.

The Lurgi system can be used for both bituminous and brown coal. The low temperature retort coal gases are flushed from the carbonization zone into the foul gas main by the purified gas produced during the carbonization itself.

The Lurgi oven (Figure 18) consists of two zones with separate heating and two gas circulation circuits. At the top is the drying zone where the coal (e.g. stable brown coal briquettes) is dried to 0–1 per cent water. The upper part of the drying zone is a coal bunker into which coal is fed periodically from the top by a charging device. The heating gas produced by a burner and compressor circulates in the drying zone. The excess heating gas escapes through a stack together with the drying vapours. The low temperature carbonization zone proper is below the drying zone, connected to it by drying tubes. Through these connecting shafts which when properly adjusted permit the passage of only a minimum amount of gas, the dried coal slides into the low temperature carbonization zone.

The temperature in this zone is kept at 600–700°C by a circulating gas stream heated by a burner. An exhaust fan removes the volatile products from the upper part of the low temperature carbonization zone through a gas duct.

The lower part of the carbonizing retort forms a cooling zone for the coke. It is closed at the bottom by a coke discharge device which is in continuous reciprocating motion. The coke drops into an airtight coke lock from which it is removed by a belt conveyor.

Openings above the coke discharge permit the entry of cold pure gas from below into the coke cooling zone where it cools the gas to about 230°C and is itself pre-heated. Throughput of the retort is controlled only by the speed of the coke discharge device. A very careful control of the pressure drop to the outside air is necessary to prevent access of air into the retort.

In large plants a whole row of these low temperature carbonization retorts is placed side by side and their bunkers are supplied with briquettes from a travelling mechanical charger.

### *Extraction of the low temperature carbonization products*

(see Figure 19)

The distillate arrives first in the primary cooler where by direct atomization of waste liquor high-boiling, dust-laden tar is separated (primary cooler tar). Other tar fractions, in particular paraffin, are contained in the gas stream in the form of mists and can be precipitated by an electrostatic precipitator. Medium oil and foul water are condensed in water coolers. Finally, crude benzole (low temperature benzole) is washed out in a gas scrubber into which is sprayed a medium oil fraction which boils at about 220–300°C. The tar yield is 85–95 per cent (see p. 38).

Part of the residual gas is used for feeding the burners of the low temperature carbonization retort. The surplus gas is required in the plant for drying and heating purposes.

Figure 19 shows the products of low temperature carbonization of brown coal and their processing.

For purposes of hydrogenation it is sufficient to eliminate from the tar its water and dust content by settling, centrifuging, or by means of filters. Failing this, a rather complex working up by distillation and refining is necessary. The diagram shows only one of many possible systems of operation. Whilst for bituminous coal tar there is usually only one method of processing, the method used for brown coal carbonization products depends on the nature of the tars.

It is, for instance, fairly simple to extract hard paraffins from the low temperature tars produced in the old Rolle ovens. Strong decomposition occurs at the oven walls whereby the wax and resin constituents are dissociated into gas and coke. In modern ovens, milder conditions of carbonization preserve the wax and resin constituents. On the other hand, crystallization of the hard paraffin is prevented by these. The old-fashioned carbonization systems permitted, therefore, the separation of paraffin directly by cooling; modern systems require a deliberately destructive distillation of the tar.

The paraffin I (see diagram) is subjected to redistillation to destroy the residual wax and resin constituents. After preliminary separation into tar, medium oil, and light oil, which occur during carbonization, these three main products are worked up separately.

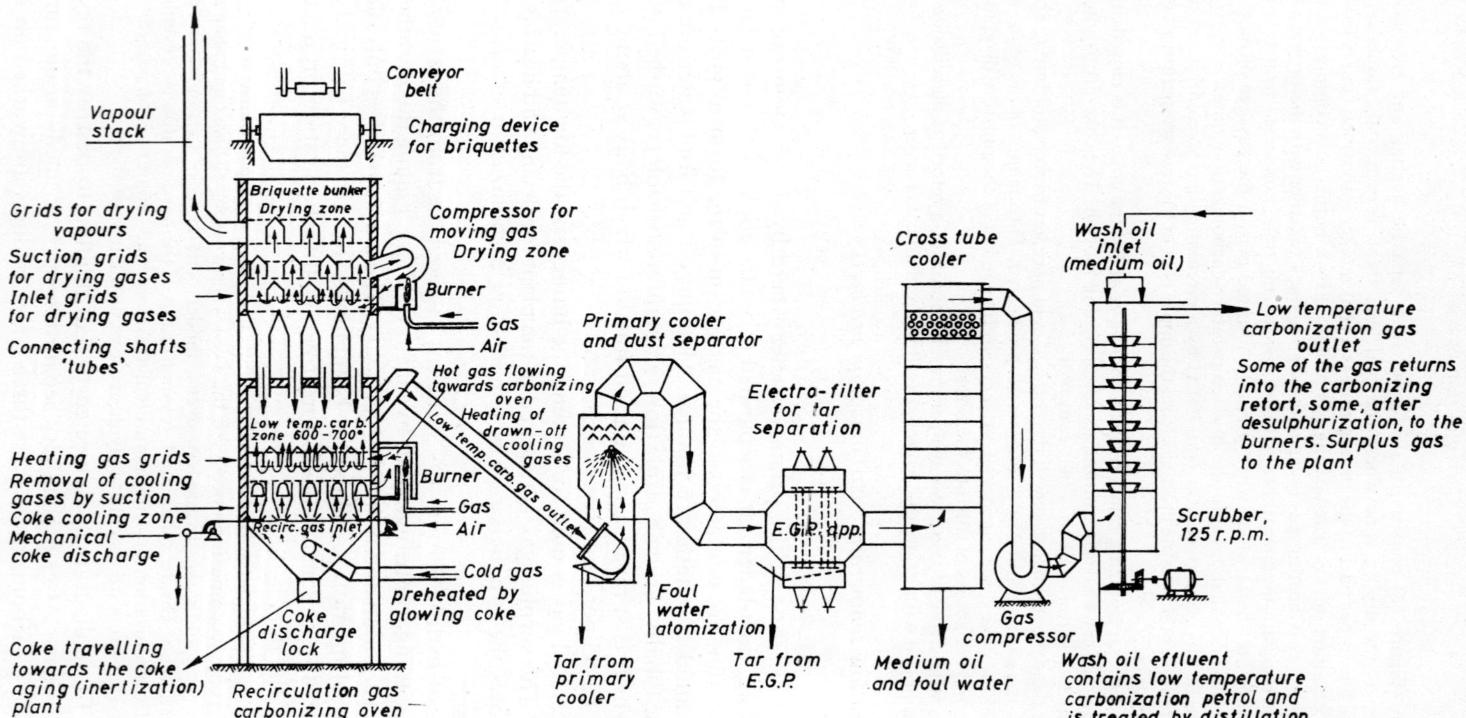


Figure 18. Diagram of a Lurgi circulating gas low temperature carbonization plant, with extraction of tar, medium oil and benzene

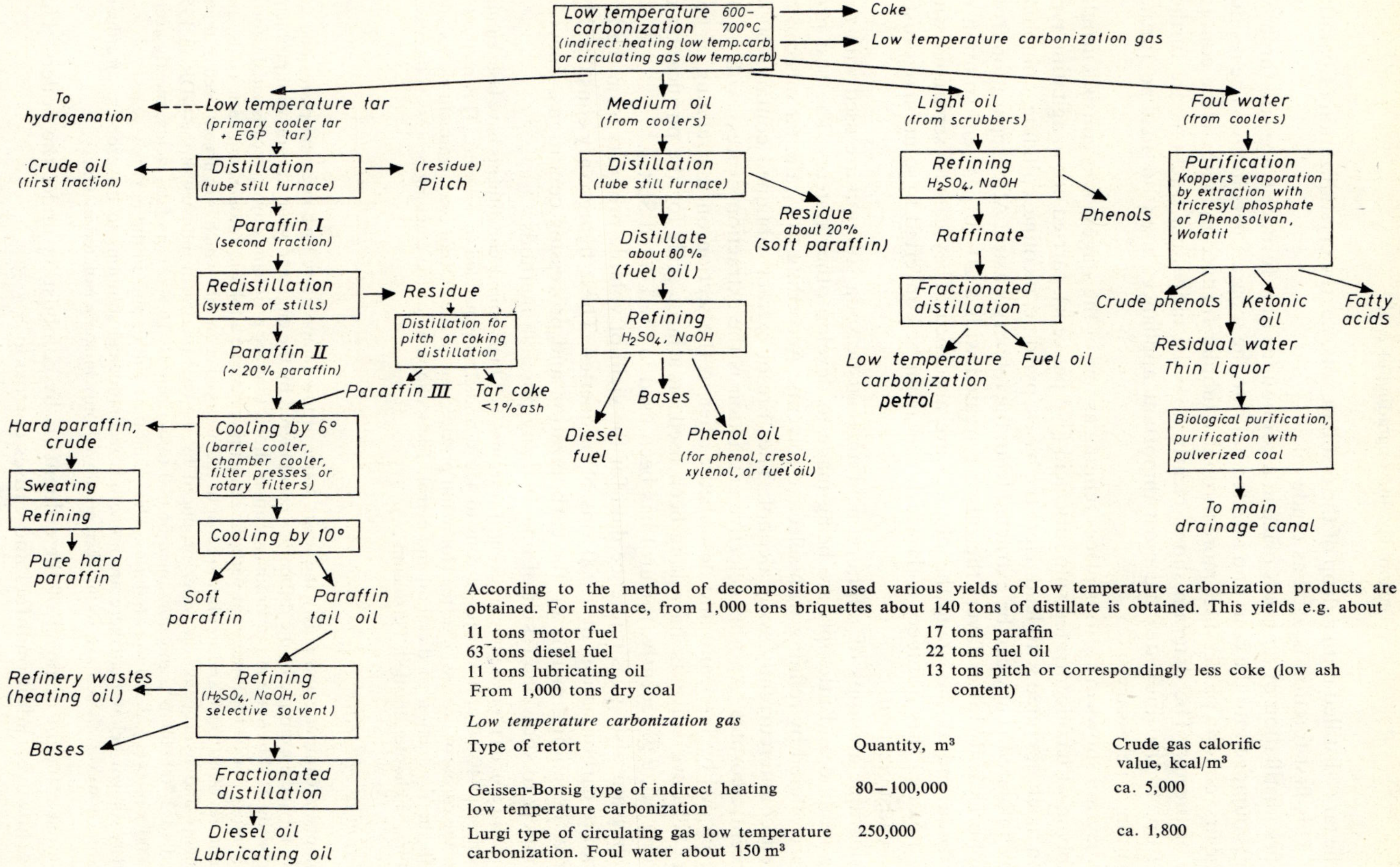


According to the method of decomposition used various yields of low temperature carbonization products are obtained. For instance, from 1,000 tons briquettes about 140 tons of distillate is obtained. This yields e.g. about

11 tons motor fuel
63 tons diesel fuel
11 tons lubricating oil
17 tons paraffin
22 tons fuel oil
13 tons pitch or correspondingly less coke (low ash content)

From 1,000 tons dry coal

*Low temperature carbonization gas*

| Type of retort | Quantity, $m^3$ | Crude gas calorific value, $kcal/m^3$ |
|---|---|---|
| Geissen-Borsig type of indirect heating low temperature carbonization | 80–100,000 | ca. 5,000 |
| Lurgi type of circulating gas low temperature carbonization. Foul water about 150 $m^3$ | 250,000 | ca. 1,800 |

Figure 19. Products of low temperature carbonization of brown coal.

First, the tar is distilled until a drop of the distillate sets, i.e. the crystallization of paraffin begins. This first fraction is used as crude oil.

The first distillation of tar takes place in a continuous tube furnace (tube still furnace). The tar is pumped into the coils and injected hot into the separation columns. The residue from the first column passes once more through a tubular furnace to increase the concentration of the tar. The diagram shows clearly how the individual fractions are further worked up. It also indicates the most important appliances used so that no further explanation is necessary.

The diagram shows in various places refining as one of the stages. This consists usually in scrubbing with concentrated sulphuric acid (3–5 per cent), carried out in agitation type flotation machines lined with lead. Here the bases, pyridine, picoline, collidine, quinoline, etc. are eliminated. They can be extracted from the refinery wastes. After separation from the sulphuric acid (lower layer) the oils are treated with a 20–30 per cent solution of caustic soda and at this stage the phenols are absorbed. They can be separated with carbon dioxide. This produces soda which can again be converted to solution of caustic soda by causticization.

Considerable amounts of sulphuric acid and caustic soda solution are used with this refining method and attempts are being made to develop other systems. The best prospects are offered by refining with selective solvents. According to Gräfe, tar oils can be extracted countercurrently with alcohol. The 'Fresol extract' obtained contains 60 to 80 per cent creosotes. Another selective extraction is the extraction process with liquid $SO_2$ developed by Edeleanu and described earlier in the section on the refining of petroleum. Selective solvents have long been used on a large scale for the purification of lubricating oil, e.g. recently, furfurol has been used in the United States.

The diagram also gives reference data for the yield of volatile products from the low temperature carbonization of 1,000 tons of briquettes. These figures vary considerably, of course, depending mainly on the carbonization and processing conditions. They are given merely to show the result of one of many possible methods.

Tar oils can also be processed by cracking. Tar from which high-boiling constituents have been eliminated is heated at 40–60 atmospheres in coils to 450°C and then expanded in columns. This method is known as the carburol process or as the Hoffmann process (e.g. in the Gölzau coal treatment works) and is used in the production of diesel oil and petrol. The technology of the cracking process is discussed in detail in the chapter entitled 'Petroleum'.

### *Purification of effluents*

The attempts to eliminate and purify the waste water of low temperature carbonization plants and gas works are a typical example of the difficulties encountered by the engineer when dealing with problems of waste products. The state of our drainage canals shows that the problem is by no means satisfactorily solved. It must not be forgotten that with the effluents, phenols, aliphatic ketones, aliphatic acids, and large quantities of nitrogen and sulphur are lost to the national economy which is in urgent need of them. The effluents and hydrogenation water produced by the coal treatment industry of the GDR in 1958 amounted to 2,700,000 $m^3$, and contained 36,000 tons phenols, 16,800 tons fatty acids, mostly acetic acid, 14,000 tons $NH_3$, 2,500 tons N-bases. These amounts will double in the next few years.

It will be shown below that there are numerous technological solutions to the problem of effluent purification; however, these have so far been employed only in some isolated instances.

The most efficient disposal of the effluents is by combustion under the boilers. This requires, however, additional fuel and causes corrosion.

In many cases the effluent from small producer gas plants is conducted into old opencast workings of brown coal mines, together with the ashes from the boiler which have an adsorbing effect on phenols. Since, however, the phenol is gradually washed out there is a danger of poisoning of the ground water. A good solution was the biological purification process of the Emscher-Genossenschaft which was first used on a large scale in England. Effluent and faeces are aerated in large basins the floors of which are lined with aerating tiles. Aerated percolating filters are also used. Thorough purification of the effluent is achieved. An improvement of the method was brought about by Nolte whose P-process involved an addition of $H_3PO_4$ which caused so strong a stimulation of the biological oxidation that faeces were no longer required.

Operation of plants for biological purification is, however, very expensive and is profitable only when a different type of purification has first been used. Moreover, the valuable phenols are lost with this method. As complete as possible recovery of the phenols is, however, of special importance for the national economy. This problem is solved by the extraction process.

After the 1914–1918 war, benzene was used as extraction solvent in countercurrent extraction of the ammoniacal liquor which contains the phenols. The working up of the benzene extract by distillation which proved too expensive was later replaced by the cheaper extraction of the benzene by caustic soda (Pott-Hilgenstock process). Phenol was separated with $CO_2$, the soda causticized with chalk, and the caustic soda thus reclaimed. The process has been adopted only in coking plants, not in low temperature carbonization plants.

The Emscher-Genossenschaft, an association of water authorities in the Ruhr, operates for their own purposes 26 dephenolization installations in coking plants, using the Pott-Hilgenstock process. Benzene is a solvent produced in the coking plants themselves. The phenoxide liquor from the 26 plants is worked up centrally by the Gesellschaft für Teerverarbeitung in Duisburg-Meiderich. These plants supply about 30 per cent of the phenols produced in bituminous coal processing works in the German Federal Republic.

To by-pass the problem of the solvents, an evaporation method was developed by Messrs. Koppers between 1935 and 1938. The phenol-containing water is evaporated in a tower with large quantities of scavenging steam in circular motion. The escaping hot phenolic vapours are extracted with caustic soda solution in a column and returned into the first tower. To prevent too much conversion of volatile fatty acids, a rather high ammonia concentration is maintained so that the fatty acids are retained as ammonium salts. Further use is made of the primary only half saturated, phenol-containing solution of caustic soda to extract phenol-containing heavy petrol or oil. In this case, too, the phenol is separated with $CO_2$ and the soda obtained is causticized. This method is still used in a few low temperature carbonization plants but is an imperfect solution of the foul water problem since only volatile phenols (phenol, cresols, xylenols) are eliminated but not, for instance, pyrocatechol and its homologues. The thin water still contains up to 8 g/l. polyvalent, and about 1 g/l. volatile phenols.

After Lurgi had developed yet another method, the adsorption process with activated coal, in which the phenol is extracted from the coal with benzene, the former IG Farben developed in about 1936 the tricresyl phosphate extraction process. Tricresyl phosphate dissolves phenol relatively well and has such a high boiling point that the phenol can be separated by heating. The process had the disadvantage that gradually a large amount of resin accumulates in the solvent which impedes the escape of phenol.

The most economic process with the widest application is the Phenosolvan process, introduced in 1940 by IG-Lurgi (see also R. Jauernik, Neuere Entwicklungen beim Phenosolvanverfahren. *Erdöl und Kohle*, 1960, **13**, 252).

Phenosolvan is a mixture of esters of the higher aliphatic alcohols, mainly butyl acetate (d. 0·88, b.p. 110–130°C). Of all the possible solvents, phenosolvan has the greatest dissolving power for phenols. For the complete extraction of foul water are required: when using benzene 220 vol. per cent, tricresyl phosphate 17·5 per cent, phenosolvan 10 vol. per cent. The basic design of a phenosolvan extraction plant in its original form is shown in Figure 20.

The effluent passes through a gas tower in which all the waste gases from the plant are collected, into a three-stage extraction, having absorbed the phenolsolvan vapours contained in the waste gases. (The extractors used now are horizontal cylinders subdivided into separate successively larger compartments

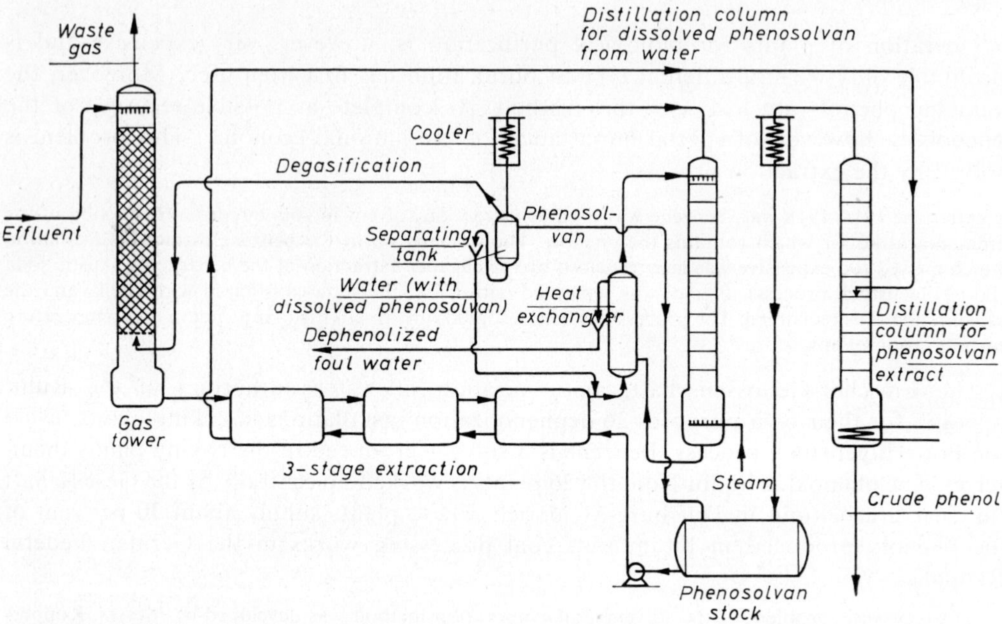


Figure 20. Operational diagram of a phenosolvan plant (system IG-Lurgi)

containing the pumps. This reduces losses by leakage, evaporation and respiration. Extraction towers with built-in rotors appear to offer even greater advantages.) There it is extracted countercurrently with 10 vol. per cent of phenosolvan. The dephenolized water is fed via a heat exchanger into a distillation column where the dissolved phenosolvan is separated with steam. Phenosolvan vapours and steam condense in a cooler and are separated in a separating tank (Florentine receiver). The phenosolvan passes into the storage tank, the water returns into the plant for the recovery of the dissolved phenosolvan. The dephenolized water, with only 0·05–0·1 g/l. phenol, still containing, however, the fatty acids as ammonium salts, runs from the sump of the column into the drains, or into a biological or adsorptive final purifier, after pre-heating of the extracted water in the heat exchanger. The phenosolvan charged with phenol is worked up by distillation. According to the nature of the carbonized coal the crude phenol contains various amount of phenol and its homologues, as well as, in many cases, large quantities of pyrocatechol and homologues.

The phenosolvan process is not suitable for the extraction (dephenolization) of oils because phenosolvan dissolves oils, but a similar process in which hydrated methanol is used in place of butyl acetate, can be utilized for refining oils. A variation is the 'Phenisol' process for coking plants, developed by Lurgi-Bayer. The selective extraction processes for the purification of tars and oils will become increasingly important in the future.

The extracted phenol water still contains all the fatty acids, about 10–15 g/l. mainly acetic acid as ammonium salts and some of the polyvalent phenols. These must not be allowed to pass into the drainage canals since their requirements of biological oxygen are very high (in effluent chemistry, the '$BOR_5$-value' is the biological oxygen requirement of a certain quantity of effluent in five days. Formerly, $KMnO_4$-numbers were used but fatty acids for instance do not react with $KMnO_4$). Various methods can be used for the final dephenolization of the thin water: oxidation, adsorption, and biological processes.

If suitable sufficiently strongl yabsorbing carbonaceous dusts and ashes are available, the residual phenols can be removed by adsorption after the bulk has been eliminated by one of the extraction or evaporation processes described by Lorenz (see below). Thus, for instance, Winkler-dust and Winkler-ash have great adsorptive power. They can adsorb 10 per cent of their weight of phenols. This method has been used for many years in the lignite works Böhlen, Leuna, and Zeitz.

The residual phenols can be precipitated by oxidation, using an ash-air treatment, if a sufficiently active ash is available and if the effluent contains mostly polyvalent, easily oxidizable phenols (A. Rieche and J. Strankmüller, *Wasserwirtschaft – Wassertechnik*, 1958, 8, **64**, 129; A. Rieche, H. Seeboth and H. Seyfarth, *Brennstoffchemie*, 1959, **40**, 194.). This process has not yet been applied industrially on a large scale.

The Emscher-Genossenschaft has for a long time been carrying out the final purification with percolating filters (i.e. towers filled with slag brick aerated from below and into which the effluent trickles from above) and an activated sludge tank. Figure 21 shows the arrangement of a two-stage biological purification plant with percolating filter tower and activated sludge tank (from K. Lorenz, Abwässer-

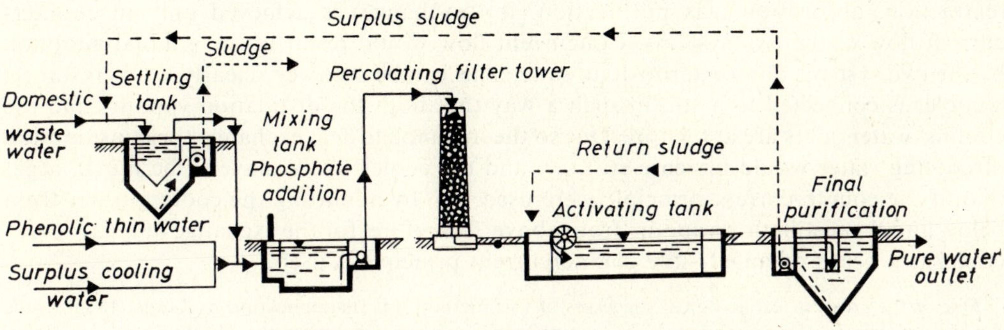


Figure 21. Diagram of a two-stage biological purification plant

probleme der Braunkohlenchemie, *Abhandlungen d. Dtsch. Akad. d. Wiss. zu Berlin*, 1957, No. 1, Akademie-Verlag Berlin). The nutritive salts required by the micro-organisms for growth must be added with the elements K, Mg, and P; N is present in adequate quantities.

For the new large brown coal processing works of the German Democratic Republic, for instance, for the 'Schwarze Pumpe' combine, plans worked out by Schulz, of the Institut für Wasserwirtschaft, Berlin, provide for a combination of percolating filter towers and activated sludge tanks (see Lorenz) for the final purification of the phenosolvan extracts.

If an adsorption, or oxidative adsorption purification process preceded the biological one, biological purification plants could be greatly reduced in size (see Lorenz). It would also make possible biological utilization of the fatty acids (e.g. in 1958 about 17,000 tons fatty acids in effluents in the German Democratic Republic) and of part of the ammonia present, by microbial protein synthesis for the production of fodder yeast, in place of the destruction of the organic matter (A. Rieche *et al.*, *Wasserwirtschaft – Wassertechnik*, 1960, **10**, 52–62; *Reinigung und Verwertung industrieller Abwässer*, Akademie-Verlag Berlin, 1962).

Another possibility is the extraction of the residual phenols by a resin exchange (e.g. Wofatit). This process, tried at the large-scale pilot project at Böhlen, proved unsatisfactory since the resin quickly became inactive and too large methanol losses occurred during the extraction and recovery of the Wofatit.

Finally, a new development: the alkaline evaporation process, devised by Dr. Grunewald, Böhlen.

Experience gained up to now has shown that none of the effluent purification processes can be used with equal success everywhere. Which process is suitable depends on the nature of the water.

*The countercurrent and concurrent principles*

The movement of two materials relative to each other can be either countercurrent or concurrent. If the materials are in direct contact so as to react mutually, e.g. in the adsorption of gases, this is called countercurrent or concurrent mass exchange. If the components are brought together for the purpose of exchanging energy, separated usually by a partition such as in coolers and heat exchangers, the system is called countercurrent or concurrent heat exchange. The extraction processes are also concurrent or countercurrent mass exchanges, for instance the transformation of a dissolved organic substance from the aqueous solution into an organic solvent, e.g. the phenosolvan process, or the exchange of ions in ion exchangers.

Depending on the technical purpose, the two materials move either concurrently or countercurrently. A complete energy exchange (heating or cooling) or mass exchange (extraction, absorption, gas purification, ion exchange) is achieved only in countercurrent flow of the two systems. Concurrent flow would result in the gradual adoption by the two systems of an intermediate state. This can be seen very clearly if, for instance, a cooler is connected to a still in such a way that both the distillation vapours and the cooling water inlets are at the top. Due to the incomplete heat exchange the consumption of cooling water would increase strongly and the cooler would have to be much larger than if the countercurrent principle were used, e.g. by admitting the cooling water from below and the distillation vapours from above. Therefore, for the exchange of heat—every cooler is a heat exchanger—the countercurrent principle is used.

One of the greatest technological successes of this method was its application by Linde (1895) to the liquefaction of air. Gases condensed by compression cool on decompression (Joule-Thomson effect). Liquefaction of air became possible when Linde used the air, cooled by decompression, in a heat exchanger (double tubular coil) to pre-cool the freshly arriving compressed air. This Linde principle was also used for the liquefaction of gaseous lower hydrocarbons to separate them by fractional distillation under pressure (see also Pressure distillation, Linde separation).

In almost all large-scale technological processes an important part is played by the heat exchange between liquids, between liquids and gases, and between gases.

Countercurrent extraction uses concentration instead of temperature. The energy exchange requires as large as possible exchange surfaces and favourable heat transfer conditions. Important factors are: size of surface, material constants, wall thickness, thermal conductivity of the materials, and convection. The problem of heat transfer is treated again in the section on Heating and heat transfer. Of decisive importance for extraction and absorption is the size of the contact area of the media. Maximum contact and fine distribution must be obtained between solid and liquid or between gas and liquid. This can be achieved by the use of fillers, by agitation, by spraying (formation of mists) and by producing thin films on rotating surfaces in separators (see section on Centrifuges). The bubble tray columns for distillation (q.v.) are ideal devices for scrubbing gases with liquids. Rectification in a column is based on the scrubbing of the components with higher boiling points by the condensate and on the artificially produced reflux on the countercurrent principle. The same principle is often used for gas circuits in hydrogenation, for many catalytic processes, and for reactions in acetylene chemistry. If the process is, for instance, the reaction of a solid with a liquid, such as wood hydrolysis, or the extraction of tanning bark, sugar beet slices, pharmaceutical products, etc., either a column is used which contains the solid and which is permeated layer by layer (percolation principle), or a series of containers with reversible inlets and outlets (extraction or diffusion battery).

When using a column the work is interrupted for the replacement of the entire contents. In extraction batteries the individual appliances are filled and emptied in turn. The most extracted appliance is filled with fresh solvent. (More about the operation of a diffusion battery under Hydrolysis of wood by the Bergius-Rheinau method.)

The concurrent and the countercurrent principles are used in drying processes (see p. 21). Temperature-sensitive materials are dried concurrently with the drying gas.

*Batch and continuous methods of operation*

Continuous heat or mass transfer on the flow principle constitutes in itself the simplest form of continuous operation. In chemical processing, batch or intermittent operation is being increasingly displaced by continuous operation. It is used for the production of most organic intermediates, dyes, pharmaceutical products, washing and textile finishing agents, plant preservatives, aromatic substances, etc. The most widely used piece of equipment is the agitator vessel (see Figure 85). It is used preferably where the quantities processed are not too large, where much space is required for the reaction where the duration of the reaction (residence time) is long, and where agitation is necessary because the systems are multi-phase. It is also used when energy and reagents are applied periodically.

The disadvantage of batch operation is the loss of time through filling, emptying, heating, cooling, and cleaning of the apparatus. A relatively large amount of manual labour is required but cannot be avoided owing to the nature of the process.

If large quantities are processed under constant conditions, as is almost always the case in basic chemical industries, only continuous operation is used. This applies to the chemical processing of coal, gases, petroleum and tar, acetylene, solvents, the simple aliphatics, the raw materials for high polymers, and to hydrogenation.

Two types of equipment are used for continuous operation: for rapid processes the flow tube, for slow ones series-connected containers, called cascades. The choice of apparatus depends on the residence time. The advantages of continuous operation are: uniform charge and discharge of the products, saving of energy by using heat transfer for heating and cooling, easy adjustability and control and, in consequence, possibility of automation, greater safety and stability, and low manpower requirements. However, the cost of the measuring and control operations required is high.

A model flow mechanism is the animal or human digestive tract, with its comminution and charging devices, mixing and emulsifying devices, addition of reagents, control of the conditions of the reaction and of the residence time, all being directed from a central 'control room'.

In flow tubes, the constituents of a reaction enter at one end. The individual particles slide one behind the other towards the exit, no mixing of the reaction zones taking place.

It is easy to operate at any required pressure by injecting the components of the reaction and fitting a throttle valve to the end of the tube.

When two materials pass continuously through an agitator vessel, some particles take the shortest path, some a medium one, and some the longest one. The residence time of each particle is, therefore, quite different from that of every other particle. This is altered when two vessels are coupled in series, for the probability that the same particle will choose an equally long path in each vessel is very small. The non-uniformity of the residence time of the particles diminishes the more vessels are connected in series. A limiting case is the flow tube where the flow takes place, as it were, through an infinite number of individual vessels. In practice 5–6 vessels are sufficient to achieve adequate uniformity, e.g. in emulsion polymerizations.

In many cases, when the reaction is not completed in one direction, continuous operation is used but deliberately with only one vessel so that all stages of the reaction are present simultaneously. An example is the production of aldol from acetaldehyde. The components of the reaction circulate and portions are continuously removed from the system for working up.

*Extraction of liquids (extraction in liquid systems)*

In chemical technology, the extraction in liquid systems increases constantly in importance. The liquids used are as sparingly soluble in each other as possible since this is the only means of achieving an ideal equilibrium of the dissolved material in the solvent. The density of the solvents must be such as to facilitate easy separation; they must not form emulsions.

The material to be extracted moves between 'donor' and 'receptor', i.e. the extracting agent. If the two solvents, one of which contains the material to be extracted in solution, were simply mixed, a certain definite distribution in accordance with Nernst's law of distribution would be obtained. This would be extraction in one 'stage'. By treating the 'donor' with fresh 'receptor' in several stages, a fresh equilibrium is continuously established and finally the extraction of the donor is completed. This corresponds approximately to the 'shaking out' of a material in a separating funnel up to the completion of the process. After each stage there must, therefore, be dissociation and separation.

Extraction follows definite laws which resemble those applying to fractionation and absorption. The amount of extracting agent and the number of stages required is calculated in the same way as the number of theoretical plates necessary in a column to obtain a certain yield in fractionation. Great uncertainty is, however, inherent in these calculations more details of which will be found in the textbooks on process technology. Essentially, one would expect to obtain the number of theoretical stages in the same way as the number of theoretical plates, when a bubble tray or a packed tower is used countercurrently for the extraction. In reality, however, the effect is much smaller. This is due to the fact that in mixing two liquids it is much more difficult to carry out the cycle: close contact, mass exchange, and separation than with a liquid and a vapour phase in a rectifying column. The diffusion velocity in two liquids is much lower; a finer splitting up into thinner layers is, therefore, necessary in extraction than in rectification.

This fine splitting up, on the other hand, impedes the separation which must follow each stage. The equipment used for countercurrent extraction must, therefore, permit continuous alternation between splitting up into layers and separation of the solvents.

In many widely used extraction columns the mixing and developing process is left to itself. A mixing vessel is often fitted before the column.

Figure 22 shows the method of operation of an extraction plant with a packed tower. Bubble tray and perforated bottom columns are also used. For large plants the output of a single column is usually not satisfactory and designs with an improved performance of the columns have been developed such as Scheibel's agitation column in which rest zones (packing material layers) alternate with agitated zones.

Because of the difficult mixing and separation in columns the extraction is often carried out in block stages with mixing and stabilizing vessels. Each stage consists of a mixer and separator pair, any number of which can follow each other. The mixer can be either an agitator vessel or simply a centrifugal pump, the separator a large vessel in which e.g. the extracting agent (receptor) is mixed approximately in the centre with the liquid to be extracted (donor), the donor and the receptor separating out at the bottom and at the top respectively. Mixing and separation are also carried out in horizontal multi-chamber cylinders (see p. 56).

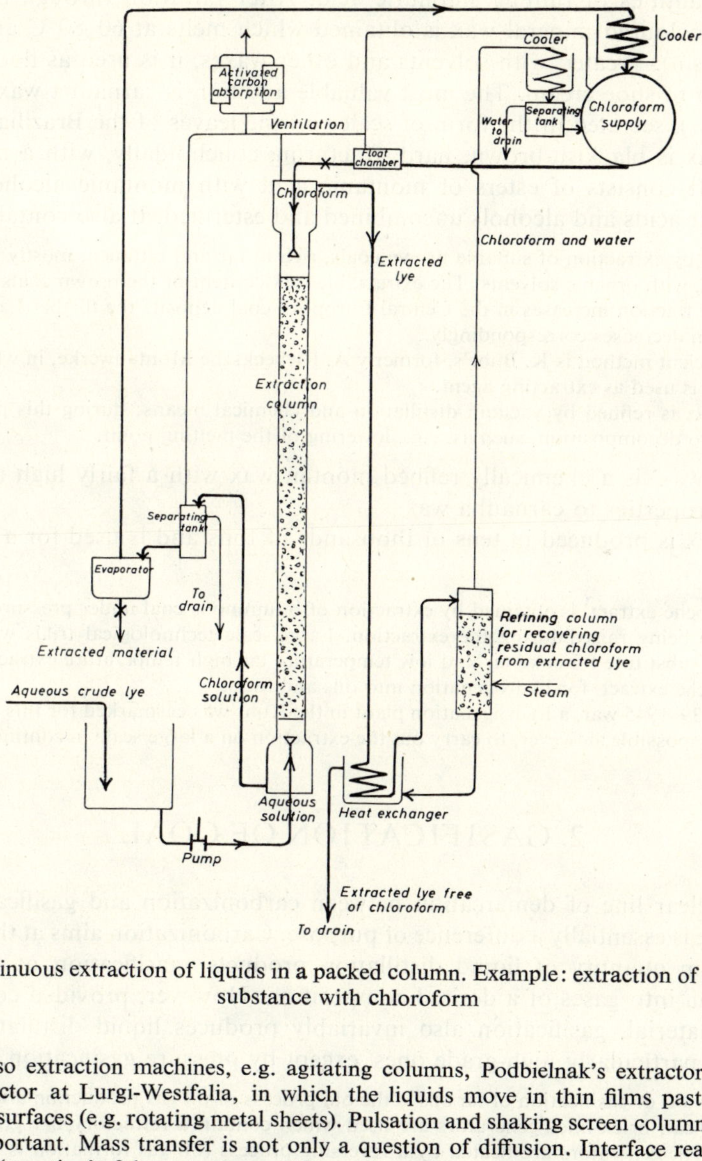


Figure 22. Continuous extraction of liquids in a packed column. Example: extraction of a water-soluble substance with chloroform

There are also extraction machines, e.g. agitating columns, Podbielnak's extractor, and Coutor's 'Luwesta' extractor at Lurgi-Westfalia, in which the liquids move in thin films past each other on rapidly moving surfaces (e.g. rotating metal sheets). Pulsation and shaking screen columns are becoming increasingly important. Mass transfer is not only a question of diffusion. Interface reactions, internal turbulence, and 'eruption' of drops to the outside accelerate the exchange since they increase the surface area. Optimum drop size during the initial formation of the drops, and the duration of drop formation are particularly important.

*Mineral wax and montan wax, extraction of coal*

Mineral wax (ozocerite) is extracted by mining and forms also the solid constituent of many petroleums. It consists essentially of non-crystallizing paraffins and bituminous resins. Rather large losses occur during purification which is carried out by smelting with large quantities of fuming sulphuric acid. After filtration through a decolorant, purified light-coloured mineral wax is obtained which melts at 60–80°C and resembles beeswax (ceresin). Treated with solvents and other waxes, it is used as floor polish and as an addition to shoe cream. The most valuable addition is carnauba wax (which produces the gloss), secreted in the form of scales on the leaves of the Brazilian wax palm.

Montan wax is blackish-brown, hard, fracturing conchoidally, with a melting point of 80–90°C. It consists of esters of montanic acid with montanic alcohol and other high-molecular acids and alcohols uncombined and esterified. It also contains resins.

It is obtained by extraction of suitable brown coals, rich in tar and bitumen, mostly from the Oberröbling coalfield, with organic solvents. The extractable wax content of the brown coals varies with the tource. The wax fraction increases in the Central European coal deposits the further East they are and she resin fraction decreases correspondingly.

The most efficient method is K. Bube's, formerly A. Riebecksche Montanwerke, in which a benzene-alcohol mixture is used as extracting agent.

The crude wax is refined by vacuum distillation and chemical means; during this process changes also occur due to decomposition, such as, e.g., lowering of the melting point.

The 'I.-G. wax' is a chemically refined montan wax with a fairly high melting point and similar properties to carnauba wax.

Montan wax is produced in tens of thousands of tons and is used for a great variety of purposes.

The 'Pott-Broche extract' is obtained by extraction of bituminous coal under pressure with solvents, the temperature being raised during the extraction. Large-scale technological trials were carried out with the aim to substitute for coke tar and low temperature tar high temperature extracts of coal, such as the Pott-Broche extract, for hydrogenation into oils and fuels.

Before the 1939–1945 war, a hydrogenation plant in the Ruhr was earmarked for this purpose. At the time it proved impossible, however, to carry out the extraction on a large scale in continuous operation.

## 2. GASIFICATION OF COAL

There is no clear line of demarcation between carbonization and gasification of coal. The difference is essentially a difference of purpose. Carbonization aims at the production of a maximum quantity of liquid distillation products, gasification at a substantial transformation into gases of a desired composition. However, provided coal, not coke, is the raw material, gasification also invariably produces liquid distillation products although not particularly high-grade ones, except by pressure gasification.

A simple method of gasification of tar oils is the old process of oil gas production after Pintsch. Two upright, cylindrical, brick-lined generators are used which contain a refractory bar grate serving as a heat reservoir. The generators are heated by a reversing process (oil gas formation is an endothermic reaction) to 1,000°C by gas or oil firing and then tar is introduced. Most of the tar is cracked into gaseous products by the heat from the reservoir; during this process, tar and coke are again produced which are burnt in the next heating period. The gas is desulphurized, condensed, and stored in liquid form. It was used, e.g. for lighting railway carriages.

By cracking cheap waste tars, e.g. generator tars, it is possible to produce gases which after separation are particularly suitable for synthesizing because in addition to the saturated hydrocarbons they are rich in ethylene, propylene and butylene ('Coker' and pyrolysis processes).

## Gasification of coal and coke

The aim is complete transformation of the solid fuel into gas with the smallest possible loss of fuel and small expenditure of energy. Possible gasification agents are: air, air and steam, oxygen-enriched air, oxygen, or oxygen with an addition of steam. According to the gasification agent used, there is generator (air) gas, water gas, mixed gas, and synthesis gas. The following table shows the most important types of gasification processes, the gasification agents, and some characteristics of the gases obtained.

Types of gasification, gasification agents, and characteristics of the most important fuel gases.

| Name of gas | Gasification agent | Percentage by volume CO | $CO_2$ | $H_2$ | $N_2$ | Characteristics | Use |
|---|---|---|---|---|---|---|---|
| Generator (air) gas | Air, or air + steam | 30 | 5–8 | 13 | 50 | Low calorific value 1,200–1,700 kcal/m³, high $N_2$-content | Industrial heating |
| Water gas | Air and steam alternating | 40 | 2–4 | 50–55 | 3–7 | Calorific value 2,500–2,800 kcal/m³ carburized with hydrocarbons up to 5,000 kcal | Heating to obtain higher temperatures, syntheses |
| Oxygen water gas, mixed gas (for synthesis gas) | 95% oxygen + steam in various proportions | 35–40 | 16–23 | 40–60 | 1–2 | Calorific value 2,400–2,800 kcal/m³ depending on gasification process, many settings | After $CO_2$ removal Fischer-Tropsch, methanol. After conversion, hydrogen |
| Synthesis gas (a) for CO syntheses | Oxygen ($CO+H_2O$ may be partly converted to $H_2$, $CO_2$ removed) | | $CO + H_2 = 85$ proportion depends on original gas and degree of conversion | | | Calorific value ~2,500 kcal, many possible settings | As above |
| (b) for hydrogenation | Oxygen water gas completely converted (hydrogen) | 0·4 | 0·3 | 97 | 1·7 | — | Hydrogenation, e.g. of coal, tar, petroleum |
| Pressure gas (town gas, rich gas, synthesis gas) | $O_2$ and $H_2O$, e.g. 10 vol.-% $O_2$, 90 vol.-% steam, 10–24 atm. (or other $O_2/H_2O$ ratios) | 16–20 | 30 | 34–50 | 2 | About 16–22% methane, many different processes. Calorific value ~4,500 kcal/m³ after removal of $CO_2$ | Town gas, turbines, synthesis gas, methane production, hydrogen |

The equipment most commonly used for gasification is the vertical body generator. The development of generators with vertical bodies is characterized by three stages: the revolving grate, the water jacket for the discharge of heat and production of steam, and suction producer gas operation (removal of the gas by a suction fan).

In addition to the body generator, the generator designed in 1957 by Gebr. Siemens, with horizontal grate and stepped grate and inclined back wall is still in use; it is particularly suitable for the gasification of low grade fine-particle fuels.

The upper part of a body generator contains a feeding device for solid fuel and the gas outlet (see Figure 23). In the lower part, a grate serves for the removal of the ash and the supply of the gasifying agents. The grate can be arranged vertically or as a revolving grate.

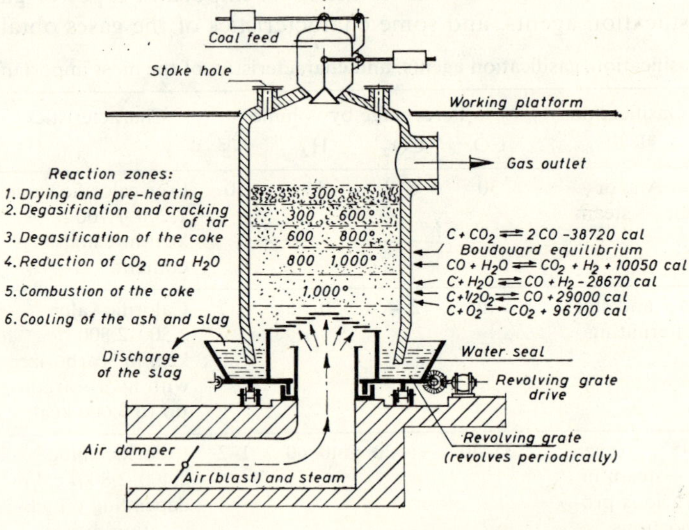


Figure 23. Diagram of a revolving gate generator for producer gas and water gas production and its reaction zones. Heat effects for 1 kg (mol) at increase of pressure: preferred: $C + 2H_2 \rightleftarrows CH_4 +$ 21,750 cal

The usual method of operation is updraught gasification. However, downdraught gasification and 'cross draught' gasification are also used. A low temperature carbonization shaft can be placed on top of the generator. It is then called a low temperature carbonization generator in which gas production is combined with low temperature carbonization, i.e. the production of liquid products. As mentioned earlier, in all generators, low temperature carbonization products are produced which must be condensed. The production of gas is usually combined with the production of steam by utilizing the heat of the generator and of the escaping gases.

Figure 23 shows the processes in a simple revolving grate generator. The chemical processes and the heat produced by them are indicated at the side of the reaction zones. The only exothermic process is in zone 5 where the fuel is burnt at more than 1,000°C; the subsequent reactions are endothermic. The composition of the gas depends on the gasification agent and the temperature at which the generator is run. The following equations show in simplified form the reactions occurring during the gasification process and the heat produced by them (after Kayser, Scharlau and Munderloh in Winnacker-Weingaertner):

Lower fuel layer:

$C + {}^1/_2\, O_2 \rightleftharpoons CO + 29$ kcal (1)
incomplete combustion
$O_2$ deficiency

$C + O_2 \rightleftharpoons CO_2 + 96{\cdot}7$ kcal (2)
complete combustion
$O_2$ excess

$C + H_2O \rightleftharpoons CO + H_2 - 28{\cdot}7$ kcal
steam decomposition
pressure dependent

$C + 2H_2 \rightleftharpoons CH_4 + 21{\cdot}7$ kcal
methane formation
pressure dependent

Middle fuel layer:

$C + CO_2 \rightleftharpoons 2\,CO - 38{\cdot}7$ kcal
Boudouard equilibrium
pressure dependent

$CO + H_2O \rightleftharpoons CO_2 + H_2 + 10{\cdot}1$ kcal
conversion reaction
pressure independent

It will be seen that a high temperature favours the $H_2O$ decomposition with C to $CO + H_2$, a low one the formation of $CO_2$ and $H_2$ from CO and $H_2O$. An increase in pressure favours the formation of $CH_4$, a factor that is exploited in the Lurgi pressure gas process. The formulae show that the formation of producer gas, being an exothermic process, does not require any additional supply of heat and can be carried on as a continuous process. The formation of water gas ($CO + H_2$), on the other hand, is endothermic and the water gas process must, therefore, be carried out intermittently. Heating up and water gas formation alternate periodically. Production of water gas by continuous operation is possible only when a mixture of oxygen and steam is used as the gasifying agent. This is the method used exclusively in large generators for the production of mixed gas and synthesis gases. In oxygen gasification partial combustion of C and the water gas reaction are, therefore, combined in accordance with the formulae:

$$\text{(a)} \quad 3C + O_2 + 2H_2O = CO_2 + 2CO + 2H_2 + 39{\cdot}4 \text{ kcal}$$

$$\text{(b)} \quad 3{\cdot}5C + O_2 + 2H_2O = 0{\cdot}5\, CO_2 + 3CO + 2H_2 + 20 \text{ kcal}$$

Both processes are, as can be seen, exothermic, and reaction (b) develops just about enough heat to make it technically feasible.

Brown coal briquettes are used as fuel for gasification; they should be as stable as possible in the degasification and fire zones; the ash must have a very high melting point so that it can be discharged in solid form. However, there are also generators with liquid discharge of the ash (wet bottom producer). Slackfree raw brown coal (so-called lump coal) can also be used because it permits uniform permeation by the gas, and so can the particularly reactive low temperature coke. In special generators, e.g. in the Winkler generator, raw brown coal, finely ground and dried, can be gasified by a suspension process (dust gasification).

The gases produced are used either directly as 'hot gases' after dust removal but without separation of the tar fractions, or as 'cold gases' after dust, tar, and oil separation and cooling. Depending on the type of fuel used and the conditions of the gasification, producer gas has an upper calorific value ($H_o$) of 1,200 to 1,700 kcal/Nm$^3$, and contains: $CO_2$ 2·5–8, CO 25–30, $H_2$ 13–17, $CH_4$ 0–4, $N_2$ 47–56 per cent by volume. The low calorific value is due to the high nitrogen and the low methane contents. Figure 24 shows the structure of a producer gas plant with gas purification, tar and effluent separation.

The crude gas produced in the generator passes through the electrostatic precipitator in which the tar mist is separated at a temperature above the dew point of the gas, and emerges as almost water-free tar. (Electrostatic gas purification (EGP) by the Siemens-Cottrell-Möller process has been described earlier, on p. 34.) The gas is then scrubbed and cooled in the gas pre-cooler and gas cooler by scrubbing with cooled foul water. During this stage most of the steam contained in the gas is condensed. The foul water heated thereby is utilized for heating and saturating the air required for gasification in the blast

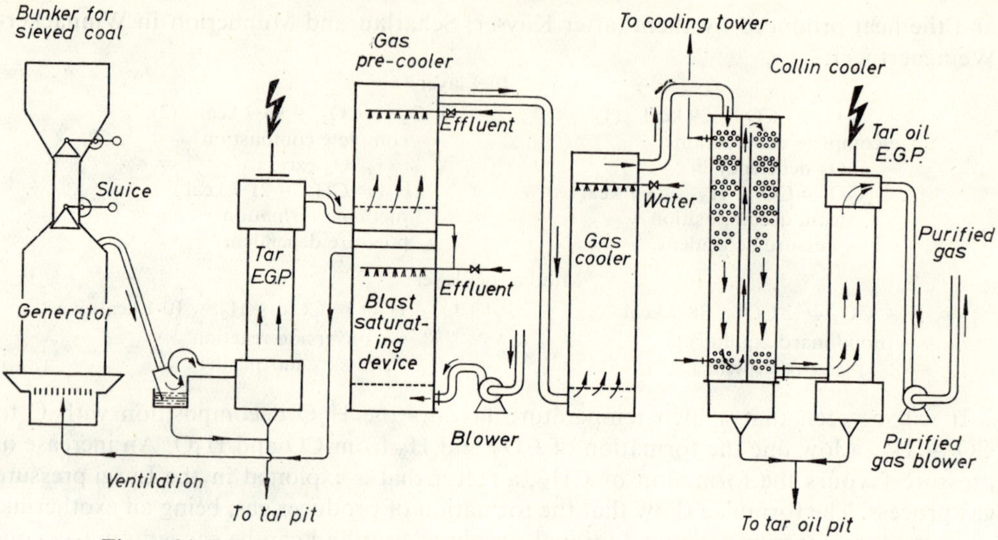


Figure 24. Diagram of a producer gas plant (water-gas-containing producer gas)

saturating device. The gas is further cooled in the indirect-action Collin cooler which causes condensation of the remaining water content and of part of the tar oil. The removal of the residual oil is carried out in the electrostatic precipitator which forms the final section of the plant. The cooled and purified gas is then delivered under pressure to the points of consumption.

By way of example, water gas from low temperature coke has a calorific value of 2,500–2,800 kcal with approximately 40 per cent by volume of CO, 50–55 $H_2$, and only 3–7 $N_2$; on the other hand, town gas produced from brown coal by the pressure gas process has a calorific value of 4,500 kcal, and 20 per cent by volume of CO, 50 per cent $H_2$, upwards of 22 per cent $CH_4$, and only 1–2 per cent $N_2$. Producer gas can, therefore, be used only for industrial purposes and not for heating purposes e.g. in the laboratory or in the home where high temperatures are required.

## Water gas, mixed gas, synthesis gas

Water gas forms the basis for important syntheses. It is too expensive for ordinary heating purposes but is used where high temperatures are required, e.g. in welding, glass melting, etc. As will be seen from the following table, it reaches its theoretical composition of 50 per cent each of $H_2$ and CO, with a calorific value of 2,800 kcal/Nm³, only at temperatures above 1,000°C.

$$C + H_2O = CO + H_2 - 28{\cdot}7\ \text{kcal}$$

$$C + 2\,H_2O = CO_2 + 2\,H_2 - 18{\cdot}6\ \text{kcal}$$

At high temperatures, more CO is formed, at low temperatures more $CO_2$.

$$CO + H_2O \rightleftharpoons CO_2 + H_2 + 10{\cdot}1\ \text{kcal} \qquad \text{(conversionreaction)}$$

Equilibrium composition of the dry gas

| °C | $CO_2$ % | CO % | $H_2$ % |
|---|---|---|---|
| 400 | 33·1 | 0·2 | 66·6 |
| 800 | 2·6 | 46·1 | 51·3 |
| 1000 | 0·3 | 49·5 | 50·2 |

For purposes of synthesis another composition is usually required, e.g. for the Fischer-Tropsch and the methanol synthesis $H_2:CO = 2:1$. The $NH_3$ synthesis and high pressure hydrogenation require pure hydrogen, free from impurities such as CO, $CO_2$ and $H_2S$. Water gas can be produced with air using a cyclic process (division into 'blowing and make periods'), in continuous operation with pure $O_2$ and heat supplied by indirect heating or by pre-heated gases (circulating gas process). Figure 25 shows the construction of a water gas plant operating on the reversing process such as is designed for instance by Messrs. VEB Gaselan (formerly Pintsch AG) (From Ost-Rassow, *Lehrbuch der chemischen Technologie*, 26th ed., p. 43).

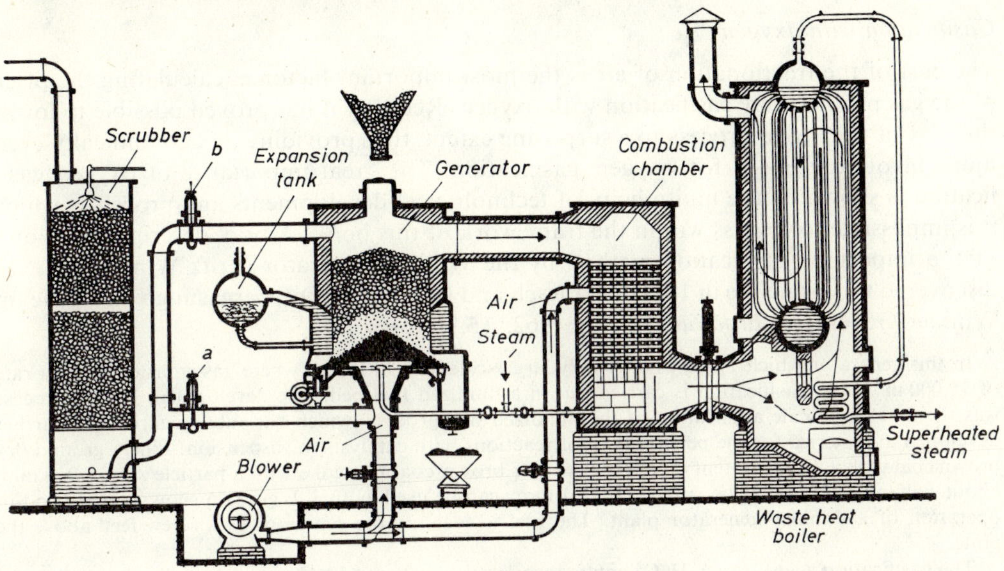


Figure 25. Design of a water gas plant

It consists of a generator, a combustion chamber, scrubber, waste heat boiler for the generation of steam, and blower. With the blast valve open, hot air is blown in and part of the fuel charge is burnt. The combustion gases, together with additional air, pass into the combustion chamber, and there they heat stone chequers which store the heat. Their sensible heat is utilized for the generation of steam in the waste heat boiler (blowing period). The steam required for the process is, therefore, produced during the process itself.

*Gas period:* The blast valve is closed, with the upper gas valve open. Steam is injected under the grate (up-steam). Water gas forms and escapes at the top from the generator into the scrubber where it is cooled and scrubbed. Then the upper gas valve is closed and the lower one opened and the steam passes through the combustion chamber where it is strongly superheated by the stored heat of the charge, impinging on the burning coke from above. The water gas then passes into the scrubber from the lower part of the generator (down gas). Once more the gases are passed upwards for a short time and then follows another blowing period.

To avoid cyclical operation in the production of water gas, processes have been developed (e.g. by Bubiag-Didier) where the necessary heat is supplied to the combustion chambers externally by producer gas. External heating requires, however, large heating surfaces so that the combustion chambers have to be combined into large blocks.

The heat required for the water gas process can, however, be supplied also directly by pre-heated gas. The gas circulates and is periodically re-heated to a temperature of about 1,300°C (Pintsch, Hillebrandt and Koppers circulating gas processes).

The conversion of coke oven or natural gas into water gas is economically very favourable. It is carried out either by cracking with steam in an externally heated tube system, or by partial combustion with limited quantities of oxygen in the presence of steam. At about 1,000–1,100°C it is possible to convert methane and the other hydrocarbons almost completely to carbon monoxide and hydrogen, with carbon dioxide as a by-product. Coke oven gas, cracked by internal combustion with $O_2$ and steam, contains about 55 per cent $H_2$, 16 per cent CO and 23 per cent $N_2$. Since cracking is carried out in the presence of nickel catalysts the original gas must be carefully desulphurized.

*Gasification with oxygen*

The cost of the fractionation of air is the most important factor in calculating the price of the gas produced by gasification with oxygen. Recently it has proved possible to lower the cost of the Linde process to a surprising extent, thus providing an economically even more favourable basis for oxygen gasification. The great importance of oxygen gasification is shown by the multiplicity of technological developments and processes which it is impossible to discuss within the framework of this book. A brief description follows of the important gasification method by the Winkler generator (Fritz Winkler, BASF, discovered this principle in 1921. W. Flesch and G. Velling, Die Vergasung von Kohle im Winkler-Generator, *Erdöl und Kohle*, 1962, **15**, 710.)

In this generator which operates on a very large scale, e.g. in Leuna where gas is produced at a rate of 75,000 $m^3$/hour, gasification is carried out in a fluidized fuel bed. The development of this process was the first large-scale application of the 'fluidized bed process' which has subsequently been further developed in America for the performance of reactions with catalysts in suspension. Finely ground dry brown coal with a water content of 6–8 per cent, or brown coal fine coke with a particle size of 0–6 mm, about half of which must be smaller than 1 mm, can be used as fuel. Figure 26 shows the method of operation of a Winkler generator plant. The fuel is fed into the generator by a screw feed above the grate.

The gasification agent ($O_2 + H_2O$) enters through the grate at 450°C at such a high velocity that the fuel is kept in suspended motion. The very high velocity of the gasification agent relative to the fuel produces a very rapid reaction which is the essence of the fluidized bed process. The gas carries of course a large quantity of flue dust most of which is precipitated in the waste heat boiler. There the heat of the gas is utilized for the generation of steam. The largest part of the flue dust is precipitated in a 'multicyclone' plant and then passes into the water scrubber. A (Lurgi) multicyclone is a group of cyclones. The process is based on the realization that several small parallel cyclones are more efficient than one large one. Finally, the remaining dust is removed from the gas in a special scrubber. The very fine ash is discharged through the horizontal grate by a water-cooled rotary arm.

The process is extraordinarily variable as far as type of fuel, gasification agent, and conditions of operation are concerned. Water gas for fuel and methanol synthesis, hydrogenation hydrogen, and mixed gas for $NH_3$ synthesis can be produced.

Various other dust gasification processes have been developed. The Flesch-Winkler process is based on downdraught gasification. The Wintershall-Schmalfeld process which has great advantages from the point of view of thermal economy but is very difficult to operate, uses circulating gas which is pre-heated by a reversing process in regenerators.

It appears that the best prospects are offered by the Koppers-Totzek process for dust gasification. The operations are carried out in two chambers (towers).

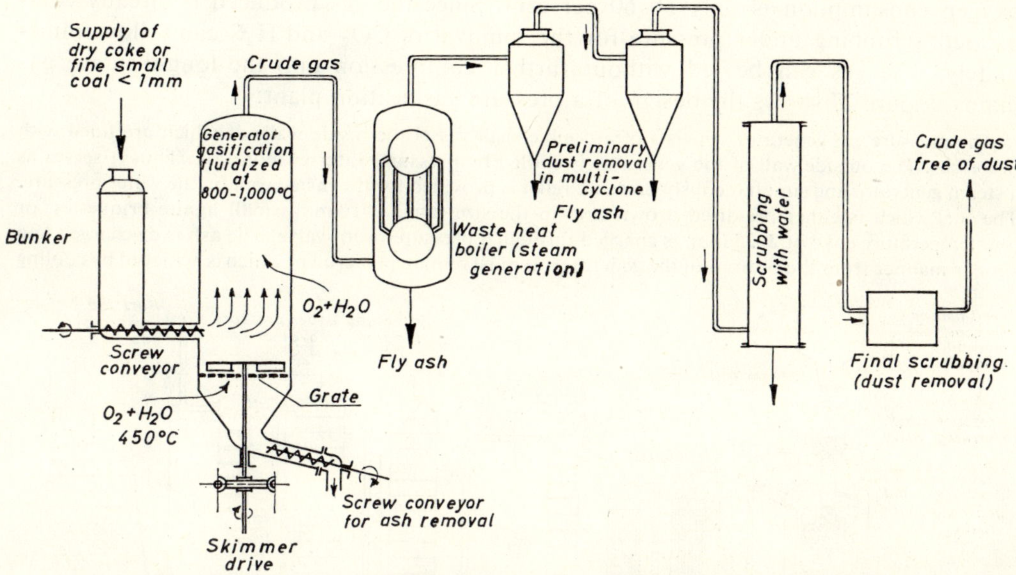


Figure 26. Diagram of a Winkler producer gas plant. Continuous gasification of fine-particle fuel by the fluidization process

The gasifier consists of a horizontal cylinder to the jacket of which is attached the gas discharge at the top and the slag discharge and scraper at the bottom. At the front are the gasifier heads into which the fuel-oxygen mixture (coal dust or fuel oil) is injected through nozzles. The super-heated steam is introduced in such a way that it envelops the flame. In this way very rapid gasification is achieved which is practically complete when the gases issue from the gasifier heads. The gases collect in the cylindrical waste heat boiler and part of the ash settles in the form of slag which is discharged at the bottom of the gasifier. The hot gas then passes through a tube boiler with vertical tubes at the lower end of which the flue dust is extracted. Further purification of the gas is then carried out in the usual manner.

Finally, mention should be made of the wet-bottom producer principle (Leuna) in which no grate is used for the fuel bed but where the gasification agent enters the fuel bed above the ash zone so that the ashes do not cool and are discharged at the high temperature of 1,700°C as liquid slags (tapped like the slags of a blast furnace). After the 1939–1945 war, cheaper oxygen production led to the gasification of high temperature coke on a large scale, to produce methane-free hydrogen, e.g. for ammonia synthesis. For this purpose high performance wet-bottom producers were developed with outputs of 60,000 $m^3$/h.

## Pressure gasification

Pressure gasification with $O_2$ and $H_2O$ was successfully accomplished by the Lurgi pressure gas process and has proved itself technically over the past twenty years. As already mentioned, the reaction $C + 2H_2 \rightarrow CH_4 + 21{\cdot}75$ kcal is a function of the pressure inasmuch as an increase in pressure creates more favourable conditions for the formation of methane and of polyatomic gases ($CO_2$ and $H_2O$). Moreover, other technological advantages are also secured by increasing the pressure: with smaller volume less space is required for the reaction and the reaction time is longer. This permits an increase in the charge in the gasifier shaft. As compared with gasification at atmospheric pressure, oxygen consumption is only 50–60 per cent. Since the gas produced is already compressed, scrubbing under pressure for the removal of $CO_2$ and $H_2S$ can follow immediately. The gas can be fed without further compression into the long-distance gas mains. Figure 27 shows the design of a pressure gasification plant.

The pressure gas generator consists of a double-shell vessel the inside walls of which are lined with firebricks. The outside wall of the generator is cooled by pressure-water evaporation. Thus it serves as a steam generator and simultaneously pressure relief is provided for the inside wall by the water pressure. The fuel which is either pre-dried brown coal in the size range 2–10 mm, small lignite briquettes, or low temperature coke of 2–25 mm, is charged through a pressure-proof valve. The ash is discharged in a similar manner from the bottom of the generator. It must remain pulverulent which is achieved by cooling

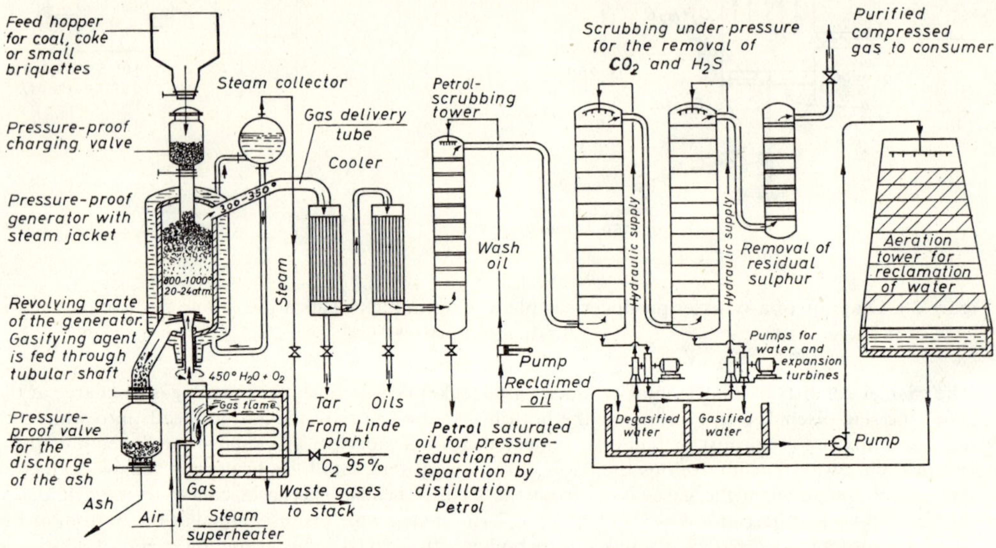


Figure 27. Diagram of a Lurgi oxygen pressure gasification system (pressure gas) for the generation of synthesis gas, town gas and fuel gas

with steam. The gasifying agent is 95 per cent oxygen, under 22–24 atm pressure, and steam, both of which are pre-heated in a gas-fired superheater to 480–500°C, and fed through the tubular shaft of the revolving grate. The gas leaves at a temperature of 250–350°C, passes first into an injection cooler (160–170°C) which is followed by the condensation of tar, oil, and steam in several coolers connected in series. The benzole is then scrubbed at 20–30°C under pressure with wash oil in the benzole scrubbing tower. Then follows scrubbing with pressurized water for the removal of $CO_2$ and most of the $H_2S$. The water is supplied to the scrubbing tower through a pressure pump and is then pressure-relieved in an expansion turbine (Pelton turbine) which is connected to the shaft of the pressure pump. This arrangement is generally used in hydraulic scrubbing plants to utilize the energy of the pressure-reduction of the water. The same principle is often applied in pressure-reduction of gases. The water containing $CO_2$ and $H_2S$ is degasified after pressure-reduction in an aeration tower. Since the gas still contains 20–50 g $H_2S/100\ m^3$, the residual sulphur must be removed, e.g. in a tower with Lauta compound (ferric hydroxide), or by scrubbing with alkazid. The latter method can also be combined with hydraulic scrubbing to extract the sulphur.

The process can be greatly varied and, in addition to town gas, synthesis gas of many types can be produced; the $CO/H_2$ ratio can be altered from 1:2 to 1:4·5. The gas produced has the very high calorific value of about 4,500 kcal. The tar yield is 70–80 per cent of the yield determined analytically. The petrol has a high octane rating. Coke is not produced since gasification is complete. A Lurgi pressure gas plant which supplies Central Germany is, e.g., the 'Otto Grotewohl' Combine, Böhlen. It is likely that the process will be increasingly used in the future (H. Just, Ferngaserzeugung durch Vergasung von Steinkohle nach dem Lurgi-Druckverfahren. *Erdöl und Kohle*, 1954, **7**, 14.)

*Conversion of water gas*

To produce synthesis gas and hydrogen, and for the removal of the toxic constituents of town gas, carbon monoxide is converted into carbon dioxide and hydrogen in accordance with the following formula (conversion reaction): $CO + H_2O \rightarrow CO_2 + H_2 + 10{\cdot}1$ kcal. This reaction was first used on a large scale when the BASF needed to provide large quantities of hydrogen for ammonia synthesis by the Haber-Bosch process. For this purpose the hydrogen sulphide must first be removed from the gas by one of the methods described below.

Since CO conversion is an exothermic reaction, a favourable equilibrium state can, by the laws of thermodynamics, be expected only at lower temperatures. The equilibrium constant $K = \frac{[CO] \cdot [H_2O]}{[CO_2] \cdot [H_2]}$ has the value 1 at 825°C. Only half of an equivalent mixture of carbon monoxide and steam is, therefore, converted, and at higher temperatures even less. $K$ rapidly decreases when the temperature is lowered, at 400°C a value of 0·07 is reached. Great credit is due to W. Wild, a colleague of Carl Bosch, who found a catalyst (iron oxide activated with aluminium oxide or chromium oxide) which works sufficiently rapidly at 400°C. Although the conversion is calculated at only 79 per cent it can be improved by using excess steam. When $[H_2O]$ is increased [CO] must decrease since $K$ remains constant. It was thus possible to convert the carbon monoxide contained in water gas and producer gas almost completely.

The conversion is carried out in converters. These are cylindrical towers in the upper part of which the catalyst in pellet form is placed on the perforated plates. The lower part contains the heat exchanger where the heat produced by the conversion is transferred to the incoming gas. This is necessary since so little heat is produced by the reaction that unless it is used for the pre-heating of the entering gases the temperature of the converter would be lowered too much. After conversion the gas is cooled in a water cooler and then passes into that part of the plant where the carbon dioxide is removed.

Three methods are usually used for the removal of the $CO_2$: scrubbing with water under pressure, scrubbing with lye, scrubbing with organic bases, e.g. triethanolamine, and alkazid scrubbing. Scrubbing with water under pressure is particularly economical when the gas is to be used as pressure gas. The method of operation of a pressure gas scrubber has already been explained in the section on the Lurgi pressure gas process.

*Purification of the gases for hydrogenation*

Removal of the sulphur compounds

Hydrogen for syntheses and hydrogenation must be purified very carefully. During gasification the sulphur compounds of coal yield mainly $H_2S$, in addition to $SO_2$. Numerous methods have been developed for the removal of the hydrogen sulphide. They are based on two principles: dry purification and wet purification. Dry purification with bog iron ore, Lux, Lauta, or Bayer compound has already been described on p. 32.

The spent compound, containing 40–50 per cent free sulphur, can be reconditioned, e.g. by roasting when the sulphur is produced as $SO_2$.

When water gas is desulphurized with activated carbon, the hydrogen sulphide adsorbed on the carbon is oxidized by catalytic oxidation with small volumes of air to sulphur which is then eliminated with ammonium sulphide as polysulphide from the carbon filters. By boiling this solution S, $NH_3$, and $H_2S$ are produced which are re-used, or processed in the Claus furnace to sulphur (see below). In this way a large part of the sulphur required in industry can be produced. If all gases were desulphurized sulphur could even be exported.

*Wet purification process*

In large gas generating plants, Bähr's alkazid pressure scrubbing process is also used which consists of absorption by an aqueous solution of amino acids (e.g. glycine or alanine). The hydrogen sulphide can be expelled by heating.

In the purification process of the Bureau of Mines the gas is first scrubbed with a 50 per cent diethanol amine solution which removes all but a small residue of $H_2S$ and $CO_2$. Dry purification follows.

The most recent gas purification process is the Rectisol process (W. Herbert, *Erdöl und Kohle*, 1956, **9**, 77–81), which permits elimination of all the impurities in the hydrogen and synthesis gas, both $H_2S$ and organic sulphur, in one operation. The gas is scrubbed under pressure in several stages with intensively cooled methanol (−40 to −70°C). Regeneration is carried out largely by simple pressure reduction in several stages (desorption); during the final stage the temperature of the methanol drops to −75°C as a result of the release of the carbon dioxide. The final purification of the gas is carried out in methanol at −62°C. The degree of purification is the best achieved so far so that the gas can be used in the Fischer-Tropsch synthesis for which a very high degree of purity is necessary.

Hydrogen sulphide produced, e.g. by heating alkacid lye, can be used for the production of elementary sulphur by the Claus method. In this process the hydrogen sulphide is burnt with the calculated volume of air on bauxite pellets, or in the Claus furnace on a $Fe_2O_3$ catalyst: $H_2S + O = S + H_2O$, the heat of combustion being utilized for the generation of steam. Final purification of gases and liquids can also be carried out with the highly active Cu catalyst (BTS of BASF) which is effective even at room temperature.

*Removal of the carbon monoxide*

In hydrogen which is to be used for general hydrogenation purposes, the CO content must be reduced to at least 0·4 per cent. This is achieved by treating the scrubbed gas with ammoniacal copper salt solution (tetrammine copper (I) carbonate) at 300 atm in high-pressure scrubbers through which the compressed gas passes. The cuprotetrammine solution is pumped under high pressure countercurrently to the gas.

The copper salt solution is reclaimed by degasification after the pressure has been reduced. Almost all the energy used for pumping the copper salt solution into the scrubber is recovered in an expansion chamber. $CO_2$ residues are also largely eliminated during the treatment. The hydrogen yield, on a 100 per cent basis, amounts to 98 per cent of the $CO + H_2$ entering the conversion. The volumetric composition of the hydrogen used for hydrogenation is: $H_2$ 97, $CO_2$ 0·3, CO 0·4, $CH_4$ 0·6, $N_2$ 1·7. This CO content is still too high for hydrogenation on a nickel catalyst since under the usual conditions nickel carbonyl may form. To remove the final traces of CO, the hydrogen is passed over a preliminary nickel catalyst at more than 180°C; this causes hydrogenation of the CO to methanol. When organic sulphur compounds are to be eliminated from a gas it is best to subject it to preliminary hydrogenation on a sulphur-resistant catalyst, e.g. an alkalized ferric oxide, at 180–280°C in the presence of a small amount of water vapour. The hydrogen sulphide formed is removed by the usual methods, or oxidized to $SO_2$ and then to $SO_3$ by the Roelen-Feist process by the addition of some $O_2$ during the hydrogenation.

The $SO_3$ is absorbed by the final purification agent (Lux compound + soda) (see Fischer-Tropsch synthesis).

The following survey summarizes the methods used for the purification of fuel gases, water gas, synthesis gas, and hydrogen:

| *Impurity:* | *Method and conditions of purification:* |
|---|---|
| Solid particle | Wet purification: in scrubbing towers or gas purifiers (Feld, Theisen, Standard, or Ströder scrubbers). Dry purification: cyclones, chambers, Siemens-Cottrell-Möller (electric gas purification) |
| $H_2S$ | Dry purification: bog iron ore, Lauta compound, Bayer compound (extraction of the sulphur by roasting $\rightarrow SO_2$). Wet purification: scrubbing with organic bases, e.g. triethanolamine, alkazid pressure scrubbing (glycine, alanine). Adsorption on carbon, partial oxidation with $O_2$ to S: solution with ammonium sulphide, boiling, extraction of S in the Claus furnace. |
| $H_2S$ and organic S compounds (final purification) | Hydrogenation to $H_2S$ and simultaneous oxidation via $SO_2$ to $SO_3$: absorption by final purification compound (Lux compound + soda), preliminary purification by active carbon. BTS Cu catalyst. |
| $CO_2$ | Scrubbing with water under pressure: final purification for hydrogenation hydrogen together with CO by tetrammine-copper (I)-solution at 300 atm. |
| CO | Conversion on a ferric oxide catalyst with steam. For hydrogenation hydrogen by scrubbing at 300 atm with tetrammine-copper (I)-solution. Final purification by hydrogenation on a nickel catalyst above 180°C. |
| $H_2S$, CO, $CO_2$ | Rectisol process. |

## LITERATURE


Scheer und Gröbner, *Die Entwicklung der Steinkohlenveredlung in den letzten 20 Jahren.* Glückauf, Essen 1941.

Ost-Rassow, *Lehrbuch der chemischen Technologie*, 26th edn, J. A. Barth, Leipzig, 1955.

Winnacker-Weingaertner, *Organ. Techn.*, I, p. 120. 2nd ed., Winnacker-Küchler, vol. 3, *Org. Techn.* I, 1953, 112.

Ullmann, *Enzyklopädie*, vol. 4.

Thau, *Leitfaden der Braunkohlenchemie*, Knapp, Halle, 1949.

Furth-Munderloh, *Braunkohle und ihre technische Verwertung*, Th. Steinkopff, Dresden und Leipzig, 1951.

Lissner-Thau, *Die Chemie der Braunkohle*, vol. II: Chemisch-technische Veredelung, 3rd ed., Knapp, Halle, 1953.

Thau, *Brennstoffschwelung*, Knapp, Halle, 1949.

Gluud, *Handbuch der Kokerei*, Knapp, Halle.

Rosendahl, *Motoren-Benzol*, Enke, Stuttgart, 1936.

Schumann, *Kokereiteer und Rohbenzol*, Enke, 1940.

v. Alberti, *Böhlener Schwelfibel*, Knapp, Halle, 1951.

Mühlert und Drews, *Technische Gase, ihre Herstellung und Verwendung*, Hirzel, Leipzig, 1928.

Schuster, *Energetische Grundlagen der Gastechnik*, Knapp, Halle, 1950.

Schöberl, Technische Brenngase und ihre Erzeugung, *Chem. Ztg.*, 1952, 76, 851.

A. Dierichs und R. Kubička, *Phenole und Basen*, Akademie-Verlag, Berlin, 1958.

# II. Natural gas and petroleum

## Natural gas

Natural gas occurs either alone or in conjunction with petroleum. It either issues from the ground spontaneously, is released by drilling boreholes, or is contained in solution in the petroleum. Natural gas, like petroleum, is held in porous reservoir rock. Many petroleum deposits have a 'gas-cap', an accumulation of natural gas in their upper part. This forms whenever more gas is present than can be dissolved in the oil. Most petroleum oils also contain, according to the conditions of deposit, pressure, and temperature, more or less natural gas in solution which is liberated when the pressure is relieved during extraction or later by pressure reduction and separation. The type and quantity of the gas dissolved in the petroleum has a very great effect on physical properties, such as density, viscosity, surface tension, compressibility, and, therefore, on the out-flow of the oil during extraction. The quantity of the dissolved gas is mainly a function of the pressure and temperature and depends on the chemical composition. According to the depth of the borehole and the thermal conductivity of the rock, temperatures up to 85°C occur. Methane-rich gases dissolve least easily; the more hydrocarbons with long chains that are present, the better the solubility.

To characterize a gas-petroleum mixture, a pressure-volume diagram and a gas liberation curve are determined on the basis of a sample from the deposit, or of a corresponding synthetic mixture produced under pressure. For this purpose the sample pressure is reduced by stages and the increase in volume is measured. Before the saturation pressure is reached the volume increases only very slightly, later when gas begins to evolve it increases greatly even for small reductions in pressure. The gas volume is calculated for standard conditions and is referred to the oil volume at the saturation pressure. The gas liberation curve thus obtained enables conclusions to be drawn as to the composition of the gas. During the liberation of the gas the liquid volume shrinks and viscosity changes; this is apparent from the curve. Other gases dissolved at normal temperature, mainly ethane, propane, and butane, are driven off by heating. In petroleum chemistry this is called 'stabilization' and the gases produced 'stabilization gases' because one of the effects is a reduction of the flash point (see also Petrol stabilization).

The two main groups of natural gases are 'dry' and 'wet' natural gases. In addition to hydrocarbons the gases often contain considerable quantities of nitrogen (e.g. the sources of natural gas at Langensalza, Thuringia, 25–69 per cent), hydrogen sulphide, some carbon dioxide, carbon monoxide, and sometimes small fractions of helium.

'Natural gas petrol' is isolated from the 'wet' natural gases. It accounts e.g. for 17 per cent of the petrol production of the U.S.A. 'Liquid gases' are those that can be liquefied at 20 atm, i.e. propane, butane, and the higher hydrocarbons. In a wider sense the olefins, although not actually occurring in natural gas, also belong to the liquid gases.

Composition of natural gases in % by volume

| Natural gas | Methane | Ethane | Propane | Butane | Pentane |
|---|---|---|---|---|---|
| Dry gas | 84·7 | 9·6 | 3·0 | 1·1 | — |
| Wet gas | 36·8 | 32·6 | 21·1 | 5·8 | 3·7 |
| Gas produced by distillation of petroleum | 5·0 | 10·0 | 30·0 | 35·0 | 20·0 and more |

Figures from: F. Asinger, *Einführung in die Petrolchemie*, Akademie-Verlag, Berlin, 1959.

The following method is used to produce the maximum amount of liquid gas from a natural gas source: The natural gas passes into a cycling plant where under its own pressure the hydrocarbons with high boiling points which are easily soluble are eliminated by scrubbing with oil; the gases with a low boiling point are pumped back into the gas pool, thus maintaining the pressure for a longer period. In the United States in 1953, 13·5 per cent of the gas produced was pumped back.

*Underground gas storage*

Empty reservoir rock of natural gas or petroleum fields can be filled again under pressure with gases, e.g. coke oven gases, cracking gases, etc., through bore holes lined with pipes. Such underground gas reservoirs have an enormous capacity. Industrial and domestic gas consumption fluctuates; it is considerably higher in winter than in summer. The total gas storage space above ground in Western Germany is the equivalent of only six hours' supply.

The extraordinarily large capacity of an underground reservoir which may reach 60 km$^3$, permits a uniform supply. In the United States, 209 underground gas reservoirs were in use at the beginning of 1960. In Germany, so far only one underground gas reservoir (near Hanover) has been established, with a capacity of 200 million m$^3$ (annual gas consumption in the Federal German Republic 10 billion m$^3$), and another reservoir is in course of construction near Darmstadt. In the GDR, a reservoir for natural gas is planned near Brandenburg.

Natural gas occurs mainly in the United States and in the Soviet Union where most of the extraction and utilization takes place. The reserves of natural gas in the Soviet Union are at present estimated at 1,500 billion m$^3$. Before the war it was utilized only at Baku and in the Carpathian Mountains (Daszawa). Numerous new sources have since been opened up, e.g. in the Ukhta and Emba regions, near Saratov, Fergana, Kuybishev, Volgograd, and Grosny. In Europe, rich deposits were found in Roumania and Italy, smaller ones in Austria and France (in the foot hills of the Pyrenees). The Federal German Republic also has several small but intensively worked fields, the most important of which are those near Rehden, Stockstadt, and Bentheim. The German Democratic Republic has a deposit in Thuringia.

In 1950 the natural gas field near Bentheim, which was opened up in 1931, supplied 51 million m$^3$ of natural gas. The gas issues under a pressure of 180–190 atm and has a calorific value of 8,600 kcal/m$^3$. It is supplied through a pipeline of 85 km length to the Chemical Works Hüls where it is converted into organic products. The gas from the recently opened up gas field near Pfungstadt is supplied over a distance of 35 km to the Farbwerke Hoechst. The BASF Ludwigshafen obtain methane from Stockstadt, mainly for the production of hydrogen for ammonia synthesis. The natural gas in Thuringia (Mühlhausen and Langensalza) is used as town gas and industrial gas in these two towns, some also in Eisenach. It is planned to connect it up with the long-distance pipeline system Magdeburg, Dessau, Leipzig.

Production of natural gas (in million $m^3$)

| Country | 1950 | 1955 | 1960 | 1964 |
|---|---|---|---|---|
| U.S.A. | 194,500 | 266,331 | 364,000 | 440,232 |
| U.S.S.R. | | 8.981 | 45,303 | |
| Italy | 500 | 3,627 | 6,429 | 7,668 |
| Austria | 470 | 748 | 1,469 | 1,764 |
| France | 250 | 266 | 2,216 | 5,088 |
| BRD | 50 | 420 | 447 | 1,457 |

Natural gas is supplied through pipelines, often over long distances. To avoid the necessity of very large pipe cross sections, the gas is compressed to 20–50 atm. Condensation is prevented by first separating out the 'liquid gas' so that practically only methane, ethane and propane are carried.

The United States have a gigantic network of pipelines. The largest and longest is the Rio Grande–New York pipeline, with a length of 3,400 km. Other pipeline networks exist in the Soviet Union, in the Ruhr, Rhineland as far as Hanover and Kassel, and in Northern Italy.

Three methods are used for the separation of liquid gas: (1) compression-condensation, (2) absorption, and (3) adsorption.

The most important method is washing with an absorption oil at a minimum pressure of 35 atm in an absorber-cooler. All hydrocarbons above propane are dissolved and methane-, ethane-, and propane-containing 'dry' gas enters the pipeline. On heating the wash oil gives off the absorbed liquid gases. These are condensed and first 'stabilized' by pressure distillation in bubble tray columns for further separation of dissolved gases, and then fractionated by distillation in several columns.

Only a small part of the natural gas is used chemically, most of it is used for the production of heat and energy. However, in several countries a chemical industry based on natural gas as raw material is in the process of construction on an ever-increasing scale. In the United States, natural gas is appreciably cheaper than petroleum and its price per calorie corresponds approximately to that of coal. It is for this reason that chemical utilization was first developed there. The conditions must be similarly favourable in the Soviet Union. Italy, though she is short of coal and raw materials, has been able to develop an important chemical industry on the basis of the natural gas found in Northern Italy.

The utilization of natural gas for the production of synthesis gas is described on p. 145, the chemical use of the hydrocarbons is described comprehensively in Section V of this book.

## LITERATURE

O. Horn, Über Erdgas und seine chemische Verwendung, *Erdöl und Kohle*, 1955, **8**, 156.

U. Nehse, Erdgas — seine Gewinnung und Verwertung, *Erdöl und Kohle*, 1957, **10**, 3, 181.

H. Laurin and G. Wedekind, Das Erdgas, die drittgrößte Energieversorgungsquelle der Erde. *Erdöl und Kohle*, 1961, **14**, 141, 221, 308.

Other literature see under Petroleum.

## Petroleum

Petroleum, also called mineral oil or naphtha, is a substance with a characteristic smell, and a viscosity varying from low when light coloured to high when dark, that burns with a sooty flame. The names are derived from the Latin *oleum* = oil, *petra* = rock, or the Greek *naphtha* = mineral oil Petroleum consists of a mixture of gaseous, liquid or solid hydrocarbons which is frequently fluorescent. Petroleum is one of the 'organic minerals' and in particular one of the zoogenous ones to which also belong ozocerite and asphalt (or bitumen). On the other hand coal, peat, and amber (a fossil resin from the Tertiary) belong to the phytogenous organic minerals. Petroleum as it is extracted today is a product converted by heat and rock pressure, a sort of distillate, from sapropelitic rock (Greek *sapros*: decaying, *pelos*: mud). According to the composition three types of petroleum can be distinguished: paraffinic which contains mainly hydrocarbons of the methane series ($C_nH_{2n+2}$) and occurs mainly in some regions of the United States, naphthenic the constituents of which are based on the formula $C_nH_{2n}$ and which is found mainly in the oil fields of the Soviet Union and of Texas, and finally a mixture of the two. Ozocerite is an oxidation product of petroleums of the methane series whilst bitumen (asphalt, mineral pitch) consists of oxidized petroleum of the naphthene series.

Petroleum does not occur in nature in 'veins' or 'underground lakes' but is absorbed in the pores of rock, e.g. sandstone, where it is absorbed like in a sponge. In conjunction with petroleum always occurs brine, 'edge water'; no explanation of this phenomenon has as yet been found. Natural gases are also almost always found in these deposits.

The specific gravity of petroleum fluctuates within wide limits from 0·71 to 0·94. Petroleum mostly has a repulsive smell and contains, in addition to the hydrocarbons, compounds of nitrogen, oxygen, and sulphur. According to its components, petroleum is more or less volatile. German petroleum is of the non-volatile type from which petrol can be extracted only by cracking or hydrogenating reformation.

Two theories have been put forward to account for the origin of petroleum, the 'inorganic' and the 'organic' one. Mendeleev, for instance, assumed that petroleum originated from metal carbides which at great depths under pressure with water produce hydrocarbons. In 1926, Franz Fischer, following his dicsovery of the direct synthesis of petroleum hydrocarbons at normal pressure, said: 'From the point of view of the origin of petroleum hydrocarbons in nature, if one can speak at all of the origin of materials other than from organisms, a new possibility has been found.'

Fischer developed the so-called 'carbide theory' for his hydrocarbon synthesis from carbon monoxide and hydrogen on cobalt and iron catalysts, and applied it to the formation of petroleum. This theory is untenable for petroleum and is also unlikely for the CO hydrogenation.

It has been proved that petroleum originated from the protein, fats, and carbohydrates of low forms of vegetable and animal life for it contains optically active compounds which cannot be produced synthetically, and the presence of chlorophyll and haemin derivatives has also been determined. Petroleum also contains substances which are not stable at high temperatures such as would have to obtain under the 'inorganic theory'. Petroleum originates, therefore, from phyto- and zooplankton, and also higher organisms. (Recapitulation of the theories on petroleum formation, E. Beerstecher, *Petroleum Microbiologie*, Elsevier Press, Houston, 1954; Stadnikoff, *Erdöl und Kohle*, 1958, **11**, 293.)

Petroleum forms where organic material collects and remains intact over long periods, such as basins protected from breakers, and shallow landlocked seas. Figure 28 (from P. Kukuk, *Geologie, Mineralo-*

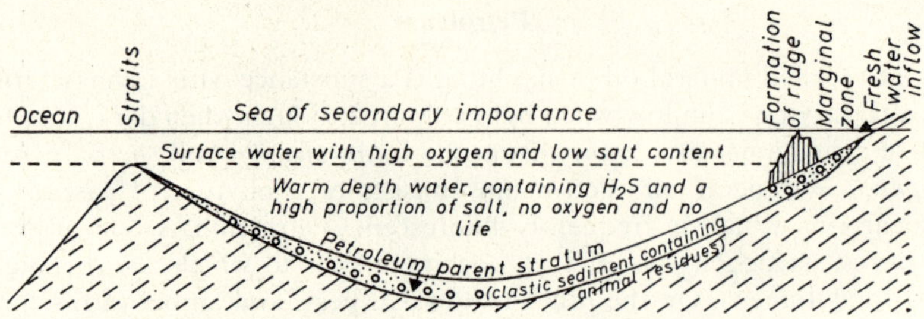


Figure 28. Formation of a petroleum parent stratum

*gie und Lagerstättenlehre*, Springer, Berlin/Göttingen/Heidelberg, 1951) shows a diagram of the formation of a petroleum parent stratum. Conditions as sketched there exist even today, e.g. in part of the Black Sea (K. Krejci-Graf, Moderne Anschauungen über die Entstehung des Erdöls. *Erdöl und Kohle*, 1960, **13**, 836.)

The conditions for the accumulation of organic material and its transformation are: surface water with high oxygen and low salt content, lively plankton formation as a result of intensive solar radiation, and deep water trapped underneath, containing $H_2S$ and a high proportion of salt and no oxygen. Together with inorganic sediment dead material constantly descends and decaying sludge forms ('Sapropel'). Owing to the lack of oxygen putrefaction cannot occur at depth, only anaerobic processes take place under the action of microbes. No doubt, extremely active enzymes play their part, eliminating $CO_2$, $H_2O$, and nitrogen under mild external conditions. It is likely that unknown anaerobic micro-organisms live in the sapropel, capable of producing higher hydrocarbons. These microbial transformations do not, however, produce petroleum but an intermediate product. This is converted into petroleum by a kind of distillation process. Petroleum is never found at the place of its formation. It 'migrates', moved by

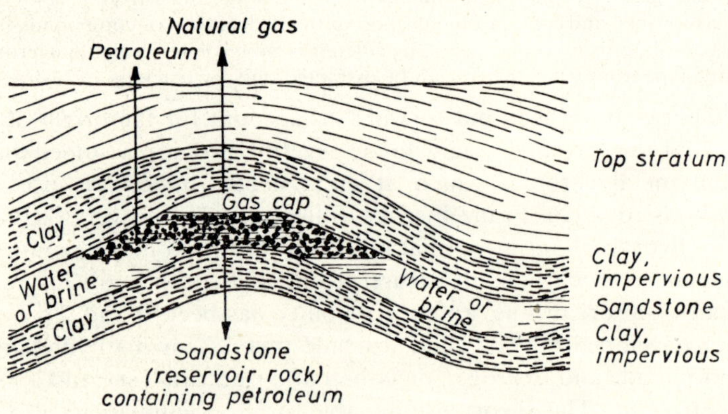


Figure 29. Accumulation of natural gas and petroleum in the crest of anticlines

tectonic pressure, gas pressure, and by water or brine, until it is absorbed by a porous rock. A further condition must be satisfied for it to accumulate and be retained in the reservoir rock: it must be prevented from further movement by impervious strata (e.g. clays, saliferous clays). The accompanying water or brine lifted the petroleum and it moved together with the gas, preferably along the line of least resistance in the rock, as far as an elevated or crest zone with impervious strata such as are often found under an anticline. Most of the fields are associated with the saddle cores of anticlines. The type of origin, migration, and accumulation of petroleum just described is, therefore, called the 'anticlinar theory'. Figure 29 illustrates the accumulation of oil in an anticline.

At the highest point of the field is the 'gas cap', underneath it the petroleum, and finally the 'edge water'. The deposits take many different forms but the petroleum always rose along an inclined stratum until it met an impervious layer. Such layers may have originated by dislocations. The German petroleum deposits are usually associated with stock shaped bodies of salt.

Petroleum was not formed in a definite geological period. Like salt and coal deposits it is present in almost all formations. However, more than half the oilfields are in the tertiary strata. They are mostly at the edge of fold mountains because there shallow seas formed most easily.

The richest oilfields are in the countries of the Middle East and North America, the Soviet Union, Venezuela and Iran, smaller ones are in Hungary, Rumania, and Austria. Recently, large quantities of oil were found in the Sahara near Hassi Messaoud, estimated to amount to 300 million tons. The German petroleum production of 600,000 tons in 1936 ranks 16th place among the producer countries of the world. In the last fifteen years production has, however, consistently increased and in 1952 it amounted to 1,755,000 tons, in 1966 to 7·8 million tons. The following table shows the petroleum production of the most important petroleum-producing countries in 1900–1966, and world oil production until 1966:

Petroleum production of the most important petroleum-producing countries, in million tons

| Country | 1900 | 1930 | 1952 | 1956 | 1958 | 1960 | 1966 |
|---|---|---|---|---|---|---|---|
| U.S.A. | 8·7 | 123·1 | 310·0 | 352·8 | 330·1 | 347·1 | 410·2 |
| U.S.S.R. | 9·8 | 18·6 | 47·0 | 83·8 | 112·9 | 148·0 | 262·8 |
| Venezuela | — | 20·1 | 95·0 | 128·9 | 138·6 | 151·6 | 174·6 |
| Kuwait | — | — | 37·5 | 55·1 | 70·2 | 81·8 | 111·3 |
| Saudi Arabia | — | — | 41·5 | 48·4 | 50·1 | 61·1 | 117·3 |
| Iran | — | 6·0 | 1·0 | 27·0 | 40·6 | 62·1 | 103·9 |
| Iraq | — | 0·04 | 19·0 | 31·3 | 35·7 | 47·9 | 67·9 |
| Canada | 0·09 | 0·2 | 8·0 | 22·8 | 22·3 | 25·8 | 42·2 |
| Roumania | 0·2 | 5·7 | 4·1 | 10·9 | 11·3 | 11·4 | 12·6 |
| Germany[1] | 0·05 | 0·2 | 1·7 | 3·5 | 4·4 | 5·5 | 7·8 |
| **World** | **19·8** | **196·5** | **618·1** | **839·0** | **907·5** | **1,060·1** | **1,629·7** |

[1] from 1952: DBR

The world oil reserves are estimated at 40–50 billion tons of which about 61 per cent are in the Middle East. If petroleum extraction continues at the present rate the reserves should last for another 30–40 years. Due to new methods of exploration the petroleum reserves have for some years been increasing at a higher rate than petroleum production. In the Soviet Union the oil under the Caspian Sea is now being extracted. Offshore oilfields exist also in the Gulf of Mexico on the Texan coast, where extraction is being carried on. The richest offshore oilfields lie off the coast of Saudi Arabia near Kuwait on the Persian Gulf (about 1,800 m below the ocean floor).

It is estimated that 60 per cent of reserves not yet worked are in Arab countries. According to 1961 estimates, the various countries share the world oil reserves in the following proportions:

| | | | |
|---|---|---|---|
| Middle East | 60·80 per cent | Venezuela | 6·03 per cent |
| U.S.A. | 11·29 per cent | Europe excl. U.S.S.R. | 0·96 per cent |
| U.S.S.R. | 10·50 per cent | | |

The most important German oil producing provinces are Hanover (the area between the rivers Elbe and Weser), Weser-Ems, the area west of the Ems and the area north of the Elbe (Schleswig-Holstein, Hamburg).

Smaller deposits are in the South German sub-alpine region and in the Upper Rhineland.

According to estimates carried out at the beginning of 1965, proven oil reserves in Western Germany amount to 111 million tons of which 87·3 million tons are considered as certain and 23·8 million tons as probable.

Asphalt, which according to recently adopted terminology means a mixture of bitumen and minerals, is rare in Germany. The largest deposit is in Holzen near Eschershausen/Vorwohle am Hils, with a production of crude asphalt of 108,000 tons in 1938 from which 4,000 tons of asphalt were extracted. Oil shale is found much more frequently in Germany. So far, however, only the deposit at Messel near Darmstadt, with an oil content of 8 per cent, has proved workable; extraction amounted to 320,000 tons in 1949.

## The search for petroleum

Formerly, drilling for oil was carried out only where there were indications on the surface of petroleum deposits.

The extraordinary successes of recent years in locating petroleum deposits are due to the large-scale employment of modern geological and geophysical methods. Before drilling starts exact investigations are carried out using, in addition to surface-geological methods, mainly seismic but also gravimetric and sometimes magnetometric methods. It is impossible to determine the presence of petroleum by geophysical methods, these indicate merely whether the geological conditions for its presence are given.

The seismic method measures the propagation of elastic waves produced, e.g., by an explosion. Sudden changes of elasticity and density of the rock, i.e. a geological change of the stratum, produce reflection and refraction of the waves.

The method resembles echo sounding. On the basis of empirical values deep mapping can be carried out. Gravimetric measurements by means of a highly sensitive torsion balance provide data about the density of the underlying strata; the magnetic balance indicates changes in the magnetic field.

## Drilling techniques

A distinctive sign of oilfields are numerous steel derricks of 30–45 m height.

The old method of rope drilling is now hardly used. The rotary method was used for the first time in Texas in 1901.

At the end of a set of pipes which are screwed into each other is a chisel with conical toothed rollers made of hard metal. The rotary drill poles are suspended from a block and tackle (*a*) so that they can be raised and lowered from the drilling box (*b*) in the derrick (Figure 30 from Ullmann, *Enzyklopädie der techn. Chemie*, 3rd edn, vol. 6, p. 573. Urban & Schwarzenberg, München/Berlin, 1955).

At the upper end of the drill poles is the rotary water swivel through which water or a clay suspension is pumped in. Momentum is supplied to the drilling bit via a cam follower on the turntable (*d*) which is driven by an electric or a diesel motor. The flushing agent forced in by pumps (*g*) drives the drillings upwards between the drill rods and the walls of the bore hole. It is collected at the edge of the bore hole, separated from the drillings over a vibrating screen, and pumped back in again through a pit, i.e. it circulates. The drill poles are extended by screwing in further rods. To exchange the drill bit it is necessary to withdraw the drill poles and unscrew them.

A new method for driving the drill bit is used in 'turbine drilling' which is particularly highly developed in the Soviet Union. A multi-stage turbine is located directly above the drill bit and is driven by high-pressure gas or flushing agent.

From time to time during drilling, pipes with a suitable diameter are cemented into the bore hole from the bottom upwards. When the first length of pipes has been secured

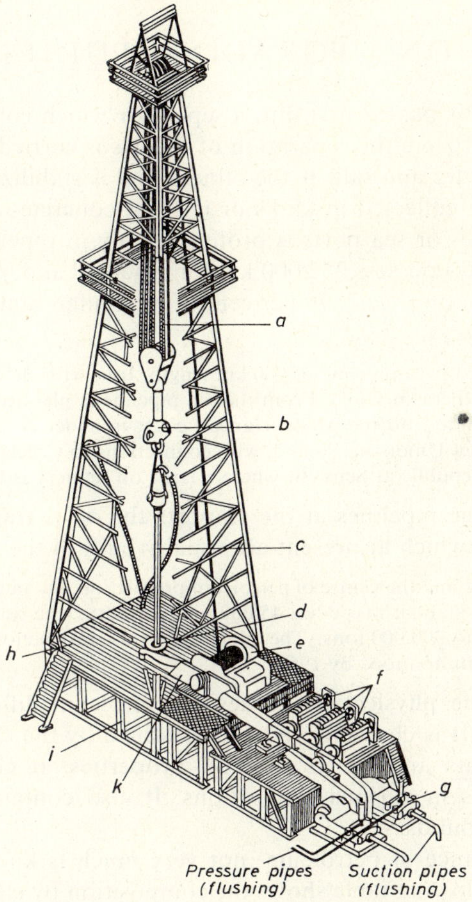


Figure 30. Rotary drilling rig

drilling is continued with the next smaller diameter. Thus a telescopic set of pipes is installed in the bore hole. To prevent gas and oil outbursts in deposits under high pressure, high-pressure slide valves which can be quickly closed are placed on the pipes.

When the oil level is reached the oil usually rises to the surface under pressure. Often gas is encountered first which may escape with great violence, sometimes mixed with oil, and even eject the drill poles (so-called 'gushers'). When the natural pressure is spent production starts with either rod or deep well pumps. The deepest bore hole in Germany is 5,956 m deep. The deepest exploratory drilling in the world is 7,620 m deep, the deepest producing well 6,880 m.

The oil flows from the deposit towards the bore hole under the action of driving forces such as water pressure, gas pressure, oil expansion pressure, and gravity. All efforts are directed towards preservation of the energy in the deposit so as to facilitate the smooth flow of the oil. If this energy drops it must be supplemented by 'secondary methods' by pumping into the deposit natural gas, air, waste gases, or edge water or fresh water. It is not usually possible to extract all the oil from an oilfield.

## 1. COMPOSITION, UPGRADING, PURIFICATION

From the bore hole the oil is passed first into a separator which consists of a container equipped with baffle plates. Here the separation of the gas is carried out. In many cases the oil is separated from water and salt at the oilfield and is stabilized to save transport costs. The crude oil is then collected in tanks or in large concrete-lined pits.

Transport to the refineries or sea ports is profitable only in pipelines of which, since about 1958, the United States possess 252,000 km for crude oil and 69,000 km for refined products (853,000 km for natural gas). 70 per cent of all United States oil is transported through pipelines.

Baku is connected with Batum by a pipeline of 900 km length, Iraq with Beirut, and the Iranian oil-fields with the port of Abadan (Persian Gulf). From there a pipeline is planned to the Mediterranean, another one from Bahrein to Sidon on the Mediterranean coast is under construction. A pipeline is being constructed from the Soviet Union via Poland, with connections to Czechoslovakia and Hungary, into the German Democratic Republic at Schwedt where a large oil refinery is being built.

From the terminals of the pipelines at the seaports the oil is transported by tankers to the petroleum refineries which at present are usually built in the consumer countries.

Since the Middle East has become the centre of petroleum production, the number and size of tankers has steadily increased. The new standard type has 45,000 tons capacity. The super tankers in operation or planned have capacities up to 300,000 tons. The global tanker construction programme provided for 1,200 new tankers with 38 million tons by 1965.

The composition and the physical properties of petroleum differ widely according to where it was produced. It is characterized and classified by the chemical constitution of its principal components and by its physical properties. It consists of paraffins, naphthenes, and aromatics in various proportions. It also contains many impurities, sulphur and oxygen compounds.

Considering the importance of petroleum, not very much is known as yet about its exact composition. The following table shows the composition by elements and the main constituents:

Composition (%) by elements

| | |
|---|---|
| Carbon | 81–87 |
| Hydrogen | 10–14 |
| Sulphur | 0–6 |
| Oxygen | 0–7 |
| Nitrogen | 0–1·2 |

*Hydrocarbons contained in petroleum*:

(a) Paraffins from $CH_4$ to $C_{36}H_{74}$, also hard paraffins
(b) Small amounts of olefins
(c) Naphthenes (cycloparaffins, cyclopentane, cyclohexane and derivatives)
(d) Aromatic hydrocarbons (traces, also polynuclear)
(e) Naphthenic acids: $C_nH_{2n-1}COOH$ (carboxylic acids deriving from the naphthenes, particularly from cyclopentane)

Petroleum contains many hydrocarbons on which little research has so far been done. Petroleum is upgraded by distillation and subsequent refining of the distillates. It was considered appropriate to precede the following chapter by some remarks on the principles of the distillation technique, the columns and coolers, and their method of operation, since the distillation technique together with the cracking technique forms the basis of petroleum processing.

*Distillation and rectification*

Distillation is the physical separation of mixtures of liquids that are completely or partially miscible with each other. The individual components must have different boiling points. Within the limits of their miscibility the liquids must form no azeotropes. An azeotropic mixture is a mixture of two or more components which at a certain boiling point has the same composition both in the liquid and in the vapour phase. This renders separation by distillation impossible. According to the number of components there are binary (two components) and ternary (three components) azeotropes.

If e.g. a two component mixture is boiled, the vapour and, therefore, also its condensate, contains more low-boiling components than the boiling liquid. When the resulting condensate is boiled again it produces more vapour and a condensate with a concentration of low-boiling components. Almost complete separation is then possible only by carrying out many individual distillations. If a column is placed on top of the distillation vessel, the vapour condenses and the condensate is then heated up by the rising vapour and vaporized. During this process the vapour produced by the reflux is enriched by low-boiling components whilst the high-boiling components are condensed from the rising vapour and flow back downwards. Thus the previously described individual distillations are carried out simultaneously within the column. This process is called rectification.

Over the length of the column the concentration both of the liquid which runs from the top to the bottom and is called reflux, and of the rising vapour undergoes changes. The vapour is enriched on its way by low-boiling constituents up to the terminal concentration, whilst the reflux is enriched by high-boiling components and the concentration of low-boiling constituents in it is reduced. Liquid and vapour are in equilibrium at every point of the column. There is also a correlation between this equilibrium and a specific temperature. The temperature drops over the length of the column from the bottom to the top.

So as to extract the distillate from this distillation process it is, of course, not possible to let the whole of the distillate reflux continuously as must be done at the beginning of the distillation to enable the condensate to reach the desired concentration at the head of the column, but part of the condensate must be extracted as a product. If, e.g. one third of the condensate produced is extracted, two thirds return into the column as reflux. The ratio reflux: extraction is called the reflux ratio and is 2 in the above example. Reflux is produced in various ways which are shown schematically in Figures 31 and 32.

(*1*) ***Dephlegmation.*** The heat exchanger which in most cases operates with water is placed directly on the head of the column and condenses part of the vapour which is immediately fed into the column as reflux whilst the remainder of the vapour is condensed in a condenser and extracted as product. The disadvantage of the dephlegmator is, however, that the reflux ratio cannot be regulated and depends solely on the temperature and quantity of the cooling water.

(*2*) *Partial condensation* (Figure 32). The dimensions of the condenser which is outside the column are such as to extract only enough heat from the distilled vapour to make the condensate correspond to the quantity of reflux whilst the remainder of the vapour is condensed in the next condenser and extracted as distillate. Both dephlegmation and partial condensation result in rectification of the distillate which represents a saving in column plates. The disadvantage of partial condensation is that the reflux ratio is fixed but can only be altered by means of the cooling water.

(*3*) *Total condensation* (Figure 31). This is the method of operation generally employed in modern distillation columns. The whole of the distillate vapour is condensed in a condenser. Part of the condensate produced is extracted as product, the remainder is fed as reflux into the head of the column. This method permits direct adjustment of any desired reflux ratio so that it is possible to control the distillation with great precision.

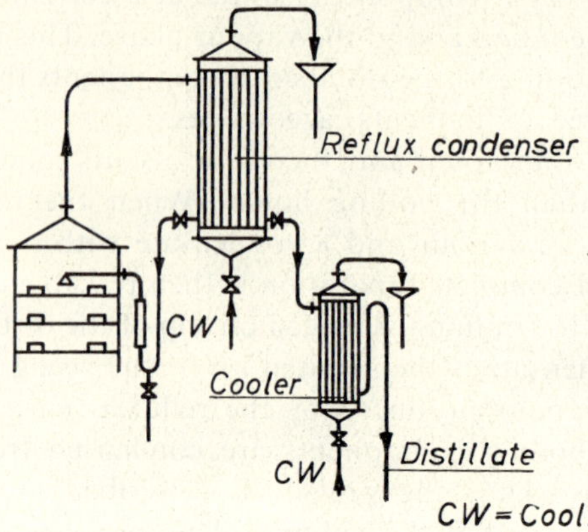


Figure 31. Column with total condensation

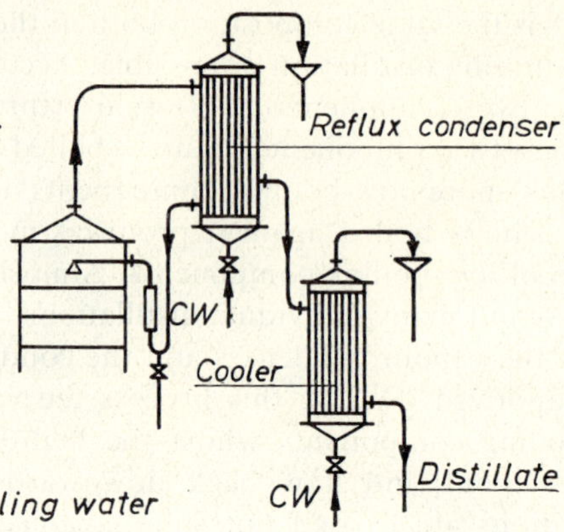


Figure 32. Column with partial condensation

In addition to the reflux produced deliberately as described above every column has another internal reflux which is produced by additional condensation at the walls of the column as a result of the temperature difference between the temperature inside the column and the ambient temperature. This internal reflux shifts the reflux ratio from the adjusted position in such a way that the amount of reflux increases continuously from the head of the column downwards because of the steadily increasing difference in temperature. In order to keep this internal reflux down to a minimum the columns are insulated. For the same reason smaller laboratory columns are surrounded by a Dewar jacket. According to the method of operation there are continuous and intermittent or retort distillation.

(*1*) *Continuous distillation* (Figure 33). It follows from the previous paragraph that the temperature inside the column drops from high at the bottom to low at the top, each temperature corresponding to a definite composition of the mixture inside the column. If the initial mixture is fed at boiling temperature into the column at that point where the temperature of the column is the same as the boiling temperature of the initial mixture, continuous distillation is obtained. The inflow of the mixture divides the column in two. Above the inlet the column is called upper, rectifier, or reinforcement column; below, lower, or separating column. The entire heat required for the separating operation is supplied to the sump of the column by steam through a heat exchanger (evaporator), and the mixture from which the lowboiling

boiling constituents have been eliminated is extracted there. With this method of operation a state of equilibrium which does not change during the distillation is established inside the column at every point.

(2) *Intermittent or retort distillation* (Figure 34). In this method the mixture to be distilled is fed all at once into the retort and then heated. When the desired concentration at the head of the column is reached, discharging starts into the first receiver. Owing to

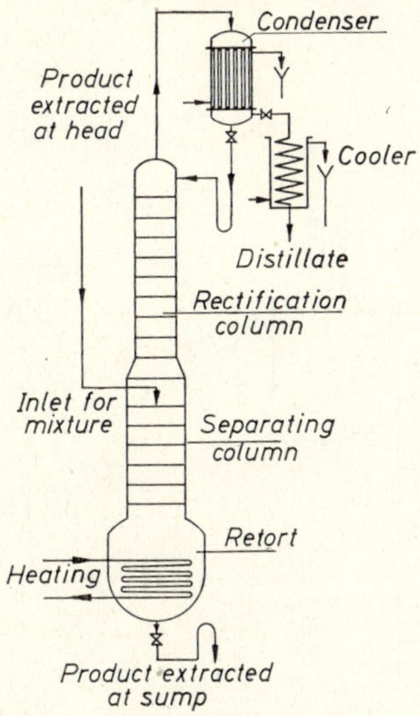


Figure 33. Diagram of a continuous rectification column

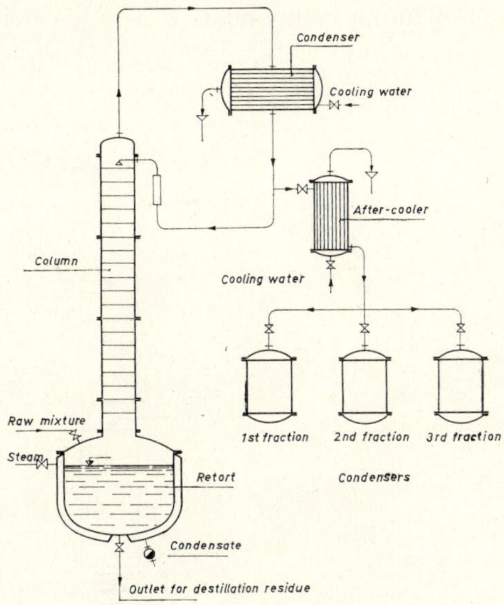


Figure 34. Retort distillation (batchwise)

the continuous extraction of the distillate the composition of the mixture in the retort changes constantly since the low-boiling components in the retort decrease continuously. As a result the temperatures inside the column-change constantly, i.e. the more low-boiling material is extracted at the head of the column the more the temperature rises both in the retort and in the column.

So as to be able nevertheless to extract a good distillate at the head it must be possible to alter the reflux ratio to obtain steadily increasing values. If the extraction becomes too small the receiver is exchanged and the remainder of the low-boiling material is then distilled into the receiver together with some high-boiling material at increasing head temperature. This is called the intermediate run. It is used again at the next filling of the retort and re-distilled. After completion of the intermediate run the highest boiling of the remaining components is distilled into the next receiver. Then follows another intermediate run and so on until the individual components of the mixture are separated.

There are several points in which continuous and batch distillation differ:

(1) In intermittent distillation several components can be extracted in succession from one column whilst with continuous operation several columns are required. The

number of columns is given by the number of individual components of the mixture minus one, i.e.:

$$Z = n - 1 \quad (n = \text{number of components of mixture})$$

Figure 35 shows the continuous distillation of a three-component mixture $A + B + C$: In the first column the extraction of the lowest-boiling component $A$ is carried out at the head whilst the mixture of the remaining components $B + C$ is extracted from the sump and fed into the next column, where the separation of the two remaining components $B + C$ is carried out.

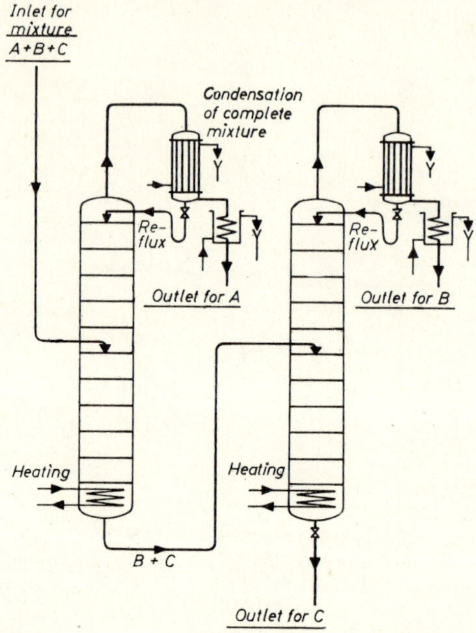


Figure 35. Separation of a three-component mixture A + B + C in two columns by continuous distillation

(2) Continuous distillation arrives at an equilibrium that does not change during the whole of the period of distillation which can be unlimited. This permits automation on a large scale so that large continuous distillation plants can be operated by a very small number of personnel.

In intermittent distillation on the other hand composition and temperature change continuously. The reflux ratio must be adjusted all the time to the changing conditions in the columns. The individual fractions and the intermediate runs must be distilled separately into different receivers. Moreover, the period of distillation is limited by the capacity of the retort which must be filled afresh for each run and emptied after completion of the distillation. This method requires a separate operator for each column.

(3) Less energy is required for continuous distillation than for intermittent distillation in which the individual intermediate runs have to be processed again at each distillation stage.

When to use intermittent distillation and when continuous distillation will be decided by the following considerations:

(1) Continuous distillation will be used where large quantities of mixtures have to be processed. Moreover, a continuous process also requires if possible continuous distillation.

(2) Intermittent distillation is appropriate for small quantities of mixtures for which continuous distillation is not practicable. Intermittent distillation must also be used where a viscous residue (pitch) impedes the passage through the column.

Technology of distillation

(*1*) *The columns.* There are two types of columns: plate columns and packed towers. Both plates and packing material serve to improve the heat and mass exchange between liquid and vapour. With plates this is achieved by running the reflux across

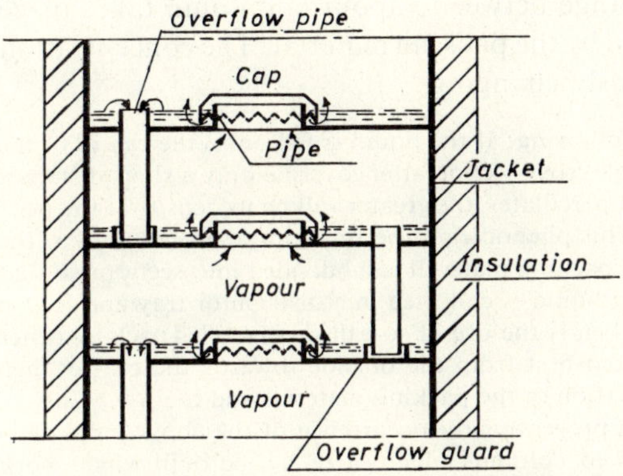


Figure 36. Diagram of a bubble tray column

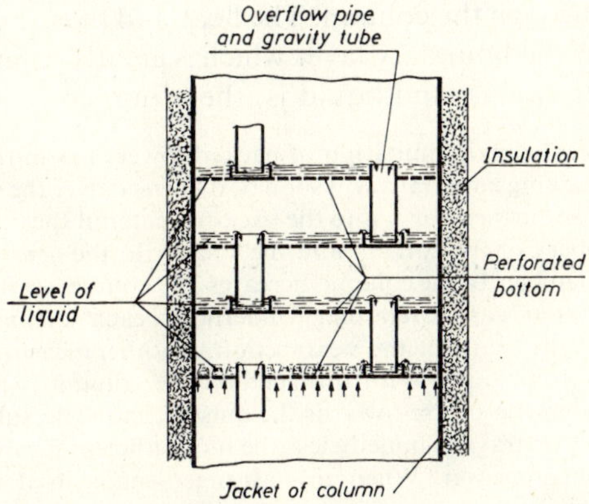


Figure 37. Diagram of a perforated plate column

the plate so that the rising vapour must pass through the liquid to reach the next higher plate. The mass and heat exchange, therefore, takes place only during the passage of the vapour through the liquid. The mutual exchange improves with increasing size of the surface of liquid and vapour. The different types of plates are designed to create as large a surface as possible.

(*a*) *Bubble trays.* These have numerous perforations for the passage of the vapour, in the shape of short tubes covered by a cap. The lower edge of the cap either has a large number of slots or it is serrated. This forces the vapour to pass through the liquid in the form of a large number of bubbles (see Figure 36).

The disadvantage of the bubble tray is that with liquids containing impurities cleaning is possible only by dismantling. There is, moreover, always a residue of liquid at the bottom even when the column is not in operation. This renders it unsuitable for intermittent distillation. Small holes are, therefore, drilled into the bottom outlet pipe which permit the liquid to run off when the column is not working.

An advantage of the bubble tray is, however, the fact that it is largely insensitive to fluctuations in load.

(*b*) *Perforated plate* (Figure 37). This consists of a perforated tray with many holes for the passage of the vapour. Just as with the bubble tray, the liquid runs across the plate, then through an overflow pipe into a gravity tube leading to the next lower plate. Owing to the design of the plate the vapour load of the plate can be altered only within narrow limits since with too small a load the liquid runs through the perforations, and with too large a load the plate is blown dry. The plate has, however, the advantage that it runs completely empty when the column is taken out of operation.

In addition to these two main types there is a large number of special designs, e.g. tunnel plates, flap plates, peripheral exchange plates, and others which, however, in almost every case are based on the principle of the two types described above.

Packed towers, on the other hand, work on an entirely different principle. The column is filled partly or wholly with packing material. The liquid is delivered at the top of the filled column and percolates through it downwards. The surface of the packing material is completely wetted by the liquid, which is thus spread over a very large surface. The vapour rising from below is also divided up into many partial currents by the packing material. Both liquid and vapour have, therefore, to traverse a considerably longer path through the column. The heat and mass exchange between vapour and liquid takes place in the boundary layer which is greatly enlarged by the packing material. The concentration of vapour and liquid is, therefore, continuously changing.

The main limitation of packed towers lies in the following: If the liquid is fed on to the centre of the packing material it will soon be dispersed over the whole cross-section after covering only a short distance. The further down into the packing material the liquid percolates, the greater will be its density at the periphery of the column and the smaller in the centre. This phenomenon becomes more pronounced as the diameter of the column increases. To counteract it the packing material is subdivided into sections so that this process is retarded. Underneath each section the liquid is collected in a distributor tray and fed on to the centre of the next section. Another measure taken is the use of so-called controlled packing (after Kirschbaum). With this method the column is packed first from the outside towards the centre, then from the centre towards the outside, etc. The subdivision of the packing material and the use of distributor trays is nonetheless the most efficient means of preventing the occurrence of the above-mentioned phenomenon. When this arrangement is used packed columns of any size can be built which work satisfactorily.

Rings of various sizes, 'Raschig' rings, are used, as packing material, also Berl saddles, and other types. All fillers can be prepared from various materials, e.g. glass, sheet metal, ceramics, graphite, plastics, etc., according to the purpose for which they are intended.

The principle of packed towers is often used also for other purposes, e.g. to achieve good distribution of a gas in a liquid. By the formation of a liquid film on the packing material and provision of a large surface, gases can be brought into close contact with liquids, or immiscible liquids with each other. Thus packed towers are used for scrubbing gases and for the countercurrent extraction in the liquid phase. These packed towers are called scrubbing, absorption, and extraction towers.

For the separation of mixtures the boiling points of which differ by only a few degrees, columns with 40–50 trays, often 80 trays, and even 120 trays are used. With multiple mixtures it is possible to tap at different heights and extract fractions with constant composition.

Figure 33 showed a diagram of a column, reflux cooler (dephlegmator) and cooler. An arrangement for the extraction of three fractions is shown in Figure 41 in which the condenser and cooler are placed above the column. The platform construction necessitated by this arrangement is very expensive and requires much material for the high and large columns used in the petroleum industry. In the design adopted more recently the columns are, therefore, freestanding. Figure 38 shows the construction and

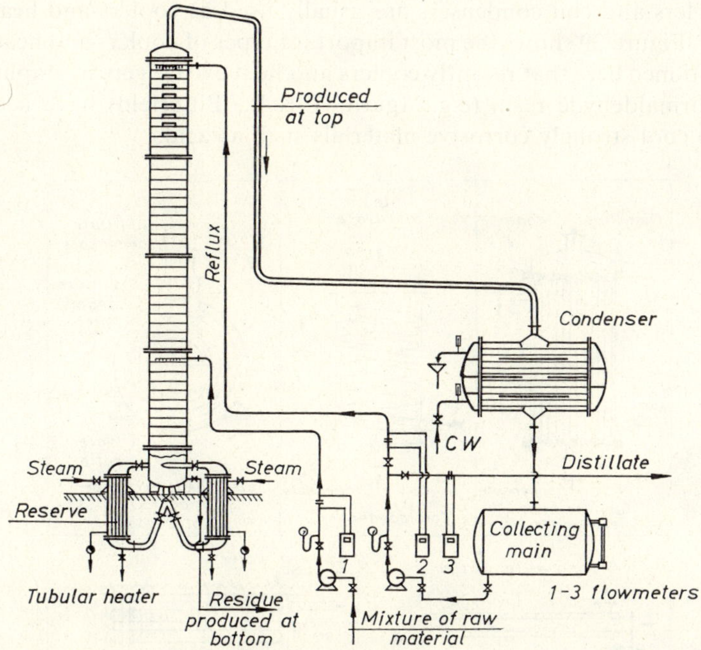


Figure 38. Method of operation of a free-standing bubble tray column for continuous distillation with recirculation heating

method of operation of such a column. To protect it from the weather the insulation is surrounded by a metal jacket with a protective coat of paint. The cooler is at the bottom. The reflux is operated by a pump, and the feed is heated either by circulation heating (e.g. in evaporators) or by pre-heating in a tubular still furnace.

### *Steam traps* (*steam separators*)

When an apparatus is heated by steam, e.g. by a double jacket or by heating coils, a quantity of steam corresponding to the heat used is condensed. If the outlet valve is left open the condensate can run off freely but it is difficult to regulate the flow in such a way that no steam escapes at the same time. If heating above 100° is required a throttle valve must be used which is so adjusted that only condensate but no steam escapes. Such an adjustment is difficult to achieve and self-adjusting steam separators, 'steam traps', are, therefore, used.

A steam trap allows the condensate produced to drain off without permitting the passage of live steam. The regulating impulses for the discharge of the condensate are given by the physical differences between steam and liquid.

When the steam trap is governed by a float, a throttle valve is opened as soon as the liquid reaches a sufficiently high level.

In other designs the regulating impulse is given by the difference in temperature between steam and water through a bimetallic regulator. The difference in the behaviour of steam and water when passing through a nozzle or a labyrinth is also utilized sometimes.

Tubular coolers and coil condensers are usually used as coolers and heat exchangers for distillation. Figure 39 shows the most important types of coolers and heat exchangers. It may be mentioned here that recently coolers and heat exchangers of graphite cemented with phenol-formaldehyde resin (e.g. 'Igurit' E.K.B., Bitterfeld) have been used very successfully to cool strongly corrosive materials such as acids.

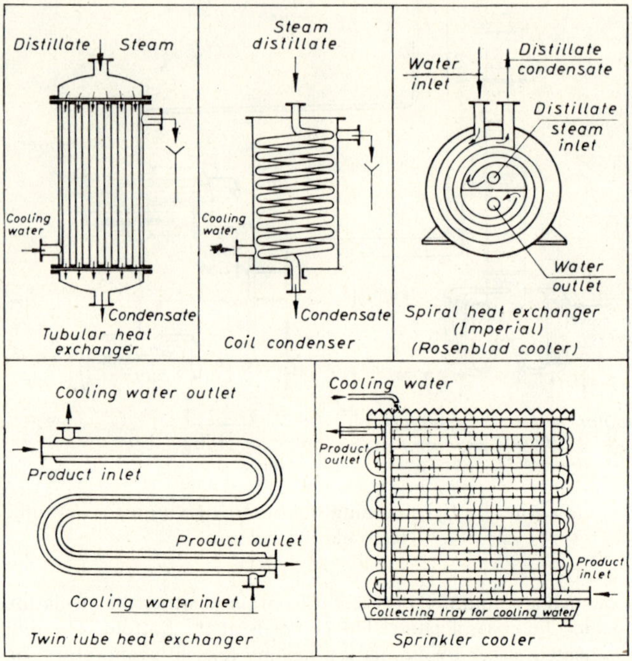


Figure 39. Diagram showing design of coolers and heat exchangers

We propose now to discuss some concepts used in distillation technology. When a column operates constantly, liquid and vapour in 'ideal scrubbing' are in equilibrium. Above each plate the product with the lower boiling point is concentrated more strongly than below it.

'Ideal scrubbing' and 'theoretical concentration' are related. Theoretical concentration is the optimum concentration for a plate at equilibrium between discharging liquid and vapour. 'True concentration' is, of course, always somewhat less than theoretical concentration. The actual concentration ratio of a plate or a tower is the ratio of true concentration to theoretical concentration.

'Theoretical number of plates' is the number of plates that would be required in a plate column at theoretical concentration on each plate to obtain the required separation at a given reflux ratio. Reflux was defined earlier.

With packed towers it is also possible to speak of 'theoretical plates', relating the effect to that of a bubble tray column with the same performance. The 'specific number of plates' is then the theoretical number of plates that corresponds to the effect of a packed tower of 1 m height. This means for instance that the specific number of plates of a packed tower with controlled packing is greater than that of one with random packing. Columns can be placed on stills, or the separating column and the still can be erected separately. The sump of the column can be heated by heating coils or by circulation heating. Finally, the mixture to be separated can be fed or injected into the column at the bottom or at a stipulated height so that fractional condensation occurs in the separating column in the various sections.

In addition to the product to be separated, steam or auxiliary liquids can be fed into the column.

Specific distillation methods have been developed for most products and there is hardly a field in process technology that has been so thoroughly worked and analysed as distillation technology, so that distillation experts can often on the basis of a laboratory distillation alone determine the type and size of a large distillation plant. For the calculations the determination of an equilibrium curve of the mixture to be separated is necessary.

This can be done only empirically in the laboratory or in the works by a trial distillation when the concentration of one of the components in both the liquid and the vapour must be continuously determined analytically during the distillation. The concentration in

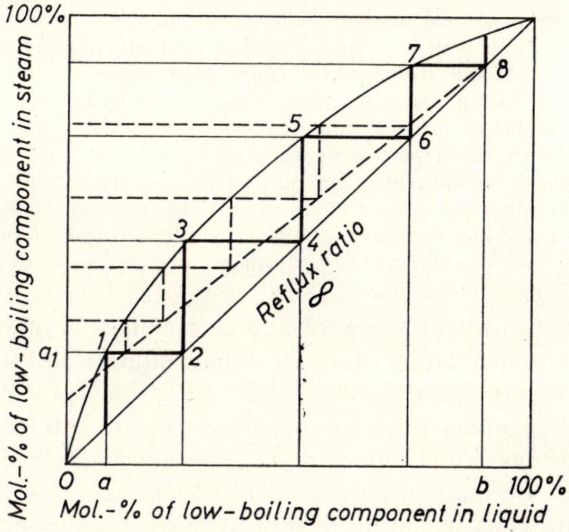


Figure 40. Equilibrium curve and stepped diagram for the determination of the number of theoretical plates

the liquid of a low-boiling component is plotted e.g. in mol.-% on the abscissa, the concentration of a low-boiling component in the vapour on the ordinate. Since in miscible materials that do not form azeotropic mixtures the component with the lower boiling point is concentrated in the vapour, a curve results such as is reproduced in Figure 40. If the mixture is azeotropic (see D'Ans-Lax, *Taschenbuch für Chemiker und Physiker*) there is a point on the curve where the composition of the vapour is the same as that of the liquid, i.e. where the curve intersects the 45° line. A well-known example is the mixture of 96 per cent alcohol and 4 per cent water which behaves as a uniformly boiling liquid. A separation can then be carried out only by distillation with an auxiliary liquid (see Azeotropic esterification, p. 208) or in certain circumstances by altering the pressure. Such azeotropic mixtures of two or more substances occur very frequently and render very difficult the rectification of mixtures, e.g. the separation of aliphatic and aromatic hydrocarbons.

From the equilibrium curve the number of 'theoretical plates' required for the separation of a mixture can be determined as follows: Let the mixture of two substances have the composition *a* from which it is intended to produce one with composition *b*. In accordance with the equilibrium curve, the concentration in the liquid of low-boiling material (mol.-%) is $a$, in the vapour $a_1$. It is assumed that the vapour is completely condensed on each theoretical plate. A line is drawn from $a$ parallel to the ordinate as far as the intersection (1) with the equilibrium curve, thence a line parallel to the abscissa as far as the diagonal (45° line). From this point (2) a line is drawn parallel to the ordinate as far as the intersection (3) with the equilibrium curve. Another parallel to the abscissa is then drawn from (3) as far as the diagonal (4) and this point is then connected with the equilibrium curve by a parallel to the ordinate as far as the intersection (5). This procedure is continued (points 6, 7, and 8) until composition $b$ is reached. Thus a stepped curve is obtained in which the number of theoretical plates is indicated by the number of steps (in our example four). The reflux in this case is $\infty$ which in practice never happens. If the reflux ratio is lower, the 45° line does not apply and one has to use another line which is displaced more at its lower end on the ordinate as the reflux ratio decreases (dotted line). On the assumption that the separting effect changes linearly, this line is called the concentration line (McCabe-Thiele). The slope of the concentration line can be calculated approximately from the reflux ratio. If a new stepped curve (dotted) is drawn in accordance with the above-described method using the equilibrium curve and the concentration line, it is immediately evident that more steps result (in our example six). The lower the reflux ratio the more theoretical plates are, therefore, required. Conversely, by a trial distillation of a known mixture, with as far as possible complete reflux, the number of 'theoretical plates' in a column of unknown performance can be determined.

For a bubble tray column the concept of the number of theoretical plates that is determined by a trial distillation of a mixture applies in general also to all other pairs of substances to be separated. In a packed tower, on the other hand, the degree of efficiency of the theoretical plates depends entirely on the nature of the mixture, and the values of the performance of a packed tower for a certain pair of substances cannot be simply applied to another.

Separation by direct distillation, e.g. of aliphatic olefins, is often very difficult. This is resolved by extractive distillation in which a third liquid is introduced as an auxiliary substance into the column.

Materials with boiling points below room temperature are distilled in pressure columns, e.g. simple hydrocarbons, vinyl chloride, and acetaldehyde. Substances with a high boiling point and reactive materials are distilled under vacuum.

The largest columns constructed up to now are used for the separation of the lower saturated aliphatic hydrocarbons.

### (a) Distillation of petroleum, conversion and refining methods

*General remarks*

Formerly, the purpose of crude oil processing was the production of the maximum amount of petroleum (paraffin, kerosene), lubricating oil, and fuel oil. This could be done merely by distillation and refining. The development of internal combustion engines necessitated the production of increasing amounts of petrol and diesel oil. Crude oil does not itself contain enough of these. To increase the yield of petrol the cracking processes were developed in which the fractions with higher boiling points are used. In addition to the straight-run petrol, petrol was also produced by cracking. During cracking, gases containing olefins are produced; olefins can also be obtained by cracking natural gas. In order to utilize these raw material sources for the production of petrol methods for the polymerization or alkylation of olefins were developed, thus adding a third type of petrol, the polymer petrol or alkylate.

Since most petroleum fractions do not have the antiknock rating (see below) required by modern car engines with high compression ratios, methods had to be developed to improve their properties. These are called the reforming processes. They consist in the dehydrogenation of hydro-aromatics to aromatics, or in the isomeration of paraffins to paraffins with branched chains and olefin groups. From these processes was developed the production of aromatics from petroleum by catalytic dehydrogenation.

*Distillation of petroleum*

To distil crude petroleum it is heated in a coil to such high temperatures that most of it vaporizes. The heating takes place in a pipe still at a pressure of, e.g. 20 atmospheres (see Figure 41). The coil is inside a furnace which is heated by atomized crude oil, coal dust, or gas, by means of a burner.

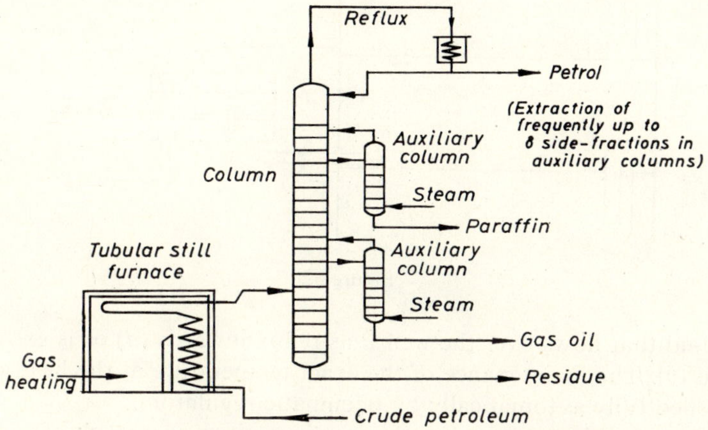


Figure 41. Diagram of single-stage distillation of petroleum

The flame is directed against an air-cooled fire bridge and does not, therefore, touch the pipes themselves. The coils in the oven are parallel and the crude oil passes through them countercurrently so that the cold oil comes into contact first with the cooled gases and only in the upper part of the furnace with the hottest gases. Heating can also be done by circulation heating shown schematically in Figure 60 (p. 126).

The superheated crude oil is injected into a bubble tower, or a system of bubble towers (e.g. a two-column, or two-tower plant). In one method the crude oil, pre-heated in heat exchangers, is injected into the middle of the first column and the petrol is taken off at its head. The hot non-volatile residual oil is drained off at the bottom and passes through a pre-heater into the second column. Some part of this residual oil is fed back, having been re-heated for a short period, and heats the sump of the first column. In the second column steam is added and fractionation takes place.

The advantage of this arrangement lies in the fact that the first column can be operated under pressure and the petrol is thus immediately stabilized. 'Stabilization' is the setting of a certain vapour pressure by fractionation with simultaneous removal of propane and/or butane in a pressure distillation. The corrosive materials, e.g. inorganic and organic sulphur compounds and acids, can also be removed directly in the first tower.

Cuts with the same boiling point are extracted from the individual plates. The fractions are adjusted by keeping the operating conditions constant, in particular the temperature of the head of the column, by regulating the inflowing and outflowing quantities. The flow from the extraction plates is automatically controlled by proportionating slide valves which regulate the flow outwards and the flow inwards to the next lower plate (Figure 42, from Winnacker-Küchler). Slide *b* which is set by the scale *e* controls the

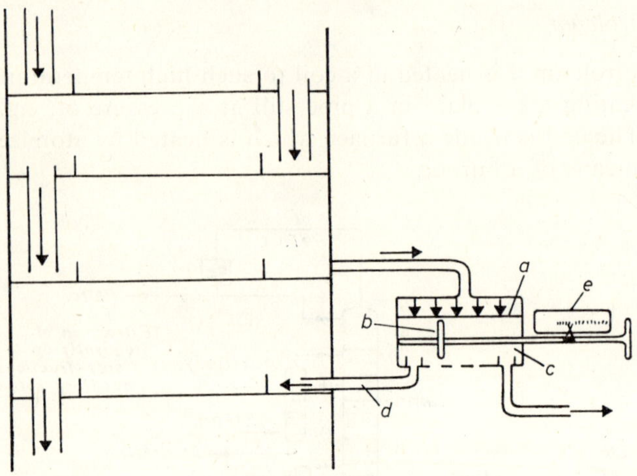


Figure 42

quantity of fluid that flows over the weir *a* as reflux inwards (*d*) or is extracted as 'fraction' outward (*c*). The maintenance of the exact temperature at the head of the column is also controlled fully automatically by pneumatic regulators.

The outflowing fractions again absorb some impurities as a result of some of the low-boiling fractions dissolving on their passage upwards in bubble-form through the plates. The extraction of the fractions is, therefore, carried out in small 'strippers' (see Figure 41). These are small bubble towers with heating coils at the bottom, into which steam is injected from below. This expels dissolved lower fractions which are then fed into the main column about two plates higher up. A similar effect is obtained by 'stripper plates' built into the column.

It is also possible to inject steam into the main column. This makes for better vaporization and conservation of the products. The steam behaves like an inert gas but is greatly superior to it because it precipitates in liquid form during condensation. Liquid water can easily be separated from petrol and oil. Steam can be used with advantage as a carrier gas also in vacuum distillation.

The following fractions are usually collected in the distillation of petroleum:

Up to 200°C petrol
200–250° kerosene
250–300° gas oils, (diesel oils)
Above 300° fuel oils, lubricating oils
Residue: bitumen.

The heavy fuel oils with boiling points above 300°C are then distilled separately under vacuum and yield heavy fuel oils, lubricating oils, hard paraffins, and vaseline. Since the consumption of lubricating oils, compared with the consumption of fuel oils, is very small (maximum 2 per cent), the fractions best suited for this purpose can be cut out. Lubricating oils for internal-combustion engines are then further refined, particularly by selective solvents.

The individual fractions can be further separated, e.g. the petrol fraction into petroleum ether, white spirit, motor fuel petrol, etc.

### (b) Refining and upgrading, dewaxing

Petroleum contains paraffinic fractions that cannot be separated by distillation, since innumerable isomeric hydrocarbons with the same boiling point are present. Dewaxing is of special importance for lubricating oils. At low temperatures paraffins tend to crystallize and raise, therefore, the solidifying point of the oil, i.e. at low temperatures it becomes viscous and finally solid. An ideal lubricating oil should have a viscosity as far as possible independent of temperature so as not to become too thin when hot and not too viscous when cold. In this respect paraffins with many side chains are best. It may be mentioned that a synthetic lubricating oil (e.g. 'Buna-Öl') produced by the VEB Chemische Werke Buna, Schkopau, by polymerization of ethylene with aluminium chloride consists mainly of branched chains. Many oils produced from brown coal, e.g. from carbonization tar, have, however, adverse properties.

The method most commonly used for dewaxing is based on cooling below 0°C; during this process the straight-chain paraffins crystallize out and the material solidifies. The crystalline constituents cannot be separated from the pulp. It is, therefore, diluted using one of a variety of technological methods with thin solvents, e.g. liquid propane, butane, benzene, acetone, 1,2-dichloroethane, $SO_2$, methylene chloride, and the crystals are separated by skimmer coolers or drum dip-cell filters.

This method is rather costly because of the expense in producing cooling and in solvent recovery in which small losses are unavoidable.

From an observation made by F. Bengen a method was developed which gives rise to the precipitation of straight-chain paraffins in the form of crystalline adducts from petroleum fractions by means of urea and their subsequent separation by filtration or centrifuging; the heat of crystallization produced is absorbed by methylene chloride cooled in reflux. The adhering oil is removed by suitable solvents and the adduct is cracked with water, the urea being recovered. Paraffin can thus be produced from petroleum. This method is used mainly for the refining of lubricating oil.

The crude petroleum fractions contain simple and complex unsaturated compounds. Started by an autoxidation process, the unsaturated compounds cause precipitation ('gumming') in cracking petrols during storage, and resinification in lubricating oil. If all unsaturated compounds are eliminated the wetting action of oil on metal surfaces is impaired. It has been shown that resinification is caused by the complex unsaturated compounds but that mono-olefins have a favourable effect on the wetting action. The refining methods aim at removing the constituents which cause the formation of resins.

A large part of the complex unsaturated constituents can be polymerized and separated out with small quantities of concentrated sulphuric acid. This effect is, however, often insufficient. In that case 'fuming sulphuric acid' containing free $SO_3$ is used. The sulphuric acid is then added on to a double bond and acid sulphuric acid esters form which are dissolved.

Separation is, however, often rather difficult. For this purpose the oils are mixed with concentrated sulphuric acid in conical lead-lined vessels. The dark acid sludge settles at the bottom and is discharged; the fluid is then washed with water, caustic soda solution, and again with water. This method consumes, however, very large quantities of reagents and causes loss of oil. Physical refining methods, e.g. adsorption with Fuller's earth, silica gel, and activated carbon, are, therefore, employed. With these it is possible to remove high-molecular coloured impurities but not the resinifying hydrocarbons.

### *Deasphalting*

Crude petroleum, distillation residues and also some distillates contain dark highly viscous substances insoluble in petrol. Because of their similarity with natural asphalt they are here called 'asphalt'. They contain oxygen and often also sulphur. Those with melting points below 100°C are called 'soft asphalt', above 100°C 'hard asphalt'. Only the latter have a harmful effect. They are precipitated with petrol and are soluble in benzene. For deasphalting liquid propane is used which precipitates the bituminous fractions ('asphalt').

### *Purification of oils by selective extraction*

Of much greater importance is the selective extraction with solvents, of which the oldest and best known is the Edeleanu process with liquid $SO_2$. Liquid $SO_2$, boiling point −10°C, dissolves aromatic and olefinic hydrocarbons more easily than paraffins and naphthenes. The oil is, therefore, scrubbed with liquid $SO_2$ under pressure, the $SO_2$ extract is separated and the solvent is recovered by expansion and distillation. The insoluble hydrocarbons are not affected by $SO_2$.

The diagram in Figure 43 shows the method of operation of an oil scrubbing plant with selective solvents, in particular $SO_2$, in a very simplified form. A similar method is used for refining lubricating oils.

Distillation and selective extraction are complementary in oil purification. Solvents used include phenol and its homologues, alcohols, esters, nitrobenzene, furfurol, and others (selective dissolving solvents), furthermore e.g. propane (precipitating solvent) and for petroleum mainly liquid $SO_2$.

$SO_2$ is liquified by compression and cooling: the petroleum fraction to be purified is pre-cooled to −10°C; the liquid $SO_2$ is fed from above into a mixer (extraction tower), the oil from below. Since $SO_2$ is heavier than the hydrocarbons the extract sinks to the bottom. The refined oil overflows from the mixer at the top, the $SO_2$ and the extract are removed by evaporation. The $SO_2$ is then fed back into the compressor.

From the economic point of view an essential factor in the Edeleanu process is an extremely exactly calculated heat, or rather cold, economy. This is achieved by the interposition of many cold exchangers (not shown in the diagram) in which e.g. the cold of the $SO_2$ evaporation of the extract, of the refined product, etc. is utilized. This is indicated in the diagram by a heat exchanger in which the utilization of the cold takes place. The Edeleanu process is used mainly for paraffin and lubricating oils.

### *The extraction of aromatics from petroleum as an example of the separation of various materials with the same boiling point*

Petroleum fractions with the same boiling point but different constitution such as paraffinic hydro-aromatic and aromatic compounds, cannot be separated by distillation. To separate the aromatics present originally the petroleum can be extracted directly. A much larger quantity of aromatics is contained in conversion products from reforming processes, e.g. pressure dehydrogenation, hydroforming, the 'DHD' (pressure-$H_2$-dehydrogenation) and platforming processes.

Liquid $SO_2$ can be used as an extracting agent; this leads, however, also to the separa-

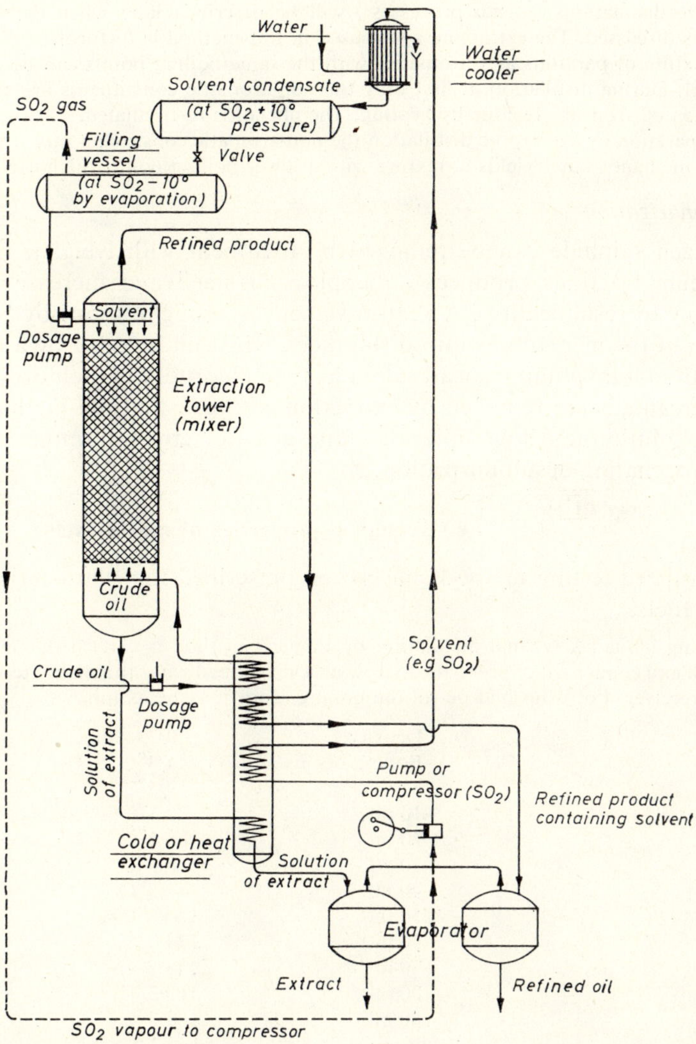


Figure 43. Method of operation of an oil scrubbing plant with selective solvents (e.g. with $SO_2$ after Edeleanu)

tion of some non-aromatic fractions. When the $SO_2$ extract is then washed with paraffin oil only aromatics are left in the $SO_2$.

Aromatics can be extracted also with selective organic solvents.

In the 'Udex process' a glycol-water mixture is employed which is so selective that a broad fraction containing benzene, toluene, and xylene can be used as a starting product. As in the Edeleanu process the extraction is carried out in a column. The aromatics can be separated from the solvent mixture by heating. Recently, a new extraction process, the Mofex process, was developed at the VEB Leuna-Werke in which monomethylformamide mixed with water is used as extracting agent.

Finally mention should be made of the extractive distillation process developed by Shell. These extractive distillations ('distex processes') will be described later when the separation of butane and butene is discussed. The extracting agent used in this method is furfurol.

A mixture of paraffins and aromatics with the same boiling points can be dissolved also in phenol or cresol. During distillation at first only the non-aromatic constituents are separated. The aromatics are driven off from the residue by heating. The phenol is re-circulated.

In separation by azeotropic distillation the non-aromatic constituent mixed with an auxiliary solvent such as methanol-water yields a mixture with a lower boiling point which can be distilled out.

*Desulphurization*

Hydrogen sulphide can be removed by treatment with lye. The mercaptan content of petroleum fractions produces an unpleasant smell and increases corrosivity and the tendency to resinification. An improvement is brought about by 'sweetening', the conversion of the mercaptans into disulphides. This can be done by a number of processes, e.g. with sodium plumbite, lead sulphide, hypochlorite, Cu(I)-chloride. In another method the mercaptans are removed by extraction with caustic soda-methanol solution, caustic potash solution and low-molecular fatty acids or ethanol amines. Yet another method is hydrogenating desulphurization.

## (c) Technical properties of motor fuels

Standardized testing methods have been prescribed for characterizing petrol and other motor fuels.

Boiling limits: A normal distillation by Engler's method is carried out in standardized apparatus, i.e. flask and condenser of prescribed sizes, at a prescribed rate, the distillate being collected in a calibrated receiver. For a normal petrol the boiling range may be as follows:

| | |
|---|---|
| Density | 0·734/20°C |
| Boiling begins at | 44·3°C |
| 5 ml | 59·3°C |
| 10 ml | 70·3°C |
| 20 ml | 85·3°C |
| 30 ml | 98·3°C |
| 40 ml | 111·3°C |
| 50 ml | 121·8°C |
| 60 ml | 134·3°C |
| 70 ml | 148·8°C |
| 80 ml | 169·3°C |
| 90 ml | 183·3°C |
| Ends at 98·3 ml | 190·3°C |

The flash point is the temperature at which a liquid can be ignited in air without a wick. The sample is heated slowly in a special test apparatus, and the temperature at which the mixture can be ignited by a flame is determined at intervals of 1°C. The temperature at which a flash occurs is called the flash point.

Various methods are used for the determination of the viscosity the degree of which is of importance particularly for lubricating oils. In the Engler viscometer, 200 $cm^3$ oil are passed through a standardized orifice at a certain temperature. The time taken is compared with that taken by 200 $cm^3$ water. The ratio of these times gives the Engler degrees. In the Ostwald viscometer, the liquid flows through a glass capillary from one side of a U-tube to the other. The reference liquid is water. In Ubbelohde's method the flow is measured at constant water-column pressure. The most reliable instrument, particularly for highly-viscous liquids, is the Höppler viscometer in which a sphere drops in an inclined tube from one calibration mark to the next. The falling time at a given temperature is a measure of the viscosity.

*Anti-knock properties*

The 'knocking' of fuels in engines is caused by premature ignition or too rapid combustion. Higher compression results in a saving of fuel, but the higher the compression the stronger the 'knocking'. Anti-knock agents are inhibitors that interrupt the chain reaction, e.g. through the surface effect of finely divided solid materials (lead tetraethyl, often mixed with ethylene bromide, or iron pentacarbonyl), or by antioxidants (amines, aminophenols) which capture the radicals produced during the autoxidation. Numerically the anti-knock properties are expressed as octane numbers. To determine the octane number, the anti-knock properties of the test petrol are compared in a test motor with mixtures of iso-octane (2,2,4-trimethylpentane) and n-heptane. A mixture is found with the same anti-knock properties as the test petrol. The octane content in per cent by volume of this mixture is the octane number. In the scale of octane numbers, iso-octane equals 100 and n-heptane zero. In order to save octane, substandard test petrols of known octane number are used. Car petrols should have octane numbers of 80–100 corresponding to a mixture of 80–100 per cent by volume of iso-octane and 20–10 per cent by volume of n-heptane. Iso-octane has almost the highest anti-knock properties of all paraffins. Triptane (2,2,3-trimethylbutane) has even higher ones, n-heptane has the lowest (zero). Di-isopropyl ether, produced from propylene, also has a very high octane number and has been used as aviation fuel. The octane number of motor fuels produced for example by cracking and reforming processes can be raised from 70–80 to 80–88 by the addition of tetraethyl lead (TEL).

The ignitability of motor fuels is expressed by the cetane number (hexadecane $C_{16}H_{34} = 100$, α-methyl naphthalene $= 0$) which is determined in the test engine. It indicates the range in which diesel fuels can be used.

Petrols must not resinify when exposed to air. The resinous residue (gumming) is determined by evaporating a sample of 100 $cm^3$. The residue must not exceed 10 mg. Gumming, too, is an autoxidation process in which metals act as catalysts. In the first stage peroxides can be detected. The process can be stopped by antioxidants, e.g. 'Stabisol' (monobenzyl-p-aminophenol).

## 2. THE CRACKING PROCESSES

In many petroleums the petrol fraction is small (this is true e.g. of those found in Germany). By superheating the higher-boiling fractions, cracking into low-molecular hydrocarbons can be achieved. Cracking can be carried out either in the vapour phase without pressure or in the liquid phase under pressure. The first cracking processes were derived from the distillation of petroleum during which cracking was already observed. The material is heated to about 450°C in a tubular heater (tube still) and then left for 15 to 20 minutes under pressure in a heat-insulated pressure vessel. It is then released into a column and the low-molecular products are separated by distillation. The uncracked petroleum can be fed back into the still. With this method a separation of coke is unavoidable. Moreover, some gases (cracking gases) of an olefinic nature are formed. Some of the cracked hydrocarbons are also olefinic. The yield of petrol produced by cracking (i.e. the processed distillates) was only 25–30 per cent with the old methods. Much more favourable, particularly as regards the separation of coke and the formation of gases which must be considered as losses, are the modern processes of catalytic and hydrogenating cracking with which the yield amounts to about 80 per cent of the petroleum used. These cracking processes are carried out either with a stationary catalyst, or, especially in the United States where this method was first developed, with a floating catalyst as a 'fluidized process'. The first large-scale fluidized process, though not for the cracking of hydrocarbons but for the gasification of coal dust, was the production of gas in the IG-Winkler Generator (see p. 69).

The following structural formulae show schematically pure cracking and hydrogenating cracking:

$$R{-}CH_2{-}CH_2{-}CH_2{-}CH_2R \rightarrow R{-}CH = CH_2 + H_3C{-}CH_2{-}R: \quad \text{cracking}$$

$$R{-}CH_2{-}CH_2{-}CH_2{-}CH_2R \rightarrow R{-}CH_2{-}CH_3 + H_3C{-}CH_2{-}R: \quad \text{hydrogenating cracking}$$

The cracked fractions in pure cracking are always partly saturated and partly unsaturated whilst in hydrogenating cracking, i.e. in the presence of hydrogen, hydrogen is again added to the unsaturated fractions so that in this process, if the hydrogen partial pressure is adequate, only saturated fractions are produced.

Cracking distillation has for many decades been of very great significance for petroleum and also for tar oils.

The method was first proposed in 1865 and was designed to increase the yield of kerosene which at that time was the product for which the demand was greatest. When town gas and electricity were introduced the experiments aiming at the cracking of petroleum were stopped. It was the development of the motor car industry and the steadily increasing demands made of the motor fuel industry that brought about a rapid development of the cracking processes. In the design of internal combustion engines the compression ratio was continuously raised in order to improve the performance of the engines, which entailed increasing demands for improvement of the anti-knock properties and it was these that were offered by the cracked petrols. Since 1936 the thermal and catalytic cracking of gaseous low-molecular hydrocarbons into unsaturated ones has been carried out to obtain the raw materials for the so-called 'polymer petrols' and 'isoparaffins'. The cracking process has also been used for the reduction of the solidifying point and of the viscosity of oils. In recent years a special cracking technique has been developed for the production of olefins such as ethylene and propylene as raw materials for plastics. These processes are described later in Chapter V.

The distillate petrols can be transformed into petrols with better anti-knock properties by a cracking process, the 'reforming process'. The reforming is carried out in a tubular heater under pressure and under accurately controlled conditions. By releasing the pressure and cooling, the process is suddenly stopped, then the material is separated by distillation and 'stabilized'. In this way the octane number of crude petrols can be almost doubled.

## Thermal cracking

Cracking takes the following course: Heavy oils undergo a certain degree of decomposition even at relatively low temperatures. Thus the viscosity of lubricating oils drops appreciably even at 200°C. At 350°C decomposition can be clearly observed and above 400°C distillation can no longer be carried out without excessive decomposition. This, then, is the upper limiting temperature for distillation in the petroleum industry. Above 400°C cracking of the hydrocarbon chain occurs with simultaneous formation of gas and of low-boiling products. At yet higher temperatures the amount of unsaturated hydrocarbons increases and even aromatics are formed. The most favourable temperature for this is between 700 and 900°C. The processes in the coke oven in which aromatic hydrocarbons are formed from hydro-aromatic and olefinic ones, are closely related to the cracking processes. Above 900°C hydrogen, methane, and coke are formed from aliphatics; in the presence of oxygen this gas mixture can, as is well known, be converted into water gas ($CO + H_2$). In thermal cracking mainly C—C— and C—H bonds are ruptured. Radicals are formed which react in a great variety of ways (D. T. A. Huibers and H. J. Waterman, *Brennstoffchemie*, 1961, **42**, 50).

Industrial cracking processes are carried out above 400° up to a maximum of 600°, the following main changes occurring:

(1) Straight hydrocarbon chains are split up into fractions, with formation of olefins.
(2) Hydrogen atoms are removed in the cracking, with formation of simple and complex unsaturated hydrocarbons.
(3) Hydro-aromatics are transformed into aromatics, possibly with reduction of the ring to a smaller size and cleaving off of side chains.
(4) Straight chains are rearranged into branched ones (isomerization).
(5) Unsaturated low-molecular products are assembled into higher-molecular ones by polymerization or condensation, sometimes with formation of rings.
(6) Aromatic materials are converted into coke by increasing condensation and dehydrogenation.

The cleaving off of hydrogen from hydrocarbons with formation of olefins is an equilibrium reaction: $C_nH_{2n+2} \rightleftharpoons C_nH_{2n} + H_2 - Q$. At low temperatures the equilibrium ought to be displaced towards the side of the paraffin, i.e. towards the left-hand side. However, at this temperature the speed of the reaction is zero. At high temperatures the equilibrium is displaced towards the olefins since heat is absorbed. The longer the chain the less thermally stable is the paraffin. It follows from the equilibrium reaction above that at high temperatures olefins are more stable than paraffins. In cracking the breaking of a C—C bond can be explained by the lower activation energy of a C—C bond than of that of a C—H bond. The activation energy required for the breaking of an aromatic C—H bond is, however, much larger and that required for a double bond even more so. In the cracking of mixtures, then, the molecules with the largest energy content are always preferred. A further consideration is that the C—C chain breaks preferably at different points according to the temperature and the pressure so that the reaction can be controlled. Naphthenes and hydrocarbons with five-membered rings are the most stable thermally. First, the side chains are split off, the breaking of the ring occurs much later. Six-membered rings yield some aromatics by dehydrogenation. At high temperatures these are also produced from olefins, e.g. from ethylene. Acetylene may also be formed which is polymerized to aromatics. Here the famous laboratory experiment should be remembered in which benzene and other aromatics were produced from acetylene in red hot tubes. Side chains are the first to be split off from aromatics, splitting occurring more easily the longer the chains are. The bonds of methyl and ethyl groups are very strong and methylated and ethylated aromatics are, therefore, found in larger quantities in bituminous tar. Aromatic rings are very stable; they tend to thermal condensation which may lead even to coke and graphite formation.

In thermal cracking the yield of petrol is about 50–65 per cent. The octane rating of petrol produced by cracking is higher than that of distillate petrols and it is particularly high in catalytically cracked petrols. Petrols produced by cracking contain 20–50 per cent olefins and 10–45 per cent aromatics.

The cracking reaction is endothermic, for instance: $C_{10}H_{22} \rightarrow C_5H_{12} + C_5H_{10} -$ 29,000 cal/mol, $C_6H_{14} \rightarrow C_6H_{12} + H_2 -$ 32,000 cal/mol. The process is, therefore, carried out in several stages with intermediate heating. The effect produced depends on the temperature, the pressure, and the duration of the reaction.

In the range 450–500°C, which is the most important one for industrial cracking, a rise in the temperature of only 15°C causes a doubling of the reaction rate. It is important for the technology of the process to separate the reaction products into different fractions immediately after the cracking of the raw material. The fractions in the desired boiling range and the gas are immediately removed in a column. The higher boiling fractions are returned while still hot. High-molecular fractions, poor in hydrogen, resins and asphalts must be removed as quickly as possible since they give rise to coke formation.

The fractions which can be vaporized but have too high a boiling point can after separation in the column be returned into certain zones of the cracking still. By prolonging the residence time in a chamber it is possible to ‘crack for coke’ deliberately.

This coke is ash-free and of a very high grade; it is used as electrode coke, like the coke produced during the processing of tar oils. The cracking is carried out thermally either in the liquid or in the vapour phase. It is thus possible to process crude oils that cannot be distilled, residues, and also distillates.

*Historical development and description of some cracking methods*

In the old Burton-Clark process of 1914 the oil is heated in a still to a high temperature. The still is connected by two ducts with a vessel above it. The oil circulates in the two vessels and the ducts on the thermosyphon principle. Thus less coke is separated than when the oil is heated simply in a vessel without circulation. The separation of coke is nevertheless very high and the still has to be cleaned after three days' operation. Accordingly, the yield of petrol is rather low (30 per cent). The method was still employed in the United States on a very large scale in 1930.

Very soon stills were discarded as heating units and tubular heaters were used in their place such as are also employed in the distillation of petroleum. Heat exchangers, columns, and condensers complete the apparatus necessary for all cracking installations. In general only distillates are used so as to make the operating periods as long as possible. It must be easily possible to open and drill the tubes of the tubular still furnace at the elbows. The return bends have, therefore, been specially developed and provided with a plug. The heaters are cleaned frequently.

*Actual working time during operational cycle*

When a plant is e.g. working for 45 days and idle for five days for cleaning purposes, the actual working time during the operational cycle is 90 per cent. This concept is important, for instance, for all catalytic installations which must be shut down for the regeneration of the catalysts, or for distillation plants in which the columns must be cleaned. The aim is always to keep the actual working time as high as possible.

*Expansion processes*

In the relaxation processes the oil is first heated under pressure, perhaps kept for a time under pressure and at a high temperature in a reaction chamber, and then the pressure is released.

The velocity of the oil in the tube still where it is pre-heated, is 3 to 5 m/sec, i.e. it flows very fast. This is the only way to achieve a favourable actual working time during the operational cycle. The next method to be considered is the Ellis process (Standard Oil of New Jersey, 1920), also called the 'tube and tank process'. The oil is heated in the tubular heater at 50–70 atm pressure to about 450°C and is then fed into a reaction chamber where it is left for about 15 minutes. It is then released into an expansion chamber where the cracked residual oil (used as fuel oil) is separated from the volatile products, mainly petrol, which is produced with a yield of 68–70 per cent by volume whilst the production of coke is only 2 per cent. If more coke is required, the residence time is extended by adding a further reaction chamber.

The 'Cross' method of the Gasoline Products Co. also uses tubular heaters, reaction chambers and expansion chambers. The reaction chamber has the remarkable length of 13 m, with a diameter of 1·30 m and a wall thickness of 125 mm, thus resembling a furnace for high-pressure hydrogenation. The processing of the cracked products is carried out in two columns. The method is hardly used any longer.

In the Winkler-Koch process (Figure 44) no reaction chambers are used, only tubular heaters. The diagram shows in simplified form the design of a Winkler-Koch cracking plant.

The crude oil pre-heated in the heat exchanger is heated further in the tubular heater I, so that in the first stage all low-boiling fractions are separated in the first column. The residue is used as fuel (bottom product of column 1). In the second column the petrol fractions are separated. The gas oil fraction (bottom product of column 2) is forced by a pump into the tubular heater II where the cracking reaction takes place under pressure. After passing through the furnace II the pressure is released in column I. The cooler products from furnace I are added after the pressure release and the sudden cooling stops

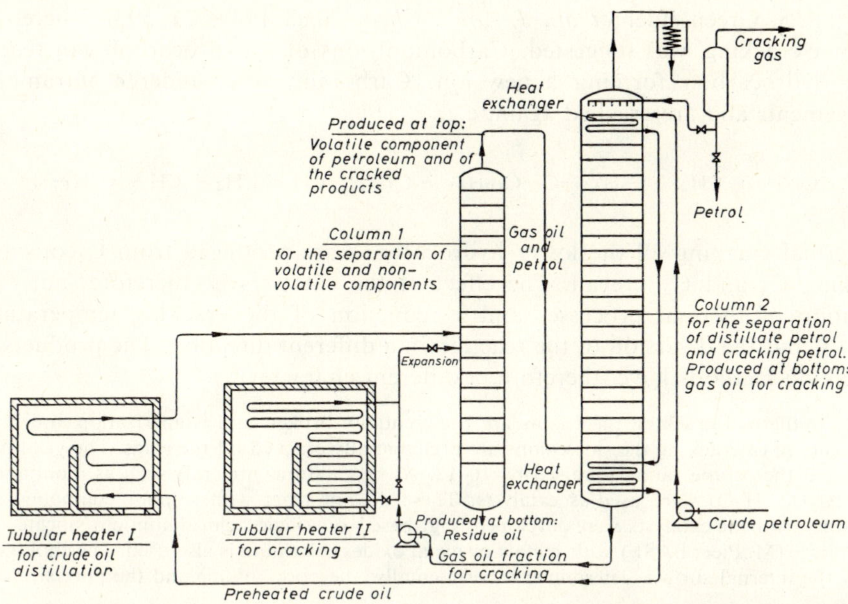


Figure 44. Simplified diagram of a petroleum cracking plant (Winkler-Koch process)

the reaction. Fractionation in two columns follows, i.e. all volatile components which are not removed as cracking gases and petrol through the top of column 2 are repeatedly returned into the process until they are cracked. In this method the distillation petrol is obtained from the first furnace and the cracking petrol from the second furnace.

Of the German methods practical significance was attained by the 'Carburol process', by H. Wolf (Rheinmetall-Borsig), originally used for tar oils. It resembles the Winkler-Koch process in that it has no reaction chambers, the cracking taking place in the tubular heater.

*Coking methods*

When the cracking period is extended more coke and gas are produced. Such coking processes include the Holmes-Manley process (1916) and the widely used Dubbs process of the Universal Oil Products Co., originally a coking process (until 1925) but later modified by the installation of expansion chambers. At the present time one tubular heater is used for heavy oil and another for light oil. The reaction period is controlled by adjusting the level in the reaction chamber.

Thermal cracking of ethylene and propylene will be described in the Section dealing with the production of olefins.

### Catalytic Cracking ('Catcracking')

Catalytic cracking is carried out exclusively in the vapour phase at low pressure. This means that only distillates can be used as raw materials.

The reaction mechanism differs completely from that of thermal cracking. In catalytic cracking, carbonium ions (a type of triphenylmethyl salt) are produced (F. C. Whitmore, *J. Amer. Chem. Soc.*, 1932, **54**, 3274; 1933, **55**, 4153; *Chem. Engng. News*, 1948,

**26**, 668; B. S. Greenfelder *et al.*, *J. Amer. Chem. Soc.*, 1949, **71**, 593). Therefore the term 'ionic cracking' was suggested. Carbonium ions of a hydrocarbon can react with another hydrocarbon forming a new ion. Carbonium ions undergo intramolecular rearrangements and are cracked again, e.g.

$$C_{15}H_{31}{-}\overset{\oplus}{C}H_2 \rightarrow C_5H_{11}{-}\underset{H}{\overset{\oplus}{C}}{-}C_{10}H_{21} \rightarrow C_5H_{11}{-}CH = CH_2 + \overset{\oplus}{C}H_2{-}C_8H_{17}$$

In thermal cracking all the lower hydrocarbons are produced from $C_1$ onwards, in catcracking $C_3$ and $C_4$ prevail. The effect of the catalysts is, therefore, not only an acceleration of thermal processes and a reduction of the cracking temperatures by 100–150°C but the diversion of the reaction in a different direction. The products of the two cracking methods have, therefore, a different character.

Whilst in thermal cracking the chains are preferentially broken and isomerization and polymerization occur, in catcracking the side chains are preferentially cracked off the rings. The yield of petrol is higher and the octane rating is raised by 10–12. At first natural minerals such as montmorillonite ($4\,SiO_2 \cdot Al_2O_3 \cdot H_2O$) were used as catalysts. These do not react with sulphur compounds. Later, innumerable synthetic catalysts were developed, e.g. based on silicic acid aluminium silicate or magnesium silicate (M. Pier; BASF) with metals or metal oxides; platinum is also used. The choice is determined by the internal surface determined experimentally, the pore volume and the pore size distribution curve.

*Principles of catalytic cracking*

Catalytic cracking methods are based on three principles: fixed bed catalyst, moving bed catalyst, and fluidized catalyst. In catalytic cracking, too, coke is separated which covers the catalyst and renders it ineffictive, so that the catalyst has to be regenerated by burning off the deposit. In the fixed bed method several catalyst furnaces are, therefore, required which are regenerated in turn.

The moving bed crackers operate continuously by circulating the catalyst through the reactor and the regenerator. From this are derived the processes with pulverulent fluidized catalyst material, fluid catalytic cracking.

The fluid process is now used most frequently, then follows the fixed bed, and finally the fluidized catalyst bed.

In the Houdry process the oil is pre-heated in a tubular heater, separated in the tar separator from the non-vaporized material, and is then fed into one of the 4–12 reactors with fixed bed catalyst each of which operates for ten minutes and is disconnected and regenerated in the following ten minutes. The temperature is kept constant at 430–480°C by heating tubes fitted in the reactors and through which flows a molten mass of potassium nitrate-sodium nitrite.

The 'Cycloversion process', used also for desulphurization and reforming, employs two alternately operating reactors and works with the addition of superheated steam.

The fluidized bed processes in which a circulating granulated catalyst is used which carries the required heat from the regenerator into the reactor, suffered at first from many disadvantages, including an unfavourable heat balance. The difficulties were eliminated by many modifications and improvements, in particular acceleration of the circulation of the catalyst by pneumatic feed and by placing the reactor directly above the regeneration chamber.

One of the most widely used methods is the Thermofor-Catalytic-Cracking process (TCC), redesigned in 1950 (TCC-Airlift), the principle of which is shown in Figure 45 (From Ullmann, *Encyklopädie*, 3rd ed., vol. 6).

The raw oil is heated to about 400°C in the tubular furnace *a* and is separated from non-volatile material (bitumen) in the tar separators *b* and *c*. The high boiling fractions in the vapour phase pass directly from *b* into the reactor *f* and the lower boiling fractions, after condensation in the cooler *d* via *e* as a liquid. The reactor contains the pre-heated catalyst which consists of 85% $SiO_2$, 15% $Al_2O_3$, and 0·003% Cr. Cracking takes place at 450–500°C. In the lower part of the reactor the catalyst is stripped of the lighter hydrocarbons using steam and drops into the regenerator *h*. There it is regenerated

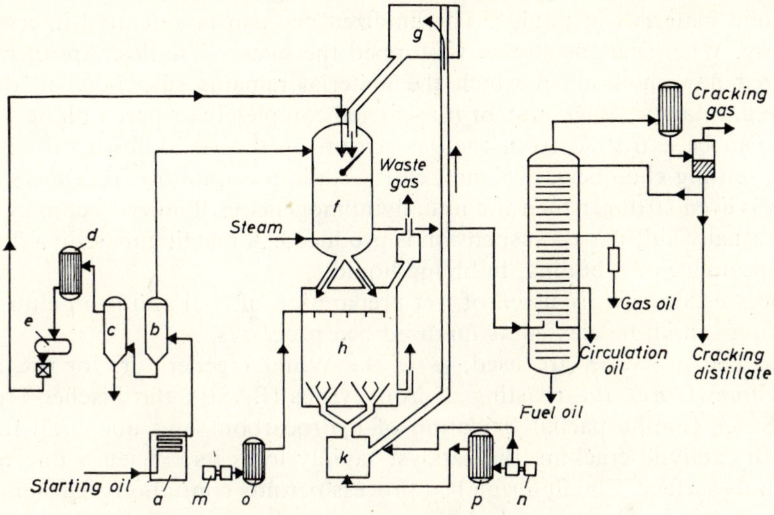


Figure 45. Thermofor-Catalytic-Cracking process (TCC)

with pre-heated air and is then conveyed pneumatically by pre-heated air from the charging container *k* upwards into the high-level tank *g*. The cracking product discharged from the lower part passes into a condensating column *i* (fractionating tower) where it is separated. The Houdriflow process is based on a similar principle but there the reactor and the regenerator form a single unit.

## *Fluidized bed processes – fluidized operations. General remarks*

(See also: Brötz, Grundlagen der Wirbelschicht-Verfahren, *Chem. Ing. Techn.*, 1952, **24**, 60)

The 'dust-flow' or fluidized bed process, used first in the Winkler generator and for various other special purposes, is now being used more generally in industry. In the American literature these processes are called 'fluidized operations'. They involve the turbulent movement of fine-particle material in a gas or liquid stream. The granular material may be a catalyst for reactions in the liquid or gas phase, it may act as heat transfer agent, or it may itself be heated, cooled or transformed in the gas or liquid stream. The fluidized bed process has recently become specially important in the technology of catalytic reactions, and its principles will, therefore, be briefly discussed in this section.

When a gas stream is directed into a particulate material the bulk material will remain at rest up to a certain velocity of the gas stream. When the gas stream is intensified the particles rise and become suspended. Although at first they are still close together they have become mobile and can move like a liquid. This property has long been used in Polysius' pneumatic conveyor chute the bottom of which is made of a porous material through which air is forced. Even at a slight inclination of the chute fine-particle material then moves like a liquid very rapidly to the lowest point ('moving bed'). Further intensification of the gas stream produces the 'fluidized bed' in which the material is held freely suspended in the gas stream or even moved against gravity. The same principles apply to solid materials in liquids. The fluidized bed can be operated intermittently or continuously. When the gas stream is stopped the material settles. Another method is the 'fluidized flow method' in which the material remains suspended in the reaction chamber (e.g. catalytic processes), or it is carried completely or partly along with the gas stream. It can be extracted from the gas stream by the well-known dust removal installations, settling chambers, cyclones, electrostatic precipitators, scrubbers, etc. If the gas stream is even stronger then the initially homogeneous fluidized bed may be broken up and eventually a dispersed suspension is produced. A reacting mass in a fluidized bed has the appearance of a boiling, bubbling liquid.

The processes already discussed of wet preparation of coal and iron, elutriation, and air separation can all be classed as fluidized bed processes.

Fluidized bed processes are used, as in the Winkler generator, for the burning of powdered lime (Dorr), the roasting of iron pyrites (BASF), the Fischer-Tropsch synthesis (U.S.A.), for the partial oxidation of hydrocarbons and above all for catalytic cracking. In catalytic cracking the catalyst rapidly loses its efficiency due to coke deposition on its surface. The fluidized bed process permits continuous regeneration of the catalyst by burning off, the catalyst being conveyed pneumatically from the reaction chamber to a regeneration chamber and then back again. The cracking process is endothermic whilst during the regeneration of the catalyst heat is liberated by oxidation. Profitable use can be made of this heat for the cracking process by heating the catalyst and utilizing the heat of the waste gases produced during regeneration. The process is carried out in reaction chambers similar to the Winkler generator (Figure 26) to which are connected the regeneration chamber, waste heat boiler, and electrostatic dust precipitator.

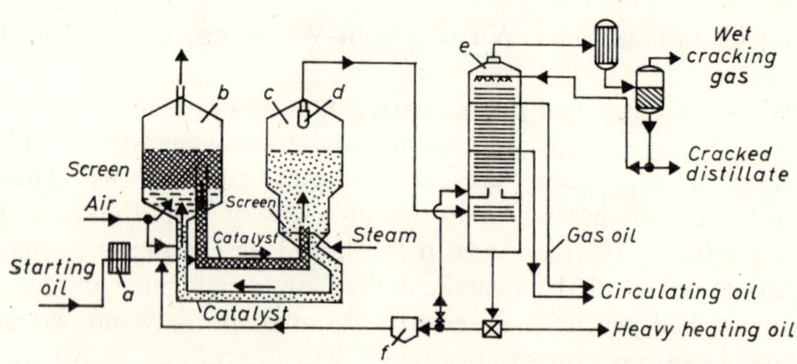


Figure 46. Cracking plant using the fluidized bed method

Figure 46 shows the principle of a cracking plant using the fluidized bed process (Fluid Catalytic Cracking) such as is being built by Standard Oil for daily outputs of 9,000 tons. The advantage of the process lies partly in the fact that it permits, by setting the catalyst-oil ratio at a predetermined point between 5:1 and 10:1, to pass all the heat through the pre-heated catalyst and to control the temperature between 470 and 530°C by adjusting the quantity of catalyst used.

Powdered catalysts with an average particle size of 70 are used. The raw oil, after heating in the preheater *a*, passes together with the catalyst (temperature 620°C) into the reactor *c* where it evaporates immediately, and where the fluidized bed is formed. The cracked products pass through a dust separator *d* into the fractionating column *e*. The catalyst, having been treated with superheated steam and carrying coke, flows into the regenerator *b* where the coke charge is burnt off with air at 620°C. The heat of combustion of the coke is sufficient to maintain the temperature in the regenerator.

## 3. REFORMING, FORMATION OF AROMATICS, POLYMERIZATION AND ALKYLATION

'Reforming' improves the motor fuel properties of the distillate petrols (straight-run petrols) and cracked products by enriching them with olefins and aromatics and by isomerization. (See also under Variations of the hydrogenation processes, p. 130). Relatively low-boiling fractions with boiling points from 100 to 230°C are used as raw material and reformed by thermal or catalytic cracking. In thermal reforming the process consists mainly in the cracking of the chains, with olefin formation; it is carried out in the liquid phase at a pressure of 50 atmospheres and a temperature of 525–575°C. In catalytic reforming, which is carried out exclusively in the vapour phase, the processes taking place are the dehydrogenation of naphthenes to aromatics, ring isomerization, such as the conversion of alkyl-cyclopentanes to cyclohexanes and their dehydrogenation to aromatics, and the isomerization of straight-chain paraffins. The catalytic process is becoming increasingly popular because of the higher octane rating, the better yields, and the more complete desulphurization.

The equipment and techniques resemble those of the cracking processes, i.e. fixed bed catalysts, moving bed catalysts, and fluidized bed processes are used here, too. However, the catalysts used are hydrogenating catalysts since it is a hydrogen circulation process; thus, for instance, molybdenum oxide on $Al_2O_3$, mixed cobalt-molybdenum catalysts, or finely divided platinum ('platforming process'), with which coke separation is particularly low, are used.

The catalysts must be able to hydrogenate and dehydrogenate, they must be 'acidic', i.e. they must be able to lose protons. As in catalytic cracking, carbonium ions are produced in the reforming processes, too. $Al_2O_3$ is, therefore, a suitable carrier since it has acidic properties. On this is based its capacity to split off water from alcohols to form olefins, and $NH_3$ from amines (e.g. in the formation of diphenylamine from aniline).

The catalytic reforming processes are also used very extensively for the production of aromatics. The methods employed for the separation of the aromatics have already been described on p. 96.

For the production of aromatics it is important to choose suitable intermediates and fractions. Most suitable are 'naphthenic petroleums' which in their crude state may contain up to 35 per cent aromatics and 55 per cent cyclo-paraffins.

In the 'hydroforming process' which is more or less identical with the 'DHD' (pressure-$H_2$-dehydrogenation) process for the production of petrol from coal (see p. 130) a molybdenum-$Al_2O_3$-catalyst is used and the operations are carried out at a hydrogen pressure of 7–20 atm and a temperature of 480–540°C. The presence of hydrogen reduces coke formation. The high reaction temperature shifts the equilibrium towards dehydrogenation. The process must be carried out in several stages with intermediate heating since the conversion of cyclohexane into a benzene residue requires 53 kcal/mole. In addition to dehydrogenation, isomerizations also take place:

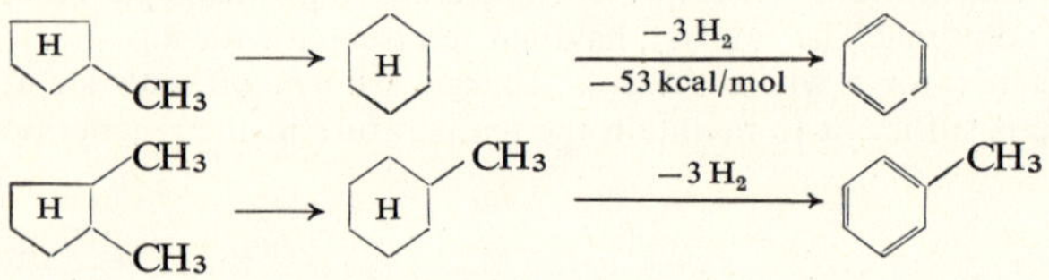


The composition of the 'hydroformate' depends to a large extent on the cyclo-paraffins used. The 'platforming process' is carried out with a platinum-$Al_2O_3$-catalyst in fixed bed, moving bed, and fluidized bed operations. In addition to the isomerizations shown above and the aromatization, paraffin-isomerization and hydro-cracking also take place, as well as hydrogenation to paraffins of the olefins produced. The hydrogen formed from the equilibrium is removed and aromatics and knockless petrols are obtained.

Polymerization is the production of polymer-petrols from low olefins and paraffins; it can be carried out either thermally or catalytically, always in the liquid phase. The raw materials are $C_3$-$C_4$-fractions. The tendency of unsaturated hydrocarbons to polymerize decreases with increasing temperature and increases with increasing pressure. The operations are, therefore, carried out at pressures of 18–25 atm. At temperatures above 500°C and low pressure, depolymerization prevails. The lowest hydrocarbons, such as ethane and propane, give better yields with thermal polymerization; higher hydrocarbons are polymerized better catalytically.

Catalysts were first introduced in 1927 by the former IG. The catalytic process has been carried out by the Universal Oil Products Co. since 1933 on an industrial scale.

Suitable catalysts are inorganic acids, such as sulphuric acid, phosphoric acid, and hydrofluoric acid, on inorganic carriers. 'Acidic film catalysts' have also recently been used, with quartz or glass as non-porous carriers. The reaction temperatures are between 170 and 220°C. Phosphoric acid tends to dehydrate even at about 150°C. To control the water content of the phosphate catalyst it is advantageous to add a small amount of steam to the gas. The catalyst is regenerated by burning off the coke at 480 to 540°C and subsequent hydration of the metaphosphoric acid formed with steam at 285°C.

Very high quality 'polymer petrols' are obtained which, when so-called 'selective polymerization' of $C_4$ mixtures is used (at 150°C), are suitable as aviation spirit. Polymer petrols contain mainly branched olefinic hydrocarbons. The yield is very good, it is possible to convert 85–90 per cent of the olefins into polymerisate.

With these processes it is possible to convert low hydrocarbons from natural gas, hydrogenation gases, the Fischer-Tropsch synthesis, and from coke oven gases after transformation into olefins by dehydrogenation, and cracking gases directly into higher hydrocarbons and petrol. This is of great importance for countries rich in natural gas since it provides them with an additional source of motor fuel. Like the conversion of these gases into synthesis gas and the production of hydrocarbons from it by the Fischer-Tropsch process, it represents another means of stretching the available petroleum reserves.

As mentioned earlier, a simple and typical example of the production of polymer petrol is the transformation of isobutene into iso-octane.

A similar process is the so-called 'alkylation' in which isoparaffins, e.g. isobutane, are condensed with olefins (propene, butene, and above) to saturated isohydrocarbons. Catalysts used are sulphuric acid and hydrofluoric acid.

## LITERATURE

P. Kukuk, *Geologie, Mineralogie und Lagerstättenlehre*. Springer, Berlin, 1951.

C. Hähnel, in Ost-Rassow, *Lehrbuch der chem. Technologie*. 26th ed., J. A. Barth, Leipzig, 1955.

Weingaertner, *Chem. Technologie, Organische Technologie*, I, p. 233. Hanser, München, 1952, 2nd ed. 1959, I, Vol. 3, p. 177.

L. F. Fieser and M. Fieser, *Lehrbuch der organ. Chemie*. Chemie, Weinheim, 1954.

W. Rühl in: Ullmann, *Encyklopädie der Techn. Chemie*, 3rd ed., vol. 6, p. 548.

F. Asinger, *Einführung in die Petrolchemie*, Akademie-Verlag, Berlin, 1959.

H. Prinzler, *Einführung in die Technologie des Erdöls*, VEB Deutscher Verlag der Grundstoffindustrie, Leipzig, 1961.

H. Ruf, *Kleine Technologie des Erdöls*. Birkhäuser-Verlag, Basel–Stuttgart, 1955.

### *Distillation technology*

E. Kirschbaum, *Destillier- und Rektifiziertechnik*, 3rd ed., Springer, Berlin, 1960.

S. Kiesskalt in: Winnacker-Weingaertner.

K. Sigwart, Destillation und Rektifikation in: Ullmann, *Encyklopädie*, 3rd ed., vol. I, p. 433, Urban & Schwarzenberg, München–Berlin, 1951.

W. Rühl, *ibid.*, vol. 6, p. 611.

E. Krell, *Handbuch der Labordestillation*. VEB Deutscher Verlag der Wissenschaften, Berlin, 2nd ed., 1960.

# III. Hydrogenation of coal, tar, and crude oil

High-pressure hydrogenation processes were developed at first exclusively in Germany. The development there was accelerated by the lack of petroleum, repeated exclusion from the world market, and by the raw materials available. The location of the hydrogenation plants was, therefore, determined mainly by the deposits of the raw materials. Later, the whole world benefited from the processing of petroleum by the hydrogenation method.

The 13 hydrogenation plants built in Germany before the 1939–1945 war which include Leuna, Böhlen, Magdeburg, and Zeitz, are situated mainly in the brown coal areas. The last three plants having brown coal low temperature carbonization installation supplield Central Germany at the samet ime with electricity. One plant was built based on Rhein brown coal, two on Ruhr bituminous coal, and one on Upper Silesian bituminous coal. One of the Ruhr plants was to have hydrogenated Pott-Broche extract but since the production of the latter could not be organized on a large industrial scale, the hydrogenation plant was converted for bituminous tar. Two plants were built for the hydrogenation of residues from petroleum refineries. It had been intended to buy these very cheap residues on the world market. During the war these plants were converted for the hydrogenation of tar or bituminous coal. Finally, there was at Ludwigshafen-Oppau, where coal was first hydrogenated, a plant for the production of aromatic petrol from petroleum, or hydrogenation petrol by the DHD process (pressure-$H_2$-dehydrogenation). Similar sections were added to five more hydrogenation plants. At the VEB Combine 'Otto Grotewohl', Böhlen, the aromatization of hydrogenation petrol is also carried out by the DHD process.

The following Table shows the German production of motor fuels and mineral oils up to the end of the war:

| Methods of production | in 1,000 tons 1940 | 1943 |
|---|---|---|
| High pressure hydrogenation | 1,503 | 1,830 |
| Fischer-Tropsch synthesis | 406 | 425 |
| Petroleum processing | 978 | 1,392 |
| Tar distillation | 575 | 945 |
| Benzene production | 378 | 355 |
| Others | 101 | 36 |
| Total | 3,941 | 4,983 |

(Figures taken from: Ost-Rassow, *Lehrbuch der chemischen Technologie*)

In the United States, high-pressure hydrogenation for the processing of crude petroleum was carried out in two plants by the Standard Oil Co., on the basis of an agreement with the former IG. At present the process is being increasingly used there for the hydrogenation of bituminous coal. In England, ICI have for a long time past been carrying out hydrogenation of bituminous coal, also under licence. Before the war, the plant at Billingham had a capacity of 100,000 tons per annum. Italy has two large plants each producing 180,000 tons per annum. Japan could not carry out large-scale hydrogenation before the war because it did not take part in the international collaboration.

### *Chemistry*

The finished products of hydrogenation contain appreciably more hydrogen than the raw materials. The difference is bridged by the supply of various quantities of hydrogen. The content of carbon, hydrogen, sulphur, nitrogen, and oxygen of the raw materials

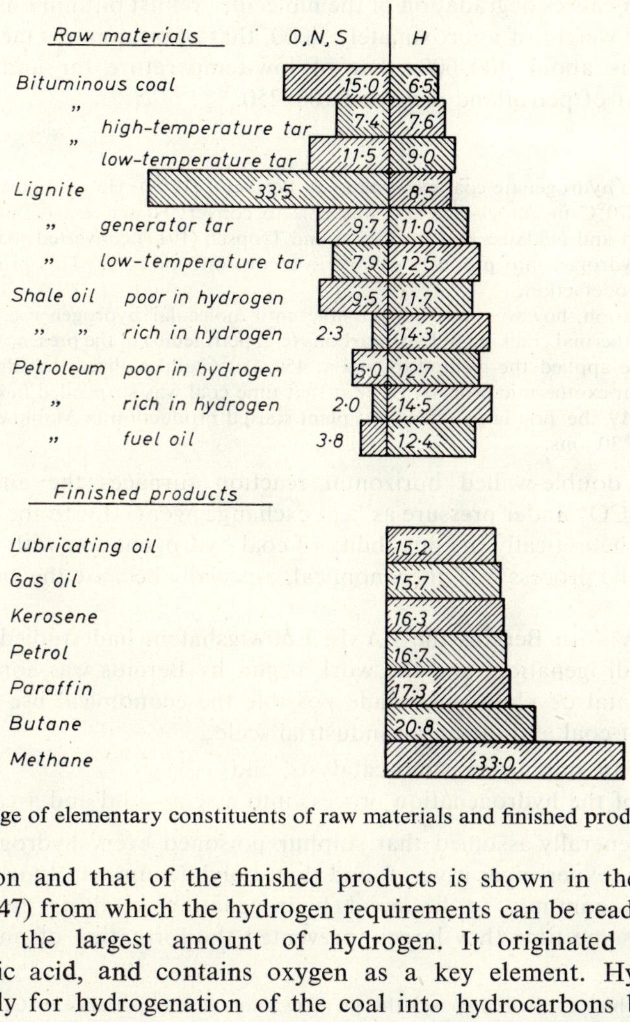


Figure 47. Percentage of elementary constituents of raw materials and finished products (after M. Pier)

for hydrogenation and that of the finished products is shown in the illustration by M. Pier (Figure 47) from which the hydrogen requirements can be read off.

Coal requires the largest amount of hydrogen. It originated from vegetable matter via humic acid, and contains oxygen as a key element. Hydrogen then is required not only for hydrogenation of the coal into hydrocarbons but also for the degradation of molecules by removal of the oxygen. Like oxygen, sulphur and nitrogen must also be converted into their hydrogenated products, i.e. hydrogen sulphide and ammonia. However, during hydrogenation splitting occurs not only by the formation of water which consumes hydrogen but also by the splitting off of oxygen-containing products, e.g. of $CO_2$ by decarboxylation, or of $H_2O$ by dehydration. As far as the finished products are concerned, the formation of gaseous hydrocarbons (e.g. methane, ethane, etc.) requires the largest amount of hydrogen; it will readily be seen that for economic reasons the formation of these 'hydrogenation gases' is kept as low as possible.

Less hydrogen is required for the conversion of cyclic, aromatic, araliphatic, and olefinic hydrocarbons into saturated ones, and least of all for the formation of short-chain hydrocarbons from long-chain ones (hydrogenative splitting). The hydrogen requirements decrease, therefore, in the order: coal → tar → petroleum.

Hydrogenation causes degradation of the molecule. Whilst bituminous coal and lignite have a molecular weight of approximately 5,000, that of bituminous tar, petroleum, and lubricating oils is about 400–600, that of low-temperature tar and paraffin about 250–400, and that of petrol and diesel oil 100–250.

*History*

The first attempts to hydrogenate coal were made by Berthelot (1869). He used nascent hydrogen from hydriodic acid at 270°C in autoclaves, and was able to convert 60 per cent of bituminous coal into distillable oil. Dafert and Niklausz (1911), Fischer and Tropsch (1917) converted coal using compounds which easily lose hydrogen into products soluble in chloroform or ether. This proved the theoretical possibility of coal liquefaction.

Industrial application, however, was not possible until molecular hydrogen was used. In 1910 Bergius carried out the thermal cracking of hydrocarbons (e.g. petroleum) in the presence of hydrogen under pressure. In 1913 he applied the method to coal at 450–480°C and realized that the hydrogenation of tar and coal was an exothermic process. Even at that time coal was suspended in oil and $Fe_2O_3$ used as a catalyst. In 1919 the first large-scale pilot plant started production in Mannheim-Rheinau with a daily throughput of 30 tons.

Bergius used double-walled horizontal reaction furnaces the annulus of which contained $N_2$ or $CO_2$ under pressure as heat exchange agent. It is to the credit of Bergius that he proved theoretically the possibility of coal hydrogenation with good yields. For various reasons the process was uneconomical, especially because the yield of petrol was too low.

Simultaneously with Bergius the BASF, Ludwigshafen, had studied the problem of tar- and coal hydrogenation, and the work begun by Bergius was continued by them.

Two fundamental developments made possible the economical use of high-pressure hydrogenation of coal and tar on an industrial scale:

(1) the discovery of sulphur-resistant catalysts, and

(2) the division of the hydrogenation process into a semi-solid and a gaseous phase.

It had been generally assumed that sulphur poisoned every hydrogenation catalyst. Contrary to these experiences it was found that molybdenum and tungsten are elements whose sulphides continue to be good hydrogenating catalysts even if sulphur is present. At the same time they largely prevented the formation of undesirable hydrocarbons.

The BASF also had at their disposal the great technological experience of C. Bosch in ammonia synthesis and experience in the use of equipment for the hydrogenation of carbon monoxide to methanol and isobutyl oil. The rapid development of the high-pressure hydrogenation process was due mainly to Matthias Pier and Alwin Mittasch, with the support of C. Bosch. The first industrial plant for the hydrogenation of brown coal tar into petrol at a gauge pressure of 200 atmospheres and a temperature of 450°C started production as long ago as 1924.

The direct conversion of brown coal tar into petrol required a very large excess of $H_2$ without which incompletely hydrogenated high-molecular weight products were deposited on the surface of the catalyst. These difficulties are avoided if hydrogenation is carried out in two phases:

(1) 'Semi-solid phase': hydrogenation of the raw material (tar or coal in suspension) in the liquid phase in the presence of a finely divided catalyst, the lighter fractions being removed as intermediates from the system and the heavier ones being returned and cracked.

(2) 'Gas phase': processing of the intermediates from the semi-solid phase and of the original light fractions of the tars, petroleums, etc. in the gaseous (vapour) phase over fixed catalyst pellets, giving the finished product.

The two phases change over at a boiling point of about 325°C. Fractions with boiling points above 325°C are hydrogenated in the semi-solid phase, those with boiling points below 325°C in the gas phase. It proved expedient to divide the hydrogenation in the gas phase into two parts:

(1) refining hydrogenation in which the oil is converted mostly into saturated hydrocarbons and N, S, and O are removed;

(2) aromatization during which mainly cracking takes place, with relatively little hydrogenation.

Hydrogenation thus includes three partial processes:

cracking hydrogenation (semi-solid phase)
refining hydrogenation (gas phase I)
hydrogenative cracking (gas phase II)

In all these partial processes reduction of the molecular weight, bridging of the hydrogen balance, and thus prevention of polymerization of unsaturated compounds on the catalyst, takes place.

## 1. HYDROGENATION IN THE SEMI-SOLID PHASE

As in the high-pressure synthesis of ammonia and methanol vertical reaction furnaces are used. At first agitators were employed but it was found that the flow of hydrogen is adequate for stirring; this also avoided the use of stuffing boxes. For agitation, excess hydrogen is used as recirculation gas which passes rapidly through the semi-solid phase and by its turbulence prevents the sedimentation of the reaction material and catalyst. The inner wall of the high-pressure furnace for the semi-solid phase should be as smooth as possible. Under pressure hydrogen dissolves in the oil but the quantity dissolved at a gauge pressure of 200 atmospheres is insufficient. The speed of dissolution of the hydrogen diffusing from the gas space is, however, so great that sufficient is always available. The solubility of hydrogen increases approximately proportionally to the pressure so that at a gauge pressure of more than 1,000 atmospheres sufficient hydrogen should be in solution for the circulation of hydrogen to be dispensed with.

The heat produced by hydrogenation is sufficient to cover the losses through radiation and some heat must even be dissipated by cooling. Cold circulating gas is used for cooling, and is fed into the reaction furnace by sunk pipes or also more recently by lateral ducts. The reaction furnace consists of seamless special-steel tubes with maximum dimension of 1,200 mm inside diameter and 18 m length.

Since 1940 furnaces have been built consisting of an interior tube of hydrogen-resistant special steel, with a wall thickness of only 20 mm, around which is wound steel tape, the 'ribbon-wrapped furnaces' (Class and Maier, Gewickelte Hochdruckkörper, *Chem. Ing. Techn.*, 1952, **24**, 184). The tape is applied at about 800°C and is then cooled so that it shrinks on tightly. Its profile is such that the teeth of the upper layer engage

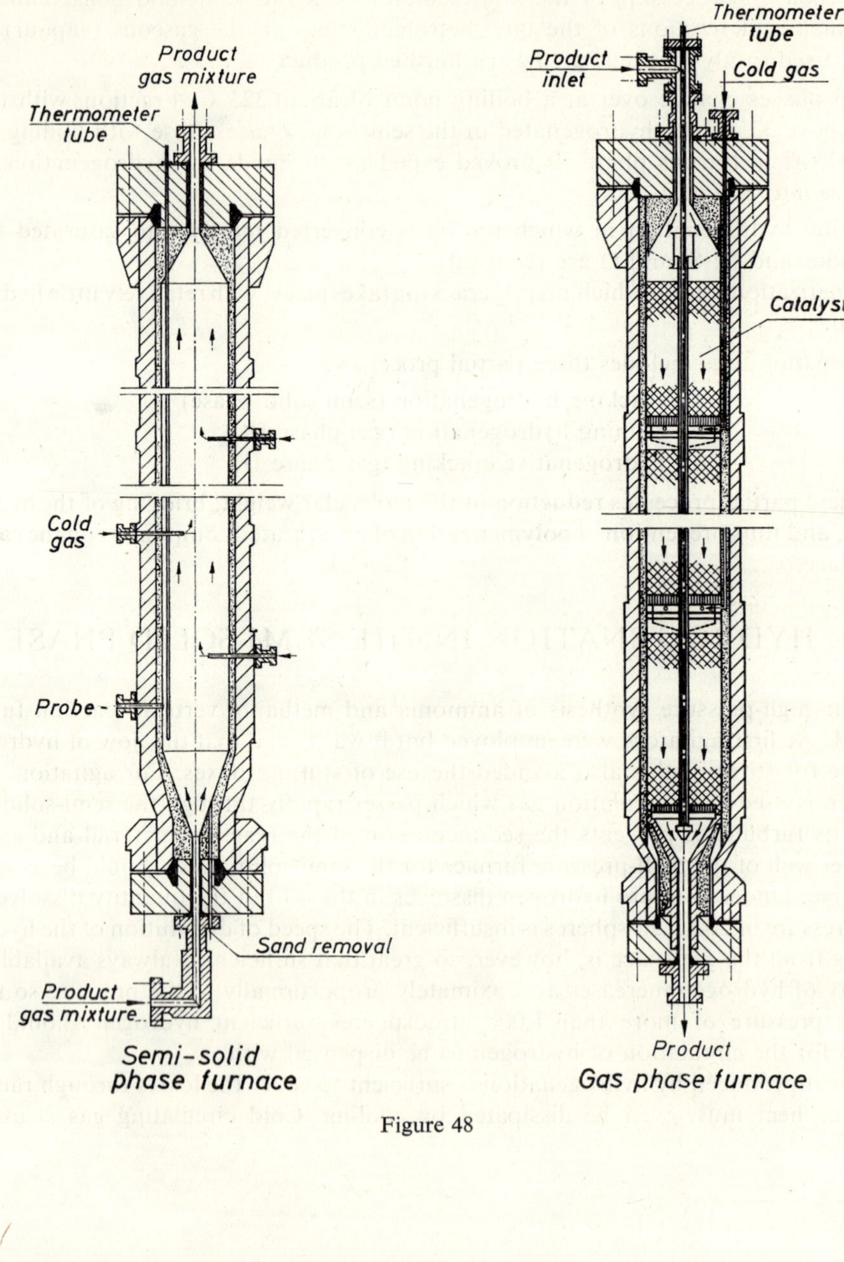


Figure 48

with corresponding recesses of the lower layer. Thus a firm indentation is obtained and the resulting bodies are of exceptional strength and elasticity so that they can withstand strong internal pressures.

The sketch in Figure 48 shows the design of standard high-pressure furnaces for semi-solid phase and gas phase hydrogenation.

The furnaces are made of low-alloy chromium steel N 1 with 2·5–3·0% Cr and 0·2–0·25% Mo. Since the wall must not be heated to more than 300°C at a gauge pressure of 300 atm, the furnace is lined with an insulating layer of asbestos cement. An internal tube made of thin $V_2A$ metal protects it against the contents of the furnace (catalyst tube). This tube and the insulating layer have a limited life and have to be renewed at intervals of one or more years. In older furnaces cold-gas tubes leading into the top of the furnace regulate the reaction temperature; however, occasionally they cause deposits. In more recent designs of semi-solid phase furnaces the cold gas is fed in through lateral inlets. The furnaces also contain thermometer wells for temperature-controlling thermocouples. Whilst the semi-solid phase furnaces are completely empty, the gas phase furnaces contain fittings for the catalyst. Since a very large amount of heat is produced during pre-hydrogenation in the gas phase furnace, cold-gas mixers are

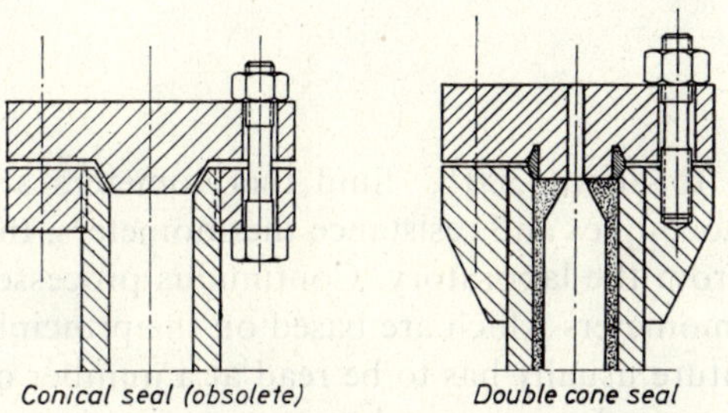


Figure 49. High-pressure seals (lid locking devices)

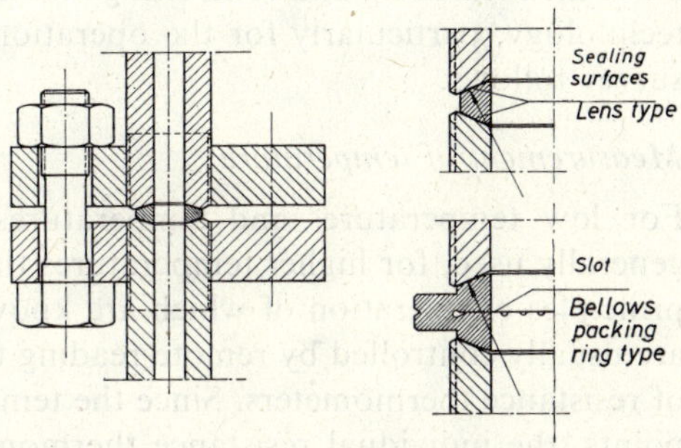


Figure 50. Seals for high-pressure tubes

fitted underneath the grates carrying the catalyst which effect a rapid mixing of the cold gas with the hot reaction gas. The cold gas introduced through the lid from above flows through an annulus from the furnace wall towards the centre mixing on the way with the hot gas and passing afterwards over a perforated bottom into the catalyst layer below.

Special care must be taken in high-pressure processes with the sealing of the apparatus and of the connections of the ducts, for the smallest leak rapidly enlarges itself and the equipment has to be taken out of service.

Formerly the lids of the high-pressure furnaces were sealed by tapered seals. A cone forged on to the lid was pressed into the conically ground end of the tube by firmly tightening the clamping screws before putting the equipment under pressure. Only a narrow strip was sealed and when it was necessary to open the furnace the conical surface had to be re-ground. Therefore, a special double conical ring with a soft iron lining is now being used, on which an increase in the internal pressure has a self-sealing effect (see Figure 49).

The sealing surfaces of the tube connections are ground conically, an iron lenticular seal is inserted and the tubes screwed together through flanges (see Figure 50). For tubes heated to a high temperature which are subjected to strong fluctuations of expansion and contraction, double divided lenses ('bellows packing rings') are used.

The furnaces are built in groups of three or four into reinforced concrete furnace chambers, together with the heat exchangers and separators; the chambers are open at the top and at the front and the frontal opening is closed by doors and gates of reinforced concrete. This open construction prevents the accumulation of explosive mixtures. The chambers must not be entered during operation. When leaks occur the mixture and the circulating hydrogen are released and ignited. The peak pre-heater and the circulating gas blowers are situated next to the chambers. All measuring and controlling devices are housed in a separate control building behind the furnace chambers so as not to endanger the service personnel. Along the row of furnace chambers a very powerful crane travels on rails by means of which the furnaces, heat exchangers, and peak pre-heaters can be changed.

Control of the hydrogenation and of the equipment for subsequent distillation consists mainly in the measurement of temperature, pressure, and quantities. These measurements are recorded so that the manager can at any time carry out a check afterwards. Since the measuring and controlling methods used here are employed generally in process technology, particularly for the operation and control of continuous processes, a brief survey follows.

*Measurement of temperature*

For low temperatures and temperatures up to about 300°C fluid thermometers are generally used, for higher temperatures thermocouples and resistance thermometers, the principles of operation of which are known from the laboratory. Continuous processes are usually controlled by remote-reading thermometers which are based on the principle of resistance thermometers. Since the temperature usually has to be read at a number of points, the individual resistance thermometers can be connected to a central galvanometer.

The recording is done by automatic temperature recording instruments which at fixed intervals connect one measuring point after another through a switch to the galvanometer and record the measured point on the graph, in a different ink for each measuring point. Thus, for example, six temperature curves can be recorded by one instrument (six-point recorder). The measuring instruments can also be equipped with alarms, operating through a relay, such as warning lights, bells, buzzers, to warn of undesirable events.

*Measurement of quantities*

Measurement of quantities of liquid and solid materials is done most accurately by weighing. No details will be given here of the various designs of weighing devices. Numerous methods have been developed for the measurement of moving materials, i.e. of mass flow. Dosing of fixed quantities of solid materials is carried out, for example, by belt conveyors or screw conveyors, rotary mixing tables, and reciprocating shoes with adjustable speeds.

A reciprocating shoe is a shoe which receives a measured quantity of a material and discharges it during the forward and reverse movements. It is suitable for coarse material. The flow of liquids is controlled by slides, cocks, valves, or perforated discs.

Their quantity can be measured, on the principle of the rotameter, by means of a spinning float with oblique grooves at the edge in a calibrated conical tube. The liquid starts the float rotating and lifts it to different levels according to the intensity of flow. The height is a measure of the quantity of flow.

The instrument must be calibrated separately for each type of liquid. Gases can also be measured by rotameters. When placed in parallel with a regulating valve the rotameter is used as a 'Frigorota measuring valve'. Instead of a floating cone an impeller is connected with a pointer and a counter for the measurement of the number of revolutions, like the milometer of a car.

The most widely used method of measuring quantities of liquids and gases is based on the orifice principle. This method can be used at any pressure and is, therefore, particularly suitable for the measurement of liquids and gases in high-pressure hydrogenation processes. The flow in the pipe is dammed by an orifice (diaphragm), nozzle, or Venturi tube. The pressure difference which varies with the flow velocity is measured, for example,

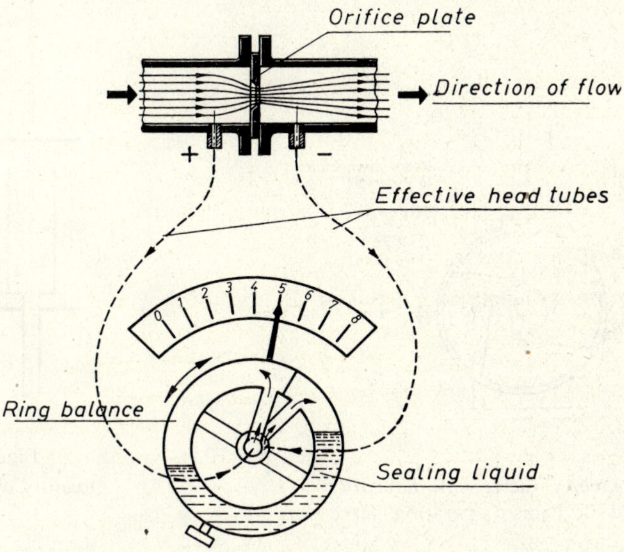


Figure 51. Diagram of the measurement of quantity of flow by means of an orifice plate and ring balance

with a differential pressure gauge or a ring balance. Of great importance is the correct selection of the sealing liquid for the pressure gauge or ring balance, which must not react with the gas or liquid being investigated. The quantity flowing through the pipe is proportional to the square root of the pressure difference. The orifice is calibrated for the liquid or the gas and kg/h or $m^3$/h can be read off directly from a suitable scale fitted to the measuring instrument. Figure 51 shows the principle of measuring quantities by means of an orifice plate and ring balance.

The ring balance consists of a smoothly revolving hollow metal ring which is half filled with a liquid, e.g. mercury, oil, water, or a high-boiling organic material. The ring is kept in its basic position by an interchangeable weight fitted at the lowest point. The smaller the balance weight the more sensitive is the balance. A partition divides the empty annular space above the fluid into two halves each of which is connected to one side of the orifice by a very long, multi-coiled flexible metal tube. The slightest difference in the pressure displaces the liquid in the annular tube causing the tube to rotate and the movement is indicated on a scale by a pointer. The measuring range can be rapidly adjusted within wide limits by means of the balance weight. The equipment of a hydrogenation plant contains a large number of such ring balances which may also be fitted with recording devices.

Gases under low pressure, e.g. from gasometers (town gas, water gas, synthesis gas), are measured quantitatively by gas meters. The mode of operation of a gas meter is shown schematically in Figure 52.

It contains in an enclosure a rotating drum with outlets, divided into four chambers by radial partitions. The inlets for the gas are at the shaft. Displacement of the liquid by the gas causes the gas-filled chamber to rise and the drum starts rotating at a speed proportional to the quantity of the gas. The number of revolutions of the drum is recorded by a counter which indicates directly the volume in $m^3$ of the gas passed through.

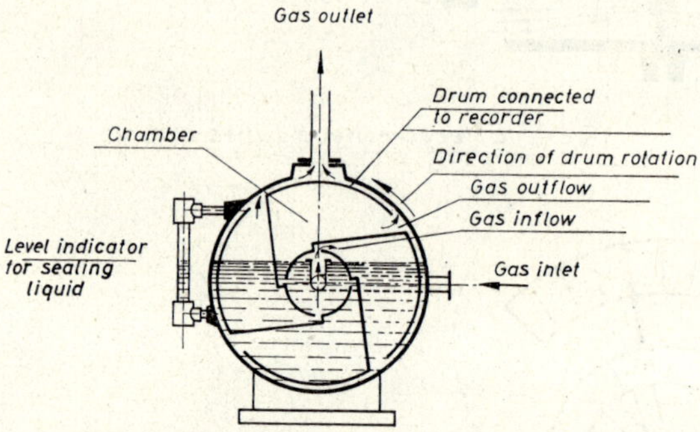


Figure 52.
Gas meter (meter with indicator and mechanism operating wet)

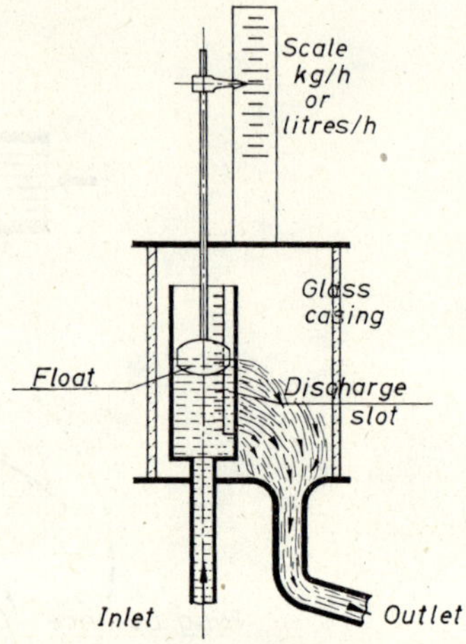


Figure 53.
Quantity measurement

A simple method for the measurement of liquids which though not very accurate is yet adequate for many purposes, is based on the following principle (Figure 53):

The liquid is fed from below into a narrow cylindrical vessel the jacket of which is slit along its whole length at the side. The level of the upper boundary of the laterally issuing jet varies According to the quantity flowing through. A calibration scale can be fitted directly on the slot. The reading is more accurate when a light float is placed on the liquid with a peg and a pointer at the top. The level of the float is read off a scale, giving a measure of the quantity of flow.

### *Measurement of pressure*

Low pressure can be measured most accurately with differential pressure gauges, higher pressures with metal gauges. Both devices also measure vacuum (vacuum gauges). The metal gauges contain a flexible membrane of a curved tube of elliptical cross-section which changes its shape under the influence or pressure (Figure 54). This change is shape is transmitted by a lever device, segmental rack, and pointer drive to a pointer which indicates by its position on a scale the pressure in $kg/cm^2$. Pressure gauges can also be connected to a recording apparatus, such as the temperature recorder (manograph). Non-corroding material must be chosen for the pressure gauge.

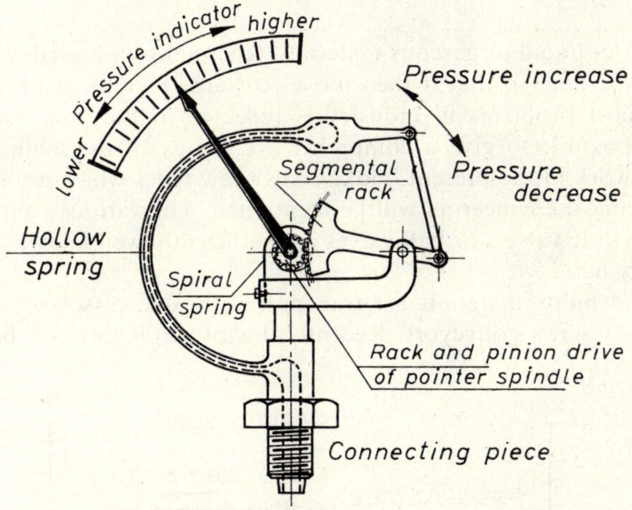


Figure 54.
Design and mode of operation of a pressure gauge

*Automatic control of chemical processes*

The measuring instruments described above can be connected to control devices for the automatic control of a process. With great ingenuity and utilizing all the possibilities offered by the science of physics, installations have been developed by specialist firms which make human supervision largely unnecessary. Mechanical and electrical control devices are used, more recently also electronic valves.

In order to maintain, for instance, a certain pressure in a duct a so-called servo-motor is used, consisting of a double-acting piston. Increase of pressure on one side moves the piston. Through a lever system the displacement sets in motion a device which counteracts the increase in pressure, e.g. by opening or closing a valve.

Automatic control devices usually consist of a sensing device, e.g. a photocell, or an instrument for the measurement of temperature, pressure, quantity, or pH. The weak signals are amplified by e.g. an electric relay or electronic valves. These produce controlling impulses which for instance actuate an electric motor or a current regulator, or release pneumatic or hydraulic forces. These in turn start the control gear proper which usually requires more power. However, even the best designed automatic control devices occasionally fail. They do not, therefore, relieve the man who supervises the operation of all responsibility. Experience has shown that with too much automation the operator easily becomes inattentive. Automatically controlled processes require, therefore, efficient checking and warning devices independent of the control and regulating instruments, programmers and 'electronic brains' (computers) of complete automation. From a psychological point of view it is also advisable to keep the operator constantly on the alert by requiring him to keep an exact record of the operation.

*Feed pumps*

For the transport of liquid or gaseous materials, feed pumps are used which increase the pressure on the material so that it then moves towards regions of lower pressure. The variety of transport problems in industry is reflected in the great number of pump designs. It is impossible to give a comprehensive survey of the main types of pumps within the framework of the present book. Only a few types which are used particularly frequently in chemical engineering will be mentioned. The various reciprocating pumps with plug valves, slide valves, or ball valves are sufficiently well known for a description to be unnecessary here.

Many solid and pulpy materials are transported, mixed, dissolved, extracted, dried, and compressed by screw conveyors. Recently this principle has also been used for the

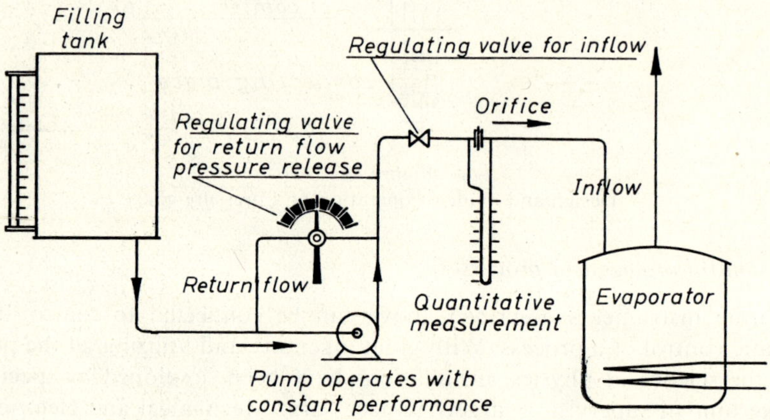


Figure 55. Diagram of the inflow regulation for an evaporator with return flow pressure release (pump return flow)

propulsion in screw pumps. Extremely precise construction and a high number of revolutions permit the use of screw pumps for high pressures (e.g. for gauge pressures of 100 atm). Screw pumps are also used for free-flowing dusts, e.g. coal dust.

Similar to screw pumps are geared pumps (see Figure 56) which are particularly reliable, and suitable for quantitative regulation, e.g. of the flow of liquids into evaporators. Any pump can regulate circulation or inflow if the arrangement shown schematically in Figure 55 is used. The pump operates at normal power and the fluid is re-circulated (adjustable return flow). The desired quantity is discharged as a partial flow through a regulating valve.

This method is called 'return flow with pressure release of liquid pumped in excess'. This principle is used for example in high-pressure hydrogenation for the oil feed. Similarly in high-pressure hydrogen compressors the compressed gas excess is returned from the pressure side to the suction side with simultaneous release of pressure.

For the quantitative regulation of smaller quantities, even against high pressure, the Bosch pumps of diesel motors are quite suitable. Reciprocating pumps with adjustable piston stroke or adjustable speed are also used.

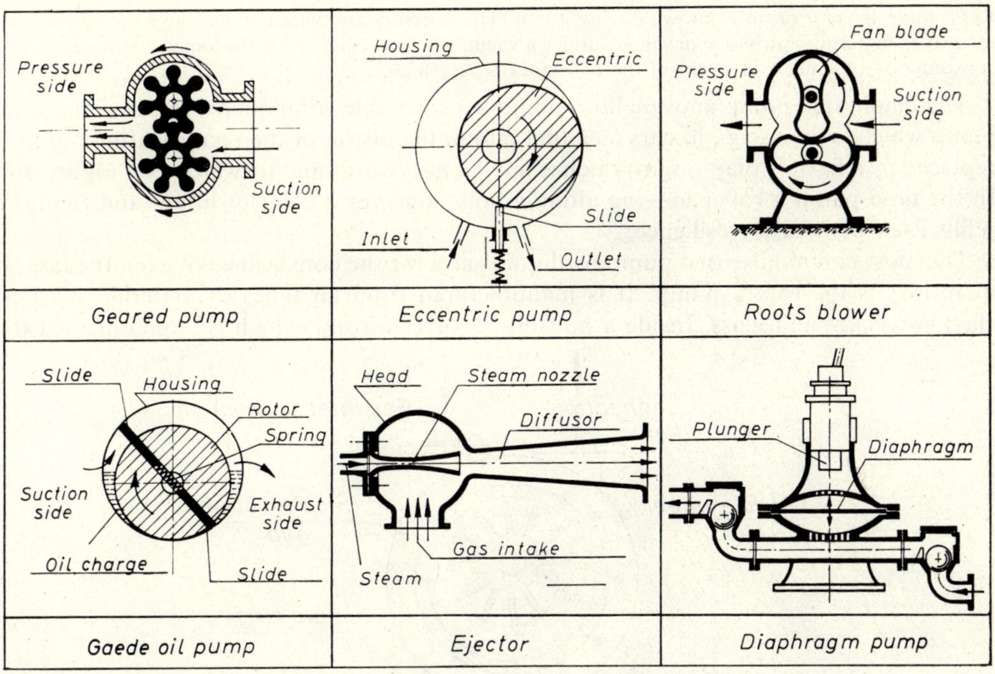


Figure 56. Feed pumps

Based on a similar principle are radial cylinder super chargers, the eccentric pump (see Figure 56) and the Roots blower (see Figure 56). In the latter, two fan blowers rotate inside a housing.

A vacuum is often produced with a Gaede pneumatic oil feed pump (rotary slide valve pump) (see Figure 56). It consists of a cylindrical housing in which an eccentric metal cylinder rotates. Two flat slide valves are flexibly fitted into the metal cylinder. The sealing fluid of the pump is oil. When the cylinder rotates the space on the suction side is enlarged and that on the pressure side reduced. This permits suction and expulsion. The principle resembles that of the water ring pump to be discussed later.

The very reliable piston pumps are also used for the production of a vacuum. Vapours penetrating inside often impair the performance of oil vacuum pumps, either by corrosion or by pollution of the oil. For this reason distillation technology uses on an increasing scale ejectors operated mainly by steam. They can be made of any desired material, have a very high performance, are fairly robust and produce a high vacuum, but consume large quantities of steam.

An ejector (see Figure 56) consists of a head with steam nozzle (driving nozzle), mixing nozzle, and diffuser. In the steam nozzle the pressure of the steam is reduced to that of the gas received, e.g. air. This gives the steam a very high speed. In the mixing valve water, steam, and gas are mixed, and in the diffuser the speed is converted into pressure. The correct dimensioning of all components is very important. Often several steam ejectors, e.g. three, are connected in series with the corresponding water-cooled condensers. The performance is improved by arranging for a pre-vacuum in front of the ejector, e.g. a water injector or an Elmo water ring pump. The injectors are placed 10 m above the level of the duct so as to permit automatic drainage, under a vacuum, of the condensate (barometric drainage). The consumption of water and steam of these ejectors is very high.

The diaphragm pump and the hose pump use a flexible component. In the diaphragm pump which is used, e.g., in cars as a fuel pump, the piston of the reciprocating pump is replaced by a flexible diaphragm which is moved backwards and forwards (see Figure 56). In the hose pump a roller moving along a hose squeezes it clear of liquid and the hose refills itself by its own resilience.

The most commonly used pump for liquids and for the conveyance of even the largest quantities is the rotary pump. It is manufactured from all types of material, plastics, clay, porcelain, and glass. Inside a housing revolves a rotor which is sealed against the

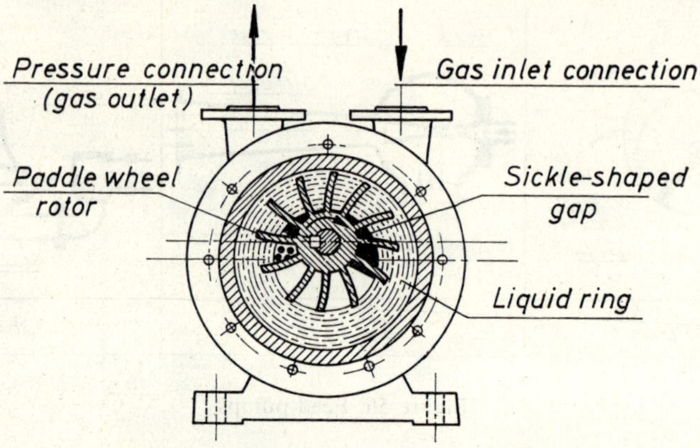


Figure 57. Siemens-Elmo water ring pump

outside. As in a turbine, the rotor carries blades. It is driven by a motor which is usually fitted on to the same shaft. Several rotors can be connected in series. A great variety of rotor designs exist that cannot be described in detail here. Rotor devices are also used as compressors.

Siemens water ring pumps (Elmo pumps) serve to exhaust large quantities of air, e.g. for suction filters, rotary filters, large-scale evaporation plant. They are built in all sizes and their performance is excellent. In practice the vacuum obtained is only 25–30 mm but this is sufficient for most purposes to exhaust large quantities of air. The mode of operation is shown in Figure 57.

A paddle wheel rotates eccentrically in a housing carrying with it a ring of liquid, which the centrifugal force causes to run concentrically along the housing. This produces a sickle-shaped gap between the rotor and the wall of the housing. Into this gap between rotor hub and liquid ring the gas from the inlet connection enters through the disc cam and the side shield. As the paddle wheel continues rotating the sickle-shaped gap narrows and the air is impelled towards the pressure connection and forced out. The liquid ring is continuously replenished from a water tank. The water also dissipates the heat generated by compression. If the vapours are acid, caustic soda can be added to the water to reduce corrosion. Water ring compressors have also been used for the past few years.

One of the pneumatic conveying devices is the air lift pump, a description of which is given on p. 318. This principle is employed also in transporting solid coarse material such as sugar beet, gravel and sand, for aerating liquids, and for stirring.

Reciprocating compressors and turbo-compressors are used for the compression of air and gases. Compressed air with a gauge pressure of 4–10 atmospheres is used a great deal in chemical plants, e.g. for emptying equipment, for blowing filter presses dry, for stirring and driving small stirrers. Because of the heavy consumption of electricity compressed air is relatively expensive. If large quantities of low-pressure air are required (up to 0·8 atm gauge pressure), e.g. for aeration processes such as yeast fermentation, oxidation, production of blast for gas generators, blowers are the most economical (e.g. Jaeger blowers).

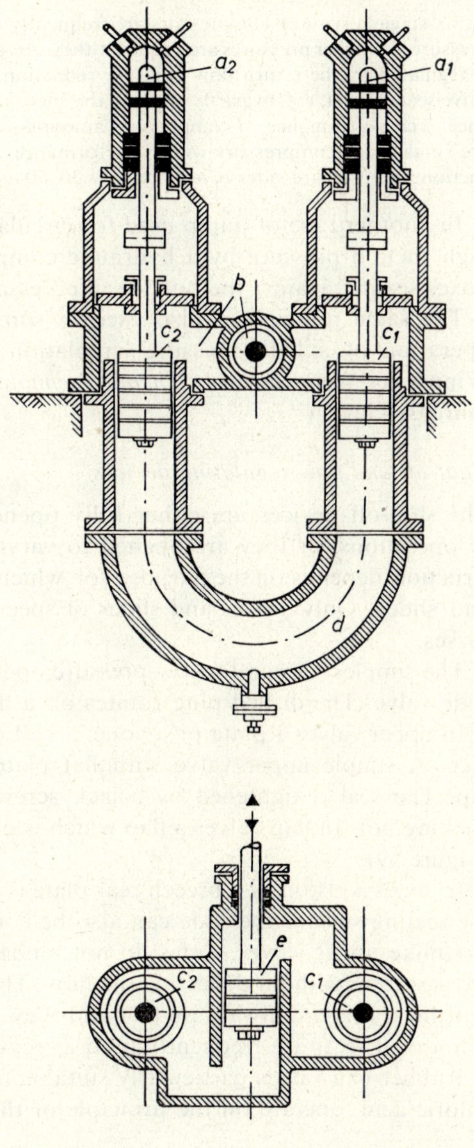


Figure 58. High-pressure gas circulation pump, Gebrüder Sulzer design, for pressure differences of 200 atm gauge pressure, operating in the range up to about 1,000 atm gauge pressure, $n = 32$–$125$ r.p.m., $N = 150$ h.p. The two pistons $a_1$, $a_2$ compress the residual gas by 200 atm gauge pressure, the hydraulic drive $b$ compensating the initial pressure through the oil immersion pistons $c_1$, $c_2$, and the hydraulic lever system $d$, which relieves the structure correspondingly. The horizontal drive piston $e$ of the valve-less hydraulic lever system also operates only against a pressure drop of 200 atm gauge pressure. Cf. *Z.V. D.I.* 1932, **76**, 117/118

In hydrogenation processes plunger pumps, usually fitted with three plungers, are used for liquids, compressors for the condensation of gases, high pressure circulating pumps and Roots rotary compressors for the circulation of gases, and hot circulation pumps for hot oils. Since the compression heat of liquids is low they can be compressed to any desired pressure in one step. The condensation of gases on the other hand must be carried out in several steps because of the large amount of heat of compression that is liberated. A large proportion of the equipment used in high pressure hydrogenation consist, therefore, of multi-stage compressors which are usually driven by slow-running Siemens-Halske synchronous motors.

Six-stage horizontal compressors are frequently used producing compression from 1 to 325 atm gauge pressure. If higher pressures are required they are produced by a booster compressor. The performance is regulated by the return flow pressure reduction described above, or by deliberately producing poor valve seat contact, e.g. by partly opening the suction valve, and by artificially enlarging the cylinder clearance. The performance of compressors amounts usually to 14,000–17,000 $m^3/h$. Gas circulation pumps are single-stage compressors with a performance of e.g. 32,000 $m^3/h$. The pressure difference between suction and pressure sides is usually only 30–50 atm.

In another type of pump used for circulation the driving mechanism is relieved of the high suction pressure by a hydraulic compensating lever system. This renders packing boxes sealing against the outside unnecessary (Sulzer design).

The same principle is also used in some refrigerator designs. Figure 58 shows the operation of a high pressure circulation pump with hydraulic lever system. (From Winnacker-Weingaertner, *Chem. Technologie I*, Kiesskalt, Allg. Prinzipien der Verfahrenstechnik.)

### *Shut-off and flow regulating devices*

The shut-off devices are either fully opened or shut at the start or the closing down of operations, or they are opened to varying degrees as regulating devices. Their construction depends on the purpose for which they are designed. There are cocks, valves, and slides. Only valves and slides of special design can be used for high pressure processes.

The simplest design for low-pressure operation is the well-known stop valve, or rotary slide valve. The drilled plug rotates on a flat, conical, or cylindrical seat.

In upper valves a plate or a cone, a bell or a sphere are pressed against a sealing surface. A simple upper valve with flat plate and leather washer is the ordinary water tap. The seal is tightened by a jack screw which is sealed off from the outside by a packing box. In flap valves a flap which is often held down by springs, is lifted off the seat (Figure 59).

In slides a piston or a breech rear plate is pushed into the seat. To give a tighter fit with the sealing surface the slide can also be conical.

Unlike upper valves, slides do not hinder the flow but when opened clear the whole cross-section without deflecting the flow. They can be made with very large cross-sections but the seal is not always perfect and they are very expensive to buy. Sleeve valves with cylindrical seats are frequently used as regulating devices for reciprocating machines.

Rubber bag valves particularly suitable for materials which attack metals, e.g. hydrochloric acid, operate on the principle of the laboratory snap valve.

Safety valves which open automatically above a certain pressure (working pressure) have a special function. They are plate or conical valves in which the seal is pressed against the seat by a spring or a weight. The shut-off devices are operated by hand, by electric motors, hydraulically, or pneumatically.

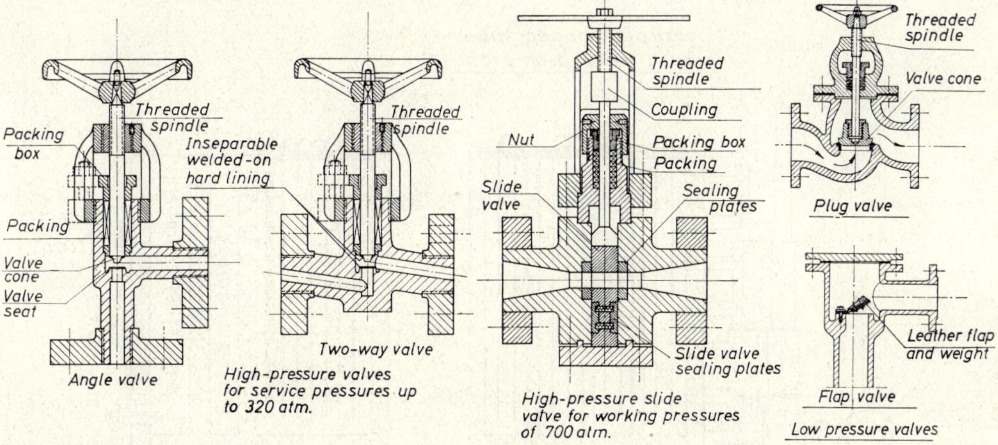


Figure 59. Shut-off and regulating devices

### *High pressure valves*

Whilst in ordinary valves and slides the sealing devices, housing, and seat are made of cast iron, high pressure valves must be shaped from a forged steel block by mechanical treatment. From this results the most commonly used shape of the upper valve. The screw thread of the spindle is very long and produces only a small lift because the counter-pressure is very large; it is operated not as usual by a small hand wheel but by a long lifting handle with a large levering effect. The packing box is particularly long and easily adjustable. The design resembles that of the well-known pressure-reducing valve of the steel cylinders for compressed gases the seal of which is shaped like a cone (needle). Large high pressure valves can also be actuated instead of by a threaded spindle by a piston operated by pressure oil. Such valves can then be remotely controlled and can be opened or closed instantaneously. The much used electrical drive corresponds in principle to the manual drive.

High pressure slide valves have also been developed which can be built into ducts more easily as angle valves. The seal consists of two plates which are forced apart by strong springs and pressed against the sealing surface since an adequate seal cannot be achieved in any other way.

Conveyors, pumps, and compressors which can be used in high pressure hydrogenation processes have already been described in another section, and so has the production and purification of hydrogen. The discussion of the hydrogenation processes will, therefore, now be continued.

Not until BASF, in cooperation with the steel industry, had developed suitable materials, was it possible to heat all the components of a reaction together in hairpin-shaped pressure tubes with welded-on chain lines, suspended vertically in gas-heated furnaces (peak pre-heaters).

*Circulation gas heating*

To heat materials moving in pipes to high temperatures circulation gas heating is commonly used, the operation of which is shown in Figure 60. It is an improvement on the pipe still furnace used widely in the petroleum distillation and cracking industries. The materials to be hydrogenated, coal slurry, tar, and petroleum, are pre-heated in peak

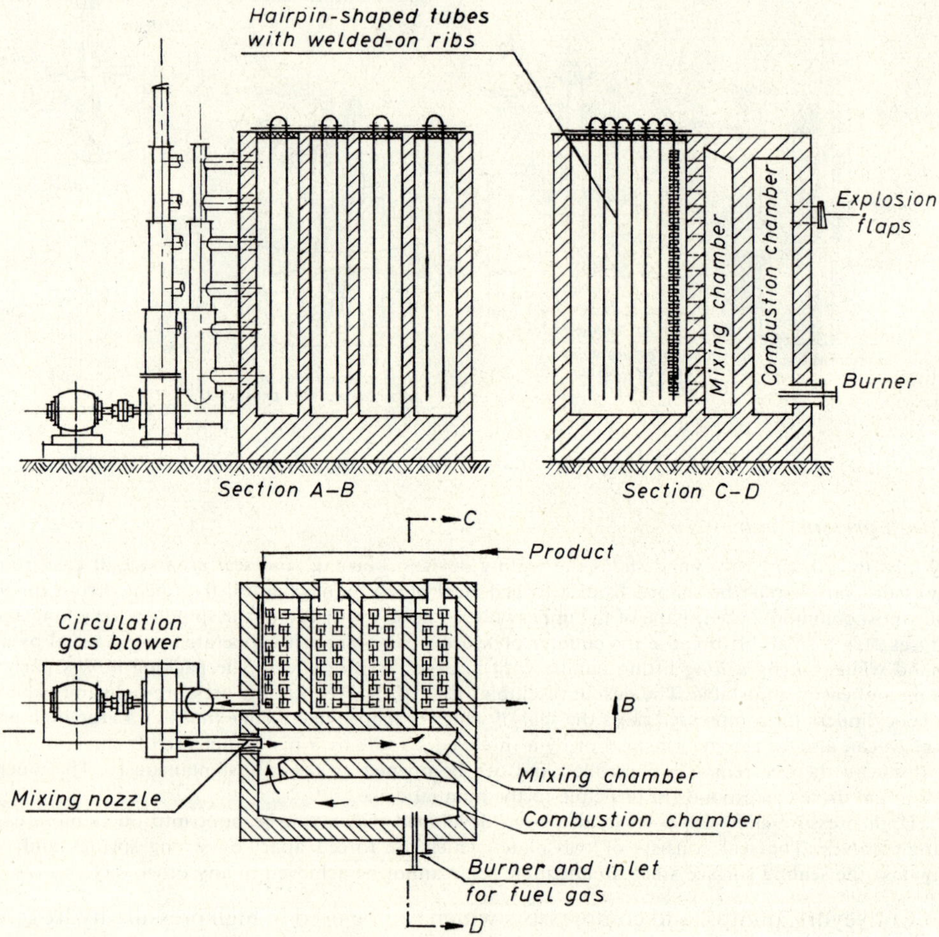


Figure 60. Circulation gas heating

pre-heaters by circulation gas. However, the pre-heating required for the distillation of hydrogenation products and of petroleum is also carried out, under pressure, in the circulation gas furnace and the pressure is then released in a fractionating column. This type of heating allows the temperature to be kept very constant and overheating is prevented.

In a circulation gas furnace the fuel gas is first burnt with air in the combustion chamber; the hot gas then flows into a mixing chamber where it mixes with the circulation gas sucked from the heating

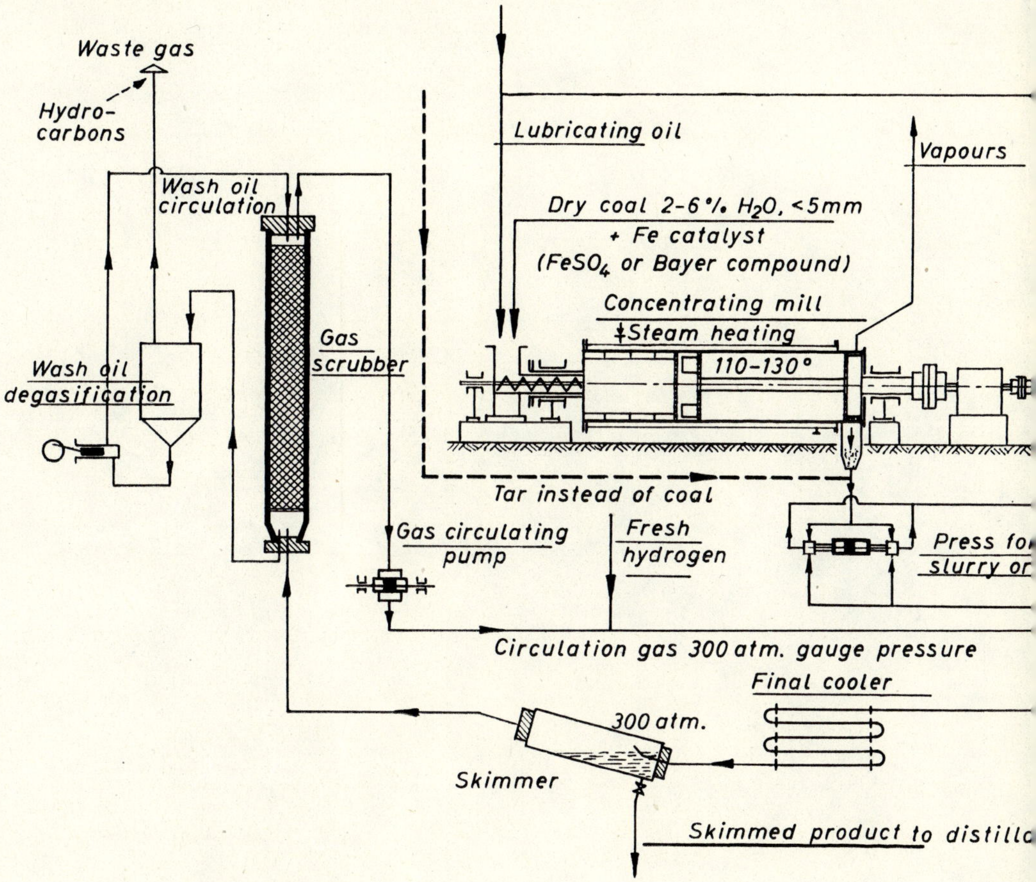


Figure 61. Simplified diagram of coal

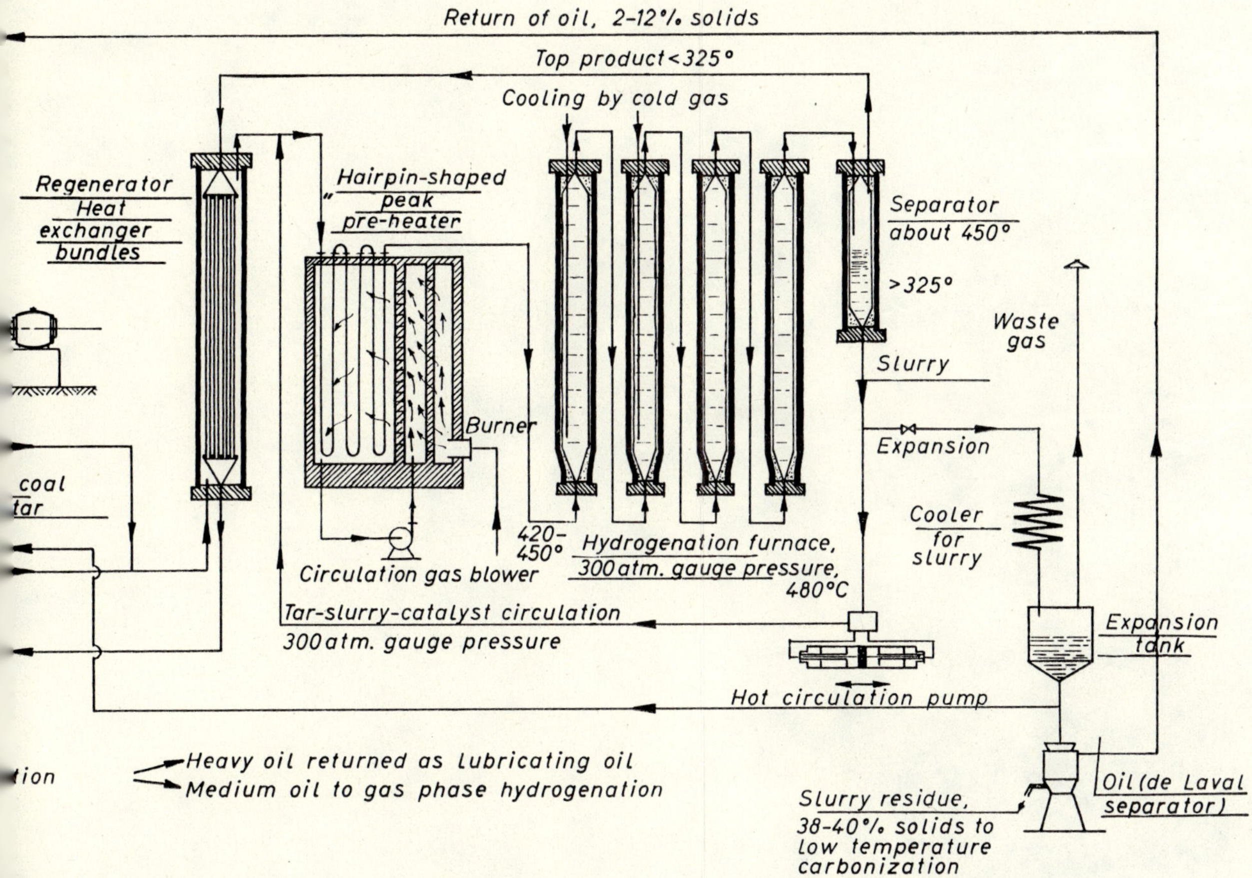


and tar high-pressure hydrogenation in the semi-solid phase

chamber proper by a blower. Only such quantities of fuel gas are added as are required to keep the temperature constant. The excess gas escapes via a stack.

So as to carry out the process as far as possible autothermally, the heat content of the gaseous and volatile reaction products is utilized for heating the components of the reaction in heat exchanger bundles. In a vertical high pressure jacket frequently more than a hundred tubes are combined into bundles. The reactiants are usually discharged through the tubes, the supply of the reactants and the circulation gas usually move on the outside. This arrangement is reversed in special cases. It is thus possible to heat the components of the reaction to 20–60°C below the reaction temperature. A peak pre-heater is then sufficient to attain the reaction temperature of 480–500°C (cf. the simplified diagram of high pressure hydrogenation, Figure 61).

Most heat is produced in the first third of the furnace, which facilitates starting. A large part of the coal is hydrogenated already in the pre-heating period from 320°C onwards. At the end of the pre-heating period the hydrogenative decomposition of the coal is almost complete and the product is largely soluble in benzene.

The coal to be hydrogenated is crushed to 5 mm, dried to a water content of 2–6 per cent, then lubricating oil is added in concentrating mills. Ferric oxide catalyst in the form of bog iron ore or Lux or Bayer compound produced in the manufacture of aluminium, or ferrous sulphate is also added here. The mills consist of revolving tubes containing chambers with grinding elements, at the end of which are slots with screens of 1 mm mesh. At 110–130°C further loss of water occurs here (heating by friction) and then complete homogenization of the coal-oil mash. The resulting slurry with about 50 per cent coal can be pumped by ordinary liquid pumps. This coal slurry is passed through suction vessels into slurry presses where it is compressed at 300 atm gauge pressure (sometimes 700 atm gauge pressure) and then fed together with the hydrogen into the heat exchangers (regenerators). (The hydrogenation of tar and petroleum is carried out by the same method, at least in principle, as that of the coal slurry.)

The tar or coal slurry is fed by a special high pressure press (slurry press) through the heat exchangers and the pre-heaters from below into the first furnace of the chamber. The circulation gas is added to the coal slurry or tar by gas circulation pumps immediately after it has passed through the press. Cracking hydrogenation of the coal or the tar up to the desired intermediate product (medium oil) is usually carried out in 3–4 furnaces arranged in series. From the furnaces the mixture is passed into separators where the temperature is 10–14° below the reaction temperature. In the lower conical part of the separator a certain liquid level is maintained which is checked by measurement. The slurry that collects at the bottom contains in coal hydrogenation up to 35 per cent solids, in tar and petroleum hydrogenation usually up to 22 per cent; it is discharged from there. From the separator the so-called top product of which about 50 per cent boil at 325° escapes in the form of gas. This top product is cooled first in the regenerators, then in the final cooler by water. The separation of gas and liquid takes place in the skimmer, a high pressure cylinder, slightly inclined against the horizontal, where a level of liquid is maintained. The circulation gas is returned by the gas circulation pump into the system, fresh hydrogen having been added. So as to maintain constantly the necessary hydrogen partial pressure of 70–80 per cent in the circulation gas it is scrubbed with oil under pressure in a scrubbing tower which reduces its hydrocarbon content. When the pressure is released the oil liberates the gaseous hydrocarbons. Circulation hydrogen is also fed directly into the furnaces for cooling (cold gas). The proportion of the skimmer product that boils below 325° has a composition similar to that of low-temperature tar or coal medium oil or petroleum medium oil. This proportion is extracted by distillation from the semi-solid phase.

Those fractions that boil above 325° (heavy oils) are returned into circulation, either by mixing them with the raw material (tar or petroleum) or by lubricating the coal with them. The liquid and solid fractions which are discharged from the lower part of the separator are further processed by centrifuging and low temperature carbonization.

Filtration is possible only after more oil has been added and at high temperatures of about 200°. Separation is most efficiently carried out by centrifuging in a de Laval separator. The slurry which in coal treatment usually contains 22–35 per cent solids, is thinned with skimmer heavy oil to 15–20 per cent solids (temperature 140–160°C). The products obtained are:

(1) centrifuge oil with only 2–10 per cent solids which may for instance be returned into the coal as lubricating oil,

(2) centrifuge residue containing 38–40 per cent solids which are processed thermally, e.g. by distillation to coke in a special furnace i.e. by a sort of carbonization process. The catalyst is not recovered. With brown coal a coil furnace is used for the carbonization where the residue, after pre-heating in the heat exchanger, is carbonized by gas heating at 550–600°C with the addition of superheated steam. In the hydrogenation of bituminous coal spherical furnaces are used consisting of a carbonization drum lined with steel plate and filled with steel balls; subsequently the oil is condensed and the coke discharged. Finished products are oils, carbonization gas, and coke.

Not all the slurry discharged from the separator is processed, only a part equivalent to the amount of coal, tar, or petroleum added. For heat economy reasons a large part is returned under pressure to the coal slurry before it enters the pre-heater so that part of the catalyst circulates.

## 2. HYDROGENATION IN THE GAS PHASE

By the conversion in the semi-solid phase of the raw materials coal, tar, and mineral oils into oils of the medium boiling range the main work of hydrogenation into fuels is already done. In this part of the process about two thirds of the total hydrogen required are used. Qualitatively the most important part of the process is the gas phase hydrogenation the success of which depends essentially on a reliable catalyst. The catalyst must be resistant to sulphur and have a long life, at least 11 months. The equipment used is similar to that employed in semi-solid phase hydrogenation. The furnaces are, however, fitted to receive the catalyst which is pressed from powder into pellets of 10–15 mm diameter and placed on perforated plates.

The technological development of hydrogenation in the gas phase is closely bound up with the discovery of suitable catalysts. Accordingly, there were four stages in the development:

(1) First, an oxide compound of molybdenum, zinc, and magnesium was used which was, however, not sufficiently reactive and produced too many gaseous products. In the course of the hydrogenation process of coals, tars, and oils, i.e. sulphur-containing raw materials, oxide catalysts are converted into sulphide ones. The catalyst compound is poisoned, like Pd-asbestos, by organic bases.

(2) A tungsten-zinc oxide compound that was converted at about 200 atm gauge pressure with $H_2$ and some $CS_2$ ($H_2S$) into the sulphide one, proved to be especially effective. It could be used at a temperature that was lower by 50° than the Mo-Zn-Mg compound referred to, with double the performance and half the amount of gas formed. It was almost insensitive to organic bases. At first, the somewhat complicated preparation was obviated by thermal decomposition of ammonium sulphotungstate $(NH_3)_2WS_4$ into $WS_2$ + ammonium sulphide which takes place at atmospheric pressure. This highly reactive tungsten sulphide has been shown by x-ray measurements to have a distorted

lattice. The lattice distortions are probably the cause of the high reactivity; opinions are, however, still very divided on this. Whilst this catalyst had a very high cracking activity the petrol produced was rich in hydrogen, with very poor antiknock properties that could not satisfy the requirements of the motor industry.

(3) The highly reactive $WS_2$ catalyst was then precipitated on to carriers that were known to have a good cracking effect, i.e. that were themselves reactive, e.g. aluminium silicate. However, bleaching earth like Terrana, a type of Fuller's earth etched with hydrofluoric acid, is better. This produced a catalyst with a good cracking effect, partly due to the carrier, with a reduced hydrogenation effect owing to the dilution of $WS_2$ in the catalyst to about 10 per cent. With it petrols with higher octane ratings were obtained. The catalyst was effective in the hydrogenation of petroleum medium oils but is sensitive to phenols, unlike the pure $WS_2$ catalyst, and its performance with brown coal medium oils and bituminous coal oils was, therefore, poor.

(4) The problem was finally solved by dividing the gas phase hydrogenation into two stages. First, phenols and bases were removed from the oils by the strongly reducing and hydrogenating concentrated $WS_2$ catalyst, and their hydrogen content was brought to the level of the petroleum medium oils. A $WS_2$ catalyst diluted with alumina and a small addition of nickel sulphide which has been used recently in industry in place of the pure $WS_2$ catalyst also proved suitable for this hydrogenating stage. In the second phase aromatization on the 10 per cent $WS_2$-carrier catalyst takes place. However, this aromatization catalyst is sensitive to $NH_3$. It is, therefore, used in a separate gas circulation connected with a pressure-water scrubber in which the ammonia is continuously washed out of the circulation gas. It is important to maintain constantly a minimum $H_2S$ residual pressure in the reaction chamber since the $WS_2$-carrier catalyst otherwise becomes oxidized. The apparatus is designed similarly to that used in the semi-solid phase; it consists of two separate circulations, one for the pre-hydrogenation and one for the aromatization.

Hydrogenation is followed by distillation of the skimmer products, leaching of the petrol, and finally removal of propane (stabilization) by distillation under pressure.

As by-products the hydrogenation process yields gases (Hy-gases) principally methane, ethane, propane, butane, and isobutane. These are used as fuel and power gases, and for chemical reactions. The saturated hydrocarbons can be transformed by cracking into olefins and form then a valuable basis for chemical syntheses.

*Processing of the hydrogenation gases*

The saturated hydrocarbons $C_1$ to $C_4$ and some pentane are constituents of the circulation gas. They do not become concentrated because under high pressure they dissolve in the liquid hydrocarbons and are thus partly removed from circulation. As already mentioned, they are also removed by the oil scrubber introduced into the circuit for this purpose. Since the solubility of the gases increases with the boiling point separation can be achieved by stepwise pressure release of the skimmer products.

This stepwise pressure release produces first a splitting up into lean gas and rich gas. The latter contains mainly the hydrocarbons $C_3$ to $C_5$. $H_2S$ and $CO_2$ are removed from the rich gas by an alkalized scrubber, and $NH_3$ by scrubbing with water. Like the separation of the natural gases in the petroleum industry, or of the cracking gases, etc., the separation can be carried out by liquefaction, e.g. by compression to 25 atm gauge pressure and subsequent fractional distillation in a pressure apparatus (Linde). The $C_5$ fraction is separated in isolation as 'gas petrol', used for the amelioration of the petrol. It is possible, by distillation under pressure, to separate also the olefins for subsequent chemical utilization.

# 3. VARIATIONS OF THE HYDROGENATION PROCESSES

## (a) Low temperature hydrogenation (LTH process) – Refining by hydrogenation

The hydrogenation process can also be used to remove from crude petroleum, tar oils, heavy oils, medium oils, etc., oxygen, sulphur, and nitrogen, and to hydrogenate completely or partly olefinic bonds and aromatic double bonds in such a way that cracking is as far as possible prevented. Low temperature hydrogenation is carried out at about 250 atm hydrogen pressure and a temperature of 350°C on the $WS_2$ catalyst in the liquid phase. Thus, for instance, heavy oils can be hydrogenated to lubricating oils, and inferior medium oils to diesel oils.

In practice this process is used principally for the transformation of brown coal tars into high-grade finished products in the VEB Hydrogenation Plant Zeitz. The high-pressure apparatus used corresponds largely to the gas phase chambers described above, in particular the pre-hydrogenation, i.e. it operates with a fixed catalyst and with excess hydrogen (circulation gas). The products are very valuable for the national economy since they are quality products, e.g. export goods such as diesel fuel and paraffins, or valuable raw materials for organic intermediates such as benzene. The extraction of aromatics requires naturally additional equipment.

## (b) High temperature processes – Aromatization, dehydrogenation

For special purposes frequently fuels with an exceptionally high octane rating are required, e.g. for aircraft engines. For this a high content of aromatics, naphthenes, and isoparaffin sin the petrol is desirable. Bituminous coal hydrogenation petrol is more suitable for conversion into special petrol than brown coal petrol since it consists of about fifty per cent naphthenes with six-ring-systems predominating. Chain branching improves the anti-knock properties of petrol.

Such converted petrols are also a source of raw material for the extraction of aromatics, i.e. benzene, toluene, the xylenes and naphthalene, a fact of special importance for the economy of the German Democratic Republic because of its very limited production of bituminous coal tar. However, separation of the aromatics directly from the dehydrogenation petrol by distillation is impossible since they form azeotropic mixtures with non-aromatic constituents which cannot be separated by distillation. For this purpose special separation methods must be used. (See Selective extraction and distillation of petroleum.)

Whilst hydrogenation is an exothermic process, aromatization and dehydrogenation are endothermic and require large pre-heaters. The dehydrogenation is carried out on catalysts, e.g. molybdic acid on alumina. It is unavoidable that gradually polymeric products settle on the catalyst surface and are there dehydrogenated to coke. The catalyst must, therefore, from time to time be regenerated by burning-off. A dehydrogenation catalyst must, therefore, operate with the least possible cracking and formation of gases and separation of coke, and it must be able to stand frequent burning with air for regeneration. The metals platinum and nickel are also excellent dehydrogenation catalysts for operation at high temperatures.

The hydroforming process (HF process) used mainly in the United States for the aromatization of petroleum products was originally developed by the former IG and in 1945 was taken over by the Standard Oil Co. The catalyst consists of 90 per cent $ZnAl_2O_4$ + 10 per cent $MoO_3$. Technologically it resembles more the cracking process than the DHD process described below.

The DHD process (pressure-$H_2$-dehydrogenation) which was also developed by the former IG and is for example used by the VEB Combine Böhlen, is carried out on a $Al_2O_3$-$MoO_3$ catalyst, recently also on a $Al_3O_3$-Pt-catalyst, with bituminous coal petrol at a $H_2$ partial pressure of 30–40 atm gauge pressure and a temperature of 500–530°C, with brown coal petrol at a somewhat lower pressure. To avoid excessive condensation, polymerization, and therefore precipitation on the catalyst it is necessary to work under a hydrogen pressure which because of the equilibrium position is, however, lower than in hydrogenation. Since the catalyst has a limited life it is called, like in cracking, a "decaying catalyst". The time to regeneration used to be 60 hours and has now been increased by various improvements, in particular after the introduction of the platinum catalyst, to far beyond 1,000 hours. Regeneration is carried out at 40 atm gauge pressure and a temperature of about 530°C, by flushing with nitrogen to which gradually 2–4 per cent oxygen are added. The reaction chamber is subdivided, and between the individual furnaces the gases are pre-heated in hairpin-shaped tubes, the peak pre-heaters. In the DHD process some olefins are always produced which have an unfavourable effect on the stability of the petrol since they are oxidizable and lower the octane rating of the fuel. They can be removed chemically, e.g. by sulphuric acid, but better by catalytic hydrogenation in the refining furnace at the end of the DHD process which also contains a $MoO_3$ catalyst. The temperature is 300–350°C so that only partial hydrogenation takes place, the aromatics being left intact. The gases from the dehydrogenation processes are largely olefinic and are, therefore, particularly suitable as raw material sources for the extraction of ethylene, propylene, and butylene, and the saturated hydrocarbons.

## LITERATURE

*Flow technology*

F. Kluge, in Ullmann, *Encyklopädie*, 3rd ed., vol. I.

*Pressure technology*

R. Berger, l.c.

A. Henglein, *Grundriss der chemischen Technik*, Verlag Chemie, 11th ed., 1963.

*Process technology*

S. Kiesskalt in Winnacker-Weingaertner, *Chem. Technologie I.* Hanser, München, 1952.

*Control technology*

E. Weiss, *Chem. Ing. Techn.* (1952), **24**, 437.

O. Winkler, Temperaturmessung in chem. Betrieben. *Chem. Ing. Techn.* (1953), **25**, 1.

*Manufacture of synthetic fuels by pressure hydrogenation*

M. Pier, Über Hydrierbenzin. *Z. Angew. Chem.* (1938), **51**, 603.

C. Bosch, *Chem. Fabr.* (1933), **6**, 127; (1934), **7**, 1.

Krauch und Pier, *Z. Angew. Chem.* (1931), **44**, 953.

G. Hähnel, in Ost-Rassow, *Lehrbuch der chemischen Technologie*. J.A. Barth, Leipzig, 1955, 26th ed., p. 134.

W. Krönig, in Winnacker-Weingaertner, *Chem. Technologie*, Organische Technologie I, 1952.

W. Krönig, *Die katalytische Druckhydrierung von Kohlen, Teeren und Mineralölen.* Springer, Berlin/Göttingen/Heidelberg, 1950.

# IV. Syntheses with Carbon Monoxide

## 1. HYDROGENATION OF CARBON MONOXIDE

Apart from some scientific experiments in other countries the hydrogenation of carbon monoxide was technologically developed at first almost exclusively in Germany. The Badische Anilin- und Sodafabrik (BASF) and its collaborators, of whom may be mentioned Mittasch, Schneider, Pier, Wietzel, Rumpff, Stein, Luther, and Winkler, recognized early the great industrial and economic possibilities of carbon monoxide hydrogenation. In 1913 a basic patent (Schneider, DRP 293,787, dated 8. 3. 1913) was applied for, which even at that time outlined all the possible developments of hydrogenation. Based on the large-scale industrial experience gained in the high pressure synthesis of ammonia by the Haber-Bosch process, the BASF was able relatively quickly to evolve an industrial method for the high pressure hydrogenation of CO to methanol on $ZnO$-$Cr_2O_3$ catalysts. This was later followed by the pressure hydrogenation of CO to isobutyl oil, a mixture of higher alcohols which introduced an entirely new epoch in industrial aliphatic chemistry in Ludwigshafen-Oppau and later in Leuna (Wietzel, 1923).

As early as 1913 hydrocarbons, in addition to oxidized products, were obtained by the BASF by the hydrogenation of CO under pressure on osmium and cobalt oxide catalysts.

In 1922, at about the same time as the first large methanol plant started production for the BASF, Fischer and Tropsch observed the formation on alkalized Fe filings, at a temperature of 400°C and at more than 100 atm gauge pressure, of higher-molecular weight oxidized hydrocarbons which they called 'Synthol'.

The year of birth of the Fischer-Tropsch synthesis proper is 1925. In the former Kaiser-Wilhelm-Institut für Kohleforschung (Kaiser Wilhelm Institute for Coal Research) at Mühlheim a. d. Ruhr the conversion of carbon monoxide into hydrocarbons, at atmospheric pressure and a moderate temperature of 220–250°C, was carried out on iron and cobalt catalysts. In the first publication with the title 'Direct synthesis of petroleum hydrocarbons at normal pressure (I)' comprising only two pages of the Berichte der deutschen Chemischen Gesellschaft, 1926, all the fundamental principles of the new synthesis are described. We shall introduce the discussion of the section 'Hydrogenation of carbon monoxide' by beginning with Franz Fischer's words from this first publication, destined to be of such far-reaching significance:

> 'It seems that carbon monoxide reduction will assume a similar importance in industry as carbon dioxide assimilation has in nature. The reason for not basing the technological-catalytic reduction on carbon dioxide is that the carbon monoxide which is poorer in oxygen can be easily manufactured directly. The reduction of carbon monoxide to methane at normal pressure which has been known for a long time, has now been joined by the catalytic production of petroleum hydrocarbons, also at normal pressure.'

## Hydrogenation of carbon monoxide

Table showing the development of the process, catalysts used, pressure and temperature conditions, and the nature of the products obtained

| Year | Authors | Catalysts | Pressure | Reaction temperature | Product |
|---|---|---|---|---|---|
| 1902 | Sabatier and Senderens | Ni pyrophorus | 1 atm | 200–280 °C | $CH_4$ |
| 1913 | Badische Anilin & Sodafabrik Mittasch, Schneider | Os- and Co-oxide | 100–120 atm gauge pressure | 300–420 °C | Hydrocarbons and oxygen-containing compounds incl. alcohols |
| 1923 | BASF-IG Wietzel et al. | $ZnO\text{-}Cr_2O_3$ | 220 atm gauge pressure | 375 °C | Methanol and higher alcohols 'isobutyl oils' |
| 1923 | Fischer-Tropsch K. W. I. Kohlenforschung Mühl. | Fe-filings alkalized | 100 atm gauge pressure | 410 °C | O-containing products 'Synthol' |
| 1925 | Fischer-Tropsch | Fe-ZnO Co-ThO $Co\text{-}Cr_2O_3$ | 1 atm | 220–250 °C | Hydrocarbons with chains of various length |
| from 1934 | Fischer-Ruhrchemie | Development of industrial synthesis plants on a large scale, operating on the principle described above | | | |
| 1936 | Fischer, Pichler Ruhrchemie „Mitteldrucksynthese“ ('Medium pressure synthesis') | Co, $ThO_2$, Fe | 5–15 atm gauge pressure | 250–300 °C | Mainly higher hydrocarbons |
| from 1937 | Kölbel-Rheinpreussen liquid phase | Fe | 11–12 atm gauge pressure | 260–270 °C | Hydrocarbons, mainly olefins |
| 1941 | Fischer, Pichler 'isosynthesis' | $ThO_2$ | 300 atm gauge pressure | 300–500 °C | Iso-paraffings, $O_2$-containing compounds |
| 1939 | IG Synolprocess (not yet operated on a large scale) | Fe fusion catalyst | 18–30 atm gauge pressure | 190–200 °C | Straight-chain, primary alcohols $C_9\text{-}C_{20}$ |
| from 1946 | Ruhrchemie-Lurgi High load synthesis | Fe | 25–30 atm gauge pressure | 220–230 °C | Variable: lower or higher hydrocarbons |
| 1941 | Ruhrchemie-IG ('Oxosynthesis') Roelen | Co (Fe) | 20 atm gauge pressure | 115–140 °C | Branched aldehydes and alcohols |
| | | Addition of CO and $H_2$ to olefins | | | |

## (a) Hydrocarbon synthesis by the Fischer-Tropsch process

### Method and technological development

The first observations were made in 1925 during experiments in Fischer's Institute designed to produce oil gas for the railways from water gas over Fe and CO catalysts. At atmospheric pressure and at a temperature of 250–300°C oily products were obtained which proved to be hydrocarbons. The method was further developed in the period 1926–1928; more reactive catalysts were discovered which were, however, very sensitive to catalyst poisons, in particular sulphur, and had, therefore, a life of only two weeks. The best behaviour was shown by a cobalt catalyst with $ThO_2$ on kieselguhr as a carrier which could be used for 6–8 weeks. When the price of cobalt on the world market dropped by about a quarter it became possible to use this catalyst industrially.

During the hydrogenation process a hard paraffin which reduced the reactivity ('catalyst paraffin') was deposited on the catalyst; it could be removed by periodic extraction and hydrogenation in the catalyst furnace itself so that the catalyst could remain in use for 2–3 months (Fischer, Koch, Roelen). Since 1934 the Ruhr-Chemie-AG in Holten has been the licensee for Fischer's patents. A rapid technological development followed. Within a few years a solution was found for the difficult problems of the design of large industrial catalyst furnaces from which the very large amounts of heat generated during the reaction could be dissipated, the manufacture of pure raw materials for the catalyst, and the manufacture of the catalyst itself, the extraction of synthesis gas from bituminous coke and brown coal, and the necessary extremely thorough purification of the gases. Production was started in the first plants in 1936 and by 1941 nine large plants had been built in Germany which in 1942 had an annual capacity of 740,000 tons primary products.

Of these plants five operated the normal pressure process, two the medium pressure process (5–15 atm gauge pressure) and two used both processes. An improvement of the catalyst by Roelen in 1938 had great industrial advantages. It consisted in a partial replacement of the Co-$ThO_2$ catalyst by a Co-$ThO_2$-MgO catalyst. When during the 1939–1945 war there was a shortage of cobalt—for in spite of regeneration of the catalyst cobalt losses cannot be avoided—attempts were made to change over to the use of an iron catalyst; these attempts were so successful in the medium pressure synthesis at Schwarzheide that the change-over to this catalyst was imminent.

After 1945 production in Germany ceased. The Schwarzheide plant soon recommenced production by the normal pressure synthesis.

In 1946 a new development took place at Ruhrchemie-Lurgi: new synthesis furnaces were introduced in which the dissipation of heat was achieved by very high gas velocities (high pressure and gas circulation). In 1952 medium pressure synthesis on an iron catalyst was restarted at the Krupp-Kohlechemie G.m.b.H. in Wanne-Eickel. In 1952–1956 the new Ruhrchemie-Lurgi synthesis furnaces started production at Kuhlmann, France, and at Sasolburg, South Africa, for the manufacture of solid paraffins. Until that time fixed bed catalysts had still been used industrially.

As early as 1930–1931 F. Fischer, O. Roelen and W. Feisst had carried out experiments with Co catalysts suspended in oil. This method was applied to iron catalysts semi-industrially in 1939–1942 by Ruhrchemie, operating with circulation gas, whilst the former I. G. Ludwigshafen (W. Michael) worked with fixed bed catalysts and circulating oil and with powdered catalysts under pressure.

At the same time successful tests were carried out of the liquid phase synthesis with iron catalysts by Rheinpreussen (H. Kölbel, P. Ackermann). In 1953 the first industrial liquid phase furnace started production there (H. Kölbel, P. Ackermann, *Chem. Ing. Techn.*, 1951, **23**, 183; 1956, **28**, 381). The synthesis from carbon monoxide and water was also developed there by H. Kölbel and F. Engelhardt, thus combining the conversion reaction with the CO-hydrogenation.

The most recent large-scale industrial development since 1943 is the use of the fluidized bed method for the Fischer-Tropsch synthesis (Hydrocarbon Res. Co., P. C. Keith, U.S.A.), and since 1946 the development of a medium pressure synthesis on finely powdered iron catalysts (Kellogg Co., U.S.A.). These produce only light liquid hydrocarbons. Both methods are used in very large plants, one in Brownsville, Texas, another at Sasolburg, South Africa.

*Chemistry of CO hydrogenation*

The chemical mechanism of CO hydrogenation has not yet been determined with certainty (Kölbel and Engelhardt, *Chem. Ing. Techn.*, 1950, **22**, 97, Zusammenstellung der Arbeiten über den Reaktionsmechanismus. See also: Anderson, Hofer and Storch, *Chem. Ing. Techn.* 1958, **30**, 560). It is a complex, heterogeneous reaction on the surface of the catalyst. The synthesis reaction consists of a number of individual reactions of various rates, some taking place one after the other and others simultaneously. Many theories have been evolved of which the 'classical theory' by Fischer will be mentioned here first. According to Fischer, carbides form first from carbon monoxide and the metal catalyst. The carbon of the carbides is supposed to form with hydrogen on the catalyst surface $CH_2$ radicals which are polymerized. This is possible only if the polymerization takes place more rapidly than the hydrogenation, otherwise only methane would be formed. The formation of methane is in any case the most likely reaction. Many investigators contest Fischer's theory, others support it. According to the Soviet authors Eidus and Pusitski, for instance, the way in which polymerization of $CH_2$ takes place is that two $CH_2$ groups which are very close together first form ethylene. This is supported by the findings of Craxford and others that in the presence of ethylene and under the same conditions an increase in the yield of liquid products can be observed as compared with CO and $H_2$ only.

For various reasons the carbide theory is untenable. Tests with the isotope of carbon $^{14}C$ have shown that on the iron and cobalt catalyst only 10 per cent of the hydrocarbons are formed from carbide (Kummer, Witt and Emmett, *J. Amer. Chem. Soc.*, 1948, **70**, 3632). The first stage in the reaction is an adduct of catalyst, CO and $H_2$ in which the carbon-oxygen bond is loosened. According to the conditions and the catalyst used the following reactions occur:

(1) breaking of the CO bond yields hydrocarbons,
(2) if the CO bond is preserved, O-containing compounds are obtained and from these olefins by separation of $H_2O$.

In the oxosynthesis which is a homogeneous catalysis carbonyls are formed. German contributions to the explanation of the reaction mechanism have come in recent times mainly from Kölbel and his colleagues, H. Pichler, W. Brötz and W. Rottig (Kölbel and Engelhardt, *Chem.-Ing. Techn.*, 1950, **22**, 97, Zusammenstellung der Arbeiten über den Reaktionsmechanismus, see also: Anderson, Hofer and Storch, *Chem.-Ing. Techn.*, 1958, **30**, 560). Basing his argument on the fact that preferentially straight-chain, unbranched paraffins are formed, W. Brötz discusses the possibility of joining together methylene radicals step by step (*Z. Elektrochemie*, 1949, **53**, 301).

It is important to preserve the crystalline metallic surface during the catalysis and regeneration of the catalyst. Oxides do not dissolve the metallic bonds but disturb them and affect the activity in a specific manner.

According to Herbert and Royen, the reaction equations are likely to be as follows:

Methane formation

$$CO + 3H_2 \rightarrow CH_4 + H_2O + 50{\cdot}4 \text{ kcal (Ni-cat. 190–250°C) (Sabatier and Senderens)}$$

$$CO_2 + 4H_2 \rightarrow CH_4 + 2H_2O + 39{,}3 \text{ kcal (Ni-cat. 230–300°C)}$$

Hydrocarbon formation

$$xCO + x2H_2 \rightarrow (-CH_2-)_x + xH_2O \text{ (Co catalyst, ideal)}$$

$$\left.\begin{array}{l} CO + 3H_2 \rightarrow CH_4 + H_2O \\ 2CO \rightarrow CO_2 + C - 38{\cdot}7 \text{ kcal)} \end{array}\right\} \text{(side reactions)}$$

Boudouard equilibrium

$$\left.\begin{array}{l} xCO + x2H_2 \rightarrow (-CH_2-)_x + xH_2O \\ xCO + xH_2O \rightarrow xCO_2 + xH_2 \end{array}\right\} \begin{array}{l}\text{(Fe catalyst)} \\ \text{conversion reaction}\end{array}$$

$$x2CO + xH_2 \rightarrow (-CH_2-)_x + xCO_2$$

The formation of $CO_2$ can be considered as a combination of the first reaction (formation of water) with the conversion:

$$CO + H_2O = CO_2 + H_2 + 10{\cdot}1 \text{ kcal}$$

The result varies according to the conditions, pressure, temperature, gas composition, type of catalyst, and residence time. From the above equation follows that each catalyst requires a definite composition of the gas, e.g. Fe : 2 CO to 1 $H_2$; Co: 1 CO to 2 $H_2$.

The unavoidable methane formation should be kept as low as possible; the 'conversion reaction' $CO + H_2O \leftrightarrows CO_2 + H_2 + 10{\cdot}1$ kcal is also undesirable.

During hydrogenation water is formed which must be removed if the reaction is to proceed smoothly. An intermediate separation of the reaction products and of the unwanted water can be achieved by carrying out the synthesis step by step, or by circulation synthesis.

According to Kölbel and Engelhardt (H. Kölbel and F. Engelhardt, *Erdöl u. Kohle*, 1949, **2**, 52; *Brennstoffchemie*, 1952, **33**, 13) the conversion reaction can be made to take place rather rapidly so that instead of carbon monoxide and hydrogen, carbon monoxide and water, and carbon dioxide and hydrogen can be converted:

$$x(3CO + H_2O) \rightarrow (-CH_2-) + {}_{2x}CO_2$$
$$x(CO_2 + 3H_2) \rightarrow (-CH_2-) + {}_{2x}CO$$

The synthesis is exothermic. In the complete conversion of $CO + H_2$, 23 per cent of the total heat of combustion contained in the reactants is liberated (600 kcal/$m^3$). Without cooling and heat dissipation this amount of heat would be sufficient to heat the synthesis gas to 1,500°C. More heat is produced here than in any catalysis and its dissipation presented one of the greatest difficulties when the method was to be applied on an industrial scale. The solution of this problem represented a great technological triumph.

*Technology of the synthesis processes*

The following basic problems had to be solved:

(1) The manufacture on a large industrial scale of serviceable catalysts that satisfied the requirements in respect of specificity, wear, activity, life, and capacity to be regenerated.

(2) The design of catalyst furnaces of sufficient size that permitted maintenance of the temperature at a constant level within narrow limits, the enormous quantities of heat produced being dissipated, and which contained as large a quantity of catalyst as possible. For the medium-pressure synthesis the catalyst furnace had to be pressure-resistant.

(3) Methods had to be developed which permitted the economical production of synthesis gas in very large quantities, and thorough purification of this gas. An analysis

of the operating costs of the Fischer-Tropsch synthesis showed that formerly about 55–60 per cent, now up to 85 per cent, of the cost was accounted for by coal and gas and gas purification, 30 per cent by the cost of the synthesis and of the catalyst and the separation of the reaction products, and about 15 per cent by the manufacture of the finished product.

For the hydrogenation of carbon monoxide the metals iron, cobalt, and nickel have a specific activity, ruthenium acts as a catalyst only at high pressure.

In all the original synthesis plants Co catalysts were used at first. Now they have gone over to the exclusive use of Fe catalysts (precipitation catalysts and fusion catalysts) which, in addition to being cheaper, are more flexible, i.e. their useful range of temperature and pressure is considerably wider. Moreover, contrary to the effect of a cobalt catalyst, the conversion, and with it the space-time yield, increase with pressure when an iron catalyst is used. By changing the gas composition and the operating conditions the composition of the synthesis products can be varied within wide limits and considerably more olefins can be obtained. The medium pressure synthesis cannot be carried out with Co.

As an illustration of the manufacture of a precipitation catalyst of exceptionally high purity, a description follows of the manufacture of the 'classical' cobalt-thorium-oxide catalyst for the normal-pressure synthesis.

From 1938 onwards the standard catalyst for the Fischer-Tropsch synthesis consisted of 100 parts Co, 5 parts $ThO_2$, 8 parts MgO, and 180–200 parts kieselguhr. It was made for all the consumers in three special catalyst plants since the manufacture requires very great care. Of specific importance are highest purity of all raw materials, and exact compliance with all the requirements such as residence times, temperatures, concentrations, and proportions mixed. The manufacture of the catalyst was carried out in four operations:

(1) preparation of the kieselguhr and other raw materials,
(2) Co- and Th-regeneration from spent catalysts,
(3) precipitation of the catalyst from nitrates with soda, filtration, washing, compression, and drying,
(4) reduction of catalyst to produce new catalyst (formation of Co).

Kieselguhr (infusorial earth) is a very good carrier particularly for active catalysts. It is natural $SiO_2$ from the remains of fossil infusoria (diatoms) and has a very large surface, very good absorptivity, uniform structure and internal cavities. Electron microscopic examination has shown that the various deposits differ in structure.

Even more uniform are dehydrated natural and synthetic zeolites. They are used as molecular sieves ('Linde sieves') for specific adsorptions and separations.

For the manufacture of the synthesis catalyst it is desirable for the kieselguhr to contain less than 1 per cent Fe (if more, methane would form), less than 0·4 per cent $Al_2O_3$ (if more, too much gel would form), and less than 1 per cent organic matter, which can be destroyed by heating at 600–700°C. To prepare the catalyst, solutions of $Co(NO_3)$, obtained from the solution of Co metal or of spent catalysts, together with $Th(NO_3)_4$ and $Mg(NO_3)_2$, are heated and precipitated with soda. Then kieselguhr is added and the material filtered and washed in a filter press or a rotary filter, compressed into small 'sausages' in an extruding press, and dried in a shelf drier with rotating arms. The substance is then screened to a coarse uniform particle size, and reduced at 390–410°C using very high purity circulated hydrogen that is free from $H_2S$ and CO. During the reduction a large amount of water is liberated which must be removed from the circulating gas. This is done by a very economical method which is frequently used in the finish-drying of gases. First, most of the water is precipitated by freezing at −12°C in a Linde cooler (the content drops from 20 g $H_2O/m^3$ to 3·5 g $H_2O/m^3$); then drying is continued with silica gel down to a content of 0·1 g $H_2O/m^3$. Drying to even lower $H_2O$ contents can be carried out with calcined zeolites.

About 40–50 per cent of the adhering paraffins is removed from the spent catalysts in the synthesis oven by means of diesel oil or petrol. Then the catalyst is dissolved in nitric acid for the recovery of Co and Th. This is followed by a difficult inorganic separation and purification process to obtain pure nitrates.

*Final purification of the gas*

So far the removal of $H_2S$ had been adequate for the hydrogenation of coal and tar; traces of organic S compounds remained in the gas. This could not be permitted for the Fischer-Tropsch synthesis since the Co metal catalyst is very sensitive to sulphur. The removal of all traces of organic sulphur compounds, such as mercaptan, RSH, COS, $CS_2$, was essential. All the methods known up to 1930 failed in this respect, and a new gas purification method had to be developed by Roelen and Feisst. When cracking gas or raw brown coal gas is used, the 'gum formers' must first be removed with activated carbon.

The organic sulphur compounds are reduced to $H_2S$ by reduction at 180–280°C on an alkalized Fe catalyst with hydrogen and a little $H_2O$ with the addition of 0·15 to 0·3 per cent by volume of $O_2$. The hydrogen sulphide is simultaneously oxidized to $SO_2$ and then to $SO_3$: $H_2S \rightarrow SO_2 \rightarrow SO_3$. This high-degree-purification material contains 30 per cent of soda for the removal of the $SO_3$ formed. It is produced from 'Lauta compound' and soda. This material absorbs acid gases up to 280°C. The final concentration is only 0·2 g of organically bound sulphur in 100 m³ of gas. The following reactions must be expected:

$$CS_2 + 2H_2O \rightarrow 2H_2S + CO_2$$
$$H_2S + 1{\cdot}5\,O_2 \rightarrow H_2O + SO_2$$
$$SO_2 + 0{\cdot}5\,O_2 \rightarrow SO_3$$
$$SO_3 + Na_2CO_3 \rightarrow Na_2SO_4 + CO_2$$
$$\overline{CS_2 + 4O_2 + 2Na_2CO_3 \rightarrow 2Na_2SO_4 + 3CO_2}$$

More recently, the Rectisol gas purification method (W. Herbert, *Erdöl u. Kohle*, 1956, **9**, 77) has been adopted in an increasing number of synthetic plants. The impurities (benzene vapour, $CO_2$, $H_2S$, gum formers, and organic sulphur compounds) are washed out with a polar solvent, such as methanol or acetone, at low temperatures (as low as −62 °C).

*Catalyst ovens (non-pressure type) for synthesis at atmospheric pressure*

In Fischer's first small experimental oven which contained only 15 g of catalyst, heat dissipation was no problem. It was achieved by enveloping the oven in a large quantity of metal. On a large scale, however, the heat liberated from 1,000 m³ of gas per hour (corresponding to about 100 kg of synthesis products ($C_3$ and higher)) produces 0·6–0·7 tons of steam per hour which corresponds to an effective use of 80 per cent of the heat of reaction. The conventional normal pressure oven for a performance of this order is 2,500 mm high, 1,500 mm wide and 5,000 mm long, and contains 10 m³ of catalyst. To carry out the process precisely it is necessary to keep the temperature constant in the range 170–210°C to ± 0·5°. As the catalyst loses its efficiency the temperature from the time of starting of the oven to the end of the period is raised by 40°C.

The former IG had experience in the construction of such large catalyst ovens with heat dissipation, e.g. for the catalytic oxidation of naphthalene to phthalic anhydride with air, and the oxidation of anthracene to anthraquinone. They used catalyst ovens with a large number of narrow tubes containing the catalyst which were cooled by water under pressure. Later they used 'lattice ovens' in which the catalyst lies on lattices embedded between closely-coiled cooled tubes.

The problem of the dissipation of the heat of reaction is solved in the standard-pressure oven by placing numerous tubes running from one end to the other in a large sheet metal box, and these are threaded through sheets of metal of about 2 mm thickness which are parallel to the front faces and are spaced 7 mm apart. This 'parcel' consists of 630 tubes and 550 sheets of metal giving a very large cooling surface of 3,700 $m^2$. The tubes pass three times through the upper part of the oven and five times through the lower part where less heat is generated, and are then combined into a collector tube (see Figure 62) filled with pressure water. Steam and hot pressure water from the collector tubes are fed into a steam collector whence the steam is extracted. The feed water has been previously softened. The space between the metal sheets is filled with catalyst to the full height of the oven. The catalyst is, therefore, nowhere more than 3·5 mm from a heat-conducting surface; this is important since due to the poor heat conductivity of the catalyst local overheating would otherwise occur.

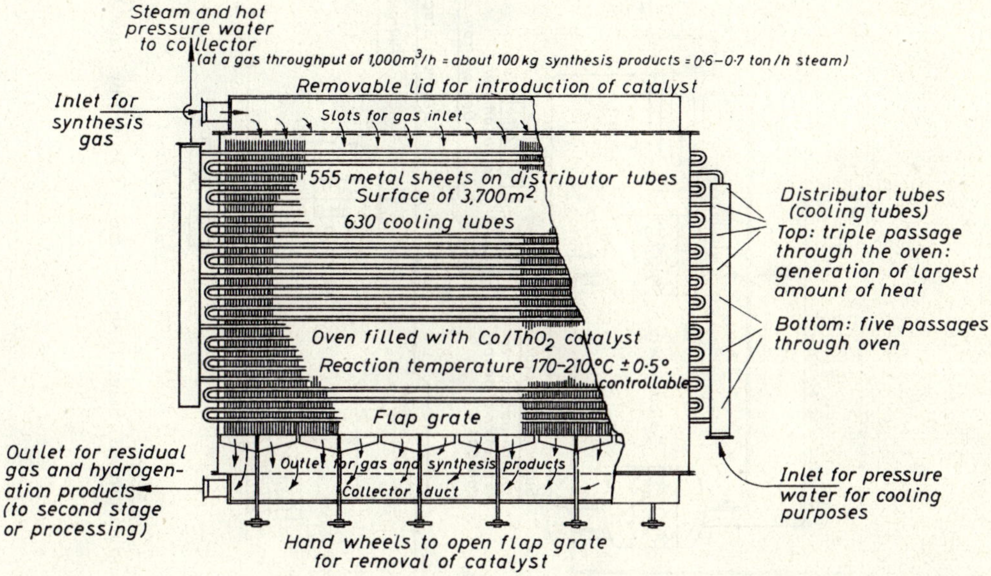


Figure 62. Diagram of a standard-pressure oven for fuel oil synthesis by the Fischer-Tropsch method. U-tube oven

In the normal pressure oven (U-tube oven) the synthesis gas enters from above through the slots in the lid, is uniformly distributed, and escapes at the bottom from the flap grate that carries the catalyst, into a collector duct. The catalyst can be removed by opening the flap grate. The lid is removed and the catalyst is fed in through the top shielded by oxygen-free gas. Since this oven, inasmuch as the maintenance of a constant temperature and heat dissipation are concerned, is unique, it is used also in other catalytic processes where large amounts of heat are generated.

*Operation*

Since only half the gas has reacted after a single passage through the oven, the synthesis is carried out in several stages, requiring an intermediate removal of the synthesis products and of the reaction water. Figure 63 shows a diagrammatic sketch of a simple two-stage synthesis scheme. The first stage is followed by the second stage using identical equipment. The end gas from the oven is cooled in a cooling tower first with circulating water and then with cold water to 20–30°C, to precipitate the oil. About half the synthesis products accumulate directly in the oil separator. Petrol and the gaseous hydrocarbons are adsorbed in an activated-carbon plant.

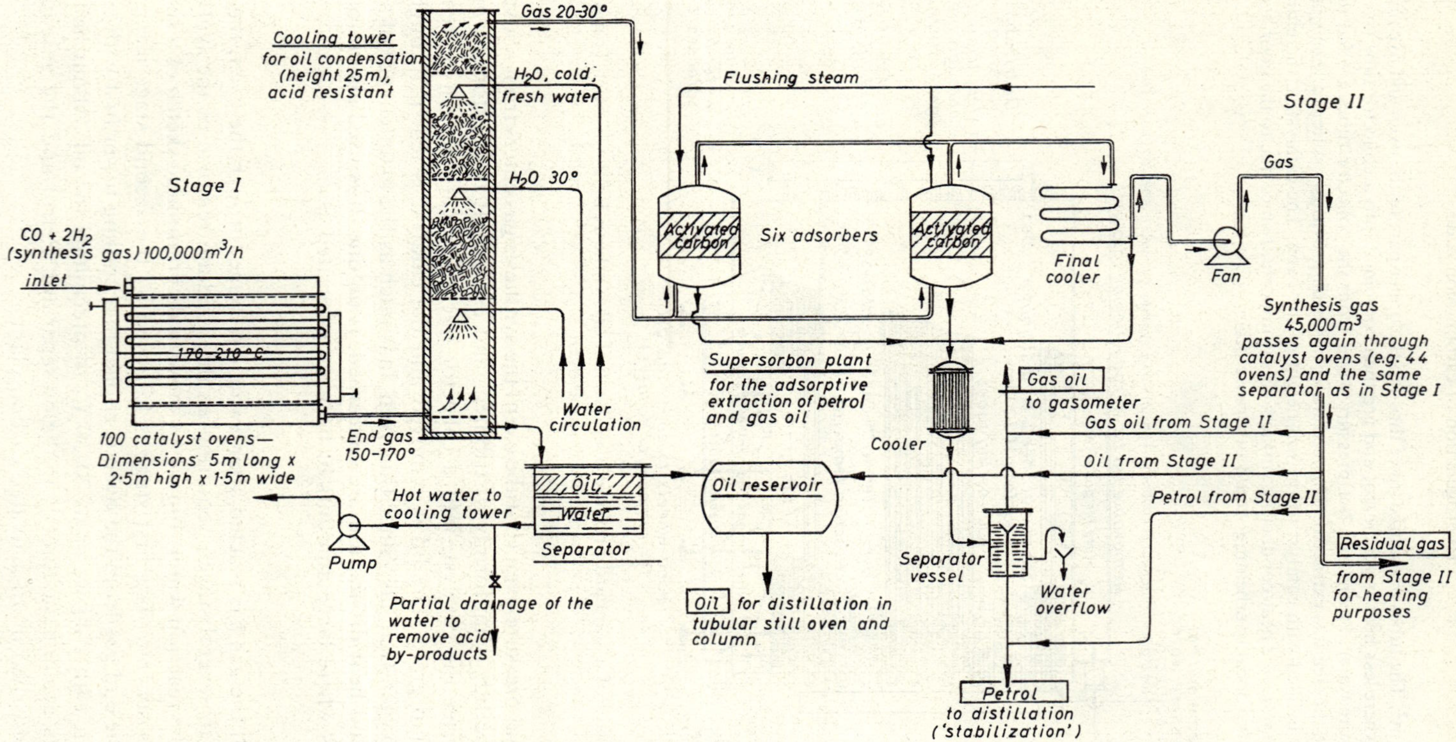


Figure 63. Simple two-stage synthesis scheme of the Fischer-Tropsch process. Stage 1, shown here, is followed by stage 2 with identical equipment

The 'Supersorbon plants' built for the Fischer-Tropsch-Works for the production of petrol and liquid gas from the end gases are the largest plants of this type ever built. They possess 7–16 adsorbers and contain 100–160 tons of activated carbon each. The principle used is the selective adsorption according to boiling point. Periodic evaporation is followed by condensation of the synthesis product in the cooler, and separation of benzene and gas oil in a separating vessel (see Figure 63).

Usually, a large number of activated carbon adsorbers is grouped together, with a four-stage working cycle:

(1) loading
(2) evaporating
(3) drying
(4) cooling ready for next charge.

Fractions not adsorbed pass through a final cooler and then, boosted by a fan, pass as 'synthesis gas 2' into the second stage which is identical with the first, except that less equipment is used. The residual gas is fed into a gasometer whence it is drawn off for heating purposes as required.

### *Medium pressure synthesis*

The shape of the ovens for the medium pressure synthesis (40 $m^3$) is different. In order to absorb the pressure they are built cylindrically and contain twin tubes sunk into the front walls of the cylinder. The catalyst is in the annular space of the twin tubes; the distance between the walls is 10 mm so that the catalyst is at the most 5 mm from the heat-conducting surface. One oven may contain e.g. 2,044 twin tubes, with a catalyst volume of 10 $m^3$. The pressure water circulates in the interior tube. Four ovens are usually combined into one bank with a common steam collector. The Fe catalyst has six times the performance and a longer life than the Co catalyst. With this design heat dissipation is more favourable than with the normal pressure oven, because of the greater heat conductivity of compressed gases ($^1/_3$ the cooling surface). When high purification of the gas by the Rectisol process is carried out no additional compression of the gases is necessary. Medium pressure synthesis, too, is carried out in two or three stages.

Medium pressure synthesis produces a far greater proportion of heavy products with a higher content of unsaturated hydrocarbons than normal pressure synthesis. The $H_2$-consumption is, therefore, higher than the $H_2/CO$ ratio of 2 : 1, e.g. 2·2–2·4 : 1. After the first stage additional hydrogen must, therefore, be introduced into the gas. Separation is easier. Owing to the pressure a smaller quantity of the products is in the gaseous phase. Paraffin and heavy oils produced on the catalyst trickle down continuously. Low-boiling products need not be removed between the stages with activated carbon. They can remain in the gas stream and will then be adsorbed at the end with activated carbon. The following table shows the various types of products obtained:

| Product | Normal pressure synthesis (per cent) | Medium pressure synthesis (per cent) |
|---|---|---|
| Gas oil ($C_3$ and $C_4$) | 14 | 10 |
| Petrol | 47 | 26 |
| Diesel oil 180–230°C | 17 | 24 |
| Kogasin 230–320°C | 11 | 13 |
| Crude paraffin | 8 | 16 |
| Hard paraffin | 3 | 10 |

Gas oil normally contains about 30–35 per cent olefins, the $C_4$ fraction 45 per cent $C_4H_8$, the $C_3$ fraction 10–15 per cent $C_3H_6$. The composition depends to a large degree on the catalyst, particularly its $ThO_2$ content. Until 1947 the Hydrogenation Works Schwarzheide have been polymerizing the $C_4$

fraction without previous dehydrogenation on a phosphoric acid catalyst with charcoal as carrier at 30 atm gauge pressure and about 200°C, to give a petrol with an octane rating of 80–85.

The gaseous hydrocarbons $C_3$ and $C_4$ can be used directly as liquid gas. They are also separated by pressure distillation and then form a valuable raw material for the synthesis of aliphatic compounds. The olefins contained in it can also be used for the production of aliphatic alcohols by the 'Oxosynthesis' process.

Petrol: The impurities contained in the petrol from the normal pressure synthesis (boiling up to 180°C) consist of only small quantities of fatty acids which are removed by a simple leaching wash. The octane number is about 50–54, i.e. very low. The problem of raising the octane number by catalytic conversion does not appear to have been solved as yet for this petrol. The only solution is the addition of anti-knock agents. The medium pressure petrols which are comparatively rich in alcohol can be converted on an alumina or bauxite catalyst at 400–450°C by separating the water, with simultaneous olefin formation, cyclization and polymerization, in such a way that the octane number is increased.

As a motor fuel, the petrols obtained by the Fischer-Tropsch synthesis have, therefore, rather unfavourable properties. On the other hand, the diesel oil (at Schwarzheide the fraction between 180 and 240°C) is a very good motor oil so that it can be blended with inferior oils, e.g. those obtained from the hydrogenation of coal and tars.

Kogasin, the fraction from 230–320°C, is a very important aliphatic raw material for the synthesis of detergents, lubricating oil, plasticizers for plastics, etc. Thus, for instance, a suitable kogasin fraction ('Mepasin', a mixture of paraffin hydrocarbons, mainly $C_{14}$ to $C_{17}$) after purification by hydrogenation is transformed by sulphochlorination in ultra-violet light into sulphonic chloride (Mersol) which after saponification becomes the important synthetic detergent Mersolat (P. Herold, Leuna). Another possibility is the sulphoxidation to sulphonic acid by the 'Hostapon process'. The plasticizer 'Mesamoll' used in the manufacture of plastics is produced by the conversion of sulphonic chloride into phenolic esters. The sulphonic acid amides of the kogasin fraction have also become very important; they consist of waxes stable in air and to light, oils for the textile industry and for the lubricating of leather, and emulsifiers for lubricating oils and soluble oils.

The paraffins satisfy all the requirements of the trade, particularly the high-melting hard paraffins with melting points of 60–66°C. It is also possible to obtain hard paraffins with melting points above 100°C. The high-boiling residue of the condensate oil, the slack wax or crude paraffin, with a melting point not exceeding 30–40°C, is the most important raw material for the manufacture of fatty acids by paraffinic oxidation.

*Recent variations of the Fischer-Tropsch synthesis*

(W. Löpmann, Betrachtungen zur Fischer-Tropsch-Synthese ..., *Erdöl und Kohle*, 1954, 7, 622).

A further development of the classical medium pressure synthesis is the Ruhrchemie-Lurgi high-pressure synthesis in which use is made of the favourable effect of an increase in pressure on the rate of the reaction, and the negative effect of the carbon dioxide and water generated is taken into consideration.

A fixed bed synthesis oven is used in which the catalyst is contained in pipes of 30–80 mm diameter; the water circulating around these as a cooling agent serves for the generation of steam. The catalyst is an iron catalyst without carrier and possessing a very large internal surface (300 $m^2$/g Fe). The synthesis takes place at 220–230°C and 25–30 atm gauge pressure. The heat of reaction is dissipated by gas circulation and introduction of fresh gas and by high gas velocity (see Figure 64, reproduced from Ullmann, *Encyklopädie*, 3rd ed., vol. 9). As compared with the former medium pressure oven, the performance is increased 25–30 fold.

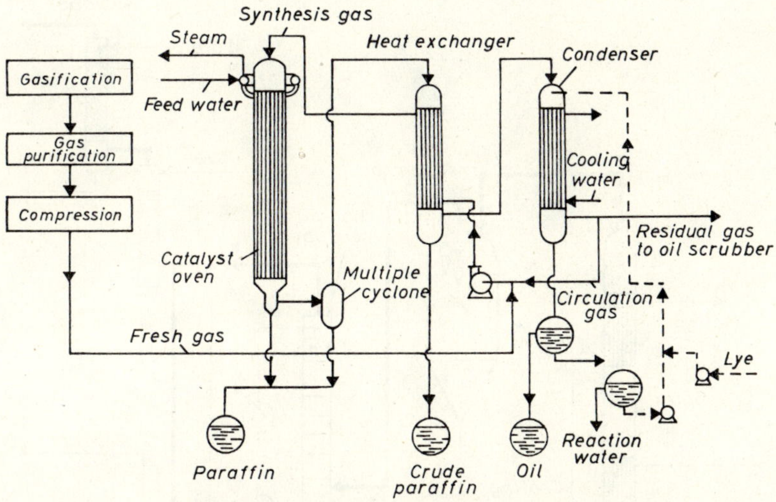


Figure 64. Ruhrchemie-Lurgi high load synthesis

Dissipation of the heat by the synthesis gas circulating through a cooler was tried experimentally on a fixed bed catalyst in the Michael process (formerly I. G. Lu.) and in the Lurgi stage oven.

This latter is the main agent for the heat transfer in the processes with circulating powdered catalysts. In the M. W. Kellogg suspension process (U.S.A.) the powdered iron catalyst passes through the reactor with the gas.

The reactor is divided into five stages, three without and two with built-in tubular cooling system. The heat carrier in the cooling system is circulating oil; the system enables the synthesis temperature to be regulated in the uncooled reactor chambers (Figure 65).

The residual gas and the reaction products pass into a dust precipitator and a two-stage cyclone. The catalyst precipitated there (less 2–4 per cent loss per day) returns into the reactor. The residues and the high-boiling fractions of the synthesis products are separated in a hot-oil scrubber. After the acid fractions have been removed by scrubbing with lye the liquid products and the reaction water are separated in coolers. Part of the residual gas is re-circulated. The process is used in the synthesis plant of the South Africa Coal, Oil and Gas Corporation, Sasolburg, South Africa, in two units with a primary output of 165,000 tons per annum (for liquid, low-boiling products) and five reactors operating on the Ruhrchemie-Lurgi high-pressure synthesis system (for higher-boiling products and paraffin).

The 'Hydrocol process' (a plant based on natural gas used to operate in Brownsville) uses the fluidized bed principle. The synthesis gas is produced at Brownsville by partial combustion of natural gas with oxygen. The synthesis takes place at 310–350°C and a gauge pressure of 25–30 atm.

The heat from the reactor (capacity 100 $m^3$) is dissipated by vertical boiler tubes suspended in it which are combined into cooling systems by collecting pipes. As in the suspended particle process, the precipitation of heavy products on the catalyst must be prevented here, too, to avoid the formation of a covering layer.

The liquid phase processes on a fixed catalyst with oil circulation (Duftschmid, BASF), with circulation and catalyst-oil-suspension (Michael-Schaum-process, BASF), and the processes with the catalyst in stationary suspension (O. Roelen, Ruhrchemie, and H. Kölbel and P. Ackermann, Rheinpreussen and Koppers) have the great advantage of

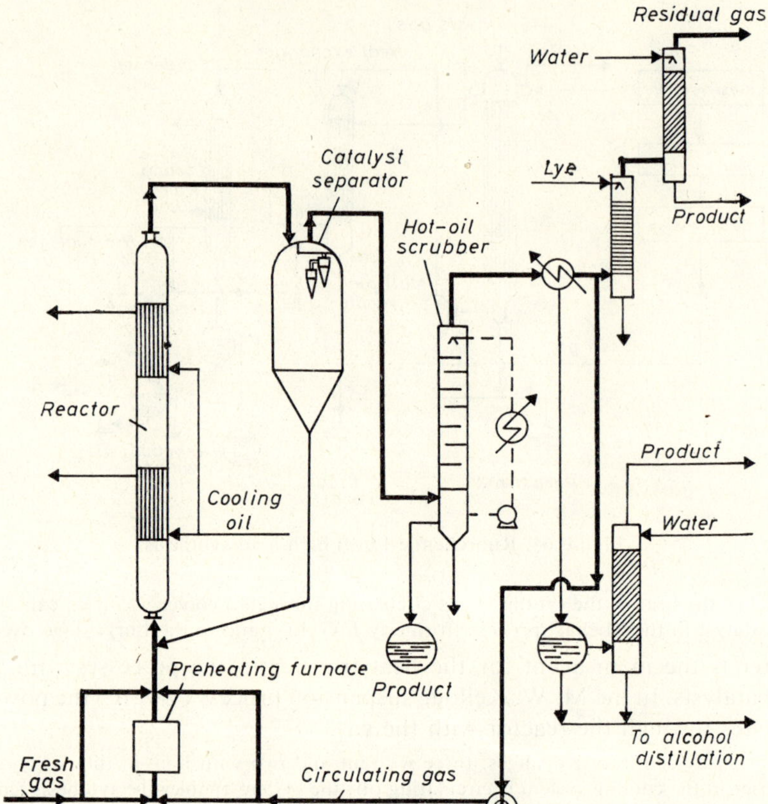


Figure 65. Principle of the Kellogg suspension process

easier heat dissipation. Medium and high pressures are used in order to have always sufficient hydrogen in solution, as in the semi-solid phase process of coal and oil hydrogenation. The principle of the plant operated by Rheinpreussen and built together with Koppers is shown in Figure 66 (after H. Kölbel and P. Ackermann, *Chem. Ing. Techn.*, 1956, **28**, 384).

The gas is raised to synthesis pressure in the compressor (1), pre-heated in the heat exchanger (2) with end gas, and fed from below into the synthesis furnace (3) which contains suspended cooling assemblies through which water circulates. The furnace temperature is regulated by the extraction of steam from the cooling assembly. Synthesis products which have already become liquefied are separated (5) in the heat exchanger (2). Medium-boiling liquid products and water of formation are condensed from the uncondensed gas in a cooler and collected in a separator; the water is separated out in the separator (5). The separation of $CO_2$, the hydrocarbons $C_3$–$C_5$ and the end gas takes place in the $CO_2$ separator (pressure scrubbing (9)). The liquid level of the furnace (3) is maintained constant by the floating governor (4) which controls the supply of suspending oil by means of a valve. The high-boiling primary products are extracted by a pump through a pressure filter (6). Part of the contents of the furnace are passed through the centrifuge (8) where the spent catalyst is separated. The separated oil can be mixed with fresh catalyst in the mixing device (7) and is then fed back into the furnace.

Liquid phase synthesis turned out to be extremely versatile. According to the initial gas and the conditions the most varied synthesis products can be obtained, including those with a high olefin content. It is possible also to use inferior gases, such as carbon monoxide from carbide furnace gas, even blast furnace waste gas if it has a sufficiently high CO content. However, only gas with a high CO content yields the particularly desirable olefins (80–85 per cent). These can be used directly for the manufacture of washing agents.

### *Variations of the process. Prospect*

In the United States, work on the Fischer-Tropsch synthesis did not start till much later and use was then made of the wealth of experience gained in Germany. This synthesis is used to convert large natural gas deposits into motor fuels. The conversion of natural gas into synthesis gas is much more economical than that of coal. Methane can be decomposed thermally with steam in the Cowper: $CH_4 + H_2O = CO + 3H_2$. The Gesellschaft für Kohletechnik (Society for Coal Technology) has developed a catalytic cracking process in which cracking is carried out in an externally heated chromium-nickel-steel cracking furnace at 900–1,000°C with the addition of a little air, on a nickel catalyst. Synthesis gas is also obtained by partial combustion of natural gas with limited amounts of oxygen. BASF have developed a special furnace for this process, using a platinum layer as a pre-catalyst and Ni as catalyst (See Bartholomé and Nonnenmacher, BASF, Erdöl und Kohle, 1959, **12**, 359).

At the beginning of the 1939–1945 war, the Ammoniakwerk Merseburg (W. Wenzel) developed the 'Synol process' as an extension of the Fischer medium pressure synthesis. In this method, fused iron catalysts are used in a similar way to the ammonia process. The process is effected at 18–30 atm gauge pressure and 190–200°C, with several circuits arranged in series, the reaction products and the carbon dioxide produced being currently removed. The products are 30–70 per cent straight-chain primary aliphatic alcohols, also branched chain ones, mainly $C_9$—$C_{20}$, and the appropriate olefin hydrocarbons. These alcohols are otherwise difficult to produce and are of some importance, for instance, for the production of textile auxiliary agents.

A further promising possibility of producing alcohols consists in the oxosynthesis, developed by Ruhrchemie and the former I.G. This involves the conversion of olefins in

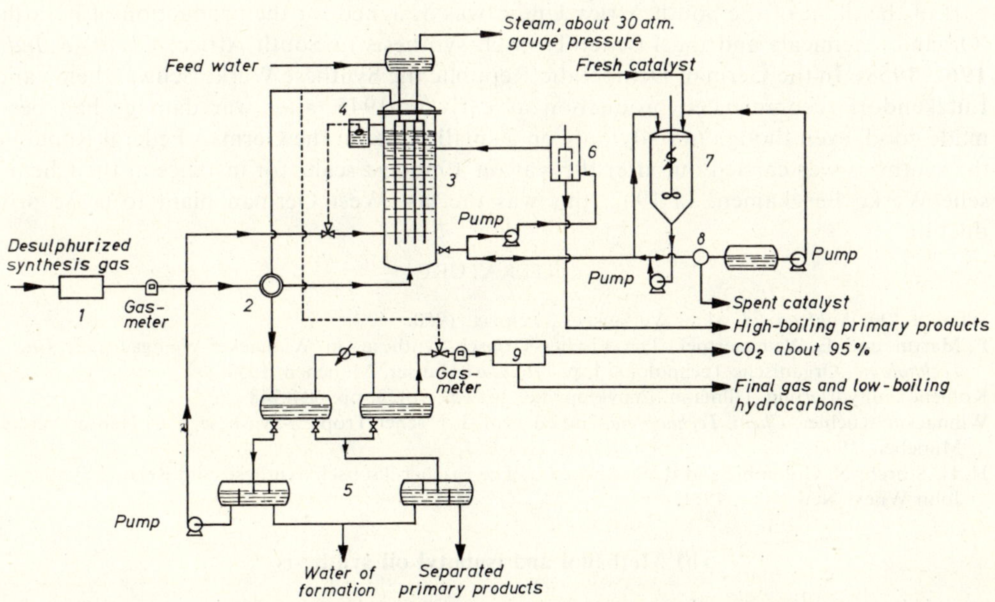


Figure 66. Liquid-phase synthesis (Rheinpreussen-Koppers)
*1* Compressor *2* Heat exchanger *3* Reactor *4* Level regulator *5* Separation of product
*6* Pressure filter *7* Catalyst mash *8* Centrifuge *9* $CO_2$ separator

the presence of CO-containing catalysts with $CO + H_2$ into a mixture, enriched in each case by one C atom, of aldehydes and alcohols, some of which are branched. In a second stage the mixture is hydrogenated to alcohols (P. W. Sherwood, Neuere Entwicklungen beim Oxoverfahren, *Erdöl und Kohle*, 1959, **12**, 20).

Finally, Fischer and Pichler developed in 1941 the 'isosynthesis' by which they obtained isoparaffins and $O_2$-containing compounds on $ThO_2$ catalysts at high temperatures and pressures. This synthesis offers possibilities of great diversity but is technically difficult.

The construction of the Fischer plants had a very favourable effect on the development of industrial methods for the production of synthesis gas and for the final purification of gases. The whole technology of gasification has benefited from these processes.

This short survey shows that the technological possibilities of CO hydrogenation are not yet exhausted. Compared with the hydrogenation of coal and tar which must be considered as upgrading, CO hydrogenation permits the production of motor fuels and aliphatic products from gases, i.e. from the simplest elements. At some future date this may become very important since world petroleum reserves will probably be exhausted much earlier than the coal reserves. Due to Germany's former lack of petroleum it was only natural that the synthesis processes based on coal and gas were so rapidly developed there. Various countries built Fischer-Tropsch plants for the same reasons. Even countries rich in petroleum such as the United States built synthesis plants for the conversion of their natural gas into liquid motor fuels.

Nevertheless, one has to be clear about the fact that pure motor fuel production by the Fischer-Tropsch process from coal, as opposed to that from natural gas and by-product gases, is born of necessity. It cannot compete economically with the coal and tar hydrogenation processes. Its future prospects lie rather in the field of the synthesis of pure aliphatic raw materials as bases for the heavy organic industry. Thus, for instance, part of the plant of the South Africa Union was designed for the production of paraffin (Organic chemicals and the Fischer-Tropsch synthesis in South Africa, *Chem. & Ind.*, **1962**, 1958). In the German Democratic Republic the Synthese Werke Schwarzheide and Lützkendorf recommenced production as early as 1945, after war damage had been made good, even though initially only on a small scale. In the German Federal Republic the synthesis was carried out after the war on a modest scale, for instance at the Chemische Werke Bergkamen. In 1962, this was the last West German plant to cease production.

LITERATURE

Schmidt, *Das Kohlenoxyd*. Akad. Verlagsges., Leipzig, 1950.

F. Martin und E. Weingaertner, Die Fischer-Tropsch-Synthese, in Winnacker-Weingaertner, *Chem. Technologie*, Organische Technologie I, p. 776, Carl Hanser, München, 1952.

Kohlenoxydhydrierung, Ullmann, Encyklopädie, 3rd ed., vol. 3, pp. 684–748.

Winnacker-Küchler, *Chem. Technologie*, 2nd ed., vol. 3, Fischer-Tropsch-Synthese, Carl Hanser Verlag, München, 1960.

H. H. Storch, N. Golumbic and R. B. Anderson, The Fischer-Tropsch Synthesis and Related Syntheses. John Wiley, New York, 1951.

### (b) Methanol and isobutyl oil synthesis

Of all the reactions possible in the hydrogenation of carbon monoxide the most favoured is the formation of $CH_4$. With 50·4 kcal per mole methane has the highest heat of reaction and its formation takes place very easily on an iron catalyst under pressure.

If metal oxides such as zinc or chromium oxide are used which are difficult to reduce, methanol and higher alcohols are obtained at pressures of about 200 atmospheres gauge pressure, together with acids, aldehydes, and ketones:

$$CO + 2H_2 \rightarrow CH_3OH + 21{\cdot}7 \text{ kcal}$$
$$CO_2 + 3H_2 \rightarrow CH_3OH + H_2O + 14 \text{ kcal.}$$

The second reaction does not include the heat of reaction of the conversion. After a comparatively short period of experimenting, the BASF (Bosch, Mittasch, Pier, Wietzel, Winkler, et al.) succeeded in producing pure methanol on a large scale, using a zinc oxide catalyst with activating additions. The 'Syntol' from Franz Fischer's catalytic pressure hydrogenation (1923) contained about 1·5 per cent methanol.

Other possibilities for the production of methanol than CO hydrogenation under pressure were also explored. Thus, the Deutsche Gold- und Silberscheideanstalt has developed a two-stage process which proceeds according to the following equations:

$$CH_3OH + CO \rightarrow HCOOCH_3$$
$$HCOOCH_3 + 2H_2 \rightarrow 2CH_3OH$$

This is CO hydrogenation by the intermediate production of formate. The process has not become technologically very important. The first stage of this process was carried out intermittently at 30 atmospheres gauge pressure and 80°C with metallic sodium dissolved in methanol as catalyst. (The BASF had earlier produced methyl formate by a similar process as the gaseous pesticide 'Areginal' (Wolfen) for the protection of silos against the corn weevil.) In the second stage the methyl formate was then hydrogenated with hydrogen at 175–185°C on an Adkins catalyst. The large-scale production by the BASF catalytic process has, however, reduced the price of methanol to such an extent that no other process can compete with it.

### *Catalyst*

The ZnO catalyst with activating additions is a model of a good catalyst with high specificity. Many catalysts for the production of methanol are described in patent applications but none has so far proved equally efficient. In general it is easier to find a good catalyst for a reaction which does not involve secondary reactions, such as the synthesis of $NH_3$ or the oxidation of $SO_2$ to $SO_3$, than to find catalysts which quite specifically promote only the formation of a certain organic substance. It is not yet possible to state with certainty the reasons for the specificity of a catalyst.

The most effective catalyst for the methanol synthesis consists of 3 parts ZnO + 1 part $Cr_2O_3$. It reacts at 380°C and is a typical 'mixed catalyst' which is more effective than can be expected from the mere sum of its individual components. It has a life of six months to one year.

The catalyst is used in the shape of cylinders of 2–14 mm diameter produced by preforming presses. As is usual with tablets, the front surfaces are curved since otherwise 'bars' would form during the filling of the catalyst furnaces by superposition of the tablets. The fresh catalyst is first formed by slowly raising the temperature during several days to 380–400°C at 225 atm gauge pressure. In the course of its use it then becomes gradually less active owing to the poisoning of its surface.

Catalyst poisons are mainly organic sulphur compounds, as in the Fischer-Tropsch synthesis. Hydrogen sulphide can be removed comparatively easily from the synthesis gas; removal of organic sulphur compounds on the other hand is more difficult. Even after final purification 1–2 mg hydrogen sulphide per cubic metre are left in the gas. The catalyst is not very sensitive to these small quantities since the following equilibrium is set up during the synthesis process: $ZnO + H_2S \rightleftarrows ZnS + H_2O$. The ZnS content depends on the S content of the gas. The $H_2S$ content causes even more damage to the copper components of the equipment than to the catalyst since the reaction $Cu + H_2S \rightarrow CuS + H_2$ is not reversible.

The role played by the $H_2O$ content of the gas with $ZnO—Cr_2O_3$ catalysts seems surprising at first sight. During operations it was found that the performance of the furnaces deteriorated temporarily each time a reducing furnace with fresh catalyst was connected to the gas circulation. The water of reduction produced during the starting of a furnace brought about a temporary poisoning of the catalyst. This is deleterious since according to a theory of Taylor's and Kistiakovski's $H_2O$ and $CO_2$ are adsorbed more strongly by ZnO than $H_2$ and must, therefore, be considered as catalyst poisons. For the same reason an $O_2$-content of the gas has a damaging effect because of $H_2O$ formation. An excess of $N_2$ in the gas leads to the formation of $NH_3$ and amines.

The alkali content of the catalyst, which is present if the ZnO precipitate was not carefully removed, causes the formation of higher alcohols; this fact is utilized in the isobutyl oil process. The formation of dimethyl ether as a by-product is unavoidable; it is caused by the $Al_2O_3$ and kieselguhr content of structural parts of the furnace which act as dehydration catalysts. The carrier problem is, therefore, always of great importance for the catalyst. There is practically no catalytic process in which the carrier material does not react with the catalyst. The spezific influence of the carrier on, for instance, the course of the hydrogenation has already been discussed in the section on the gas phase hydrogenation of middle oil.

Iron directs the reaction towards the formation of methane and hydrocarbons. Occasionally, a catalyst furnace would 'run away' owing to the great quantity of heat liberated during the methane formation, which can be countered only by rapid relaxation and filling with inert gas or by closing the outlet so as to 'suffocate the reaction in the product'. This proved to be due not to the smooth walls of the equipment but to finely divided reactive iron which precipitates in certain parts of the hydrogenation apparatus. This iron is formed from the decomposition of $Fe(CO)_5$ which above 250°C has only limited stability and results from CO + Fe above 100°C under pressure, $Fe + 5\,CO \rightleftarrows Fe(CO)_5$ (pressure-dependent). During the thermal decomposition of iron pentacarbonyl highly reactive iron is formed. Favourable conditions for the formation of pentacarbonyl exist in the higher stages of the compressors and on the iron parts of the heat exchangers; it decomposes in the catalyst space.

The gas is, therefore, purified in an activated carbon filter after the compressor stage. The formation of pentacarbonyl on colder parts of the equipment can be prevented by lining with copper. 1·2–1·8 per cent manganese are added to render it malleable.

### *Starting gas*

According to the formula $CO + 2H_2 \rightarrow CH_3OH$, the $CO/H_2$ ratio of the gas must be 1 : 2 or, in practice, 1 : 2·2. It is produced by the process described in the chapter on 'Gasification' from water gas by partial conversion with steam over an iron oxide catalyst.

During the conversion, decomposition of the organic sulphur compounds takes place giving $H_2S$ which is then removed, together with $CO_2$, by scrubbing with water under pressure, for the production of ammonium carbonate with aqueous $NH_3$, or by the Alkazid process.

### *Method of operation*

Since with longer residence times of the reaction products in the synthesis furnace there is a risk of secondary reactions (e.g. increased formation of dimethyl ether), the conversion of the synthesis gas on an industrial scale cannot be carried out in a single run over the catalyst. Also, too much heat would be produced. Based on experience gained in the ammonia process, the gas is, therefore, circulated and only part of it converted at a time with the addition of fresh gas.

A smaller flow velocity in the equipment entails also stronger dehydration, i.e. formation of dimethyl ether.

Carbon dioxide and hydrogen sulphide having been removed by scrubbing under pressure, the gas, converted in the ratio 1 CO : 2 $H_2$, is further compressed, according to the method of operation, to 200–700 atm gauge pressure and passes through an oil cup into a pressure vessel filled with activated carbon in the shape of cylinders of 4–5 mm length. There the harmful iron pentacarbonyl is removed by decomposition. The gas is then fed by the shortest possible route through a heat exchanger into the synthesis furnace. In the heat exchanger, the gas entering the furnace is heated by the hot gas leaving it, as near as possible to the reaction temperature. If this temperature is not fully reached, additional heating can be carried out in a peak pre-heater. This is important when, for instance, during the decline of the catalyst the reaction heat produced is no longer adequate. In the furnace the reaction takes place on the catalyst which is placed in layers on built-in shelves. The furnace is of a similar design to the gas phase furnace illustrated in Figure 48. The optimum temperature is between 350 and 390°C. Originally, the furnace designs of the ammonia process were used, later special designs were developed for the methanol synthesis.

Heat exchangers used are those with built-in electrical peak pre-heaters, or tube bundle heat exchangers such as those used in the hydrogenation of coal and tar. In that case the peak pre-heater is built directly into the synthesis furnace. This so-called 'burner' is heated electrically. From the heat exchanger the gas passes through a central tube inside which the burner is fixed to the top lid, into the upper part of the furnace. At the top the gas flow is reversed into an annular space containing the catalyst. The annular space is separated from the pressure-proof jacket by an insulating material. The catalyst is arranged on screens in several layers. The temperature is controlled by thermocouples. At intervals of 1–2 m annular tubes, so-called cold gas annuli, are installed, by means of which cold gas can be led between the individual catalyst layers for cooling purposes. After cooling in the heat exchanger the reaction product is fed into a high-pressure terminal cooler which is composed of bundles of tubes. The separation into liquid products and gas takes place in a separator. Part of the gas is returned into the process by a recirculating pump.

The methanol from the high-pressure cooler contains a large amount of gas in solution. Thus, for instance, at 225 atm gauge pressure 35–40 $m^3$ gas are dissolved in one $m^3$ of crude methanol. This gas is liberated when the pressure is released. Liquid impurities include: dimethyl ether, propanol, isobutanol, isohexyl alcohol, trace acids, ketones, amines, olefins, some water and iron pentacarbonyl. The dimethyl ether, boiling point −24·8°C, is rectified at a gauge pressure of 10 atmospheres when it is liquid (1–2 per cent of the methanol yield). It has a variety of uses and is employed in industry for the introduction of the methyl group, e.g. for the conversion of aniline into dimethyl aniline, for the production of dimethyl amine and trimethyl amine, and also for the production of dimethyl sulphate in a very elegant continuous process in which $SO_3$ and dimethyl ether converge in a reaction tube and yield directly pure dimethyl sulphate:

$$SO_3 + CH_3 - O - CH_3 \rightarrow (CH_3)_2SO_4.$$

Methanol is distilled with an addition of 30–40 per cent water. Fractionation is carried out in a column of 60–70 trays. Before the distillation the semi-solid phase from the column is treated with a 1 per cent solution of potassium permanganate until discoloration ceases. A small amount of fore-runnings is taken off at the head. Pure methanol accumulates several trays below the head, the higher alcohols on one of the lower trays. The quantity is relatively small. The yield by distillation is about 90 per cent.

*Isobutyl oil synthesis*

With the same pressure and a temperature of about 30–40°C more higher alcohols are obtained on an alkalized ZnO— $Cr_2O_3$ catalyst. The gas must contain a somewhat larger quantity of CO. The flow velocity is set to about one third that of the methanol process. Consequently, only one third of the heat exchange surface is required, with the same catalyst space, for the circulating quantity of gas is correspondingly smaller. The equipment used is the same as that used in the methanol process but the methanol must be separated; it is then re-injected into the circuit. From this and from the fact that injection of propanol also produces a larger yield of isobutyl oil, it follows that in this process the lower products precede as preliminary stages. The more alkaline the catalyst, the higher the temperature, and the smaller the flow, the more high-molecular products are obtained. There is, however, a limit since many inseparable and useless products result if these measures are taken to extremes.

Since in the isobutyl oil process more carbon is separated the catalyst has a short life. The most favourable method of operation is one in which only about 11–14 per cent of the isobutanol fraction is obtained. The reaction product then still contains about 50 per cent methanol, 2 per cent propanol, 1–3 per cent amyl alcohol, and 2–3 per cent of the alcohols $C_6$ and $C_7$. The main product of the isobutyl oil process is, therefore, methanol. However, since the methanol, together with the other lower by-products, is returned into the synthesis, it does not actually appear outside the process. It would not pay to upgrade the mixture with 50 per cent methanol to obtain commercial methanol since it is of very low quality and cannot be used as a solvent or for synthesis purposes.

The upgrading of the isobutyl oil is limited to obtaining certain purified fractions. In addition to the normal alcohols mainly iso-compounds are present since the thermodynamic conditions for their formation are somewhat more favourable than those for the formation of unbranched ones with the same C-content. The separation and purification of isobutyl oil is a combination of various difficult distillations, partly with the addition of water for the formation of azeotropic mixtures with a hydrogenating after-treatment.

*Use of the products*

As a result of the development of the synthesis, methanol has become one of the cheapest solvents and intermediate products. Because of its low price and many uses it has been called 'organic water'. Apart from its employment as a solvent for the production of esters for antifreeze, a large part is used in the production of formaldehyde.

In 1928, the production of methanol in Germany amounted to 18,000 tons. The 100,000 tons mark was passed in 1937. The record production was reached in 1943 with 293,000 tons. In the United States, production rose from 22,900 tons in 1939 to 196,000 tons in 1943 and reached 247,000 tons in 1947, the price dropping by about half during the same period. In 1959 the B. R. D. produced 296,000 tons pure methanol, the USA 797,947 tons.

The isobutyl oil process formed the basis of an entirely new chemistry and technology concerning synthetic solvents. The alcohols $C_4$ and $C_5$ are marketed as 'Intrasolvan', the alcohols $C_6$ and $C_7$ as so-called HS-alcohols 'Intrasolvan HS'. Their esters with acetic acid and propionic acid are the 'Polysolvans', used also for the selective extraction of phenols, e.g. from carbonization liquors. Of some importance also are the pure alcohols used as solvents for lacquers, for extraction processes, etc. Esters, e.g. with phthalic acid, are called 'Palatinols'. n-propanol can be converted into di-n-propyl ether which is an excellent solvent.

Primary alcohols can be converted easily into carboxylic acid by treatment with alkali at higher temperatures, if necessary under pressure. Isobutanol is converted on an alumina catalyst at 360–400°C and 3–6 atm gauge pressure into isobutylene which separates under pressure in liquid form from the water of reaction and is then extracted. Polymerization of isobutylene, carried out continuously with $BF_3$ as a catalyst on conveyor equipment, yields poly-isobutylene, a valuable synthetic material, used e.g. for the lining of vessels, acid-resistant coatings, etc., under the name of 'Oppanol' (BASF-Oppau). Isobutylene can also be dimerized or trimerized with sulphuric acid or phosphoric acid. Dimeric isobutylene in its hydrogenated form gives the motor fuel iso-octane, used as a test substance because of its anti-knock properties. Iso-octane was produced during the war on a large scale as aviation fuel. However, tryptan (2,2,3-trimethylbutane) even improves on the anti-knock properties of iso-octane. Polymerized isobutylenes with their terminal double bonds can be used, for instance, for the alkylation of phenols. If such alkylated phenols are reacted repeatedly with ethylene oxide, to give alkylphenolpolyglycol ethers, important materials are obtained which are marketed as dispersing and emulsifying agents (e.g. Igepal, Peral).

*Formaldehyde*

A large part of the methanol production is used for the manufacture of formaldehyde. With the growth of the synthetics industry, in particular the phenol-formaldehyde resins, urea-formaldehyde resins, and other condensation resins, the consumption of formaldehyde has increased in the past 15 years to an enormous extent.

In 1960, the production of formaldehyde in the B. R. D. amounted to 153,000 tons, in the United States to 230,000 tons. Large quantities of formaldehyde are also used in the production of pentaerythrite. In recent years the way has been prepared for a further use of formaldehyde: Du Pont de Nemours have succeeded in polymerizing a pure monomeric formaldehyde into the thermoplastic synthetic material 'Delrin'. For the production of formaldehyde by catalytic methods, methanol of very high purity is required. Even small quantities of iron pentacarbonyl are capable of poisoning the catalysts used in this process.

As early as 1867 A. W. Hofmann found that formaldehyde was produced when methanol-air mixtures were passed over a glowing platinum coil; however, the yield was low. Much more suitable are the metals in Group Ib of the periodic system: Ag, Cu, Au. The optimum temperature for the formation is between 550 and 600°C. Without oxygen dehydrogenation takes place, with limited quantities, partial oxidation:

dehydrogenation $$CH_3OH \rightleftharpoons CH_2O + H_2 - 20 \text{ kcal/mole}$$

partial oxidation $$CH_3OH + {}^1/_2\, O_2 \rightarrow CH_2O + H_2O + 38 \text{ kcal/mole}$$

If the operation is carried out with a deficiency of air, it is possible to let partial oxidation and dehydrogenation take place in about equal amounts on the Ag and Cu catalyst. The problem of the dissipation of heat which normally presents itself with oxidation reactions does not arise in this case owing to the coupling of the endothermic and exothermic reactions.

In the process developed by the former IG and in general use in Germany, steam is added as an inert gas to the air-methanol mixture. This increases the quantity of air to such an extent that almost complete conversion of the methanol is possible and its recovery is not necessary.

In the United States the following method is widely used (Figure 67). A mixture consisting of about equal parts of methanol and air is employed. The methanol is evaporated in 'saturators' which are designed like a column, by steam heating and admission of air into the lower part of the evaporator. The methanol-air mixture escaping at the top passes through an electrically heated silver wire gauze serving as a preheater into the catalyst furnace, which also contains silver wire gauzes or catalyst layers of loose silver crystals. The temperature rises spontaneously to 520°C and the heating can be switched off. The mixture which now consists of $N_2$, $H_2O$, $CH_2O$ and $CH_3OH$ is fed directly into a column from the base of which formaldehyde in a 40 per cent concentration with 12–14 per cent methanol is drawn off ('Formalin').

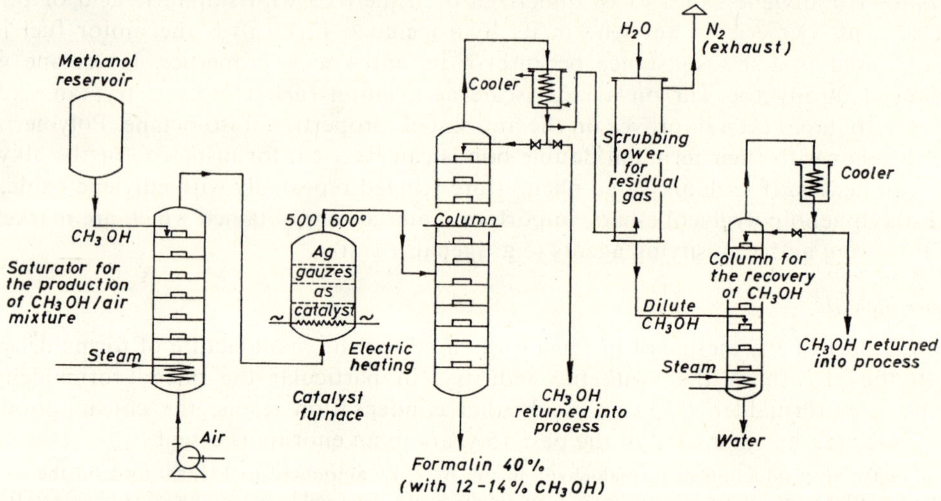


Figure 67. Diagram of a plant for the production of formaldehyde by catalytic methods

The methanol escaping through the head of the column returns into the process. From the escaping gas consisting mainly of nitrogen the methanol is removed in a scrubbing tower with water. The aqueous methanol is distilled to 100 per cent methanol in a distillation column.

The process for the production of formaldehyde by the partial oxidation of methane has not so far been used industrially because of the high oxidation temperature of methane (600°C); at this temperature formaldehyde decomposes into CO and $H_2$. The addition of small quantities of nitric oxides to the oxygen greatly reduces the reaction temperature.

## 2. VARIOUS SYNTHESES WITH CARBON MONOXIDE

One of the syntheses with carbon monoxide is the production of formic acid.

Berthelot had discovered as early as 1855 that the reaction of carbon monoxide with caustic potash at 100°C produces formate:

$$CO + KOH \rightarrow KOOCH$$

The reaction of CO with alkaline substances has since formed the basis for most of the industrial processes for the synthesis of formic acid.

There are two basic processes, the wet and the dry. The carbon monoxide required can be produced, for instance, in generators by incomplete combustion of coke with air. The gas containing, in addition to CO, about 70 per cent nitrogen is carefully purified, in particular $CO_2$ is completely removed with milk of lime. The gas does not need to be dried, however, since a small amount of moisture is necessary for the reaction with caustic soda.

Caustic soda in 97 per cent concentration in walnut size is treated with CO in horizontal autoclaves with horizontal stirrers at 115–120°C and 8 atm gauge pressure. The autoclaves are provided with a jacket for the initial heating at the beginning of the reaction and for cooling in the course of it later. The residual nitrogen (unless pure carbon monoxide is used) must be continuously distilled off. Finely-powdered sodium formate in 95 per cent concentration is obtained from which are screened the larger pieces that still contain caustic soda.

A process developed by R. Koepp & Co., Oestrich am Rhein, uses soda in a liquid medium, and finally cheap sodium sulphate can be used in the presence of milk of lime:

$$Na_2SO_4 + Ca(OH_2) + 2\,CO \rightarrow 2\,HCO_2Na + CaSO_4.$$

In this case the 'wet' process is employed. The operation can be carried out continuously in a battery of autoclaves at 25–30 atm gauge pressure, temperature about 150°C. The calcium sulphate precipitated is filtered out of the formate solution and the latter is evaporated under vacuum after purification with lime.

Formate can also be produced by the reaction of calcium hydroxide with CO to calcium formate and subsequent treatment with sodium sulphate (EKB Bitterfeld, BASF, Koepp & Co.).

To produce free formic acid the formate must be decomposed by stronger acids. The difficulty is that strong acids decompose formic acid, weaker ones, however, lead to a diluted formic acid which cannot be concentrated to more than 76·5 per cent since this mixture has a minimum vapour pressure. A great many technological suggestions have been made for ways round this difficulty. One possibility is to decompose the finely powdered solid sodium formate with concentrated sulphuric acid, the mixture being simultaneously cooled and vigorously stirred, and to draw off the liberated formic acid under a vacuum into a heated drum.

Condensation is carried out in stoneware coolers. Commercial formic acid in 85–90 per cent concentration is obtained. The sodium sulphate is used again for the production of formate. Acid in 100 per cent concentration can be obtained by concentration with acids combining with water and distillation, by freezing or azeotropic dehydration with auxiliary liquids. Sodium formate is also obtained as a by-product of the manufacture of pentaerythrite:

$$CH_3 - CHO + 4HCHO + NaOH \rightarrow C(CH_2OH)_4 + HCOONa.$$

Sodium formate is the raw material for the production of oxalic acid (see below).

Aliphatic formic acid esters, e.g. formic acid methyl ester, are obtained according to the equation:

$$CH_3OH + CO = HCOOCH_3$$

by stirring in an autoclave at 70°C, with some sodium methylate as a catalyst. Carbon monoxide is introduced until saturation occurs. Formic acid methyl ester is used as a gaseous pesticide against corn beetles in silos and flour mills (Areginal). The ester is an important intermediate product; it is easy, for instance, to produce formamide from it:

$$HCOOCH_3 + NH_3 \rightarrow HCONH_2 + CH_3OH$$

The methanol is recovered.

With dimethyl amine, dimethyl formamide, $(CH_3)_2NCHO$, is formed, an important polar solvent, e.g. for spinning polyacrylonitrile. Another formic acid process is based on formamide as a raw material:

formamide is saponified with sulphuric acid in 70 per cent concentration. Distillation in a rotary tubular furnace yields formic acid in 90 per cent concentration.

The reaction of formaldehyde, carbon monoxide and water in the presence of acids at 200°C and under very high pressure produces glycolic acid $HOH_2C—COOH$ which can be hydrogenated by the Du Pont process to ethylene glycol.

An industrial process for the production of hydrocyanic acid is based on the catalytic separation of water from formamide:

$$HCONH_2 \xrightarrow[-H_2O]{} HCN.$$

The industrial production of oxalic acid is also based on carbon monoxide. Originally, oxalic acid was produced by alkaline fusion of sawdust at 270–280°C, at the present time frequently by oxidation of carbohydrates with nitric acid in the presence of sulphuric acid. A large amount of oxalic acid is also obtained as a by-product of the microbiological production of citric acid.

Sodium oxalate has long been produced by the Koepp & Co. process from sodium formate by heating. The process has three phases:

(1) $2HCOONa \rightarrow NaOOC—COONa + H_2$ beginning at a temperature of 290°C and ending at 360°C.

(2) Sodium oxalate is reacted with milk of lime: $Na_2C_2O_4 + Ca(OH)_2 \rightarrow CaC_2O_4 + 2NaOH$.

The oxalate is precipitated in insoluble form. The sodium hydroxide in solution is reacted in a single operation introducing steam and gaseous CO at a gauge pressure of 15–20 atm according to the equation.

$$2NaOH + 2CO = 2HCOONa.$$

After separation of the oxalate the formate solution (filtrate) is evaporated and returned into the process.

(3) The calcium oxalate is converted into free oxalic acid with sulphuric acid:

$$CaC_2O_4 + H_2SO_4 \rightarrow HOOC—COOH + CaSO_4.$$

### *Production of higher carboxylic acids using CO*

With carbon monoxide and olefins carboxylic acids can be obtained by two processes:

(1) By oxosynthesis (see p. 145,) with aldehydes by oxidation.

(2) By Reppe and Kröper's method from olefin, CO and water on a nickel salt at 250–300°C and 200 to 300 atm gauge pressure.

$$H_2C = CH_2 + CO + H_2O \xrightarrow{\text{Ni-salt}} CH_3—CH_2—COOH.$$

H. Koch devised a new method for the synthesis of aliphatic carboxylic acids by addition of CO and water to olefins at 0–30°C and moderate pressure in the presence of sulphuric acid (*Brennstoff-Chemie*, 1955, **36**, 321). Since the mechanism is an ionic one, branched carboxylic acids are obtained according to Markovnikov's rule:

$$H_3C—CH = CH_2 \xrightarrow[H_2O]{CO} H_3C—\underset{\displaystyle CH_3}{\underset{|}{CH}}—COOH$$

LITERATURE

J. Giesen and H. Hanisch, Die Methanol- und Isobutylölsynthese, in Winnacker-Weingaertner, *Chem. Technologie*, Organische Technologie I, p. 19. Hanser Verlag, München, 1952.

Ullmann, *Encyklopädie der techn. Chemie*, 2nd and 3rd ed.

O. Horn and H. Bestian in Winnacker-Küchler, 2nd ed., vol. 3, Hanser Verlag, München, 1959.

# V. Production and processing of simple paraffins and olefins

## Sources and production

The most important sources of the $C_1$–$C_4$ saturated hydrocarbons, apart from coke oven and carbonization gases, are natural gas (p. 74), the residual gas from the Fischer-Tropsch synthesis (pp. 141, 142), the hydrogenation gases (p. 129), gases from petroleum distillation, and cracking gases.

Olefins, particularly ethylene and propylene, are found in small quantities in coke oven and carbonization gases. The short-chain paraffins and olefins from coke oven and carbonization gases are produced mainly by the cracking of side chains during the formation of aromatics. The residual gases of the 'DHD' and hydroforming aromatization processes are, therefore, particularly rich in olefins (pp. 99, 107, 130).

A large part of the gaseous products of the cracking and pyrolysis of tar oils and petroleum, and of reforming gases consists of olefins. The thermal or catalytic dehydrogenation of the lower paraffin hydrocarbons, such as ethane, propane, butane, etc., which occur in large quantities in petroleum and in the hydrogenation gases, offers particularly favourable conditions for the production of the lower olefins.

For the economy of the German Democratic Republic the hydrogenation gases could be a rich source for the production of olefins but this possibility is little used as yet. The petroleum refinery planned at Schwedt is intended to supply a large part of the olefin requirements of the German Democratic Republic. Almost all the gases mentioned also contain hydrogen which at present is largely lost. Its recovery and utilization would be of the utmost significance for the economy.

Ethylene can be produced also by the partial hydrogenation of acetylene (p. 165).

## Industrial Processes for the Concentration and Production of Hydrocarbons in Pure Form

### *Fractionation of coke oven gas into its components by the Linde-Bronn process*

The separation of gaseous hydrocarbons was originally based on the processes developed by the Gesellschaft für Linde's Eismaschinen A.G. and J. Bronn. Further development led to the fractionation of coke oven gas in order to obtain hydrogen for synthesis purposes. Even before the 1939–45 war a considerable part of the hydrogen required for the production of ammonia was obtained from coke oven gas. In this process nitrogen and hydrogen are produced in about the desired proportions and at the pressure required for the synthesis. It was also possible to obtain hydrocarbons but at first not much interest was shown. It was not until 1930 that a start was made on the extension of the process to the production of hydrocarbons.

Not all the gas mixture can be liquefied at once; this would cause disturbances by precipitating some of the components as solids. Cooling is, therefore, carried out in steps, utilizing carefully the cold in the heat exchange for the pre-cooling of the next fraction. At $-10°C$ benzene and its homologues are separated from the coke oven gas, as well as water, the separation of which is completed at $-45°C$. This is the point at which fractionation starts in benzene-free hydrocarbon gases, natural gas, and cracking gas. At $-110°C$ to $-115°C$ liquefaction of ethylene $C_2H_4$, propylene $C_3H_6$, and ethane $C_2H_6$ (ethylene fraction) occurs, that of methane $CH_4$ (methane fraction) between $-150$ and $-190°C$.

In order to obtain hydrogen, the gas which at this stage contains only $H_2$, $N_2$, and CO, is scrubbed at $-195°C$ with supercooled liquid nitrogen, causing CO to be dissolved (Linde, Messer, Claude). This yields a fraction rich in ethylene and methane. These are separated in rectification columns operating under pressure into the pure hydrocarbons and are thus made available for synthesis purposes. The separation of the pure hydrocarbon gases, natural gas, cracking gases, and Hy-gases is similar but simpler.

*Separation of natural, hydrogenation and cracking gases*

The first aim is the separation of hydrogen and methane and separation by C-numbers. Figure 68 (From F. Asinger, *Einführung in die Petrolchemie*, Akademie Verlag, Berlin, 1959) shows a simplified diagram of the most important technological possibilities for the fractionation of gas mixtures.

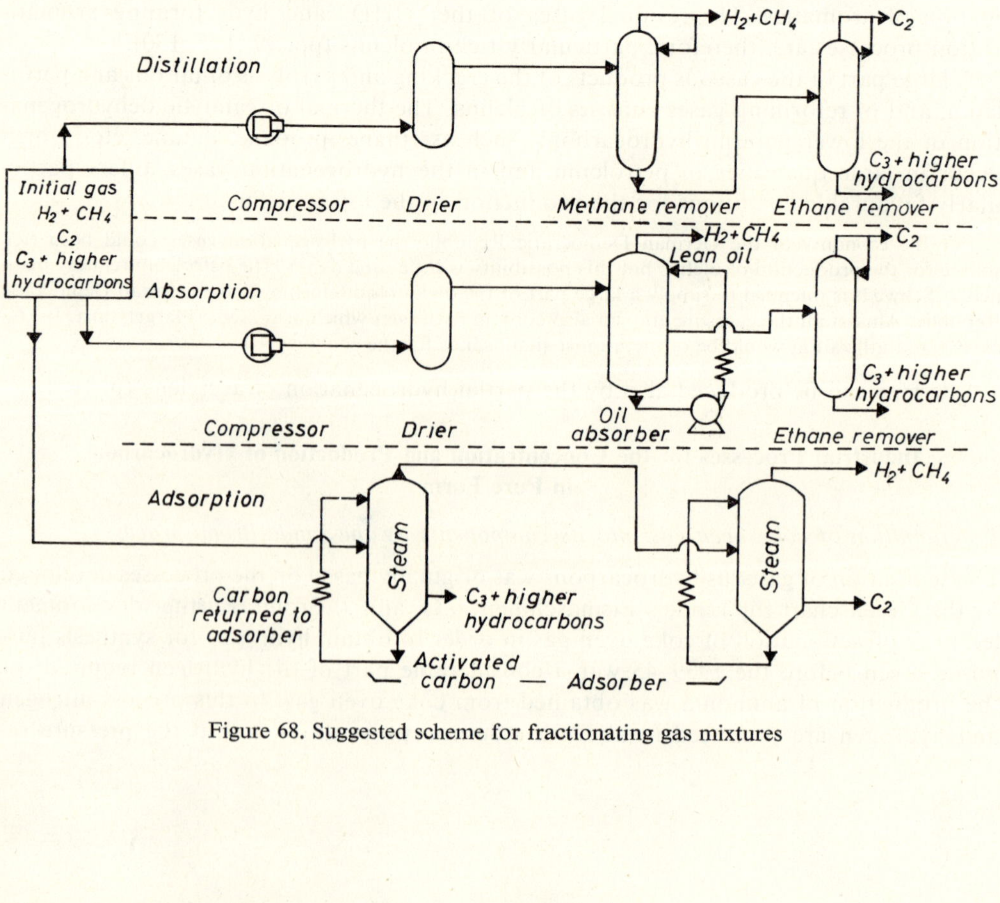


Figure 68. Suggested scheme for fractionating gas mixtures

(1) Separation by pressure distillation in columns.

(a) Drying in pressure column, sometimes followed by silica gel treatment.
(b) Separation of $H_2$ and methane as 'head product'. ('Head product' is the fraction escaping from the upper part of the column, the 'head'. The 'bottom product' or 'sump' is extracted from the bottom of the column).
Methane can be liquefied only under pressure with simultaneous cooling. The dried gas mixture is, therefore, fed at $-18°C$ into the methane column which must contain a methane reflux to render the $C_2$ mixture amenable to separation.
(c) Separation in a column into $C_2$ (head product), $C_3$ and higher (bottom product).

(2) Separation by adsorption in oil.

(a) Drying in a pressure column.
(b) Absorption in oil under pressure. $C_2$ and higher are dissolved, $H_2$ and $CH_4$ are not absorbed (head product).
(c) Separation of $C_2$, $C_3$ and higher. $C_2$ and higher are expelled from the oil and separated by subsequent pressure distillation.

(3) Adsorption-desorption method.

(a) By means of activated carbon adsorption of $C_3$ and higher hydrocarbons takes place in the first stage up to saturation. Then separation with steam is carried out. $C_2$, $CH_4$, and $H_2$ are not adsorbed.
(b) In a second adsorber $C_2$ is adsorbed whilst $H_2$ and $CH_4$ pass through. $C_2$ is separated with steam.

(4) Hypersorption process.

The hypersorption process corresponds to a continuous application on an industrial scale of gas chromatography. In a column the gases to be separated meet countercurrently a stream of recirculated carbon and are adsorbed more or less rapidly according to their molecular weight. The carbon is fed into the upper part of the column (cooling section), the gas below.
The non-adsorbed gases ($H_2$ and $CH_4$) pass through the carbon layer and escape at the head. The carbon layer migrating downwards (charged with $C_2$ and higher) enters a 'rectification section' where, as in a distillation column, a fraction of heavy hydrocarbons which has been heated in the lower part of the column (evaporating section) is forced upwards. The adsorbed hydrocarbons are fractionated and separated and escape from the column at different heights as side gas streams. The lower the molecular weight of the hydrocarbon, the higher the part of the column from which it escapes. The desorbed activated carbon leaves at the bottom and is transported to the top by a conveyor. Saturated and unsaturated hydrocarbons with the same number of carbon atoms cannot be separated by this method, either.

Pyrolysis and cracking gases can contain acetylene, which before it is used in a separating process must be selectively hydrogenated to ethylene (see p. 164).

*Separation of paraffins and olefins*

The following table shows the formulae and boiling points of the simple saturated and unsaturated hydrocarbons. When the hydrocarbons are classified by C numbers, the $C_2$ and $C_3$ products can be separated by pressure distillation. The boiling points of ethane and ethylene differ by about 15°C (see Table), those of propane and propylene by about 5°C. The boiling points of butane and butene are so close that separation by simple fractionation is impossible, but can be carried out by 'distillation extraction' ('Distex process'). The presence of an auxiliary solvent, e.g. furfurol, changes the volatility relationships so strongly that it is now possible to separate by distillation substances with boiling points which are practically identical (see pp. 93, 96).

| Name | Formula | Boiling point in °C |
|---|---|---|
| Methane | $CH_4$ | −164 |
| Ethane | $CH_3—CH_3$ | − 88·5 |
| Ethylene | $CH_2=CH_2$ | −103·9 |
| Propane | $CH_3—CH_2—CH_3$ | − 44·5 |
| Propylene | $CH_3—CH=CH_2$ | − 47 |
| n-butane | $CH_3—CH_2—CH_2—CH_3$ | + 0·6 |
| iso-butane | $(CH_3)_3CH$ | − 10·2 |
| 1,2-n-butene (butylene) | $CH_3—CH_2—CH=CH_2$ | − 6·1 |
| cis-butene-(2) | H H<br>$\rangle C=C \langle$<br>$H_3C$ $CH_3$ | + 3·72 |
| trans-butene-(2) | H $CH_3$<br>$\rangle C=C \langle$<br>$H_3C$ H | + 0·88 |
| Butadiene-(1,3) | $H_2C=CH—CH=CH_2$ | − 4·4 |
| Carbon monoxide | | −190 |
| Hydrogen | | −252·8 |

The following Table shows the pressure and temperature conditions in a group of four columns for the separation of $C_1$–$C_3$ (see p. 85).

| Type of column | Gauge pressure atmospheres | Temperature in °C | | Escaping at head |
|---|---|---|---|---|
| | | Head | Bottom | |
| Methane column (intake at −18°) | 40 | −90 | +38 | $CH_4 + H_2$ |
| Ethylene column | 27 | −18 | +43 | $C_2H_4$ |
| Ethane column | 24 | + 4 | +65 | $C_2H_6$ |
| $C_3$-column ($C_4$ and $C_5$ bottom product) | 15 | +93 | +38 | $C_3$-fraction |

The very economical Linde Low Temperature Separation Process is becoming increasingly important for cracking and pyrolysis gases. It is based on liquefaction by pressure and low-temperature cooling. The $C_3$ and $C_4$ fractions are separated and the cooling produced by the Joule-Thomson effect during the pressure release of the raw gas permits the liquefaction of $CH_4$ and the $C_2$ fraction. The liquefaction takes place by stages. The heat of condensation is utilized at each stage for the evaporation of the condensate with the next lower boiling point. The utilization of the heat exchange in highly efficient cross countercurrent heat exchangers with which a very small difference in the temperature between the incoming and outgoing gases can be achieved, yields an optimum of economy (for a short description and diagram of the process see K. Schoenemann, *Chem. Z.*, 1959, **83**, 435).

# 1. TECHNICAL PROCESSING OF METHANE

Methane is present in all the above-mentioned gases and cán easily be produced directly from coal by the Lurgi pressure gas process (see p. 69).

In nature methane occurs wherever organic matter decays with limited access of air sewer gas, e.g. in the sludge of sewage farms whence it is extracted and utilized commercially. In bogs and swamps it occurs as marsh gas and in coalmines as mine gas (firedamp) where it may cause explosions. It is the main constituent of natural gas. The 'dry natural gases' contain up to 100 per cent methane. It is a very cheap raw material but is still mainly used as a fuel gas. Its production from various sources was described in the preceding section.

Figure 69 shows the most important industrial uses of methane.

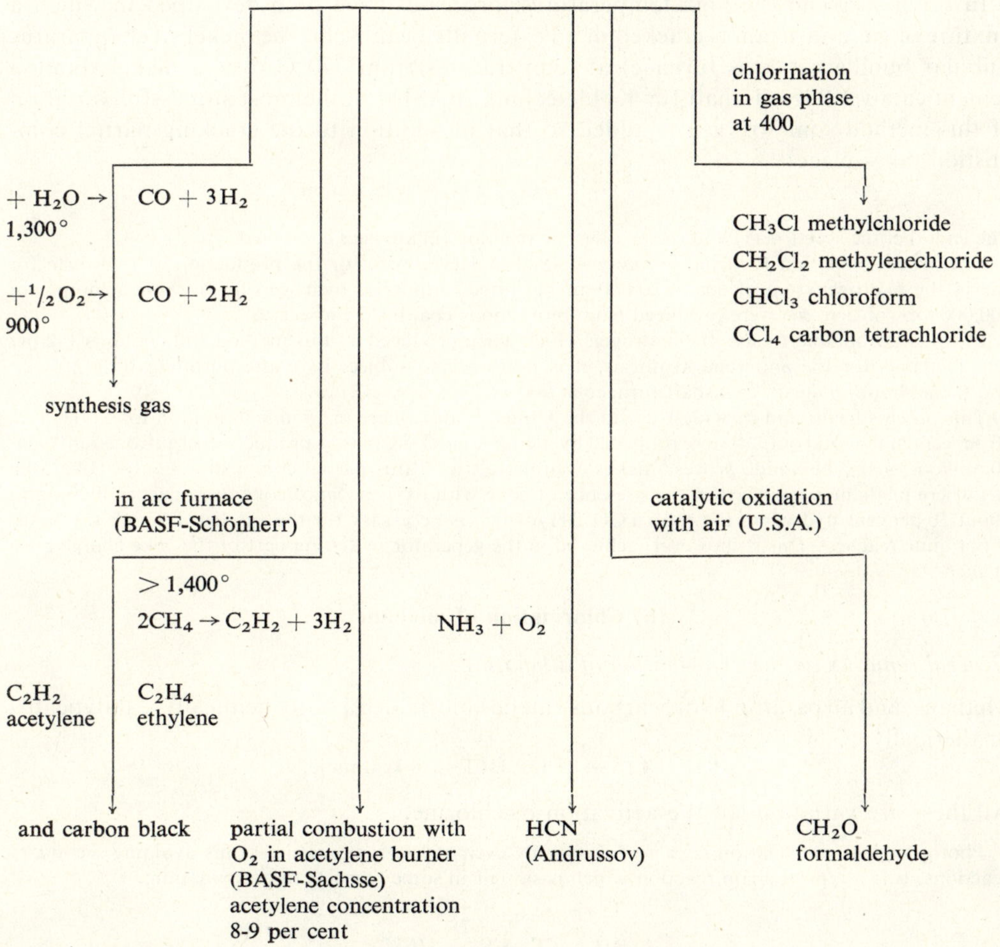


Figure 69. Industrial uses of methane

### (a) Synthesis gas from methane

The production of synthesis gas and hydrogen from methane has already been mentioned briefly in the section on Gasification. Several methods have been developed. In both the Kuhlmann and the Koppers processes, coke oven gas and hydrogen pass over stone surfaces heated to a high temperature (regenerators, Cowpers) in a reversing process and have gas added to them. At a temperature of 1,400°C or 1,300°C, 90 per cent methane is separated, according to the equation:

$$CH_4 + H_2O = 3H_2 - 50\ \text{kcal/mole} \qquad (1)$$

Since the process is strongly endothermic, a large amount of heat of reaction, at high temperature, must be supplied to the system. This problem can be solved either by external heating, e.g. in the tube bundle catalysis furnace, or by internal heating, e.g. by the combustion of part of the hydrocarbons with oxygen.

In order to avoid the high temperatures, processes have been developed in which a mixture of gas and steam is cracked in an externally heated chrome nickel steel apparatus (tubular bundle catalysis furnace) at temperatures from 750°C over a nickel-alumina cement catalyst (Gesellschaft für Kohletechnik, BASF). In the most successful variation of this method some oxygen is added so that in addition to the cracking partial combustion takes place:

$$CH_4 + {}^1/_2\,O_2 \rightarrow CO + 2\,H_2 + 9\ \text{kcal/mole} \qquad (2)$$

The endothermic reaction (1) and the exothermic reaction (2) are here combined.

The process was first used in Germany in 1940 at Heidebreck for the production of hydrogen for use in the hydrogenation of coal. 120,000 $m^3$/hr (rated output) of hydrogen for an annual output of 400,000 tons of benzene were produced from bituminous coal in 12 furnaces.

In order to be able to use the hydrogen which when produced by this method still contains 1–2 per cent methane, for the ammonia synthesis, it is necessary to subject it to afterburning with a little air and the addition of steam in a shaft furnace at 900°C.

This process has found its widest use in the United States where in 1956 3·36 million tons $NH_3$, i.e. 80 per cent of the total output, were produced by methane cracking (world production of nitrogen in 1958: 10 million tons). The Linde process makes even more use of this partial combustion e.g. of coke oven gas where methane is oxidized in the presence of coke with oxygen. Since coke oven gas itself contains about 50 per cent hydrogen, the ratio $1CO:2H_2$ which is necessary for the Fischer-Tropsch synthesis, is not quite reached. This is, however, achieved in the generator by letting part of the coke charge burn at each stage.

### (b) Chlorination of methane

*General remarks on the chlorination of aliphatics*

Methane, and all paraffin hydrocarbons can be chlorinated photochemically, catalytically, or thermally:

$$RH + Cl_2 \rightarrow RCl + HCl + 24\ \text{kcal/mole.}$$

All these are variations of the activation of chlorine.

Photochemical chlorination is carried out easily even at low temperatures, thus avoiding secondary reactions. It is a typical chain reaction which is similar in some respects to autoxidation.

$$Cl_2 + h_\nu = 2Cl^\bullet$$
$$RH + Cl^\bullet \rightarrow R^\bullet + HCl$$
$$R^\bullet + Cl_2 \rightarrow RCl + Cl^\bullet$$

With oxygen the chains break easily (reaction of the radical R˙ to ROO˙). With the purest raw materials the quantum yield is 30,000–40,000, with industrial ones 2,000. This figure indicates the number of hydrocarbon molecules converted by a Cl-atom once formed, before the chain breaks.

Catalytic chlorination is carried out by means of chlorine carriers, such as iodine, sulphur, phosphorus, and antimony, i.e. elements which can react with chlorine in at least two valency steps. The necessary chlorine atoms may also be formed by radical-forming compounds such as diazomethane, tetraethyl lead, or organic peroxides (homogeneous catalysis).

In heterogeneous catalysis in the gas phase, activated carbon, pumice, alumina, saturated with metallic salts, are used as catalysts. Chlorine is ionized on the catalyst. These reactions are insensitive to chain-breakers such as oxygen. It is even possible to proceed in such a way that the hydrochloric acid formed is oxidized with oxygen to chlorine which is fully utilized (e.g. Deacon process).

Industrially, the most important is the thermal chlorination of the paraffin hydrocarbons in the absence of light and catalysts which is also a chain reaction, inhibited by oxygen. The chlorine atoms are formed by thermal dissociation. The activation energy at 20 kcal/mole is about 8 kcal higher than in catalytic chlorination.

One of the most interesting chemical treatments of methane is its direct thermal chlorination, developed by the former IG Hoechst in 1923. The Chemische Werke Hüls also developed a methane chlorination process on the basis of Bentheim natural gas after 1945. At Hoechst methane and chlorine in the gas phase are reacted at 400°C in a furnace, the methane being kept in excess. Only a little above this temperature carbon black forms; this must be prevented. By varying the quantities and conditions one obtains mainly methylene chloride or methyl chloride, in addition to chloroform and carbon tetrachloride.

In the chlorination furnace a cold mixture of 5 parts $CH_4$ and 1 part $Cl_2$ is first injected through a nozzle in a circulation furnace into a diffusion tube whence it passes into an empty chamber. Here fresh gas is diluted with reacted gas. Nevertheless, autogenous heating to 400°C occurs through the heat of reaction; the heat is dissipated as the gas passes over a cooler. Owing to the loss of heat in this manner it used to be necessary continuously to supply additional heat to the furnace by gas heating. Now, the fresh gas is heated before being fed into the furnace so that the heating of the furnace can be dispensed with (see O. Horn and H. Bestian in Winnacker-Küchler, *Chemische Technologie*, 2nd ed., vol. 3, p. 729).

The hydrochloric acid is first scrubbed with water in earthenware scrubbing towers. This yields directly highly concentrated commercial hydrochloric acid. Whilst hydrochloric acid is often produced by 'combustion' of chlorine with hydrogen, this process is a sort of 'combustion with methane'. After scrubbing with a solution of caustic soda the gas is dried in a refrigeration cooler and then liquefied by deep freezing. The excess methane is not condensed and returns into circulation. The chlorination products are fractionated in two columns by pressure distillation.

Distillation is also continuous. The yield is, for instance, 35 per cent methyl chloride, 45 per cent methylene chloride, and 20 per cent chloroform and carbon tetrachloride. The methyl chloride can be returned into the process if there is not sufficient demand for it.

Formerly, methyl chloride used to be produced by esterification of methanol with hydrochlorid acid:

$$CH_3OH + HCl \rightarrow CH_3Cl + H_2O.$$

Since 1937 the process described above has been preferred. Some methyl chloride, and also ethyl chloride, is produced at atmospheric pressure in the gas phase from the alcohols with HCl on an activated carbon catalyst which contains some $HgCl_2$. Methyl chloride boils at −23·7°C, is marketed in pressure bottles or air pressure tanks, and is used as a filler for cooling plants, and for methylation purposes.

*Chemical use of methyl chloride*

Reaction with Na-cellulose → methyl cellulose
(an important thickener used in textile chemistry and in cosmetics)

Production of dimethyl chlorsilane (silicone)

Reaction with NaSH → methyl mercaptan

Reaction with $Na_2S$ → dimethyl sulphide $\xrightarrow{+O_2}$ dimethyl sulphoxide (a new, universal, polar solvent)

Several thousand tons of methyl chloride are produced annually.

Methylene chloride, with a boiling point of +39·8°C, is an important solvent for lacquers, polyvinyl chloride, for spinning PC fibres, and is a very efficient non-inflammable extraction solvent. It is the main constituent of PC-adhesive for polyvinylchloride. The demand continues to increase.

Chloroform (boiling point 61·2°C) is a by-product. It is also produced by a well-known process from trichloroacetic acid and alkali: $CCl_3 - COOH + NaOH \rightarrow CHCl_3 + NaHCO_3$. Chloroform for anaesthetic purposes is stabilized with 0·5 per cent ethanol (dephosgenization).

Carbon tetrachloride (boiling point 76·7°C) is also produced by the process described above. Since there is a very large demand for it for a great variety of purposes, e.g. as a solvent for the extraction of fats and oils, as a cleaning material and fire extinguishing agent, and for fluorochloromethane, supplies must also be obtained from other sources. Carbon tetrachloride is produced additionally by the old process from carbon disulphide and chlorine (e.g. EKB Bitterfeld).

PC is a per-chlorinated polyvinylchloride. This renders it soluble in acetone or methylene chloride so that it can be spun directly from the solution. There are also PC varnishes.

*Carbon tetrachloride, carbon disulphide*

Carbon disulphide is heated with excess chlorine, at first at 60°C, later at boiling temperature, in the presence of iron powder as a catalyst in lead-lined boilers. After completion of the reaction fractional distillation is carried out. In this reaction sulphur and $S_2Cl_2$ (sulphur monochloride) are separated and returned again into the process:

| | |
|---|---|
| Carbon disulphide: | $2\,S + C \rightarrow CS_2$ |
| Carbon tetrachloride: | $CS_2 + 2\,S_2Cl_2 \rightarrow CCl_4 + 6\,S$ |
| | $4\,S + 2\,Cl_2 \rightarrow 2\,S_2Cl_2$ |
| (total equation): | $CS_2 + 2\,Cl_2 \rightarrow CCl_4 + 2\,S$ |

The industrial production of carbon disulphide $CS_2$ (boiling point 46°C) is a very evil-smelling and extremely dangerous work. The reaction of carbon—formerly only hard charcoal was used, now also coke—and sulphur takes place in cast iron cylinders which at the start are filled with coal and pre-heated by gas or electricity to 800–1,000°C. Liquid sulphur runs in at the bottom; its vapours permeate the coal charge and combine with it in an endothermic reaction to $CS_2$. Because of the exceptional volatility it is necessary to use very large condensation plants and to take special measures for the prevention of explosions and fires. After condensation the residual gases contain $H_2S$ and COS. The ignitability of carbon disulphide is very high, e.g. on steam-heated pipes at about 130°C.

Carbon disulphide is used mainly for the production of viscose (p. 381) and for carbon tetrachloride. By adding amines, dithiocarbamic acid compounds are obtained which are used as vulcanization catalysts and as fungicides for the protection of crops. The great toxicity and explosivity of carbon disulphide limits its usefulness.

### (c) Hydrocarbons containing fluorine

Brief mention must be made of the simple hydrocarbons containing fluorine which are produced from halogenated hydrocarbons. Certain compounds containing fluorine are the strongest known respiratory and contact poisons, e.g. $CH_3—SO_2F$, others, particularly the simple hydrocarbons containing fluorine, are physiologically completely neutral and extremely stable. They can be produced from halogenated paraffins with antimony pentafluoride, or with anhydrous hydrofluoric acid in autoclaves with antimony pentafluoride as catalyst. Of these, dichlorodifluoromethane, $CF_2Cl_2$, boiling point $-21°C$, has become especially important for the generation of cold, i.e. as fillers for refrigerators, under the name 'Freon' or 'Frigen'. Aliphatic fluorine compounds are non-toxic, non-inflammable, and non-corrosive, and are, therefore, much more suitable for refrigerators than the widely used methyl chloride.

Aliphatic fluorine compounds are also used as solvents and atomizing agents for pesticides and in cosmetics. About 14 fluoromethanes and -ethanes are produced industrially.

Tetrafluoroethylene, $F_2C = CF_2$, boiling point 76·3°C, is becoming increasingly important. It can be polymerized by the Du Pont de Nemours process at normal temperature under pressure in a methanol silver nitrate solution to produce the plastic 'Teflon'. This is heat-resistant and resistant to all chemicals (see chapter on Plastics p. 421).

### (d) Hydrocyanic acid from methane (Andrussov process)

On catalysts used for the combustion of ammonia to nitric oxides, e.g. platinum, a mixture of methane or other low hydrocarbons with air is converted to hydrocyanic acid in the presence of ammonia at 1,000–1,200°C (L. Andrussov, *Chem. Ing. Techn.* 1955, **27**, 469):

$$CH_4 + NH_3 + 1{\cdot}5\,O_2 \rightarrow HCN + 3H_2O + 114{\cdot}9 \text{ kcal.}$$

The gas issuing from the combustion furnace contains 8 per cent by volume of HCN. It is immediately quenched to 95°C, the excess ammonia is absorbed with acid ammonium sulphate lye, and the hydrocyanic acid is scrubbed with water at 5°C in a scrubbing tower, separated in a column, and distilled to 100 per cent.

The Andrussov process is also the basis for the 'ammonoxidation process', e.g. the direct conversion of propene into acrylonitrile with $NH_3 + O_2$, the reaction taking place via aldehydes.

F. Endter has reported a process (*Chem. Ing. Techn.*, 1958, **30**, 305) in which hydrocyanic acid is obtained without combustion, i.e. without addition of oxygen, from ammonia and methane only over a platinum catalyst.

### (e) Carbon black from methane and other hydrocarbons

Active carbon black is of the greatest importance for the rubber industry. Synthetic tyre rubber contains 40 kg carbon black per 100 kg rubber. The addition of carbon black lengthens the life of the tyres sixfold. World production is 1 million tons annually.

Carbon black can be produced for instance by cracking acetylene, but this process is not economical. It is mostly obtained by incomplete combustion of hydrocarbons with air, or by thermal cracking into carbon and hydrogen. $CH_4 \rightarrow C + 2H_2 - 22$ kcal/mole.

One distinguishes gas black (active) and lampblack from tar oils (printer's ink). For Germany the combined combustion of coke oven gas with tar oil which yields active carbon black is very important.

Carbon black is produced by one of two processes, the 'channel process' and the 'oven process'. In the channel process the hydrocarbon (e.g. natural gas) is burnt in numerous small burners with an inadequate air supply (luminous flames). The sooty flames impinge from below on cooled iron channels from which the carbon black is removed by scrapers.

In the oven process gas or light gas oil is burnt with an inadequate air supply. The liberated heat cracks the excess hydrocarbon into carbon and hydrogen. The separation of the carbon black is difficult; it is carried out with cyclones and in Cottrell installations (see p.p. 33–34).

A small proportion of the production of carbon black is also carried out by thermal cracking on the regenerator furnace principle. Some carbon black is also obtained as a by-product in the production of acetylene, from gases in the electric arc furnace, or by partial oxidation of hydrocarbons at 1,400°C; it is, however, not active.

## 2. PROCESSING OF ETHYLENE

### *Production methods*

At the present time ethylene is the most important material in the manufacture of industrial aliphatic products. In countries possessing natural gas with $C_2$ fractions or an extensive petroleum cracking industry it is one of the most abundant and cheapest raw materials. In 1957 the production of ethylene in the United States amounted to 1·8 million tons, of which half was obtained as a by-product in refineries, the other half by cracking of low molecular weight hydrocarbons. The position used to be different in Germany where the demand for ethylene had always been very high (e.g. in 1944 200,000 tons) and was further greatly increased by the introduction of the plastic polythene.

Almost half the production of ethylene in 1944 was obtained by partial hydrogenation of acetylene although the cost is nearly three times the cost of the production from cracking gases or coke oven gases by the Linde-Bronn separation described earlier. Even today ethylene is produced in the German Democratic Republic mostly from acetylene whilst in the Federal Republic most of the production is from gases and more recently by cracking and pyrolytic processes. Even more expensive and uneconomical for Germany was its production by dehydration of ethanol which in 1934 supplied about four fifths of the demand in four plants. In 1944 the production of ethylene by the cracking of ethane was of about the same order of magnitude. The price of ethylene produced by this process was about half that of ethylene produced from acetylene or from ethanol.

Ethanol is produced over an $Al_2O_3$ catalyst of strictly defined structure in a tubular furnace in the gas phase at 400°C at atmospheric pressure. About 90 per cent is converted. Ether is obtained as a by-product. Both ether and non-reacted alcohol are separated from the ethylene. At that time the Wolfen plant, for instance, operated with alcohol from the wood hydrolysis plant of the Dessau Sugar Refinery.

In order to dissipate the large amount of heat generated in the production of ethylene from acetylene, this hydrogenation reaction has to be carried out with a large hydrogen excess in the circulation.

$$CH{\equiv}CH + H_2 \rightarrow CH_2{=}CH_2 + 35 \text{ kcal.}$$

The best catalyst for partial hydrogenations of triple bonds, e.g. also for vinylacetylene → butadiene, is Pd, e.g. on $BaSO_4$.

The hydrogenation furnace, consisting of three parts, contains the catalyst (palladium) which is spread out on boards. The hydrogenation temperature is 270°C. In order to avoid overheating, superheated pressure water is injected for cooling between each layer of catalyst. For an oven which produces 200 tons of ethylene per month, 300 $m^3$ acetylene and 15,000 to 20,000 $m^3$ hydrogen circulate per hour. Some part of the gas circulation is branched off and after drying with silica gel and separation by deep freezing and distillation a yield of 85–87 per cent of the reacted acetylene is obtained in the form of ethylene. In 1944 this process was used in four plants in Germany. The partial hydrogenation method is also used to separate pyrolysis and cracking gases from acetylene. Catalysts used include chromium-nickel catalysts.

## The production of ethylene by dehydrogenation, cracking and pyrolysis processes

### *General remarks on dehydrogenation and cracking*

By heating to 850–900°C ethane can be cracked purely thermally into ethylene and hydrogen. Purely thermal dehydrogenation of the higher hydrocarbons is difficult because the cracking reaction predominates. The production of olefins from the saturated hydrocarbons by dehydrogenation must, therefore, be carried out by catalytic methods. Even with these the proportion of cracking reactions increases with the length of the chains.

Heat must be supplied to both dehydrogenation and cracking since both these reactions are endothermic. The dehydrogenation reaction requires a considerably greater supply of heat (—30 kcal/mole) than the cracking reaction (—17 kcal/mole). Heating of hydrocarbons with more than 2 carbon atoms tends, therefore, to produce cracking rather than dehydrogenation. Pyrolysis is cracking by heating to high temperatures so that $CH_3$, $CH_2$, and CH radicals are formed. With cracking processes it must be borne in mind that a considerable increase in the volume of gas takes place. The heat required for the cracking reactions can be supplied in various ways:

(1) Passing through externally heated tubes.

(2) Passing over pre-heated regenerators (reversing process), regenerative furnace.

(3) Pre-heating to a high temperature under pressure and releasing the pressure.

(4) Supplying heat by superheated steam (perhaps combining points 3 and 4) which acts simultaneously as an inert gas and reduces coke separation.

(5) Cracking over pre-heated solid bodies (stone or coke spheres, catalyst particles) which after separation are burned off and recirculated.

(6) Supplying the heat required for dehydrogenation by supplying limited quantities of oxygen and partial combustion.

All these processes resemble closely the gas phase cracking processes discussed in the chapter on Petroleum. In those the aim was the production of petrol or of aromatics. For the production of ethylene a minimum of liquid products and a maximum of gaseous olefins are aimed at. However, there are also methods for the production of olefins which are coupled with the production of aromatics.

Of the multiplicity of processes developed for the production of ethylene and olefins only a few typical ones will be mentioned here:

### *Dehydrogenation of ethane*

Ethylene can be obtained by thermal dehydrogenation or pyrolysis of ethane (e.g. of gases containing ethane) in a tubular furnace made of silicon chromium steel at about 900°C with a residence time of one second. In a single passage about 50 per cent of the ethane is cracked.

The so-called autothermal cracking of ethane to ethylene, i.e. without external heating, is carried out in a ceramic cracking furnace filled with porcelain spheres and into which is fed a pre-heated mixture consisting of 100 parts per volume of ethane and 30 parts per volume of oxygen (H. Klein) (Bartholomè and Nonnenmacher, *Erdöl u. Kohle*, 1959, **12**, 359). The first experimental plant started in Leuna in 1936, originally based on propane, later on ethane. The process was further developed by BASF. The gas mixture formed contains, in addition to other gases, carbon monoxide, carbon dioxide, methane, and higher hydrocarbons.

*Ethylene produced by cracking of gases*

The thermal cracking of gases, e.g.

$$CH_3—CH_2—CH_3 \rightarrow CH_2 = CH_2 + CH_4 - 16{\cdot}1 \text{ kcal/mole},$$

is carried out in a manner similar to the dehydrogenation. Fresh gas and circulation gas from which the olefin has been separated are heated for a short period in tubular heaters to about 800°C. The mixture is then cooled as rapidly as possible ('quenched') e.g. by injection of water or of oil in order to prevent secondary reactions. Hydrogen and methane can be used, after separation of the olefins by the compression-condensation processes discussed earlier, as fuels or for the production of hydrogen or synthesis gas (methane cracking).

Coke deposits in the pyrolysis coils are unavoidable. From time to time they must be burnt out with air and steam.

In order to prevent heating of the gases in the tubular heaters processes have been developed in which the thermal cracking takes place over heated surfaces.

In the 'Pebble Heater' process stone pellets are pre-heated by combustion gases in a refractory-lined heater. They then slide through a refractory tube into the reactor below into which the gas is fed for cracking. The pellets then drop into the lower part of a tube conveyor where they are lifted by hot air and returned from above into the heater (see also Hoechster-Coker).

The regenerator furnace principle (see p. 40) can also be used for cracking gases. In the Koppers-Hasche furnace, for instance, 85–90 per cent of propane can be converted into ethylene and methane at 870–1,110°C.

*Pyrolysis of liquid hydrocarbons*

The fact that ethylene is produced during the manufacture of petrol by cracking has already been mentioned.

In countries that do not possess sufficient gases for dehydrogenation and cracking, pyrolysis processes have been developed for the conversion of tars and petroleum into ethylene. For reasons already discussed these processes are usually carried out in the presence of steam. Heat is supplied by one of the methods described on p. 165. The aim is to obtain a maximum of gaseous olefins.

Light petroleum fractions can be cracked by the Shell pyrolysis process in a tubular furnace in the presence of steam at 600–700°C. The quenching of the pyrolysis gas is coupled in a waste heat boiler with the generation of high pressure steam. The yield is: $C_2$ 26 per cent, $C_3$ 15 per cent, $C_4$ 10 per cent, aromatics 20 per cent, fuel oil 26 per cent.

In the Kellogg process pyrolysis is carried out by feeding both a mixture of gas oil and steam, heated to 650°C, and steam heated to more than 900°C, into a pyrolysis tube.

A mixture of acetylene and ethylene is obtained from liquid gas or light petrol by a method developed by Farbwerke Hoechst and Koppers-Totzek, in use since 1957. The hydrocarbon is injected into a gas flame at 2,000°C (pyrolysis burner). Heat is produced by the combustion of the residual gas (methane and hydrogen) with oxygen. Separation of the acetylene from the mixture of cracking gases is possible by scrubbing under pressure.

The cracking of very-high-boiling oils (tars, crude oils, petroleum residues) during which the formation of considerable amounts of coke is unavoidable, is carried out in a heat carrier-travelling bed (see Pebble-Heater process). In order to be able to extract the carbon residue of oil, coke pellets are used as heat carriers in the 'Hoechster Coker' (K. Winnacker, *Chem. Ing. Techn.* 1955, **27**, 400). The coke formed is deposited on the spheres thus steadily increasing their size. According to the oil used as raw material 50–60 per cent olefins are obtained. In the Farbwerke Hoechst a plant for the processing of 50,000 tons of crude oil annually has been in operation since 1955. Figure 70 shows the method of operation of the Hoechster Coker.

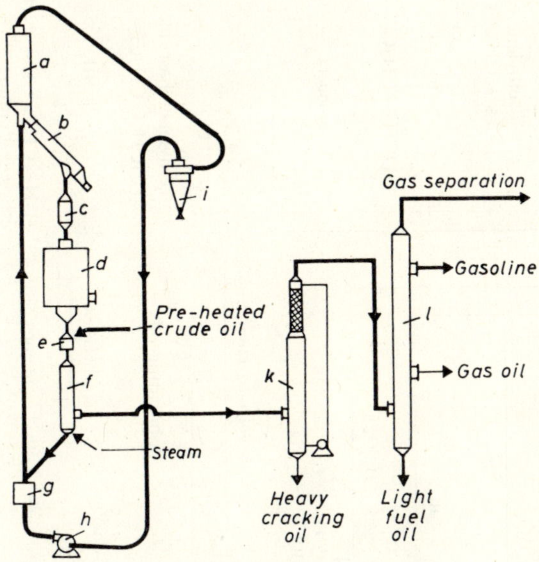


Figure 70. Hoechster Coker

*a* Coke trap *b* Coke separator *c* Intermediate vessel *d* Coke heater *e* Mixing zone *f* Reactor *g* Lift pot *h* Hot gas blower *i* Cyclone *k* Quenching and scraping *l* Fractionating column

In the tubular heater *d* coke spheres are heated indirectly by gas to 650–750°C. The pre-heated oil is sprayed on to the spheres in the reactor *f* where cracking takes place.

The pyrolysis gases pass into a quencher *k* where they are cooled with heavy oil. Then follows separation in the column *l*. By means of a lifting device the coke is returned via *g*, the coke trap *a*, and the screening device *b* into the heater *d*.

A plant built for 'Erdölchemie' by B. P. and Bayer in 1958 at Dormagen where mainly olefins are produced from petroleum products, is the Lurgi Ruhr gas sand cracker (*Erdöl und Kohle*, 1961, **14**, 537; 1962, **15**, 799). In this cracking plant sand is used as heat carrier in a fluidized bed reaction, the sand being heated and regenerated whilst being moved pneumatically.

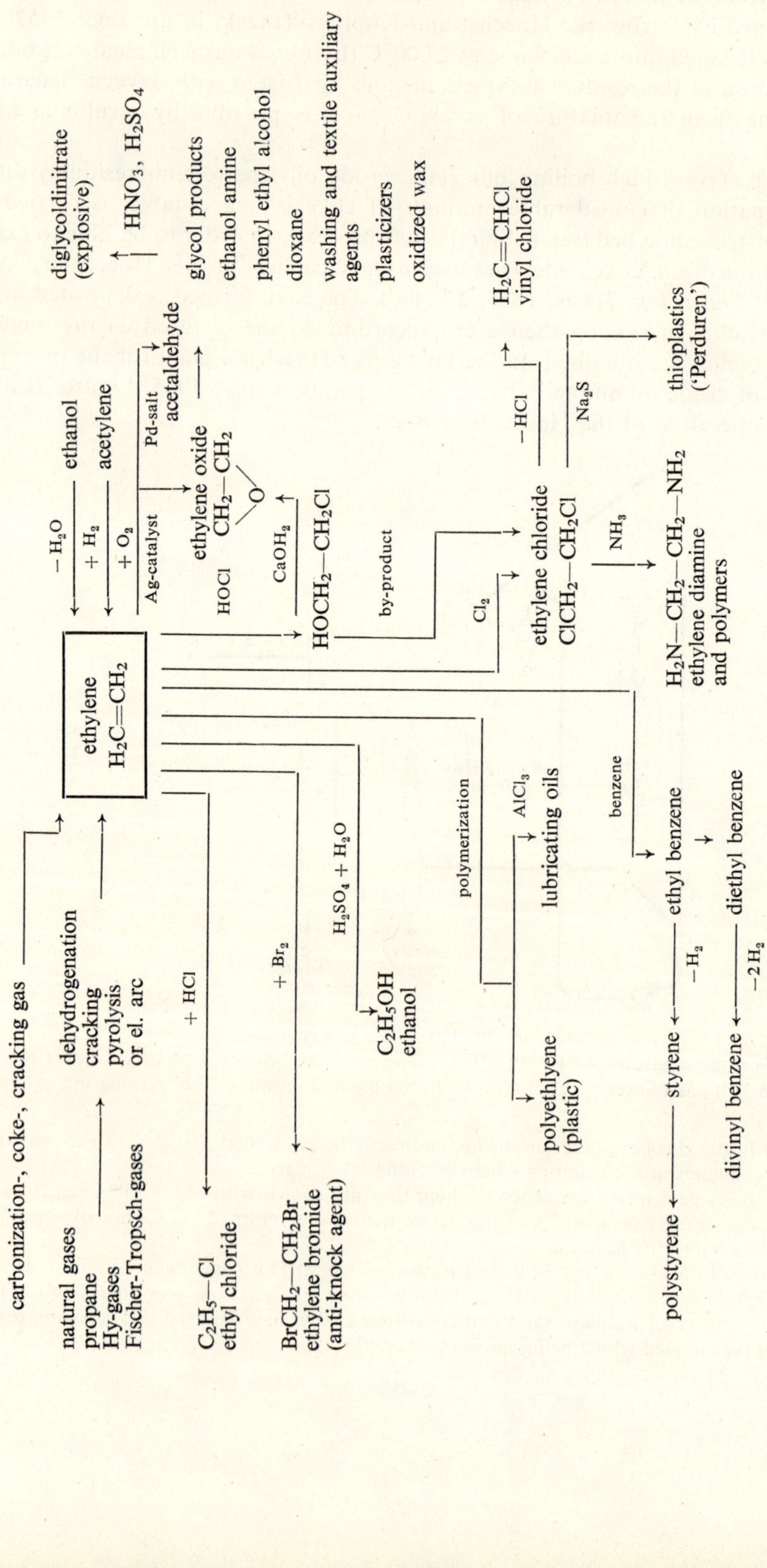


Figure 71. Production and industrial use of ethylene

Ethylene forms the raw materials for very many aliphatic compounds (see Figure 71). In the United States the bulk of it used to be processed to alcohol and ethylene oxide, more recently to polyethylene. The large new ethylene plants in Western Germany have been built mainly for the production of polyethylene.

Very pure ethylene is required which at gauge pressures of 500 up to as high as 2,000 atm is transformed into the high-polymer form. With 'Ziegler catalysts' polymerization without pressure is also possible (for details see under Plastics, p. 418). It is one of the thermoplastic materials with the brightest future.

A medium polymerization of ethylene is carried out with aluminium chloride. This yields valuable lubricating oils with a low setting point which are produced at Leuna and Schkopau.

More than half of the very large American production of alcohol is based on ethylene. In Germany and other countries where alcohol is produced by fermentation, ethylene oxide is the most important intermediate product of ethylene processing.

### (a) Halogen derivatives of ethylene

Foremost among the halogen derivatives of ethylene is ethyl chloride, boiling point 12·2°C (U.S.A. production in 1956: 245,000 tons). Apart from its use in medicine for local anaesthetics by cooling it is important mainly for the introduction of the ethyl group, e.g. for the manufacture of ethyl cellulose and of tetraethyl lead, which is a well-known anti-knock agent in common use: $4\ PbNa + 4\ C_2H_5Cl \rightarrow Pb\ (C_2H_5)_4 + 4\ NaCl + 3\ Pb$. K. Ziegler and H. Lemkuhl have developed a new method for the production of tetraethyl lead via aluminium triethyl, i.e. on an ethylene basis (*Angew. Chemie*, 1955, **67**, 424).

Ethyl chloride used to be produced mostly by esterification of ethyl alcohol with anhydrous hydrochloric acid under slight pressure in the presence of $ZnCl_2$ as a catalyst, or at atmospheric pressure over a carbon-$HgCl_2$ catalyst: $C_2H_5OH + HCl \rightarrow C_2H_5Cl + H_2O$. More attractive is the production of ethyl chloride from ethylene directly by combining it with hydrochloric acid in the liquid phase in the presence of aluminium chloride as a catalyst: $CH_2{=}CH_2 + HCl \rightarrow C_2H_5Cl$. Since it is necessary to cool the mixture the only materials available formerly were platinum and tantalum. The BASF had very expensive equipment for the manufacture of ethyl chloride consisting entirely of tantalum.

During the 1939–45 war a new method was developed at Leuna for the production of ethyl chloride from ethane by chlorination, in which chlorination took place over a catalyst in suspension, coke or ceramic material saturated with metallic chloride: $CH_3{-}CH_3 + Cl_2 \rightarrow C_2H_5Cl + HCl$. The method resembles the chlorination of methane (see p. 160).

By further chlorination of ethyl chloride the Dow-Chemical Corp. produces 1,1,1-trichloroethane (methylchloroform): $CH_3{-}CCl_3$. It is used a great deal in 'chemical laundries'. It has only one fifth the toxicity of carbon tetrachloride. By HCl-cracking with milk of lime vinylidenechloride is obtained from it which is the monomer of the plastic 'Saran'.

Ethylene chloride (boiling point 84°C) is produced easily by the addition of chlorine to ethylene: $CH_2{=}CH_2 + Cl_2 \rightarrow CH_2Cl{-}CH_2Cl$. Chlorides of iron, copper, and antimony are used as catalysts. It is, however, also a by-product in the manufacture of ethylene oxide via chlorhydrin. If the manufacture is on a large enough scale sufficient ethylene chloride is produced.

1,2-dichloroethane is used for the production of vinyl chloride $H_2C{=}CHCl$, e.g. by cracking hydrochloric acid at 400°C over an alumina catalyst. It is also an intermediate product in the manufacture of ethylene diamine $H_2N—CH_2—CH_2—NH_2$, of polyethyleneamine, and of chloroethane sulphonic acid which is important as an intermediate product in the manufacture of washing and textile auxiliary agents. Ethylene chloride used to be important for rubber-type plastics, 'thioplastics', e.g. 'Perduren', obtained by reaction with alkali sulphides.

By the addition of bromine by the well-known method ethylene bromide (boiling point 131–132°C) is obtained which is much used for synthesis purposes and as an addition to anti-knock agent mixtures.

### (b) Ethylene oxide

*Production of ethylene oxide by the chlorhydrin process*

Ethylene oxide (boiling point 12·5°C) is one of the most important aliphatic products of the large-scale chemical industry. Almost all glycol derivatives are produced from it. The oldest method for the production of ethylene oxide is carried out in two stages.

First, hypochlorous acid is added to ethylene, giving the intermediate product ethylene chlorhydrin which is converted into ethylene oxide, without being first isolated, by boiling with milk of lime:

$$CH_2{=}CH_2 + HOCl \rightarrow CH_2OH—CH_2Cl$$

$$2\,CH_2OH—CH_2Cl + Ca(OH)_2 \rightarrow 2\,\underset{\diagdown O \diagup}{CH_2—CH_2} + CaCl_2 + 2\,H_2O$$

The reaction is carried out in water in a 15 m spray tower, filled with Raschig rings, at 70°C. Chlorine and ethylene are introduced from below. The undissolved ethylene escapes at the top, is separated from the liquid in a separating vessel, compressed again and returned into the tower. After the pressure has been released the aqueous solution of chlorhydrin passes into a reaction chamber lined with rubber and brick where it is brought to boiling heat with milk of lime. During this process crude ethylene oxide is liberated, together with ethylene chloride and $\beta,\beta'$-dichloroethylether. These two impurities are produced to the very considerable amount of 20 per cent. The yield of ethylene oxide is 75 per cent of the theoretical. Part is immediately processed further, the rest is dispatched by tank wagon and in steel cylinders.

Strict safety must be observed for the storage of, and work with, liquid ethylene oxide: Before sufficient experience had been gained of the behaviour of liquid ethylene oxide, it repeatedly occurred that pressure vessels exploded spontaneously. It was found that certain substances with a catalytic effect, e.g. acids, can cause spontaneous polymerization and the generation of a large amount of heat if they pass accidentally from another part of the equipment into the ethylene oxide. Because of its great toxicity, ethylene oxide, mixed with $CO_2$, has been used as a pesticide, under the name of 'Catox'. It also forms very easily ignitable mixtures with air.

In 1943, production in Germany amounted to nearly 67,000 tons, in the United States in 1951 to 247,884 tons, in 1960 to 669,190 tons. Because of the large consumption of chlorine the chlorhydrin process can only be used where electrolytic chlorine is available in sufficient quantities.

The large quantities of chlorine required in the chlorhydrin process are due to the fact that all the chlorine put in is lost as calcium chloride lye. To produce 100 kg of ethylene oxide, 209 kg of liquid chlorine and 70 kg CaO are used. The $CaCl_2$ formed places a great strain on the sewerage system. For these reasons, and also because of a shortage of chlorine, modern development tends towards the direct oxidation method.

*Ethylene oxide produced by the catalytic oxidation of ethylene (Direct oxidation method)*

The direct oxidation of ethylene with atmospheric oxygen was first published by the Société Française de Catalyse Generalisée in 1931 as a patent, and was carried out over an Ag catalyst by T. E. Lefort (Carbide and Carbon Chem. Corp., USA) in 1936. On the basis of the patents of this firm an industrial process for the direct production of ethylene oxide from ethylene and air was developed several years ago by the former IG plant at Ludwigshafen who built a large-scale works for this purpose. Twenty years after its invention the reaction is in general use on a large scale. Plants have been built in all industrialized countries and new installations of ethylene oxide works are based exclusively on this process.

The formation of ethylene oxide from ethylene can be considered as a partial combustion process. 56 kcal per mole are liberated, whilst in complete combustion of ethylene 631 kcal are generated. It follows that certain measures must be taken in order to hold the reaction at the labile oxide stage and to avoid a 'running away'. These measures include the use of a catalyst with maximum specificity (silver on $Al_2O_3$), the prevention of any local overheating which requires the smallest possible distance of each reaction point from the cooling surfaces, and rapid discharge of the ethylene oxide formed. This is achieved by the use of tubular catalyst furnaces in which the catalyst is charged into a large number of tubes as uniformly as possible to ensure uniform admission of the gas to each tube. These tubular catalyst furnaces had proved their worth long ago in the catalytic oxidation of naphthalene to phthalic anhydride. The fluid process can also be used but does not offer any special advantages. The operations are carried out at a gauge pressure of 20 atm and at 240–260°C. The temperature must be controlled exactly; it rises as the catalyst is consumed. The ethylene must be very pure and completely free of acetylene which is always contained in cracking ethylene (see for gas phase oxidation: F. Broich, *Chem. Ing. Techn.*, 1962, **34**, 52).

The ethylene content is kept low and the excess of air very high if only to operate below the lower explosion limit. This is of importance in all catalytic oxidations. The ethylene concentration amounts, therefore, to 3–5 per cent by volume. Since 40–45 per cent are converted, the concentration of ethylene oxide with a single passage through the catalyst is low and the isolation of the oxide would not be economical. The process is, therefore, carried out as a circulation process and, if necessary, in several stages, i.e. the gas passes through the catalyst furnace several times and the ethylene consumed is made up in the gas so that finally a concentration of 7–8 per cent of ethylene oxide is obtained. It would be uneconomical to try to attain higher concentrations.

Part of the gas circulation is continuously bled off and the ethylene oxide extracted. The total yield amounts to a maximum of 60–62 per cent of the theoretical, calculated on an ethylene basis, the rest is lost. In this respect direct oxidation is, therefore, less favourable than the ethylene chlorhydrin process.

Isolation of the ethylene oxide itself is difficult. It is possible to adsorb it on carbon and then to desorb it again but it tends to polymerize on the carbon. The most practical method is the extraction of the ethylene oxide from the cooled gas with water. It is expelled from the aqueous solution by heating and then distilled in a pressure column.

The ethylene oxide can be introduced into diluted acid, e.g. sulphuric acid or oxalic acid, and processed directly to glycol or di- and triglycol.

A direct glycol synthesis process has been developed by Du Pont de Nemours, U.S.A. Carbon monoxide and water are reacted with formaldehyde at 200°C and 700 atm gauge pressure, using sulphuric acid or boric acid as catalysts. The glycolic acid formed is esterified with methanol. The glycol and methanol are recovered by catalytic reduction.

$$HC\begin{matrix}\nearrow O \\ \searrow H\end{matrix} + CO + H_2O \rightarrow HOCH_2\text{—}COOH$$

Direct oxidation of ethylene to acetaldehyde is possible by the new process developed by the Consortium für die elektrochemische Industrie G.m.b.H., München, at the Farbwerke Hoechst (J. Smidt et al., *Angew. Chem.*, 1959, **71**, 176; G. Berger and R. Mittag, *Erdöl u. Kohle*, 1962, **15**, 699), using palladium copper chloride catalysts in the aqueous phase at atmospheric or slightly higher pressure, at temperatures around 100°C. Noble metal-olefin complexes are produced as intermediates:

Total reaction $C_2H_4 + {}^1/_2\,O_2 \rightarrow CH_3\text{—}CHO + 52\text{ kcal}$

Partial reactions $C_2H_4 + PdCl_2 + H_2O \rightarrow CH_3\text{—}CHO + Pd + 2\,HCl$

$Pd + 2\,HCl + {}^1/_2\,O_2 \rightarrow PdCl_2 + H_2O$

The method can be used generally for the oxidation of olefins to carbonyl compounds and has great possibilities.

### (c) Syntheses with ethylene oxide

For reasons discussed earlier it is necessary, when working with ethylene oxide, to interpose an empty chamber between the pressure vessel and the reaction vessel and to make absolutely certain that no reflux of liquids and vapours is possible. By heating with water in the presence of acids ethylene oxide yields glycol and diglycol as well as a little triglycol which are separated by distillation.

During the 1939–1945 war the production of diglycol was very important for the production of diglycoldinitrate which has properties similar to nitroglycerine. Glycol has a variety of uses as a substitute for glycerine, e.g. to prevent the freezing of cooling water.

By cyclization of two molecules of ethylene oxide in the presence of acid catalysts dioxane is produced which is an excellent though slightly toxic solvent. With alcohols in the presence of acids glycol ethers are formed, which are used as plasticizers for plastics. Specially pure glycol ethers, e.g. phenyl glycol ether and benzyl glycol ether, which are almost odourless and non-volatile, are used as 'fixing agents' in the perfumery industry. They prevent too rapid evaporation of volatile perfumes and give them a certain permanence. Glycol esters are obtained with carboxylic acids. The benzoic ester of diglycol is a non-toxic plasticizer for polyvinylchloride. The polymeric glycol ester of terephthalic acid is becoming increasingly important as a raw material for fibres (see p. 389).

Formerly the addition of hydrocyanic acid to ethylene oxide was of industrial significance. Ethylene cyanhydrin was obtained according to the equation:

$$2\,\underbrace{CH_2\text{—}CH_2}_{O} + HCN \rightarrow CH_2OH\text{—}CH_2CN$$

By the splitting off of water, acrylonitrile is produced which is important for plastics, and in an alcoholic solution, by the simultaneous splitting off of water and saponification, acrylic ester is obtained:

$$CH_2OH\text{—}CH_2\text{—}CN \rightarrow CH_2\text{=}CH\text{—}CN + H_2O$$

$$CH_2OH\text{—}CH_2CN + ROH + H_2SO_4 \rightarrow CH_2\text{=}CH\text{—}COOR + NH_4HSO_4$$

Hydrocyanic acid adds on to ethylene oxide at 52–55°C in the presence of water and diethylamine and with the addition of NaOH, heat being generated in the process. Acrylonitrile (boiling point 78°C) is formed from cyanhydrin by heating, the temperature being raised from 200° to 280°C, in the presence of magnesite or bauxite in powdered form as catalyst. Water and acrylonitrile distil out of the reaction vessel whilst cyanhydrin can be continuously fed into the catalyst so that the process takes place continuously for a period. (Acrylonitrile from acetaldehyde, see p. 212.)

Acrylonitrile polymerizes easily and is, therefore, distilled in an enamelled column, with the addition of methylene blue which acts as an antioxidant. It can also be distilled in iron equipment with the additions of 500 g of hydroquinone per ton.

Acrylonitrile can also be produced from acetylene and hydrocyanic acid. Formerly it was used mainly for Buna N. From its polymers is produced the American 'Orlon' fibre, which is called 'Wolcrylon' by the VEB Film- und Chemiefaserwerk 'Agfa', and 'Dralon' by Bayer; this fibre had originally been developed by the former IG under the name of PAN fibre (see Synthetic fibres).

The reactions of phenols, alcohols, and acids with ethylene oxide can also take place in such a way that during the formation of the glycolether of the phenol or alcohol, or of the glycolester of the acid, a further number of ethylene oxide molecules are added so that derivatives of polyethylene oxide are produced.

The polyethylene oxides themselves have the characteristics of waxes and are often used as synthetic waxes. The reaction products with fatty alcohols are important emulsifiers. Alkylphenolpolyethylene oxide ethers are washing and emulsifying agents. The distinguishing feature of these detergent and surface-active compounds is the fact that they are free of cations (see p. 436).

The oxidation of glycol in the gaseous phase yields glyoxal OHC—CHO which is marketed in aqueous solution as a partly polymeric hydrate. It reacts with the OH groups of cellulose and with the amine groups of proteins. Cross-linking increases the resistance to water (textile finishing agent).

The reaction of ethylene oxide with ammonia yields mono-, di-, and triethanolamine for which there is a variety of uses. Here only their capacity to combine with $SO_2$ and $H_2S$ shall be mentioned, use of which is made in the purification of town gas. With amines, alkylaminoethanols are obtained, a group which is used e.g. as diethylaminoethanol residue in 'Procain' (amino benzoic ester of diethylaminoethanol) and other local anaesthetics. 'Morpholin' is obtained by the splitting off of water from diethanolamine with sulphuric acid of 70 per cent concentration:

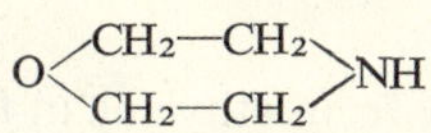


Starting from monoethanolamine, ethyleneimine $H_2C—CH_2$ (N, H) is produced by boiling with a solution of caustic soda, with the intermediate stages ethyl chloride amine-hydrochloride, obtained with $SOCl_2$ ($ClCH_2—CH_2—NH_2 \cdot HCl$), or amino ethylsulphuric acid ($HO_3S—O—CH_2—CH_2—NH_2$).

Ethyleneimine is processed to polyethyleneimine which is used increasingly for syntheses. Various cytostatics contain several ethyleneimine residues.

Ethylene oxide can also be used in the Friedel-Crafts reaction. From benzene and ethylene oxide, in the presence of $AlCl_3$, phenylethylalcohol is obtained:

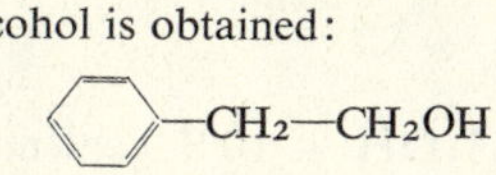


which in its pure form has the odour of roses and is one of the most important synthetic perfumes. Phenylethylalcohol can also be produced by the following method:

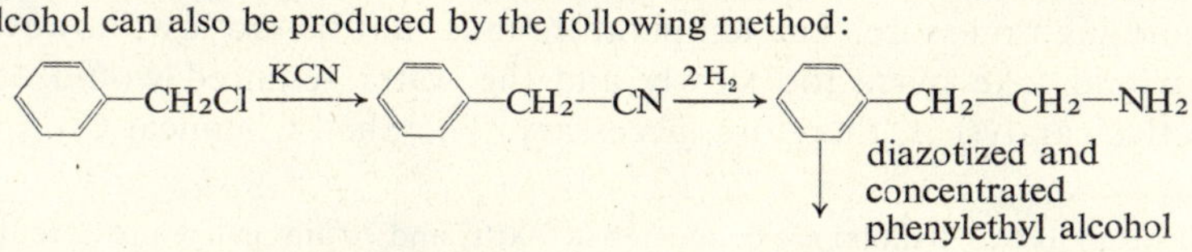


### (d) Hydration of olefins to alcohols

This reaction is of great importance. In 1955, the United States produced 620,000 tons ethanol, 80 per cent of this from ethylene and of these four fifths by hydration with sulphuric acid, $^1/_4$ directly. Moreover, 450,000 tons of isopropanol were produced from propene exclusively with $H_2SO_4$ (in the same year, the production of methanol amounted to 630,000 tons). Both in the United States and in the Soviet Union

acetaldehyde is being produced from ethanol whilst in Germany it is produced from acetylene. Like methanol, ethanol is oxidized catalytically with atmospheric oxygen to formaldehyde.
(Direct method ethylene → acetaldehyde p. 171).

*Indirect hydration*

In indirect hydration, addition of $H_2SO_4$ takes place first followed by hydrolysis of the alkyl sulphate:

$$R{-}CH{=}CH_2 \xrightarrow{H_2SO_4} R{-}\underset{\displaystyle OSO_2OH}{\underset{|}{CH}}{-}CH_3 \xrightarrow{H_2O} R{-}\underset{\displaystyle H}{\underset{\displaystyle O}{CH}}{-}CH_3 + H_2SO_4$$

Sulphuric acid always migrates to the point where there is the least amount of hydrogen: Propene → isopropanol, isobutene → tertiary butanol. The addition must be carried out with rather concentrated sulphuric acid, the hydrolysis in diluted $H_2SO_4$. For re-use, the diluted sulphuric acid must be concentrated again which requires a large amount of energy.

The ease of addition of $H_2SO_4$ to olefins increases from ethylene to the hexanes; iso-olefins add particularly easily. A concentration of the acid of 98 per cent, pressure, and a temperature of 80°C are, therefore, required for ethylene, for isobutene a concentration of 65 per cent is sufficient at 0°C, propane and n-butene require 85 per cent. Sulphate formation and hydration depend on the various degrees of ease of formation and stability of the carbonium ions produced as intermediates (see also Separation of butene, p. 179).

Ethanol is produced continuously from ethylene in three stages:

(1) Absorption of the ethylene in $H_2SO_4$ of 98 per cent concentration, 35 atm gauge-pressure, 80–90°C. 50 per cent mono- and 50 per cent diethylsulphate are formed, 75 per cent of the $H_2SO_4$ being utilized. Heat is liberated and cooling is, therefore, required.

(2) Dilution and saponification under pressure.

(3) Separation of the alcohol by injection of steam ($H_2SO_4$ in a concentration of 50–60 per cent is formed, also ether by ethylation of alcohol with diethylsulphate).

The mixture of ether and alcohol is scrubbed and distilled. Indirect hydration is possible in the presence of saturated hydrocarbons since these are not absorbed by the $H_2SO_4$.

*Direct hydration*

The reaction $C_2H_4 + H_2O \rightleftarrows C_2H_5OH + 10{\cdot}9$ kcal/mole is an equilibrium reaction. The ethylene used must be fairly pure. Theoretically, the optimum equilibrium is at low temperature and high pressure. The temperature must not be too low, however, because the reaction would take place too slowly and the water required would condense. A sufficiently active catalyst is, therefore, necessary. The Shell Chemical Co. process is as follows:

Ethylene and steam (0·6 : 1·0 mole) are conducted at 300°C and 70 atm gauge-pressure through phosphoric acid into diatomaceous earth. The ethylene must be at a concentration of 97 per cent. The reaction is only 4·5 per cent and ethylene circulation is, therefore, necessary. Since vitiated gas accumulates during this process, part of the ethylene is extracted by pressure release (85 per cent) and is concentrated back to 97 per cent in the Linde apparatus.

Some of the alcohol separates out in liquid form, the steam is removed from the gas by scrubbing with water. Since the distilled alcohol contains acetaldehyde and croton aldehyde (permanganate test) additional hydration on a Ni catalyst is carried out.

# 3. PROPANE AND PROPYLENE

Figure 72 shows the most important industrial uses of propane and propylene (propene). Ethylene can be obtained by cracking of propane thus presenting a further source of this important olefin. The nitration of propane will be discussed briefly later.

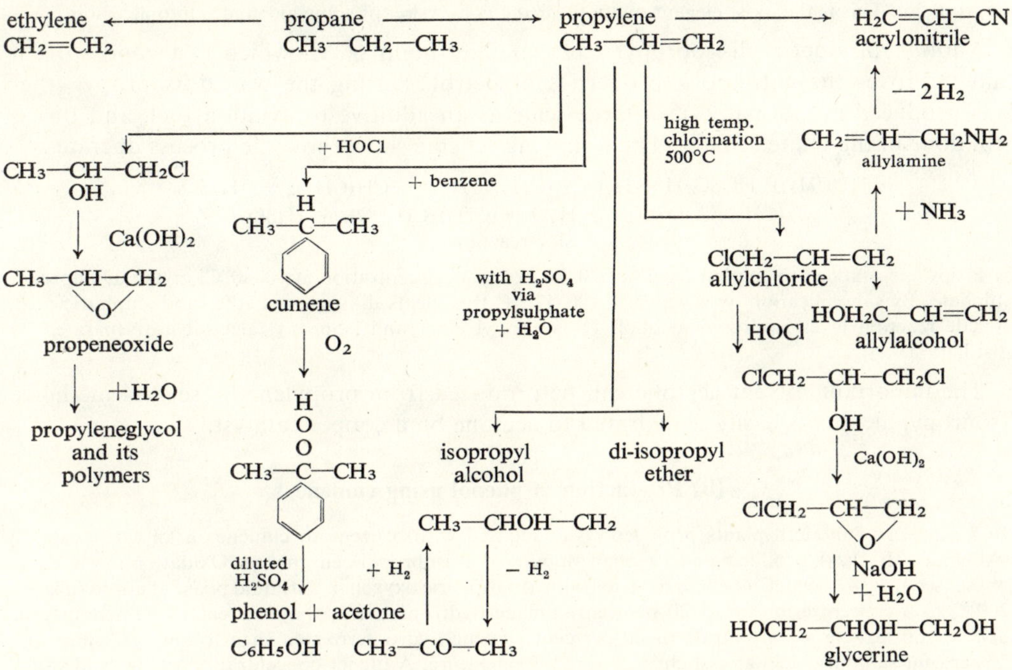

Figure 72. The use of propane and propylene in industry

## (a) Propanol and di-isopropyl ether

One of the most important uses of propylene is its conversion to isopropanol. In this process, propylene is fed countercurrently, at a gauge-pressure of 8–10 atm and at normal temperature, into towers into which is trickled sulphuric acid of 92 per cent concentration. Propylene reacts with concentrated sulphuric acid much more easily than ethylene. The acid with its large propylene content (propyl sulphonic acid) is diluted with water to about 20 per cent and the isopropyl alcohol is cracked by heating ('strong acid process'):

$$CH_3{-}CH{=}CH_2 + H_2SO_4 \rightarrow (CH_3)_2\,CHOSO_3H$$

$$(CH_3)_2\,CHOSO_3H + H_2O \rightarrow CH_3{-}CHOH{-}CH_3 + H_2SO_4$$

In the so-called 'weak acid process' propene is fed into sulphuric acid of 70 per cent concentration at 65°C and a gauge pressure of 25 atm. The acid then passes into a vacuum column with a little steam. Isopropyl alcohol is distilled off and the 70 per cent acid is returned into the propane absorption tower. Isopropyl ether rather tends to autoxidation.

Isopropyl alcohol can also be obtained by catalytic hydrogenation of acetone produced by different methods (see p. 223). Isopropyl alcohol (boiling point 82·5°C) is a fully equivalent substitute for ethyl alcohol as a solvent, particularly in the perfume and cosmetics industries. Its disinfecting effect is stronger than that of ethyl alcohol. Its esters are important solvents, and it is used in many syntheses.

Mention must be made of the reduction, by Meerwein-Ponndorf, of higher aldehydes and ketones to alcohols using isopropyl alcohol in the presence of Al-isopropylate, in which acetone is converted to isopropanol. This is the most elegant method for the conversion of some aldehydes into alcohols.

Another product is di-isopropyl ether, boiling point 66°C, which is a good solvent and improves the anti-knock properties of petrol. During the war, di-isopropyl ether was produced in Germany on a large scale as an additive to aviation fuel, and this is still done in the United States. The following scheme shows how the process operates:

$$2(CH_3)_2CHOSO_3H + H_2O \rightarrow (CH_3)_2CH{-}O{-}CH(CH_3)_2 + 2H_2SO_4$$

$$(CH_3)_2CHOSO_3H + H_2O \rightarrow (CH_3)_2CH{-}OH + H_2SO_4$$

side reaction

Propylene is added to sulphuric acid of 70–80 per cent concentration at 10–30°C, giving di-isopropyl sulphate. By saponification with water at 100–125°C this yields di-isopropyl ether and sulphuric acid. A side reaction gives isopropyl alcohol. Di-isopropyl ether and isopropyl alcohol are separated by distillation.

The important solvent acetone can be produced from propylene by several methods. Isopropyl alcohol is easily dehydrated to acetone on a copper catalyst.

### (b) Production of phenol using cumene

In a number of modern plants propylene is condensed with benzene to cumene on an $AlCl_3$ catalyst (Al filings, HCl gas), producing also a small amount of di-isopropyl benzene (see Oxidation in the liquid phase, see pp. 389–390). Cumene is oxidized with atmospheric oxygen in the liquid phase at approximately 200°C to a concentration of 15–20 per cent cumene hydro-peroxide, the non-reacted cumene is then distilled off leaving a concentrate of 70 per cent cumene hydro-peroxide. Then follows cracking with 70 per cent sulphuric acid at a slightly elevated temperature. A higher concentration of the acid would produce formation of diphenylolpropane ('Dian'). The sulphuric acid circulates. Finally, the non-reacted cumene, acetone, and phenol are separated in three columns. The process is due to H. Hock, and the technology was developed in the U.S.A. which was the first country to use it on a large industrial scale.

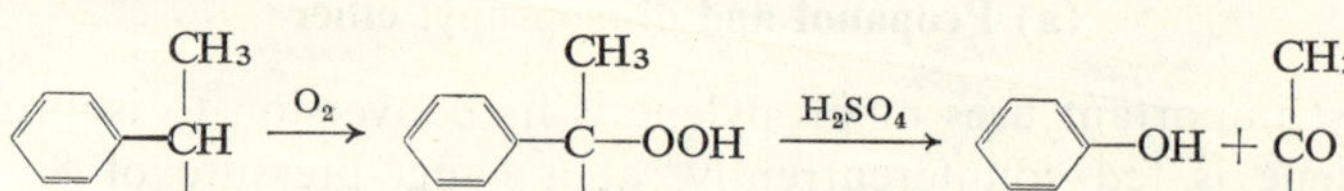


This new phenol process is very economical for countries disposing of adequate supplies of propylene since it saves the sulphuric acid normally used for the production of phenol from benzene, as well as the alkali used in separation by melting, or the chlorine and alkali in the chlorobenzene process. There is a continuously rising demand for the acetone produced as a by-product (e.g. for 'Dian', p. 412). A new process has been developed for the production of phenol from toluene via benzoic acid and its phenol-esters (*Brennstoffchemie*, 1960, **41**, 71; 1962, **43**, 134).

### (c) Glycerine

From propylene there is also a route to glycerine which was developed industrially by Shell (H.P.A. Groll) in 1935. Owing to the consequences of the war the process could not be used satisfactorily on an industrial scale in Germany (Ludwigshafen and Heydebreck). In this process propylene is chlorinated in the so-called 'cold flame' at 500°C and with a residence time of 2–3 seconds. Allylchloride (b.p. 46°C) is formed directly, with a yield of 85 per cent of the theoretical.

$$H_2C{=}CH{-}CH_3 + Cl_2 \rightarrow H_2C{=}CH{-}CH_2Cl + HCl$$

By-products are 2-chloropropene, propylenechloride, and dichloropropene. As in the classic ethylene oxide process, hypochloric acid is added to allyl chloride, then hydrochloric acid is cracked off with milk of lime. Epichlorhydrin is obtained which yields glycerine when saponified with caustic soda solution.

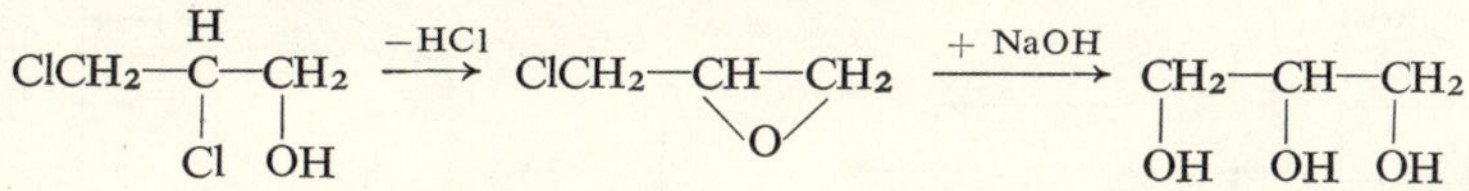

A technical difficulty was presented by the fact that it was very difficult to obtain glycerine free of chlorine which is essential for the manufacture of explosives.

Epichlorhydrin is an intermediate with a great variety of uses. It has a very mobile chlorine atom and the very reactive epoxide ring. Phenoxypropenoxide, a stabilizer for halogen alkyls, is obtained by the addition of sodium phenolate:

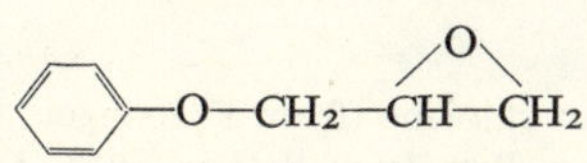

At present, the most important use for epichlorhydrin is in the manufacture of epoxy resin (see Plastics, p. 426). For this purpose it is usually reacted with 'Dian' to diandiglycidether.

By saponification with allylchloride, allyl alcohol is obtained, $H_2C{=}CH{-}CH_2OH$, which is also an important intermediate. Esters of allyl alcohol, e.g. with phthalic acid, have polymerizable vinyl groups, are bi-functional, and can be used for polymerization and cross-linking in the manufacture of plastics.

Another glycerine process (Shell), based on allyl alcohol, is less favourable than that via epichlor hydrin:

A 5 per cent aqueous solution of allyl alcohol with chlorine is converted at 15°C with circulation cooling into chlorhydrin:

$$H_2C{=}CH{-}CH_2OH + HOCl \rightarrow ClCH_2{-}CHOH{-}CH_2OH.$$

Chlorhydrin can be saponified with a mixture of sodium hydroxide and soda at 150°C in 30 minutes. The rather concentrated NaCl solution is evaporated, the salt is filtered off, the crude glycerine is distilled and extracted with xylene for purification.

### (d) Direct syntheses of allyl alcohol and glycerine from propene

It is possible to hydroxylate allyl alcohol directly to glycerine with hydrogen peroxide using a $WO_3$ catalyst, via cyclic tungstic acid esters.

Utilizing this reaction, the Shell Co. have branched off in an entirely novel direction by producing allyl alcohol (and acetone) by autoxidation of propene with oxygen, using acrolein. In a separate process hydrogen peroxide is obtained by autoxidation of isopropanol, acetone again being a by-product (see Figure 73) (B. P. 73/238, 758,967, 760,605; P. W. Sherwood, Chem. Ing. Techn., 1960, **32**, 459). No chlorine is used in this process, in contrast to those described above, which represents in a way an analogue of the ethylene oxide production by direct oxidation.

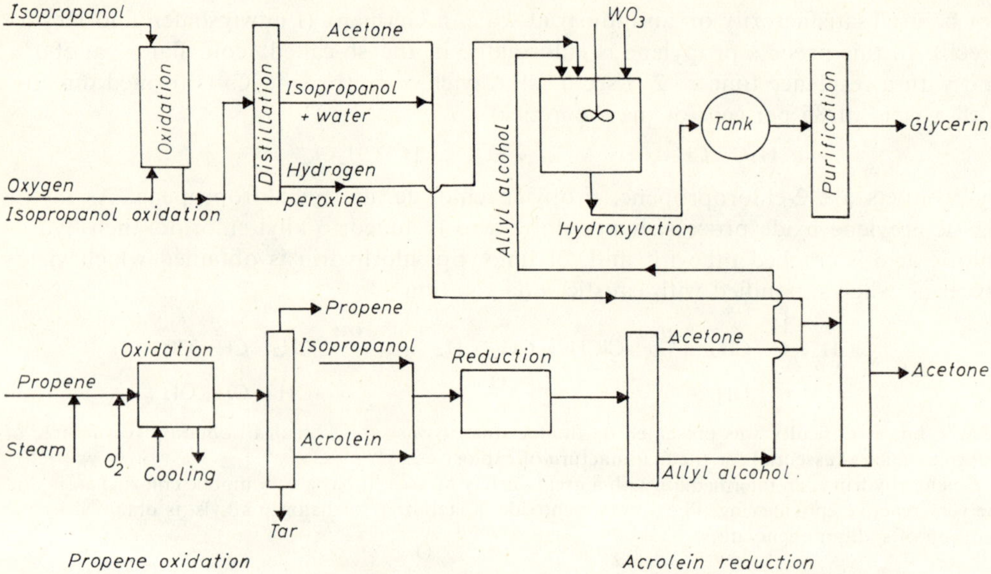


Figure 73. Glycerine synthesis without chlorine (from F. Asinger, *Einführung in die Petrolchemie*, after T. P. Forbeth and B. J. Gaffney, *Petroleum Refiner*, 1955, **34**, (12) 160, slightly modified)

(1) Allyl alcohol: At 300–400°C and at a gauge pressure of up to 10 atm propene is first oxidized to acrolein in the presence of steam on copper oxide-silicon carbide, giving a yield of 85 per cent:

$$CH_3—CH=CH_2 + O_2 \rightarrow O=\underset{H}{C}—CH=CH_2 + H_2O$$

Isopropanol is added as the $H_2$ source and allyl alcohol (77 per cent yield) is obtained on a MgO-ZnO catalyst at 400°C, with acetone as a by-product:

$$H_2C=CH—CHO + H_3C—\overset{H}{\underset{OH}{C}}—CH_3 \rightarrow H_2C=CH—CH_2OH + H_3C—CO—CH_3$$

(2) Hydrogen peroxide: At a gauge pressure of 2·5 atmospheres, 90–140°C, in the liquid phase, with the addition of $H_2O_2$ (to set off chain reaction) part of the isopropanol is oxidized with $O_2$ ($H_2O_2$ is the first reduction product of the oxygen):

$$CH_3—CHOH—CH_3 + O_2 \rightarrow H_3C—CO—CH_3 + H_2O_2$$

(3) Glycerine: An aqueous $H_2O_2$ solution (2 n) is reacted with allyl alcohol and a 0·2 per cent tungstic acid solution at 60–70°C in two hours. After evaporation and vacuum distillation glycerine is obtained, with a yield of 85 per cent, calculated on a propene basis.

When reacted with ammonia allyl chloride yields allyl amine from which acrylonitrile, an important raw material in the production of 'Orlon', can be obtained.

Very important is a direct method for the conversion of propene into acrylonitrile using ammonia and $O_2$ on a bismuth-Foster-molybdate catalyst by ammonium oxidation (Du Pont de Nemours, Sohio, Distillers Co. Ltd.) (D. J. Hadley, *Chem. & Ind.*, 1961, 238). Finally, it should be mentioned that, as the 'chemistry of ethylene oxide' was developed for ethylene, so the 'chemistry of propylene oxide' was developed for propylene, so that propylene glycol and its derivatives can be produced cheaply. With allyl alcohol allyl chloride yields diallyl ether from which acrolein is produced by pyrolysis. Thus acrolein has become easily obtainable and can be used for many syntheses (In Asinger's *Einführung in die Petrol-*

*chemie*, p. 243, there is a list of the uses of acrolein). With KCNS, allyl isothiocyanate, $H_2C{=}CH{-}CH_2{-}N{=}C{=}S$, is obtained from allyl chloride. Propene is tetramerized to duodecylbenzene, a detergent.

## 4. BUTANE AND BUTENE

Based on butane and the butenes an entirely novel chemistry of aliphatic products used on a large scale in industry was developed which was first applied industrially in the United States, then in the Soviet Union, and which is beginning to be used in other countries including Germany. Large quantities of butane are obtained in the production of natural gas, the processing of petroleum, and the hydrogenation of coal, tars, and petroleum. It is easily produced in pure form. The butane fraction must be separated by distillation into n-butane and isobutane (difference in boiling points about 10°C). By catalytic dehydrogenation a mixture of the n-butenes with isobutene and butadiene is obtained. Mixtures of butenes are also produced in the cracking processes, in various quantities and compositions according to the process used. Since the boiling points lie very close together (see table on p. 158) their separation is very difficult. Butadiene can be separated under pressure with a solution of copper oxide in ammonia. In other cases extractive distillation is used. In many industrial syntheses the butane-butene fraction can be used directly. Isobutene reacts very easily with sulphuric acid of 65 per cent concentration and can be extracted with it. From tertiary butyl sulphate either isobutene is recovered or, if desired, tertiary butanol, di-isobutylene or tri-isobutylene, the latter for interface-active materials, are produced.

First, the products from n-butylene will be considered: n-butene can be converted into butadiene. In the United States and in the Soviet Union n-butylene is dehydrogenated catalytically to butadiene on a very large scale, and recently this process has also begun to be used in West Germany.

In the United States, 'Buna' is produced almost exclusively from butadiene manufactured by this process and not, as formerly in Germany, from acetylene. More recently, the Buna Works Hüls have also based their Buna production on n-butane and n-butene since this is a more economical method than the acetylene process. The butadiene obtained is not as pure as that obtained from acetylene by the classical four-stage-method. (The methods for the production of butadiene for Buna manufacture are discussed in Chapter VII (Syntheses with acetylene) p. 214).

From butadiene one of the raw materials for the super polyamides is produced, e.g. by chlorination, a mixture of 1,2- and 1,4-dichlorobutene is obtained which can be readily separated by distillation. The undesirable 1,2-dichlorobutene can be isomerized catalytically to a mixture of the two so that in the end all the butadiene can be converted into 1,4-dichlorobutene. With KCN 1,4-dichlorobutene yields 1,4-dicyanobutene-2 and by hydrogenation adipic acid dinitrile. From adipic acid dinitrile, hexamethylenediamine is obtained by hydrogenation and by saponification of the nitrile, adipic acid. Adipic acid dinitrile is also obtained from 1,4-dichlorobutane with NaCN (or KCN) (see p. 387). Finally, there is also the possibility of converting it into ε-amino caproic acid. Thus a possibility exists to produce monomers for superpolyamides from cheap butane, and by-passing acetylene or phenol.

By polymerizing isobutylene which, as mentioned earlier, can also be obtained from isobutylalcohol by separation of water, the plastic 'Oppanol' can be produced by the addition of boron trifluoride. Isobutylene can also be dimerized, trimerized, etc. and thus all the derivatives of isobutanol be obtained which have already been discussed. The possibility of dimerizing and aromatizing isobutene to p-xylene is of great importance (Review of the production from petroleum: H. W. Haines jr., J. M. Powers and R. B. Bennett, *Chim. et Ind.*, 1956, **75**, 75).

The requirements for p-xylene for the production of terephthalic acid for Terylene continue to increase. The most important uses of butane and the butenes are shown in the diagram in Figure 74.

The oxosynthesis of olefins discussed earlier offers great possibilities for the production of valuable aliphatic alcohols. It consists in the 'addition' of carbon monoxide to the olefin under hydrogenating conditions, the product being aldehydes or alcohols. Thus, the oxosynthesis of propylene yields butanolene, that of butylene amylalcohols, and that of di- or higher polymerized butylene interesting aliphatic alcohols.

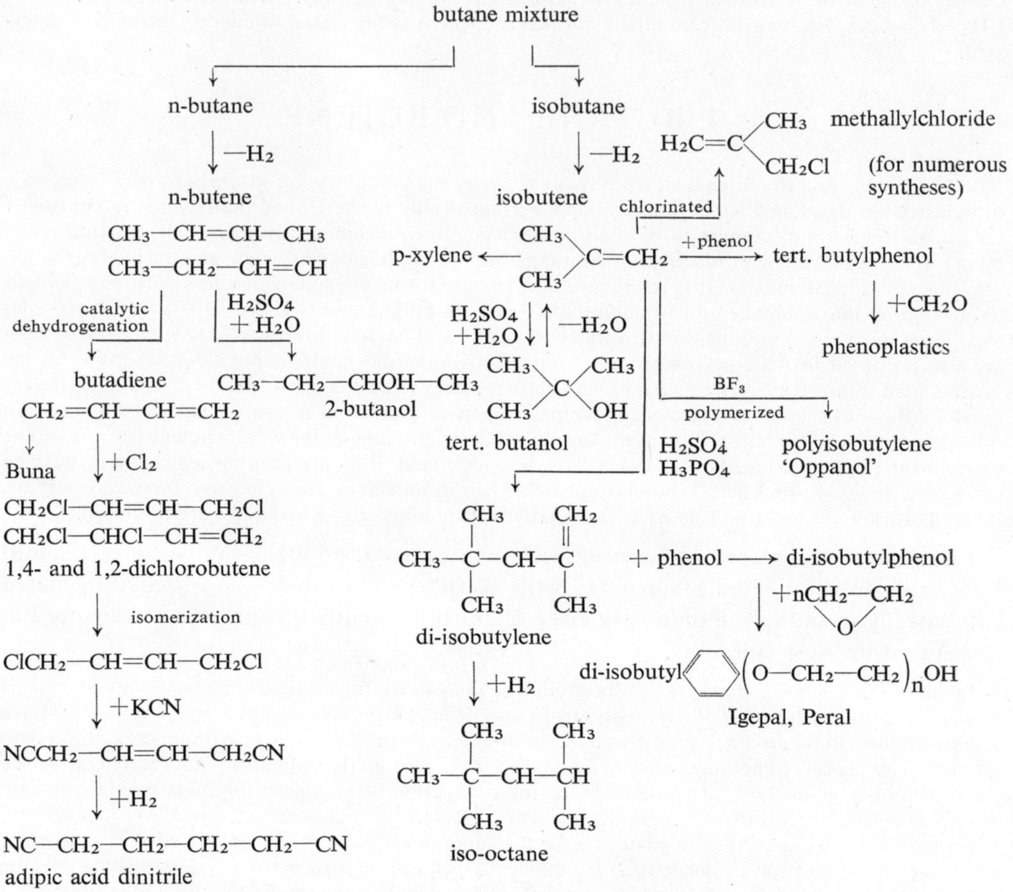


Figure 74. The use of butane and butene in industry

## 5. HIGHER MOLECULAR WEIGHT OLEFINS

Of the olefins with more than four C atoms which also occur in cracking gases, renewed interest has recently centred again on isoprene (2-methyl-1,3-butadiene)

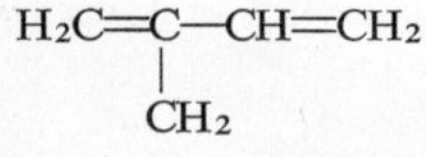

By means of Li or with 'Mülheimer polymerization catalysts' (aluminium trialky and titanium tetrachloride) it can be polymerized to a product that is at least equivalent to natural rubber. The $C_5$ fraction obtained by pyrolysis of petroleum fractions has a very complex composition. It can contain 15–25 per cent isoprene which is obtained in its pure form by extractive distillation.

Higher cracking olefins ($C_{10}$–$C_{18}$), too, are produced industrially on a large scale. They can also be obtained by polymerization of lower olefins; such olefins are highly branched. Di- and tri-isobutene have already been mentioned, similarly tripropylene (isononylene) and tetrapropylene (isododecylene) are produced on an industrial scale. They are added to benzene or phenol for the manufacture of synthetic detergents. More uniform products are obtained by catalytic dehydration of fatty alcohols.

It has recently become possible to obtain α-olefins by controlled polymerization of ethylene with triethylaluminium, using Ni as a catalyst (Ziegler) (*Angew. Chem.*, 1952, **64**, 323; *Brennstoff-Chem.*, 1954, **35**, 321). Ethylene is thus dimerized to 1-butene, propylene to 2-methylpentene-1. There is thus no displacement of the double bond.

## 6. CHLORINATION OF THE HIGHER PARAFFINS

(See also p. 160)

The chlorination products of pentane are used for the manufacture of amyl alcohols, by hydrolysing under pressure at 170–180°C with alkali, in the presence of sodium oleate as emulsifier. The amyl alcohols produced contain 60 per cent primary and 40 per cent secondary alcohols. Diamylether is obtained as a by-product. A certain amount of pentene formation is unavoidable.

Chlorination takes place statistically like autoxidation or sulpho-chlorination. This means that the chlorine attacks every C atom in the chain (F. Asinger).

Higher molecular weight halogen paraffins are of importance in the manufacture of detergents and textile auxiliary materials, lubricants, pour-point depressants, oils for dressing leather, etc. They are obtained:

(1) By condensation of chloroparaffin with aromatics (benzene, phenols, naphthalene), using $AlCl_3$.

(2) Using $AlCl_3$ or only Al and HCl, thus obtaining HCl separation and polymerization.

'Paraflow', an important pour-point depressant, is a chlorinated hard paraffin containing 14 per cent Cl, condensed with naphthalene using $AlCl_3$ at 30 to 60°C in ethylene chloride. The addition of 1 per cent reduces the pour point of lubricating oil from −1 to −20°C. Alkyl-aryl sulpho acids (from $C_{12}H_{26}$) are important detergents. Aliphatic sulpho acids are obtained directly by sulphochlorination with $Cl_2$ and $SO_2$ (see Mersol process, p. 450).

## 7. OXIDATION OF PARAFFINS

Like chlorination, the autoxidation of paraffin hydrocarbons is a radical chain reaction. In industrial processes activation is produced thermally. The first stages are in every case hydroperoxides. Once small quantities of it are formed, its decomposition produces radicals which promote the formation of chains. The decomposition into radicals of peroxides added for the purpose accelerates autoxidation. The accelerating effect on autoxidation of metallic salts consists mainly in the fact that they accelerate the decomposition of peroxides formed and so create active centres for the start of chains.

$$R^{\bullet} + O_2 \rightarrow ROO^{\bullet}$$
$$ROO^{\bullet} + RH \rightarrow ROOH + R^{\bullet}$$

F. F. Rust and W. E. Vaughan, working at Shell (*Ind. Engng. Chem.*, 1949, **41**, 2595), made the important observation that HBr greatly facilitates autoxidation ('induced oxidation'). Bromine atoms formed by $O_2$ set off bromination (as a chain reaction) which is transformed into an oxidation chain. The speed of reaction of the hydrogen atoms drops from the tertiary to the secondary to the primary so that a rising reaction temperature is required.

At first, in industrial processes (e.g. at the Celanese Corp.) operations were carried out with an air deficiency, under pressure, sometimes in the presence of steam. Propane required a temperature of 330–370°C, with a residence time of about one second. Then rapid cooling was brought about by the injection of water. One of the concurrent reactions with oxidation is the formation of olefins, preferably at higher temperatures. Later, oxidation of gaseous paraffins with oxygen (3–6 per cent) was carried out at a reaction temperature of 430°C and a gauge pressure of 7–11 atm. The quantities of oxygen-containing products increased, atmospheric nitrogen was not required and the size of the plant could, therefore, be reduced, and the isolation of the products was easier. The product of the reaction is a mixture of alcohols, aldehydes, acids, acetone, $CO_2$, and CO. Butane yields similar oxidation products.

In the Shell process, using the HBr catalyst, with paraffins containing tertiary H atoms, the hydroperoxide $R_3COOH$ or a dialkylperoxide could be obtained, $R_3C{-}OO{-}CR_3$. From hydrocarbons with secondary H atoms ketones are obtained, with hydro-peroxide as an intermediary:

$$R_2C\begin{matrix}\diagup H \\ \diagdown OOH\end{matrix} \rightarrow R_2CO + H_2O$$

from those with primary H atoms, acids:

$$R{-}CH_2{-}OOH \rightarrow R{-}CHO \xrightarrow{{}^1\!/_2 O_2} RC\begin{matrix}\diagup\!\!\diagup O \\ \diagdown OH\end{matrix}$$

It may one day be possible to utilize the oxidation of propane and ethane for the production of $H_2O_2$. At 470°C and a residence time of only a few seconds and with rapid cooling the condensate contains $H_2O_2$ (Kooijman).

A new method for the petrochemical production of hydrogen peroxide is the Shell process described above. Isopropanol is oxidized in the liquid phase at 90–140°C and a gaugepressure of 15–20 atm (Hydrogen peroxide production using petrochemical raw materials. P. W. Sherwood, *Chem. Ing. Techn.* 1960, **32**, 459). The 'non-electrolytic processes' for the production of $H_2O_2$, e.g. using 2-ethylanthrahydroquinone or hydrazobenzene with $O_2$, the reduction of the anthraquinone or the azobenzene being repeated ('cyclic process'), are being more and more widely used (O. v. Schickh, *Chem. Ing. Techn.*, 1960, **32**, 462).

The oxidation of butane in the liquid phase in a solvent, e.g. acetic acid, takes place more uniformly than the gas phase oxidation. Metallic salt catalysts (Mn, Co, Ni) are used at temperatures of 165–170°C and a gauge pressure of 60–80 atm. (Celanese Corp., Distillers Co. Ltd., Chem. Werke, Hüls) (F. Broich, Oxydationsreaktionen in der Petrolchemie, *Chem. Ing. Techn.*, 1962, **34**, 45).

The liquid phase oxidation process can be controlled more easily than the gas phase oxidation. Another liquid phase oxidation is the 'paraffin oxidation' yielding saponaceous fatty acids (see p. 445, 447).

## 8. NITROPARAFFINS

The first nitroparaffins were produced by Victor Meyer in 1874, then by Konowalov in 1893. Industrial application of the method became possible only by nitration in the gaseous phase at high temperatures (U.S.A., 1936). The gaseous phase nitration has been used since 1955 by the Commercial Solvents Corporation and recently also in England. Small quantities of nitric acid in 70–80 per cent concentration are injected at several points into the flow of pre-heated hydrocarbons (e.g. propane). The temperature is between 390 and 550°C. In order to prevent secondary reactions, very rapid cooling must be arranged. Nitric oxides, water, and carbon dioxide are removed and separation by distillation is carried out. Propane yields a mixture of 25 per cent nitromethane, 10 per cent nitroethane, 25 per cent 1-, and 40 per cent 2-nitropropane. There are two methods for the nitration of higher paraffins: At 170–180°C with superheated $HNO_3$ vapour (Grundmann), or with dinitrogen tetroxide $N_2O_4$ (Geiseler).

From nitroethane, 2-nitropropanol $HOCH_2—CH(NO_2)(CH_3)$ can be obtained with formaldehyde and diluted alkali from 1-nitropropane 2-nitrobutanol:

$$CH_3—CH_2—CH_2—NO_2 \xrightarrow{+CH_2O} CH_3—CH_2—CH(NO_2)(CH_2OH)$$

$$CH_3—CH_2—CH_2—NO_2 \xrightarrow{+H_2O} CH_3—CH_2—COOH + H_2NOH$$

These can be reduced to the corresponding amino-alcohols.

1-nitropropane can be cracked with sulphuric acid into propionic acid and hydroxylamine. A great variety of uses has been found for the nitroparaffins produced industrially (*Chem. Engng*, 1955, **62**, (10) 136, 142; *Chem. Z.* 1956, **80**, 139).

# VI. Industrial processes for the production of various aliphatic compounds

Following the 'petrochemical' production of aliphatic compounds a description is given of a whole series of 'classical' methods for the production of important aliphatic intermediates, solvents, etc. which are still of great importance.

## 1. DIMETHYL ETHER

Dimethyl ether is a gas (b.p. −23·65°C). It is transported in liquid form in steel cylinders and tank wagons. It can be produced from methanol by splitting off water with concentrated sulphuric acid or by catalytically splitting off water over an $Al_2O_3$ catalyst at 300–340°C. Adequate quantities of it are, however, produced as a by-product during the methanol synthesis.

Dimethyl ether can be used directly as a methylating agent in pressure reactions. However, it is mainly used in the production of dimethyl sulphate. The ether is combined in the liquid phase in a reaction tube with $SO_3$:

$$CH_3OCH_3 + SO_3 \rightarrow O_2S(OCH_3)_2.$$

Chlorodimethyl ether $ClCH_2—O—CH_3$ can be obtained by chlorinating dimethyl ether but more simply from methanol, formaldehyde and HCl. Chlorodimethyl ether promotes the introduction of chloromethyl groups into aromatics ('chloromethylation'), a particularly easy reaction with toluene and naphthalene:

$$CH_3OH + CH_2O + HCl \rightarrow CH_3—O—CH_2Cl + H_2O$$
$$RH + CH_3—O—CH_2Cl \rightarrow R—CH_2Cl + CH_3OH$$

By chlorinating chloromethyl ether further a symmetrical and an asymmetrical dichloromethyl ether are produced which can be separated. The latter, $Cl_2CH—O—CH_3$, formic acid ester dichloride, is a universal chlorinating and condensing agent and is suitable for syntheses, e.g. for the introduction of the aldehyde group (A. Rieche and H. Gross, *Chem. Techn.*, **10**, 515; *Chem. Ber.*, 1960, **93**, 88; H. Gross, *Chem. Techn.*, 1958, **10**, 659).

## 2. DIETHYL ETHER AND DI-ISOPROPYL ETHER

Diethyl ether is an important solvent, e.g. for the celluloid film industry, an extracting agent, and is also used as an anaesthetizing agent (1957 production U.S.A. 36,500 tons, Federal German Republic 1991 tons). In the United States and other countries that produce ethanol from ethylene the total requirements are probably covered by the by-products from this reaction. In both parts of Germany diethyl ether is produced from ethyl alcohol, by one of two possible methods: splitting off water with concentrated sulphuric acid, and splitting off water catalytically in the gaseous phase over an alum catalyst.

In the old method, alcohol is reacted with sulphuric acid in lead-lined vessels at 130–140°C. The sulphuric ether (ethyl ether)' formed is distilled off. During the reaction ethylsulphonic acid is formed.

By the modern method ether is produced by splitting off water catalytically from alcohol in a catalyst tube at 180–200°C. This temperature must not be exceeded to avoid excessive ethylene formation. A similar method of producing ethylene, but at a considerably higher temperature and over an $Al_2O_3$ catalyst, is well known. Catalysts used for the production of ether are alumina or dehydrated alum, i.e. acid catalysts.

The alcohol vapours pass from a vaporizer from below into the furnace and are condensed in a cooler. The condensate is distilled in a column in the still of which aqueous ethyl alcohol (reaction water) remains as a residue which is concentrated by distillation and used again. The ether can be purified with bisulphite and by scrubbing with a solution of caustic soda; potassium permanganate is also used. Ether to be used in anaesthetics must be of the highest degree of purity and completely free of any traces of aldehydes and peroxides. Non-denatured alcohol is also used for the process and ether plants are, therefore, under continuous strict supervision by the Excise authorities.

Di-isopropyl ether is produced in large quantities as a by-product in the production of isopropanol from propene (see p. 175).

Both di-isopropyl ether and diethyl ether are liable to autoxidation. First, ether hydroperoxides are formed which when left standing for extended periods of time, and particularly when heated, are easily transformed into multimolecular weight alkyl peroxides which are highly explosive (A. Rieche, *Angew. Chem.*, 1958, **70**, 251). These have often given rise to violent explosions. Peroxide-containing ether must, therefore, never be distilled from the residue. The peroxides can be destroyed, e.g., by treatment with a solution of ferrous sulphate and subsequently with strong alkali. Impure ether and ether which has previously been oxidized have a particularly strong tendency to peroxide formation ('oxidation affinity'). What has been said above applies also to the cyclic ethers tetrahydrofuran and dioxane.

## 3. GLYCERINE

Glycerine is a basic substance of the natural fats and oils and is obtained during their saponification. Its production by this method will, therefore, be discussed in the section on fats and oils. Glycerine is mainly used in the manufacture of explosives, but also in cosmetics, e.g. in toothpaste, in the form of its esters, as a plasticizer for plastics, and for many other purposes. In wartime there is, therefore, a great shortage in all countries and efforts have been made to produce glycerine independently of the production of fats. The syntheses from propylene have already been discussed.

Worth noting is the interesting process for the production of glycerine on a large industrial scale by the fermentation of sugar. The method was developed by Connstein and Lüdecke and was used successfully during the 1914–1918 war in the Dessau Sugar Refinery. The operative principle is the addition of sodium sulphite which intercepts the acetaldehyde formed as an intermediate during the fermentation so that the intermolecular oxide-reduction leads not to alcohol but to glycerine. A large part of the organic matter is, however, lost as $CO_2$.

The former IG Works Hoechst developed a technically most interesting process for the production of glycerine-type products by the hydrogenative cracking of sugar under high pressure. In addition to glycerine, the product contained propylene glycols, ethylene glycols, erythrite, hexite and other polyvalent alcohols and was marketed under the name 'Glycerogen'. This product, like glycerine, can be used for many other purposes, in addition to the manufacture of explosives. In contrast to the production of glycerine by fermentation the sugar is used in the hydrogenation process with little loss.

## 4. SYNTHETIC SUBSTITUTES FOR GLYCERINE

Since in many cases, particularly for plasticizers, it is not absolutely necessary to use glycerine derivatives, glycerine substitutes have been developed, e.g. by the addition of formaldehyde to propionic aldehyde or butyric aldehyde, giving trimethylolethane and trimethylolpropane:

$$CH_3\cdot CH_2\cdot CHO + 3H_2CO + H_2O \rightarrow CH_3{-}\underset{\displaystyle CH_2OH}{\overset{\displaystyle CH_2OH}{\overset{|}{\underset{|}{C}}}}{-}CH_2OH + HCOOH$$

trimethylolethane

Another glycerine substitute is hexantriol, a by-product of the 1,3-butylene glycol production for Buna:

$$\begin{matrix} CH_3{-}CHOH \\ CH_3{-}CHOH \end{matrix}\!\!>\!CH{-}CH_2OH$$

Glycol, diglycol, and triglycol are often used as glycerine substitutes; they are slightly toxic. Their monomethyl ethers are more toxic.

## 5. ALIPHATIC AMINES

In addition to the reduction of nitroparaffins discussed earlier, aliphatic amines can be obtained mainly by the alkylation of ammonia and the catalytic hydrogenation of nitriles. Since during the hydrogenation $NH_3$ is easily cracked off and a mixture of primary, secondary and tertiary amines forms, the hydrogenation is often carried out in the presence of gaseous $NH_3$ on a nickel or cobalt catalyst at 200–210°C, obtaining mainly primary amines. Nitriles can be produced easily, as is well known, from the ammonium salts of the fatty acids by catalytic dehydration.

The vapours of carboxylic acids can be conducted, together with excess circulating ammonia, at 320–400°C over catalysts such as silica gel, phosphate, and aluminium oxide, a particularly effective catalyst being boron phosphate. Thus acetonitrile is produced from acetic acid, stearic acid nitrile from stearic acid and adipic acid nitrile from adipic acid. The yield of the latter is, however, only 75 per cent. Better methods for the production of adipic acid nitrile have already been discussed, e.g. from acetylene. Amines can also be obtained directly from acids and ammonia over special catalysts which have simultaneously a dehydrating and a hydrogenating effect. The reactions can also be carried out with aldehydes instead of with carboxylic acids. Under these conditions benzaldehyde with ammonia yields benzylamine.

The simple aliphatic amines can also be produced from alcohols with ammonia over a dehydrating catalyst. Thus a mixture of methyl amine, dimethyl amine, and trimethyl amine is produced from methanol in a catalyst furnace containing a pelletized $Al_2O_3$ catalyst, at 450°C and a gauge pressure of 14–15 atm. In addition to unchanged methanol and ammonia the mixture contains 13–14 per cent $CH_3NH_2$, 7–8 per cent $(CH_3)_2NH$ and 10–11 per cent $(CH_3)_3N$ which are separated by distillation.

Of great importance are ethylene diamine and polyethylene amines, diethylene triamine, triethylene tetramine, and polyethylene imine. Ethylene diamine is obtained in autoclaves by the reaction of ethylene chloride with excess ammonia at 180°C under pressure. The quantity of monomeric amine obtained increases with the excess of ammonia. The yield of monomeric diamine is, however, in any case only 40–50 per cent, the rest are polymers of ethylene diamine:

$$ClCH_2—CH_2Cl + 4\,NH_3 \rightarrow H_2N—CH_2—CH_2—NH_2 + 2\,NH_4Cl$$

Ethylene diamine is a raw material for the important textile auxiliary ethylene diamine tetra acetic acid ('Trilon B', 'Komplexon'), obtained by means of chloroacetic acid:

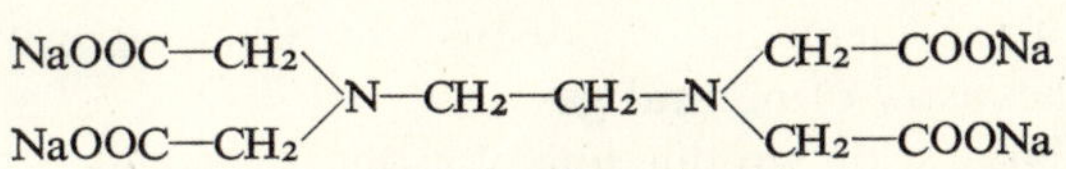

'Trilon B', 'Komplexon'

Ethylene diamine tetra acetic acid forms soluble complexes with calcium, iron, and other cations. It can be used for instance as a water softener, and wherever traces of metals have disturbing effects. Recently it has also become important for the analysis of metals and there seem to be prospects of its use in pharmaceutics (dissolution of stones in the kidney). Ethylene diamine and its polymers are also used in the manufacture of condensation products with fatty acids which are important textile auxiliaries (e.g. 'Sapamine', 'Soromine').

The higher diamines are, of course, important for polyamides. Hexamethylenediamine-(1,6) is used in the manufacture of 'Nylon'. It is also an intermediate in the production of di-isocyanate ('Desmodur H'). To produce isocyanates, amines are reacted with phosgene, e.g. in o-dichlorobenzene as a solvent. First, the polycarbaminate is produced with carbon dioxide from hexamethylenediamine at 90°C and then phosgene is introduced, initially at 0°C (formation of carbamic acid chloride); further phosgene is reacted at 150°C:

$$H_2N—(CH_2)_6—NH_2 + 2\,COCl_2 \rightarrow OCN—(CH_2)_6—NCO + 4\,HCl$$

The production of the amines of the higher fatty acids as disinfectants should also be mentioned. The first of these was 'Zephirol', made by Bayer, Elberfeld. Since then a large number of such disinfectants has been developed. They are odourless and physiologically harmless. In the field of textile auxiliaries (q.v.) higher amines are of great importance as raw materials.

The polyamines are used in the rubber industry to accelerate vulcanization. As fatty acid amides they are of great importance in the textile industry as cation-active auxiliaries.

Yet another product is the N-analogue of the ethylene oxide, the ethylene imine $CH_2—CH_2$ which is used in industry (see p. 173).
$\diagdown NH \diagup$

It can be obtained from chloroethylamino-chlorohydrate with dilute NaOH:

$$Cl—CH_2—CH_2—NH_2\ HCl + 2\,NaOH \rightarrow \begin{matrix} CH_2 \\ | \\ CH_2 \end{matrix} \!\!>\! NH + 2\,NaCl + 2\,H_2O$$

Like ethylene oxide it polymerizes easily and is very toxic.

## Retrospect and prospect of the technology of aliphatic compounds

Up to the end of the 19th century, industrial organic chemistry was concerned almost exclusively with the aromatic compounds, based on the products of the bituminous coal tar industry. The aliphatic industry on a large scale did not begin to develop till after the turn of the century. The first beginnings were the production of acetaldehyde and acetic acid from acetylene, and the production of formic acid and oxalic acid. Biological methods, such as glycerine fermentation, were also introduced under the pressure of war conditions. A new era began with the introduction of the synthesis of methanol and the higher alcohols from carbon monoxide in 1923 by the BASF. Synthetic fuels followed and formed the start of the chemistry of gaseous hydrocarbons, with its parallel in the United States in the chemistry of natural gases.

In the U.S.A. the large-scale production of isopropylalcohol from propylene was started in 1922 and from then onwards the aliphatic industry in the U.S.A. developed consistently faster than the aromatic industry so that in 1940 it produced quantitatively about four times as much. Added to this was the development in the U.S.A. of the fermentation processes on the basis of carbohydrates for the production of solvents, taken up also by the Soviet Union. The problem of the supply of solvents became increasingly urgent with the introduction of synthetic materials, plastics, and lacquers, and the production of synthetic fibres. Formerly, only natural resins and oils, natural rubber and gutta-percha, were available for the production of lacquers and resins. The necessity of producing monomeric vinyl compounds for the manufacture of plastics on a large scale led to an extraordinary growth of the aliphatic industry. For these plastics, lacquers, etc. large quantities of plasticizers, solvents and auxiliaries were also required. In addition there were the requirements of the textile industry for ever more up-to-date detergents and cleaning agents, for textile auxiliaries with specific action for the finishing, impregnation, and processing of fibres, which formerly had been manufactured only on the basis of soap and natural oils. Everything combined, therefore, to give the aliphatic industry an enormous impetus. The causes of these steadily increasing requirements of aliphatic products are to be found in the rapid development of traffic, in the displacement of the railway by the car and the plane, in the development of electrification and telecommunications, of radio and television, all demanding ever higher quality of insulating materials. Increasing mechanization in trade, agriculture, and in the household, such as preservation, storage, and transport of foodstuffs, has led to a steady increase in the demand for plastics. They are becoming increasingly important also in the construction of dwellings and in the manufacture of furniture and fittings for houses, trade premises, laboratories, and hospitals, and recently also for external wall panellings and, with foamed materials, in large structures built from prefabricated components.

This development is still in full flood. In the United States, the Soviet Union, and many other countries the production of plastics and the auxiliary products required for them increases year by year on an enormous scale. The European countries were rather backward until a few years ago but are now catching up rapidly. Another reason compelling this development is the growing shortage of non-ferrous metals. Meanwhile the German Federal Republic has become the second largest producer of plastics in the world.

Owing to their wealth in natural gases and the development of the cracking industry, the United States had at their disposal at a very early stage gaseous hydrocarbons as raw materials at very low prices. There the large-scale aliphatic industry developed, therefore, first.

In Germany gaseous hydrocarbons were available on the whole only as mixtures and had to be purified or separated, or the derivatives had to be produced in difficult syntheses based on the expensive acetylene. However, with the development of the hydrogenation processes, in connection with the discovery of important petroleum fields in Western Germany, the natural gas fields e.g. at Bentheim, and above all by the construction of petroleum refineries and cracking plants, and of pyrolytic cracking plants, great industrial possibilities opened up also in Germany. Similar plants are in course of construction also in the German Democratic Republic, where above all large quantities of hydrogenation gases are still not fully utilized for synthesis purposes. If they are to be used for such purposes it will be necessary to create substitutes where they are at present used for heating purposes, in the form of producer gas, power gas, pressure gas, etc.

The acetylene chemistry has to carry the high initial costs of electric power and in the long run it will be necessary to limit the use of acetylene to those products which cannot be produced from other sources. As shown earlier, the possibility exists in principle. In the German Democratic Republic, too, it would be possible to construct an aliphatic industry on the basis of the gases produced as by-products in hundreds of thousands of tons, e.g. the hydrogenation gases. This will, however, mean much and protracted development work. A very large expenditure of capital for the construction of new plants will then be required which will first have to be created. This can be achieved by a nation that has lost a war and has experienced the total collapse of its economy only by great sacrifices and extraordinary diligence.

## LITERATURE

O. Nicodemus, Die aliphatischen Chemikalien und Zwischenprodukte, in Winnacker-Weingaertner, *Chem. Technologie*, Organische Technologie, I, pp. 611. Hanser, München, 1952.

O. Horn and H. Bestian, in Winnacker-Küchler, *Chem. Technologie*, 2nd ed., vol. 3, p. 650–783, 1959.

Ullmann, *Encyklopädie*, 3rd ed., vols. 3 and 4. Urban & Schwarzenberg, München–Berlin, 1953.

Ost-Rassow, *Lehrbuch der chemischen Technologie*, 26th ed., J. A. Barth, Leipzig, 1955.

H. Sönsken, Die Erdölchemie und ihre Bedeutung für die Kunststofferzeugung, *Kunststoffe*, 1953, **43**, 490.

Schiemann, *Die organischen Fluorverbindungen in ihrer Bedeutung für die Technik*. Steinkopff, Darmstadt, 1951.

H. Gnamm, *Die Lösungsmittel und Weichmachungsmittel*. 6th ed., Wissenschaftliche Verlagsgesellschaft, Stuttgart, 1950.

F. Asinger, *Chemie und Technologie der Paraffinkohlenwasserstoffe*, Akademie-Verlag, Berlin, 1956.

F. Asinger, *Chemie und Technologie der Monoolefine*. Akademie-Verlag, Berlin, 1957.

F. Asinger, *Einführung in die Petrolchemie*. Akademie-Verlag, Berlin, 1959.

# VII. Acetylene production and syntheses based on acetylene

The diagram on p. 36 showing the natural raw material sources for aliphatic base materials and their production indicates the position of acetylene. Petroleum, coal, and tar form the basis for both the aliphatics, olefins, and their reaction products, and the aromatic heterocyclic base materials and phenols. The syntheses with carbon monoxide also lead to the paraffins, olefins, and their simple derivatives such as methanol and the higher alcohols. They form equally the basis for derivatives of long-chain hydrocarbons. Acetylene has become the basis for an entirely different type of chemistry. There are, of course, connecting lines; thus, ethylene for instance forms a bridge from the chemistry of petroleum, coal, and tar to acetylene chemistry. A large number of compounds can be obtained starting from either ethylene or acetylene. (The United States production of acetylene in 1958 amounted to 300,000 tons, that of ethylene in 1957 to 1·8 million tons.)

Acetylene permits the production of a large number of aliphatic compounds with two and three carbon atoms or a multiple thereof, so that syntheses based on acetylene had an extraordinarily fertilizing effect on the chemistry of the solvents, the vinyl monomers for plastics, the plasticizers, and many other fields. All these processes and reactions are based in the final analysis on the great capacity of acetylene to add other atoms and groups of atoms. Of central significance for these syntheses is the addition product of water and acetylene, acetaldehyde. The possibilities for syntheses using acetylene with aliphatic substances, but also with araliphatic, cyclic, heterocyclic, and aromatic compounds, appear almost unlimited; in spite of the extraordinarily comprehensive work of W. Reppe (Ludwigshafen), we are only at the beginning of the development, particularly in the scientific application of acetylene syntheses to the building up of many new compounds.

In this connection mention must be made of the very wide-ranging work in the field of acetylene chemistry in the research institutes of the Academy of Sciences of the U.S.S.R. (I. N. Nasarov *et al.*). A very impressive survey of the industrial possibilities of using acetylene in all fields of chemistry is afforded by the 'acetylene tree' first designed by Nicodemus, Hoechst, to whom many developments in the fields of acetylene chemistry, particularly solvents, are due ('Acetylene tree' with surveys of the many uses of acetylene see Ullmann, *Encyklopädie der techn. Chemie*, 3rd ed., vol. 3, 59–64. Urban & Schwarzenberg, München–Berlin, 1953).

The particularly great and rapid development of acetylene chemistry in Germany, like that of the hydrogenation processes and the syntheses with carbon monoxide is due to the lack of adequate aliphatic raw material resources. In other countries many of the products which in Germany are obtained from acetylene can be produced from natural gas, petroleum, cracking gases, etc. One of the most important mass products of acetylene chemistry, Buna, is also manufactured in other countries but from different raw materials, e.g. in the Soviet Union its manufacture is based directly on alcohol from agricultural products and in the United States on alcohol and natural butane.

All products derived from acetylene carry the large energy cost of about 11 kWh/m$^3$ acetylene, and although in the German Democratic Republic it is at present necessary to produce even ethylene and the glycol products from acetylene this can only be a temporary expedient. The supply of raw materials is, therefore, the reason why in no other country acetylene chemistry has the same industrial importance as in Germany, and the development of the acetylene processes, like that of the hydrogenation processes and the Fischer-Tropsch synthesis, is among the greatest pioneering achievements of German scientists and technologists. It is nevertheless quite possible that a gradual shift from acetylene as the raw material basis to other sources will take place in Germany. A new economic factor favouring the production of acetylene via carbide has arisen in recent years: carbide furnaces can be used for levelling out the supply of electricity by consuming excess off-peak capacity.

The modern modifications of the Fischer-Tropsch synthesis can be expected to yield quantities of by-products which have so far been obtained from acetylene, such as acetic acid, acetone, and acetaldehyde (already realized in a plant at Brownsville, Texas). Since the Fischer-Tropsch synthesis can be carried out using cheap gases like coke oven gas, natural gas, cracking gas for synthesis gas production, some of the aliphatic compounds produced so far from acetylene will become available from this source.

## 1. PRODUCTION OF ACETYLENE

Four processes are used for the production of acetylene on a large industrial scale (see also: H. Kehlen, Probleme und Entwicklungstendenzen der Acetylengewinnung, *Chem. Techn.*, 1961, **13**, 386).

(1) Decomposition of calcium carbide with water

$$CaC_2 + 2\,H_2O \rightarrow C_2H_2 + Ca(OH)_2 + 31 \text{ kcal.}$$

(2) Thermal conversion of methane into acetylene in the arc furnace (former IG-Ludwigshafen, Chemische Werke Hüls)

$$2\,CH_4 \rightarrow C_2H_2 + 3\,H_2 - 89{\cdot}97 \text{ kcal (heat of reaction for 25°C).}$$

(3) Incomplete combustion of hydrocarbons (Sachsse process, BASF).

(4) High-temperature cracking and pyrolysis.

### (a) Carbide and acetylene from carbide

The formation of carbide from carbon and lime is a reversible process which is endothermic and requires very high temperatures in order to fuse CaO. These high temperatures are achieved in the arc furnace and by combustion.

$$CaO + 3\,C \rightleftarrows CaC_2 + CO - 111 \text{ kcal.}$$

This reaction is reversible so that solid carbide in a CO atmosphere at a gauge pressure of 1 atmosphere below 1,630°C is reconverted into lime and carbon. In the carbide furnace the equilibrium shifts towards $CaC_2$ until a $CaO : CaC_2$ ratio is achieved which at a certain furnace temperature corresponds to the CO partial pressure.

The melting-point diagram in Figure 75, based on measurements carried out by Pinkas (from Wildhagen, Calciumcarbid und die Carbidindustrie, Akademische Verlagsgesellschaft, Leipzig, 1953) shows that the eutectic mixture of 68 per cent $CaC_2$ and 32 per cent CaO has a definite melting point of 1,800°C. The softening temperatures of mixtures of different compositions are higher. Pure carbide melts at

2,300 °C, pure CaO at 2,500°C. When using freshly calcined lime and low-ash coke in the carbide furnace, the product contains 80–85 per cent carbide and melting must, therefore, be carried out at more than 2,000°C. If residual lime (from the carbide gasification) and lower-quality coke are used, temperatures of 1,900–2,100°C (see Figure 75) give a product containing 75–80 per cent carbide.

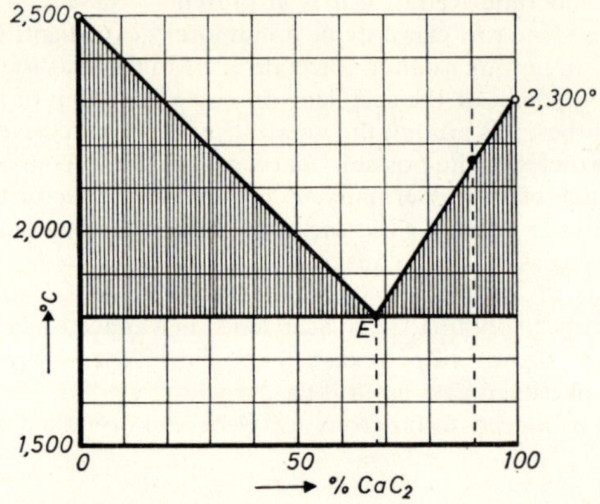


Figure 75. Melting-point diagram for $CaC_2$ and CaO, after Pinkas

On account of the large current consumption carbide plants have been built either in brown coal fields, e.g. Piesteritz, Schkopau, and Knapsack near Cologne, or where cheap water power is available, e.g. in Bavaria and in South Tyrol.

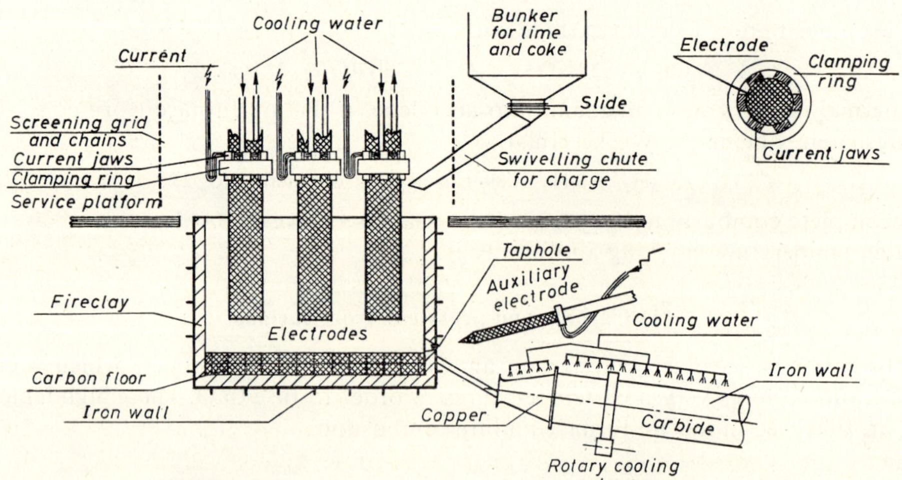


Figure 76. Diagrammatic sketch of a carbide furnace

A furnace for the production of carbide (20,000 to 40,000 kW) consists of an iron box with a volume of 120–130 m³ (see Figure 76). The floor has a fireclay lining of 1 m thickness on which is placed a double layer of carbon blocks of 1 m sides and 30 cm thickness. The outer walls have a thinner fireclay lining. Between the iron walls and the fireclay lining there is an insulating layer of sand or slag. Current is supplied through copper jaws which are pressed by an annular clamp with spring pressure against the suspended carbon electrodes. The furnace has three electrodes because three-phase a.c. is used. The electrodes, with a diameter of 1–1·5 m, are pressed from ground electrode coke, a by-product of tar processing (see p. 53), with pitch as a binder and are pre-burnt at Siemens-Plania. Only the lowest part of the carbon is converted into electrically conductive graphite. Since the electrodes are being continuously consumed they are lowered in their movable mountings several centimetres every few hours.

When the electrodes are consumed, the remainder of the assembly with clamping and current supply are removed. A new electrode assembly is ready, the current supply is reconnected and the new electrodes are lowered into position. This operation takes only about 20–30 minutes; it does, of course, entail an interruption of the melting process. The exchange of the electrodes is particularly difficult when the carbon monoxide formed in the process is to be extracted. For this purpose the upper part of the furnace contains water-cooled, fireclay-lined gas collecting funnels which are liable to be damaged during the electrode exchange.

To permit continuous operation the 'Söderberg electrode' was developed which can be added continuously from above the furnace. It consists of an iron jacket of special shape (see sketch) into which is fed the hot Söderberg compound which has been premixed at 90°C. The compound consists of a mixture of graphitized anthracite, coke (1 : 1), tar, and pitch.

When a sufficiently large part of the electrode and iron jacket has been burnt away, an extension of the iron jacket of about 1 m height is welded to the upper end and filled with electrode compound. In order to obtain a maximum contact surface between iron wall and carbon compound metal fins are welded inside (see sketch). Originally, the Söderberg electrode was even made up of individual circular segments. Since the electrodes are very heavy the clamping chucks must be very strong and when the assembly is lowered they have to be loosened and tightened each time. It is not easy to lower the electrodes uniformly, but the Söderberg Company has developed a method for doing this safely and evenly. A long iron strip (Wisdom Strip) is welded to both sides of the iron jacket and is held by a shoe brake. When the electrode is to be lowered the shoe brake is released and the electrode slides slowly down through the spring-clamped current jaws. Compressed air can be used on the current jaws to facilitate the sliding of the electrode. The lowering takes place at full current load. If the electrode should 'stick' a 10 ton weight can be placed on top of it by means of a strong overhead crane which travels the whole length of the furnace battery. This crane is also used to fit and weld the box extensions for the electrodes and to lift the Söderberg compound in iron buckets for the re-fill.

The operation of the furnace is controlled and regulated from a central control panel.

In principle, the electrodes are arranged in one of two ways by which the shape of the furnace is determined: the triangular furnace with three round electrodes arranged in a triangle, and the square furnace with three plate electrodes in a row (see Figure 77). Formerly the triangular furnace was considered more economical because of the uniform distribution of the current passage. The design and operation of the square furnace have, however, been so much improved that it is now considered the equal of the triangular one. The current is supplied through large water-cooled copper sockets on the contact surfaces of which the current density is at least 4–7 A/cm$^2$.

All the structural elements inside the furnace are water-cooled. The current is supplied through copper tubes in which water flows. The cooling tubes of the current-carrying parts are connected to the supply by rubber tubes, for insulating purposes. The cooling system of a carbide furnace is extremely complicated and requires constant control. The control valves and drains for the control and regulation are arranged in a row on one side of the furnace to facilitate inspection. Minor leakages are not dangerous. When larger quantities of water penetrate into the furnace near the melting hearth gas eruptions with powerful jets of flame may occur. The hydrogen content of the furnace gas (CO) is, therefore, checked continuously. It has proved possible to prevent furnace eruptions by strict control of the $H_2$-content.

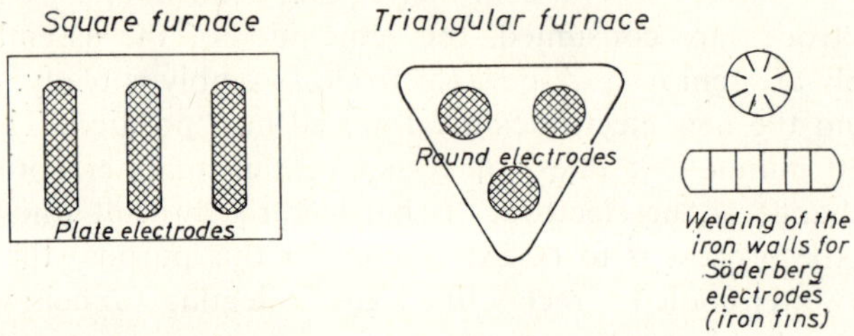


Figure 77. Arrangement of the electrodes in the carbide furnace

The furnace is started by placing the three electrodes on top of iron drums filled with coke. Up to mid-height the drums stand in a coke bed which provides the electrical connection between the electrodes. The mixture of 100 parts of lime and 65 parts of walnut size coke is charged only above this level. The new Söderberg electrode is formed in the furnace in 24 hours. First, an electric arc is struck. The mixture melts and more lime and coke in the above-mentioned proportions can now be charged. The supply comes from storage bunkers equipped with elevators through a manually swivelled chute which permits uniform distribution over the entire surface. The electrodes are then lifted and the heating by electric arc turns into resistance heating by molten carbide. Sometimes another arc forms at various points during operation. This is highly undesirable since a local rise in temperature causes dissociation of the carbide already formed. Clogging and gas eruptions follow. The service crew intervenes with long scrapers, protected from the radiant heat and spattering glowing particles by chain curtains and perforated metal sheets, and redistributes the material in the furnace.

At Knapsack, a completely covered furnace (40,000 kW) with cooled copper lid started operating a few years ago. The supply of powdered lime and coke is fully automatic.

About once every hour a taphole is melted into the bottom of the furnace wall by means of a swivelling auxiliary electrode. This permits the periodic discharge of the viscous carbide melt which solidifies when cooled by contact with the air. Carbide has a crystalline structure and its fresh fracture surface is reddish. On the outside it is black due to the exudation of carbon which takes place during the formation at the surface of calcium cyanamide: $CaC_2 + N_2 = CaCN_2 + C$.

Formerly the carbide was left to solidify in pans. After many vain attempts the continuous solidification of the molten carbide in rotary drums over the outside of which trickles water was successfully accomplished.

A patent for the first rotary tapping drum was taken out in 1918 by Karl Burkart. In 1923 F. W. Arnet was granted in Knapsack an additional patent under which a reciprocating or pendulum-type movement of the drum was protected. The first tests in the Bayerische Stickstoffwerke were unsuccessful since no material capable of resisting for long periods the high temperatures and material stresses could be found.

It was finally realized at Piesteritz that copper was a suitable material for the feed head of the drum which could withstand the enormous strain for a year or more although the temperature of the carbide melt is almost 1,000°C above the melting point of copper. To the copper head is connected a long iron drum placed obliquely on rollers which rotates fairly fast and is sprayed on the outside with water. The solidified carbide passes into crusher, grinders, and silos, and is fed from there into the gasification apparatus.

The carbon monoxide formed during the process is exhausted and can be used, after dust removal in Theisen scrubbers, e.g. as fuel gas (calorific value 2,000–2,200 kcal/$Nm^3$). It contains only about 65 per cent CO and 10–15 per cent hydrogen and nitrogen. The hydrogen is derived directly from the coke or anthracite, and partly from the water content of the charge (water gas reaction). When anthracite is used the gas also contains some methane. The CO concentration is not always high enough for chemical syntheses, but it is adequate for the Kölbel-Rheinpreussen liquid-phase synthesis (see p. 145).

Gas is exhausted from all parts of the operating carbide furnace; it contains a very large quantity of dust and must, therefore, be conducted away through very high stacks. Finally, it should be mentioned that it is very difficult to extinguish a carbide fire. Only hydrogen can be used for this since carbide reacts with oxygen, nitrogen, and carbon dioxide and particularly easily with steam.

### *Oxy-thermal carbide process*

Recent experiments (C. Wurster, *Chem. Ing. Techn.* 1956, **28**, 1) at the BASF and in the Soviet Union have shown that the temperatures necessary for carbide formation can be achieved also by combustion of coke with oxygen and that, with the addition of lime, carbide can be produced economically. With this process the carbon consumption is, of course, correspondingly higher. On the other hand, large quantities of carbon monoxide are produced as a by-product which must be used industrially. The reaction furnace is a combination of a carbon monoxide generator and melt bed with a carbide furnace. The carbide produced is not of such high quality as from an electric furnace because it contains ash from the coke and impurities from the lime. The coke should, therefore, contain as little ash as possible. Large experimental plants exist at the BASF, Ludwigshafen, and in the U.S.S.R.

The following comparative figures per 1,000 kg carbide show the consumption of current and coke for the two methods (*Chem. Ing. Techn.* 1956, **28**, 759):

| | Electro-thermal process | oxy-thermal process |
|---|---|---|
| kWh | 2,940 | 600 |
| coke kg | 550 | 2,000 |
| $Nm^3$ gas | 350–380 | 2,300 |

### *Acetylene from carbide*

$$CaC_2 + 2\ H_2O \rightarrow C_2H_2 + Ca(OH)_2 + 31\ \text{kcal/mole}$$

As this formula shows the decomposition of carbide with water liberates a large quantity of heat. Account must be taken of this in the gasification equipment for the production of acetylene. Fundamentally there are three processes in general use for the decomposition of calcium carbide with water:

(1) carbide-to-water system,
(2) displacement system,
(3) dry gasification.

Only lump carbide of 25–80 mm particle size can be used for the carbide-to-water system (wet gasification). Carbide is thrown into large quantities of water which dissipates the heat. Fine-particle carbide must be charged continuously into a gasifier fitted with an agitator. This is a great disadvantage of wet gasification. The process is still used in the Dissous gas works (manufacture of welding acetylene in cylinders) because the gas produced is very pure and maintenance is simple. In order to produce a sufficiently thin milk of lime a tenfold quantity of water must be used in wet gasification. This thin milk of lime is pumped into large settling ponds where the mass gradually solidifies; it is cut and can either be used in building or dumped. There is, however, always the risk of a sudden liquefaction of parts of the dump.

Smaller plants use the displacement system which is based on the principle of the Kipps apparatus in which the water is also dammed up by turning off the gas stream. The generation of acetylene can thus be interrupted at will. The principle is used in the design of carbide lamps and acetylene welding appliances.

The large scale production of acetylene from carbide is carried out exclusively on the principle of dry gasification. The quantity of water used is kept down to a minimum. Only the amount required for the reaction and for the dissipation of the heat by eva-

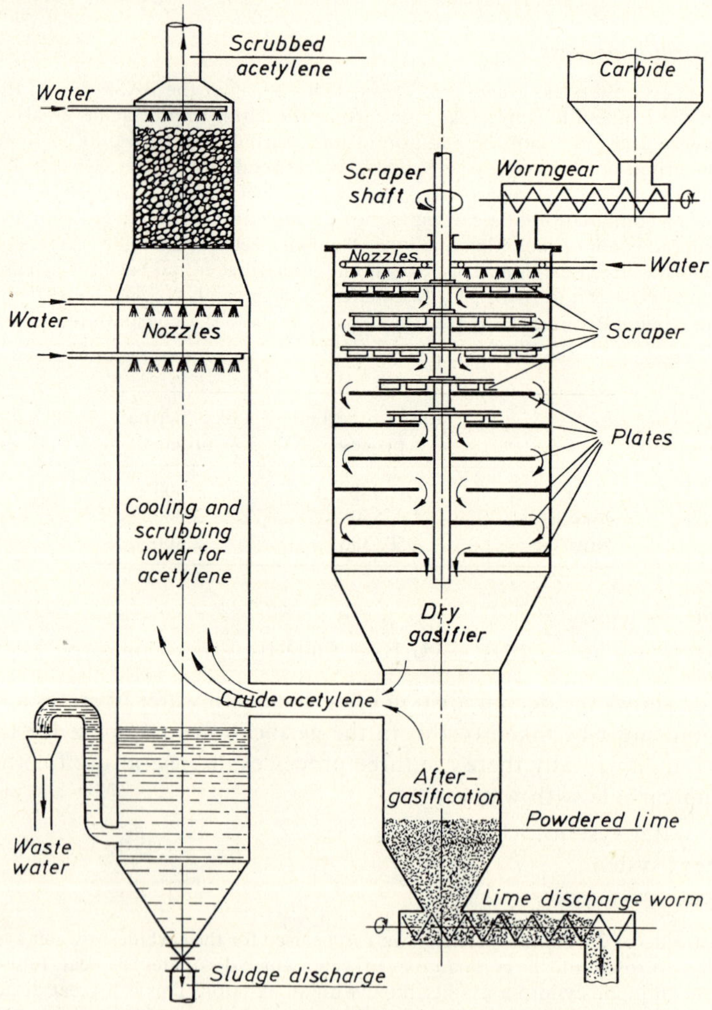


Figure 78. Dry gasifier (Knapsack)

poration of the water is added. The gas obtained consists of 80 per cent steam and 20 per cent acetylene; a by-product is calcium hydroxide with 7–8 per cent free water for which, in the form of a finely-dispersed solid, 'slaked lime', there exists a variety of uses (Buna lime). At too high gasification temperatures, particularly during gasification of carbide with steam, yellow to dark brown, cuprene-type polymerizates and hydrogen are formed.

In order to ensure complete gasification very thorough mixing is essential. Special equipment has been designed for this purpose, the so-called dry gasifiers which are either similar in design to the well-known pyrites roaster, or work on the principle of the rotary tube furnace.

A rotary tubular gasifier consists of a revolving iron cylinder, inclined downwards, into which granular carbide is fed and water is injected. If sufficient water is present the temperature can never exceed 100°C.

Figure 78 illustrates the principle of the 'Knapsack dry gasifier', designed by the AG für Stickstoffdünger.

The carbide is fed into the gasifier from a silo equipped with a charging valve. The gasifier has 20 round plates on which scrapers attached to revolving agitators perform a circular movement in such a way that the material is alternately pushed from the periphery towards the centre on one plate and back again on the one below. The water required is sprayed on to the top plate through atomizer nozzles. Owing to the great velocity of the agitators gasification takes place on all plates. A small residue of carbide is gasified during its residence in the lime discharge area where the powdered lime which is free of carbide is discharged by a wormgear. The gas is ready for use after it has passed through a scrubbing and cooling tower equipped with water spray nozzles and in which the steam that has been carried forwards is condensed.

The purification of the acetylene (removal of hydrogen phosphide and organic sulphur compounds) is carried out in scrubbing towers with chlorine water (1 g Cl/litre) or with a NaOCl solution, pH 8, which have to be measured out very carefully since for example small quantities of halogen-containing hydrocarbons disturb the catalytic conversion of acetylene into acetaldehyde. Smaller quantities of gas, e.g. from portable generators, can be purified easily with chromic acid.

A difficult problem is the utilization of the large quantities of powdered lime (tens of thousands of tons per month in a large acetylene-producing plant). The conversion into CaO by calcining and returning it to the carbide furnace is more expensive than using fresh limestone. Moreover, there is the silicate and aluminate content which comes mainly from the coke ash. Only 25 per cent of the lime can, therefore, be re-used. For this purpose the lime must be regenerated at such a high temperature that it sinters ('sinter lime'). It would be desirable if all the lime that is not used in the building industry could be used as a fertilizer in agriculture. However, there remain difficult problems of explanation and organization, transport and packing to be solved.

### *Properties of acetylene*

Acetylene has a density of 0·906 (air = 1), b.p. −83·8°C. It liquefies at 0°C and a gauge-pressure of 21·5 atm. At higher pressures it tends to decompose into its elements: $C_2H_2 = 2\,C + H_2 + 60$ kcal/mole. It is difficult to carry out reactions under pressure because spontaneous decomposition occurs very easily giving rise to explosions. When mixed with inert gases, e.g. $N_2$, or dissolved in acetone, pressures up to 15 atm can be applied so that steel cylinders with an acetylene solution in acetone, absorbed in a porous material such as pumice concrete, can be handled quite safely.

### (b) Production of acetylene by the electric arc process

The production of acetylene from gases is an endothermic pyrolysis process the heat being supplied by an electric arc. The higher the C number the less energy is consumed. When methane is used 11–22 kWh per kg of acetylene are required, with hydrogenation gases 8–9 kWh. The by-product is always ethylene.

Methane is the most stable hydrocarbon during pyrolysis. Decomposition only begins at 700°C. At about 1,400°C cracking off of $H_2$ produces the radicals $—CH_3$, $>CH_2$, $\geqslant CH$ which dimerize. Acetylene-forming methine radicals increase with increase in temperature; the acetylene must be rapidly cooled to $<150°C$.

The reaction $2\ CH_4 \rightarrow C_2H_2 + 3\ H_2 - 95{\cdot}5$ kcal takes place satisfactorily only above 1,400°C. However, at these temperatures the unstable acetylene decomposes in accordance with the equation given above, and carbon black is formed. In order to obtain

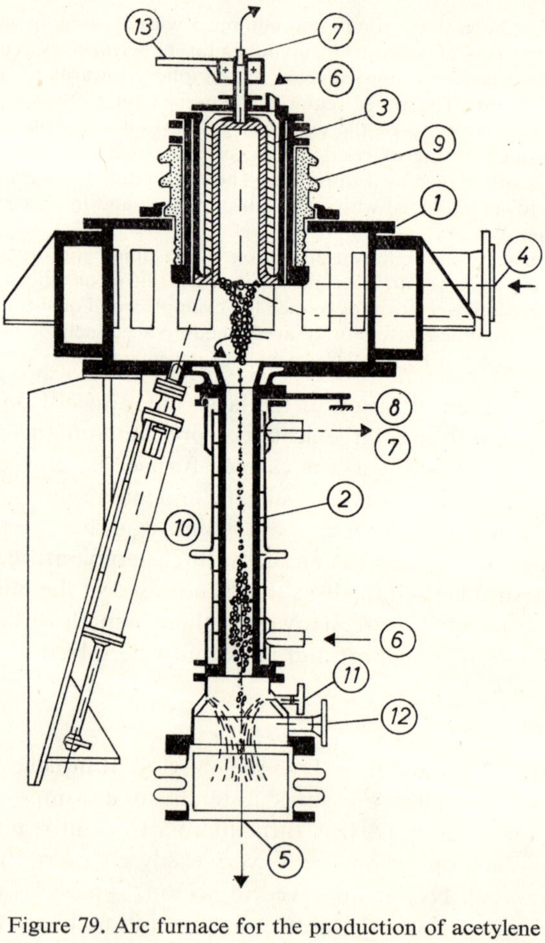


Figure 79. Arc furnace for the production of acetylene
*6, 7* Cooling water *8* Earth *9* Insulator *10* Ignition rod *12* $N_2$ *13* High tension

undecomposed acetylene from the arc, rapid heating to 1,400°C must be carried out, followed by cooling as rapidly as possible, 'quenching'.

Methane is cracked in a d.c. arc furnace. The Chemische Werke Hüls were the first large industrial undertaking to produce acetylene from coke oven gas. The hydrogenation gases from Scholven were formerly available at Hüls which because of their high C-content were particularly suitable for the production of acetylene. During the treatment of the gases in the arc furnace acetylene is formed, in addition to ethylene and hydrogen, some diacetylene and hydrocyanic acid. After separation of the gases pure hydrogen remains which is used for hydrogenation purposes. Acetylene becomes thus in a way a 'by-product' of the hydrogen production. Formerly, 40,000 tons of butadiene were produced annually at Hüls from the Scholven waste gases. The production of Buna at Hüls which has recently been re-started is based partly on the natural gas sources of Bentheim, and partly on butane-butene.

Figure 79 (after Nicodemus in Winnacker-Weingaertner, *Chem. Technologie*) shows the design of a 7,000 kW arc furnace for the production of acetylene.

The velocity of the gas in the arc furnace is very high to prevent dissociation of the acetylene formed. It is about 1,000 m/s. The feed gas enters tangentially into the twisting guide (1) and passes spinning through the flame tube (2) of 2 m length. The arc burns downwards from the insulated copper electrode (3) (negative pole). The other parts of the furnace are made of iron and are earthed. When the gas has passed through the arc the equilibrium between feed gas and acetylene is 'frozen' by injection of water at (11) which causes sudden cooling to 150°C. The steam generated escapes with the gases, at (5).

About 50–60 per cent of the energy is utilized and about 50 per cent of the hydrocarbons are reacted. The acetylene is separated from the gases by scrubbing with pressure water. At a gauge pressure of 18 atm acetylene dissolves fairly well in water whilst hydrogen, ethylene and ethane remain undissolved. When the water is depressurized acetylene of 98 per cent purity is obtained directly. The principle of scrubbing by water under pressure, used in the scrubbing of carbon dioxide and hydrogen sulphide from gases, is discussed on p. 70. Here an apparatus of similar design is used.

The separation of carbon black is carried out in cyclones or cloth filters, that of tarry and polymer products by oil scrubbing, at a gauge pressure of 18 atm, leaving $H_2$ and low hydrocarbons undissolved, that of HCN by water, and that of $H_2S$ by iron oxide.

### (c) Acetylene produced by incomplete combustion of hydrocarbons

The first process of this type was developed by Sachsse in the former IG plant at Oppau. It is based on the autothermal pyrolysis of methane to acetylene. The heat required for the pyrolysis is generated by the combustion of part of the gas with pure oxygen.

This is technologically the simplest process for the production of acetylene. It is used in the United States with natural gas by the Monsanto Chemical Co., the Carbide and Carbon Chemical Co., and the American Cyanamid Co., and in Northern Italy. It is intended to produce 1·3 million tons of acetylene in the United States by 1975, of which 75 per cent will be from natural gas by the Sachsse process.

Figure 80 shows the principle of the process: Oxygen and methane are pre-heated separately to 500°C and burnt in a special burner in the molar ratio of 1:0·65.Two thirds of the methane are burnt and one third is converted into acetylene. The gases must be ideally mixed and after burning they must be rapidly quenched with water in order to prevent dissociation and formation of carbon black. A temperature of 1,500–1,600°C is achieved, with a contact time of 1/1,000–1/100 s. After cooling the mixture is compressed to a gauge pressure of 10·5 atm, higher hydrocarbons are removed by scrubbing under pressure with oil, and the acetylene is absorbed in dimethylformamide as a selectice solvent. Then follows distillation under pressure of the dimethylformamide solution ('stabilization'), and finally extraction of the acetylene.

The 'burnt' fraction of the gas yields mainly hydrogen and carbon monoxide (suitable for conversion or as synthesis gas). The process is particularly suitable in combination with the production of ammonia since the nitrogen required becomes available as a by-product of the extraction of $O_2$ from the air.

The yield from 1,000 kg methane is 230 kg acetylene and 1,160 kg synthesis gas. For 1 ton acetylene: 4 tons $NH_3$.

The energy yield of the acetylene processes is

| | |
|---|---|
| via carbide | 50 per cent |
| electric arc | 66 per cent |
| Sachsse process | 77 per cent |

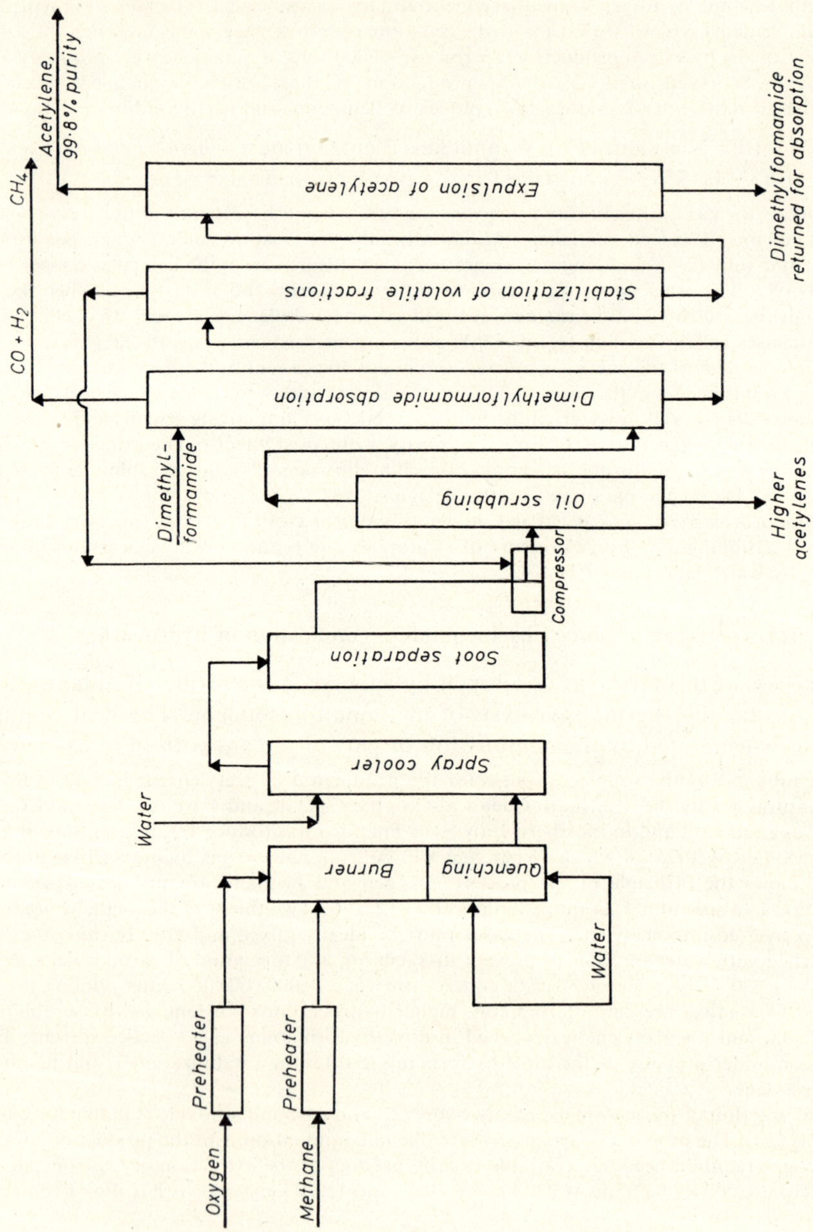


Figure 80. Diagram of the production of acetylene by the Sachsse-BASF process (from F. Asinger, *Einführung in die Petrolchemie*)

### (d) Acetylene produced by pyrolysis

Recently acetylene has been produced frequently together with ethylene, e.g. in the Hoechst pyrolysis process (see also p. 167). As in the Sachsse process, the heat for the endothermic pyrolytic process is generated by the combustion of gases with $O_2$. For this purpose, instead of sacrificing some of the raw material, the waste gases of the process are burnt. Lower and medium petroleum fractions and gases with more than 2 C atoms can be used as raw material.

Pyrolysis residual gas and oxygen are burnt in a water-cooled burner (Figure 81), (From Winnacker-Küchler, *Chem. Techn.* 2nd ed., vol 3, p. 665) and, e.g., light petrol is injected into the very hot flame. Then follows quenching with water. The acetylene/ethylene ratio is influenced by the conditions. Pyrolysis gas contains about 30 per cent by volume acetylene + ethylene. Further treatment is similar to that of the Sachsse process.

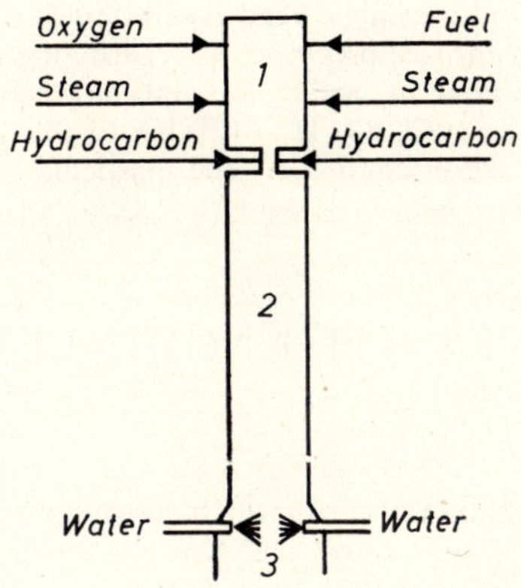


Figure 81. Hoechst pyrolysis burner

The Eastman pyrolysis process is based on a similar principle. Natural gas or gas petrols are pre-heated to 600°C and burnt, together with pre-heated oxygen and air, in a furnace chamber.

The flame temperature is 2,000°C. The material to be pyrolysed is also pre-heated with steam to 600°C and combined with the combustion gases in a mixing zone.

*Acetylene by the pyrolysis process using the regenerative principle*

The heat for the cracking of natural gas can also be supplied on the regenerative principle using checker brick work heat storers (Wulff process). First, the furnace is heated in a reversing process by gas, e.g. waste gas of the process itself, and then methane is introduced, producing acetylene formation. Each cycle lasts 60 sec, the contact time of the gases is 1/10 sec. Quenching is done by water and treatment is similar to that of the Sachsse process.

*Prospect*

The peak of acetylene production in Germany was reached in 1943–44 when all the Buna plants were in operation. 440 million $m^3$ were produced; 86 per cent of this quantity was produced by the carbide process and 14 per cent by the arc process, or by incomplete combustion of methane.

In 1959, the Federal German Republic (BRD) produced 341,000 tons of acetylene, 76·5 per cent from carbide, 23·5 per cent from petrochemicals; the U.S.A., 379,000 tons, 71·5 per cent from carbide, 28·5 per cent from petrochemicals. In 1956, the German Democratic Republic (GDR) produced 190,000 tons, of 85 per cent were used for Buna and plastics manufacture. In 1960, carbide production in the GDR was 923,000 tons, the equivalent of 316,000 tons acetylene; BDR production was 1,101,000 tons. Most acetylene is, therefore, still produced everywhere from carbide.

As a result of the increase in production at the VEB Chemische Werke Buna,Schkopau, production of acetylene from carbide has also considerably increased.

It is true that in recent decades newly developed processes for the production of acetylene have undermined the monopoly of carbide; nevertheless, in spite of all its disadvantages (large consumption of limestone, by-product of slaked lime or dry lime) it will maintain its position in countries lacking petroleum but possessing reserves of energy in the form of coal or hydraulic power. The purification and concentration of the crude acetylene produced by all the other acetylene processes also require large investments in capital equipment. Carbide production has the advantage of being elastic and able to utilize off-peak current thus permitting a more evenly spread generation of current and cheaper supply. (This high elasticity of the carbide and arc furnace processes as regards the consumption of current is of great importance for the Chemische Werke Buna, Schkopau, and the Chemische Werke, Hüls. See in this connection K. Schoenemann, Chem. Z., 1959, **83**, 437.) It is hardly likely that the new methods will bring about an appreciable reduction in the price of acetylene, particularly in countries that have to import petroleum or petroleum products.

## 2. SYNTHESES WITH ACETYLENE

*General rules for working with acetylene*

Certain experience has been gained in working with acetylene (W. Reppe, Ludwigshafen).

In view of the ease with which auto-decomposition of acetylene takes place, operations with undiluted acetylene should never be carried out at pressures exceeding a gauge pressure of 2 atm. Large spaces should be filled with packing material which inhibits the chain reaction of the dissociation by the 'wall' effect.

At pressures exceeding a gauge pressure of 2 atm, the concentration should not exceed 70 per cent, the rest should be an inert gas, e.g. nitrogen.

Since an acetylene explosion causes the pressure to rise 12 to 13-fold, vessels are designed to withstand pressure increases to 12–13 times the operating pressure. Oxygen must be excluded by purging with nitrogen which is completely free of oxygen to prevent the formation of peroxides. The acetylides of, for instance, Cu, used as catalysts, are very sensitive to impact and heat in their dry state. These catalysts must, therefore, never be allowed to dry. The apparatus should be equipped with rupture diaphragms and if possible be installed in the open air. It must be operated from separate protected service platforms.

### (a) Additions to acetylene

Within the framework of the present book it is impossible to discuss the large number of products which are produced from acetylene and the processes of their manufacture. Only a few of the industrially most important products will be selected, representing only a small part of the 'acetylene tree'. The following table shows the most important additions to acetylene:

| Acetylene + | Product | Catalyst | Temperature °C | Pressure |
|---|---|---|---|---|
| HCl | Vinyl chloride $H_2C{=}CHCl$ | 10 per cent $HgCl_2$ on active carbon | 120–150 | Atmospheric |
| $H_2O$ | Acetaldehyde $H_3C{-}CHO$ | 15 per cent $H_2SO_4$ with $HgSO_4$, $Fe^{\cdot\cdot}$ + $Fe^{\cdot\cdot\cdot}$ sulphate | 98 | Atmospheric |
| $CH_2O$ (ethylination) | $CH{\equiv}C{-}CH_2OH$ propargyl alcohol | Cu acetylide | 90–100 | Gauge pressure 3–4 atm |
| $CH_3{-}COOH$ (vinylation) | Vinyl acetate $CH_3{-}COO{-}CH{=}CH_2$ | Active carbon with zinc acetate (in gas phase) | 180–220 | Atmospheric |
| $C_2H_5OH$ (ROH) | Vinyl ether $C_2H_5OCH{=}CH_2$ | $C_2H_5ONa$ (RONa) | 150–180 | Gauge pressure 3–15 atm |
| HCN | Acrylonitrile $CH_2{=}CH{-}CN$ | CuCl, $NH_4Cl$, pH 3·5 (in liquid phase) | 80 | Atmospheric |

The manufacture of the most important products containing chlorine which are derived from acetylene is shown in the following table:

| Name | Chem. formula | From | Reagent | Remarks |
|---|---|---|---|---|
| s-dichloro-ethylene | $ClCH{=}HCCl$ | Acetylene + $Cl_2$ deficiency | 40°C active carbon | b.p. 48–60°C (cis-transisomers) |
| s-tetra-chloroethane | $Cl_2CH{-}CHCl_2$ | Acetylene + 2 $Cl_2$ added to tetrachloro-ethane at 80°C | $SbCl_5$ or $FeCl_3$ | b.p. 146°C. Corrosive effect. Used for $AlCl_3$ syntheses as solvent. With strong alkali produces explosive dichloro-acetylene |
| Trichloro-ethylene 'Tri' | $ClCH{=}CCl_2$ | Tetrachloroethane − HCl | With $Ca(OH)_2$ 50°, or pyrolytically $BaCl_2$, $CuCl_2$, carbon 220–330°C | b.p. 85°C (important solvent, stable, non-toxic) |
| Tetrachloro-ethylene 'per-chloroethylene' | $Cl_2C{=}CCl_2$ | Pentachloroethane (trichloroethylene + $Cl_2$) | With $Ca(OH)_2$ | b.p. 121°C (important solvent, very stable) |
| Hexachloro-ethane 'perchloro-ethane' | $Cl_3C{-}CCl_3$ | Tetrachloroethylene + $Cl_2$, 100–140°C | $FeCl_3$ | Sublimates at 180°C, melting point 187°, (camphor substitute) for fluorine compounds |

| Name | Chem. formula | From | Reagent | Remarks |
|---|---|---|---|---|
| Vinyl chloride | $CH_2{=}CHCl$ | Acetylene + HCl or 1,2-dichloroethane −HCl | 10 per cent $HgCl_2$ on activated carbon at 120–150°C | b.p. 13·9°C (monomers for plastics) |
| 1,1-dichloroethylene (vinylidenechloride) | $H_2C{=}CCl_2$ | Vinylchloride + $Cl_2$→ 1,1,2-trichloroethane HCl | 20 per cent $Ca(OH)_2$ 50°C | b.p. 32°C monomers for plastics ('Saran') |

Those chlorine-containing derivatives of acetylene that are stable and are not corrosive are very important solvents for industry. Thus, the s-dichloroethylene, the boiling point of which is similar to that of methylene chloride $CH_2Cl_2$, is used as a solvent for rubber and as an extracting agent. It is inflammable but the flame can be extinguished very easily. Tetrachloroethane cannot be used as a solvent for laquers, etc. because of its corrosive properties and because it causes atrophy of the liver. It is, however, very suitable as a solvent for aluminium chloride syntheses since it does not react with anhydrous aluminium-chloride. Trichloroethylene is not combustible and does not react even with diluted alkalis. When it is pure it has no corrosive effects. Physiologically it is comparatively harmless, its heat of evaporation is low, and it is, therefore, widely used as a solvent, purification and extracting agent. Perchloroethylene (or tetrachloroethylene) is obtained from trichloroethylene by the addition of chlorine to pentachloroethane and cracking off hydrochloric acid with milk of lime. Its dissolving properties are as good as those of trichloroethylene but it has a higher boiling point. The most suitable material for transporting and storing trichloroethylene and other chlorinated hydrocarbons is lead-coated iron, in the stabilized form iron or galvanized iron is used. Note that stabilized trichloroethylene is stored covered with a layer of soda solution, thus rendering aluminium unsuitable as a container material. Saturated chlorinated hydrocarbons are less stable than the unsaturated ones.

### *Acetaldehyde*

The most important intermediate in all acetylene chemistry is acetaldehyde which in Germany is produced exclusively from acetylene. In other countries where ethylene and, therefore, also ethyl alcohol are cheap, such as in the U.S.A., or in countries where ethyl alcohol can be produced particularly cheaply from ethylene or other agricultural or forest products, such as the Soviet Union or the United States, acetaldehyde is produced by oxidation from ethyl alcohol:

$$CH_3{-}CH_2OH + \tfrac{1}{2}O_2 \rightleftharpoons CH_3{-}CHO + H_2O.$$

This can be done in a catalytic process in which the vaporized alcohol is preheated with air in a heat exchanger; the mixture is then, as in the production of formaldehyde from methanol (see p. 151), passed over a silver wire gauze. The reaction is strongly exothermic and takes place at a temperature of 500 to 570°C. The gases are scrubbed in a scrubber, the aldehyde and the alcohol forming a solution. The mixture is then rectified and with a 50–55 per cent conversion the yield is 85–95 per cent of the theoretical. In some respects the process resembles the catalytic production of ethylene oxide from ethylene + oxygen in which 3 per cent ethylene is reacted with 97 per cent air at 200–240°C on an Ag catalyst on pumice (see p. 170).

In Germany acetaldehyde has up to now been produced mainly from acetylene by hydration over a mercury-iron catalyst. Working with mercury is rather difficult and the losses of mercury are considerable. A second process was, therefore, developed (Reppe) which is based on the reaction of acetylene with methyl alcohol. This gives vinyl methyl ether which is easily saponified by diluted phosphoric acid to acetaldehyde and methanol.

$$HC{\equiv}CH + CH_3OH \rightarrow CH_2{=}CH{-}OCH_3$$

$$CH_2{=}CH{-}OCH_3 + H_2O \xrightarrow[\text{diluted } H_3PO_4]{} CH_3{-}CHO + CH_3OH$$

A new process for the production of acetaldehyde from ethylene has been developed by the Consortium für Elektrochemische Industrie, München, (J. Smidt, *Angew. Chem.*, 1959, **71**, 176) which is discussed in the section on ethylene, p. 171.

Acetaldehyde is isomeric with vinyl alcohol (enol form) $CH_2{=}CH{-}OH$. It is not stable in its monomer form but is in the form of its polymer ether which can be obtained by polymerization of the vinylmethylether mentioned above.

The hydration of acetylene was first carried out in the Chemische Fabrik Griesheim-Elektron and was later developed by the Gesellschaft für Stickstoffdünger in Knapsack and the Alexander Wacker Company, München. The reaction used to take place batchwise in stirrers, but is now carried out continuously in a tube (generator) containing the 'catalyst acid' at 98°C.

At the bottom of the tube acetylene is pumped in continuously. At the top the acetaldehyde, diluted with excess acetylene, is removed by steam from the catalyst solution. Water condenses in the first cooler, acetaldehyde in the second. Then scrubbing is carried out with water and the excess of acetylene is returned to the generator. The catalyst acid consists of sulphuric acid of 15 per cent concentration, containing $HgSO_4$ and, in order to maintain a certain Redox potential, $FeSO_4$ and $Fe_2(SO_4)_3$ in the proportion of 60:40 per cent. The acid draining from the generator contains about 1 g/l of dissolved mercury. The mercury sediment formed by reduction is collected in a settling vessel and recycled. The aldehyde also carries mercury vapours. Because of the great toxicity of mercury and its relatively high volatility great efforts were made to find a substitute for mercury as a catalyst; these do not appear to have been successful as yet. The losses of mercury are rather high. With the very large production of acetaldehyde of 340,800 tons in 1943, 172·4 tons of mercury were lost. Mercury is found in all parts of the equipment, particularly in the acetaldehyde storage tanks; from time to time it is extracted and used again.

The daily output of a generator is about 60 tons of acetaldehyde, with a throughput of 1,200 $m^3$/h of crude acetylene and 2,400 $m^3$/h of circulation gas. To prevent the formation of explosive gas mixtures which may occur in closed spaces owing to leakages, and to reduce the danger of poisoning by mercury vapours, the generators for acetaldehyde are built as 'open air plants'. The equipment is suspended in a steel construction, open all round and only protected above by a roof. Mercury poisoning is, therefore, no longer found among the workers.

The aqueous acetaldehyde is finally distilled in a very high column under pressure to 99·6 per cent aldehyde. For pressure distillation water cooling is adequate. Pressure-less fractionating such as is carried out, e.g. at Knapsack, requires cooling with brine which is expensive. The yield is 96 per cent of the theoretical. Pure aldehyde contains only traces of crotonaldehyde and acetic acid. The chlorine content is only 4 mg per litre, originating from the acetylene purification.

Acetaldehyde is an intermediate for a variety of important aliphatic compounds. By oxidation acetic acid is obtained and from it many derivatives, such as the chlorinated acetic acids, and acetone and its derivatives. Oxidation can also lead directly to acetic acid anhydride which is important for many acetylations, e.g. for the production of acetyl cellulose. Dimerization of acetaldehyde leads to aldol from which 1,3-butandiol is produced, the raw material for butadiene and hence synthetic rubber. By dehydration aldol also yields crotonaldehyde which is hydrated via butyraldehyde to butanol. Butyraldehyde can also be oxidized to butyric acid. The seven works operating in Germany in 1943 produced3 40,800 tons of acetaldehyde, a fivefold increase in production as compared with 1937. In 1960, the Federal German Republic produced 244,000 tons p.a.

### *Acetic acid*

Right through history acetic acid has been produced by biological oxidation of alcohol. The production of acetic acid is the oldest of all biological oxidations and was probably invented at the same time as the manufacture of alcoholic drinks from sugar. Even today a particularly tasty vinegar is produced from wine. Another possible method to produce acetic acid is the carbonization of wood, in which wood vinegar is obtained from grey lime. This method will be discussed later.

Here only the most important large-scale industrial process for the production of acetic acid by the oxidation of acetaldehyde will be described. It is based on the simple equation: $CH_3{-}CHO + {}^1/_2\,O_2 \rightarrow CH_3{-}COOH$. The mechanism of the reaction is, however, much more complicated. (See in this connection A. Rieche, Organische Peroxyde und ihre Bedeutung für die Autoxydation organischer Verbindungen. *Wiss. Ann.*, 1953, **2**, 721–723. Akademie-Verlag, Berlin. H. Jokusch, *Chem. Z.*, 1956, **80**, 241).

During oxidation aldehydes add one mole oxygen by 'interposition of oxygen between the CH bond', giving per-acids. Per-acid reacts with one mole aldehyde via an acetal-type addition product, acetaldehyde-monoperacetate (I), which is stable at low temperature. By oxidizing acetaldehyde in acetone a solution of peracetate is obtained which can be used for epoxydization. At room temperature, particularly in the presence of catalysts, it decomposes into two molecules of acid. The autoxidation of acetaldehyde follows, like almost all autoxidation reactions, a radical-chain mechanism:

$$\mathrm{R^{\bullet} + O_2 \rightarrow ROO^{\bullet} + RH \rightarrow ROOH + R^{\bullet}}\ \text{(chain)}$$

$$\mathrm{CH_3{-}CHO + O_2 \rightarrow CH_3{-}\overset{O}{\overset{\|}{C}}{-}OOH}$$

$$\mathrm{CH_3{-}\overset{O}{\overset{\|}{C}}{-}OOH + \underset{H}{\overset{O}{}}\!\!\gg\! C{-}CH_3 \rightarrow \underset{I}{CH_3{-}C{-}\overset{O}{\overset{\|}{C}}{-}OO{-}\underset{H}{\overset{OH}{\overset{|}{C}}}{-}CH_3} \rightarrow 2\,CH_3\cdot COOH}$$

Autoxidation used to be carried out formerly in cylindrical horizontal aluminium tanks, equipped with agitator and cooling coils, oxygen being introduced under pressure. At first, acetic acid containing 0·1 per cent manganese acetate was obtained. Manganese acetate is used as a catalyst to speed up as much as possible the reaction of the aceto per-acid, formed as an intermediate, with the acetaldehyde. If the reaction is not rapid enough, perhaps because addition of the catalyst had been omitted, the aceto per acid accumulates and the mixture may explode. Acetaldehyde which in view of the dangerous acetylene peroxide formation must be completely free from acetylene is added to the mixture and simultaneously oxygen is introduced at a temperature of 60°C. This method produced directly acetic acid of 97–98 per cent concentration (glacial acetic acid) containing a small quantity of acetic anhydride which during distillation reacts with the residual water to acetic acid. This observation merits recording since its exploitation and systematic further investigation led eventually to the Knapsack process for the direct production of acetic anhydride from acetaldehyde. The side reaction became the main reaction by changing the catalyst and the reaction conditions.

The batchwise process is hardly used any more. In more modern plants operations are carried out continuously in vertical reaction columns with built-in cooling coils, or with circulation via a cooler. The column is filled with acetic acid containing 0·1 per cent manganese acetate in solution. Oxygen and aldehyde are supplied continuously together with fresh catalyst. A reaction temperature of 50–70°C is maintained. The acetic acid formed runs off at the top of the column. The waste gas which must not contain more than a few per cent of oxygen, is diluted with nitrogen thus preventing explosions in the gas-filled space at the upper end of the column. Aldehyde and oxygen must be introduced in such quantities that they are consumed completely. A generator produces about 1,000 tons per month.

The direct oxidation of propane and butane from natural gas also produces acetic acid, in addition to other products (Celanese Corp., U.S.A.).

*Acetic anhydride*

By using a catalyst consisting of copper and cobalt acetate, and in the presence of ethyl acetate as a diluent, it is possible, when the water formed is removed quickly from the reaction products, to obtain directly a mixture of acetic acid and acetic anhydride in the proportion 44 : 53. The compound I (p. 206) decomposes into anhydride and water. The process can be represented simply as follows:

$$2\,CH_3CHO + O_2 \rightarrow CH_3{-}\overset{\overset{\displaystyle O}{\|}}{C}{-}O{-}\overset{\overset{\displaystyle O}{\|}}{C}{-}CH_3 + H_2O$$

The reaction temperature is 50°C, the oxygen gauge pressure is 4 atm.

The anhydride can be separated by distillation. Compared with the former dehydration methods this was a great technological and scientific advance, particularly in view of the fact that with the rising manufacture of acetyl cellulose, e.g. for safety film, the production of acetic anhydride had to be constantly increased. There is also a number of petrochemical routes to acetic anhydride (see F. Asinger, *Einführung in die Petrolchemie*, Akademie-Verlag, Berlin, 1959, p. 218).

Acetic anhydride can be obtained also from ketene by adding acetic acid:

$$CH_3{-}COOH \rightleftharpoons H_2C{=}C{=}O + H_2O$$

$$H_2C{=}C{=}O + HOOC{-}CH_3 \rightarrow CH_3{-}CO{-}O{-}OC{-}CH_3$$

One of the processes for the production of ketene is the Wacker process in which acetic acid is reacted with a small addition of triethyl phosphate at 700–730°C at slightly reduced pressure in a Si-Cr-Al steel tube. The cracking gases escaping from the furnace must be quenched as rapidly as possible. Small quantities of bases are added to stabilize the gases.

In the Anglo-Saxon countries the production of ketene from acetone at 600–700°C in Cr-Ni steel tubes, with a contact time of 0·3 sec, has been adopted. (Reaction 25 per cent per pass, yield 82 per cent.)

Ketene is an intermediate with a variety of uses. It is used mainly in the production of acetic anhydride, but many other syntheses can also be carried out with it. For instance, at Hoechst the important non-toxic preservative sorbic acid is produced from ketene and croton aldehyde:

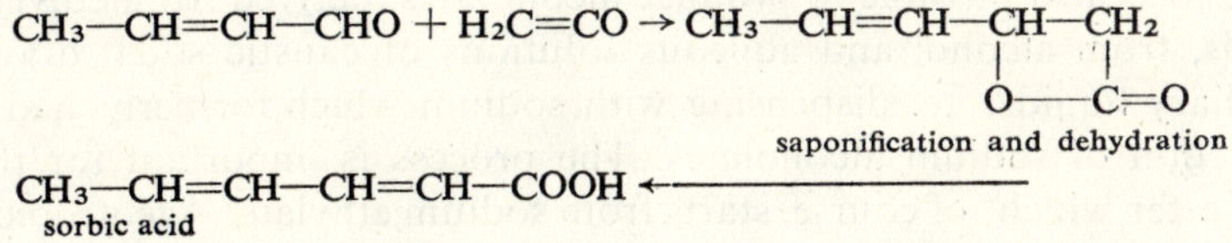

Vinylidene cyanide $H_2C{=}C\langle{}^{CN}_{CN}$ is produced from ketene and HCN; its copolymerizate with vinyl acetate is the 'Darlan' fibre (Goodrich Chem. Co.).

Dimerization of ketene, easily carried out in spray columns, leads to diketene which adds alcohols to acetic acid ester and aromatic amines to the important acetoacetic acid arylamides:

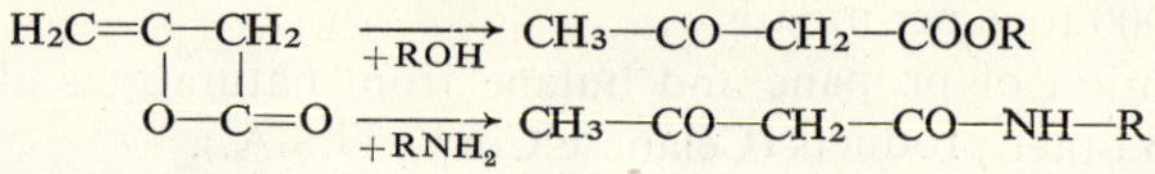

*Esters of acetic acid, azeotropic esterification*

Industrially, mainly methyl acetate (with methanol content), ethyl acetate, propyl acetate, butyl acetate, and amyl acetate are used. They are excellent universal and non-toxic solvents.

The esters of acetic acid, like all esters of aliphatic carboxylic acids, can be produced by the usual method of boiling the acid with alcohol in the presence of a catalyst (e.g. $H_2SO_4$) or of an acid ion exchanger:

$$R{-}COOH + ROH \rightleftharpoons R{-}COOR + H_2O$$

Esterification is a time reaction and reversible. If the water or the ester can be removed continuously the equilibrium is shifted towards the ester side in accordance with the law of mass action.

The choice of the method depends on the boiling points. The alcohol often forms with water constant boiling binary mixtures and the water can be distilled off with the alcohol, separated in a separating vessel, and the alcohol can be recycled. However, if the boiling points are unfavourable, auxiliary liquids are used which form ternary mixtures with alcohol and water, e.g. benzene, carbon tetrachloride, toluene, xylene, or, with high-boiling esters, chlorobenzene. An example of an azeotropic esterification of a carboxylic acid with ethyl alcohol in the presence of benzene will be described.

The apparatus used is a still with an efficient column, cooler, and separating vessel from which the aqueous layer is extracted at the bottom, the benzene layer at the top and returned into the still. During the distillation of a mixture of benzene-ethanol and a carboxylic acid the first fraction distilled is a ternary mixture with a boiling point of 64·86°C. It contains 74 per cent benzene, 18·3 per cent ethanol, 7·7 per cent water. After condensation and cooling a wet layer of benzene and spirit forms at the top of the separating vessel (Florentine receiver), and at the bottom a water layer which contains spirit and benzene. The latter is extracted and later processed by distillation whilst the upper benzene-spirit layer is returned into the still as long as the ternary mixture distils over. After completion of the esterification the second fraction distilled is a mixture of ethanol (32·4 per cent) and benzene (67·6 per cent) at 68·02°C which can be used for the next esterification. Finally, the third fraction is absolute spirit with a boiling point of 78·3°C, leaving the ester in the still.

This method of azeotropic esterification is generally applicable. The azeotropic dehydration of spirit with auxiliary liquids (proof-increasing process) is based on the same principle. It is also possible to produce alcoholates, derived from ethyl alcohol and higher alcohols, from alcohol and aqueous solutions of caustic soda, drawing off the water by auxiliary liquids, i.e. dispensing with sodium which formerly had to be used for the production of sodium alcoholates. The process is important for the synthesis of acetoacetic ester which, of course, starts from sodium ethylate. Azeotropic distillation can be carried out continuously, e.g. for the production of acetic esters. Three columns with coolers are required for this process.

Technologically more important than esterification is a direct ethyl acetate process based on a reaction of Tishchenko's and used by the Alexander Wacker Company: 2 $CH_3$—CHO → $CH_3$—$COOC_2H_5$. The operative principle is the disproportionating effect of aluminium ethylate $Al(OC_2H_5)_3$ with the addition of a little $ZnCl_2$ or $AlCl_3$. Under the influence of this catalyst anhydrous acetaldehyde is converted directly into acetic ester, with simultaneous excessive generation of heat. The reaction resembles the reduction of carbonyl compounds developed by Meerwein-Ponndorf:

$$RCHO + R'CH_2OH \underset{\text{Al-alcoholate}}{\rightleftharpoons} RCH_2OH + R'CHO$$

This reaction takes place concurrently with the esterification. The Meerwein-Ponndorf reduction is used in industry in various reductions of carbonyl compounds, isopropyl alcohol which is converted into the very volatile acetone being particularly suitable for this purpose.

Specially important among the acetic acid esters is n-butylacetate (b.p. 125°C) which finds use as a solvent for laquers and resins of many kinds.

The most important of the chlorine derivatives of acetic acid is monochloroacetic acid. It can be produced by chlorination of acetic acid in the presence of 1 per cent sulphur as a catalyst, some di- and trichloroacetic acid always being produced also. Less chlorine and acetylene is used in this process than in the saponification of trichloroethylene as carried out for example in the VEB Chemische Werke, Buna:

$$ClCH{=}CCl_2 + H_2SO_4 \rightarrow CH_2Cl{-}CCl_2{-}OSO_3H \xrightarrow{+H_2O} CH_2Cl{-}COCl + H_2SO_4 + HCl$$

$$CH_2Cl{-}COCl + H_2O \rightarrow CH_2Cl{-}COOC + HCl$$

The sodium salt of trichloroacetic acid is a weed killer.

## (b) Vinylations

Vinylation is the addition of groups of atoms or molecules to acetylene, establishing a triple bond and maintaining a double bond. The vinyl group is very reactive, and these acetylene derivatives are used in a great variety of processes for the production of high-polymer compounds. The first industrial vinylations were carried out in the Chemische Fabrik Griesheim-Elektron (Klatte).

### *Vinyl acetate*

Vinyl acetate, too, is an important intermediate in the production of plastics. It is used mainly in mixed polymerizates with vinyl chloride. By interchange of ester radicals with methanol the polymer vinyl acetate can be saponified to polyvinyl alcohol ('Vinarol'). Vinyl acetate and vinyl chloride must be considered as esters of the hypothetical vinyl alcohol:

$$[H_2C{=}CHOH], \quad H_2C{=}CH{-}O{-}CO{-}CH_3, \quad H_2C{=}CH{-}Cl$$

Vinyl acetate can be produced by addition of acetic acid to acetylene in the presence of Hg salts at 40–50°C, with the quickest possible removal of the acetate formed from the liquid reaction zone. The catalyst is produced from $SO_3$ and glacial acetic acid (aceto-sulphuric acid) and dissolution of mercury in this mixture. In the liquid phase it is possible also to work with boron trifluoride or mercury phosphate catalysts. These methods are no longer used in industry.

The only process used at present is the gas phase process in various forms, developed by the Consortium für elektrochemische Industrie. The catalyst furnace at Hoechst is the Fischer-Tropsch furnace which permits particularly good heat dissipation. Operations are carried out on a carbon catalyst impregnated with zinc acetate, at 160–205°C.

Vinyl acetate boils at 76°C. Its polymerizate is marketed under the name 'Movilith' and other names, also as a mixed polymerizate with vinyl chloride. In the Reppe process vinyl esters with more than four carbon atoms can be produced at 180°C on a zinc oxide catalyst in autoclaves in the liquid phase. Recently vinyl propionate has become important under the name of 'Propiofan'.

### *Vinyl chloride*

In its polymer form vinyl chloride has become one of the most widely used and most important plastic materials both in Germany and abroad. The industrial production of the monomer vinyl chloride is, therefore, of the greatest importance. Whilst it is produced in Germany mainly from acetylene, it is obtained in the United States from ethylene via ethylene chloride and subsequent catalytic cracking of hydrochloric acid at 400–450°C on activated carbon. The anhydrous hydrochloric acid formed can be added to acetylene so that all the chlorine used reappears in the vinyl chloride.

Hydrochloric acid can be added easily to acetylene in the gas phase at 150–220°C on a $HgCl_2$ catalyst on activated carbon: $HC{\equiv}CH + HCl \rightarrow H_2C{=}CHCl + 44$ kcal. For this process originally developed by Klatte the former IG built an experimental works at Rheinfelden for a monthly production of 50 tons of vinyl chloride. A large-scale plant was later built at Schkopau where in 1943 about 3,000 tons per month were produced in twenty furnaces. The yield is very good; calculated on an acetylene basis it is 99 per cent of the theoretical.

In this plant the acetylene is first thoroughly dried in drying towers with solid KOH. In a tank filled with activated carbon which acts at the same time as an adsorbent for traces of chlorine the gaseous hydrochloric acid which has been dried previously with sulphuric acid is mixed with acetylene. The hydrochloric acid is present in excess. The mixture then passes into a tubular furnace with 800 catalyst tubes filled with granular activated carbon to which 10 per cent $HgCl_2$ has been added as a catalyst. Oil which can be heated and cooled circulates in the furnace. To initiate the reaction the oil is heated to 80°C, when the reaction has started the temperature is maintained at 150°C and the system is switched over to cooling. A great disadvantage is the fact that $HgCl_2$ is continuously sublimed off and the catalyst is gradually spent so that the reaction temperature is bound to increase and finally goes as high as 250°C. The gas escaping from the catalyst oven contains some acetaldehyde and ethylidene chloride as by-products, together with small amounts of non-reacted acetylene and HCl, and is, therefore, scrubbed first with water and then with alkali by means of a brine cooler in which the water is separated; then the gas is dried and finally subjected to a second drying process with solid KOH. It is then liquefied in cooling brine and distilled in a column, the condensation taking place in a low-temperature iron cooler at −40°C. When the pressure is released the by-product gases (acetylene, $H_2$, $CO_2$, $N_2$) escape first from the condenser, and the remainder is then extracted by boiling in a second column. Vinyl chloride is stored and transported in pressure vessels.

### *Vinyl ether*

These preliminary products which because of their great versatility are of so much interest to the chemistry of plastics were produced in Ludwigshafen by W. Reppe first from vinyl chloride and alcoholates, or alcohol and alkali. The reaction takes place quite smoothly in autoclaves at 80–100°C in the presence of alcohol as a solvent:

$$H_2C{=}CHCl + NaOR \rightarrow H_2C{=}CHOR + NaCl$$

A great many alcohols, including polyvalent alcohols and phenols, (the latter giving, however, low yields) can be reacted in this way. During these experiments an important observation was made which was to become the starting point for acetylene syntheses under pressure. It turned out that acetylene present in excess is by itself capable of reacting with the alcoholates when added to vinyl ether.

It was found that alcohols can be added generally to acetylene, using alkalizing agents as catalysts, e.g. 0·5–1 per cent K or Na as alcoholate. The most favourable reaction temperature is 150–180°C. Methanol reacts briskly even at 120°C. The intermediate is probably the metallic salt of vinyl ether which regenerates the alcoholate with alcohol:

$$R—OK + HC{\equiv}CH \rightarrow R—O—CH{=}CHK$$
$$R—O—CH{=}CHK + ROH \rightarrow R—O—CH{=}CH_2 + ROK$$

Figure 82 illustrates schematically a plant in which vinylations are carried out, e.g. the production under pressure of vinyl ether from low alcohols. The operation must be carried out under pressure be-

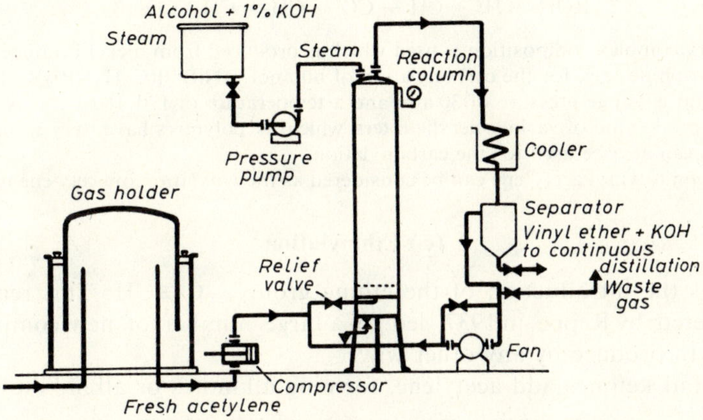


Figure 82. Vinylation under pressure

cause the vinyl ethers have very low boiling points, e.g. vinyl methyl ether −9°C, and it is necessary to work in the liquid phase. Higher ethers can be produced without pressure. (Figure 82 after W. Reppe, *Neue Entwicklung auf dem Gebiet der Chemie des Acetylens und Kohlenoxyds*, Springer-Verlag.)

There are so many possible applications of the vinylation reaction that it is out of the question to discuss here the industrial processes based on it, and even less so the many possibilities for the production of vinyl compounds. The reader must be referred to the specialized literature. Mention will be made of only a few more reactions:

Amines can also be vinylated, the heterocyclic amines with 'acid H on nitrogen' (e.g. carbazole) with particular ease. Solid caustic potash is used as catalyst and acetylene is introduced at a gauge pressure of 10–20 atm and a temperature of 180°C. Vinyl carbazole can be polymerized and then forms the very high-melting plastic 'Luvican'. Of interest is a product obtained from α-pyrrolidone. α-pyrrolidone can be produced via acetylene. 1,4-butandiol which will be discussed later is oxidized to γ-butyrolactone and the latter reacted with ammonia. Vinyl pyrrolidone can be polymerized to chains of various length ('Collidon'). A polymer of a certain viscosity and exactly-defined osmotic properties has found a surprizing application. Under the name of 'Periston' (Reppe, Hecht and Weese) it is used in solution as a blood substitute, and has a similar effect to the glycogen-type carbohydrate Dextran based on the fungus *Leuconostoc mesenteroides*.

In a process developed by the former IG Leverkusen hydrocyanic acid is added to produce acrylonitrile:

$CH{\equiv}CH + HCN \rightarrow CH_2{=}CHCN$. A cuprous chloride-ammonium chloride solution, pH 3·5, is used as catalyst.

The production of acrylonitrile from ethylene oxide and HCN is discussed elsewhere. Recently, the acrylonitrile process from acetaldehyde via lactic acid nitrile has regained some importance after success had been achieved in stabilizing the acrylonitrile during dehydration by the addition of large quantities of phosphoric acid (see p. 172):

$$CH_3{-}CHO + HCN \rightarrow CH_3{-}CH(OH){-}CN \xrightarrow{-H_2O} H_2C{=}CH{-}CN$$

Water is split off by the direct effect of hot combustion gases. The process via allylamine has already been discussed (p. 178).

Finally, mention must be made of the 'carbonylation' discovered by Reppe. This reaction resembles the production of esters from alcohols and carbon monoxide discussed earlier. Here, acetylene is added, according to the equation:

$$ROH + HC{\equiv}CH + CO \rightarrow CH_2{=}CH{-}COOR$$

A catalyst of very complex composition is used which is produced from nickel bromide, alkyl bromide, and triphenylphosphine, e.g. for the carbonylation of butanol: $\{NiBr_2[(C_6H_5)_3P]_2\}C_4H_9Br$. The reaction takes place at a gauge pressure of 30 atm and a temperature of 170–180°C. It is important industrially for the production of various acrylic esters which as polymers have very valuable properties. Metal carbonyls can also be used for the carbonylation.

The dimerization to vinyl acetylene can be considered as the vinylation of acetylene itself (see below).

### (c) Ethinylation

Ethinylation is the introduction of the ethine group $-C{\equiv}CH$. This reaction of acetylene, discovered by Reppe in 1937, led to a large number of new compounds which were difficult to produce by any other way.

Aldehydes and ketones add acetylene, forming alkinoles or alkindioles:

$$R{-}CHO + HC{\equiv}CH \rightarrow \underset{\text{alkinole}}{R{-}CHOH{-}C{\equiv}CH}$$

$$2\,R{-}CHO + HC{\equiv}CH \rightarrow \underset{\text{alkindiole}}{R{-}CHOH{-}C{\equiv}C{-}CHOH{-}R}$$

Copper acetylide is an effective catalyst. Formaldehyde can be added once and twice to acetylene, forming propargyl alcohol or 1,4-butindiole. In Ludwigshafen, a second Buna process was based on this reaction.

$$HC{\equiv}CH + CH_2O \rightarrow HC{\equiv}C{-}CH_2OH \quad \text{propargyl alcohol}$$

$$HC{\equiv}CH + 2\,CH_2O \rightarrow HOCH_2{-}C{\equiv}C{-}CH_2OH \quad \text{butindiole}$$

The reaction is exothermic. The carrier for the Reppe catalyst red copper acetylide, $Cu_2C_2$, is silica gel. It is important to see that the catalyst is not disintegrated and 'peptized' since in fine dispersion it would be floated off. Figure 83 (from W. Reppe, Neue Entwicklungen etc.) shows the pressure apparatus for the ethinylation reaction.

The reaction is carried out in a spray column filled with catalyst. For this purpose silica gel is soaked with pure copper nitrate which is converted into CuO at 400–500°C in a retort furnace. At 70°C and a gauge pressure of 3–5 atm acetylene diluted with nitrogen is pumped through the column concurrently with the 30 per cent formaldehyde solution. The acetylene is recycled. Before the reaction begins the 'formation of the catalyst', i.e. the formation of acetylide, must take place. Aldehyde and acetylene combine at a reaction temperature of 90–100°C. At the bottom of the column is a pressure separator in which

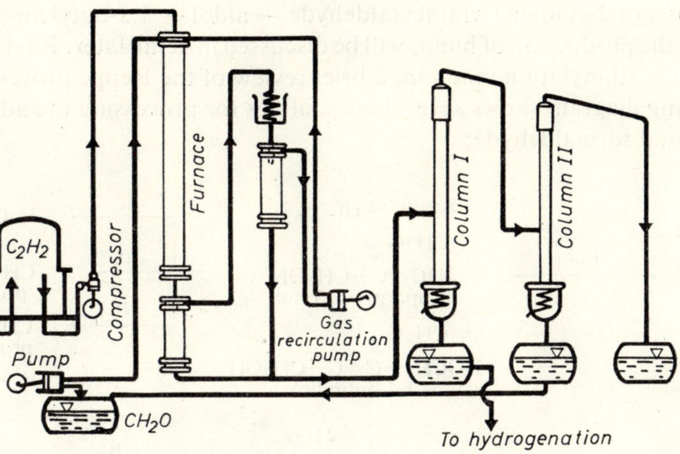


Figure 83. Ethinylation under pressure

the liquid products are separated from the residual acetylene. After cooling the acetylene is returned into the cycle. The crude addition product containing propargyl alcohol, 1,4-butindiole, methanol from formaldehyde, and water, is processed in two columns. In column 1, 1,4-butindiole is obtained as a 30–35 per cent aqueous solution which is hydrogenated to 1,4-butane-diole for the production of butadiene. At the top of column 2 methanol is separated, and at the bottom progargyl alcohol which in the 1,4-butindiole process is recycled. In this method the heat of reaction is dissipated by the steam entrained in the circulating acetylene-nitrogen flow. The yield is 98 per cent of the theoretical. Up to 1945 the BASF operated six columns 18 m/high and with a diameter of 1·5 m, each containing 2 tons of copper acetylide, which produced 150 tons of butindiol per day, b.p. 105–108°C at 1 mm.

Propargyl alcohol and even more so butindiole have become the raw materials for an entirely novel chemistry of aliphatic compounds. Hydrogenation can lead to the most varied products according to the catalyst used and the reaction conditions. Propargyl alcohol can be partly hydrogenated to allyl alcohol with Pd as a catalyst (allyl alcohol from propylene: see p. 177). The esters of allyl alcohol, e.g. allyl phthalate, are of importance in the field of plastics. A route via allyl alcohol and glycerine chlorhydrin leads to glycerine, or directly by the addition of $H_2O_2$ to allyl alcohol in the presence of $OsO_4$. With Cu or Ni in a neutral or slightly alkaline medium n-propanol is obtained, in acid medium almost exclusively propionaldehyde. Using the catalytic effect of $CuCl_2$, two molecules of propargyl alcohol can be oxidized with atmospheric oxygen to hexadiene-2,4-diol-1,6: $HO{-}CH_2{-}C{\equiv}C{-}C{\equiv}C{-}CH_2{-}OH$. From this, with nickel as a catalyst and hydrogen, 1,6-hexandiol is obtained at 40–60°C which is a useful intermediate in the production of high-molecular weight polyesters. It presents, however, also a new route to the important intermediate class of the initial products for superpolyamides. By oxidizing 1,6-hexandiol with nitric acid adipic acid is obtained which can also be produced by oxidizing cyclohexanone.

Propargyl alcohol can, however, be oxidized also with potassium permanganate to propargyl aldehyde (Leverkusen). This can be reacted with guanidine to yield the sulphonamide component 2-aminopyrimidine. Thus a large number of important syntheses is based on the intermediates produced with the aim of obtaining butadiene by a novel route.

The synthesis of butadiene via acetaldehyde → aldol → 1,3-butylene glycol, used in Schkopau for the production of buna, will be discussed in detail later. First follows, in connection with the ethinylation reaction, a brief review of the Reppe process via butindiol.

The following diagram shows several possibilities for processing the addition products of acetylene and formaldehyde:

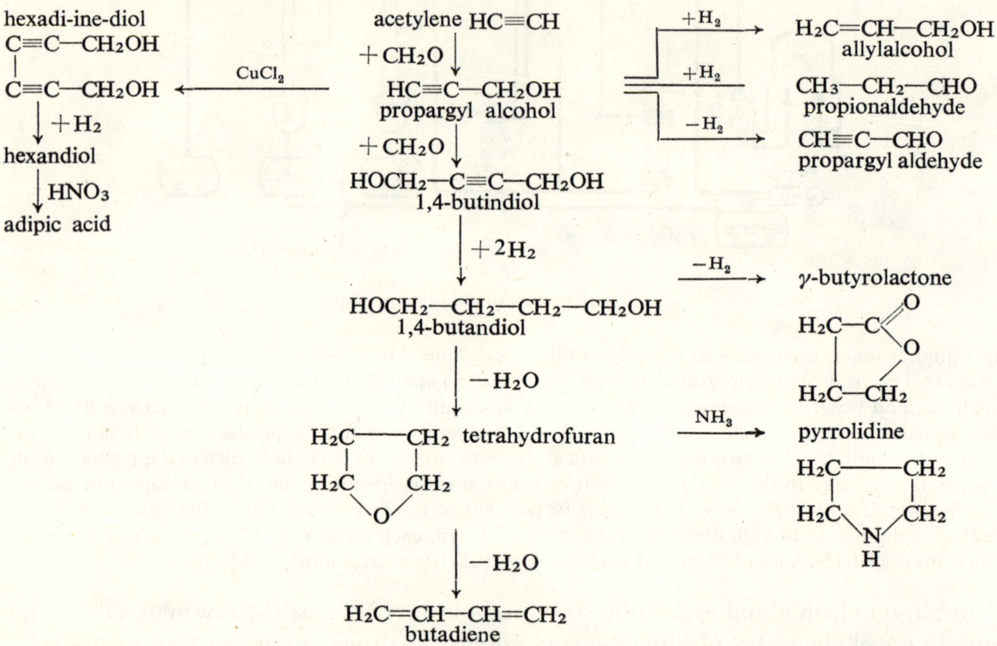


## (d) The butadiene process

*Butadiene produced by the Reppe process*

The first stage in the production of butadiene is the catalytic hydrogenation of the 35 per cent aqueous butindiole solution to 1,4-butylene glycol at a temperature of 180–200°C and a gauge pressure of 280 atm. The catalyst is a mixture of 90 per cent nickel, 9 per cent Cu, and 1 per cent Cr on silica gel. Water can be cracked off from the aqueous solution at 260°C and a gauge pressure of 70 atm with free phosphoric acid in a high-pressure tube. In the pressure distillation following, 94 per cent tetrahydrofuran is obtained almost quantitatively. Tetrahydrofuran is a very good solvent which boils at 64–65°C. A disadvantage is that it oxidizes relatively easily in the air, forming peroxide, and above all that physiologically it is not quite harmless.

Water is again split off from tetrahydrofuran at 260–270°C in the presence of steam over a sodium phosphate catalyst on coke or granulated graphite; this process yields butadiene. It is possible also to obtain butadiene directly from 1,4-butylene glycol by cracking off 2 moles of water on a phosphoric acid catalyst, but there are side reactions and the yield is not as good as when the process is carried out in two stages.

It is well known that the yield is not satisfactory with the cracking off of water from 1,3-butylene glycol which leads to the production of butadiene by the 'Four-stage process'. The various routes to butadiene are shown in the following diagram.

Various routes leading to butadiene

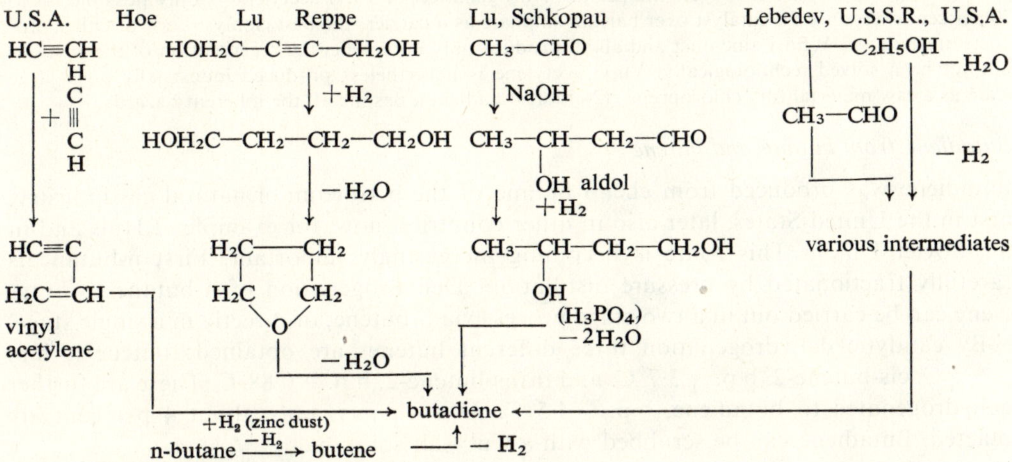

*The Lebedev process*

The diagram shows also the butadiene process developed by the Soviet scientist Lebedev which derives directly from ethyl alcohol and is used a great deal in the U.S.S.R.

In the original process vapour is passed at 400°C and a gauge pressure of 0·25 atm over an aluminium oxide-zinc oxide catalyst, or a silicate catalyst which have both dehydrogenating and dehydrating properties.

Acetaldehyde is obtained as an intermediate. The yield is 60 per cent, but all by-products, e.g. diethyl ether, must be carefully separated and recycled. This is so expensive that the otherwise simple process loses much of its value. The plant required is more or less the same as in the Four-stage process.

The process is carried out more easily with acetaldehyde and ethyl alcohol. This modification has been adopted both in the U.S.S.R. and the U.S.A. For this process, a catalyst consisting of 2 per cent tantalumpentoxide and 98 per cent silica gel is used. The reaction temperature is 330°C. The acetaldehyde is first subjected to aldol condensation and subsequently to dehydration to croton aldehyde:

$$2H_3C{-}CHO \rightarrow H_3C{-}CHOH{-}CH_2{-}CHO \xrightarrow{-H_2O} H_3C{-}CH{=}CH{-}CHO$$

Croton aldehyde is reduced by alcohol to crotyl alcohol and then follows another dehydration to butadiene:

$$H_3C{-}CH{=}CH{-}CHO + C_2H_5OH \rightarrow H_3C{-}CH{=}CH{-}CH_2OH + H_3C{-}CHO$$

$$H_3C{-}CH{=}CH{-}CH_2OH \rightarrow H_2C{=}CH{-}CH{=}CH_2 + H_2O$$

During the war, the United States were able, owing to the abundance of the available spirit supplies, to build up an enormous manufacture of Buna within a very short time.

*The Nieuwland process*

The diagram shows also the process via vinyl acetylene developed by Nieuwland at Du Pont, U.S.A. It was not possible to exploit it on a large industrial scale in spite of its attractive apparent simplicity. This was due mainly to two causes: In addition to vinyl acetylene, the extremely dangerous divinyl acetylene is formed during the dimerization of acetylene. Divinyl acetylene is highly autoxidizable and forms very dangerous and almost indestructible peroxides of a resinous nature which have on several occasions given rise to explosions. However, the partial hydrogenation of vinyl acetylene is only possible up to 70 per cent even on a Pd catalyst over barium sulphate as a carrier, the best catalyst known. Other products are butenes. Whilst zinc dust and alkali produce only butadiene, the regeneration of the zinc has not yet been solved technologically. Vinyl acetylene is nevertheless produced industrially on a large scale as a raw material for 'chloroprene' (2-chlorobutadiene), despite all the inherent hazards.

*Butadiene from butane and butene*

Butadiene was produced from cheap butane of the petroleum or natural gas industry, first in the United States, later also in other countries, now, for example at Hüls and in the Soviet Union. This route is becoming increasingly important. First n-butane is carefully fractionated by pressure distillation. Dehydrogenation of n-butane to butadiene can be carried out in a two-stage process via n-butene, or directly in a single stage.

By catalytic dehydrogenation three different butenes are obtained: butene-1, b.p. $-6{\cdot}2$°C, cis-butene-2, b.p. $+3{\cdot}7$°C, and trans-butene-2, b.p. $+0{\cdot}88$°C. These are further dehydrogenated to butadiene, b.p. $-4{\cdot}5$°C. At each pass only about 4 per cent are reacted. Butadiene can be scrubbed with a $Cu^{II}$-salt solution.

*Two-stage process for the dehydrogenation of butane (Phillips Petroleum Co.)*

The raw material is n-butane from natural or refinery gases, or n-butene mixtures from cracking plants. Isobutene is removed from the latter with 65–70 per cent sulphuric acid, and they are then purified by extractive distillation ('Distex process') to 90 per cent butene mixtures.

The catalytic dehydrogenation of n-butane to butene can be carried out for example by the Standard Oil process in the presence of superheated steam over MgO-, $Fe_2O_3$-, $K_2O$-, and CuO-catalysts. Only about 20 per cent are converted, the non-dehydrogenated butane is recycled.

During the dehydrogenation of n-butane not only butenes and hydrogen are formed but also butadiene, isobutene, isobutane, $C_3$ hydrocarbons and those higher than $C_4$. The dehydrogenated gas mixture is compressed and freed by scrubbing with oil under pressure of hydrogen (which does not dissolve) and of the lower hydrocarbons. A complicated rectification follows, carried out in two columns with 100 trays each, by extractive distillation with furfurol. Cis- and trans-n-butene-2 and n-butene-1 are dehydrogenated to butadiene (stage 2).

In the Phillips Petroleum Co. process dehydrogenation is carried out on a chromium oxide-$Al_2O_3$ catalyst at 670–680°C. The catalyst is contained in a tubular furnace, with the heat supplied from outside. The conversion is again not more than 22 per cent.

From the furnace issue: n-butene, butadiene, and $H_2$.

By-products from the cracking and isomerization are: isobutene, isobutane, acetylene homologues, e.g. dimethyl acetylene.

Processing by distillation is again a combination of rectification and extractive distillation. After final distillation in a 120-tray column 98·71 per cent butadiene is obtained which still contains small quantities of butene.

*The Houdry single-stage process*

This process is operated in the Chemische Werke, Hüls, and new plants are also under construction in the United States.

Butadiene is obtained in a concentration of 8–12 per cent on a chromium oxide-aluminium oxide catalyst at 600°C and 100–150 mm Hg, i.e. pressure below atmospheric. Heat cannot be supplied by steam since this catalyst reacts with steam, unlike the catalyst used in the Standard Oil process.

During dehydrogenation much carbon is precipitated on to the catalyst which has to be burned off in periods of 8–10 minutes (regeneration). In the catalyst area, catalyst pellets are mixed with aluminium spherules which have no catalytic effect themselves but great heat capacity. During the burning off the temperature rises up to 700°C. The spherules dissipate the heat during dehydrogenation. The catalyst furnace can be considered as a kind of regenerative furnace which also contains the catalyst. The butadiene is extracted with a solution of copper ammonium acetate. The escaping n-butane-butene mixture (circulation gas) is recycled. With small butadiene concentrations the extraction method is superior to extractive distillation. The Dow Chemical Corp. also developed a butadiene process on the basis of selected $C_4$ fractions. This process is used in 10 plants with a total output of 350,000 tons per annum.

*Isoprene*

Recently isopropene, $H_2C{=}C(CH_3){-}CH{=}CH_2$, 2-methyl-1,3-butadiene, has become very important. It is the structural element of natural rubber. Using 0·1 per cent lithium or 'Mülheim catalysts' (aluminium trialkyl and titanium tetrachloride, see p. 180), isoprene has been successfully polymerized to a synthetic rubber at least equal to natural rubber.

The first industrial isoprene process was also based on acetylene chemistry: acetone-sodium and acetylene were reacted together:

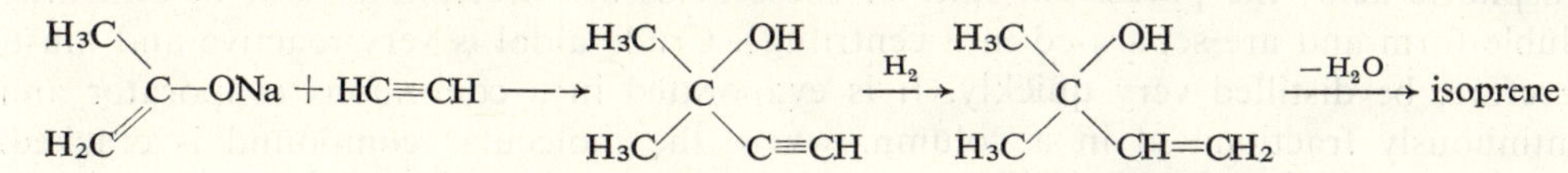

3-methylbutinol 3-methylbutenol

Isoprene can also be produced from isobutylene and formaldehyde by the Prins reaction:

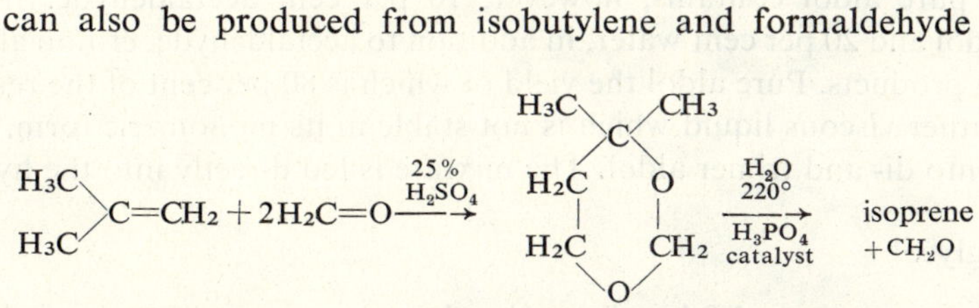

The main raw material for the petrochemical production of isoprene is the $C_5$ fraction of the high temperature pyrolysis of petroleum fractions at 650–760°C which contains about 15–25 per cent. Purification is carried out, similarly as with butadiene, by extractive distillation, using acetone with 5 vol.-per cent water as an extraction agent.

*Butadiene from acetaldehyde, the Four-stage process*

The process was developed for the production of Buna. Natural rubber consists of polymeric isoprene and the first attempts to produce synthetic rubber were aimed at imitating natural rubber (F. Hofmann).

The production of isoprene is, however, more costly than that of butadiene and, in addition, the polymerization of isoprene was technically particularly difficult. The five technological routes to butadiene have already been shown in the diagram on p. 215, as have the individual reaction stages of the Four-stage process (Hofmann, Meisenburg, Delbrück).

## Aldol condensation

Whilst acetaldehyde to which a large amount of alkali has been added resinifies, aldol is obtained by the addition of a very little KOH (0·2 per cent) in aqueous solution at a temperature of +15°C. The reaction takes place with simultaneous generation of heat:

$$2\,CH_3\text{—}CHO \rightarrow CH_3\text{—}CHOH\text{—}CH_2\text{—}CHO + 300 \text{ kcal per kg acetaldehyde.}$$

The reaction is possible only with aldehydes possessing, in addition to the aldehyde group, a reactive hydrogen atom. The aldol formation is not complete. Aldol also forms with unaltered acetaldehyde acetal-type molecular compounds which, however, crack during distillation. The industrial process is based on a reaction of only 60 per cent of the acetaldehyde. If the reaction rate is higher, and also at higher temperatures, increased formation of croton aldehyde, dehydration of aldol, and resin formation occur. Very thorough cooling with water is, therefore, essential. The reaction was developed further into a continuous reaction by Hoechst which made possible the exploitation of the process on a large industrial scale.

It is carried out in a system of flow tubes consisting of horizontal double-walled tubes in a water jacket (see p. 90). Agitator vessels were used up to 1936. Ready-made aldol is used and acetaldehyde and diluted caustic potash solution are forced through the system of tubes by means of a circulating pump.

Since a Cannizarro reaction takes place simultaneously, some of the KOH is consumed in the reaction and must be replaced. The contact time is several hours. Crude aldol is continuously extracted from the tubular system and neutralized with acetic acid or phosphoric acid; the potassium salts of these acids are precipitated out in difficultly soluble form and are separated in a centrifuge. Crude aldol is very reactive and must, therefore, be distilled very quickly. It is evaporated in a continuous evaporator and continuously fractionated in a column, where the molecular compound is cracked. Acetaldehyde escapes at the top of the column and is returned into the aldolization process. The pure aldol contains, however, 10 per cent acetaldehyde. It consists of 65 per cent aldol and 20 per cent water, in addition to acetaldehyde, croton aldehyde, and higher boiling products. Pure aldol the yield of which is 80 per cent of the reacted acetaldehyde is a rather viscous liquid which is not stable in its monomeric form. It is rapidly transformed into di- and trimer aldol. The mixture is fed directly into the hydrogenator.

### 1,3-butylene glycol

Aldol is hydrogenated to 1,3-butylene glycol at a gauge pressure of 300 atm and 100–120°C over a catalyst consisting of Cu, Ni, and Cr, on granular silica gel which must be of a well-defined structure:

$$CH_3\text{—}CHOH\text{—}CH_2\text{—}CHO + H_2 \rightarrow CH_3\text{—}CHOH\text{—}CH_2\text{—}CH_2OH$$

The catalyst is contained in hydrogenation furnaces of conventional design, made of copper-lined steel or V2A steel, up to 18 m long and with a diameter of 0·8 m. The largest hydrogenation furnaces contain about 8 m³ catalyst. The hydrogen is circulated by a circulation pump; the gauge pressure on the inlet side is about 280 atm, on the pressure side 300 atm. These special compressors work on the principle of the reciprocating pump and are driven directly by slow-running synchronous motors. The aldol is compressed in a single stage by a plunger pump to a gauge pressure of 300 atm and is injected into the circulation in measured quantities (see Figure 84).

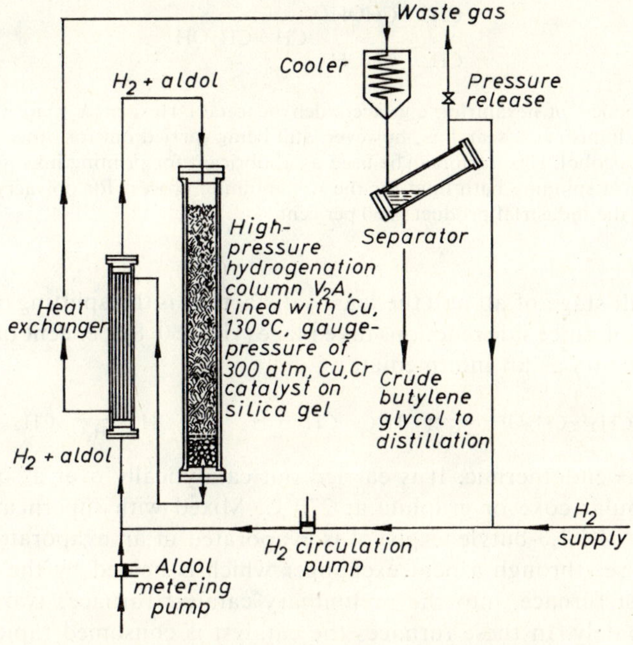


Figure 84. Hydrogenation of aldol to 1,3 butylene glycol

Aldol and hydrogen pass first through a heat exchanger which is heated by the escaping hydrogenation products and then enter the catalyst furnace from above concurrently with the circulation gas. About 4,000 litre aldol are reacted every hour in this tube. The gases leave at the bottom of the column through a V2A high-pressure hydrogenation column lined with copper Raschig rings. The 1,3-butylene glycol passes via a heat exchanger and a cooler into a separator where separation of the excess hydrogen takes place. The reaction temperature must be maintained very accurately and this is achieved by the circulating hydrogen. A small amount of waste gas is allowed to escape continuously in order to prevent an undesirable accumulation.

The hydrogen must be as free as possible from CO since carbonyl formation on the nickel catalyst would occur. The usual purification with ammoniacal Cu-oxide under pressure is inadequate. To be purified the hydrogen is passed over a nickel catalyst above 180°C, which causes the carbon monoxide still contained in the gas to be hydrogenated to methane. A very careful check of the CO concentration is carried out by means of recording infra-red analysers.

Finally, pure 1,3-butylene glycol is obtained by distillation in a system of columns, operating continuously, the last under reduced pressure. It has a b.p. of 204–205°C at 760 mm. (For method of operation and design of columns see p. 87.) The yield is 96–97 per cent of the theoretical.

**In addition to 60–65 per cent of 1,3-butylene glycol, 3·5–5 per cent of ethanol are contained in the crude butylene glycol, produced by hydrogenation of the admixture of acetaldehyde, together with**

4–5 per cent n-butanol stemming from croton aldehyde, and 4–5 per cent higher alcohols from condensation products of croton aldehyde, which are obtained as by-products of the distillation. With an annual production of 200,000 tons butylene glycol the yield of ethanol, butanol and higher alcohols, particulary hexantriol, is considerable:

$$\begin{matrix} CH_3—CHOH \\ CH_3—CHOH \end{matrix} \rangle CH—CH_2OH$$

The reaction products of hexantriol, e.g. acetaldehydeacetal ('Hexalat A'), are valuable plasticizers used in the plastics industry. A search is, however, still being carried out for other applications of this interesting trivalent alcohol. Hexantriol can be used as a lubricant for printing inks in place of glycerine. It is also suitable as a spinning bath fluid for the wet spinning process for polyacrylonitrile ('Prelana' fibre). The yield of the industrial product is 60 per cent.

## Butadiene

The most difficult stage of all and the least satisfactory is the splitting off of water from 1,3-butylene glycol, since side reactions take place (yield 80–82 per cent of the theoretical). Allyl carbinol occurs as an intermediate:

$$CH_3—CHOH—CH_2—CH_2OH \xrightarrow[-H_2O]{} H_2C{=}CH—CH_2—CH_2OH \xrightarrow[-H_2O]{} CH_2{=}CH—CH{=}CH_2$$

The reaction is endothermic. It is carried out catalytically over a sodium phosphate catalyst on granular coke or graphite at 270°C. Mixed with superheated steam in the proportion 1 : 1, the 1,3-butylene glycol is evaporated in an evaporator at 300°C. The mixture first passes through a heat exchanger which is heated by the crude butadiene from the catalyst furnace, into the preliminary catalyst furnace; two of these can be connected alternately. In these furnaces the catalyst is consumed rapidly since volatile butyl phosphate is formed which removes the phosphoric acid. It must be replenished frequently and a large catalyst plant produces exclusively for this stage of the process. When the catalyst in one furnace is spent, the other furnace is connected.

Mainly allyl carbinol is formed in the preliminary catalyst furnace. The vapour mixture then passes into the main catalyst furnace which consists of separate rings with horizontal coils heated by high-pressure steam. Between the coils is the sodium phosphate catalyst which is precipitated on to granular coke or graphite. The gaseous crude butadiene passes through a heat exchanger and a terminal cooler into a separator where the liquid products and water are separated.

The crude gas still contains $CO_2$, $H_2$, $CH_4$, $CH_2{=}CH—CH_3$, $CH_2O$, $CH_3CHO$. It is treated with NaOH to remove the resinifying constituents, dried, condensed by low-temperature cooling, and distilled under pressure in two columns. From the first column propylene is extracted, from the second 99 per cent butadiene, b.p. −4 to −5°C, is obtained with a total yield of 80–82 per cent calculated for the reacted butylene glycol.

The admixtures of propylene, formaldehyde, etc. show that side reactions occur, e.g. $H_2C{=}CH—CH_2—CH_2OH \rightarrow CH_2{=}CH—CH_3 + HCHO$. By splitting off water between carbon atoms 2 and 3 crotyl alcohol is formed: $H_3C—CH{=}CH—CH_2OH$. Butyraldehyde and higher molecular weight oils are also obtained.

The processing of butadiene to Buna is discussed in the section 'Fully synthetic plastics.'

The four-stage process described above was operated in 1945 in Germany in five Buna plants with a total capacity of 150,000 tons of which about half was produced at the Schkopau works. At Ludwigs-

hafen the Reppe process via 1,4-butandiol was used on a large industrial scale. In 1939 the Soviet Union had four Buna plants where butadiene was produced from alcohol by the Lebedev process. After the war butadiene was produced in the Soviet Union also from acetaldehyde and alcohol. In 1944 the United States had a total production of 626,320 tons, from ethyl alcohol, acetaldehyde and ethyl alcohol, butene, butane, and cracking gases (data taken from O. Nicodemus in Winnacker-Weingaertner, *Chem. Technologie*). Production from alcohol is economical only where adequate quantities of cheap hydrocarbons are available, such as in the U.S.S.R., U.S.A., Canada, or where ethyl alcohol is produced from ethylene, coke oven gas, hydrogenation gas, cracking gas, or natural gas. In the Scandinavian countries with their large forests sulphite spirit is also used as a raw material.

*n-butanol from aldol*

It has been shown above that n-butanol is formed from croton aldehyde as a by-product of the hydrogenation of aldol. Since the output from this source is inadequate (butanol is one of the most important solvents) it is also produced directly. When the acidulated aqueous aldol is heated in the copper still of a distillation column, an azeotropic mixture of croton aldehyde and water distils over. After condensation 86–88 per cent croton aldehyde is separated from the water. The croton aldehyde is purified to 96–97 per cent (b.p. 104°C) by distillation. Butyraldehyde is obtained by careful hydrogenation in an agitator autoclave with nickel as catalyst in the liquid phase. In a high-pressure spraying apparatus with hydrogen circulation only butanol is obtained at 200°C on a stationary catalyst consisting of copper of a certain structure, the so-called 'violet copper'. The prepared catalyst is produced by precipitating copper as a carbonate on pumice stone and reducing carefully with hydrogen at 180–200°C.

Pure butanol boils at 117°C. As such or as an ester, e.g. butyl acetate, it is widely used in the lacquer industry and as a solvent for extraction purposes. It is, for instance, an excellent solvent for nitrocellulose lacquers and has the advantage that it is physiologically harmless. In other countries, e.g. in the U.S.S.R. and the U.S.A., butanol is produced also by bacterial fermentation of carbohydrates. Thus the Commercial Solvents Corporation has for a long time been manufacturing butanol, propanol, acetone and ethanol using butyl bacteria for the fermentation of grain. In Germany there are not sufficient agricultural products available.

### (e) Dimerization of acetylene

It has been known for a long time that when acetylene is polymerized, e.g. by feeding it into hot oil in the presence of a little copper, a brown, loose, cork powder-type material, 'cuprene', is formed. It occurs as an undesirable by-product in all acetylene reactions where acetylene and copper are brought into contact. R. Meyer succeeded in trimerizing acetylene to benzene, albeit with a low yield. It was not until 1931 that Nieuwland, of Dupont's, started producing definite di- and trimerized acetylenes. The acetylene dimer, monovinylacetylene, $CH_2{=}CH{-}C{\equiv}CH$, is abbreviated to 'Mova', the trimer, $CH_2{=}CH{-}C{\equiv}C{-}CH{=}CH_2$, to 'Diva' (divinylacetylene). These compounds are very reactive, decompose readily and adsorb oxygen, forming peroxides. In acetylene reactions they can initiate an ignition.

The dimerization process invented in the U.S.A. was further developed, and operated, in the former IG Works Hoechst. The dimerization takes place in an acid, highly concentrated copper (I)-chloride solution at temperatures below 100°C. Acetylene is circulated in great excess in order to minimize the formation of higher polymers. The reaction heat of 1,210 kcal per kg of acetylene is dissipated by steam with the acetylene flow and by external cooling of the reaction column which is enamelled on the inside. It is important to ensure complete freedom from oxygen, and thus nitrogen of the highest degree of purity is used. The separation of water, higher boiling by-products, and dissolved acetylene has been solved by a very elegant technique; descriptions can be found in the original literature (*Ind. Engng. Chem.*, 1935, **27**, 850; Schema d. Mova-Herstellung: Winnacker-Küchler, *Chem. Technologie*, **3**, 709). Mixtures containing polymers of acetylene can be distilled at atmospheric pressure only in the presence of thinning agents. After final distillation the yield of vinyl acetylene is 75 per cent of the acetylene consumed. Further treatment as early as possible is essential since even in the absence of air unexpected ignition and decomposition from unknown causes may occur.

Monovinylacetylene can take part in many interesting reactions. For instance, as with acetylene, water can be added with $HgCl_2$ as catalyst, obtaining methyl vinyl ketone. The most important reaction, however, is the addition of hydrochloric acid, using CuCl/HCl in aqueous solution as catalyst, obtaining chloroprene $CH_2{=}CH{-}CCl{=}CH_2$, i.e. chlorobutadiene, first produced by the American Carothers. By-products are the addition product with water, methyl vinyl ketone, and that with 2 moles HCl, 1,3-dichlorobutene-(2): $ClCH_2{-}CH{=}CCl{-}CH_3$, which must have been formed by 1,4 addition of HCl to chloroprene. Other interesting 1,4 additions (Diels-Alder reaction) of chloroprene are possible. Chloroprene (b.p. 59·2°C) has a much greater tendency to polymerization than butadiene. To initiate polymerization, heat, traces of metallic salts or oxygen are sufficient. It is prevented by introducing continuously into the distillation column small amounts of a polymerization inhibitor, in this case e.g. thiodiphenylamine:

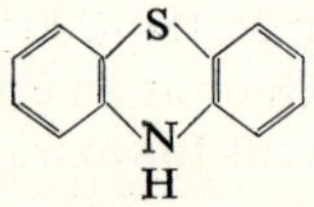

Polymerization inhibitors, like anti-oxidants, break the chain and prevent the continuation of an oxidation or polymerization chain reaction. This expedient is often used in the distillation of vinyl compounds and other oxidizable materials, adding for example a small amount of hydroquinone to styrene.

The chloroprene polymerizate is very important in industry ('Chloroprene' of the former IG, 'Neoprene', formerly 'Duprene' of Dupont de Nemours, 'Sovprene' of the U.S.S.R., and 'Mustone' of Japan).

### (f) Acetone

Acetone is one of the most important solvents. It has the great advantage of being both soluble in water and soluble in, or miscible with, nearly all organic solvents. It is used to gelatinize gun cotton, to dissolve and spin acetate silk, as a low-boiling lacquer solvent, as a solvent for acetylene in pressure cylinders, and it is also an important organic intermediate. Efforts have, therefore, always been made to find cheap processes for the manufacture of acetone. A number of these processes are listed in the following summary. The most economical at the present time is probably the manufacture by the processes listed under (2) and (3) from propylene via isopropanol if the latter is available from cracking gases, coke oven gases, or low temperature carbonization gases.

In 1956 the production of acetone in the United States amounted to about 250,000 tons of which 230,000 tons were produced from isopropanol, in 1961 it amounted to 327,000 tons.

For the catalytic dehydrogenation of isopropanol (or secondary butanol for methyl ethyl ketone) equal parts by volume of alcohol vapour and hydrogen are fed at 400°C over a catalyst into a tubular furnace. The reaction is endothermic:

$$CH_3—CHOH—CH_3 \rightarrow CH_3—CO—CH_3 + H_2 - 15{\cdot}9 \text{ kcal/mole.}$$

After cooling, scrubbing is carried out countercurrently with water. The yield is hydrogen of 99 per cent and a 20 per cent acetone in water solution which is separated by distillation.

With air the dehydrogenation can be carried out in a similar manner to that used for formaldehyde production from methanol (described on p. 151).

Noteworthy is the Degussa process from alcohol on an alkalized iron oxide catalyst at 450°C with excess steam. Acetone can also be produced from acetylene with steam directly on a zinc oxide catalyst at 400°C by a process of the former IG; the processes for the manufacture of acetone are, therefore, listed briefly below.

Processes used for the manufacture of acetone

(1) The old 'grey lime process': distillation of $Ca\langle{}^{OOC—CH_3}_{OOC—CH_3}$

(2) Dehydration of isopropanol

$$H_3C—\overset{H}{\underset{OH}{C}}—CH_3 \xrightarrow[250\,°C]{Cu} H_3C—CO—CH_3 + H_2$$

(3) $H_3C—\overset{H}{\underset{OH}{C}}—CH_3 \xrightarrow[+\text{air } 300\,°C]{Cu} H_3C—CO—CH_3 + H_2O$ (like $CH_2O$ from $CH_3OH$)

(4) $2\,C_2H_5OH + H_2O \xrightarrow[450\,°C]{\text{alkal. } Fe_2O_3} CH_3—CO—CH_3 + CO_2 + 4H_2$

*(Degussa)*

(5) $2HC{\equiv}CH + 3H_2O \xrightarrow[400\,°C]{ZnO} CH_3—CO—CH_3 + CO_2 + 2H_2$
(also diluted acetylene)

(6) $2CH_3—COOH \xrightarrow[500\,°C]{\text{cerium carbonate}} CH_3—CO—CH_3 + CO_2 + H_2O$

Hoechst since 1917

When discussing the production of acetylene a third possibility was mentioned, that of partial combustion of hydrocarbons. The acetylene obtained contains other hydrocarbons and is unsuitable for many syntheses for which pure acetylene is used. This impure acetylene is, however, eminently suitable for the production of acetone over a zinc oxide catalyst.

There is yet another route from acetylene via acetaldehyde and acetic acid to acetone which has been used since 1917 by the Hoechster Farbwerke and later by the Gesellschaft für Stickstoffdünger at Knapsack. It is a variant of the old grey lime process. As equation 6 shows, acetic acid can be converted over a cerium carbonate catalyst on pumice stone or charcoal at 500°C directly into acetone, $CO_2$ and $H_2O$ being split off. The acetone furnaces are made of iron or chrome alloy non-scaling steel, e.g. NCT 3. Traces of non-reacted acetic acid are neutralized with a solution of caustic soda before the acetone is processed further.

Finally, the process via cumene and cumene hydrogen peroxide must be mentioned which yields phenol and acetone (see p. 176). An increasing amount of acetone is also produced as a by-product of the oxidation of gaseous hydrocarbons.

## LITERATURE

U. Wildhagen, *Calciumkarbid und die Karbidindustrie.* Akad. Verlagsges. Geest & Portig, Leipzig, 1953.

Taussig, *Die Industrie des Calciumkarbids.* Knapp, Halle, 1930.

A. Henglein, 75 Jahre Calciumkarbid. *Chem. Z.* 1937, **61**, 684.

O. Nicodemus, Die aliphatischen Chemikalien und Zwischenprodukte, in: Winnacker-Weingaertner, *Chem Technologie, Organische Technologie I*, p. 611. Hanser Verlag, München, 1952.

W. Reppe, *Neue Entwicklungen auf dem Gebiete der Chemie des Acetylens und Kohlenoxyds.* Springer, Berlin/Göttingen, 1949.

W. Reppe, *Chemie und Technik der Acetylen-Druck-Reaktionen.* Verlag Chemie, Weinheim, 1952.

F. Walter, Acetylen, in Ullmann, *Encyklopädie*, 3rd ed., vol. 3, p. 43, 1953.

A. Rieche, in: *Handbuch der Katalyse*, 7/1, 136. Springer, Wien, 1944: Peroxyde als Katalysatoren.

O. Horn and H. Bestian, Winnacker-Küchler, *Chem. Technologie*, 2nd ed., 3. Hanser Verlag, München, 1959.

W. Breuers and H. Luttropp, Buna, *Herstellung — Prüfung — Eigenschaften.* VEB Verlag Technik, Berlin, 1954.

# VIII. Aromatic intermediates

## General remarks on aromatic intermediates

The production of aromatic intermediates for dyestuffs, pharmaceutical products, weed killers, plastics, washing and textile auxiliaries, tanning agents, and perfumes, is the oldest, most versatile, and most highly developed field in industrial organic chemistry. Until about 1930 the majority of organic chemists and technologists, including those working in the dyestuffs industry, had been engaged in research on intermediates. The large modern plants of the organic chemical industry stem originally from works producing dyestuffs and intermediates, a fact which in many cases is still apparent from their names. Bituminous coal tar is the basis of intermediates chemistry. Early on in the United States, petroleum also became the basis for the production of aromatic compounds. At the present time, more aromatics are produced there from petroleum than in coking plants. In Germany it is only in the last few decades that aromatics production (aliphatic base materials, gas-carburation and diesel oils, oils, solvents, and plastics) from petroleum has overtaken that from coking plants, though only quantitatively and probably not yet by value.

The year 1841 may be considered as the year in which the intermediates chemistry was born when the Russian Zinin produced aniline from nitrobenzene. Unverdorben had produced it first in 1826 from indigo by lime distillation; Runge discovered it in bituminous coal tar in 1834. A. W. v Hofmann, the grand old man of aniline chemistry, proved the identity of all these products. Intermediate and dyestuff chemistry have always been closely connected. The discovery of a new intermediate was almost invariably followed in the coal-tar dye industry (as the industry was called that produced synthetic dyestuffs) by a series of new dyes. Thus, for instance, the new derivatives of benzene or naphthalene were reacted with a great variety of compounds and the most surprising results were obtained. The first dyes were produced by accident. At that time nothing was known of the principles governing the composition of a compound and its nature as a dyestuff. The year 1856 inaugurated a new era of industrial organic chemistry when the Englishman Perkin – who had actually intended making quinine – produced the first coal-tar dyestuff 'Mauveine', by oxidizing aniline with potassium dichromate. Fuchsin followed, obtained in 1858 by both Nathanson who heated aniline with ethylene chloride, and A. W. v. Hofmann from aniline and carbon tetrachloride; industrially, however, it was produced for the first time in 1859 by the Frenchman Verguin by heating technical aniline containing toluidine with anhydrous stannic chloride. After this came the discovery of a large number of triphenylmethane and azine dyes. The industrial development of the intermediate and dyestuff chemistry was at first completely in the hands of English and French factories. This became very obvious at the Great Exhibition in London in 1862 when six-and-a-half million visitors from all parts of the globe admired for the first time aniline dyes in quantities which until then had been unimaginable. This may have been one of the causes for the start of coal-tar dyestuff production by German chemists, technologists, and industrial pioneers.

The German coal-tar dyestuff works were founded at that time. One of the first was Oehler's at Offenbach, later Griesheim-Elektron (1842). In 1861 followed Weiler ter Meer, Ürdingen. Approximately simultaneously in 1863 followed Bayer, Elberfeld, Meister Lucius & Co., Hoechst, Kalle, Biebrich. The Badische Anilin- und Sodafabrik, Ludwigshafen, started production in 1865, Casella, Mainkur, in 1870, and the Aktiengesellschaft für Anilinfabrikation (Agfa), Berlin, in 1873. At the next World Exhibition, in Paris in 1867, German dyestuff firms showed a large number of new dyestuffs.

Of great importance for the development of synthetic dyestuffs and for systematic research was the creation of clear ideas on the nature of benzene and of aromatic compounds by August Kekulé, Justus v. Liebig's great disciple, in 1865. 'It brought order into what already existed, and it projected a beam of light on to what was yet to come', said Caro in 1891, referring to Kekulé's theory (*Ber. dtsch. chem. Ges.*, 1892, **25**, R. 955) in his great retrospect of the development of the coal-tar dyestuff industry. These new ideas, and above all A. W. Hofmann's research into aniline chemistry liberated the chemists' work from the shackles of mere empiricism.

Two other landmarks in dyestuff chemistry must be mentioned: alizarin and synthetic indigo.

At that time medicines were produced almost exclusively from natural materials. Large-scale production of synthetic drugs followed about twenty years later. Drugs like salicylic acid (Kolbe), antipyrine (Knorr), phenacetin (Duisberg and Hinsberg), pyramidone (Stolz) became the starting points for the synthesis of drugs, and their introduction into medicine. The manufacture of intermediates is closely interlinked with that of pharmaceutics.

## 1. THE MAIN OPERATIONS IN THE PRODUCTION OF INTERMEDIATES AND THE EQUIPMENT USED

Intermediates are semi-finished products, produced from aliphatic and aromatic raw materials, which by comparatively simple further conversions can be transformed into finished products (e.g. dyes and drugs).

Within the framework of this textbook it is impossible to give a complete survey of the extensive field covering the industrial production of intermediates. It is one of the largest fields in industrial organic chemistry. In the production of intermediates in many cases certain basic operations are used, however, e.g. those starting from benzene, its homologues, from the phenols and from naphthalene, as well as certain types of apparatus, which it is quite feasible to discuss briefly in this connection. In the factories these basic operations are usually combined in individual plants since the raw materials and equipment for the reactions, for separation and purification, are very similar. Plants, or parts of plants, have been equipped for the following operations:

(a) halogenation,
(b) nitration,
(c) sulphonation,
(d) reduction,
(e) separation by melting, alkali fusion,
(f) carboxylation,
(g) amidation.

Distillation and fractionation, grinding and mixing of the finished product are also concentrated in certain sections of the plants. First, a survey will be given of the equipment used in these operations and some problems concerning materials will be discussed.

### (a) The equipment used in intermediate chemistry; problems concerning materials

When a process is to be scaled up to the industrial level the question of materials is of paramount importance.

Classification of materials of construction:

(1) metallic materials (ferrous and non-ferrous metals),
(2) inorganic (non-metallic) materials,
(3) organic materials.

(1) Metals have especially favourable physical properties (e.g. thermal and electrical conductivity). In contrast to many other inorganic materials they can withstand abrupt changes of temperature. 'Frigidal' is a FeNi alloy with low thermal conductivity, which is important for some special applications. The strength of metals, particularly as far as tension, impact, and bending are concerned, is very high. They are easily worked and can be made to adopt any shape, e.g. by bending, casting, forging, rivetting, welding, and soldering. In the combustion of ammonia the gauzes, 2m diameter, used as catalysts, have 3,600 apertures of 0·04 mm diameter per $cm^2$, which shows their great mechanical strength.

Spinning nozzles made from tantalum can have 3,000 drillholes in a few $cm^2$. In many metals chemical strength is lower (except in Au, Pt, Pd, Ag, and Ta). Aluminium must be considered as a very base metal, it is attacked even by water; it is, nevertheless, resistant to air and water as well as to oxidizing substances since it is protected by an oxide film. When this oxide film is destroyed, e.g. by amalgamation caused by the breaking of a mercury thermometer in an aluminium apparatus, it is rapidly corroded. The stability of all materials capable of forming an oxide film is good. Recent investigations (Schwabe) have shown that the passivity of metal surfaces is based not so much on the formation of a protective layer but is rather a question of the crystalline structure and the formation of certain types of surfaces.

(2) With inorganic non-metallic materials the position as compared with that of metals is reversed. Some are universally chemically stable. Their disadvantages consist in the fact that they cannot be worked like metals and that they are poor conductors of heat and electricity.

(3) Organic materials like wood, rubber, leather and, above all, in recent times, plastics, are in many cases chemically fairly stable and can be worked easily. They suffer the disadvantage of having only limited thermal stability. The upper limit is 130°C except for special plastics.

In chemical plants copper used to be the most widely used material apart from iron. The part played by the coppersmith was very important. Copper was used for reaction vessels, particularly stills. Copper is resistant to acetic acid and many organic acids. Nowadays aluminium has replaced it in many applications, especially pure aluminium (with an Al content above 99·5 per cent) which is resistant to organic acids, acetic anhydride, and concentrated nitric acid.

Lead has been used for a very long time, e.g. in sulphuric acid plants and sulphite cellulose works. It is resistant to sulphuric acid of many concentrations below 78 per cent, and to sulphurous acid and bisulphites. Lead-lined stills are very important in modern sulphonation plants and for many operations carried out in acid media which attack iron, e.g. in chlorination. A lead-lined apparatus is easy to heat or to cool; this is rather more difficult in brick-lined equipment unless heating or cooling coils of lead or acid-resistant steel are used. The apparatus is lined with sheet lead or lined 'homogeneously' with lead. In homogeneous lead-lining, molten lead is applied to the iron surface which has first been polished by sandblasting or turning, and may also have been tinned. The lead lining then adheres very tightly and does not lift off when a still is evacuated. The chemical stability of lead depends to a large degree on its content of admixtures of other metals. Thus some lead linings produced from scrap lead often have only a short life. Hard lead contains 10–12 per cent antimony.

Silver conducts heat best of all metals. Its corrosion resistance is excellent, particularly to hydrochloric acid, provided the concentration is not too high. Only tantalum is resistant to highly concentrated hot hydrochloric acid. Silver is also resistant to all types of acid chlorides. Recently, titanium has been adopted increasingly for the construction of acid-resistant equipment.

Steel can be plated with nickel silver (FK silver) in a hot-rolling process, a silver layer of 1 mm thickness being adequate. Such plates can be welded easily which permits the design of apparatus of any desired shape and size.

Equipment is also made of nickel. Nickel is resistant to alkalis at high temperatures. Particularly resistant to heat are alloys of chromium-nickel and iron-chromium-nickel. Bronze, an alloy of copper and tin, with possibly some additions, is used wherever great strength is important, e.g. for valves, bearings, kneading machine arms, worm gears, and for non-sparking tools.

For many purposes lead and noble metals have been rendered obsolete by special steels. The chemical stability of steel, particularly to acids, can be greatly increased by the addition of chromium, nickel, molybdenum, aluminium, and copper. However, not only the composition is important but also the crystalline structure. Alloys of iron and silicon (cast silicon) have no great mechanical strength and are difficult to work but are very resistant to acids except to hydrofluoric acid and hot hydrochloric acid. Acid pumps and stills for the concentration of inorganic acids are made of cast silicon.

Extraordinarily resistant to heat and corrosion at elevated temperatures is sichromal steel, an alloy of iron and varying quantities of Cr, Al, and Si. It is used for high-pressure steam boilers, preheaters, catalyst tubes, e.g. for cracking plants, for Geissen low temperature carbonization ovens, etc.

The most important material has always been iron in its wrought and cast forms. No other material can be used in so many applications. Apart from its great mechanical strength it is also the cheapest material for the construction of chemical equipment. It is impossible to discuss here iron-carbon alloys and their properties. Iron is resistant to alkaline media of not too high concentrations but not to acid media. Strong alkalis, particularly when hot, cause iron to become brittle by the destruction of the carbide fraction, the 'alkali brittleness'.

Whilst the resistance of lead to sulphuric acid decreases with increasing concentration owing to increasing solubility of the sulphate, that of iron increases. In general, the degree to which metals are attacked by chemical substances depends on the reactivity of the covering layers formed. The effect of inhibitors is based on the, at least temporary, inactivation of the metal surface (protective layer) or the prevention of the formation of local reaction sites. One of the causes of corrosion is often oxygen; protection can be obtained by the addition of reducing agents. In halogenized hydrocarbons stabilizers prevent the splitting off of hydrochloric acid. Acidification is started by an autoxidation process against which antioxidants are often effective.

Iron can be used with sulphuric acid at a concentration of more than 70 per cent, therefore also for oleum. Cast iron is more stable than wrought iron steel. For all processes using acid media cast iron is a more suitable material. Cast iron should be rich and steel poor in carbon. Iron is an unsuitable material for gaseous hydrochloric acid in the presence of water, but in the absence of water it can be used perfectly well, particularly at temperatures above 100°C.

Ammonium chloride in its liquid and vapour states is more corrosive than would correspond to its dissociation into $NH_3$—HCl. Corrosivity decreases with increasing excess of $NH_3$ so that it is quite possible to carry out conversion of halogenated hydrocarbons with ammonia to amines in iron even at high-temperatures (e.g. p-nitrochlorobenzene → nitraniline at 170–180°C) if a sufficiently high ammonia excess is maintained.

Whilst nitric acid of all concentrations corrodes iron very strongly, mixtures of nitric acid and concentrated sulphuric acid can be stored and reacted in iron, e.g. for nitration. This is of great industrial importance. A mixed acid composed of 11 per cent $H_2SO_4$ and 88 per cent $HNO_3$ (1 per cent $H_2O$) has little corrosive effect. Mixed acids that do not contain water, with 30–40 per cent $HNO_3$, balance $H_2SO_4$, can be freely stored in iron. Mixed acids are almost always used for nitration. The protective effect of sulphuric acid is based on the reduction of the water content of $HNO_3$, and the formation of the stable $Fe_2O_3$ form and of a protective layer consisting of insoluble sulphates. If a residual acid after nitration still contains 70 per cent $H_2SO_4$ iron can be used. With easily nitratable aromatics, such as toluene, xylene, and naphthalene, the concentrations are below 70 per cent and the nitration must, therefore, be carried out in lead-lined or brick-lined apparatus.

Iron, even cast iron, is rather reactive to strong hot alkali lye which causes it to dissolve as a ferrate, particularly in the presence of air.

In intermediate product chemistry enamelled iron vessels are very extensively used. Formerly, only cast iron stills could be enamelled since the enamel was not sufficiently elastic to follow the movements of relatively thin sheet steel. There are, however, types of enamel (e.g. Pfaudler enamel) which are suitable for the enamelling of steel stills. The advantage of such stills lies in their better thermal conductivity for heating or cooling as compared with cast iron ones. Enamel is resistant to all acids, even strong hydrochloric acid, but is attacked by alkali. It may be possible to work in enamel with a dilute hot soda solution but never with a concentrated one, or with caustic alkali. Enamel stills suffer from the disadvantage that they are very vulnerable to mechanical damage and rapid heating and particularly to rapid cooling. Their upper temperature limit is in general about 200°C. Heat must be supplied by steam or by heating baths.

To provide protection for iron equipment it is often lined. Ceramic plates are the most commonly used linings e.g. for agitator vessels, suction apparatus, condensers, and storage vessels. According to the cement used for the joints the vessels are either 'acid-lined' or 'alkali-lined'. It is not possible to change the medium subsequently. Bonding and pointing agents are cement, acid cement (water-glass cement), Asplit (cold-hardening phenol-formaldehyde resin), and lead glycerate. Finely-powdered silicates, e.g. kieselguhr, finest sand, or glass powder are added. For linings that are to be specially resistant to acids two layers of plates are used, often on top of a homogeneous lead-lining or rubber-coating. A good and lasting lining requires much skill and experience on the part of the craftsman and takes time.

Phenol formaldehyde resins are largely resistant to acids, particularly hydrochloric acid. The vessels are lined with the untreated resins which is then hardened in the furnace ('baking'). Bakelized or phenytalized vessels are resistant to HCl, $SO_2$, and $Cl_2$ but not to alkalis. Haveg consists of asbestos and pheno-plastics.

To resist hydrochloric acid and other dilute acids, rubber, polyisobutylene, polyethylene, teflon, and polyvinylchloride linings can also be used. The lining which gives the best protection against mechanical strain is rubber-coating. It is used for agitator vessels, storage vessels, tank wagons, pipe ducts, pumps, valves, rotary filters, power presses, suction vessels, coils, in short, almost all the equipment used in the chemical industry. The difficulty in rubber-coating is due to the fact that for the vulcanization heated pressure vessels are required which for large items must be of considerable size. Rubber-coatings can be used only for pure aqueous solutions. In the presence of even small quantities of organic materials the rubber swells and disintegrates. Oppanol behaves similarly with solvents. The behaviour of bakelite is more favourable. On the other hand, as mentioned earlier, rubber-coating, particularly with soft vulcanizates, offers great resistance to mechanical strain. When solid, abrasive material is suspended in the liquid to be agitated, all the materials are quickly attacked mechanically and both agitator and wall are soon abraded. A soft rubber-lined still and rubber-coated agitator on the other hand have a very long life.

Equipment made of polyvinylchloride, such as Vinidur, PVC, Ekadur (hard), Ekalit (soft), either made of welded sheets or lined with foil, is widely used in industry. This material is resistant to almost all acids and alkalis, to chlorine and $SO_2$. Pipes, valves, and fittings of all sorts are made of PVC. It can, however, be used only for temperatures up to a maximum of 70–80°C, it is also rather sensitive to mechanical shock, and parts made of Vinidur must be handled very carefully.

With all plastic materials their poor thermal conductivity and their sensitivity to heat must be taken into account. They are not very suitable for coolers and heat exchangers.

In a special position are apparatus, linings, coolers, and heat exchangers made of graphite saturated and covered with a hardenable synthetic resin, e.g. phenolformaldehyde resin (Igurit, Karbate, Durabon, Bascodure). Heat transfer is better than with plastic and ceramic materials so that it is possible to design coolers and heat exchangers with very high performances. The material is resistant to hydrochloric acid and other inorganic and organic acids and very insensitive to thermal shock.

Ceramic products will be discussed next. Clay vessels for distillation, coolers and condensers were used earlier by the alchemists and they used to be the only possible material for reactions in acid media. Clay apparatus of remarkable strength must have been available to judge for example by the fact that in 1526 in Nordhausen am Harz 'vitriol oil' was produced in clay vessels by bringing iron sulphate to white heat. Even today many fittings, suction apparatus, pipes and pumps are made of clay, however only where the temperature is low (up to 80°C) and no thermal shocks occur. Storage vessels of up to several thousand litres capacity, clay condensers, absorption plants for hydrochloric acid, absorption and scrubbing columns, agitator vessels and ball mills, are all made of clay. Its high mechanical sensitivity limits its usefulness. The same applies in principle to the ever-improving equipment made of porcelain. It is now possible to use it at quite high temperatures if it is heated uniformly and above all if it is protected against rapid and non-uniform cooling. Agitator vessels, stills, columns, and coolers made of porcelain are resistant to almost all chemical agents.

Great progress has been made in recent years by the development, e.g. by Messrs. Schott, of industrial glass equipment in ever larger sizes so that many industrial processes, in particular continuous processes such as photochlorinations, can be carried out with great advantage in glass.

Finally, mention must be made of the oldest material, wood, still widely used in the chemical industry. Its resistance to chemicals is often unexpected and remarkable. Acids and alkalis attack it superficially, of course, but it often has a very long life. Frames and chambers of 'acid' filter presses, suction filters, agitators, and, above all, vats are made from it. Wooden vats, used in the cellulose industry, tolerate for decades hot sulphite spent liquor containing bisulphite. Storage and fermentation vats are preferably made of wood, and wooden casks are still the most extensively used packing in the chemical industry. Equipment and vats are made preferably from pine wood and pitchpine wood, i.e. woods rich in resin, since they best resist water and decay.

### *Sealing materials*

An important part is played by sealing materials for pipe joints, flanges, lids for agitator vessels and autoclaves, and packing materials for stuffing boxes of agitators, autoclaves, pumps and valves. The most suitable of these is the rubber seal because it is elastic and always seals effectively without being compressed too strongly. Rubber is, therefore, the natural sealing material for enamelled, ceramic, and glass apparatus. A disadvantage of rubber is its sensitivity to organic solvents and to higher temperatures. In general, rubber seals can be used only up to 100°C.

Much more stable is a rubber-asbestos vulcanizate called 'Klingerit'. It is suitable for sealing flanges and lids even at higher temperatures and high pressure. For even higher temperatures, e.g. for the sealing of lids of autoclaves, flow pipes, etc., asbestos alone can be used which can be impregnated for example with water glass. Asbestos can also be coated with thin copper or silver sheets and used as a seal (e.g. for cylinder heads of car engines). Lead seals are also used for many purposes, particularly for packing stuffing boxes of agitator autoclaves and pumps. Great advantages are offered by the so-called asbestos-graphite braids which are saturated with high-boiling mineral oil (e.g. superheated steam cylinder oil) for sealing stuffing boxes of agitator shafts and pump axles. Superheated steam cylinder oil is often forced into such packings by means of a Bosch pump in order to reduce friction and to replace constantly small quantities of oil that are being lost.

For many high pressure ducts metal seals are used, e.g. silver and copper, and for high pressure hydrogenation, mild steel. Plastic seals are gaining wider acceptance for many purposes, e.g. polyvinylchloride which, however, can be used only at temperatures below 80°C and must contain very high boiling and chemically stable plasticizers.

Especially good is the chemical and thermal resistance of polytetrafluoroethylene and polytrifluorochloroethylene ('Teflon' and 'Hostaflon' respectively).

When a process has been designed in the laboratory in glass apparatus the problem of the material to be chosen for the industrial process arises. Many otherwise promising processes have come to grief over this problem. In order to discover a suitable material corrosion tests are made in the laboratory with small specimens of certain dimensions which are suspended in the reaction vessels. The loss is determined by exact weighing and the reduction, e.g. in thickness, is calculated for a longer period or for a large number of applications. Microcrystalline changes in the surface are also determined.

The task which the Deutsche Gesellschaft für chemisches Apparatewesen 'Dechema', Frankfurt a. M., has set itself is to promote the development of chemical apparatus and of chemical technology. The exchange of information in the field of chemical technology takes place mainly at an exhibition held every three years, called 'Achema', by means of an annual 'Achema-Jahrbuch', and other publications. Information on problems concerning materials is found in the 'Dechema-Werkstofftabellen' ('Dechema material tables') supplements to which are published regularly.

*Accessory equipment*

Still the most widely used standard appliance in the manufacture of intermediates by batch operation is the agitator vessel which should be pressure and vacuum proof. The National Technological Control Office classes it with pressure drums and designates it as a 'pressure vessel'. A 'pressure drum regulation' lays down design characteristics and maintenance standards of these vessels and also provides for inspection. It may be

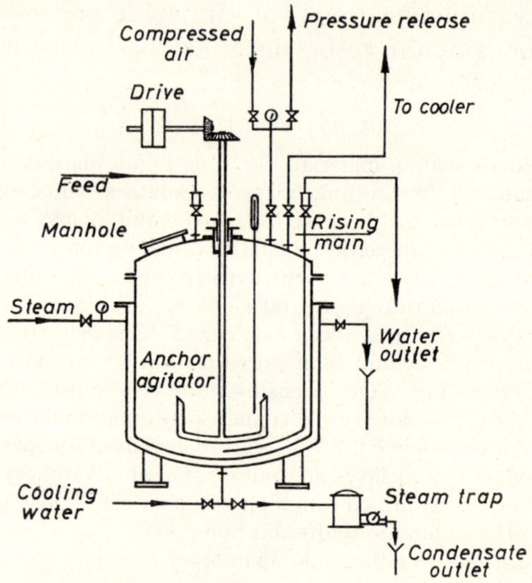


Figure 85. Diagram of an agitator vessel with heating or cooling jacket

mentioned in this connection that almost all industrial equipment is subject to constant technical supervision by experts whose reports are made under oath.

The agitator vessel is an enlarged form of the laboratory stirrer. Its design is shown schematically in Figure 85.

It is equipped with agitator, manhole, thermometer, pressure gauge, a rising main for emptying by blowing, feed with inspection glass, compressed air and vacuum feed, pressure release and coil or jacket for heating or cooling, and sometimes with a reflux cooler. For high temperatures the jacket may be manufactured by Samesreuther double wall welding or be equipped with Samka pipes. An agitator vessel used as a vacuum evaporator is shown on p. 324.

Stirring is very important for many reactions. A large number of agitator designs have been developed, the most important of which are shown in Figure 86. When transferring laboratory results to large-scale processes one must remember, as far as stirring is concerned, that the stirring effect depends not so much on the number of revolutions as on the peripheral speed of the agitator. The larger the agitator the slower can it run with equal efficiency. It is impossible to exaggerate the importance of the stirring

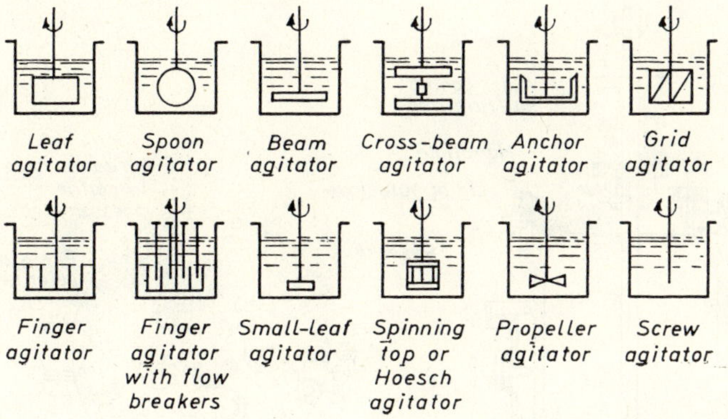


Figure 86. Agitator designs

problems for reactions in heterogeneous phases, but stirring is usually indispensable also in the homogeneous phase.

Agitators used to be driven mostly by belt drives. Nowadays direct drives are used almost exclusively, often with variable drives.

Pressure reactions are carried out in agitator autoclaves. A widely used design is shown in Figure 87. For safety reasons the test pressure is fixed at double the operating pressure. The design characteristics are shown in the sketch.

Since the stuffing box, which seals the agitator shaft into the lid, and the stuffing box packing often cause breakdowns as a result of leakage, designs of autoclave agitators have been developed which eliminate this source of breakdowns. The most widely used design is the electromagnetic stirrer which does not rotate but is moved up and down by a magnet.

*Means of heat transfer and heat dissipation*

Heat up to 100°C can be supplied and dissipated (cooling) by water (water bath). For higher temperatures high-pressure steam at various pressures is used. The usual operating steam pressure of 4 atm gauge pressure is sufficient for most purposes, e.g. boiling of aqueous liquids, distillation of liquids with boiling points up to about 100°C, heating to initiate chemical reactions, running of drying chambers, roller driers, and evaporating plants. Many reactions, however, require a higher temperature. Particularly frequently a temperature of 170–180°C is desirable for the generation of which there is usually available in the factory a steam supply with 10 to 12 atm gauge pressure. If even higher temperatures are required individual heating plants are mostly provided which are operated either by electricity using hot air with resistance or induction heating, or by coal or coke and producer gas, the heating being either direct (see p. 103) or indirect with the interposition of a heating medium. In direct heating by gas a circulation heater (see Figure 60) which makes it possible to maintain a very constant temperature has proved very advantageous.

Heat is transferred either by feeding in the steam directly or through worms, or by heating jackets, tubes cast into the still wall (Frederking principle), or heating plates (e.g. for drying chambers). Since for high pressures the walls of heating jackets would have to be very thick, the double walls are sometimes fastened by spacer bolts welded to

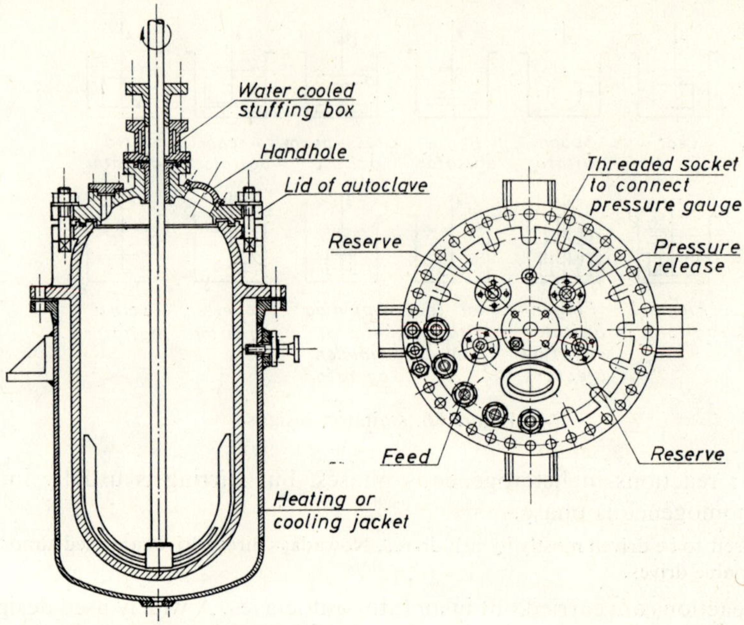


Figure 87. Agitator autoclave made of cast steel, with anchor agitator and heating jacket

| | Autoclave | Heating jacket |
|---|---|---|
| Operating pressure | 40 atm | 10 atm |
| Test pressure | 80 atm | 13 atm |
| Volume | 1,600 litres | |

the inner wall (Samesreuther welding) or by heating pipes (Samesreuther's Samka pipes), channels made from half-tubes, or steel sections, welded from the outside on to the walls of the vessel (Figure 88).

For the past twenty or thirty years one of the most common forms of heating above 200°C, or of dissipating heat in reaction plants, has been the high pressure hot water heating developed by Maschinenfabrik Sangerhausen, the 'Sangerhauser hot water heating'. Heat is transferred much more easily from liquid to solid than from gas to solid, and smaller heating surfaces will, therefore, be sufficient. It is most widely used as 'Perkins heating' in the steam ovens of modern bakeries.

Forced circulation of the pressure water by means of a pump permits easy control of the heating and ensures the maintenance of a highly constant temperature. Any desired lower temperature can be quickly obtained by feeding-in cold water. This is important for catalytic processes, e.g. for Fischer-Tropsch synthesis ovens, in reaction furnaces for the catalytic oxidation of naphthalene to phthalic anhydride, of anthracene to anthraquinone, the oxidation of ethylene oxide, for catalytic continuous hydrogenation, e.g. phenol to cyclohexanol, or aniline to cyclohexylamine, and many other processes. In these processes at the beginning heating is required, with cooling later. Figure 88 shows the mode of operation of a pressure water heating installation.

*Heating baths and heating liquids*

For smaller installations, however, a pressure water heating plant is too complicated and simpler heating baths are, therefore, used which are operated either by gas or by electricity.

Salt baths are little used, on the whole only for temperatures above 300°C. Circulation is difficult since salts set at relatively high temperatures, e.g. a zinc chloride melt. One which has proved to be very good is a nitrite-nitrate melt consisting of 40 per cent $NaNO_2$, 7 per cent $NaNO_3$, and 53 per cent $KNO_3$, with a setting point of 143°C, which at high temperatures has the viscosity of water and is then an excellent heat conductor. Such baths can give rise to dangerous explosions if they are contaminated by organic substances. A mixture of 47 per cent KOH and 53 per cent NaOH, with a melting point of about 185°C, can also be used.

Of the metals used as material for baths, lead has too high a melting point; more suitable are Wood's metal or Rose's metal, with melting points down to 70°C. Disadvantages are the high weight and the poor thermal conductivity. In a special position of its own is mercury with a melting point of −38·8°C and a boiling point of 375°C. In the United States heating plants are operated on mercury vapour. Apart from its high price and weight, its toxicity renders mercury hardly suitable as a heating liquid.

In recent times organic liquids have become increasingly popular for heating chemical equipment. Heating by mineral oil circulation which has been known for a long time and which permits temperatures up to 300°C will not be discussed in detail. All oils crack gradually above 200°C and are highly viscous at low temperatures which renders starting and circulation difficult. Glycerine and polyglycols decompose even more easily. There are, however, organic materials which are stable even at about 400°C and, as compared with water, have the advantage of a very much lower vapour pressure. A mixture of 73 per cent diphenylether and 27 per cent diphenyl ('Diphyl', 'Dowtherm') has proved to be particularly advantageous. Above 12·3°C this mixture has a low vis-

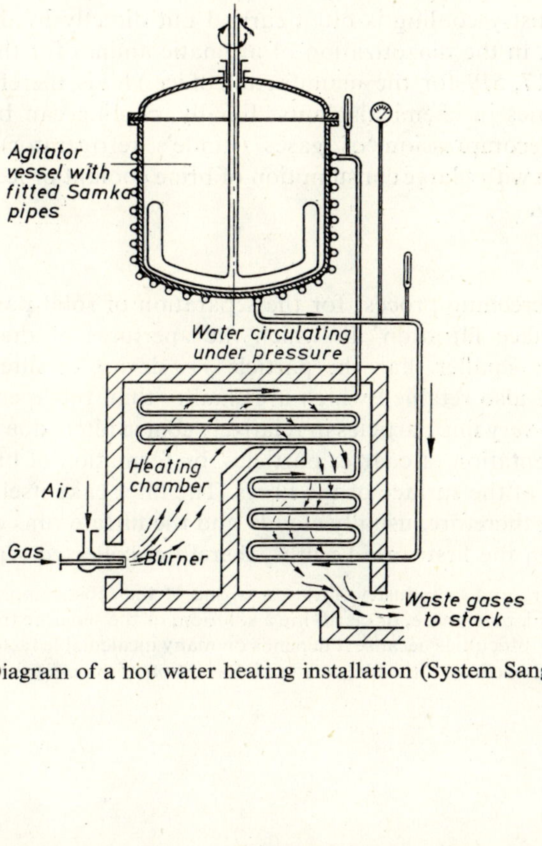


Figure 88. Diagram of a hot water heating installation (System Sangerhausen)

cosity and boils as an azeotrope at 225°C. The mixture can be used as a liquid or as a vapour with or without pressure, is non-corrosive, will keep at 350°C and above for years, and is non-toxic but flammable. The following table shows the vapour pressure of 'Diphyl' at various temperatures:

| Temperature (°C) | 250 | 275 | 300 | 325 | 350 | 375 |
|---|---|---|---|---|---|---|
| Vapour pressure (atm. abs.) | 0·88 | 1·57 | 2·57 | 4·03 | 5·96 | 8·51 |

A chlorinated diphenyl is used under the name of 'Aroclor'. It is non-combustible.

Recently silicic acid esters have begun to be used in heating technology, e.g. cresyl silicate, an isomeric mixture $(RO)_4Si$, 'H.T.ÖlC', and a tetra-arylsilicate, 'TAS'. The solidifying point of these oils is at about −30°C, they are stable up to 360°C, can be ignited only with difficulty, have a very high boiling point but are somewhat sensitive to humidity.

Even more stable are certain silicone oils and it can be assumed that these will find ever-increasing use.

### *Coolants*

The cooling agent for indirect cooling at temperatures up to about 25°C is water which can be cooled in cooling towers. The cooling effect of cooling towers is produced by draught (flue effect or fans). Water losses by evaporation are large. At low temperatures the consumption of water is very high, particularly in summer when river water often has temperatures exceeding 20°C, well water up to 15°C. In order to obtain lower temperatures, cooled salt solutions (brine coolers) are used as heat carriers, e.g. sodium chloride solution, calcium chloride and magnesium chloride solution, to which small quantities of a solution of caustic soda are added to prevent corrosion. The brine coolant is circulated by a circulation pump, it is cooled by compression or absorption refrigerators. In the chemical industry cooling is often carried out directly by dropping ice into the reaction vessels, e.g. in the diazotization of aromatic amines for the production of azo dyes. (See also p. 517, 519 for the manufacture of ice.) Ice is, therefore, manufactured in considerable quantities in chemical plants. Finally, cooling can be brought about by compression and decompression of gases (Linde's refrigerator), i.e. by electricity. Processes carried out with a large consumption of brine coolant or ice are, therefore, large consumers of energy.

### *Filtration*

Filtration is a wet screening process for the separation of solid particles from a liquid. However, only 'surface filtration' in which the apertures of the filter (tissue, sieve, porous material) are smaller than the particles, is sieving or filtering proper. During filtering particles are also retained which are smaller than the apertures of the filter. It is possible to retain very fine particles in relatively coarse filters due to constriction of the pores, e.g. by sedimentation of coarser particles, by formation of bridges, or by adsorption of fine particles at the surface of the fibres. The filter cake itself acts as a filter. The first filter product is, therefore, usually cloudy and the filtrate runs clear only gradually; in industrial filtration the first turbid run is, therefore, collected separately.

Both in the laboratory and in industry filtration is one of the most important and most common operations for obtaining a precipitate, or clarifying a solution, or the separate treatment of both. Filtration often presents great difficulties because it depends on many incalculable factors. It can be completely prevented by adsorption, swelling of the filter material, formation of minute gas bubbles, or unfavourable electrostatic phenomena. The most favourable conditions for filtration can be determined only by semi-industrial tests. Many processes failed because of the inadequate filtering capacity of one of the intermediate stages. By-pass processes are centrifuging and sedimentation. It is possible to influence the filtering capacity of a suspension. It is most important first to obtain a suitable particle size and shape. The separation conditions must be chosen so that these favourable conditions are attained. This frequently requires a long series of tests.

Preflocculation can often be obtained by suitable temperature conditions, addition of electrolyte, leaving to stand, addition of gels and powders as filter aids (e.g. kieselguhr), powerful mechanical movement, or by changing the degree of concentration.

In order to prepare a solution which contains a very finely dispersed precipitate for filtration for the purpose of clarification, another precipitate is often produced in the solution, e.g. $Fe(OH)_3$ or $Al(OH)_3$, or finely dispersed substances that can be filtered are added which adsorb the slurry. Filtration is often facilitated by a layer of charcoal spread over the filter surface.

*Suction filters*

The filter equipment used depends on the type and the quantity of the precipitate. For easily filterable, coarsely dispersed, or crystalline materials which permit the formation of a very thick filter cake, stationary suction filters are mostly used (Figure 89). These

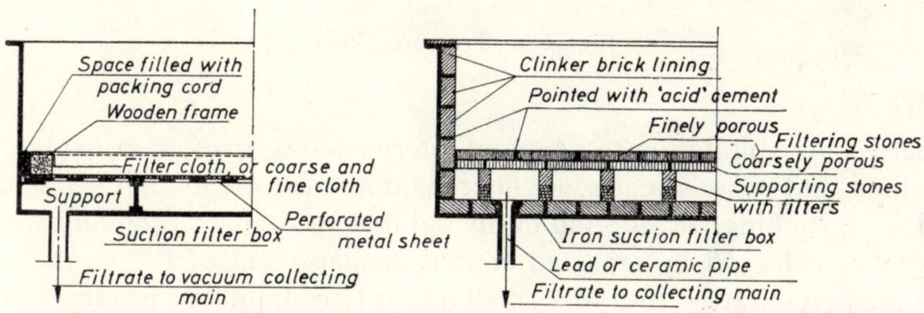


Figure 89. Left: Iron box suction filter with filter cloth. Right: 'Acid lined' box suction filter with filtering stones

can take the form of box suction filters, agitator suction filters, i.e. agitator vessels with filter tray for suction, or pressure filters (Figure 90). Pressure filters permit the use of a greater pressure drop than suction filters. The pressure medium is air or inert gas. With pressure filters it is also possible to separate precipitates from volatile, combustible, or harmful solvents since the suspension does not come into contact with the outside atmosphere. They are, however, less easy to operate than box suction filters since they have to be opened for emptying and then be resealed airtight for the next run whilst the box suction filter is simply emptied with shovels.

To facilitate emptying, pressure filters have been developed which are equipped with agitators that can be raised and lowered, and which have the manhole at the side. For small quantities, stoneware suction filters with perforated bottoms are used. For larger quantities, iron or wooden flat boxes are used with built-in perforated metal bottoms on which is placed a frame covered with filter cloth. The space between frame and wall is packed with a thick cord. For acid materials suction filters are brick-lined. Porous ceramic plates are often used as filtering media (Figure 89). In filtering by suction, sealing of cracks and 'displacement scrubbing' by covering with a little detergent are important.

Each suction filter has a feed tank, pressure filters, a pressure vessel, one or more collecting mains for vacuum filters for the suction, and a vacuum pump.

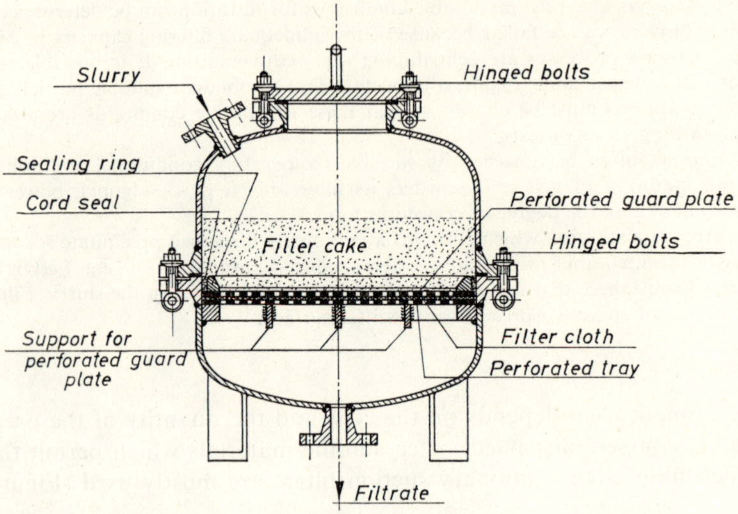


Figure 90. Pressure filter

*Filter presses*

Finely dispersed precipitates, such as many intermediates, dyes, iron oxide precipitates from reductions, and baker's yeasts, are filtered under pressure in filter presses. Filtration is followed by scrubbing. In filtering easily soluble substances economy of use of the washing agent is achieved by repeated careful displacement of the mother liquor and covering of the precipitate with a very small quantity of liquid. It is very convenient to scrub also in filter presses if they are specially designed for the purpose (wash presses). In contrast to suction filters, in wash presses the washing liquid does not pass through the filter cake in the same direction as the filtrate but countercurrently through special wash ducts, a much more advantageous system. Figure 91 shows the mode of operation of a filter press in the filling phase and in the washing phase.

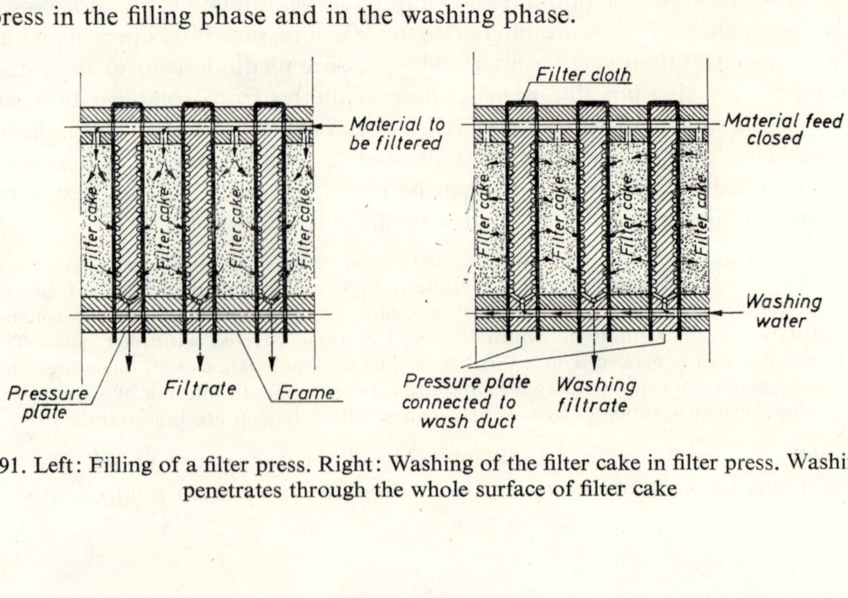


Figure 91. Left: Filling of a filter press. Right: Washing of the filter cake in filter press. Washing water penetrates through the whole surface of filter cake

A filter press consists of a frame, and grooved, slotted, or perforated alternating plates. The plates are of iron, wood or other materials, and are fitted with handles and stays which can be moved along the two spars on which they rest. The filter cloths which must have circular apertures in the appropriate places for the washing fluid channels are placed across the plates so that they overlap on both sides.

The plates and frame are pressed very tightly together by means of a jack screw or by hydraulic pressure acting on a head piece; the seal is effected by the filter cloth (the 'press is closed'). Usually, a coarse undercloth is used, e.g. jute, and a closely woven top cloth, e.g. cotton. The suspension is pressed through the material feed into the chambers formed by the frame and the cloths ('filling of the press'). The filtrate passes along the grooved plates and leaves through the ducts which usually discharge through taps in the bottom part of each plate. The taps are arranged above a collecting channel so that the proper working of each chamber can be easily observed. If the liquid issuing is cloudy the tap can be closed and the chamber is thus disconnected. This prevents a leaky cloth from spoiling the whole filtrate. The filtrate can also be drawn off through a collecting channel which passes through all the plates and the frame.

For the washing operation (see Figure 91) the taps of the chamber or the channel for the filtrate are closed and the washing liquid is forced through alternate plates which for this purpose are drilled with special boreholes. The washing liquid, evenly distributed over the whole surface, penetrates through the filter cloth and the filter cake and runs off through a special channel at the opposite plate. In the same way the filter cake can be pre-dried by injecting compressed air ('drying of the press by air injection').

When the filtration is completed the press is 'opened', the frame and the plates are slid apart manually and the filter cake is dumped from the frame into a truck underneath ('emptying of the press').

There are many different designs of filter presses. Some for example are equipped for heating or cooling, others are enclosed to exclude outside air. There are two basic types of filter presses, the chamber filter presses, suitable for clarifying liquids with low solid content, and frame filter presses which can hold very large quantities of filter cake.

Filter presses are designed in frame sizes up to $1{\cdot}50 \times 1{\cdot}50$ m and capacities up to $1{\cdot}5\ m^3$. They have the following advantages: large filter area (upwards of $100\ m^2$), high performance, particularly for slurries with great compressibility, small bulk, and the possibility to choose among many different materials for the plates, chambers, and filters. Thus there are 'acid presses' made of wood (pitch pine, pine, larch), rubberized, or of ceramic material with wool cloths or acid-resistant synthetic-fibre cloths such as PC-fibre, polythene, and Orlon, and 'alkaline presses' made of cast iron with cotton cloths or cloths made of vinyl chloride polymerizate mixtures, polythene, or Terylene.

The disadvantages of filter presses are the great amount of manual labour necessary for charging, closing, opening, emptying, and cleaning, i.e. batch-wise operation, and the often quite high consumption of filter cloths. Recently, fully mechanized or automatic filter presses have come into use on an increasing scale. Some of the disadvantages are thus obviated.

### *Special filters*

It is impossible here to discuss in detail the many designs of batch-operated filters. Only a few can be mentioned.

Scheibler filters have very large filter areas. They are bag filters. A very large filter bag, closed all round, is folded and pulled concertina-like over a collapsible frame. The frame is hinged at the four corners and covered with chains or knotted elastic cords to keep the bag extended. It is collapsed, pushed through a narrow opening in the bag and then spread out. The upper part of the side frame is also the discharge pipe for the filtrate. A number of such individual units are combined in a box and the discharge pipes are joined together. The liquid enters from outside through the bags into the frames. Scheibler filters are used when small quantities of very fine slurries are to be extracted from large quantities of liquids. The filter is cleaned by back washing and the solids are lost.

Sweetland filters and Kelly filters are shell-and-leaf filters. The filter elements are contained in a pressure housing which in the Sweetland filter has a longitudinally hinged top for the discharge of the filter cake. In the Kelly filter the carriage with the filter leaves, which runs on rails, is withdrawn through the front of the housing. Another type is the dip filter designed by Moore in which the evacuated filter leaf batch is dipped into, and lifted out of, the material by a hoisting crane, and the solids are removed by compressed air.

Whilst all these designs have only a limited range of applications, the continuously operating rotary filters described below can be used for almost any purpose.

*Rotary filters*

Rotary filters are designed as drum-, disc-, or plane filters. The most widely used and most versatile filters are the drum filters which may be either external filters (deposition of the filter cake on the outside of the drum) or internal filters (deposition on the inside of the cylinder jacket).

The simplest designs operating without a vacuum, such as the sifting drum machines used in the paper and cellulose industry, and the Strindlund filters, used by the same industries to desiccate the fibres, will not be discussed here. The simplest vaccum drum filters are for example the suction drums of papermaking machines and the non-cellular type vacuum filter designed by the Maschinenfabrik Sangerhausen.

The correct choice of filter cloth and a suitable method for the discharge of the filter cake are of decisive importance for the use of rotary filters. This can be ensured only by preliminary tests, first on a suction filter plate covered with the cloth to be tested and dipped into the slurry, then on a small experimental rotary filter. The performances determined in this way can usually be applied directly to filters of very large size (70 $m^2$ filter area). In the modern chemical industry drum chamber filters are used almost exclusively which permit separate discharge of filtrate and wash water; the filter cake is dried by a short period of suction, and dumped by releasing briefly the vacuum or replacing it by compressed air.

This is achieved by dividing the rotary drum which is covered with a sieve and the filter cloth, on the inside, into parallel chambers with radial partitions (see Figure 92). Each chamber is connected by pipes with the fixed 'control head' at the front of the drum which is carefully ground and past which slide the openings of the pipes from each chamber, giving an almost airtight fit. The control head is subdivided and has individual chambers and separate suction connections for the filtrate, the wash water and the air and often also a short zone for the compressed air feed. The drum dips into a trough containing the slurry and is charged during the passage with filter cake which on leaving the liquids is dried by suction. As the drum continues to rotate the filter cake passes into the wash zone where water is sprayed on to it or where a felt washing strip which is continuously sprinkled with water moves along with the drum. Sometimes rollers compress the washing strip in order to increase the density of the filter cake. A short suction zone follows, then the release of the vacuum, and finally at the point of removal a slight internal pressure.

The manner of removal of the filter cake depends on the characteristics of the suspension to be filtered (see sketches in Figure 93). Crystalline materials forming thick layers can be removed simply with a scraper. For many inviscid materials, e.g. hot-separated plaster of Paris, 'cord removal' is used in which string wound closely round the whole drum is stripped off by a comb. Viscous materials which are often produced in thin layers are best removed by a rubber roller running at the same speed which in turn is cleaned by a scraper. In order to achieve good filtering performance in this case it is necessary for the removal of the filter cake from the cloth to be as complete as possible so that a practically clean cloth dips into the material. For this purpose, in addition to the removal by a single roller, various complicated methods are used, e.g. roller combined with removal roller, double rollers, and perforated rollers.

In future the drum chamber filters are likely to become the most widely used filtering device.

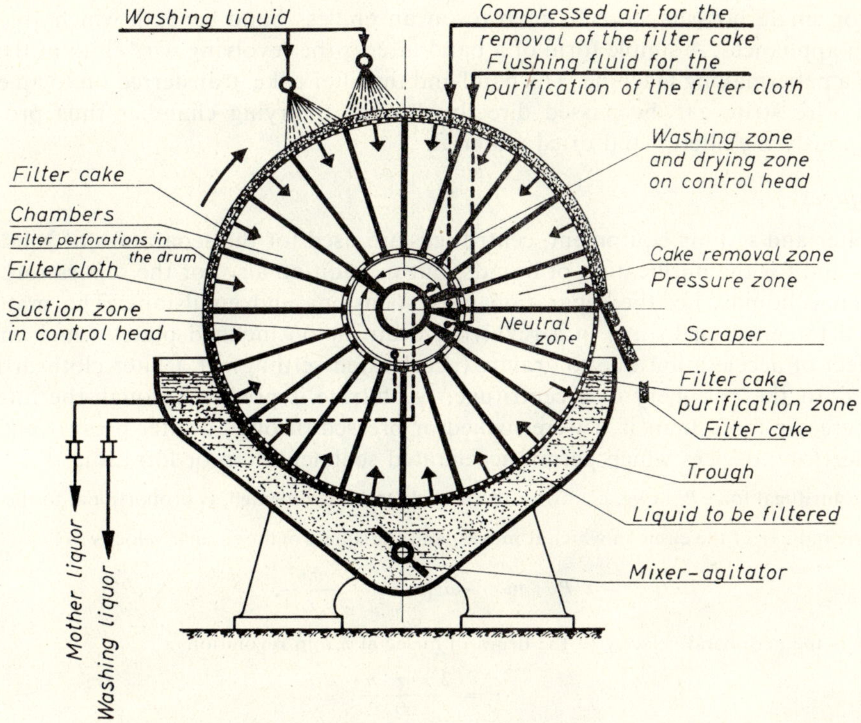


Figure 92. Rotary filter (plate filter), Schematic diagram

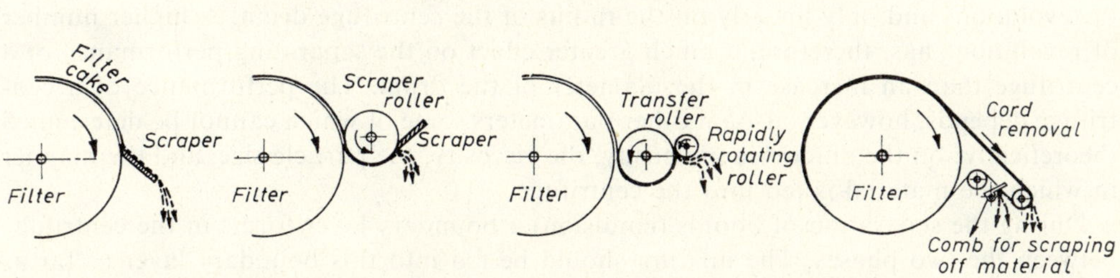


Figure 93. Removal of filter cake in rotary filters

Disc filters have filter discs divided into easily replaceable segments. These segments, which are covered by filter cloth, are fitted on to a hollow shaft which also receives the filtrate. Many discs are arranged side by side on a shaft, each with its pair of filter cake removal rollers which must be conical, or a pair of scrapers. The advantage of disc filters is the possibility of creating large filter areas in a small space. Filtration and particularly the removal of materials which are difficult to filter are, however, neither as reliable nor as adaptable as with drum filters and they cannot, therefore, be used as universally as the latter.

Other filters in this group are the vacuum band filters or band cell filters (e.g. designs by Imperial, Wolf-Buckau, Lurgi) in which filter cloths revolve on perforated rubber tapes or single box-shaped cells revolve on an endless band beneath which are placed suction appliances. A simple form of a band filter is the revolving wire cloth at the input end of a papermaking machine. The band and the filter cake, transferred on to an endless woven wire strip, can be passed directly through a drying chamber thus producing continuously the filtered and dried material.

*Centrifuges*

Like filter and settling equipment, centrifuges are used for the separation of solids from liquids and for the purification of liquids, but in addition also for the separation of two immiscible liquids, i.e. the separation of suspensions and emulsions. The material is exposed to centrifugal force in a revolving drum or on inclined plates which multiply the effect of acceleration due to gravity (accelerated settling). If a filter cloth or a sieve is fitted to the periphery of a centrifuge, the filtrate is pressed through the filter with much greater force than if it were sucked or pressed through a filter press. Centrifuges are, therefore, devices which permit accelerated sedimentation or filtration.

The centrifugal force $P$, in kg, acting on the particle to be sedimented, is proportional to the mass $m$ in kg, the radius $r$ of the circle in which it moves, and the square of the angular velocity $\omega$ $\left(\omega = \frac{2\pi n}{60}\right)$

$$P = m \cdot r \cdot \omega^2, \text{ or } P = \frac{m v^2}{r},$$

where $v$ is the peripheral velocity of the drum in m/sec at $n$/min revolutions:

$$v = \frac{2\pi \cdot r \cdot n}{60}.$$

The acceleration of the sedimentation depends, therefore, on the square of the number of revolutions and only linearly on the radius of the centrifuge drum. A higher number of revolutions has, therefore, a much greater effect on the separating performance of a centrifuge than an increase in the diameter of the drum. The performance of a centrifuge depends, however, also on other parameters some of which cannot be determined theoretically: on the difference in density, the viscosity, the particle size, and the manner in which the material is fed into the centrifuge.

During the separation of liquids (emulsion) a boundary layer forms in the centrifuge between the two phases. The mixture should be fed into this boundary layer as far as possible in laminar flow. The performance of a small centrifuge is usually a little better in laboratory tests. The greater the thickness of the boundary layer, the greater is the likelihood of the formation of turbulence, the danger of advance or retarding of liquid layers. Though attempts are made to prevent this by fitting fins into the centrifuge, only a large-scale test can show whether this has been successful.

The most important types of centrifuges are shown schematically in the diagram in Figure 94: screen centrifuge (sugar centrifuge), solid jacket centrifuge (peeling centrifuge) with vertical shaft, screw discharge centrifuge, and plate centrifuge (separator).

Screen centrifuges are designed with vertical and horizontal, rigid or suspended (flying) shafts. The perforated drum jacket is covered with a fine screen or filter cloth. The drum is placed inside a vessel to receive the filtrate which runs off through a drain valve. The solids can be removed manually or mechanically by a knife or spoon which are introduced very slowly at a lower revolution rate and are driven hydraulically. The centrifuges operate at 500–1,200 r.p.m. In the special design with automatic bottom

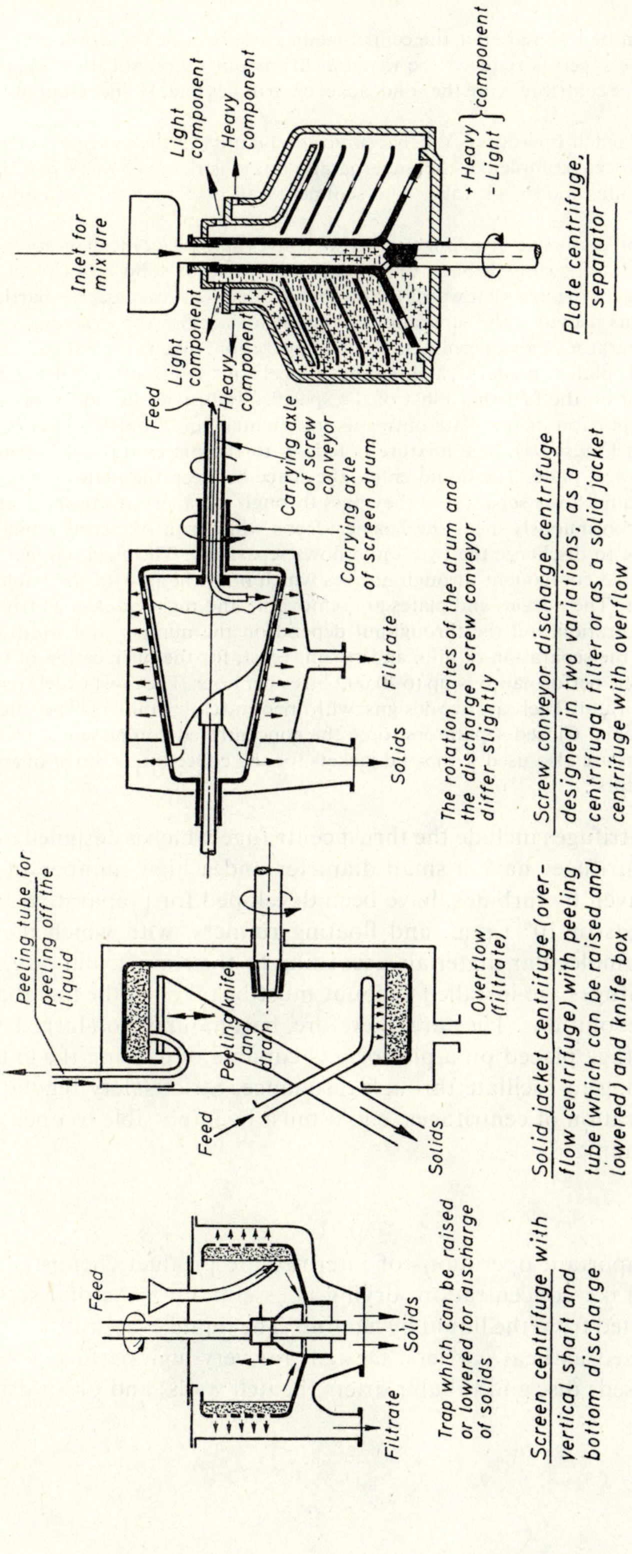


Screen centrifuge with vertical shaft and bottom discharge

Solid jacket centrifuge (overflow centrifuge) with peeling tube which can be raised and lowered) and knife box

Screw conveyor discharge centrifuge designed in two variations as a centrifugal filter or as a solid jacket centrifuge with overflow

Plate centrifuge, separator

Figure 94

discharge the bottom can be lowered after the centrifuge has stopped, or the drum is conical and open at the bottom. When the speed is reduced the material drops out automatically (sugar centrifuge).

The most powerful large centrifuges are the solid jacket centrifuges (e.g. Heine, Haubold, Escher-Wyss, VEB Nagema).

They are suitable for batch operation. When a sufficient quantity of the mixture has passed through, i.e. after charging has been completed, the supernatant liquid is removed via a peeling tube and the solids with a peeling knife, spoon, or tube. The sediment can also be removed continuously via a peeling tube.

In the screw conveyor discharge centrifuge, a screw conveyor or a conical scraper with a slightly different number of revolutions runs through the conical perforated or solid bowl drum, conveying the solids continuously along the drum wall towards the discharge opening. This design is particularly suitable for very large throughputs provided the separation requirements are not too exacting.

Plate centrifuges ('separators') also permit continuous operation. The rotor consists of a large batch of slotted plates (40–100 plates) made of high strength steel and running at 5,000–6,000 r.p.m. The separating effect is based on the fact that most of the specifically heavier liquid moves underneath the inner wall of the outer portion of the plate outwards, the lighter liquid on the outer wall of the plate inwards (see diagram in Figure 94). The mixture is fed on to the plates through a central feed pipe, discharging on to the lowest plate. The liquid enters the space between the plates through a slot in the lowest plate. The two liquid flows separate as they pass through the apertures in the plates. The lighter liquid flowing inwards continuously sheds the heavier drops which run outwards along the plate wall above. It is thus possible to discharge the two liquid flows separately. The essential factor is the forced run-off of the concentrated component through nozzles which limit the flow of the liquid and maintain the level in the separator. The nozzles and plates are removable and must be cleaned from time to time.

The degree of concentration and the throughput depend on the number and width of the nozzles. Separators are used for the separation of milk and cream, yeast, for the purification of tar oils, and for many other purposes. The performance is up to 40,000 litres per hour. The most widely used types are the Laval and Westfalia separator. There are two designs: with open and closed outlets. The latter is particularly suitable for foaming media. Closed separators force the concentrated component directly towards the desired point. Separators are also used with solid jackets for the collection of small quantities of solids, the 'clarification separators'.

Other types of centrifuges include the thrust centrifuge which is designed for continuous operation. Tube centrifuges have a small diameter and a high number of revolutions. Ultracentrifuges, driven by turbines, have been developed for preparative and analytical purposes, with speeds of $10^6$ r.p.m. and floating runners, with which it is possible to separate even macromolecular materials according to their molecular size.

When large centrifuges are installed account must be taken of the fact that they have a critical number of revolutions. They are, therefore, best mounted on large beds at ground level. A large centrifuge placed on a platform is capable of causing the entire steel construction of a building to oscillate through resonance. Strict safety regulations must be observed in the operation of centrifuges, e.g. it must be impossible to open the lid whilst the centrifuge is running.

*Drying*

One of the most important operations of intermediate product chemistry is the drying of solid materials. Like concentration, drying is essentially a thermal separation since the solids are separated from the liquid by heating. The liquid evaporates. Heated walls or hot gases can be used as heat carriers. Designs for very high performance driers have already been discussed: disc and tubular driers (heated walls) and drum driers (hot gas), see p. 21.

The material can be placed on a fixed support (shelf drier) or it can be continuously stirred and revolved as in the disc and drum driers, the pulverizing drier (illustrated in Figure 8), and the screw drier and paddle drier. In a class of its own is the atomizing drier which is particularly suitable for heat sensitive materials. (For details see p. 305).

In intermediate product plants the shelf drier is used most. In order to ensure better utilization of the heat it is designed for air circulation, e.g. Schilde-Haas's shelf drier heated by circulating air.

It has a great many applications and is particularly suitable for granular material such as filter press cake, crystalline materials, dyes, vegetables and fruit. The material is placed on metal sheets resting on mobile hurdles; these are removed from the drier for filling and emptying. The air circulation drier contains heating elements which heat the air circulated past them by a fan. Baffle plates direct the air on to the material. Part of the air current is discharged through an adjustable damper as exhaust air. This exhaust air current can be so adjusted that it carries a maximum amount of humidity. Drying times are between 10 and 24 hours. Under favourable conditions the consumption of steam is 2–3 kg for 1 kg evaporated water.

For temperature and air sensitive materials vacuum shelf driers are used. In these the heat is supplied by heated plates on which rest metal sheets supporting the material. Their design is identical with that of the well-known laboratory vacuum shelf driers.

Shelf driers require a large amount of attention. Dust is raised when the material is removed from the metal sheets. In tunnel-drying ovens the metal sheets move continuously through a tunnel through which hot air passes countercurrently. In the conveyor drying machine the metal sheets are replaced by a continuous perforated strip. The conveyor drier is sometimes combined with a rotary filter; the filter strip from the filter drum is then passed through a drying oven.

In the paddle drier the material moves through a heated tube in which it is continuously turned by blades. Drying can also be carried out under a vacuum; in this case the material is introduced into, and removed from, the drier through vacuum valves.

The design of ring shelf driers is similar to that of disc driers. The material on the shelves is moved on by scrapers and trickles from the top downwards against a stream of hot air. This drier can be considered as a variation of the continuous air circulation drier with moving material.

The type of drier chosen depends on the nature of the material, its sensitivity, and above all on the quantities involved.

### *Roller driers*

Roller driers are extremely efficient drying devices with minimum steam consumption and dimensions. Many delicate materials, such as foods (yeast, potato flakes), can be dried in them without being damaged since the drying process is very rapid. They are suitable for liquid, pasty and pulpy materials. Approximately 1·7 kg of steam is consumed per kg of evaporated water. Over revolving, heated, polished steel rollers is spread a thin liquid film which dries quickly and is removed with a scraper (thin film drying). The dry material is produced in the form of paper-thin plates, scales, or flakes. The driers differ according to the method of charging, i.e. the production of the film. There are single and double roller driers.

In order to utilize as far as possible the whole area of the roller, the points of charging and removal must be as close together as possible. This aim is achieved best by the immersion roller drier (see Figure 95). The roller dips directly into the dry material which is evenly distributed by a distributor bar. In place of the distributor bar it is also possible to use a charging roller which in turn dips into a trough. This type exists also as a double roller design with heating areas of 20 $m^2$. In roller driers the scrapers must be pressed against the rollers firmly but elastically and they must be carefully sharpened from time to time. The dried material must be scraped off very cleanly.

A widely used type is the double roller drier with top feed (see Figure 95). The rollers are pressed firmly together by sprung bearings to prevent any leakage of the liquid material. Between the rollers is the pouring sump in which the liquid is kept on the boil and is thus thoroughly preheated. The rollers revolve in opposite directions and the formation and distribution of the film is carried out by the rollers themselves. These driers occupy less space than any other type and are very efficient; they require, however, a somewhat higher steam pressure and especially careful servicing.

A particularly useful design for careful drying is the Escher-Wyss's spray roller drier. It combines the advantages of the spray drier (careful drying) with those of the roller drier (large heating area in a small space; see Figure 95). Instead of a charging roller which transfers the material on to the drying roller, discs tapering at the edge (spray discs) are used here which are mounted on a common rapidly rotating shaft. They dip into a channel containing the material and are coated with a liquid film. Air nozzles (spray nozzles) blow this liquid finely and evenly on to the drying rollers. Since these revolve rather rapidly (12–20 r.p.m.) and a very thin layer of the material is being dried, the process takes only a short time and the material is not damaged.

The heating steam is fed into the drying rollers through stuffing boxes. The condensate is discharged at the other end also through a stuffing box with a steam trap connected to it. The pressure to be used depends on the performance required, the thickness of the film and mainly on the permissible maximum temperature. It is between 2 and 6 atm.

In addition to food, chemicals of all sorts, organic intermediates, washing and textile auxiliaries, and dyes are dried in roller driers. They are the most widely used continuous drying equipment in industry. Vacuum roller driers also exist. In these the drying rollers, feeding arrangement, and scraper are contained in an airtight housing. The dried material is flushed out. In this type air sensitive materials can be dried.

A special type of roller drier is the cylinder drier such as is used for example in paper making (q.v.).

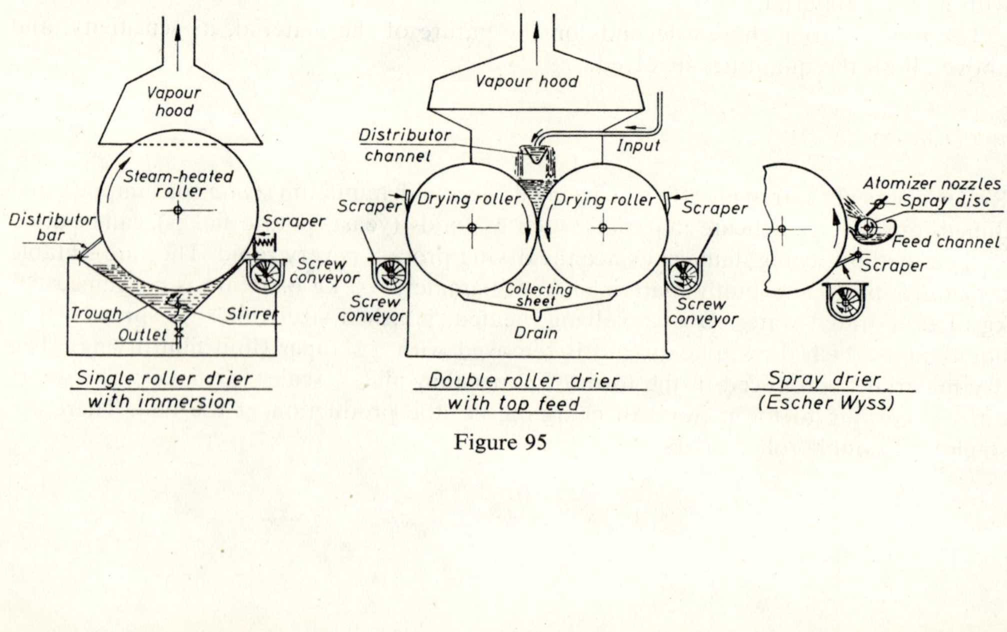


Figure 95

Similar to driers are cooling rollers but instead of being heated by steam they are cooled by a refrigerant, e.g. water or cooling brine. The roller dips into a melt (e.g. $\beta$-naphthol, naphthalene, paraffin). The liquid film adsorbed sets and is scraped off in flakes.

LITERATURE

F. A. Henglein, *Grundriß der chemischen Technik*. 9th ed., Verlag Chemie, 1955.

S. Kiesskalt, in Winnacker-Weingaertner, *Chem. Technologie*, I., Hanser Verlag, München, 1950, 2nd ed., Winnacker-Küchler, I, vol. 1, 1959.

Ullmann, *Encyklopädie der techn. Chemie*, vol. 1, Chemischer Apparatebau- und Verfahrenstechnik. Urban & Schwarzenberg, München–Berlin, 1951.

B. Matthias, *Chemisch-technische Arbeitsgänge und Apparaturen*. Verlag Dr. Bodenlander, Berlin, 1945.

H. E. Fierz-David and L. Blangey, *Farbenchemie*, 8th ed., Springer, Wien, 1952.

H. Travinski, Zentrifugen. *Chem. Ing. Techn.*, 1954, **26**, 189.

A. Eucken and M. Jakob, *Der Chemie-Ingenieur*. Akad. Verlagsges. m. b. H., Leipzig, 1935: Trocknung, vol. I, Part 3; Zentrifugen, Vol. I, Part 2; Filtern, Vol. I, Part 2.

Erdmenger and Neidhardt, Das Rühren von Flüssigkeiten. *Chem. Ing. Techn.*, 1952, **24**, 248.

W. R. A. Vauck and H. A. Müller. *Grundoperationen chemischer Verfahrenstechnik*. Verlag Th. Steinkopff, Dresden, 1962.

## (b) Basic operations in intermediates chemistry

### *General remarks*

The raw materials for the production of most aromatic intermediates are the simple basic hydrocarbons, e.g. benzene, toluene, xylene, phenol, naphthalene, anthracene, pyridine, carbazole, and pyrene, such as they are separated from bituminous coal tar and frequently converted in the plant itself from the crude products of tar distillation plants into pure intermediates. (Pure aromatics are now also supplied by tar distillation plants). The sequence of operations in intermediate product plants begins, therefore, with the distillation of benzene, the separation of carbolic oil by distillation, and the purification of higher aromatics such as naphthalene, carbazole, and anthracene. The first stages in the further processing are usually quite simple operations which recur time and time again, e.g. chlorination, nitration, sulphonation, reduction, oxidation, hydrolysis, fusion, carboxylation, alkylation, amidation, diazotization, and hydrolysis of diazonium salts to phenols esterification, saponification, and condensation. In only a few cases do these lead specifically to a definite product. Due to the different orientation of substituents, mixtures of reaction products are usually obtained. These have to be separated and the process is preferably chosen so that the substitution products are obtained in such quantitative proportions as they are required in industry. This is not always possible and where there are not sufficient uses for an intermediate it becomes a by-product or a waste product. Often there are several routes to an intermediate. The choice depends on technological-economic factors, the raw material position, the availability of auxiliary chemicals, equipment, energy consumption, labour costs, purity of product, waste and effluent problems. These conditions vary greatly in different countries and factories and the same product is, therefore, often produced by different methods and its cost price is also different.

### *Halogenation*

Halogens, mostly chlorine, less frequently bromine, and very rarely iodine, are usually introduced by direct action of the elements on the molecule which is to be substituted.

Fluorine compounds are mostly obtained from chlorine derivatives with anhydrous hydrofluoric acid.

Catalysts, almost exclusively ferric chloride, are used for nuclear substitution by chlorine. (In nuclear chlorination it is probable that it is the $Cl^+$ cation produced by the halogen carrier from $Cl_2$ that reacts, in nitration it is the nitryl ion $NO_2^+$, in sulphonation the cation $SO_3H^+$.) This is an ionogenic reaction. The substitution of the side chain must be carried out in the complete absence of iron and other carriers. The side chain chlorination is a radical chain reaction like that of the aliphatics (see p. 160). In the anthraquinone series the sulpho groups can also be substituted by chlorine. For this purpose the chlorine can be produced from hydrochloric acid and chlorate. Nascent chlorine is also obtained from hypochlorite and mineral acid. Hypochlorite in an alkaline medium produces a particularly mild form of chlorination whilst the most active chlorinating agent is the oxidation of hydrochloric acid with potassium permanganate or sodium chlorate. The reaction is facilitated by heat or light. If the halogen does not substitute at the desired site, or if for other reasons direct halogenation is impossible, it can be introduced into aromatics via the Sandmeyer reaction, i.e. by substitution of the amine group. This method has been developed in diverse ways in intermediate product chemistry and is used in numerous variations. In some cases it is possible to transform sulphochlorides into halogen compounds by splitting off $SO_2$. Sulphuryl chloride, $SO_2Cl_2$, is also yielding preferentially certain isomers depending on the method chosen (chlorine, hypochlorite, or sulphuryl chloride).

The chlorination reaction is exothermic and adequate cooling must, therefore, be ensured.

Usually during chlorination monosubstitution products and higher ones are formed. From benzene the two dichlorobenzenes, trichlorobenzene, and some tetrachlorobenzene are obtained in addition to chlorobenzene. Penta- and hexachlorobenzene are formed only under the most rigorous conditions. It is impossible to influence the o- and p-dichlorobenzene ratio. In side chain chlorination toluene always yields benzal-chloride and benzo-trichloride, in addition to benzyl chloride. The composition of the mixture depends on the degree of chlorination.

Selective monosubstitution is possible with gas phase chlorination, the monochlorine product being continuously extracted. The product to be chlorinated (benzene, toluene) is recycled (*Ber. Dt. Akad. Wiss.*, 1961, **3**, 643).

Side chain chlorination must be carried out in iron-free vessels, nuclear chlorination in iron or lead-lined ones. Cast iron is more durable than wrought iron.

For smaller products the large Jena glass reaction vessels are used which permit the introduction of fluorescent tubes or illumination from the outside. Formerly, mercury pressure lamps were used. Now, fluorescent tubes with maximum intensity in the ultra-violet range, which is perfectly adequate for the activation of the chlorine molecule, are becoming increasingly accepted (for instance, the RFT N 11/350 tube. *Elektrotechn.*, 1952, **6**, 471).

During chlorination hydrochloric acid is produced which is absorbed in water in scrubbing towers and marketed as crude technical hydrochloric acid. It must be borne in mind that the hydrochloric acid contains traces of benzene from the benzene chlorination, and attacks, for instance, rubber.

The chlorination products are separated by distillation in iron bubble cap tray columns or where there is a risk of HCl being cracked off, in acid-resistant packed towers. The separation of the solid products takes place, after they have been concentrated to the greatest possible degree by distillation, by crystallization in refrigerators, tube crystallizers, or on cooling rollers. The crystals can be purified with suitable fractions of the chlorination products or by means of solvents; for in general the crystals are perfectly pure and only their surface is covered by eutectic (in crystallization from melts) or by mother liquor. The intermediate chlorobenzene and the dichlorobenzenes form the basis for a large part of the whole intermediate product chemistry (see below). Under very rigorous conditions the chlorination can be taken as far as hexachlorobenzene.

*Nitration*

Nitration is the substitution of a hydrogen atom by the $-NO_2$ group. The reaction is carried out almost always with nitrating acid since if suitably composed it does not corrode iron, as already discussed in the section on materials and corrosion. Iron equipment can, therefore, be used provided care is taken to see that the remaining waste acid (residual acid) does not contain more than 30 per cent water. During nitration water is liberated which combines with the sulphuric acid: $RH + HONO_2 \rightarrow RNO_2 + H_2O$. The degree of the reaction of the nitric acid depends on its concentration, the water content of the residual acid produced, and the kind of substituents present. If, therefore, sufficient concentrated sulphuric acid is present which also acts as a good solvent for nitration, the nitric acid can be fully utilized. Owing to its good thermal capacity and its thermal conductivity, sulphuric acid permits the dissipation of the considerable heat of reaction by cooling. If it is desired to introduce several nitro groups an adjustment can be made to a higher residual acid, or an excess of nitric acid can be used. Addition of $SO_3$ is also commonly employed. The composition of the nitrating acid depends completely on the requirements of the nitration. Before the reaction is carried out the composition of the residual acid to be obtained must, therefore, be calculated. During the process a check is kept on the concentrations of nitric acid (with Lunge's nitrometer) and of sulphuric acid.

Thorough agitation and cooling are the main requirements. The compound to be nitrated can be introduced in solution in sulphuric acid and the nitrating acid added subsequently. It is very dangerous to introduce the nitrating acid and then to add the material. It may happen that nitration does not start at once. Often it begins spontaneously only after a larger amount of material has been added and an explosion results. Even when the nitrating acid is added to the dissolved material the nitration can become dormant and may start only after some delay. The heat of reaction thus suddenly liberated can then be contained no longer by cooling and an explosion occurs which has often led to grave accidents. It is, therefore, important to check in all nitrations that the reaction has actually started right from the beginning. This can be seen most clearly from the generation of heat. A delayed beginning of the reaction can be corrected by addition of nitrite, or increasing the content of nitrous gases in the nitrating acid. The nitration can also be arrested by a stoppage of the agitator. If the agitator then starts moving again, nitration begins spontaneously and will take an explosive course if more nitrating acid has been added in the meantime. A stoppage in the inflow of cooling water is also very dangerous. In nitrating plants the agitators are, therefore, coupled mechanically with the inlet control of the acid mixture in such a way that the inflow can be turned on only when the agitator is running. When the agitator stops the inflow is automatically cut off and acoustic and optical alarms are set off. Automatic warning devices also operate when the temperature rises too high.

The consumption of sulphuric acid in a nitrating plant is very large. The residual acid cannot be disposed of simply by diluting it with water, precipitating the nitro compound and allowing the acid to run off. Apart from economic considerations, strong corrosion of the waste water ducts, mainly by nitrous gases, would result. The isolation of the nitro products is carried out in various ways which cannot be discussed here. The aim is, however, always to recover the maximum amount of acid, including the excess nitric acid. The residual acid is denitrated in separate plants and is restored to a usable concentration by the addition of oleum or by evaporation in a Pauling plant.

In some cases the nitration is carried out with concentrated nitric acid with or without a solvent. Acetic acid is a particularly suitable solvent. Such nitrations are, however, very expensive and can be carried out only in acid-resistant vessels. The same applies to nitration with diluted (e.g. 10–20 per cent) nitric acid, which is used for materials that can be nitrated particularly easily, e.g. toluene-sulphonated amines. For this method alcohols, e.g. methanol and ethanol, are also suitable as solvents.

### *Sulphonation*

Sulphonation (sulphuration) is the introduction of a $SO_3H$ group. This renders the compounds soluble in water. For this reason almost all intermediate products for water-soluble dyes contain sulphonic acid groups, mostly in the form of their sodium salts. The sulphuric acid groups effect in acid dyes the bond with the organic fibre and in tanning agents the bond with the collagen. Sulphonic acid groups are introduced into aliphatic or araliphatic compounds, particularly long-chain ones, in order to render them surface active (wetting agents) or wash-active.

Aromatic sulphonic acid compounds are intermediates for phenols (e.g. phenol, resorcinol, $\beta$-naphthol) into which they are converted by alkaline fusion. Sulphonic acid groups are also introduced temporarily in order to carry out certain directed substitutions e.g. nitration, or in order to render atoms or groups mobile. They are subsequently cracked off by heating with diluted sulphuric acid under pressure (hydrolysis). The ease with which this cracking-off can be carried out varies and this fact is utilized for separations, e.g. of m-xylene from o- and p-xylene, because m-xylene-sulphonic acid is most easily hydrolysed.

In easily sulphonated compounds, such as naphthalene, the sulphonic acid group is introduced with ordinary concentrated sulphuric acid (95 per cent) or with 100 per cent sulphuric acid (monohydrate). For materials that react less easily, fuming sulphuric acid (oleum) with a concentration of usually 3–65 per cent, is used. '20' and '65' oleum are preferred because they do not solidify in the storage vessels owing to their low melting point. Temperatures vary from 0 to 160°C. In the naphthalene series the position of the sulphonic acid group depends on the temperature. Isomerizations can also be carried out by changing the temperature. Sulphonation is possible, too, with chlorosulphonic acid in concentrated sulphuric acid or in inert organic solvents. A large excess of chlorosulphonic acid results in sulphochlorides.

Various amines, e.g. aniline, yield aminosulphonic acids (e.g. sulphanilic acid) when their acid sulphates are heated in the dry state to a high temperature ('baking'). Baking can also be carried out by heating in a high-boiling, inert solvent, e.g. in o-dichlorobenzene. Sometimes halogen atoms are so mobile that they interchange with sulphite producing sulphonic acid groups.

The introduction into phenolic materials of the —$CH_2$—$SO_3H$ group with formaldehyde-bisulphite is particularly easy. Use is made of this property in the solubilization of dyes, tannins ('sulphitization') and textile auxiliaries. Under rigorous conditions it is also possible to obtain sulphones from sulphonic acids, e.g. from phenol via phenolsulphonic acid the important dihydroxydiphenylsulphone. Sulphones occur as by-products also in most sulphonation reactions.

Halogen- and nitro-groups, and any sulphonic acid present, render sulphonation more difficult, methyl-, hydroxy- and amino groups make it easier.

As in nitration, water forms during sulphonation:

$$RH + H_2SO_4 \rightarrow RSO_3H + H_2O$$

It retards the progress of the sulphonation. A large excess of sulphuric acid or the presence of $SO_3$ is, therefore, necessary.

The isolation of sulphonic acid is carried out in a variety of ways. Particularly favourable are the cases where the sulphonic acids separate during dilution, such as happens with the difficultly soluble sulphates or the ionic salts of many amines. Some sulphonic acids separate as free acids from sulphuric acid. They can then be removed by suction in acid-resistant suction filters. The tanks must be brick- or lead-lined.

The most common methods are salting out or 'unliming'. Many sulphonic acids precipitate as salts, frequently acid salts, when the sulphonation mixture is diluted by the addition of NaCl or $Na_2SO_4$, or of KCl, or $K_2SO_4$. The salts can then be filtered off, centrifuged, or pressed out directly. These sulphonic acids then always contain salt which, however, does not usually interfere during the manufacture of dyes.

Where direct precipitation is impossible the diluted sulphonation mixture is mixed with milk of lime, powdered lime (e.g. Buna lime, a by-product of the production of acetylene), or calcium carbonate (foaming!). The calcium sulphate is removed by suction filters, presses, or rotary filters and the filtrate is evaporated, the Ca salt of the sulphonic acid being a possible by-product. If it is desired to obtain the Na salt, the mixture, from which the calcium sulphate has been removed, is mixed with soda, separated from the precipitated $CaCO_3$ by filtering, and evaporated. Special separation methods have been developed for each sulphonic acid. For processing, the sulphonic acids must often be separated and these separation stages are amongst the most difficult operations of the intermediates industry. The decisive factor for successful 'unliming' is the easily filtrable form of the calcium sulphate produced. This is achieved by providing certain well-defined conditions in which precipitation takes place. It is known from experience that a precipitation temperature of more than 70°C is favourable. Detailed preliminary studies of the problem of the filtrability of the sulphonic acid salts and of the calcium sulphate on the laboratory and semi-industrial scale are necessary for working out efficient processes. Special methods are also required for each process by which the progress of the sulphonation can be checked and the end of the reaction determined. Since the sulphonation products are rarely homogeneous substances, very special and often very complicated analytical methods have been developed for their characterization. Sulphonation can be carried out in iron but the iron is corroded by the water used for dilution. The sulphonation mixture must, therefore, be diluted in lead-lined, acid-resistant brick-lined, or rubber-lined vessels. Very large amounts of calcium sulphate are produced from the acid waste liquor in sulphonation plants.

Pure free sulphonic acids containing no salt can be produced by means of exchange resins.

Sulphonation with gaseous $SO_2$ has been adopted quite recently, e.g. for the production of alkyl benzene sulphonic acids, large quantities of which are used as detergents.

### *Problems of waste water*

The wet calcium sulphate is loaded on mud trucks and transported into refuse pits or on to waste dumps. The acid waste liquors must be specially purified by unliming and then clarified by sedimentation or filtration. Intermediates and dyestuff plants usually have a central waste water purification plant. The waste waters converge into large tanks, acid and alkaline portions reacting with each other, and precipitation often takes place even at this stage. It is completed by the addition of milk of lime. Agitation is effected by the injection of compressed air. The water clarified by sedimentation or filtration, though usually coloured, is allowed to seep into suitable terrain, the ground-water stream of which must be entirely separate from any wells in the area, or it drains into a worked-out pit, or into the main sewer. The disposal of the waste water is often a difficult and costly problem for works that are not situated on large rivers.

### *Reductions*

Most reductions in intermediates chemistry are used for the conversion of the nitro-, sometimes also of the nitroso group into the amino group. Aromatic amines are the most important constituents of azo dyes, the largest dyestuff group, which are obtained by the diazotation of amines and subsequent coupling with phenols or other amines. Amines enter into the production of naphthol-AS-dyes at two stages: for the priming as 2-hydroxy-3-naphthoic acid arylide, and the developer (naphthol-AS-bases) which as diazotized amines are bonded on the fibre with the priming. However, amines are also the most important raw materials for triarylmethane dyes, quinonimine dyes, and many vat dyes.

Most reductions are carried out with iron in a wrought-iron reduction vessel equipped with a highly efficient agitator. Except for copper, other materials, including e.g. glass, are unsuitable. This applies even to laboratory experiments. In only a few cases so much acid is required that the iron is completely dissolved as ferrous salt. Any type of scrap iron can be used for reductions. For industrial reductions the method used almost exclusively is Béchamps' in which iron is suspended in water with very little acid. The development of this method by Béchamps in 1854 greatly contributed to the rapid development of the aniline dye industry. (Zinin had produced his aniline with ammonium sulphide from nitrobenzene.) Within a few years of its development industry used cast iron for the manufacture of aniline and other bases. However, only cast iron (grey iron) of a certain type and grain size is suitable, and this must in each case be tested by experiment. Intermediates plants possess iron grinding mills where cast iron, usually scrap, is ground to powder (not dust) in ball mills. The scrap iron used must be free from oil and fat and if necessary it must be cleaned by washing with solvents and then freed from dust. The iron powder is separated by screens into various classes of particle sizes.

In carrying out reductions the following procedure should be followed: Hot water is stirred by a high-speed agitator in a heated reduction vessel, usually of cast iron, with the total quantity of iron (an excess of 15–50 per cent) which is first cauterized with hydrochloric acid or acetic acid. During the whole reduction process the medium must be slightly acid. The nitro compound is then introduced in batches at the same rate as it is reduced, the temperature being kept constant (80°C, mostly 100°C). The progress of the reduction is checked by spot tests on filter paper, the colour indicating whether consumption of the nitro compounds is complete and whether there is always iron in solution. The latter can be recognized by applying a spot of a solution of ammonium sulphide next to the test spot. Where the spots touch a strip of black iron sulphide must appear. It is important that the phenylhydroxylamine stage of the reaction should be completed as quickly as possible. Phenylhydroxylamines condense with themselves to form azo compounds, or with nitroso compounds, formed at an intermediate stage, to form azoxy compounds recognizable by their yellow colouring. These are recognizable by their yellow colouring. The reduction may then stop. After completion of the reduction the excess iron is removed with soda. During the reduction heat is liberated. In the presence of stoichiometric quantities of acid the reaction takes place according to the equation:

$$R{-}NO_2 + 3\,Fe + 6\,HCl \rightarrow R{-}NH_2 + 3\,FeCl_2 + 2\,H_2O.$$

According to Béchamps, however, about 3 per cent of the theoretical quantity of acid is sufficient if the iron is utilized up to the ferric stage. The ferric chloride formed at an intermediate stage is decomposed by hydrolysis into ferric hydroxide; the hydrochloric acid liberated forms more ferrous chloride:

$$R{-}NO_2 + 2\,FeCl_2 + 2\,Fe + 4\,H_2O \rightarrow R{-}NH_2 + 2\,FeCl_2 + 2\,Fe(OH)_3.$$

This is, however, merely a stoichiometric explanation which does not correspond to the true course of the reaction.

At the end of the reaction mainly $Fe_3O_4$ is present, in addition to the excess iron. The position at the completion of the reduction is best represented by:

$$4\,RNO_2 + 9\,Fe + 4\,H_2O \rightarrow 4\,RNH_2 + 3\,Fe_3O_4.$$

The hydrogen is, therefore, derived from the water.

Most aromatic amines are produced on this basis. If the amines are not sufficiently soluble in water, aqueous methanol, ethanol, or butanol can be used as solvents in the reduction. For further processing it is best if the base which is either liquid or dissolved in the aqueous filtrate from the iron slurry, or liquid is precipitated directly or as a salt. The amines can be expelled from the iron slurry also by evaporation. Amines with boiling points which are too high for evaporation are obtained by extraction with solvents, e.g. with benzene. From the waste water, too, the amines must be removed since they are toxic; with aniline this is done by extraction with nitrobenzene.

In some plants which produce aniline from nitrobenzene the iron oxide produced is used for the manufacture of iron oxide pigments (red, yellow, black) by annealing.

A reducing agent frequently used in industry for the conversion of nitro groups into amino groups is hydrogen sulphide in the form of sodium sulphide, sodium hydrosulphide, or ammonium sulphide. With these, partial reduction of materials with several nitro groups is possible, e.g. of m-dinitrobenzene to m-nitroaniline, which is not possible with iron. 1·8–2 mol. of the hydrosulphide group must be used per nitro group.

Zinc and zinc dust are used for reductions in acid and alkaline solutions, particularly when it is desired to retain intermediate stages, e.g. that of phenylhydroxylamine in a slightly acid reaction in the presence of $NH_4Cl$. Scrap aluminium can also be used for this purpose. With zinc dust in an alkaline medium hydrazobenzenes are obtained which can be rearranged into benzidines, important dyestuff intermediates.

Special reductions can be carried out with sodium sulphite and bisulphite, e.g. quite generally the production of arylhydrazines from diazo compounds. Hydrazines are, of course, very important for the manufacture of pyrazolones which are widely used for organic pigment dyes and for the synthesis of pharmaceutical preparations.

The methods of catalytic hydrogenation were discussed earlier. In intermediate products chemistry they play an important part in the manufacture of aniline, and for nuclear-hydrogenated compounds, e.g. cyclohexylamine, cyxlohexane, cyclohexanol, cyclohexanone, and many other products.

### *Alkali melts, phenols*

(Survey of the phenol processes: *Chem. Engng.*, 1954, **64**, (11), 157; *Chem. Z.* 1955, **79**, 707).

Alkali melts of sulphonic acids for the conversion of the $—SO_3H$ group into the OH group require high temperatures, in some cases up to 350°C. Without pressure in open vessels such a temperature can be achieved only by the use of anhydrous caustic alkali. Aqueous alkali can also be used but in order to obtain a sufficiently high temperature an autoclave is required. A 50 per cent solution of caustic soda produces at 300°C a gauge pressure of 50 atm. Melting in autoclaves is preferred for sensitive materials where atmospheric oxygen is to be excluded and where it is important to maintain a certain concentration of alkali and a constant temperature. Caustic potash melts more easily, and reacts more strongly than caustic soda, and is, therefore, used for some melts although it is more expensive. Mixtures of caustic potash and caustic soda also have a lower melting point than the latter alone and are used for alkali melts. In the great majority of cases, however, caustic soda is sufficient, e.g. for the manufacture of $\beta$-naphthol from the sodium salt of $\beta$-naphthalene sulphonic acid ($\beta$-salt), the most widely used alkali melt in chemical engineering. During the alkali melt sodium sulphite forms which is usually isolated:

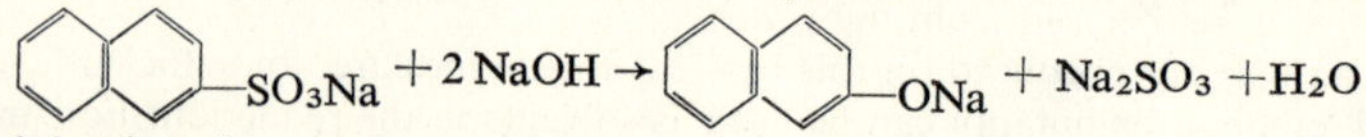


Di- and poly-sulphonic acids can be melted partially and then yield hydroxy-sulphonic acids. From amino-sulphonic acids amino-phenols are obtained. The alkali melt does not always follow smoothly the above equation. The melting alkali may have an oxidizing effect, and the sulphite present a reducing one. Thus, 1,2-di-hydroxy-anthraquinone (alizarin) is obtained directly by alkali melt from anthraquinone-2-sulphonic acid. The reducing effect of the sulphite on the nitro group becomes obvious when an attempt is made to melt nitro-sulphonic acids; melts of nitrosulphonic acids are, therefore, dangerous and may lead to explosions. For the production of m-aminophenol (e.g. for PAS) the m-amino benzene sulphonic acid is, therefore, melted.

Not only aromatics with sulphonic acid groups are converted into phenols by alkali melt but also those with chlorine. The best known method is the manufacture of phenol from chlorobenzene at a gauge pressure of more than 300 atm and at about 350°C (Dow process), first used at Chem. Fabrik Heyden, Radebeul. Copper has proved to be a suitable catalyst for the splitting off of halogen with alkali, and also with ammonia, for the production of amines (aniline from chlorobenzene).

At a high temperature chlorobenzene can also be cracked with steam into phenol and hydrochloric acid. The hydrochloric acid is then converted by the Deacon process into chlorine which in turn is utilized for the chlorination of benzene (Raschig process).

The halogen is particularly easily split off when a nitro group is in the o- or p-position. Nitrophenol and dinitrophenol are produced more economically from o- or p-nitrochlorobenzene and 2,4-dinitrochlorobenzene than by the nitration of phenol, particularly when it is desired to go beyond the dinitro stage (picric acid, picramic acid).

The conversion of chlorobenzene into phenol is carried out continuously in a flow tube (in Germany e.g. at Bayer, Leverkusen, formerly also at v. Heyden, Radebeul). The course of the reaction is not so uniform as in the melting of benzene-sulphonic acid. Nuclear condensations occur, i.e. formation of diphenyl, o- and p-hydroxydiphenyl and diphenyl ether. Phenol is also being produced by oxidizing cumene with oxygen (see p. 176).

The best materials for alkali melts are copper and nickel. However, in industry cast iron agitator vessels are usually used for melting. In the design by Frederking these can be heated up to 350°C with high-pressure steam (see Figure 88), for higher temperatures gas is used. (Effect of alkali on iron: see p. 228.)

To separate the phenols the phenolate melt is dissolved in water after cooling.

Sodium sulphite is precipitated at this stage and is filtered off. The filtrate is acidified and the $SO_2$ formed is expelled by boiling. After cooling, the salt solution is separated from the crystallizing phenol, or it is separated in the molten state. With $\beta$-naphthol this method would, however, be too expensive.

After sulphonation of the naphthalene, the sodium salt of the β-naphthalene-sulphonic acid is precipitated with NaCl or with the sodium sulphite-containing waste liquor of the β-naphthol production. The escaping sulphurous acid is fed into the naphtholate solution and precipitates the β-naphthol. The waste liquor now contains sodium sulphite which is used for the precipitation of β-salt from the sulphonation mixture. Since β-naphthol always contains some α-naphthol, the latter must be removed and this can be done by careful extraction with a diluted solution of caustic soda. A better method is, however, to remove it earlier by cracking back the naphthalene-1-sulphonic acid in the sulphonation mixture.

The crude β-naphthol is distilled under vacuum. All the ducts and vessels of the distillation equipment used must be heated (melting point 123°C) to prevent clogging. The melted β-naphthol is converted into flake form by a cooling roller.

In a similar manner α-naphthol can be obtained from α-naphthalene sulphonic acid. This and several other phenols can be produced better from the appropriate amines if these are easily obtainable from the nitro compounds. The amino group of many aromatic compounds can be easily split off with diluted sulphuric acid. For this, temperatures of about 200°C are required and it is, therefore, necessary to use acid-resistant (brick- or lead-lined) autoclaves which may be heated by injection of steam, i.e. by 'direct steam'.

α-naphthol is obtained by this process from α-naphthylamine. Resorcinol can be produced either by alkali melt from the sodium salt of benzenedisulphonic acid ('bi-salt') or from m-phenylenediamine, or from m-aminophenol by acid hydrolysis. Amino-ε-acid is converted by acid hydrolysis into epsilon acid (see p. 267), and 1,8-aminonaphthol-4-sulphonic acid (S-acid) into 1,8-dihydroxynaphthalene-4-sulphonic acid. Many more examples could be cited. Conversion of the amino group into the OH group by diazotization and boiling is usually too expensive. This method is used for the production of guaiacol from o-anisidine.

*Carboxylation (Kolbe-Schmitt process)*

Many phenols in the form of their alkali salts can be converted with carbon dioxide into carboxylic acids. Particularly important in industry is the manufacture of salicylic acid, p-hydroxybenzoic acid, cresotinic acid, 2-hydroxynaphthalene-3-carboxylic acid, and 2-hydroxycarbazol-3-carboxylic acid (for naphthol-AS), 2-hydroxynaphthalene-6-carboxylic acid, and p-aminosalicylic acid (PAS). Salicylic acid is one of the oldest and most important synthetic pharmaceuticals and dyestuff intermediates. Up to recently it had been assumed that the reaction took the following course:

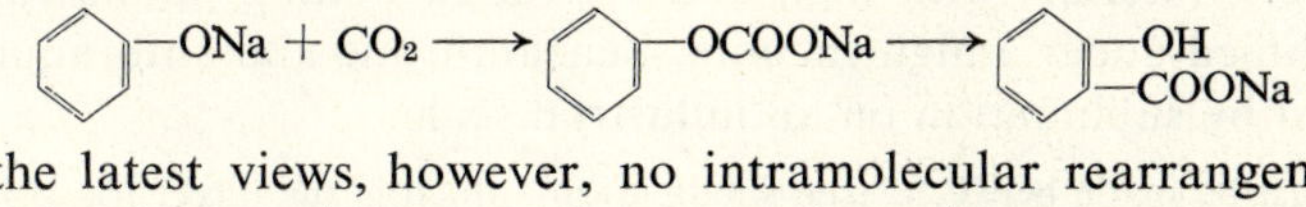


According to the latest views, however, no intramolecular rearrangement occurs but the reaction consists in the addition of the aromatic nucleus to one of the two carbonyl groups of the carbon dioxide:

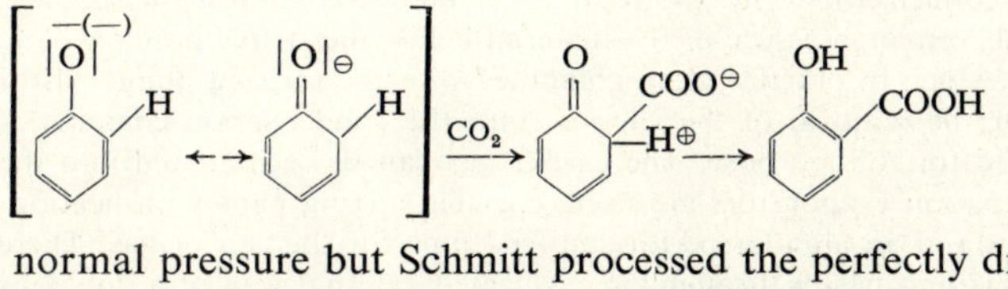


Kolbe worked at normal pressure but Schmitt processed the perfectly dry phenolate by agitation in autoclaves with gaseous carbon dioxide under pressure. To produce salicylic acid carbon dioxide is first introduced at room temperature which is gradually increased to 125°C, the gauge pressure being maintained at 5–7 atm. The phenol is almost completely converted. The small amount of unconverted phenol is distilled off under vacuum. The salicylate is dissolved in water and salicylic acid precipitates in almost pure form during acidification.

The position of the carboxyl group depends also on the carboxylation temperature. At higher temperatures p-hydroxybenzoic acid is obtained in place of salicylic acid. In the form of its ester this is a widely used preserving agent ('Nipagin').

Carboxylation is especially easy with aromatic polyhydroxy- and amino-hydroxy compounds. o- and m-dihydroxy- and hydroxy-amino compounds can be partly carboxylated in aqueous solution even without pressure. p-aminosalicylic acid (PAS) has gained special importance as a remedy for tuberculosis. It is obtained from m-aminophenol, produced by melting metanilic acid, in aqueous solution with potassium bicarbonate and carbon dioxide at 85–90°C and a gauge pressure of 30 atm.

The conversion of $\beta$-naphthol to 2-naphthol-3-carboxylic acid is much more difficult. A conversion temperature of up to 260°C is required. Unlike phenol and m-aminophenol only half of the naphtholate charged is converted to carboxylic acid. The remaining $\beta$-naphthol must be distilled off under vacuum.

Exceedingly important for the success of many carboxylations is the total exclusion of water. Since the gaseous carbon dioxide reacts with the solid phenolate, the latter must be very finely dispersed in powder form, it must not cake, and the surface must be continuously renewed. The agitator autoclaves used must have very strong, peripheral agitators. Atmospheric oxygen is harmful (resinification). The nature of the alkali is also important. Thus 2-naphthol-6-carboxylic acid and the 2-hydroxycarbazole-3-carboxylic acid which is much used for naphthol-AS-dyes are obtained from potassium phenolates.

Salicylic acid must be sublimed for the purpose of purification for pharmaceutical use (DAB 6 manufacture). Since sublimation is often carried out as a purification operation, its technology will be briefly discussed.

### *Sublimation*

For many substances sublimation produces a higher degree of purity than crystallization or distillation, and for many compounds it is the only possible purification method. During sublimation usually a very reactive, finely crystalline form is produced and it is, therefore, used for pharmaceutical products, such as salicylic acid, Resorcinol, camphor, benzoic acid, etc. Materials with high and very high boiling points can also be very successfully sublimed under a high vacuum. Benzanthrone and anthraquinone, even vat dyes, are purified by sublimation on an industrial scale.

A distinction must be made between 'true sublimation' which is the direct transformation from the solid state into the vapour phase and subsequent condensation directly into the crystalline product, and 'pseudo-sublimation', which is the transformation of the softened or molten product into the vapour phase and condensation into the solid sublimate. A condition for the use of true sublimation is a sufficiently high vapour pressure of the material below the 'triple point' where all three phases are in equilibrium and which in practice often coincides with the melting point. Also important is the use of a carrier gas for the removal of the vapours into the condensation chamber. Air, carbon dioxide, or nitrogen are used for this purpose. The carrier gas can be recirculated into the evaporator. Sometimes special sublimation evaporators are used, e.g. flat stirring pans with heating jacket, heating coil, or Frederking pipes and an annular perforated feed pipe for the carrier gas. There are also designs in which the carrier gas first passes through the gas-heated external wall of a double-wall boiler, heats the inner wall, and then through perforations in the inner wall reaches the material which is spread out on a perforated bottom, penetrates it and carries the vapours to the condenser. Some materials can be evaporated without a vacuum and can be precipitated either in a large empty chamber containing suitable cloth-covered frames for the precipitation of the crystals, or in a water-cooled vessel. Specially suitable are benzoic acid, salicylic acid, and phthalic acid.

High-boiling or sensitive materials are sublimed under vacuum. A low gas circulation towards the vessel is provided. The material to be evaporated is spread out as thinly as possible. Figure 96 shows the diagram of a vacuum sublimation apparatus for high-boiling materials. The material is spread out on metal sheets and is evaporated by means

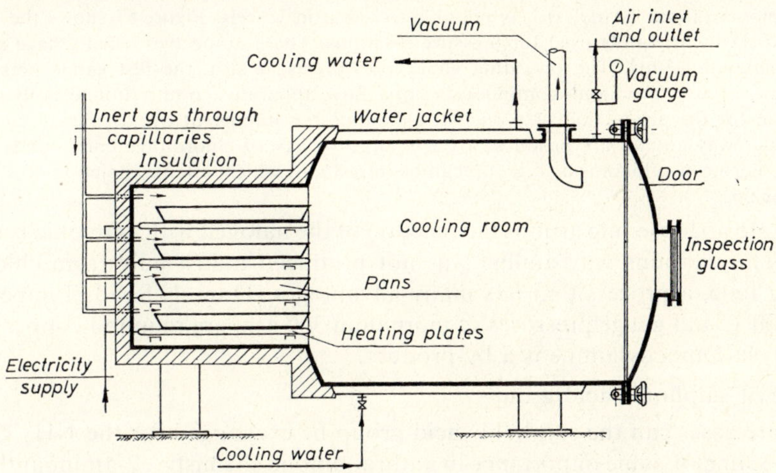


Figure 96. Vacuum sublimation apparatus for high-melting materials

of electrically or steam-heated plates. The crystals are deposited at the water-cooled wall of the cooling chamber. The process can be observed through an inspection glass. Recently the fluidized bed process has also been used for sublimation. Evaporation is effected by a hot inert gas flowing concurrently.

*Amidation*

Most aromatic amines are obtained by the reduction of nitro compounds. In many cases this is, however, impossible. Various other methods are, therefore, used in industry. The most important ones will be briefly discussed.

Exchange of halogen

The exchange of halogen for ammonia or amines is usually fairly easy if the halogen is activated by certain groups, 'negative substituents', e.g. nitro- or sulphonic acid groups. Whilst, for instance, the reaction of o- and p-nitrochlorobenzene with aqueous ammonia has to be carried out at a temperatures up to 170–180°C and a gauge pressure of up to 40 atm, it is possible to react 2,4-dinitrochlorobenzene and 4-nitro-1-chlorobenzene-2-sulphonic acid without pressure. The important intermediates o- and p-nitrochlorobenzene are obtained by nitrating chlorobenzene and separating the two isomers by distillation and crystallization (about one third o-, and two thirds p-nitrochlorobenzene). The boiling points are close together and the two isomers can be concentrated by distillation only to such a degree that they crystallize out in pure form on cooling (e.g. in tube crystallizers). The eutectics are re-circulated. At present more p-nitrochlorobenzene is produced than is required but the consumption fluctuates.

The conversion to o- and p-nitroaniline is carried out in autoclaves at 170°C and a gauge pressure of 35–40 atm in ten hours with a large excess of 25 per cent ammonia. The pressure is then released and most of the excess ammonia is re-absorbed in the reaction water. After cooling to 150°C the residual $NH_3$ pressure empties the contents of the autoclave into a filter press. The filtered nitroaniline is passes into cold water and is extracted in the form of fine crystals.

Although ammonium chloride formed during the reaction strongly corrodes iron with the high excess of ammonia (about 10 moles) it is possible to use iron vessels. Figure 87 shows the sketch of an agitator autoclave commonly used for pressure reactions. These autoclaves usually have a capacity of 1·5–2 $m^3$, maximum 3 $m^3$. The operating gauge pressure is 40 atm, the test gauge pressure 80 atm. p-nitroaniline is also produced continuously in a flow apparatus. p-nitroaniline is used mainly for azo-dyes and for the production of p-phenylenediamine (an important dye for furs, Ursol D).

In a similar way amines are produced from a large variety of chloronitro compounds. In order to loosen the chlorine, a sulphonic acid group can be introduced which, after completion of the amidation, is split off again.

If there are no loosening groups the reaction of the halogen with ammonia is much more difficult. This explains why aniline was not produced industrially from chlorobenzene for a long time, a process that has many advantages (Dow, U.S.A.). Temperatures exceeding 300°C and gauge pressures of more than 300 atm are required copper being used as a catalyst. Diphenylamine is a by-product.

### Exchange of sulphonic acid groups

Only in rare cases can the sulphonic acid group be exchanged for the $NH_2$ group. This method became of some importance in anthraquinone chemistry. 2-aminoanthraquinone and 1-aminoanthraquinone are obtained from the sulphonic acids by reaction with ammonia in autoclaves at temperatures of 200°C and 175°C respectively. Since the sodium sulphite formed interferes with the reaction an oxidizing agent e.g. sodium arsenate or m-nitrobenzene sulphonic acid is added which converts the sulphite into sulphate.

### Exchange of OH groups, Bucherer reaction

The direct exchange of OH groups for —$NH_2$ or RNH— is possible in only a few cases. β-naphthol can be converted according to an old BASF patent into β-naphthylamine above 150°C with gaseous ammonia, 1,3,5-hydroxylenol catalytically at high pressure into 1,3,5-hydroxylidine. Another important example is the industrial production on a large scale of phenyl-β-naphthylamine from aniline and β-naphthol, large quantities of which are used as age resistors for rubber. Phenyl-β-naphthylamine is produced from β-naphthylamine and aniline in the presence of sulphanilic acid, in a similar way as diphenylamine is formed from aniline and a small amount of aniline hydrochloride in autoclaves at about 280°C. The presence of quaternary ammonium salts facilitates the reaction when an OH group is exchanged for the residual aryl amide. A special case is the easy conversion of 2,2′-dihydroxy-1,1′-dinaphthyl into dinaphthocarbazole with ammonia under pressure.

Lepetit (1896) and shortly after him Bucherer (1899) found that the substitution of aromatic OH groups by —$NH_2$, —NHR, or $N\langle^{R}_{R}$ is greatly facilitated by the presence of $SO_2$ or bisulphites. This important discovery was of the greatest significance for intermediate and dyestuff chemistry. Whilst the reaction can be used only to a limited extent and is not yet being applied industrially for the benzene and anthracene series, it is a very widely used operation in the naphthalene series, especially for the production of $\beta$-naphthylamine derivatives. It is possible to convert many naphthols with ammonium bisulphite into amines, with amines and sodium bisulphite into secondary and tertiary amines, and inversely primary amines with sodium bisulphite into naphthols. Until recently the mechanism of the Bucherer reaction was not fully understood. Naphthols and naphthylamines yield identical addition products when heated with sodium bisulphite, $NH_3$ being split off the latter. So far it was assumed that $\alpha$-hydroxysulphonic acids or $\alpha$-aminosulphonic acids formed by addition of sodium or ammonium bisulphite to phenols and amines (to the keto form or the double C,C-bond). These bisulphite adducts-many of which have been isolated are capable of the following conversions:

(1) Bisulphite adducts of amines with alkalis yield the corresponding hydroxy-compounds (phenols).
(2) Bisulphite adducts of naphthols with excess ammonia or with amines yield the corresponding primary, secondary, or tertiary amines.
(3) By treatment with arylhydrazines they form carbazoles.

The bisulphite adducts formed by heating many phenols and arylamines with sodium- or ammonium bisulphite, with or without pressure, are very easily soluble in water. This can be used as an indication of the progress of the reaction. That conversion of the amine is complete is recognized from the fact that it can no longer be diazotized. The capacity for coupling with a diazotized base is a measure of the amount of phenol present.

*The mechanism of the Bucherer reaction*

Rieche and Seeboth (*Angew. Chemie*, 1958, **70**, 52, 312; *Liebigs Ann. Chem.*, 1960, **638**, 43—101; *Mber. Dt. Akad. Wiss.*, 1961, **3**, 43) have shown that the existing theories were incorrect. According to a theory developed earlier by Bogdanov, the hydrogen sulphite addition compounds of the naphthols are tetralon-m-sulphonic acids, e.g. for the $\alpha$-naphthol 1-tetralone-3-sulphonic acid. The latest results of investigations show the course of the Bucherer reaction to be as follows: With $\alpha$-naphthols and $\alpha$-naphthylamines the first step is the electrophilic addition of a proton to a C atom 4:

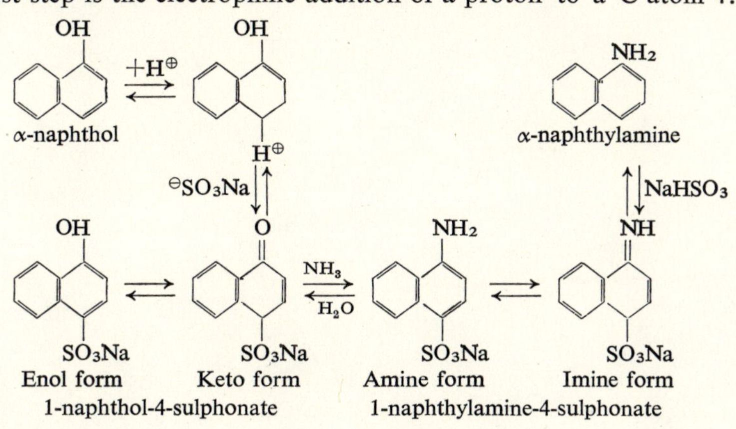

The essential reaction is, therefore, the conversion of tetralonesulphonic acid into tetraloneiminosulphonic acid and vice versa. The equilibrium point is determined by the degree of concentration of the constituents of the reaction and by the tendency of the ketimine or enaminosulphonic acid either to split off or to hydrolyse $NaHSO_3$. Accordingly the carbazole formation takes the following course:

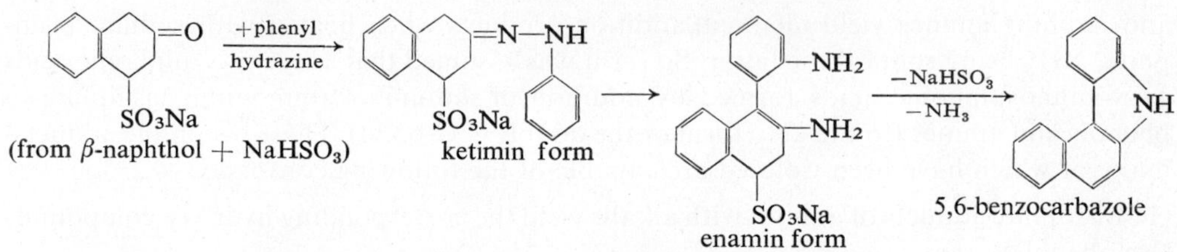


The possible uses of this reaction have been very thoroughly investigated by Bucherer and by industrial laboratories. Certain regular features were found. There is, e.g., no reaction in the $\alpha$-series of naphthalene if a sulphonic acid group is in the o- or m-position to the $NH_2$- or OH group. In the $\beta$-series a sulphonic acid group in the m-position has an interfering effect, which will be readily understood after what has been said above. With dihydroxy- and diaminonaphthalenes only one of the groups reacts, with aminonaphthols an OH group or an amino group may react according to the position.

By the reaction of bisulphites with sulphonic acids of dihydroxynaphthalenes and aminonaphthols important dyestuff intermediates are obtained (e.g. I-acid, M-acid). The reaction takes place always at that group which is not affected by the $SO_3H$-group.

The conversion of naphthylamines into naphthol is of importance only in the $\alpha$-series, e.g. the conversion of naphthionic acid (1-naphthylamine-4-sulphonic acid) into 1-naphthol-4-sulphonic acid (Nevile-Winter acid). This process is much more convenient than the diazotization and boiling of naphthionic acid. It should be mentioned that the conversion of naphthionic acid into 1-naphthol-4-sulphonic acid with sulphites had been discovered even before Bucherer by Lepetit (1896) and handed over to Bayer, Elberfeld, for exploitation, without, however, Bucherer being aware of this fact.

In the $\beta$-series the bisulphite adducts react with ammonia and with aliphatic and aromatic amines, those of the $\alpha$-series on the other hand do not react with aromatic amines.

The behaviour of 2,3-hydroxynaphthionic acid is interesting. It splits off $CO_2$ with bisulphite and amines and yields the same compounds as $\beta$-naphthol but reacts more easily than the latter.

# 2. DIAGRAMS OF THE PROCESSES FOR THE PRODUCTION OF IMPORTANT AROMATIC INTERMEDIATES

### (a) Benzene derivatives

Chlorine derivatives of benzene and its secondary products

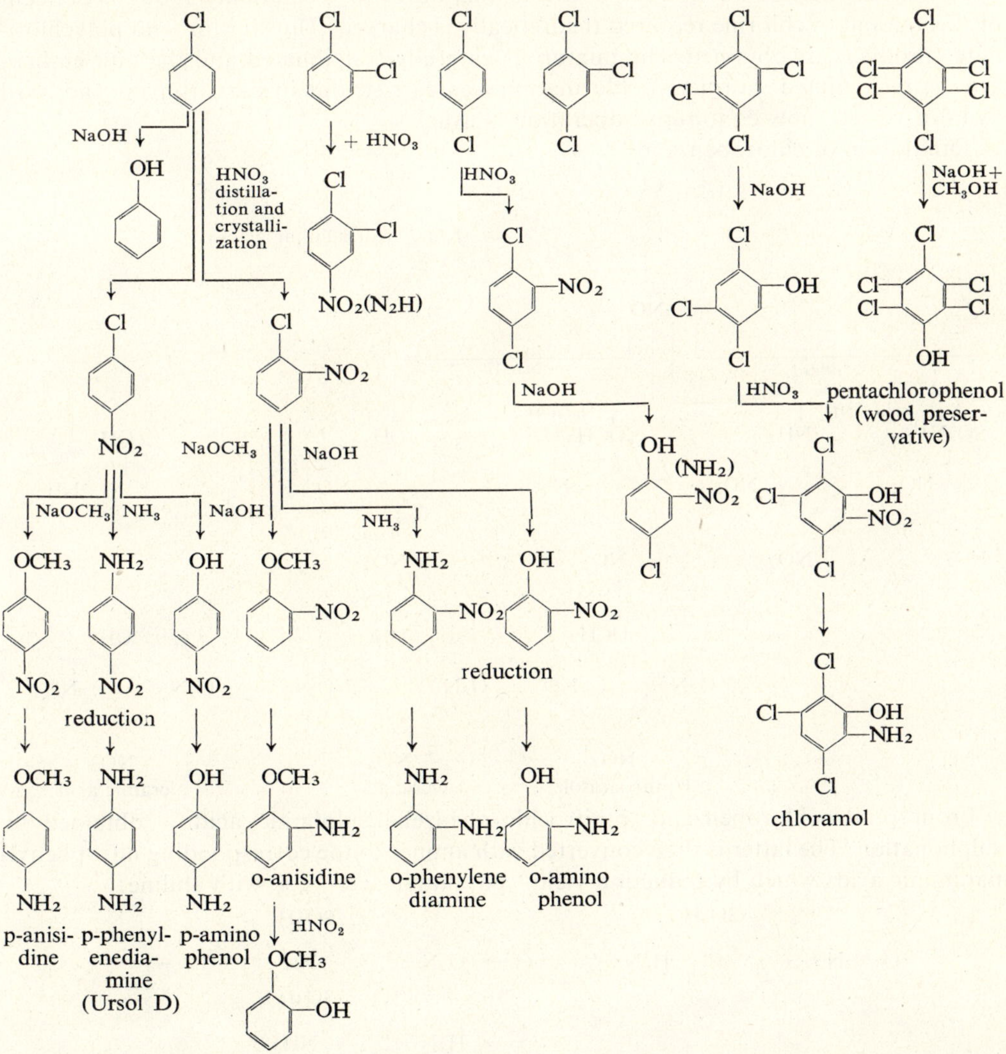


The most important raw material for the production of intermediates is benzole from bituminous coal tar. It is supplied to the coal tar processing industry as pre-purified '90' benzole. Tank wagons or tankers transport it to the benzole storage tanks. As a precaution against fire the storage vessels have thick protective walls. Pure benzene, toluene, and the xylenes are produced by fractional distillation in tall bubble tray columns (see p. 247).

*Chlorinated benzenes and their secondary products*

The nuclear chlorination of benzene takes place only in the presence of halogen carriers, e.g. ferric chloride. Since the most important intermediate product is monochlorobenzene, operations are carried out at a rather low temperature (30°C) and only about 60 per cent of the amount of chlorine required theoretically is charged. Thus few di- and polychlorinated products are obtained. The mixture is carefully fractionated and the unused benzene is recirculated. Formerly, chlorination was carried out in cast iron or lead-lined agitator vessels, now continuous operation is usual.

Dinitration of chlorobenzene

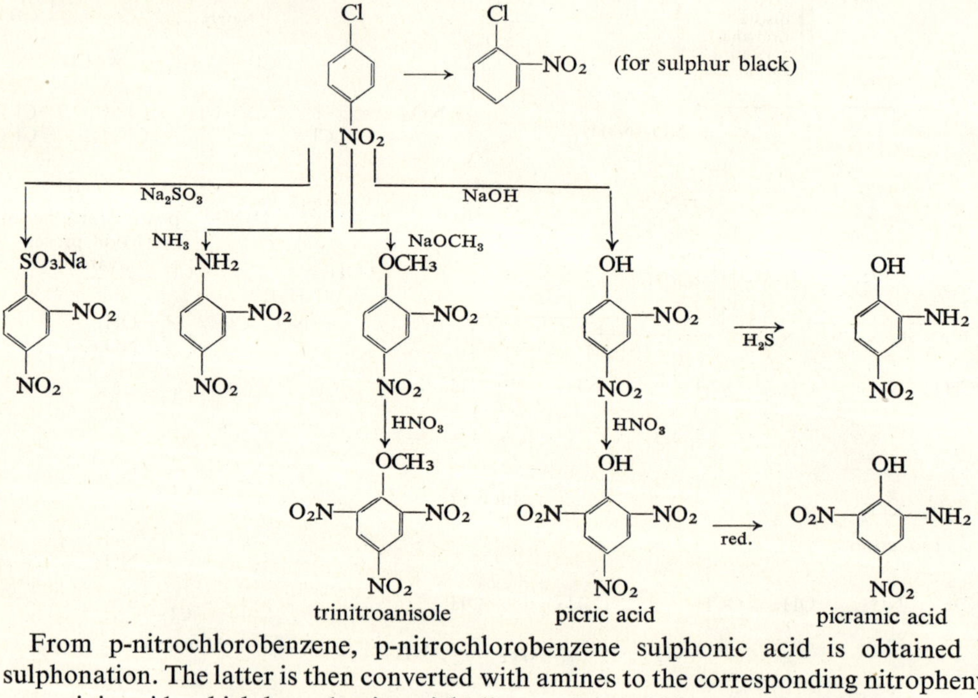


From p-nitrochlorobenzene, p-nitrochlorobenzene sulphonic acid is obtained by sulphonation. The latter is then converted with amines to the corresponding nitrophenylparaminic acids which by reduction yield "paraminic acid" e.g. with aniline:

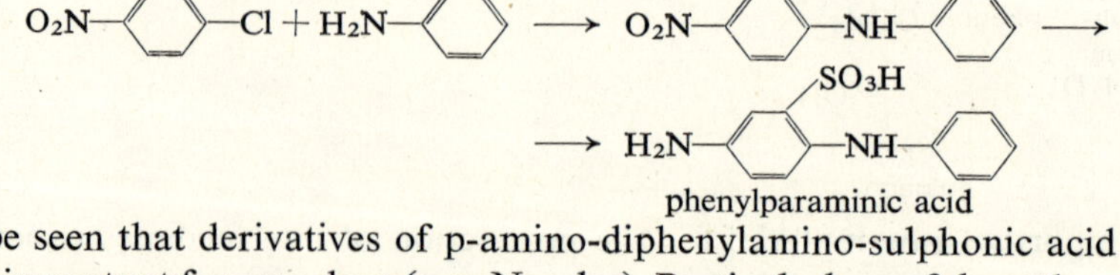


It will be seen that derivatives of p-amino-diphenylamino-sulphonic acid are formed which are important for azo dyes (e.g. Neroles). Particularly useful are the reaction products with p-toluidine, p-phenylenediamine, o-anisidine. With phenols p-nitrochlorobenzene sulphonic acid yields derivatives of diphenyl ether sulphonic acid.

## Nitration of benzene and its secondary products

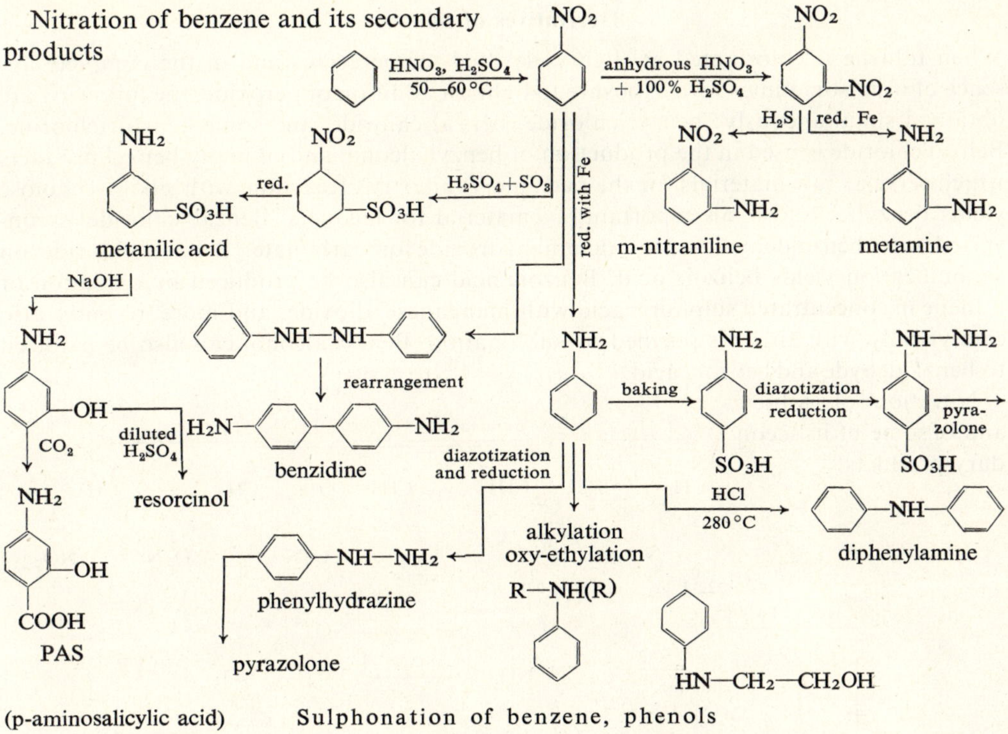


## Sulphonation of benzene, phenols

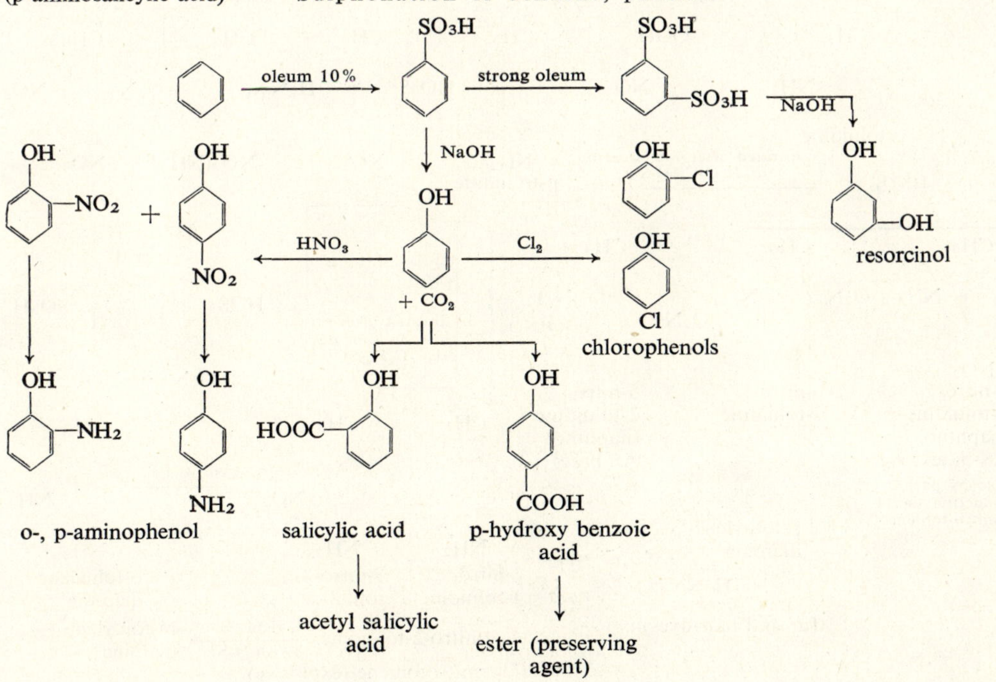

## Derivatives of toluene

When toluene is chlorinated at slighty elevated temperatures and in the complete absence of iron, sometimes with exposure to light, or addition of peroxide, the following are obtained simultaneously: benzyl chloride, benzal chloride, and some benzotrichloride. Benzyl chloride is used in the production of benzyl alcohol and of many benzyl products which serve as raw materials for the perfumery industry. Reaction with alkali cellulose yields benzyl cellulose, an important raw material for lacquers. Benzal chloride is converted into benzaldehyde with calcium hydroxide or carbonate. Benzotrichloride on saponification yields benzoic acid. Benzoic acid can also be produced by oxidation of toluene in concentrated sulphuric acid with manganese dioxide, and more recently also catalytically with air. It is purified by sublimation. Benzyl alcohol can also be oxidized to benzaldehyde and benzoic acid.

Nitration of toluene and of some of its secondary products

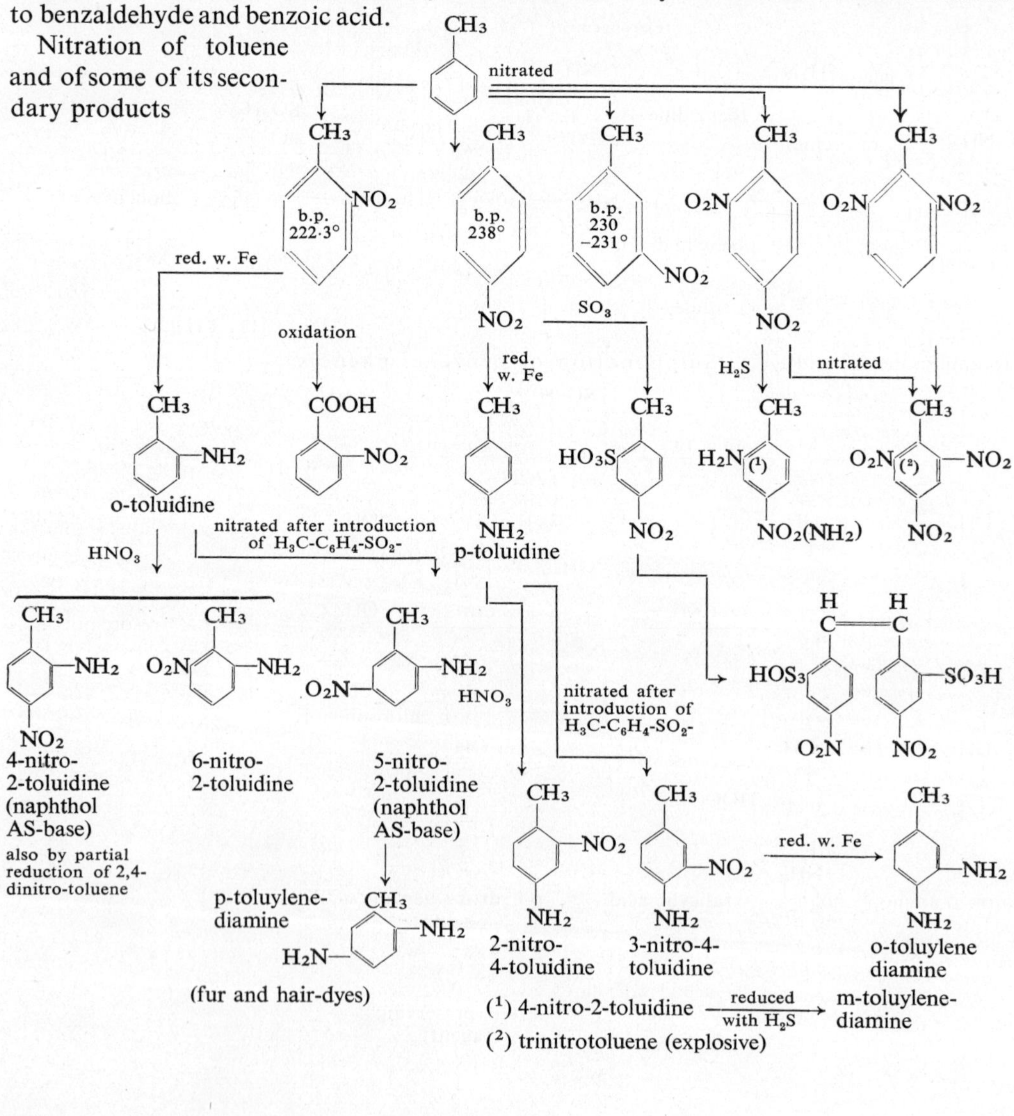

Sulphochlorination with chlorosulphonic acid yields a mixture of o- and p-toluenesulphochloride which is separated by crystallization.

In the mononitration of toluene the low orientating effect of the $CH_3$ group produces an isomeric mixture of about two thirds o- and one third p-nitrotoluene and 3 per cent m-compound. By dinitration mainly 2,4-dinitrotoluene is formed, some 2,6 dinitro- and a little 3,5-dinitrotoluene. The separation of nitrotoluenes is very difficult because their boiling points are close together. Very efficient distillation columns are required. Distillation is carried out under vacuum. The crystalline p-nitrotoluene is precipitated from the highly concentrated fractions. m-nitrotoluene is also produced since it is used for special purposes. Toluidines which are important for azo dyes are produced from the nitrotoluenes by reduction with iron. They can be further nitrated after acetylation or toluenesulphonation of the $NH_2$ group. With o-toluidine the $NH_2$ group moves mainly into the p-position to the amino group, with p-toluidine into the o-position. Thus 5-nitro-2-toluidine and 3-nitro-4-toluidine can be obtained which by reduction with iron yield p-toluylenediamine (an oxidizing dye for fur and hair dyes) and o-toluylenediamine. By direct nitration of toluidines, 2-nitro-4-toluidine is obtained from p-toluidine and 4-nitro-2-toluidine and 6-nitro-2-toluidine from o-toluidine.

By oxidation of nitrotoluene with potassium dichromate o- and p-nitrobenzoic acids are obtained and from these aminobenzoic acids. o-aminobenzoic acid (anthranilic acid) is produced more easily, however, from phthalic anhydride via phthalimide.

Almost all the reactions carried out with nitrobenzene can also be carried out with nitrotoluenes. Due to the presence of the $CH_3$ group they can be performed more easily. Hydrazobenzenedicarboxylic acid is obtained from o-nitrobenzoic acid by reduction with zinc dust and alkali and is converted with hydrochlorid acid into benzidine-3,3′-dicarboxylic acid, an important diazotization component for diazo dyes. On oxidation with air p-nitrotoluenesulphonic acid forms 4,4′-dinitrostilbene-2,2′-disulphonic acid.

### (c) Derivatives of naphthalene

*Nitration of naphthalene and its secondary products*

Naphthalene is much more reactive than benzene. Nitration with a weak mixture of nitric and sulphuric acid takes place easily even at 50–60°C. The nitro group moves almost exclusively into the α-position (4–5 per cent β).

Since the nitration is carried out in a suspension the naphthalene must be very finely powdered to avoid the production of too many dinitro products and not to leave some naphthalene unreacted.

Reduction by the Béchamp method yields α-naphthylamine which can be converted to α-naphthol with diluted sulphuric acid at 200°C. By baking α-naphthylamine sulphate under vacuum, or by stirring in boiling o-dichlorobenzene the important l-naphthylamine-4-sulphonic acid (naphthionic acid) is produced which, as mentioned earlier, can be easily converted to l-naphthol-4-sulphonic acid (Nevile-Winther acid) by Bucherer's method.

Phenyl-α-naphthylamine is obtained from α-naphthylamine and aniline by heating in the presence of a little sulphanilic acid, phenyl-β-naphthylamine from β-naphthol and aniline in the presence of sulphanilic acid (age resistor for Buna).

Nitration of naphthalene and its secondary products

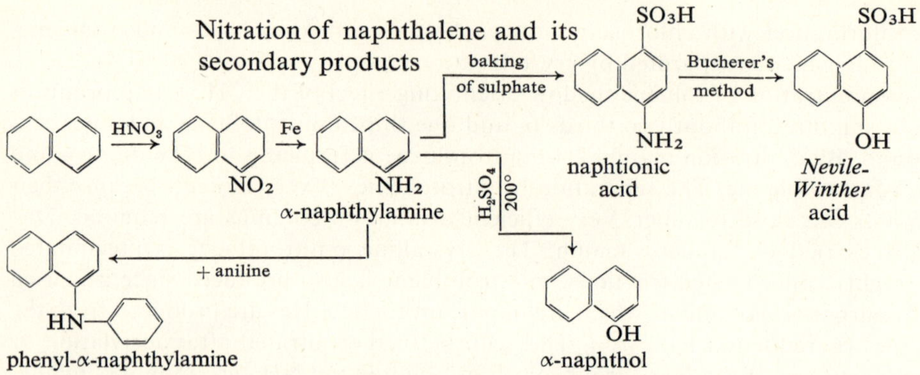


*Monosulphonation of naphthalene and its secondary products*

The point of entry of the sulphonic acid group depends on the temperature. Cold sulphonation leads mainly to α-substitution, hot sulphonation to β-substitution. β-sulphonation is carried out at the maximum temperature possible without decomposition or disulphonation, i.e. 160–170°C. About 15 per cent α-products are formed nevertheless. For the β-sulphonation sulphuric acid of 94 per cent concentration is introduced dropwise into molten naphthalene and water and the unreacted naphthalene is distilled off. (Reserve cracking of α-acid to naphthalene at 180°C, see pp. 5 and 7). After cooling the mixture is neutralized with soda and separated by addition of NaCl, the sodium salts of β-naphthalenesulphonic acid (β-salt) being precipitated.

Monosulphonation of naphthalene, β-naphthol, β-naphthylamine and their sulphonic acids

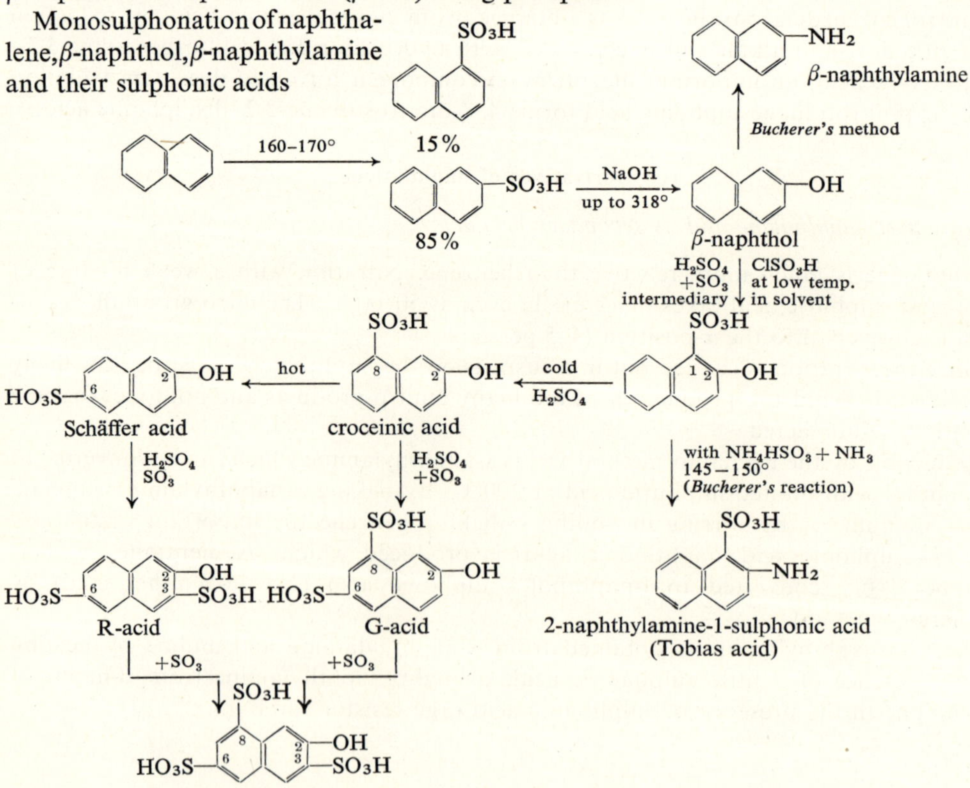

To produce β-naphthol, the β-salt is heated in an alkali melt starting at 270°C up to a temperature of 318°C. The naphtholate-sodium sulphite mixture is acidified with sulphuric acid up to the sulphite stage (almost neutral), β-naphthol being precipitated. (For the industrial process, see pp. 7 and 251.

According to the conditions sulphonation of β-naphthol yields different 2-naphtholsulphonic acids. In very mild conditions in solution (e.g. nitrobenzene) at a low temperature with chlorosulphonic acid, 2-naphthol-1-sulphonic acid is obtained.

Naphthalene disulphonation, nitration, reduction followed by separation of naphthylaminedisulphonic acids (S = $SO_3H$)

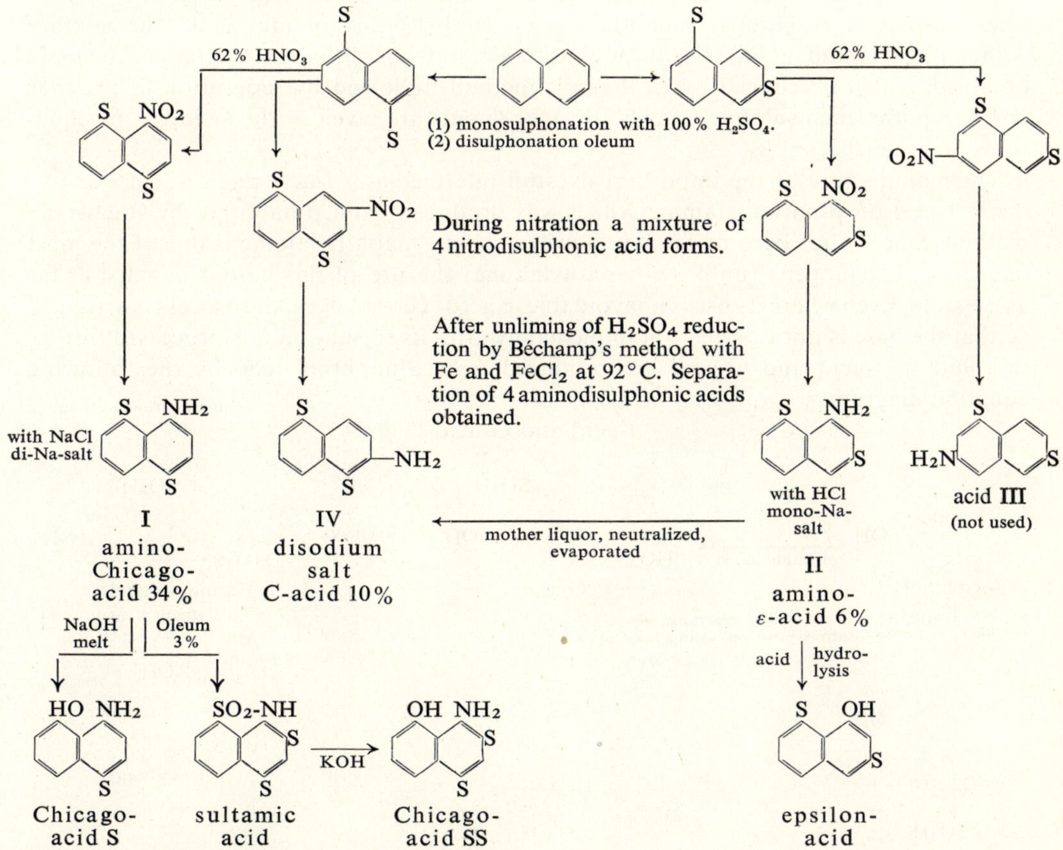


The latter forms also in other sulphonation processes as an intermediate but immediately converts in the cold to 2-hydroxynaphthalene-8-sulphonic acid (croceinic acid) and in a warm atmosphere to 2-hydroxynaphthalene-6-sulphonic acid (Schäffer acid). The rearrangement of the l-sulphonic acid is catalysed by $H_2SO_4$, and the production of 2-naphthol-1-sulphonic acid is, therefore, carried out in the absence of sulphuric acid. Schäffer acid may also be obtained from β-naphthol with sodium sulphate and sulphuric acid by baking at 160°C. A particularly suitable apparatus for baking and alkali fusion, i.e. for reactions in the solid or plastic phases, is the 'baking drum' (Virck,

Farbenfabrik Wolfen). It consists of a gas-heated rotary drum containing grinding elements. A gas inlet or outlet pipe and stuffing box may be fitted to the shaft.

On further sulphonation Schäffer acid yields R-acid and croceinic acid yields G-acid. Both together with a third sulphonic acid group form eventually 2-hydroxynapththalene-3,6,8-trisulphonic acid. The acids are separated by utilizing the different solubilities of their alkali salts.

*Disulphonation of naphthalene*

First naphthalene is monosulphonated with 100 per cent sulphuric acid. Sulphonation is continued by adding oleum to the mixture and two disulphonic acids are formed (see synopsis of equations): naphthalene-1,5- and -1,6-disulphonic acid. The mixture is then nitrated and a mixture of four different nitronaphthalenedisulphonic acids is obtained, which is reduced by Béchamp's method, followed by separation of the four aminonaphthalenedisulphonic acids. Further details are given in the synopsis of equations on page 267.

The manufacture of the important dyestuff intermediates I-acid and G-acid was formerly based on β-naphthylamine which was produced from β-naphthol by Bucherer's method. Since it has become known, however, that β-naphthylamine is one of the most dangerous carcinogens (unlike α-naphthylamine) the use of this base is avoided as far as possible. Even where its use is unavoidable, e.g. for certain dyes, the process is arranged so that the base is not isolated but immediately after its formation it is processed further in solution. I-acid and G-acid are produced from α-naphthol, e.g. by the following equation diagram:

I-acid and G-acid

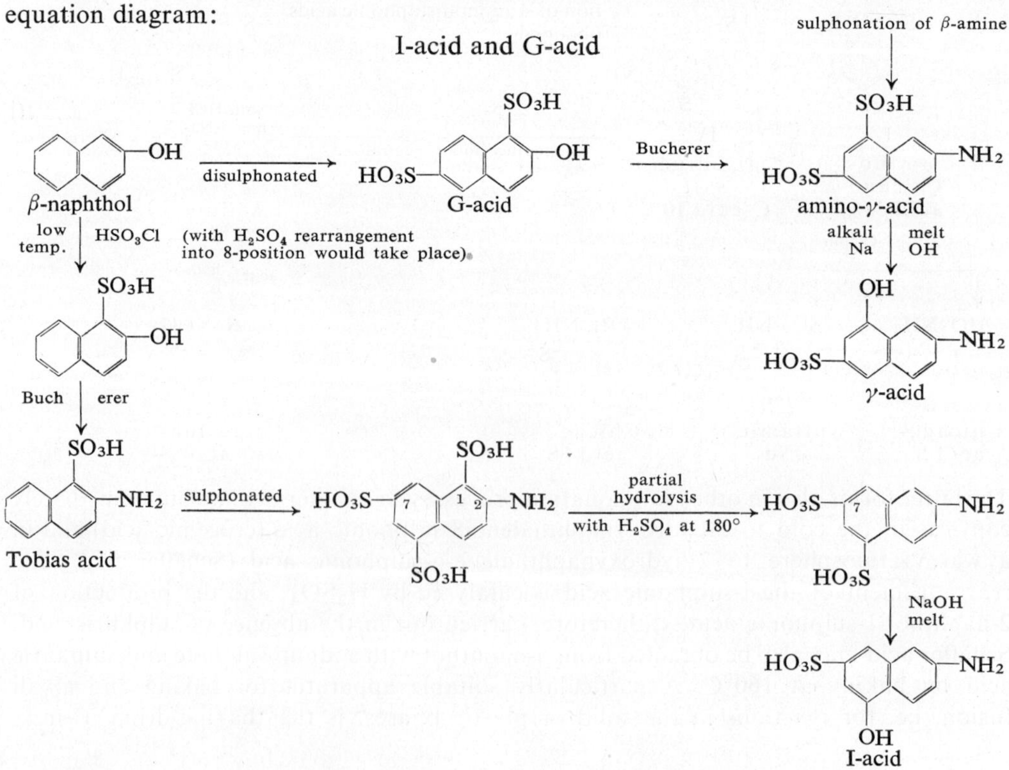

The conversion of amino-G-acid (2-aminonaphthalene-6,8-disulphonic acid) into $\gamma$-acid (2-amino-8-naphthol-6-sulphonic acid) is an example for a partial alkali fusion process. It is based on the fact that a substance with an —$SO_3H$ group in the $\alpha$-position is fused much more easily than one with a —$SO_3H$ group in the $\beta$-position. This method is also used for alkali fusion of 2-aminonaphthalene-5,7-disulphonic acid to I-acid (2-amino-5-naphthol-7-sulphonic acid) which is an isomer of $\gamma$-acid. 2-Amino-naphthalene-5,7-disulphonic acid is obtained from 1,5,7-trisulphonic aicd by acid hydrolysis with sulphuric acid. In aromatic systems sulphonic acid groups have bonds of various strength, according to their position. This fact can be utilized for some separations. Thus, for instance, m-xylenesulphonic acid cracks off most easily during the heating of the three monosulphonic acids of the xylenes with diluted $H_2SO_4$ to 170–180°C, under pressure. When the pressure in the pressure vessel is continuously released during the reaction over a cooler, pure m-xylene is distilled off directly.

One of the most important intermediates for azo dyes is H-acid (1-amino-8-naphthol-3,6-disulphonic acid). It couples in two positions: acid in the 2-position beside the $NH_2$ group, and alkaline in the 7-position beside the OH group. Thus interesting diazo dyes can be synthetized, triazo dyes with benzidine.

The following diagram shows the manufacture of H-acid from naphthalene.

1,8-aminonaphthol-3,6-disulphonic acid (H-acid)

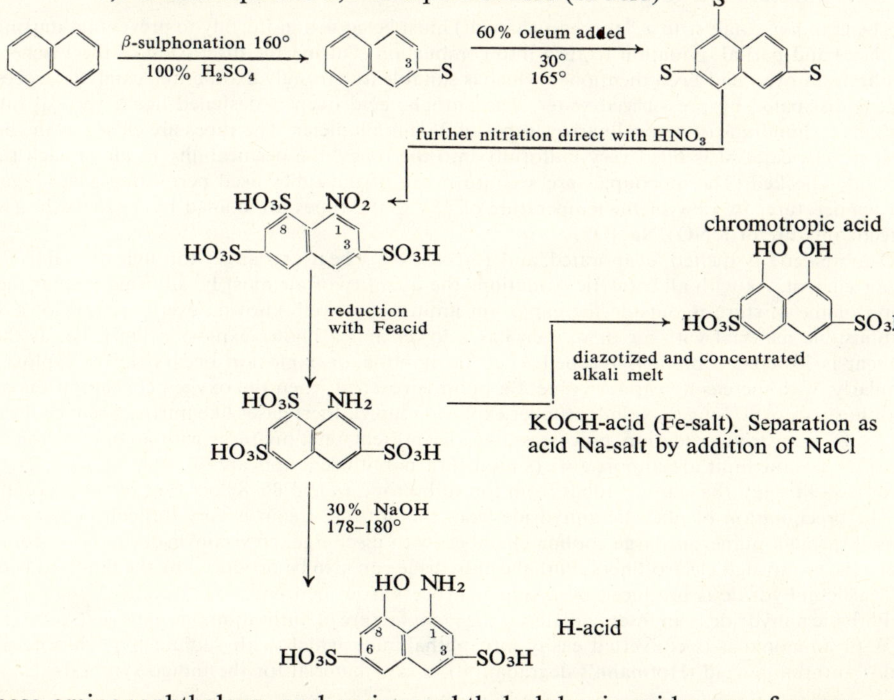


All these aminonaphthalene- and aminonaphthalsulphonic acids are, of course, not absolutely pure compounds. The solubility of a sulphonic acid is a decisive factor for its assessment. Each acid has a certain solubility value. The purer it is the less soluble. Sulphonic acids are purified only to the extent that they yield 'type-conforming' dyes, i.e. dyes which in a test produce shades and fastness corresponding to a standard dye recognized as 'type'. The amino acid content is given as 'nitrite value' indicating the amount of nitrite required for a given quantity for diazotization. The dye regulations, therefore, lay down '... kg nitrite H-acid'. The yields of amino acids is thus determined by diazotization.

*Phthalic anhydride*

Formerly, phthalic anhydride (for table see Chem. & Ind., 1961, 1143) used to be produced by a process developed by BASF Ludwigshafen, by oxidation of naphthalene with fuming sulphuric acid and a mercury salt as a catalyst. Based on Wohl's invention (Friedl. 14; 450 DRP 347 610, 379 822) BASF developed a new process which is now generally used:

Naphthalene or o-xylene (U.S.A.) is oxidized in the vapour phase at a high temperature (450°C) on a vanadium pentoxide catalyst with air. The solution of the problem of heat dissipation and of the condensation of the phthalic anhydride from the large amounts of gas oxidized with air was particularly difficult. When 1 mole naphthalene is converted into 1 mole phthalic anhydride by oxidation theoretically 784 kcal are liberated, i.e. more than half of the total heat of combustion of naphthalene. A small proportion of naphthalene is oxidized beyond the phthalic acid stage, e.g. to maleic anhydride. Another product formed is 1,4-naphthaquinone marked by an extremely pungent smell of the oxidation waste gases.

The granular catalyst (e.g. 'torpedo shaped') must be cooled uniformly to prevent its starting to glow in places and partial oxidation to turn into combustion. During the discussion of the Fischer-Tropsch synthesis an oven had been mentioned which is suitable for strongly exothermic catalytic processes. The heat is dissipated by pressurized water. The phthalic acid oven is designed like a vertical tube cooler with many hundreds of vertical narrow tubes of 25 mm diameter. The tubes are closed at the bottom by screens. The catalyst is filled very uniformly into the tubes, the permeability to air of each tube being carefully checked. Thermocouples are set into many of the tubes used permitting exact regulation of the temperature. In view of the temperature of 450°C the tubes are cooled by a salt bath, a very fluid eutectic mixture of $KNO_3/NaNO_2$.

Naphthalene is melted, evaporated, and forced over a heat exchanger through the catalyst oven by an air current. As with all catalytic oxidations the quantity of air must be adjusted so that the composition of the mixture is outside the explosion limit. As is well known, every mixture of a vaporous combustible material with air or oxygen has a lower and a higher explosion limit, i.e. as the vapour content is lowered a point is reached when no ignition or explosion occurs (lower explosion limit). Similarly, with increasing vapour content a point is reached when the oxygen concentration is so small that the mixture no longer explodes (upper explosion limit). Inert gases like nitrogen reduce the explosive range. In industrial oxidation processes, like the present one, mixtures with a composition below the lower explosion limit are almost always used. For naphthalene this are 30–40 g for 1 $m^3$ air. The mixture flows through the catalyst tubes from top to bottom. (Yield 85–88 per cent of the theoretical.)

The precipitation of phthalic anhydride from the hot air stream is very difficult. The air current is passed through numerous large cooling chambers with discharge screw conveyors at the bottom, through cyclones, or through electro-filters. Phthalic anhydride can also be produced by the fluidized bed process.

Maleic anhydride is produced by a similar process from benzene.

Phthalic anhydride is an intermediate for the manufacture of anthraquinone vat dyes (see next chapter).

With ammonia it is converted easily into phthalimide which with sodium hypochlorite and alkali yields anthranilic acid (Hofmann's degradation), a raw material for the indigo synthesis:

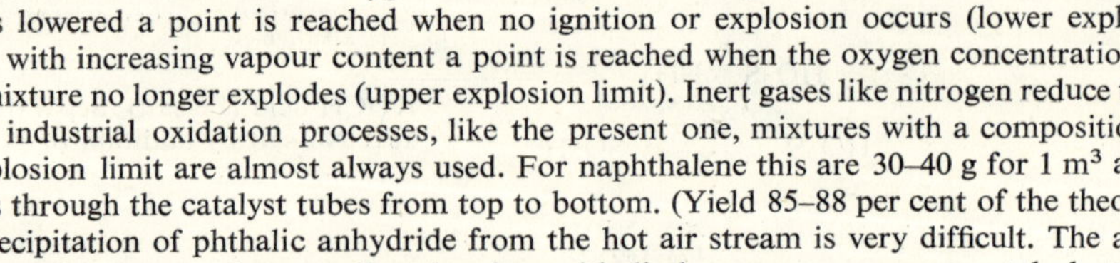

Phthalic acid esters are very important plasticizers for plastics (q.v.). Phthalic anhydride is converted to phthalodinitrile, the base material for phthalocyanin dyes.

About half the production is used in the manufacture of alkyd resins. It is also used as a bi-functional acid for polyester resins.

In 1959 phthalic anhydride production in the U.S.A. amounted to 222,000 tons, of which 95 per cent was from naphthalene and 5 per cent from o-xylene derived from petroleum. A new plant has been built where the three xylenes are oxidized in the liquid phase and the three isomeric phthalic acids are separated (P. Sherwood, Modern methods for the production of phthalic anhydride, petroleum, and coal. *Erdoel und Kohle* 1959, **12**, 890–894). The production of phthalic anhydride from o-xylene is more difficult than from naphthalene.

### (d) Anthracene derivatives

*Anthraquinone*

This intermediate, important for vat dye manufacture, can be obtained by three industrial processes:

(1) Phthalic anhydride and aromatics yield o-benzoylbenzoic acids with $AlCl_3$ which are then condensed with concentrated sulphuric acid to anthraquinones (important for the production of substituted anthraquinones).

(2) Anthracene from bituminous coal tar is oxidized with chromic acid. This is a particularly economic process because the chromium sulphate lyes are used for the production of chromium oxide green, or as chrome tanning agent. The oxidation is carried out in an aqueous suspension in the presence of a wetting agent.

(3) Catalytic oxidation of anthracene with atmospheric oxygen similar to the conversion of naphthalene to phthalic anhydride.

Evaporation of anthracene for the oxidation is carried out with an air/steam mixture to give 20 g anthracene in 1 $m^3$. The catalyst oven contains a granular ferro-vanadate catalyst spread out on shelves. It is embedded between tubular coils which are cooled by pressurized water so that the temperature in the reaction zone of the oven is 390°C and no local over-heating can occur. After passing through a heat exchanger the anthraquinone is precipitated in cooling towers or chambers and dust filters. Its purity is better than 99 per cent. The process is also suitable for a fluidized bed process using a suspended catalyst.

The methods for the production of the mono- and polysubstituted anthraquinone derivatives differ in many respects from those used in benzene and naphthalene chemistry.

Whilst chloroanthraquinones can be obtained by chlorination of anthraquinone in oleum, the exchange of the sulphonic acid group for chlorine by means of sodium chlorate in hot hydrochloric acid (A. Fischer) is more important because sulphonation results in less complex products than chlorination. Thus, for instance, 1-chloroanthraquinone and 1,5- and 1,8-dichloroanthraquinone are produced from sulphonic acids.

2-chloroanthraquinone is obtained from phthalic anhydride and chlorobenzene in the same way that the important 2-methylanthraquinone is obtained from toluene (attack only in p-position).

The halogen in the anthraquinone molecule is fairly mobile and can be interchanged easily.

Nitration of anthraquinone does not take a uniform course.

Dihydroxyanthraquinones, important for mordant dyes, are obtained by three methods: oxidative alkali fusion of sulphonic acids (1,2-dihydroxyanthraquinone, alizarin), condensation of phthalic anhydride with dihydroxyphenols (1,4-dihydroxyanthraquinone, quinizarin), and by exchange reactions, e.g. of sulphonic acid groups by calcium pressure fusion.

The amino-anthraquinones are of great importance as initial and intermediate products for many dyes of the alizarin series and the vat dye group, as well as for acetate silk dyes. Whilst 1-aminoanthraquinone is produced from the potassium salt of anthraquinone-1-sulphonic acid with aqueous ammonia in the presence of the sodium salt of m-nitrobenzenesulphonic acid, 2-aminoanthraquinone is produced from 2-chloroanthraquinone with ammonia. 2-aminoanthraquinone is the raw material for indanthrene (indanthrene blue RS) and flavanthrene.

In many cases further rings are added to the anthraquinone system. With sulphuric acid, iron powder, and glycerine benzanthrone is obtained (via anthrone), another important intermediate for vat dyes, which can also be produced from phenyl-α-naphthyl-ketone.

## LITERATURE

H. E. Fierz-David and Blangey, *Farbenchemie*, 8th ed., Vienna, Springer, 1952.

G. Jung and H. Petri, Die aromatischen Zwischenprodukte, in: Winnacker-Weingaertner, *Chem. Technologie*, vol. 4, München, Hanser Verlag, 1954.

K. Venkataraman, *The Chemistry of Synthetic Dyes*, New York, Academic Press Inc., 1952.

N. N. Vorozhtsov, *Osnovi sinteza promezhutochnykh produktov i krasatelei*, Moskva, Goskhimizdat, 1950.

# *B. The chemical processing of raw materials produced in agriculture and forestry*

Unlike coal, petroleum, and mineral raw materials which were no longer formed during historic times, the products of agriculture and forestry are a perpetually renewed and inexhaustible source of raw materials. This may at some future date become of the utmost importance when all fossil carbon sources are exhausted. Then the chemical utilization of the products of agriculture and forestry will have to be relied upon in increasing measure. At the present time agricultural and forestry raw materials are processed chemically only to a small extent compared to coal and petroleum chemistry; they are nevertheless important and indispensable. The following synopsis shows the most important possibilities of the chemical utilization of the constantly regenerated organic raw materials:

Utilization of wood:

Carbonization and distillation of wood,
Mechanical wood pulp, paper,
Bark tannins,
Cellulose (by-products: sulphite spirit, yeast, turpentine),
Wood saccharification.

Utilization of straw and reed:

Cellulose (utilisation of waste sugars), straw hydrolysis products, and yeast.

Foods and beverages
Sugar industry:

Beet sugar, cane sugar (by-products: molasses, spirit, yeast).

Starch and starch sugar
Fermentation industry:

Wine, beer, spirits, vinegar.

Fats and oils; processing:

Hardening of fats,
Fatty acids,
Soap and glycerine.

# I. Wood as a chemical raw material

The forest is the largest perpetually renewed source of raw materials on earth. Without forest and wood no cultural and industrial development would have been possible. Wood was the first raw material that man could work and shape and it made possible his first artistic creations. In the last analysis shapes and styles of construction are derived from constructions in wood. If so little of all this has been preserved this is due to the transient nature of wood. For Germany which is poor in other raw materials wood has always been an important source of material for the manufacture of chemical products. Before the last war forests covered 12,6 million hectares = 27 per cent of Germany's total area. They supplied about 50 million festmetres of wood, i.e. almost 1 festmetre per head of population, used partly as timber and partly as firewood. (1 festmetre (fm) = 1 $m^3$ of stacked logs). The large amount of firewood represents a reserve of raw materials which it will be possible to utilize as soon as firewood can be replaced by other fuels.

Despite this large amount of timber felled which used to be well balanced with the new growth, 15 million fm were imported annually, for use mostly in the chemical industry and as pit props. Since 1945, only about 9,8 million hectares of forest area remain and new growth amounts to only 27 million fm annually. Nevertheless, 69 million fm were felled recklessly in 1947 on the territory of the undivided Germany. Serious damage was thus caused to this important raw material source which can only be preserved if no more is felled than is replaced by new growth, and if it is carefully cultivated and not recklessly exploited.

This basic principle for orderly forest management was recognized in Germany for the first time 200 years ago. In other countries it was formerly unknown and was adopted only gradually.

Thus large forest areas of the globe were turned into desert by reckless felling, with grave results for the climate. Mention may be made here only of large parts of the savannahs of the former German East Africa which were laid waste by mismanagement of the forests by the mandatory power, Northern Norway, the Karst (Yugoslavia) and large areas of the U.S.A. Even today in some countries the primeval forests are being exploited like the mines.

The following survey shows the population, area of forest, and the amounts of new growth and of timber production in the various timber-producing countries of the world. The population figures are those of 1939, the timber production is that of 1946 (the figures are from Gläser in: Winnacker-Weingaertner, *Chem. Technologie*)

Population figures, forest area, and timber production

| Country | Popul'n 1939 millions | Forest area, million hectares | New growth million fm | Production 1946 million fm |
|---|---|---|---|---|
| World (total) | 2,167 | 3,650 | 2,406 | 1,280 |
| U.S.S.R. | 178 | 960 | 630 | 250 |
| British Empire | 548 | 762 | 400 | ? |
| U.S.A. | 131 | 296 | 334 | 361 |
| French Union | 106 | 170 | 180 | ? |
| Netherlands and colonies | 78 | 151 | 130 | ? |
| Germany 1939 | 69 | 12·6 | 40 | 56 |
| Germany 1946 | 66 | 9·8 | 27 | 61 |

It is obvious that in all thickly populated countries with a highly developed industry there is a shortage of timber whilst there is a surplus in the undeveloped areas of the world. It can be estimated that only about half the new growth of timber in the world is utilized, the other half rots in the primeval forests. The following survey shows very impressively the extraordinarily high percentage of timber used in Germany as compared with the rest of the world (after Gläser in Winnacker-Weingaertner)

Consumption of wood classified by use (1937)

| Use | World % | Germany % |
|---|---|---|
| Construction, saw mills | 22 | 44 |
| Mechanical processing | 8 | |
| Cellulose, paper, textile raw material | 7 | 14 |
| Pit props | 2 | 11 |
| Sleepers and masts | 2 | 2 |
| Other timber industries | 6 | 10 |
| Total timber | 47 | 81 |
| Firewood | 53 | 19 |

The growing need of timber in Germany strongly influenced afforestation policy. In past decades an increasing number of pine trees were planted, with a corresponding reduction of deciduous trees. The latter were used mainly as fuel, apart from veneers and furniture. From the point of view of forest management this development of mono-cultures was unfavourable. The policy was changed when before the 1939–1945 war hardwood was used for the first time at Wolfen for the large-scale production of cellulose. The use of hardwood, especially beech, has since been generally adopted in the cellulose industry. The constantly rising consumption of wood in the chemical industry has had a favourable effect on the increased planting of deciduous and mixed forests.

Classification of wood according to its use (after Trendelenburg: *Das Holz als Rohstoff*, 1st ed., München, I. F. Lehmanns Verlag, 1939)

(1) Wood used in the building industry and as industrial raw material

(2) Wood used as chemical raw material

| Wood fibre | Saccharification | Carbonization | Fuel |
|---|---|---|---|
| (cellulose, mechanical wood pulp) | (wood sugar, wood alcohol) | (charcoal, wood distillation) | (domestic wood, distillation) |

For a better understanding of the processing of wood described in the following section it is necessary to discuss briefly the structure of the ligneous tissue in the tree, and the microstructure of cellulose and of lignin. The structure of wood is the outward expression of its tasks:

(1) Wood is a supporting structure of ideal strength and elasticity. The tree often has to support enormous loads of snow, is exposed to storms, and lasts for very long periods.

(2) Wood ensures the transport of water and nutrient salts from the root sometimes to heights of up to 100 m and into the tip of the furthest leaf.

(3) Wood must provide space for the storage of nutrient salts, reserves for the winter and for the new growth in spring.

The origin of trees dates back 100–300 million years, the age of gymnosperms 200–300 million years, that of angiosperms (deciduous trees) 100 million years, as the often very impressive remains of wood found in bituminous coal and brown coal deposits show. After all coal also originates from wood.

Figure 97 shows the microstructure of ligneous tissue in a section through the bark, the cambium, and the sparwood of a pine. The illustration shows clearly the cambium during division. This is the group of cells which contains the mysterious germ of growth and of the whole structure of the trunk. The ring, consisting of only one row of cells, is dividing continuously both inward and outward. Inside wood cells are formed, outside bark cells. In this ring zone the cambium cells divide by longitudinal walls. Obviously wonderful and very accurately controlled enzymatic processes permit the cambium

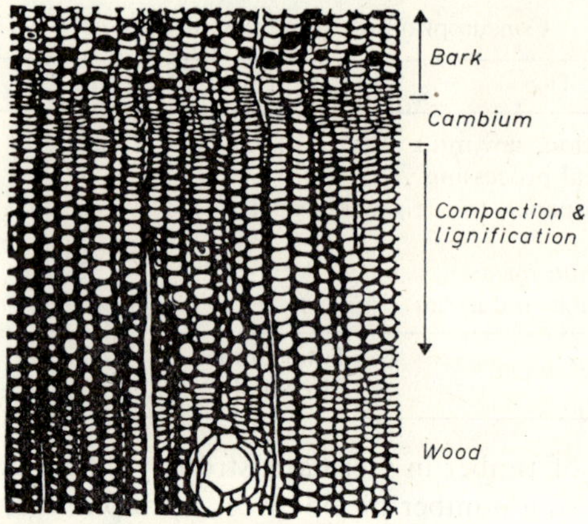


Figure 97. Microstructure of ligneous tissue

cell to produce in succession the most varied types of cell: strength cells, water transport cells, storage cells, and medullary rays. Clearly, special relationships exist between neighbouring cells since they produce simultaneously, though often varying in kind, a certain type of cell. The cambium lives as long as the tree is alive. The illustration also shows the tracheids, arranged in radial rows. They are the most important type of cell of pine trees (90 per cent), giving strength to the structure. They are 3–4 mm long. In deciduous trees it is the hard fibres (sclerenchymatous fibres) that determine the strength; these are 0·3–1 mm long. The proportion present in beech varies between 30 and 80 per cent. The ducting cells form a system of interlinked capillary tubes carrying the sap. The size of the cells varies greatly, their length being between 0·1 and 1 cm. The cellulose fibres of ramie are up to 25 cm long.

## Chemical composition of wood

The composition of wood is best considered from the point of view of growth. Initially the young cell developing in the cambium contains only cellulose. At this stage the very thin walls are cemented together only by pectin so that changes of shape are still possible. Then follows a thickening of the cell walls, micelle being deposited next to micelle and lamella next to lamella. Simultaneously lignification progresses which consists of the incorporation of lignin and other substances into the cell structure developing from the living plasma of the cell.

With lignification (solidification) the growth of the cell is complete. K. Freudenberg compared the solidification to the setting of reinforced concrete, the cellulose corresponding to the tensile iron rods, the lignin to the concrete. The compressive strength of wood is a function of its lignin content. In pine trees the proportion of lignin is larger. As compared with deciduous trees they have a greater lateral buckling strength. The lignin content of beech is about 20 per cent, that of pine about 30 per cent. Their cellulose contents of about 40 per cent are approximately equal. Some tropical deciduous trees contain up to 40 per cent lignin and can withstand extremely large mechanical stresses. Figure 98 (from Trendelenburg) shows the composition of pine and beech wood.

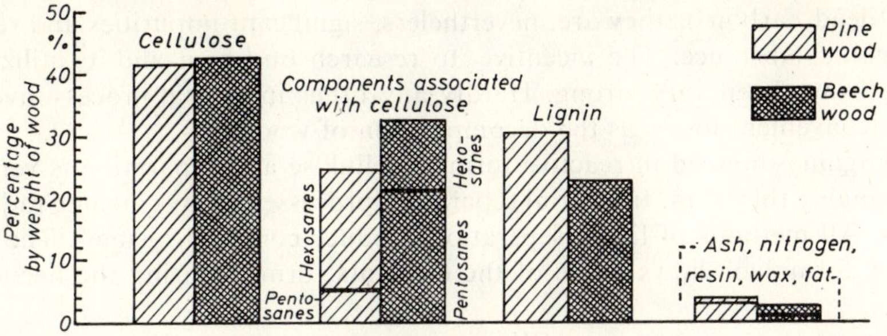


Figure 98

Cellulose is the commonest organic material and the main constituent of wood. It is formed from atmospheric carbon dioxide. About $10^{12}$ tons $CO_2$ are estimated to be fixed on earth in the form of cellulose.

X-ray diagrams show that celluloses of very different origins are very similar. Their differences consist essentially in the arrangement of the groups of molecules in the fibres. Cellulose is a polymer anhydride of d-glucose which is formed during complete hydrolytic decomposition (wood sugar). 40 per cent cellobiose can be obtained as a defined decomposition product during acetolysis. The formulae in Figure 99 show the structure of $\beta$-glucose, $\beta$-cellobiose, and cellulose. A genuine cellulose molecule contains upwards of 2,000 glucose molecules. The molecules are arranged regularly, a proportion having a higher degree of order (crystallites) than the rest. The transition is fluid, there being no strictly defined phase limits. The proportion with a higher degree of order determines certain properties, e.g. the mechanical ones, the remainder has a higher speed of reaction. The macromolecules form micelles (crystallites), compact groups with a similar structure in all fibres. The micelles combine into fibrils. These fibrils contain intermicellar spaces which have a strong effect on the properties of the cellulose fibre, for instance on their density. Wood also contains hemicellulose which can either be dissolved out of the less readily hydrolysable cellulose with alkali, or it can be hydrolysed with dilute acid. A certain amount of these hemicelluloses which include the pentosanes and hexosanes remains, however, in the cellulose. The sugars contained in the sulphite waste liquor and the acid pre-hydrolysis lyes of wood are formed mainly from the associated carbohydrates, the hexosanes and pentosanes. In hard wood the pentosanes predominates and thus beech wood sulphite waste liquor contains mainly pentosanes.

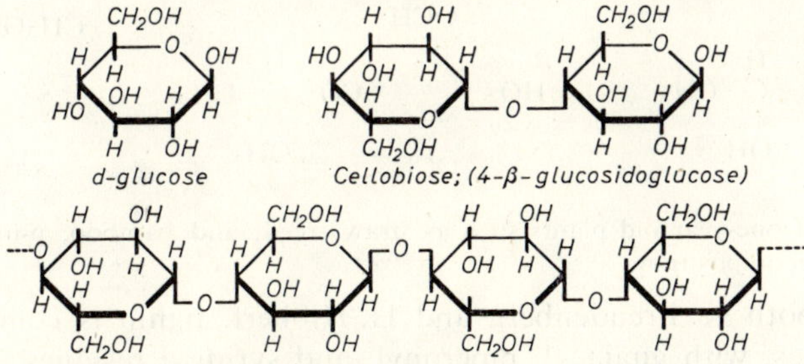


Figure 99. Cellulose (Haworth, Freudenberg)

## Lignin

During the decompositon of wood one third of the lignin is lost. Whilst the sugar-free transformation products of lignin do not cause any loss of oxygen in the drainage mains, (they are 'dead carbon'), they are, nevertheless, significant impurities and represent a loss of organic substance. The incentive to research on lignin and its utilization has, therefore, always been very strong. Hardly any other industrial process involves such high and inconvenient losses as the decomposition of wood.

In wood lignin is bonded in reactive form to cellulose and separation is very difficult without changing this form. Industrial separation processes always include radical transformations. All methods of lignin seperation produce coniferyl systems. The following synopsis of formulae shows some of the elements formed during the decomposition of lignin:

alkali fusion: pine: protocatechuic acid

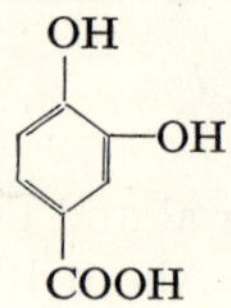

beech: gallic acid

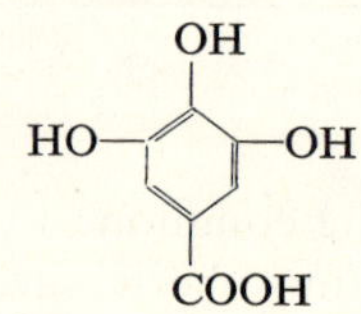

zinc-dust distillation: guaiacol

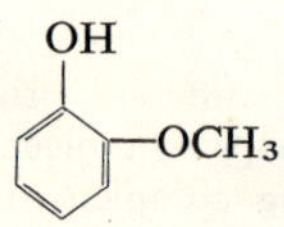

propyl-guaiacol

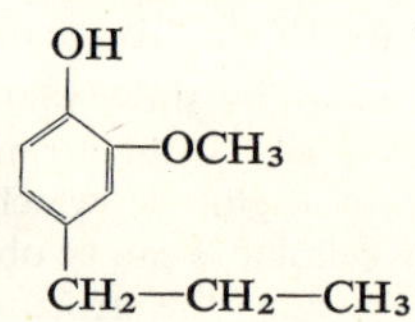

Oxidation with nitrobenzene: pine: up to 22 per cent vanillin, beech: vanillin + syringic aldehyde. Extraction with alcoholic HCl:

pine:

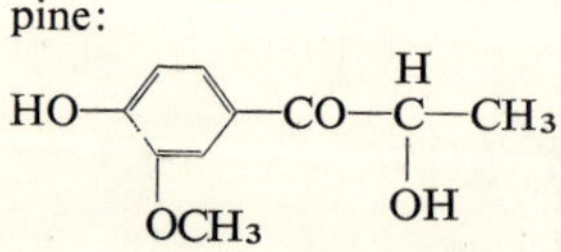

maple:

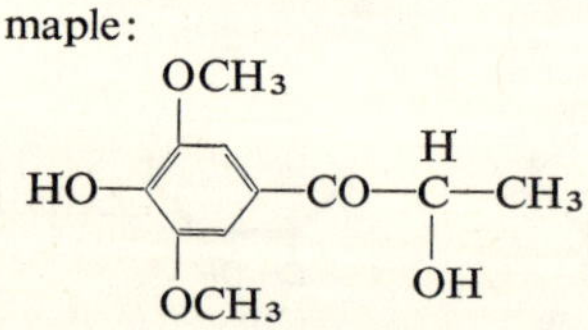

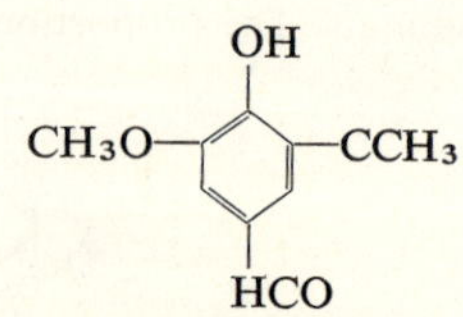

syringic aldehyde

From the lignin of one-year old plants such as straw, reed, and bamboo, p-hydroxybenzaldehyde is obtained during decomposition.

According to both K. Freudenberg and H. Hibbert, lignin is composed of phenyl propane derivatives, with guaiacyl, piperonyl, and syringyl residues. Primary units of lignin are (see e.g. Freudenberg, *Naturwiss.*, 1955, **42**, 29–35):

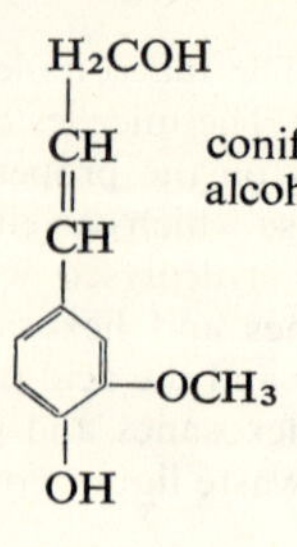

coniferyl alcohol

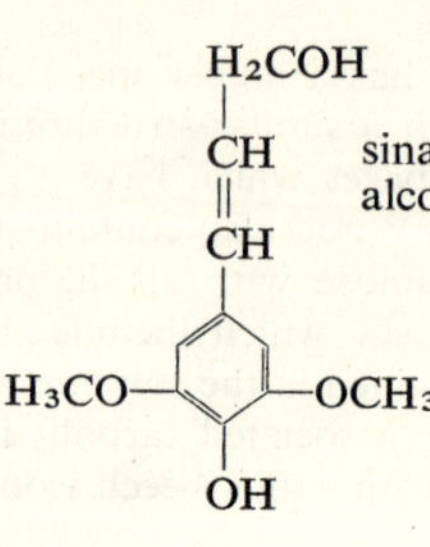

sinapin alcohol

secondary units:

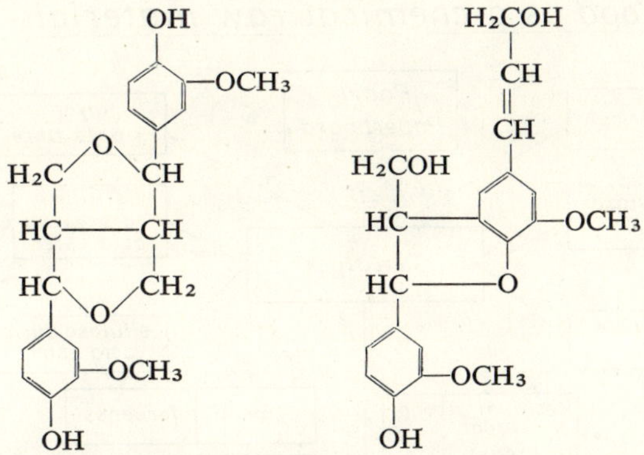

pinoresinol dehydro-diconiferyl alcohol

The units shown in the chart are joined together in the lignin molecule but nothing is known about the sequence of the individual residues. Native lignin is bonded in the wood to the cellulose part. Freudenberg succeeded recently in demonstrating elegantly the correctness of the lignin formula established by him a long time ago: coniferyl alcohol and fungus dehydrases yield a lignin-type substance. Dehydroconiferyl alcohol was detected as a preliminary stage of lignin. Like the fungus enzyme, cell-bound pyrocatechin dehydrogenases and peroxidases contained in the cells newly developing in the cambium are able to convert coniferyl alcohol into lignin-type dehydrogenation polymerizates (K. Freudenberg, Biochemische Vorgänge bei der Holzbildung. *Angew. Chem.*, 1959, **71**, 43. Summary: Fortschritte d. Chemie organ. Naturst., 1962, **20**, 41). Tiemann and Haarmann had, of course, found that coniferyl alcohol occurs in the cambium as glycoside. Lignification consists, therefore, in the decomposition of this glycoside and the polymerizing dehydrogenation of coniferyl alcohol via dehydro-diconiferyl alcohol to lignin. In hard woods a similar reaction takes place with sinapin-alcohol. The essential biochemical problem is the question how the phenylpropane unit is formed in the plant. There is no doubt that this takes place via shikimic acid and prephenic acid which are formed from assimilation products.

This means that lignin is probably ultimately also based on glucose.

## 1. CHEMICAL PROCESSING OF WOOD

(Figure 100)

The chemical processing of wood can be divided into six main groups:

(1) Mechanical wood pulp,
(2) Cellulose and by-products,
(3) Wood sugar and by-products,
(4) Wood distillation,
(5) Resin production,
(6) Tannin production.

The first method to be discussed will be the oldest one which at the same time involves the most radical action affecting the structure of wood: wood distillation and carbonization. Its origin can be traced back to the times of the early lake dwellers who produced charcoal by pile charring.

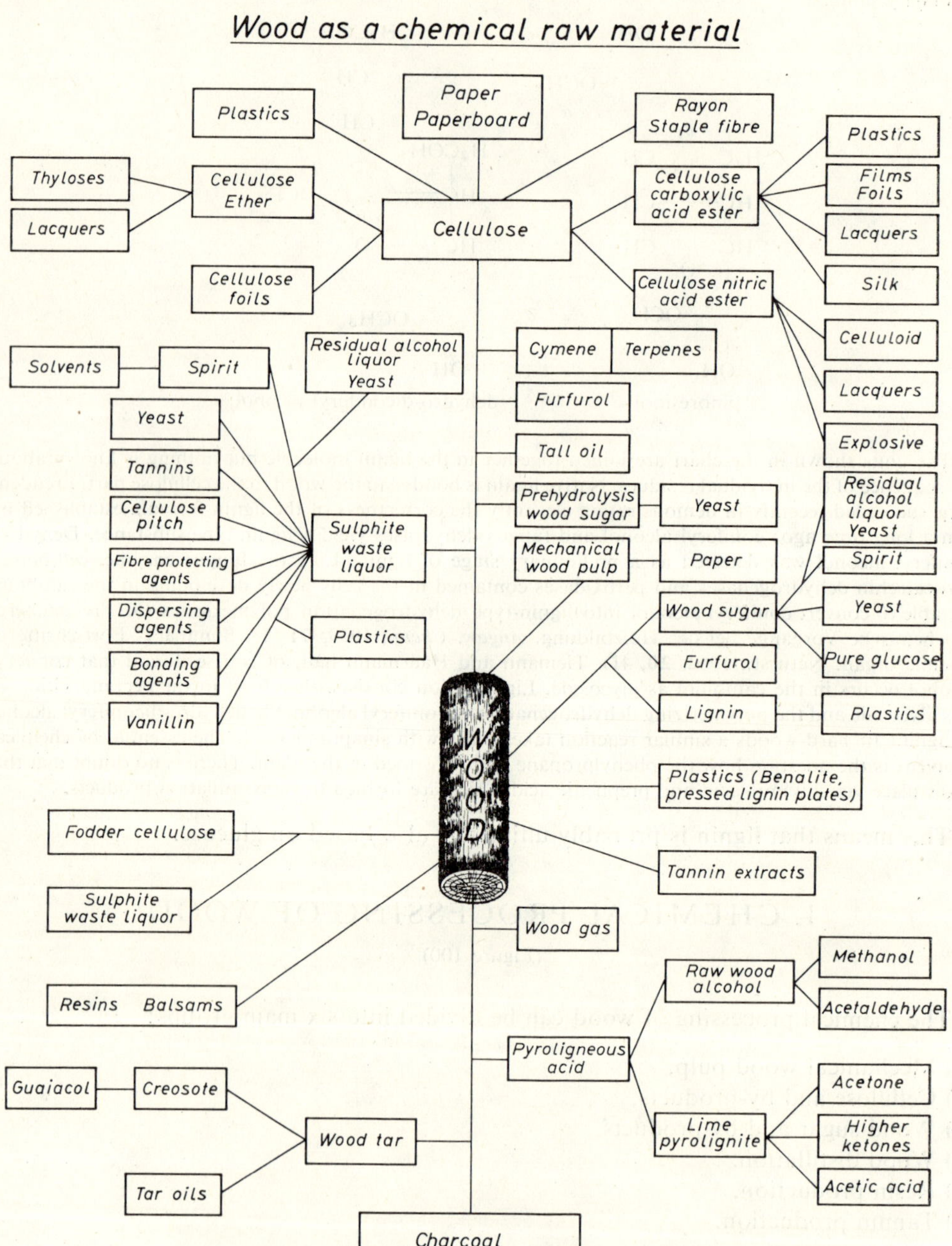


**Figure 100.** Chart showing the industrial products obtained by chemical transformation of wood and wood constituents

### (a) Wood distillation, wood carbonization

Formerly, the only purpose of wood distillation was to produce charcoal. This was used mainly for the production of metals and was later replaced in metallurgy by coke. Charcoal burning for this purpose was carried on since time immemorial, particularly in the Harz, in Sweden, and in the Urals.

To build a charcoal pile, the sticks of wood are piled into a flat cone with a cut-off top. A shaft in the centre permits the combustion gases to escape. The pile is covered from the outside first with faggots, then with loose earth, clods, etc. At the bottom there is a number of air inlets. A charcoal pile usually burns for about eight days.

In pile charring all gaseous products burn or escape and are lost. It was only the carbonization in iron retorts, heated externally, that made possible the utilization of all the products. Wood distillation is true destructive distillation in the absence of air. The yield of charcoal and distillates varies greatly, depending on the type of equipment used, the wood, and the conditions. Wood distillation in a retort (wood carbonization) takes the following course:

(1) At 150–170°C water distils off (drying)
(2) At 170–270°C gas ($CO + CO_2$) begins to form and some acetic acid and methanol distil over.
(3) At 270–280°C, i.e. rather suddenly, an exothermic reaction takes place. Less CO and $CO_2$ are formed, instead hydrocarbons are liberated (wood gas). During this period most of the acetic acid, wood spirit, and tar is produced.
(4) At 280–400°C the carbonization process is terminated without an external supply of heat. During this period spontaneous heating takes place and all the gases are evolved from the charcoal with simultaneous splitting off of hydrocarbons.

The following figures show the proportions of the products of the German wood carbonization industry as an average over several years:

| | |
|---|---|
| 34·7% charcoal | 5·7% acetic acid |
| 27·0% water | 2·4% wood alcohol |
| 21·0% wood gas (for heating retorts) | 9·2% wood tar |

The many designs of retorts and furnaces developed in the course of centuries will not be discussed here. 1–2·5 fm can be charged to a simple distillation plant with horizontal carbonization retorts. In larger plants long chamber retorts are used into which wood (e.g. 10 fm) is brought on trucks and is carbonized directly on the trucks. All these methods use an external heat supply which is, of course, limited. In the Reichard process of the 'Hiag Verein für Holzverkohlung' the heat is supplied internally direct by the carbonization gases. 40 fm of wood can be charged to the retorts in one batch. Such plants exist at Konstanz and Mannheim. Continuous processes are also operated. In the Gröndahl process the wood passes on a truck through an oven of about 100 m length with compartments designed as sluices. The heat of the charcoal produced is utilized to pre-heat the heating gases, and the charcoal is cooled.

Apart from charcoal, methanol, acetone, and acetic acid were for a time very important products of the wood carbonization industry since large-scale industrial syntheses of these raw materials did not then exist. Cheaper ways for the production of these compounds have since been found and the wood distillation industry has a hard struggle.

Raw wood vinegar consists of 82 per cent water, 8–10 per cent acetic acid, some formic acid and higher homologous acids. It used to be produced by the so-called three-still system, illustrated in Figure 101.

The gaseous components issuing from the retort oven are cooled in a cooler and pass into the first still where the tar is retained in the still residue. The vapours then pass through hot milk of lime which removes the acids. The wood alcohol passes beyond the hot milk of lime into a cooler where it is condensed and then into a third still and column where it is fractionated. The calcium solution is concentrated by evaporation and dried on roller driers. The product is crude calcium acetate from which acetone can be obtained directly by an old process. To produce the acids, lime pyrolignite is decomposed with concentrated sulphuric acid and the crude vinegar with a concentration of 75–80 per cent is distilled off under a slight vacuum to prevent coking and the evolution of $SO_2$. In Germany auxiliary solvents are used for the purification by distillation of the crude vinegar, azeotropic mixtures being obtained. Acetone and higher ketones are also

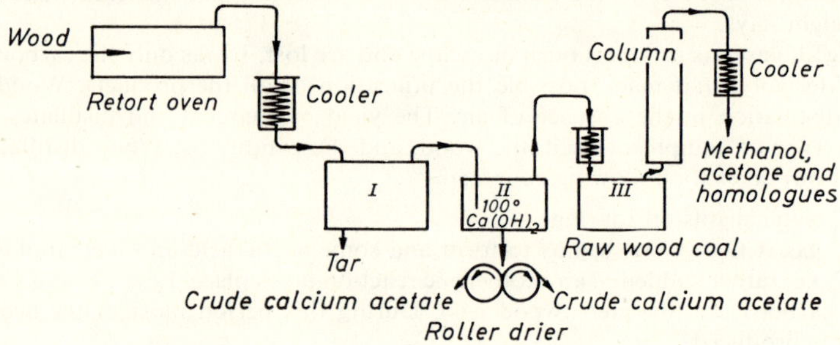


Figure 101. Simplified scheme of wood distillation by the 'three stills process'

produced by fractional distillation. Separation is very difficult because the lower ketones form azeotropic mixtures with methanol. Wood tar contains many phenolic compounds: cresols, xylenols, guaiacol and creosol (methyl guaiacol). By alkaline extraction under certain conditions 'creosote' is obtained which consists essentially of guaiacol and methyl guaiacol.

Charcoal is used for special heating purposes and for the production of gas in generators. Its most important use in chemical engineering is based on its large internal surface area and its excellent adsorption capacity for a large variety of materials. Thus it is used as activated carbon for gas adsorption, as decolourizing carbon for liquids, for the deodorization of e.g. alcohol, for the purification of water, and for medical purposes. Because of its adsorptive effect in the digestive tract of domestic animals it is often mixed with their food. The use of charcoal for the production of carbon disulphide was discussed earlier. Its importance as a raw material now is, however, less than it used to be, partly because peat is now also used for the production of activated carbon.

Wood tar which is decomposed by distillation into three fractions, light tar oil, heavy tar oil, and residual tar, is used for the production of flotation agents, as fuel oil, for impregnation and rot-proofing, and for the production of guaiacol and creosote used in pharmaceutics.

### (b) Wood saccharification

Wood saccharification on a large scale can be justified economically only if waste wood is used. The waste of timber in saw mills and cabinetmaking factories (saw dust, wood shavings, and cut-off pieces) amounts to at least 25 per cent of the wood used, but is probably even higher.

Collection from the numerous small producers of waste timber would be uneconomical. There are, however, areas where it should be possible to transport the waste timber to favourably situated wood saccharification plants, e.g. in the Scandinavian countries, U.S.A., Canada, and the Soviet Union. At present wood saccharification is no longer carried out in either of the two parts of Germany although experience shows that wood sugar, apart from molasses and sulphite waste liquor, is the cheapest raw material for the further processing of fermentation products, and for the production of pure glucose. Wood sugar, together with straw- and reed hydrolysate, forms the basic raw material for the production of fodder yeast.

As early as 1819 Braconnot found that water-soluble sugar could be obtained by the reaction of strong sulphuric acid with saw dust. This experiment was not made to produce wood sugar; the object of the investigation at the time was to study the effects of all available chemicals on wood. The hydrolysis of wood with concentrated hydrochloric acid has been known since 1880 and Dangvilliers took out the first patent. Willstatter and Zechmeister (1913) used hyperconcentrated hydrochloric acid, obtaining cellulose solutions of about 15 per cent concentration but with only 3–4 per cent monosaccharide. Of importance was Hagglund's discovery that spent hydrochloric acid which had lost its dissolving capacity still had some dissolving effect when acting on a fresh supply of wood. When the charge of wood was replenished equilibrium was always re-established so that finally up to 30 g sugar were dissolved in 100 g hydrochloric acid. This made possible the economical recovery of the hydrochloric acid by evaporation under vacuum.

The possibility of hydrolysing wood by hydrofluoric acid on a large experimental scale was also explored. This process invented by Fredenhagen and Helferich was developed industrially by the former IG Oppau. It was found, however, that with larger quantities it was impossible to dissipate the liberated heat from the wood. Schlubach and Darboven attempted to obviate this difficulty by using inert organic solvents which were intended as heat carriers for the dissipation of the reaction heat. It proved impossible, however, to develop an industrial process.

Three methods have so far been used on a large industrial scale, the 1914–1918 war process, the Bergius-Rheinau process, and the Scholler process with modifications. The 'war process' which employed diluted acids in autoclaves was entirely uneconomical and was abandoned in 1919. This gave rise to the assumption that wood saccharification with diluted acids was uneconomical. It must be borne in mind that in wood saccharification, in addition to hydrolysis, there is also a building up of saccharides (reversion).

*Bergius-Rheinau process* (*Bergius-Hägglund*)

Similarly as with hydrofluoric acid, a large amount of heat (about 36 cal per g wood) is liberated during the hydrolysis of cellulose with hyperconcentrated hydrochloric acid (40 per cent concentration), and a further quantity of heat by the dilution of the hydrochloric acid by the water contained in the wood. In the laboratory the dissipation of this heat causes no difficulties owing to radiation. On a large scale, however, the temperature of the dry cellulose which is a poor conductor of heat rises to an undesirable level. This excessive rise of temperature which causes losses of sugar can be avoided only by the use of relatively large amounts of liquid which acts as a heat absorber. However, this involves a considerable circulation of concentrated hydrochloric acid which has to be recovered by distillation under vacuum. The advantage of the method lies in the fact that the operations are carried out without pressure increase at room temperature.

Only partial saccharification is achieved with 40 per cent hydrochloric acid owing to the lack of water. When the decomposed cellulose is to be used as animal feeding stuff this is acceptable since the flora of bacteria in the animal's stomach can complete the decompensation. For biological syntheses, however, acid after-treatment in a thin solution is necessary for complete saccharification. In the Bergius process, owing to the low temperature at which the hydrolysis is carried out, the yield is better than in the process in which diluted acids are used under pressure. The cellulose is converted almost quantitatively into glucose. In order not to dilute the hydrochloric acid too much from the start thorough drying of the wood is necessary. The value of Bergius's and Hagglund's work lies in the fact that they applied these fundamental observations on a industrial scale, first in collaboration with Messrs. Th. Goldschmidt A.G. and later with the Deutsche Bergin A.G. Despite the extreme difficulties involved in working with hyperconcentrated hydrochloric acid it eventually proved possible to carry out the hydrolysis process on a large scale without trouble. The diagram in Figure 102 shows in simplified form the method of operation of the Bergius-Rheinau process. Since it shows all the essentials of the process no detailed description will be given.

The concentration of the recovered hydrochloric acid is too low for immediate re-use. The water must, therefore, be removed from the acid. This was done by a very original method with a concentrated calcium chloride solution in a subdivided evaporator of special design, with heating tubes made of tantalum.

In recent years, the Rheinau process has been radically changed and improved by the Rheinauer Holzhydrolyse GmbH and the Chemische Holzverwertung GmbH Mannheim-Rheinau. This was the

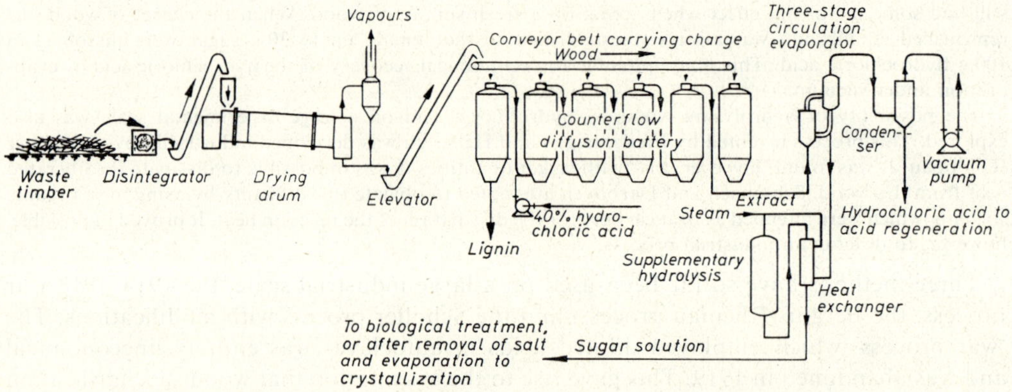


Figure 102. Bergius-Rheinau process of wood saccharification (simplified flow diagram) (old Rheinau process)

result of intensive research on the method of operating the hydrolysis (division into preliminary hydrolysis and main hydrolysis) and on simplifying the recovery, involving less loss, of the hydrochloric acid (only 6 per cent loss relative to total sugar).

In the U.S.A. the Madison process is used for the production of fodder molasses from wood (Ind. Engng. Chem., 1952, **44**, 1712). In the U.S.S.R. wood hydrolysis has for a long time been carried out for the production of alcohol for rubber synthesis and more recently for the large-scale production of pure glucose and of yeast for animal fodder.

### *Method of operation of an extraction or diffusion battery working on the counterflow principle*

For the extraction of solid materials by solvents battery extraction on the counterflow principle is used, with periodic filling and emptying of each individual extractor. In the Bergius-Rheinau process the hydrolysis of wood with concentrated hydrochloric acid is carried out in a diffusion battery. In the diffusors it is possible to extract beet slice with hot water, oily fruit with organic solvents, tan-bark and crushed tanniferous plants with water, drugs with alcohol, etc. A large number of extraction vessels are connected in series in which the crushed material to be extracted is placed on perforated bottoms or filter cloths and through which the solvent flows consecutively. In fat extraction the extractors are equipped with agitators. After saturation of the solvent the solution is clarified, e.g. by filtration (suction filter, filter press) and then evaporated, usually under vacuum. The dissolved material is precipitated by crystallization (sugar), the concentrated extract is used directly (tannin), or the dry goods may be produced by evaporation with an atomizing drier. The extraction is so arranged that maximum concentration of the dissolved material is obtained (low cost of evaporation). During processing the solvent, unless it is water, must be recovered with as small a loss as possible. The desired high concentration of the solution is achieved by use of the counterflow principle. The extractors are connected so that the fresh charge always meets the most saturated extract, and the fresh solvent meets the material that has been extracted most. This is achieved by connecting the battery in the manner illustrated in Figure 103. Only four extractors are shown whilst in reality six, eight, or more are used.

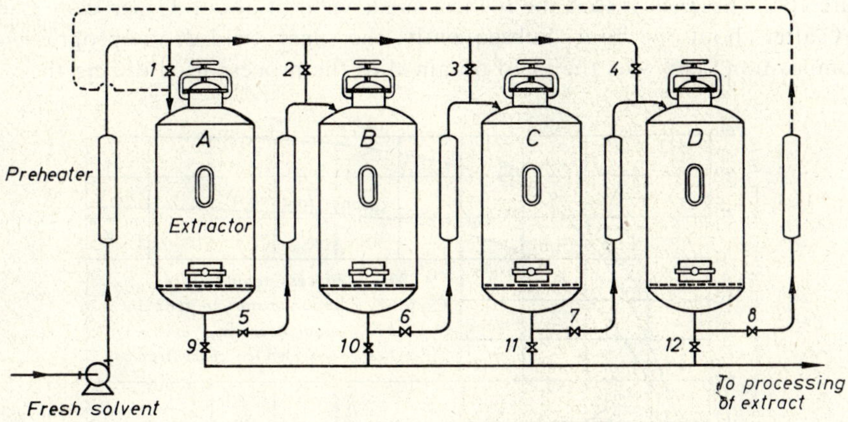


Figure 103. Diagram showing the connections in an extraction battery (counterflow extraction) for the extraction of solid materials by solvents

First taps 1, 5, 6, 7, 12 are opened allowing the solvent to flow through the extractors A. B. C. D. When a spot sample shows that the content of A is exhausted, taps 1 and 5 are closed. The fresh solvent now flows into B. When A has been emptied and freshly filled it is connected to the last in the series by opening taps 8 and 9 and closing tap 12. The content of B is exhausted next. Now taps 2 and 6 are closed and fresh solvent flows through tap 3 into C which is now the first extractor.

If organic solvents are used in place of water they have to be expelled from the material after extraction, usually by steam. For this purpose the extractors are equipped with a steam inlet; they can be connected to a cooler followed by a water separator (Florentine receiver). The solvent recovered from the material is then combined with that recovered from the extract, and re-used.

Organic solvents used are usually: petrol, benzene (toxic!), ether, ethanol, acetone, trichloroethylene, carbon tetrachloride, methylene chloride, ethylene chloride.

For the extraction of bulk materials, for instance of oil seeds, continuous extraction is also used (Hildebrandt process). The extractor has the shape of a communicating tube. The material is moved by two vertical and one horizontal screw conveyors (see section on oils and fats).

*Scholler wood hydrolysis process*

Working with diluted acids has the advantage that there is no need for the expensive and technically difficult recovery of the acid, and that undried waste timber of any size, in particular wood shavings, can be used. For these reasons Scholler and his collaborators returned to the use of diluted sulphuric acid. Scholler managed to obtain a four-fold increase in yield from the process used during the war. The following observation led him to success:

During simple heating of wood in autoclaves in diluted acid, strong sugar decomposition, as well as saccharification, takes place. Moreover, the extraction of the lignin is incomplete. The course of saccharification in closed autoclaves can be seen in the graphs in Figure 104. These show that the highest sugar yield of about 33 per cent is reached at 170°C after about one hour. Subsequently, the sugar yield drops progressively due to decomposition. This was the yield obtained in the process used during the war.

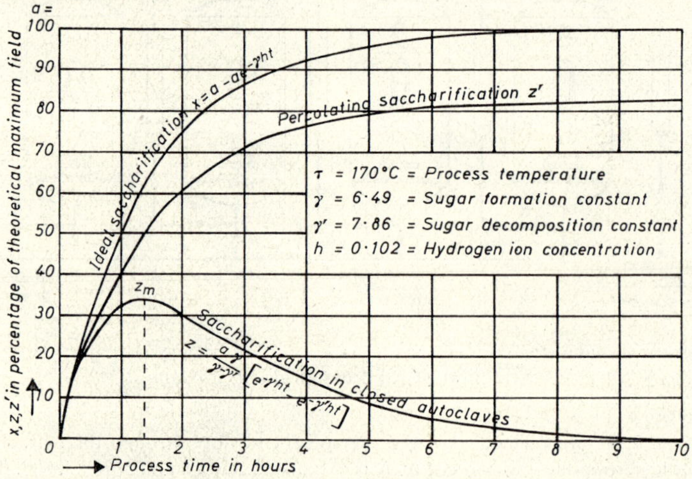


Figure 104

If on the other hand percolating saccharification is carried out, i.e. by forcing the acid through the wood under pressure, the sugar solution being continuously removed, much higher yields are obtained. Figure 104 shows the course of percolating saccharification. At 170°C it is complete in about six hours. The yield is a little over 80 per cent. The topmost curve shows the course of ideal saccharification with a 100 per cent yield which, however, cannot be reached since even with percolating saccharification some decomposition is unavoidable.

Figure 105 shows the method of operation of a wood saccharification plant using the Scholler process. The individual percolators of the battery are filled successively with waste timber by a conveyor belt. The wood is tamped down by steam impact pressure which raises the temperature to about 130°C. The 0·5 per cent sulphuric acid percolates batchwise via a so-called piston feeder in 15–20 thrusts per cycle. The sugar solution issuing at 170°C from the percolators passes through a filter into the pressure release vessels. The vapours escaping during the pressure release pass via condensers, which act at the same time as heat exchangers to pre-heat the water, into the further processing stage. Additional scrubbing in the percolator follows and then expulsion of the lignin from the percolator by the sudden opening of a wide sectional area. The sudden expansion of the hot water (170°C) in the percolator disrupts the lignin cake and pushes it into a cyclone where the steam is separated from the lignin. A sugar solution with a concentration of 3·5–4 per cent is obtained which can be processed directly into alcohol or yeast. The yield of sugar is 50–55 per cent of the dry weight of the wood. It is also possible to produce pure glucose from it.

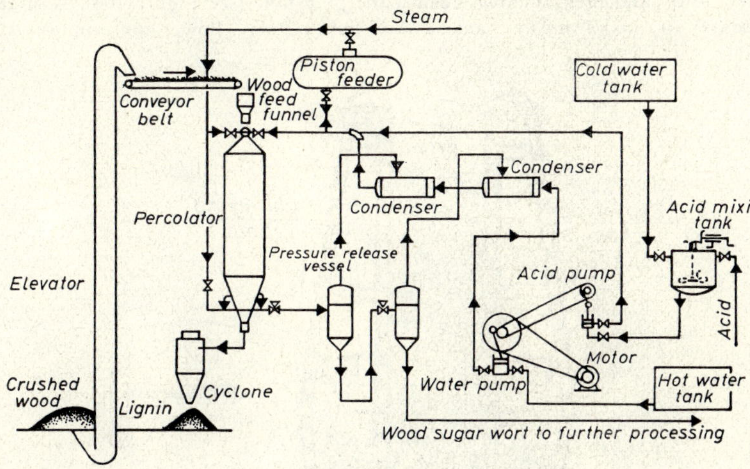


Figure 105

During saccharification furfurol is formed as a by-product from part of the pentoses: $C_5H_{10}O_5 \rightarrow C_5H_4O_2 + 3H_2O$. This can be obtained from the pressure release steam condensates by fractional distillation.

Lignin which is produced by the Bergius and the Scholler processes with water contents of about 75 and 50 per cent respectively finds only little use in industry. By acid treatment it is largely denatured and becomes chemically inert. Attempts have been made to press lignin into briquettes and to carbonize these, yielding tar of the wood tar type. Its best prospects, however, are in its replacement of phenol in phenolformaldehyde resins (see below).

### (c) Mechanical wood pulp and paper

The production of mechanical wood pulp and paper is not a chemical transformation of the wood but an upgrading process which is carried out essentially by mechanical means. Since the methods and equipment used are of general importance in the chemical industry, the processes will be discussed here at least briefly.

The principle of the production of mechanical wood pulp for paper is to separate the individual wet fibres, measuring a few millimeters in length, from the absciss layer consisting of decomposable materials. The individual fibres obtained can be formed into thin layers and felted, producing the cheaper types of paper. Better papers are produced from cellulose or waste textile fibres. The process of mechanical wood pulping for the manufacture of paper was invented in 1843 by Friedrich Gottlob Keller. Pulping consists of the mechanical separation of the fibres by pressing 1 m long pieces of round wood against rapidly rotating grindstones. Various grinders have been designed for this purpose.

In the three-pocket-grinder the chambers or pockets are filled manually with wood which is forced against the pulp stone by spring-loaded props. When the contents of the chamber are exhausted it has to be re-filled. In the magazine grinder a high shaft above the pulpstone is filled with roundwood which slides down automatically. Props at the side of the pulpstone move slowly towards it. It is sufficient to pull the props back for a fresh supply to slide into the chamber.

Continuous grinders are the most efficient. The feed shaft is formed by toggle chains which grip the wood fed in at the top and press it down against the pulpstone (see Figure 106). A large continuous grinder produces 60–70 tons of dry mechanical wood pulp per day. Power consumption is very high.

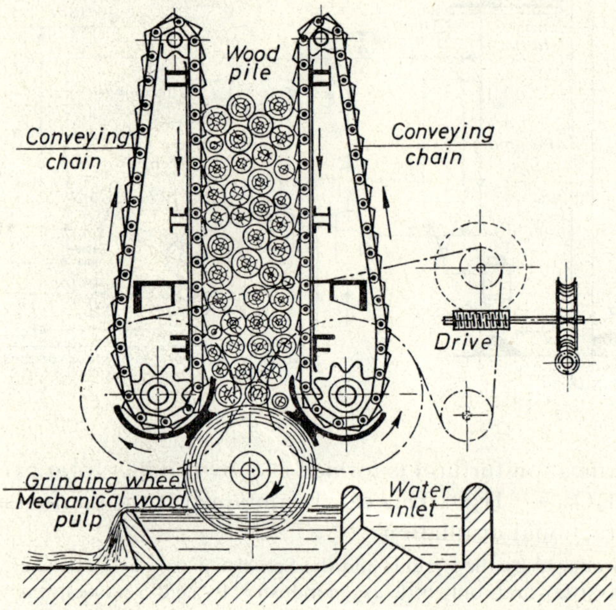


Figure 106. Continuous grinder (design of a vertical grinder)

During grinding single fibres and bundles of fibres are pulled out of the wood substance. Owing to the frictional heat dry wood would quickly ignite, and it is, therefore, constantly sprayed with water. Simultaneously, the water softens the gluey resinous substances, a process facilitated by the rise in temperature. If sufficient water is used the temperature does not rise above 25–30°C. The pulp then remains very light in colour. This 'white mechanical pulp' gives rather short fibres and a little cellulose must, therefore, be added to give strength to the paper manufactured from it. 'Brown wood pulp' gives much longer fibres which are much more pliable and are, therefore, suitable for the manufacture of wrapping paper and paperboard. For this purpose the wood is steamed in digesters before grinding so as to loosen the structure.

The pulp, consisting of a thin paste with a fibre content of 3 per cent, passes into purifiers, refining, and sorting machines. The fine pulp is separated from the coarse pulp by screens. Fibres incompletely divided are split in the refiner. All these operations are carried out in the presence of large quantities of water. Finally, the pulp passes into a wet classifying drum where the fibres, first by draining, then by compression during their passage between the rollers, aggregate more and more and are dried so that eventually a strip of pulp emerges.

In the United States a treatment of wood fibres has been developed which makes possible the manufacture of cheap plastics of reasonable quality; this method is used industrially in the 'Masonite process'. It is also used in Sweden. Finely pulped wood is exposed for a short period to rather high temperatures under pressure by means of steam and then the pressure is suddenly released (exploded wood). The decomposition products of lignin and cellulose floating to the surface act as bonding agents in the subsequent hot compression of the material giving a mechanically very strong compound that is resistant to water. Thus very strong boards are obtained, as well as plates and bars, which are widely used as substitutes for wood and plastics. The heating and the pressure-release can be applied only to very small batches; a special machine was designed for this purpose, resembling an internal combustion engine, which permits continuous working.

A further important use of wood fibres was found in the manufacture of hardboard-sheets for the building and cabinet-making industries. Waste timber, e.g. the shavings from cellulose and paper factories, is disintegrated into fine fibres by heating with steam under pressure in a screw conveyor, with the addition of small quantities of caustic soda solution followed by pressure release and unravelling in a special grinder. The pulp produced is then mixed with a very dilute alkaline solution of a hardenable phenolformaldehyde resin ('Resoltyp'), acidified so as to precipitate the resin on the fibres, and washed on a sort of paper machine, dried, and made into large boards. These are then pressed in a multilayer press between heated plates, and hardened. The hardboard sheets are non-expanding and very strong.

### *Paper*

Paper is made from mechanical wood pulp, cellulose, waste paper, and, for special papers, from waste cotton- and linen rags. Paper manufacture was invented about 2,000 years ago by the Chinese who produced a fibrous paste by crushing ramie (China grass). The plant fibres disintegrated into single fibres which were then scooped in thin layers from a thin aqueous suspension by ladles with perforated bottoms, producing an irregular fibrous tissue which dried in the sun into a thin sheet. Hand-made paper is still produced by this method. The old Egyptian papyrus made 5,000 years ago and from which the name paper is derived, was manufactured by an altogether different process. It was made from the fine medullary substance of rushes which was carefully placed crosswise and glued together and then beaten into thin sheets.

From China the art of papermaking spread to other countries, arriving eventually in Germany where the first paper mill was built in Nuremberg in 1389. There waste linen and cotton were beaten into single fibres. Later this crushing was performed in 'rag engines'. The crushed fibres were scooped out with sieves, piled on felt sheets in presses, and after the water had been squeezed out, dried in sheets. The modern paper machine, of which the first primitive hand-driven model was built in 1789 in Paris, still includes all the elements of the old manual equipment. The sieve of the scoop is replaced by an endless wire cloth running on rollers. A rubber pad moving at the same speed forms the frame and prevents the pulp from running off at the sides. Further dewatering and thickening takes place on a felt strip which runs with the web of paper between rollers where it is pressed, smoothed, and finally dried.

The disintegration into single fibres, the first operation in the 'conditioning the furnish' process, is usually carried out in beaters or, more recently, in conical refiners ('jordans') (see Figure 107). A grooved roller loaded with weights rotates in a tub against a beater plate carrying knives. The material to be ground is shredded and squeezed between the roller and the knives. Beyond the roller the bottom of the tub rises, forming the 'breasting', over which the pulp is thrown. The tub is approximately oval with a central partition, and the pulp circulates in it until it is ground sufficiently which takes upwards of 12 hours. The necessary additions are also made in the rag engine, e.g. fillers which improve the closeness, smoothness, and whiteness of the surface. Size is also added at this stage. Preferably rosin size, colophony, is used since its main constituent, abietic acid, when reacted with alkali or ammonia, yields rosin soap which is then fixed on the fibre in the mass with aluminium sulphate as aluminium soap. This renders the paper smoother and impermeable to ink.

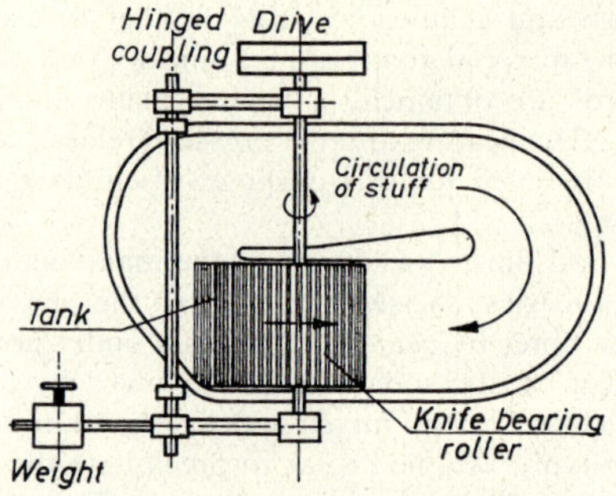


Figure 107. Beater for paper manufacturer

From the rag engine the pulp paste passes into the vat. The pulp is maintained homogeneous at a concentration of 2–4 per cent by stirring or recirculation pumping. The vat is the storage vessel for subsequent further processing in the paper machine.

The pulp runs into the paper machine through a button catcher, a pulp strainer consisting of a rotary sieve with fine slots which catches the felted fibres (knots), or special centrifuges. From the forward end of the paper machine the stock, diluted to 0·2–0·5 per cent, flows over a weir on to an endless, very fine metal sieve (wire cloth) running on rollers. Most of the water is drained off here, and more is removed by several suction boxes. Usually the sieve is shaken transversely as was done formerly by the paper-maker's hand using the scoop, in order to prevent too much longitudinal alignment of the fibres which would reduce the all-round strength of the paper. The method of operation of a paper machine is shown in the schematic sketch in Figure 108. For the sake of simplicity the felt strips stretching over almost the

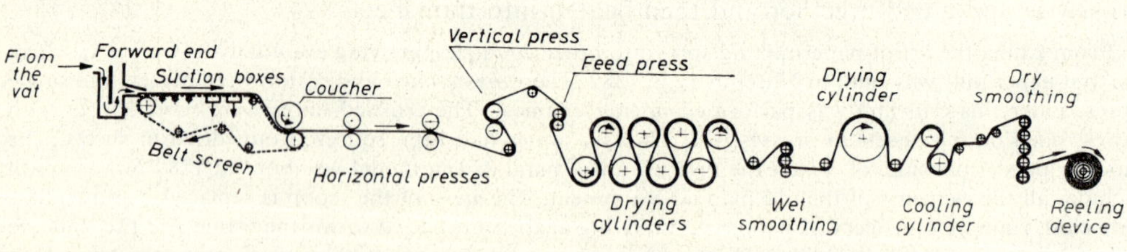


Figure 108. Scheme of the operation of a paper machine (the endless felt strips moving concurrently are omitted)

entire length are omitted. After the belt screen a fine fibrous fleece remains which is pressed and strengthened by a 'coucher press' to such a degree that it can be transferred from the wire cloth on to the endless felt strips which from then on take the paper almost to the end of the machine.

Next follow the 'wet presses', pairs of rollers pressed together which squeeze the water from the paper and the felt (wet felt). The web of paper is by now so strong that it can be fed freely via a pair of rolls on to the steam-heated drying cylinders. On these the web runs again on felt which in turn is dried on special drying cylinders during its revolution. In large machines there are upwards of 15 drying cylinders.

The paper is smoothed first on 'wet smoothing rolls' and after a further pass over a drying cylinder and cooling by cooling cylinders on 'dry smoothing rolls'. Then it is reeled. Paper which is not to yellow when exposed to light must not contain any mechanical wood pulp.

For every type of paper there is a different design and mode of operation of these machines. Thus, for example, in the manufacture of absorbent papers (filter and blotting paper) any strong pressure has to be avoided. For the production of paper twine the longitudinal orientation of the fibres must be enhanced. The pulp or stuff produced by different manufacturing methods and the amount of mechanical wood pulp added must also be adjusted to the type of paper made. For strong wrapping paper and paper bags sulphate pulp (Kraft pulp) is used. For the production of synthetic cellulose fibre the pulp is fabricated in the paper machine into an absorbent pasteboard which is cut into large sheets and baled.

For the production of parchment paper the web of paper undergoes a dipping process in strong sulphuric acid, followed by washing, neutralization, and drying.

The paper machine is used for various purposes in chemical engineering, e.g. for the processing, after polymerization, of aqueous Buna emulsion into crude Buna, and for the production of hardboard sheets.

### (d) Cellulose, and processes for the pulping of wood

Lignin is a cementing substance for the cellulose fibres which strengthens them against lateral pressure. When this cement is carefully removed from the wood fibres pure cellulose fibres remain which have high tensile strength, are pliable and ductile, similar to cotton fibres which are, of course, pure natural cellulose fibres. Cellulose is not produced from cotton because of the high cost of cultivation. The annual yield of pure cellulose per acre from wood is many times that from cotton. For pine, for instance, the yield is five times as large. Moreover, forests will grow under much more unfavourable conditions than does cotton. All countries where the cultivation of cotton is impossible, either because of lack of suitable agricultural land or for climatic reasons, must, therefore, consider the production of rayon via wood pulp. Thus the increase of the rayon industry in the whole world has brought about an expansion in the cellulose industry. Rayon is 'regenerated cellulose', i.e. cellulose dissolved in a suitable solvent and then precipitated in insoluble form during 'spinning'. Cellulose derivatives, e.g. acetyl cellulose, are also used as artificial silks.

Pulp always contains some hemicellulose and lignin. If these are only partially removed, the product is called half-stuff. The production of pure pulp aims at their maximum possible removal. Pulp is assessed according to its $\alpha$-cellulose content, i.e. the fraction which is insoluble in caustic soda of 17·5 per cent concentration. $\beta$- and $\gamma$-cellulose (hemicelluloses) are soluble in caustic soda of this concentration. Ordinary pulp, e.g. for papermaking, contains 85–88 per cent $\alpha$-cellulose.

For the production of viscose rayon and staple fibre, pulp with 90–92 per cent $\alpha$-content is adequate. For cuprammonium rayon and acetate rayon a high-grade pulp with 98 per cent $\alpha$-content has to be produced.

The degree of depolymerization of the pulp during digestion and upgrading is of the utmost importance for the production of synthetic fibres. It is characterized by the viscosity of the solutions, e.g. in ammoniacal copper oxide (Schweizer's reagent).

Pulp that has been only slightly digested and contains, therefore, an appreciable amount of lignin, is called 'hard pulp': it was boiled 'hard' i.e. simmered. The degree of digestion and the lignin content of sulphite pulp are closely interrelated:

| Degree of hardness | Very soft | Normal | Very hard |
|---|---|---|---|
| Lignin content in % | 1·0–1·2 | 3·2–4·4 | 7·2–9·2 |

The yield depends on the method of digestion used, i.e. the intensity and duration of the cooking. This changes the nature of the spent liquors ('black liquors'). In the sulphite waste liquor for instance the type and quantity of sugar is a function of the calcium bisulphite digestion of the wood and the nature of the lignin sulphonic acid depends on the type of cooking, a fact that is important for any further processing of the liquor.

Kraft pulp produced by the sulphate process still contains fair amounts of lignin and hemicelluloses. The fibres are treated very carefully and the product after processing in the paper machine is, therefore, a very strong, cohesive paper which is very suitable as wrapping paper and for paper bags, but it has a brownish colour. According to G. Jayme an optimum hemicellulose content must be present in Kraft pulp for the attainment of maximum paper strength. The lignin and hemicellulose contents facilitate the cementing together of the fibres during the manufacture of paper.

The shape of the fibres and nature of the pulp vary according to the wood used and the source of the pulp can be determined easily under the microscope. The fibre cells (tracheids) in deciduous wood are astonishingly short and thin. Deciduous wood pulp cannot, therefore, be made into strong papers; it is used only for filter and blotting paper and for surgical dressings.

*Pulping processes*

Four methods are used for the digestion of wood:

(1) The alkali process (soda and sulphate cellulose). The digestion is carried out in 5–6 hours using a solution of 25 per cent $Na_2S$ and 75 per cent NaOH at a gauge pressure of 6–7 atm and a temperature of 170°C. In 1949, about 10 per cent of all requirements were produced by the alkali process. The wood used is mostly pine. Turpentine and resin production is possible.

(2) The acid process (sulphite cellulose). The cooking liquor consists of calcium bisulphite, $Ca(HSO_3)_2$. According to the type of cooking used, the temperatures vary between 115 and 150°C, the gauge pressure is about 4 atm and the time 12–14 hours.

(3) The chlorine process and digestion with sodium chlorite ($NaClO_2$).

(4) Nitric acid digestion (N-cellulose). Oxidation of lignin to alkali-soluble compounds and subsequent alkali extraction (no longer used).

A special process is the Italian 'Pomilio process' which is used for straw and reed. First, diluted caustic soda is used which facilitates the action of chlorine acting essentially as an oxidizing agent. Then extraction with a stronger alkali is carried out. For this process, chlorine and caustic soda are required in the proportions produced during the electrolysis of NaCl. It is, therefore, used to advantage where cheap electricity is available.

*Bleaching of wood pulp*

Digestion is usually followed by bleaching. The methods used have been developed from those applied in the textile industry. They are designed to lighten the colour, to concentrate α-cellulose, to stabilize the pulp against discoloration in air and light, and to depolymerize it. The bleaches used are: hypochlorite solution of NaOH + $Cl_2$ or milk of lime + $Cl_2$, giving NaOCl or $Ca(OCl)_2$. In order to protect the fibres and to achieve a controlled digestion, bleaching is carried out in several stages, the soluble reaction products being removed by intermediate scrubbing; the initial treatment is with chlorine water: $Cl_2 + H_2O \rightarrow HCl + HOCl$; $HOCl \rightarrow O + HCl$. Then follows alkaline bleaching with hypochlorite.

The excess active chlorine is removed by $SO_2$ or bisulphite, or sodium thiosulphate. The tendency is towards multiple-stage bleaching. The equipment used is a bleaching engine, with a design similar to the rag engine discussed earlier but with an impeller replacing the roller and the knives to keep the pulp and the bleach circulating.

For high-grade pulp and cotton fibres, chlorine dioxide bleaching (E. Schmidt, 1921) is becoming increasingly popular. Chlorine dioxide can be produced easily from sodium chlorite, $NaClO_2$, by acidification. It is, therefore, possible to bleach with a sodium chlorite solution. However, it is much more economical to prepare a chlorine dioxide solution from sodium chlorate in situ by one of the technologically very sophisticated processes and to use it directly (E. Kesting, Eigenschaften und Herstellung von Chlordioxid. *Das Papier*, 1952, **6**, 155).

*Sulphate digestion, alkali pulp*

The advantage of the 'sulphate process' lies in the fact that it permits the digestion of highly resinous woods, e.g. fir wood. During cooking, turpentine oil is distilled off (about 10 kg per ton of pulp). The other constituent, fir resin, rosin, is contained, partly denatured, in the spent liquor as rosin soap (sulphate soap), obtained as tall oil.

Before cooking, the bark is removed and the wood is shredded into chips in a chipper, and when knots and sawdust have been removed in screening drums, it is fed into the digester. Chopped straw can be processed by the same method. Owing to the high silicic acid content of straw, sulphite digestion is impracticable. Only alkaline decomposition is possible. The production of straw pulp is of great importance in countries devoid of forests.

Digesters for sulphate reduction can be made of iron, which is an advantage as compared with acid decomposition. They have capacities of 70–160 $m^3$ and are heated up to 170°C, with a gauge pressure of 6–8 atm. The liquor is circulated by pumps. Heat can be supplied directly by steam or indirectly by coils, or by heating units introduced into the liquor circulation. One litre of cooking liquor contains e.g. 60 g NaOH, 22 g $Na_2S$ (essential for the digestion), 15 g $Na_2CO_3$, 4 g $Na_2SO_4$, 2 g $Na_2SO_3$. After removal from the digester the pulp is freed of liquor in rotary filters and then washed in diffusors with water in counterflow.

The principle of counterflow washing is used in order to obtain as concentrated a liquor as possible for the process is economical only when the alkali can be recovered and a minimum amount of steam is consumed for the evaporation of the liquor. The procedure in counterflow washing is that the cleaner pulp is washed with fresh water and the contaminated pulp is contacted by the spent washing water.

In order to recover the alkali, the liquor is evaporated to a solids content of 55 per cent in multiple-stage evaporators. These are based on the principle that the vapours of the first stage are used to heat the next stage. This is possible only when the evaporation takes place under vacuum which increases from stage to stage (see p. 325).

The alkali is regenerated by burning the liquor which is injected through atomizers into a melting furnace. Enough lignin and other organic components are present to obtain a sufficiently high combustion temperature (> 1,000°C) and to produce enough additional heat to cover almost the entire energy requirements in modern plants by combustion of the black liquor, in other words of the lignin. The loss of alkali is made good by the addition of sodium sulphate before combustion.

A reductive process takes place in the melting furnace so that $Na_2S$ is formed from the sulphate and soda from the alkali which are discharged together as a fused mass. The latter is dissolved in water and is causticized with milk of lime, producing NaOH. The additional consumption per ton of pulp is 100–130 kg sodium sulphate and 230 kg calcined lime. Since sodium sulphate is used in this process, it has been called sulphate digestion.

The pulp is ground in an edge mill and then—as described earlier for mechanical wood pulp—dewatered and brought to about 50 per cent solids content. Its subsequent processing was discussed in the section on paper manufacture.

*Sulphite digestion (sulphite pulp)*

The sulphite digestion process was first used by a Swede, Ekmann. Independently from him, Mitscherlich peveloped a process in Germany that has been used industrially since 1870. Mitscherlich reduced wood by a calcium bisulphite solution in an indirect digestion process. The Austrians Ritter und Kellner used direct digestion.

The wood is first carefully barked and decorticated; special machines have recently been developed for this purpose. Choppers with knives rotating at 200 r. p. m. cut the wood into discs of 2–3 cm thickness which exfoliate in the direction of the fibres. Then follows size-grading in conical screening drums. The larger pieces are cut again (see Figure 109). The calcium bisulphite liquor required for the digestion is prepared from $SO_2$

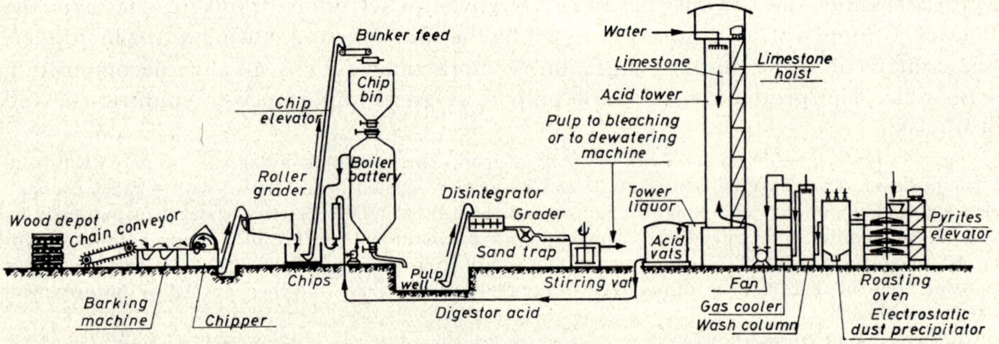


Figure 109. Method of operation of a sulphite pulp works

and lime. $SO_2$ is obtained by combustion of sulphur in specially designed sulphur ovens, or by roasting of iron pyrites or of gas-purifying material. Dust is removed from the gases by the Siemens-Cottrell-Möller system or in gas scrubbers where the interfering $SO_3$ is also removed. The gases contain 8–11 per cent $SO_2$. The purified and cooled gas enters from below into the 'tower' which used to be constructed from wood and is now made of concrete with an acid-resistant lining. This tower is continuously filled from above with lumps of limestone transported to the top in small skips by an elevator. Water is sprayed from above on to the limestone. $SO_2$ reacts with the limestone dissolving it as calcium bisulphite, and producing the 'tower liquor'. $SO_2$ escaping during the pressure relief of the pulp digester is re-circulated into the tower.

The digestor liquor produced contains per litre 40–45 g total $SO_2$ a quarter of which is fixed by the lime. The $SO_2$ excess is necessary since without it $CaSO_3$, which does not dissolve easily, would be precipitated during the digestion process. For one $m^3$ wood, 1,350–1,500 litres digestor liquor are required.

*Reduction in the digestor*

Digestors are made of iron and are lined with acid-resistant brick, pointed with a double layer of cement and water-glass cement. Lead glycerate can also be used for pointing. In modern plants pulp digestors made of V4A (an acid-resistant steel) are used, with capacities of up to 300 $m^3$. Heat can be supplied either directly by injecting steam, or indirectly with coils or external heating elements. If steam is injected directly it is necessary to take into account the fact that the digestor liquor will be diluted by the condensate. The most modern form of heating is by liquor circulation with indirect heating via a steam heated liquor preheater in which the liquor circulates. The pumps must be made of V4A. Digestors with liquor circulation and direct steam injection are also used. Figure 110 shows a vertical digestor with indirect heating and liquor circulation.

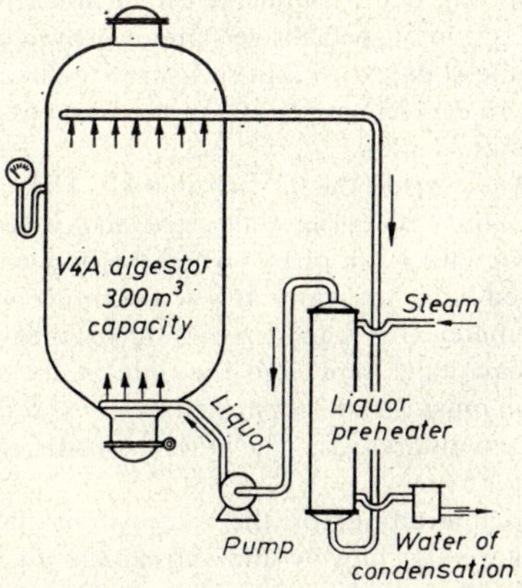


Figure 110. Vertical sulphite digestor with liquor circulation and indirect heating

According to Hägglund, two processes take place during cooking:
(1) Formation of an insoluble compound of lignin sulphonic acid with a carbohydrate.
(2) Hydrolysis of this acid into carbohydrate and water-soluble lignin sulphonic acid.

When sulphurous acid reacts with lignin under the conditions of sulphite digestion, first of all the free secondary phenylcarbinol groups such as are present in a guaiacyl glycerine ether are fastened. Then aliphatic ether groups are split off. Such ethers are found in the pinoresinal fraction of lignin.

According to modern theories, no rupturing of the hydrofuran ring, which occurs e.g. in the dehydrodiconiferyl alcohol fraction, takes place (formulae p. 281). With a sulphur content of 5·6 per cent there are about 3 to 4 phenylpropane units for one sulphonic acid group. Sulphite spent liquor contains 7–12 per cent lignin sulphonic acid.

Indirect cooking by the Mitscherlich method is carried out for 20–35 hours at a gauge pressure of 2·5–3 atm and a temperature of 115–130°C, direct cooking by the Ritter-Kellner method for 7–15 hours, at a gauge pressure of 4–6 atm and a temperature of 140–150°C. Duration, temperature, and pressure vary greatly in the pulping plants according to the type of wood used and the type of pulp required.

When the pressure in the digestor is released 'free sulphurous acid' escapes and is fed into the tower. 'Fixed sulphurous acid' remains in the liquor. After the pressure has been released the bottom of the digestor is opened and the pulp drops out. By this time the chips are thoroughly softened and are then pounded and pulped in separators and carried with large quantities of water through the knot drums, the sand trap, and the pulp strainer into the dewatering machine where they are processed into pulp paste of about 90 per cent solids content. The water requirements of a pulping plant are very high.

The distinctive sign of a sulphite pulp plant is the acid tower and the high digestor assembly where batteries of at least six digestors are placed side by side, and the large wood yard.

### *Nitric acid digestion, 'N-cellulose'*

This process is no longer used.

Wood digestion with nitric acid and with chlorite is based on their oxidizing effect on lignin. Originally, the process was considered uneconomical and foundered on the problem of materials of construction. This problem was solved by the introduction of special steels and it proved possible to reduce the consumption of chemicals, nitric acid and caustic soda, by skilful measures to such an extent that the process became economical. It was developed on an experimental scale in the Farbenfabrik Wolfen (Petersen and Bayerl) and was used in the Agfa Works at Wolfen.

The treatment with nitric acid is carried out in two phases. The 'impregnation phase' is carried out below 50°C and the minimum quantity of 14 per cent nitric acid necessary for the oxidation of lignin is used. This phase is followed by the 'digestion phase' when the hogged chips, after removal of the excess nitric acid by pumping, are treated with hot water at a temperature which slowly rises to 100°C, and at a gauge pressure of 1 atm; the material is kept circulating by means of pumps. At the end the acid concentration is very low, less than 1 per cent, and the pulp is not attacked. Digestion takes 7 hours.

The oxidation products of lignin must then be separated from the pulp fibre. This is done in the digestor with hot dilute solution of sodium hydroxide. The liquor produced is blackish-brown, still slightly alkaline, and is called 'black liquor'.

Extraction is followed by cold upgrading for the removal of the last lignin residues and of the hemicelluloses, carried out with gaseous chlorine and subsequent alkalization and treatment once more of the raw pulp with 10 per cent sodium hydroxide solution at 25°C. The $\alpha$-cellulose content rises from 87 to 98 per cent.

The high-grade pulp obtained is used as linters substitute directly in the form of 50 per cent wet flakes for the production of cuprammonium rayon. Alternatively, it can also be processed as crepe paper to cellite- and cellulose ether. The waste liquor is reconditioned, the soda being recovered.

As soon as a suitable type of linters became available again after the war the process was abandoned.

### *Straw pulp and other pulps*

Corn straw and grasses become ligneous when they mature. In addition to cellulose they contain large quantities of polyoses (hemicelluloses), particularly large amounts of pentoses, also lignin, proteins, and uronic acids. They frequently have silicic acid contents of upwards of 4 per cent. Digestion with calcium bisulphite liquor is, therefore, impossible, only alkaline digestion can be carried out. By cooking straw with milk of lime under pressure pulp containing impurities is obtained which is nevertheless very suitable for the production of yellow straw board for roofing paper. Cooking under pressure similar to sulphate digestion with NaOH, $Na_2S$, $Na_2SO_3$, yields a pulp which can be bleached well and is even suitable for the production of good-quality writing paper.

Straw pulp was produced by the former "Phrix" at Wittenberge. Before cooking under pressure with alkali the straw was subjected to acid pre-hydrolysis. With straw, alkali digestion alone does not produce results. After digestion there still remain hemicelluloses which are insoluble in 17·5 per cent caustic soda and thus give the impression of a high $\alpha$-cellulose content. They are removed by acid pre-hydrolysis and then alkaline digestion can be carried out successfully. Sugar liquor is a by-product which is eminently suitable for yeast manufacture and can also be processed with advantage into furfurol. In many cases the failure of straw pulp production was caused by difficulties with transport and storage of the bulky material.

In spite of many efforts the use of the potato plant for pulp production was an industrial failure because of the difficulties of collection and storage. The position is similar with stalk plants, e.g. topinambur. These pulps are not of good quality. In Southern Europe fine writing papers are produced from esparto and alpha grasses. In Italy where there is a permanent shortage of wood, *Arundo donax* which grows in many parts of Northern Italy, has been successfully used for the production of pulp. It is digested by the Pomilio method described earlier which is also suitable for the production of straw pulp. The large quantities of reed growing in the Danube delta are also used for this purpose.

# 2. BIOLOGICAL PROCESSING OF WOOD SUGAR SOLUTIONS AND SULPHITE WASTE LIQUORS

## General remarks

The simplest biological utilization of wood sugar solutions and sulphite waste liquor from soft wood is fermentation to alcohol. When the raw material is wood sugar the well tried methods of the fermentation industry (see the chapter on fermentation) can be used but for the fermentation of sulphite waste liquor experience had to be gained first. It must not be forgotten that not all wood sugar consists of fermentable hexoses. From the wood pentosans in which hardwoods are especially rich, pentoses reach the liquor during hydrolytic cracking; some of these are converted to furfurol but the majority are found after fermentation as 'residual sugars' in the residual alcohol liquor. This applies to an even higher degree to sulphite waste liquors. Hardwood liquors contain mainly pentoses and cannot, therefore, be fermented to alcohol. For the residual sugars of the wood sugar worts (residual alcohol liquor) Scholler and his colleagues developed a yeastification process using growth yeasts which was practised at Tornesch and in the Dessauer Zuckerraffinerie for the production of yeast for animal fodder. The residual sugar from the residual alcohol liquor produced during the manufacture of alcohol based on sulphite waste liquors (which because of their oxygen consumption placed a great strain on the sewage system) can be very conveniently processed into fodder yeast by a process developed by the former IG Farbenfabrik Wolfen who made use of the experience gained by Scholler and his colleagues. Large plants were built for this process by the Cellulosefabrik Attisholz, Switzerland, and by Enquist, Tammerfors, Finland. It is an example of the perfect utilization of all the biologically decomposable materials in the sulphite waste liquors, i.e. residual sugar, sugar conversion products, and acetic acid (A. Rieche, Über die mikrobiologisch-technische Eiweiß- und Fettsynthese auf der Basis von Zellstoffablaugen. *Wiss. Ann. d. DAdW*, **3**, (12), 1954. *Chem. Ing. Techn.*, 1957, **29**, 818. Reinigung u. Verwertg. industrieller Abwässer, Akademie-Verlag, Berlin 1962).

The only means of utilizing biologically the hardwood sulphite waste liquors is yeastification by growth yeasts, or bacterial fermentation when butanol, propanol, acetone, and ethanol are obtained. Both processes were first developed by the former IG Ludwigshafen, Farbenfabrik und Filmfabrik Wolfen (Vierling, Rieche, Hilgetag, Grützner).

Since the wood sugar worts are relatively pure sugar solutions, they can also be used for special fermentation processes, e.g. glycerine-, citric acid fermentation or other biological syntheses.

Biological utilization of sulphate waste liquors is impossible because in this process the sugars are completely destroyed. If the hemicelluloses of wood are to be used, acid hydrolysis before digestion is indispensable, as explained in the section on straw pulping. Acid pre-hydrolysis of hardwood would yield liquors suitable also for the production of furfurol. This gives, however, a somewhat smaller yield of pulp after digestion (Voss, Agfa, Wolfen). (W. Voss, Holz als chemische Rohstoffquelle. Tagungsbericht der Chemischen Gesellschaft in der Deutschen Demokratischen Republik. Deutscher Verlag der Wissenschaften, Berlin, 1954).

Even after complete biological utilization of sulphite waste liquors lignin still remains in solution as lignin sulphonic acid.

### (a) Alcohol from wood sugar

Before fermentation the wood sugar worts have to be partially neutralized. This is done hot with milk of lime, limestone powder, or other calcareous waste materials. After clarification in a filter press and cooling, the wort, with a pH of 5, is fed into the fermentation apparatus. The process is continuous. In order to speed up fermentation by increasing the yeast concentration part of the yeast which has been separated by separators after discharge from the fermentation apparatus, is re-circulated.

The carbon dioxide escaping during fermentation is very pure and can be collected and used for the production of solid carbon dioxide. The fermented wort is distilled in one of the usual alcohol distillation units.

Formerly a large part of the Soviet Buna manufacture was based on the production of alcohol from wood sugar.

### (b) Alcohol from sulphite waste liquor, sulphite alcohol

Sulphite waste liquors used to be drained into the rivers where because of their high sugar content strong oxygen absorption took place resulting in fishes dying due to lack of oxygen. Simultaneously the growth of algae increased enormously. Cellulose works could, therefore, be built only near rivers carrying large quantities of water. In 1891 Mitscherlich took out a patent for the production of alcohol from sulphite waste liquors. In 1907 the first sulphite alcohol plant was built in Sweden by Eckström and Wallin. In Germany the first sulphite alcohol works on the Swedish model began operation in 1916. Until then alcohol had been produced exclusively from potatoes. The process has since been considerably changed and improved and has been adopted by almost all pinewood pulp works. For some countries, such as the Nordic countries which lack raw materials, sulphite alcohol has temporarily become a basis for the supply of motor fuel and a raw material for the development of an organic industry (particularly in Sweden).

Sulphite waste liquor contains 20–35 g/litre reducing sugars (Fehling); two thirds consist of hexoses and are fermentable, one third is not. On the basis of the hemicellulose content of wood a much higher sugar content should be expected in the sulphite waste liquor. Only about one third of the quantity of hemicellulose is available for fermentation, two thirds are converted during boiling into non-fermentable conversion products. However, these can be partly utilized during the yeastification of the distilled residual alcohol liquor and, together with acetic acid, they account for the high yield of yeast during the yeastification of residual alcohol liquor (see p. 303). The liquors are strongly acid, and even after $SO_2$ has been expelled by aeration of the hot liquor its pH value is still 2·2–2·6 due to its content of lignin sulphonic acid and organic acids such as formic and acetic acid formed during boiling.

The production of alcohol makes use of both the original liquors and the pulp washing liquors so that about 90 per cent of all waste sugars are utilized. The diagram in Figure 111 shows the course of sulphite alcohol production.

After acid degasification which is best carried out in an acid-resistant plant under vacuum the liquor is neutralized with milk of lime or finely powdered limestone and aerated in neutralization vats at a temperature which should if possible be above 90°C, the mixture being vigorously agitated with compressed air. For this purpose acid-resistant tower-type vessels are also used. pH value is about 6, duration 2–4 hours. Calcium sulphate and calcium sulphite are then allowed to settle. Calcium sulphite does not easily dissolve in the liquor at pH 6. Its removal is essential for the smooth progress of the fermentation. The hot liquor is then cooled to 34°C in a heat exchanger, the most suitable being a Rosenblad spiral Cooler made of V4A steel (see p. 90). The cooling water is used for the hot water supply for which there is constant need in a pulp works.

Fermentation can be carried out batchwise in 3–4 days in fermentation vats made of wood or concrete, with capacities of 150–200 $m^3$. When fermentation is complete the

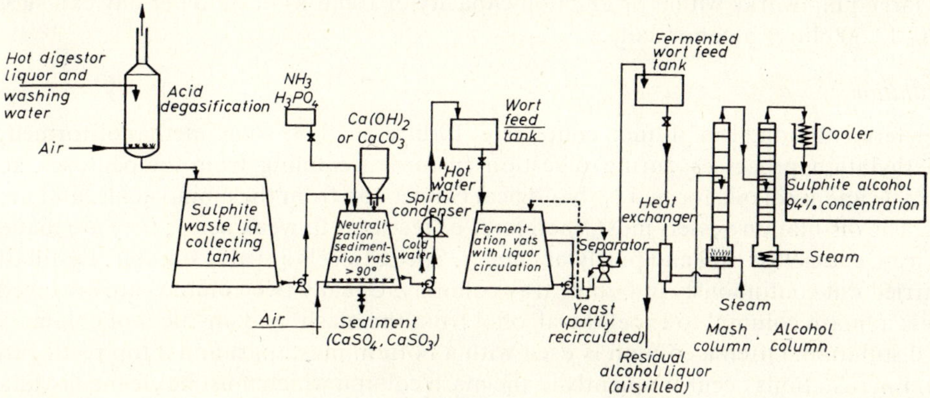


Figure 111. Sulphite alcohol production

contents of the vats are pumped out but for a residue which contains the yeast for the new charge. Yeast needs nitrogen, phosphoric acid, potassium, and magnesium for growth.

Potassium and magnesium are usually present in the liquor to a sufficient degree but the other two nutrients are added in the form of ammonia, ammonium phosphate, ammonium sulphate, or superphosphate.

Fermentation is now mainly carried out continuously in vats arranged in series. Since in this method the yeast is continuously removed special measures must be taken to maintain the yeast concentration at an adequate level. This can be done e.g. by the "fixd yeast process" developed by the Deutsch-koloniale Gerbstoffgesellschaft, Karlsruhe, which took out a patent for it in 1915. The process consists in the introduction of wooden structures of large surface area into the fermentation vats on which the yeast deposits. It is used, for instance, by the Zellstoffabrik Waldhof in Mannheim. In the process practised at the Usines de Melle and the Merck Company, Darmstadt, the distilled mash is separated in separators into a part containing little yeast which passes forward for distillation, and a yeast concentrate which is returned into the fermentation vats, thus raising the yeast concentration. The usual *saccharomyces* types are used as yeast stock which, however, have first to get used to the impurities, mainly $SO_2$ and furfurol, of the sulphite waste liquor. Yeast originating from molasses pre-cultures does not, therefore, at first perform satisfactorily. On the other hand, the $SO_2$ content of the sulphite waste liquors is an advantage since yeast is less sensitive to $SO_2$ than bacteria. The danger of infection with sulphite liquors is, therefore, considerably less than in the fermentation of molasses. This is further improved by a slightly acid adjustment of the liquor used for fermentation.

Control is exercised mainly by determination of the residual sugars and of the alcohol yield. Under the most favourable conditions a degree of fermentation of 65–80 per cent of the sugars, determined by Fehler's method, is achieved (the remainder consists of pentoses). With "hard" liquors the yield is considerably lower because the polysaccharides are decomposed less.

One kilogram of glucose yields 0·643 litre alcohol which on the basis of the equation $C_6H_{12}O_6 \rightarrow 2C_2H_5OH + 2CO_2$ corresponds to a yield of 96 per cent of the theoretical.

According to the type of digestion used this means that from 1 ton of cellulose 50–125 litre alcohol, calculated for 100 per cent, are produced. The yield of alcohol calculated for hexosans in the wood is only 20–25 per cent. This is due to the fact that during the digestion of wood a large part of the sugars form non-fermentable conversion products. The alcohol yield varies therefore with digestion. However, the sugar conversion products can be partly utilized by growth yeasts in an aeration process.

A large pulp works with a production capacity of 100 tons of pulp per day can supply 5,000–12,500 litres alcohol daily.

*Distillation*

After fermentation, 1 $m^3$ liquor contains 6–10 litre alcohol, some methanol formed by demethylation processes during digestion furfurol stemming from the pentoses, acetaldehyde and fusel oils formed by biological conversion from the amino acids, and acetic acid. The distillation vessels must, therefore, be resistant to weak acids; they are made of cast iron, enamelled iron and partly of copper. Special steels are also suitable. Distillation is carried out continuously in bubble tray columns. Usually two columns are arranged in series: a 'mash column' to expel the alcohol from the mash, and an 'alcohol column' for fine distillation. Often a column is used with a bottom mash part and a top rectification part. Encrustations occur frequently in the mash column which must be cleaned regularly (see p. 85, Figure 33).

In addition to water, the 94 per cent sulphite alcohol contains about 1·5–2·5 per cent methanol, 0·1 per cent acetaldehyde, and traces of fusel oil, and is only used for chemical purposes and as a motor fuel. Unless carefully purified it is unsuitable for drinking purposes since it has an unpleasant taste and smell. In Sweden sulphite alcohol has been successfully purified to such an exent that it has become drinkable, and the "Norland Aquavit" is a very tasty brandy, produced from the purest sulphite alcohol. Activated corbon is very suitable for the purification of alcohol.

### (c) Biological protein and fat synthesis (yeastification)

*General remarks on biological protein production*

Protein occupies a special position among the three main nutrients, protein, carbohydrates, and fat. It cannot be replaced by any other component in nutrition. In the food of animals, and of course of man, too, a certain ratio between protein and carbohydrates must be maintained. If there is a deficiency of protein and an excess of carbohydrates in animal and human food, such as always occurs in Germany in wartime, the carbohydrates are not fully ultilized, i.e. they are wasted. The vegetable products of Central European agriculture contain on the whole less protein than is required for the optimum ratio. The correct proportions of protein and carbohydrates can, therefore, be maintained only by the addition of highly concentrated protein. Whilst the organism can live for long periods on its reserves when there is a protein deficiency, the high-grade amino acids gradually disappear and serious biological damage and psychological disturbances may by caused which are known as protein deficiency phenomena. The products of the industrial biological protein synthesis, in particular yeast nutrient, are often used for the treatment of such cases.

Recent results in the field of yeast culture have made possible economically and industrially on a large scale the biological synthesis of protein. Two facts above all have advanced the development in this field: the discovery of rapidly-growing types of yeast which are capable of propagating on a purely mineral basis, and the realization that the 'unfermentable sugars', principally the pentoses, can be used for the synthesis of protein.

Protein yeast was produced for a time from molasses, as long ago as the 1914–1918 war, using growth yeast, *Torula utilis*, by Delbrück's process (Institut für Gärungsgewerbe, Berlin). Because of the shortage of raw materials the process was, however, abandoned. It was recognized in many quarters that it was possible to yeastify the waste sugars of the cellulose industry by using growth yeasts. Here be mentioned only Krohn's work in Finland in 1919. Heijkenskiöld, Sweden, showed in 1925 that sulphite waste liquor can also be used for the production of baker's yeast; this process is operated so far only by Rosenlev in Finland. In 1936 Fink and Lechner published a paper on the yeastification of pinewood sulphite waste liquor with pure inorganic nutrients. Yeast from wood sugar liquor has been produced by Scholler and his colleagues and by the Bergingesellschaft.

In recent years certain findings in the field of animal nutrition have become generally known, particularly in Germany, the United States and in the Soviet Union, and have led to an increase in the production of yeast for animal feeding purposes. In the German Democratic Republic 85 per cent of the sulphite waste liquors are now used for the production of alcohol and yeast. Molasses, too, is increasingly utilized for the production of protein fodder. This is due to the fact that yeast contains many active substances, particularly vitamins, which are very useful for the health and the rapid growth of domestic animals. Fodder yeast is added to the commercial mixed fodder for various species of animals. Another factor is the constantly increasing necessity in many countries to fight river pollution by preliminary purification of industrial effluents. Although microbiological purification methods are used mainly for this purpose, it is possible in many cases to replace these to advantage by microbial processes. (A. Rieche, Untersuchungen über Wuchshefen und Mycelpilze im Hinblick auf die Ausnutzung industrieller Abwässer. *Wasserwirtschaft-Wassertechnik*, 1960, **10**, 52). Various effluents containing fatty acids, such as subsequently dephenolized foul water and waste waters from paraffin oxidation, are suitable for this purpose and are as yet unused substrates for the production of protein fodder.

### *Yeastification of wood sugar worts*

The culture of growth yeasts from wood sugar solutions is older than that based on sulphite waste liquor. The process was developed by Scholler and his colleagues, and the first plant started operation in 1933 at Tornesch, at first with the aim of testing the suitability of growth yeasts for producing animal fodder. The favourable result of the feeding experiments hastened the development of the production of yeast, on the basis of sulphite waste liquor, for animal and human consumption. Wood sugar solution which does not contain inhibiting impurities presents considerably less difficulties in the biological treatment than sulphite waste liquor.

The preparation of the liquor resembles that for the production of alcohol. As in the production of baker's yeast, yeast cultures from molasses must be aerated as intensively as possible since yeast obtains the energy required for the protein synthesis by combustion of some of the carbohydrates. A continuous fermentation apparatus was designed for this purpose in which a large number of circulation tubes are fitted on the sides of a long box which act as air-lift pumps (see p. 318). The liquid jets from the circulation tubes impinging on the surface of the liquid shatter and suppress the foam.

By Scholler's method via wood saccharification and subsequent yeastification of the residual alcohol liquor, it is possible to produce from 100 kg pinewood 4–5 kg dry yeast, with a protein content of 50 per cent, in addition to 20–22 litres alcohol.

### *Yeastification of cellulose waste liquors*

A further incentive was given to the development work for the industrial production of protein by the pentose-containing liquors obtained as by-products in the production of cellulose. This process, operated continuously, was used on a large industrial scale for the first time by the Filmfabrik Wolfen; earlier beechwood cellulose had been produced temporarily elsewhere. At first, no use was found for these pentoses since it was not then known how to grow yeast on pentose-containing substrates. Even eminent scientists were at that time still of the opinion that pentoses could not be utilized by yeast.

Credit is due to Vierling, Ludwigshafen, for having shown that the unfermentable residual sugars in sulphite liquors could be utilized by *Oidium lactis*, a surface fungus, for the synthesis of protein (DRP 688 632). In 1936, Seidel, a colleague of Scholler's, filed a patent for the yeastification of wood sugar residual liquor. Induced by the large amounts of beech waste liquor available the former IG conducted a countrywide simultaneous investigation of the problem of pentose yeastification. This collective work produced the basic German Patent DRP 705 763, filed in 1937, for the yeastification of beechwood sulphite waste liquor (Vierling, Rieche, Hilgetag, Grützner).

Beechwood liquor gives surprisingly high yields because materials other than sugars are also yeastified. The 'unfermentable' sugars of the hardwood liquors and residual alcohol liquors represent the cheapest raw material for protein synthesis.

### *The Waldhof cellulose process*

Earlier, as far back as 1940, the Zellstoffabrik Waldhof, and in particular Claus, developed the first industrial yeastification process for beechwood sulphite waste liquor using *Torula utilis*, on a large scale. The first large plant was built at Mannheim.

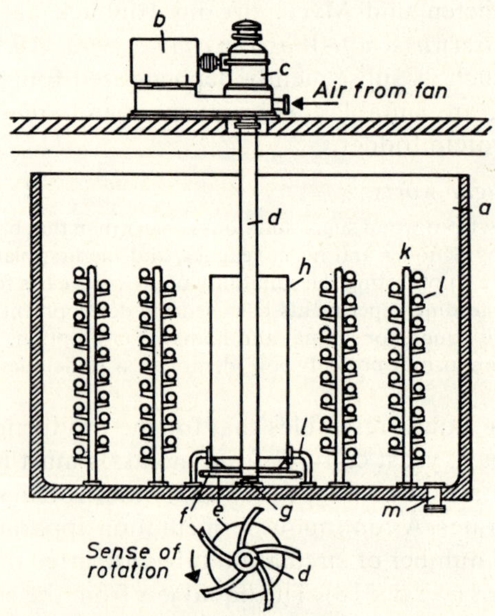


Figure 112. Waldhof vat (from Ullmann, *Encyklopädie*, 3rd edn, vol. 8)

The process utilizes the fact that it is extremely easy to produce foam from sulphite waste liquor; the operation is carried out in a fairly stable air-liquid emulsion produced by a special agitator known as 'Waldhof agitator' (Figure 112). The design of the agitator is best compared to a rapidly driven, i.e. 'inverted' Segner water wheel (see G. Butscheck, in Ullmann, *Encyklopädie der techn. Chemie*, 3rd ed., vol. 8, Urban & Schwarzenberg,

München, 1957, 477). Through a tubular shaft which serves also as a drive shaft, air is drawn in which divides into extremely small bubbles due to shearing in the liquid, at the pointed and slightly bent ends of the agitator. Simultaneously circulation takes place in the fermentation vat, the spent air being expelled and fresh air being continuously admitted, without increasing the volume of the liquid.

The fairly considerable amount of heat of about 3,000–4,000 kcal/kg dry yeast has to be dissipated by cooling, to which must be added the frictional heat of the rapidly moving agitator. Cooling coils are, therefore, fitted in several layers into the vat. The entire equipment must be resistant to acid.

For the further treatment it is necessary to convert the foam emulsion back into liquid. In the Waldhof process this is done on the centrifuge principle, a design by Claus. Right up to the periphery of a normal solid-jacket centrifuge, into which the foam enters centrally, a streamlined extraction 'sword' is fitted radially from the inside, with a fine aperture at the end on the side opposite the direction of rotation. This 'sword' has a central bore through which the foam, thus re-converted to liquid, is discharged outside by a throttle valve. The extraction of the liquid from the centrifuge is regulated by means of this valve.

The 'de-foamed' liquor which contains yeast is separated in separators, washed, if necessary, drained on a rotary filter, plasmolysed by heating to 80°C, evaporated in a circulation condensor to 22–25 per cent solids content, and finally carefully dried on a roller drier or in an atomizing drier.

*Atomizing driers*

(See also E. Kirschbaum, Grundsätzliches und Neues über die Zerstäubungstrocknung. *Chem. Ing. Techn.*, 1952, **24**, 3).

Atomization of solutions, emulsions, or suspensions into fine droplets increases the surface area to such an extent that in a warm air stream drying takes place in a fraction of a second. This type of drying least affects the substance to be dried so that even highly sensitive products, such as milk, eggs, coffee extract, sera, are preserved in their original state in the form of fine powders with a diameter of 0·1–0·3 mm and in an easily soluble form. But even less heat-sensitive materials, such as yeast, gelatine, soaps, detergents, tannins, are often dried by the atomizing method in order to obtain a powder of maximum fineness and good solubility. Materials dried in an atomizing drier can be seen under the microscope as small burst hollow spheres.

Atomization is carried out in a large empty chamber through which passes, usually from the top downwards, i.e. concurrently, hot gas, or air pre-heated by steam or electrically. The coarser solid particles which drop to the floor are removed by a mechanical reamer from the drying chamber into a screw conveyor. The powder carried along with the air stream is precipitated by cyclones or cloth filters which are mechanically shaken at short intervals. The heat consumption is rather high. 3·5 kg steam per kg of water to be evaporated must be supplied, which is much more than is consumed by a roller drier (about 1·7 kg) or a drying cabinet with circulating air (about 2–3 kg). Atomizing driers are also expensive to buy and require much room. On the other hand, the pulverizing operation is saved so that the process is economical even for so cheap a product as household detergents.

Atomizing driers are classed mainly according to the type of atomization and the design of the drying chamber. Atomization of the liquid is carried out directly by atomizing nozzles (Siccatom nozzles, Schlick nozzles, Lecher nozzles), or with the aid of compressed air (Nubilosa nozzle, Kirschbaum nozzle). In the Krause-, Lurgi-, and Zahn driers the liquid is atomized by a rapidly rotating disc. Various designs of discs and plates have been developed for this purpose, all based on the principle that the liquid which enters centrally is forced outwards through channels or tuyere pipes. Owing to the high relative velocity of the liquid compared with the ambient air in the drying tower it is broken up into a broad mist layer. The dry air introduced tangentially can have temperatures up to 180° for delicate materials, for many others upwards of 200°. The material to be dried is nevertheless heated up only very little owing to cooling by evaporation (about 30–40°C), particularly if it is removed quickly from the warm air region.

*IG-Scholler process*

Using Scholler's and Seidel's experience gained in the field of wood sugar yeastification, the Farbenfabrik Wolfen developed a yeastification process for beechwood liquor and residual alcohol liquor (literature, see p. 299). The liquor is circulated in circulation tubes by the air-lift pump principle. (The French process of Lefrançois (Messrs. Sorice) also uses an air-lift pump but it is fitted inside the yeastification vat.)

Figure 113 shows the method of operation of this fermentation apparatus as well as the preparation and processing of the yeastified liquor.

The great advantage of the Rieche-Hilgetag process is that cooling coils are not necessary. The tubes can have cooling jackets and this cooling, coupled with cooling by evaporation, is adequate for dissipation of the heat. The yeastification vat contains, therefore, no movable parts. The apparatus is set up in the open air. Cylindrical or pear-shaped bodies with many small holes are used for the aeration of the air-lift pumps. The fine dispersion of the air is due to the shearing of the emerging air bubbles into extremely fine bubbles by the liquid flowing past with great speed. The principle resembles the "Vogelbusch" aerator agitator which was developed for the baking yeast industy and is much used in fodder yeast production (E. Bergander, Biochemie und Technologie der Hefe, Verlag Th. Steinkopf, Dresden, 1959). This is a hollow agitator flattened at the ends, with a tubular shaft for the air supply, and perforations at the top and bottom through which the air escapes. Towards the ends of the agitator vanes the perforations increase in size, in accordance with the increasing peripheral speed, thus ensuring uniform shearing of the bubbles into tiny bubbles over the whole area of the agitator vane.

The hot acid beechwood sulphite waste liquor is aerated in the starting vats, neutralized with milk of lime and the nutrient salts are added. The nitrogen nutrient is aqueous ammonia or ammonium phosphate, the phosphoric acid nutrient can also be superphosphate. The Mg content is adequate but not the K content so that this has to be added in the form of KOH. The cooled liquor is then yeastified at 30–34°C. After removal of the foam which is carried out by the Waldhof principle or with a chemical defoaming agent, the yeast is first separated in separators. The concentrate is then washed once or twice and separated again in separators. After plasmolysis the separator concentrate can then be dried directly on a roller drier. It is, however, more advantageous to concentrate first by means of an overflow centrifuge (solid jacket centrifuge). Since the overflow from the centrifuge is often not quite clear, it can be passed through a separator, or it can be added to the second or third separation. Rotary filters have not proved suitable because of the frequent change of cloth necessitated by clogging. After plasmolysis (heating to 80°C) the plasmolysate is dried on roller driers.

With beechwood liquor the degree of yeastification is 85 per cent. The yield is up to 60 per cent of the reducing substance introduced (dry yeast weight), with almost 100 per cent utilization of nitrogen. The organism used (*Candida arborea*, mixed with *Torula utilis* and others) is larger than *Torula utilis* but is not a mycel fungus. The formation of large compounds of yeast germ can facilitate the separation.

*Yeastification of alcohol residual liquor*

The waste liquors obtained after fermentation which even today are partly discharged unused represent an important source of sugar. These residual liquors contain only 0·5–1 per cent of reducing substance which consists of a maximum of two thirds pentoses.

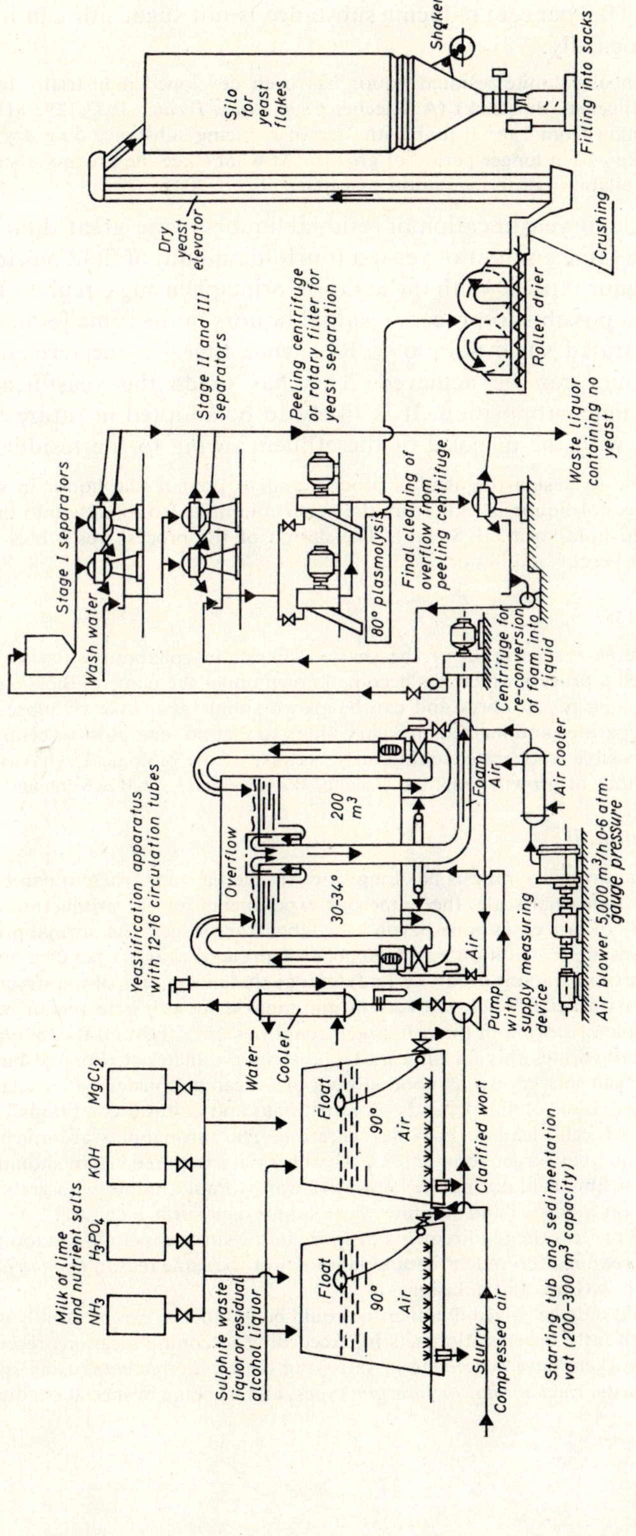


Figure 113. Sulphite waste liquor with aeration by the air-lift pump principle

The remainder of 0·3 per cent reducing substance is not sugar nor can it be decomposed any further biologically.

The yeastification of sulphite residual liquor has been developed industrially by the Farbenfabrik Wolfen (Rieche, Hilgetag, Butschek) (A. Rieche, *Chem. Ing. Techn.*, 1957, **29**, 818). By this process it is possible to obtain from 1 $m^3$ liquor with 7 kg of reducing substance 5 kg dry yeast with a rapid throughput, and 7 kg with a longer period of growth. At what stage the process is stopped is a question of economy and availability of fermentation space (see also p. 304).

The main obstacle to yeastification of residual liquors is the great dilution of the sugar solution. For the same amount of yeast a fourfold amount of fluid has to be put through when residual liquor is used. With the aeration principle using circulation tubes as shown in Figure 113 it is possible to process residual liquors in the same fermentation chamber as more concentrated sugar solutions. Residence times in the fermentation vat of as little as $2\,^1/_2$ hours can be achieved. This has made the yeastification of residual liquors an economic proposition. It is likely to be adopted in future wherever special difficulties arise with the disposal of the effluent owing to the residual sugar content.

It is now possible to design the alcohol process and to prepare the liquor in such a way that the distilled residual alcohol liquor issuing from the mash column is fed directly into the fermentation vat (Cellulosefabrik Attisholz vorm. B. Sieber). The design of the process resembles completely that of the yeastification of beechwood liquor.

### *The Biosyn process*

In order to achieve easy separation of the yeast, Peukert, in collaboration with the Zellwollefabrik Schwarza, developed a process which has become known under the name of Biosyn process. The mould fungus used forms long mycel fibres and can be grown submerged. Like cellulose, this mycel fungus can be very easily separated and washed on rotary filters so that no separators or centrifuges are required. However, it is impossible to operate continuously because of the biological behaviour of mycel fungus which differs from that of growth yeast (A. Rieche, *Wasserwirtschaft-Wassertechnik*, 1960, **10**, 52).

### *Biological fat synthesis*

Whilst the biological protein synthesis has long since reached a stage where it can be satisfactorily carried out on a large industrial scale, the numerous experiments for the production of fats with micro-organisms have only in rare cases gone beyond the laboratory stage. The normal protein yeasts or protein mycel fungi contain, in addition to 45–50 per cent protein, also 5–8 per cent lipoids. All micro-organisms contain a certain amount of fats or lipoids since these form part of the structure of the cell. The quantity present must never fall below a certain minimum if the cell is to remain viable. Some micro-organisms are capable of storing in the cell larger quantities of fat beyond this minimum.

Like certain carbohydrates this fat appears to be a reserve material. For the biological production of fat only micro-organisms known as good storers of fat can be considered. As detailed investigations have shown, the formation of fat depends on nutritional and cultural conditions. Thus nitrogen deficiency and ageing of cells lead to fatty degeneration. The throughputs are much smaller than with protein formation so that larger fermentation chambers are required. Here should be mentioned the first detailed work in this field during the 1914–1918 war by Paul Lindner who grew the hyphen fungus *Endomyces vernalis* on molasses and sulphite waste liquor as a surface culture.

For an industrial process the requirements of space in the surface method are too large and the handling of the cultures requires too much manual labour. For the same reason *Oidium lactis* was unsuitable for industrial synthesis (Fink and colleagues).

The problem could only be solved if a microbe could be found that can be rapidly grown in submerged culture and in which fatty degeneration sets in. According to communications received during the last ten years, several workers have succeeded in satisfying these requirements, using special strains of *Torula utilis*, nectar yeasts, *rhodotorula*, or *fusarium* types, and working in special conditions which are usually kept secret. This means that the biological synthesis of fat with selected heterotropic organisms has in principle been successfully carried out.

The best solution appears to have been found so far by the Zellstoffabrik Waldhof. In the special aeration equipment developed there a continuous process produced during a generation period of 10 hours, i.e. double the throughput period of the protein yeast, from 100 kg of RS (= reducible substance) introduced, 40 kg fat yeast with a fat content of 18–20 per cent and a probable protein content of 25–30 per cent.

The process, details of which have not so far been published, appears to have been carried out industrially only for a short period. Probably all difficulties have after all not yet been overcome, or the fat yeast production proved to be uneconomic as compared with the protein production.

It appears more advantageous to utilize the available waste sugars for the biological protein synthesis and not for the fat synthesis.

The synthesis could, however, become important for the production of lipoids or sterones which are difficult to obtain by other means. A large proportion of the ergosterone requirement is nowadays satisfied from yeast (Zellstoffabrik Waldhof).

### (d) Butanol-acetone fermentation of sulphite waste liquors

Whilst starch hydrolysates and molasses have for a long time in a number of countries been fermented with butyric bacteria to butanol and acetone on an industrial scale, this was not possible with sulphite waste liquors. According to the German Patents DRP 635 572 and 659 389, Vierling (Ludwigshafen) found that this was quite feasible when most of the lignin sulphonic acid was eliminated from the sulphite waste liquor by precipitation with milk of lime. The fermentation processes with butyric bacteria used so far had always started from a pure culture which can only be augmented in the transformation ratio of 1 part incubation medium to 10 parts mash. Five stages were, therefore, required for a large-scale charge.

Vierling found that by using a preliminary culture produced from spores of butyric bacteria, the inoculation of 2,000 times the quantity of mash is easily possible. Since the spores of butyric bacteria are largely heat resistant, the incubation medium containing them can be used comparatively hot, and fermentation starts during cooling. They are thus well protected against infection by other bacteria. There was, however, no incentive to use the pinewood sulphite waste liquor for bacterial fermentation. The situation changed when beechwood sulphite waste liquors which cannot be fermented to alcohol became available.

At first the fermentation of the beechwood sulphite waste liquor by bacteria on an industrial scale at the Farbenfabrik Wolfen proved highly unsuccessful until it was realized that this was due to the varying concentrations of formic acid and furfurol in the liquors. After these had been removed by hot aeration, fermentation of the beechwood sulphite waste liquor could be carried out without difficulties (Rieche-Hilgetag). The process was initially used on a semi-industrial scale only.

The concentration of the fermentation products in the liquor is as low as 0·7 per cent. They are extracted by distillation which, because of the low concentration, has to be carried out by exploiting all possible means of utilising the waste heat. The solvent mixture obtained consists of 2 per cent acetone, 12 per cent ethanol, 36 per cent isobutyl alcohol, and 50 per cent n-butanol. Its composition is, therefore, different from that obtained by the fermentation of hexoses which contains 75 per cent butanol. In large plants the cost price appears to be no higher than for the synthetic products.

Unlike the processes based on starch products and molasses continuous operation of the bacterial fermentation of sulphite waste liquor is likely to prove possible since it is to a lesser degree a nutrient medium for foreign infection.

## 3. INDUSTRIAL USE OF LIGNIN

During the chemical processing of wood about one third of the organic substance is lost as waste lignin. There is hardly any other industrial process which operates with such losses. It has been estimated that in Germany alone about 1·5 million tons lignin were lost annually during the years up to 1939.

To judge by the very voluminous patent literature the problem has long since been solved. According to this literature, almost anything imaginable can be made from lignin. In reality, however, only very few large-scale uses have been found for it which are negligible measured against the total supply of lignin available, particularly of lignin sulphonic acid.

The most important use of concentrated sulphite waste liquor is probably as a binding agent in the briquetting of bituminous coal, lignite, and coke dust. In foundries lignin sulphonic acid is also used as a core binder for moulding sand. Decalcified lignin sulphonic acid is a good foaming agent for fire extinguishers ('Tutogan'). It is also a very cheap adhesive but suffers from the great disadvantage that it is very dark, always penetrates paper to some extent, and never dries completely. Lignin sulphonic acid in the form of its sodium salt has very good dispersing properties and can, therefore, be used as a protective colloid, e.g. for spraying agents in the protection of vegetation, for the wetting and emulsifying of dyes, as a textile auxiliary and as a fibre-protecting agent (Decol, Protectol). It is also used as a plasticizer for both, the material which is burnt to clinker in the manufacture of cement (saving of water), and the liquefaction of concrete mixes so as to render them suitable for pumping. The Zellstoffabrik Waldhof produced the filler for curd soaps 'Zewa-Pulver' which is sulphite waste liquor that has been freed of iron and lime and dried in an atomizing drier and for which many other applications exist.

Sulphite waste liquor is carbonized at low temperature, activated carbon is made from it and it is also possible to produce cation exchangers from it by decomposition under certain conditions.

In the last few years many uses have been found in tanning for 'lignin extracts', purified sulphite waste liquors which have been acidified and contain mainly the ammonium salt of sulphonic acid, in addition to free lignin sulphonic acid. Lignin sulphonic acid has a certain tanning effect which, however, is small and does not constitute true tanning. On the other hand, these lignin extracts could well be used as tannin diluents and fillers; they behave similarly to tanning auxiliaries, having a dispersing effect on true tannins and a solvent effect on phlobaphenes (e.g. 'Hansa D', made by Zellstoffabrik Waldhof).

Many processes for the production of lignin tanning extracts have been described in the literature. Purification is usually carried out with sulphuric acid, calcium sulphate being precipitated and free sulphurous acid separated, or by precipitation with excess milk of lime, calcium sulphite being separated. The excess lime is removed with sulphuric acid. Iron can be precipitated with potassium ferri- or ferrocyanides, or be removed by means of Wofatit. The material is finally concentrated in multi-stage evaporators.

Much more convenient than lignin sulphonic acid are the tannins produced by condensation of phenols using nucleus-linking agents such as the $CH_2$ or $SO_2$ groups, and subsequent partial condensation of lignin sulphonic acid. These are true tannins. The first of this type was Tanigan extra A, made by Bayer-Leverkusen, the demand for which during the last war was so great that its manufacture used all the sulphite waste liquor produced in a large cellulose factory. Such exchange tannins based on sulphite waste liquor are, or were, produced also in the German Democratic Republic, e.g. 'Boemotan' or 'Pellutan' at the VEB Fettchemie- und Fewawerke, Karl-Marx-Stadt, and 'Wofagan M' at the Farbenfabrik Wolfen, from beechwood sulphite waste liquor, the latter via the basic lime salt of lignin sulphonic acid. It is possible that more use could be found for sulphite waste liquor in the tanning field. Mauthe observed that alkaline cracking of lignin sulphonic acid under very rigorous conditions produces polyhydroxy-compounds the tanning action of which resembles that of vegetable tannins (e.g. Quebracho). The production of vanillin from pinewood lignin sulphonic acid is also based on cracking with alkali.

### Vanillin from sulphite waste liquor

For more than sixty years it has been known that traces of vanillin occur during certain reactions of waste liquors of the cellulose manufacture. Grafe in 1904 was the first to isolate small quantities of vanillin by alkaline treatment of the waste liquor from the sulphite decomposition of wood by lime. Quite recently there have again been reports from America of the decomposition of lignin sulphonic acid with lime to produce vanillin. In 1928 Kürschner reported that he had succeeded in obtaining abundant quantities of vanillin by cracking sulphite waste liquor with alkali, at first by passing air through. He found later that alkali treatment alone also leads to vanillin. Pauly and Feuerstein obtained considerable yields of vanillin using chemical oxidizing agents, e.g. dichromate, though only from sulphite waste liquors of certain origins. Freudenberg and Lautsch showed that vanillin yields of up to 27 per cent could be obtained from purified lignin compounds by combined alkaline treatment and oxidation. Messrs. Schimmel & Co., Miltitz, obtained similar large yields from wood or sulphite waste liquor by oxidation with nitrobenzene in an alkaline medium. A favourable effect on the yield appears to be produced also by short-term treatment with oxygen during the alkaline cracking of lignin sulphonic acid, and by the addition of Cu as an oxidizing agent. First the alkali causes separation of the sulphonic acid groups of the lignin sulphonic acid, then cracking. In addition to vanillin a mixture of alkali-soluble phenolic condensation products is produced which contain different numbers of phenol-propane residues and are appropriately called 'lignophenol'.

The industrial production of vanillin from sulphite waste liquor, based on Kürschner's results, was first developed in the United States and in Canada. A large plant is also operating in the U.S.S.R. and an experimental one in Poland. In principle there are two processes for the industrial production: crude sulphite waste liquor can be heated under pressure with caustic soda, a process developed by Hibbert and Tomlinson and used by Howarth Smith & Co., or lignin can first be precipitated as a basic lime salt which is cracked with caustic soda, thus achieving a certain saving of alkali as compared with the use of unpurified liquor (G. O. Howard, of Marathon Paper Mills).

The alkali salt of vanillin can be extracted directly from the alkaline solution with butanol. The extracted alkaline liquor is used, after butanol has been extracted, as a binding agent for wood meal, paper, etc., for the production of building boards and sections which, however, are not very waterproof.

The problem of the industrial production of vanillin from sulphite waste liquor was tackled independently also by Farbenfabrik Wolfen, C. F. Boehringer & Söhne, Mannheim-Waldhof, and Hoffmann-La Roche, Berlin and Basle. A semi-industrial pilot plant was operated and preparations were made for large-scale production. However owing to the outbreak of war actual production was never started. A 'wood vanillin' plant is reported to be in operation in Poland.

The economics of wood vanillin production are largely influenced not only by the vanillin yield calculated for lignin but also by the amount of chemicals, particularly caustic soda, required and the possibility of recovering valuable chemicals and of utilizing the lignin slurry produced. The Farbenfabrik Wolfen (Rieche and Gnüchtel) found that desulphonated lignin from the vanillin production is an excellent phenol substitute for bakelite moulding material.

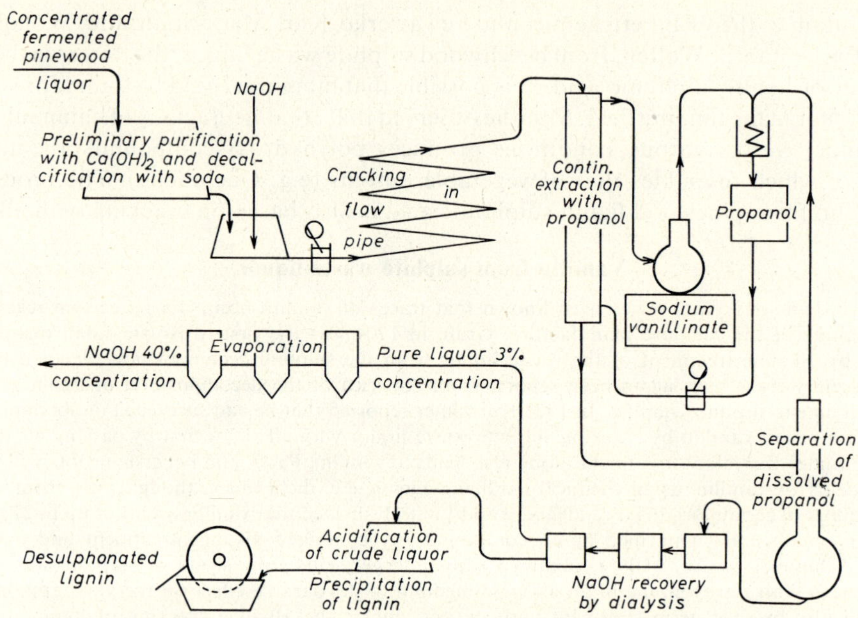


Figure 114. Production of vanillin from sulphite waste liquor

Wood vanillin is considerably cheaper than vanillin produced from guaiacol or oil of cloves.

The mode of operation of a plant for the production of vanillin from sulphite waste liquor is illustrated in Figure 114.

## Lignin used in the production of plastics

Like desulphonated lignin ('lignophenol') obtained by alkaline cracking of lignin sulphonic acid, acid lignin can also be used for the production of hardenable resins. Acid lignin from wood saccharification plants can be reacted with phenol in the presence of sulphuric acid and with simultaneous splitting off of water; the condensate obtained can be further condensed in the presence of excess phenol with formaldehyde. This mixture of lignin-phenol-formaldehyde-resin and phenol-formaldehyde-resin is thermoplastic and can be processed in the usual way with wood meal, hexamethylenetetramine (hardener), etc. into moulding materials (Rieche, Tagungsberichte der Chemischen Gesellschaft, p. 157. Deutscher Verlag der Wissenschaften, Berlin 1954). Under normal pressing conditions pressed products are obtained from these the properties of which fully correspond to those of bakelite S or type 31.

Bergius lignin is used by the Holzhydrolyse-Ges., Rheinau, for the production of core binders for foundry moulding sands (for the Croning moulding process) by condensing a preliminary condensate of phenolformaldehyde with acid lignin. Such resins are soluble in alcohol.

Alkali lignin from the decomposition of wood with caustic soda a process which at present is used only by the VEB 'ORWO' Filmfabrik Wolfen, can be condensed in an acid medium with phenol and subsequently with formaldehyde. Very versatile resins of the Novolack- or Resol type can be produced (see p. 424).

Without doubt the greatest possibilities for the use of lignin lie in the field of plastics. It is possible to use desulphonated lignin, acid lignin, and also alkali lignin for the production of building boards by utilizing the resins produced from them as binders for wood waste, paper, straw, textile waste, etc. By using it in this way lignin would after various transformations by man be used again for the purpose which it had originally in nature: as a cementing material for substances with a fibrous structure.

## 4. NATIVE BARK EXTRACTS USED AS TANNINS

(See also Section BV, Tanning and leather manufacture)

The production of native tan barks has contributed greatly towards rendering Germany independent of the world market in tannins. In practice, the only barks used for tanning at present are oak and pine barks; the use of fir, willow, birch, elder, and chestnut barks is still in the experimental stage. Ground oak bark has long been known as a tannin. Formerly it was the principal tanning agent.

Oak bark is now harvested by the 'bark-forest system' in which the trees are allowed to reach an age of only 10–20 years when they are felled and barked. The tannin content of the bark is then at its maximum. Pine bark is processed almost exclusively into aqueous tanning extracts, the 'pine bark extracts'. The method of production and drying of the pine bark decides the yield of tannin during extraction. The felled trees are barked during the sap period from May to the beginning of August, the bark is rolled inside out and stacked in the forest, protected from rain and sun. Native barks contain about 15–18 per cent extractable fractions, of which about two thirds are tannins and one third non-tannins. The extraction of pine bark is carried out on the countercurrent principle by the usual process in a diffusion battery with water to which some sodium sulphite and sodium bisulphite have been added. The extract is then evaporated under vacuum and acidified for tanning. The extractable tannin content of the bark fluctuates according to the place where the tree grew, its age, the season and time of the barking. After felling changes caused by enzymes occur in the bark; thus if it gets damp during storage the extractable fraction is greatly reduced. Obviously condensation (phlobaphen formation) occurs. The Studiengesellschaft der Deutschen Lederindustrie [Research Association of the German leather industry], Dresden, (Grassmann) developed a process by which the tannin in fresh barks is conserved in extractable form by poisoning the enzymes by heat treatment or with enzyme poisons. Increased utilization of the insoluble bark tannins is possible by sulphitizing extraction. Additional yields of up to 45 per cent can thus be obtained.

Whilst oak bark produces tanned leather of very high quality, pine bark extract has certain disadvantages. It tans for instance rather slowly and yields flat dark leather of low 'rendement' (weight-increasing properties). Good-quality leather can be obtained by the addition of suitable auxiliary and exchange tannins. In 1943, 20,000 tons, calculated on a pure tannin basis, of native barks and bark extracts were used in Germany annually.

# 5. OIL OF TURPENTINE AND ROSIN

The formation of turpentine and rosin in the wood of conifers often renders it highly rot- and water-resistant. Vats and agitators for the chemical industry are made of pitch pinewood. Those hardwoods which have a high tannin content (e.g. oak and mahogany) are particularly rot-resistant. Oil of turpentine which consists mainly of pinenes of various modifications can be obtained from the waste gases of the alkaline decomposition of fir wood. During sulphite decomposition the oil of turpentine of pinewood is converted largely into cymene, i.e. it is aromatized. It can be isolated from the digester waste gases. With steam ground conifer wood which is rich in resin yields wood oil of turpentine. In the United States 'Yellow Pine Oil' is produced from fir rhizomes. During alkaline decomposition of wood the separation takes place from the black liquor of the difficultly volatile fractions of fir resin, the resinic acids (pimaric acids) and fatty acids (tall oil) which in the form of their alkali salts are used for cheap soaps.

The bulk of tree resin in Germany is produced by cutting the cambium layer of living firs which stimulates the tree to exude resin. This method is being adopted increasingly and in old fir woods one can often see the characteristic herringbone incisions with resin welling from them which slowly runs downwards through the centre rib into a collecting vessel. During the resin period from April to October one tree yields 3–4 kg of crude resin. Crude fir resin is melted, poured into casks and processed in distillation plants.

There the large lump of resin is dropped into a heated melting pot, melted, and passed in the liquid state through a strainer which retains needles, pieces of bark and mechanical impurities into a vacuum still where the volatile oil of turpentine is separated from non-volatile resin fractions. The oil of turpentine of the *Pinus silvestris* consists mainly of D-$\alpha$-pinene apart from $\beta$-pinene and carene. From other species of trees, e.g. *Pinus nigra*, black fir, mainly L-$\alpha$-pinene is separated.

Oil of turpentine is an excellent solvent and cleansing agent for fats, lacquers, and resins used in painting and decorating. It is autoxidizable and promotes the drying of oils. It must not be forgotten that when it is spread on a large surface of an organic material, e.g. cleaning rags, it oxidizes so rapidly when exposed to air that spontaneous ignition may occur if the dissipation of the heat of reaction is impeded. Purified pinene is used for the production of one of the most important perfumes, terpineol, and of camphor. Terpineol is produced via the intermediate terpineol hydrate.

When oil of turpentine is distilled off rosin, a hard resinous substance, remains as a residue. It consists mainly of resinic acids. Rosin is one of the cheapest raw materials for lacquers but it lacks good film-forming properties. The finish is brittle, and where the water is hard light-coloured spots appear due to the formation of calcium salts. This defect shows particularly in cheap floor varnishes. The properties of rosin can be improved by neutralizing the acid with alkali or alkaline earth hydroxides or by making resinic acid anhydride with acetic anhydride.

The price of rosin depends on its colour which varies from dark brown to almost colourless. In order to purify it rosin can also be distilled under a high vacuum. Light-coloured pure rosin is converted to resin ester by esterification with glycerine. If these improved resin products are boiled with wood oil or linseed oil, lacquers with remarkably good qualities are obtained, which can be used even for outside decoration. The best lacquers the properties of which resemble those of natural copal are obtained from rosin by condensation with phenolformaldehyde resins of the Resol type ('Albertols') (see section on plastics). Rosin is also used in resin esters of the Glyptal type.

Rosin can be hydrated either as itself or in the form of its esters. Pure hydrated abietic acid ester was used as fixative for perfumes (Agfa).

Rosin soaps and soaps made from hydrated rosin are widely used emulsifiers in the polymerization of butadiene for the manufacture of Buna. Their advantage is that they precipitate during the coagulation of latex and remain in the rubber.

## LITERATURE


Winnacker-Weingaertner, *Chem. Techn.*, Organ. Techn., I. Hanser Verlag, München.

R. Trendelenburg, *Holz als Rohstoff*. J. F. Lehmann Verlag, München, 1939. New edition revised by H. Meyer-Wegelin, Hanser-Verlag, München, 1955.

H. Vogel, *Die Sulfitzellstoffablauge und ihre Verwendung*. Enke, Stuttgart, 1939.

K. Freudenberg, Neuere Ergebnisse auf dem Gebiet des Lignins und der Verholzung. *Fortschr. d. Chem. organ. Naturst.*, 1953, **11**, 43; 1963, **20**, 41.

E. Hägglund, *Chemistry of wood*. Academic Press Inc. New York, 1951.

F. E. Brauns, *The chemistry of lignin*. Academic Press Inc. New York, 1952.

E. Schmidt, *Angew. Chem.*, 1947, **60**, 16; *Papierfabrik.*, 1943, 43.

H. Scholler, Die Gewinnung von Zucker, Spiritus und Zuckerhefe aus Holz als Rohstoff. *Chem. Z.*, 1936, **60**, 293.

A. Rieche, Über die mikrobiologisch-technische Eiweiß- und Fettsynthese auf der Basis von Zellstoffablaugen. *Wiss. Ann. d. DAdW., Berlin*, 1954, **3**, 705–728.

A. Rieche *et al.*, *Reinigung und Verwertung industrieller Abwasser*. Akademie-Verlag, Berlin, 1962.

K. R. Dietrich, *Ablaufverwertung und Abwasserreinigung in der biochemischen Industrie*. Dr. Alfred Hüthig-Verlag G. m. b. H., Heidelberg, 1960.

E. Bergander, *Biochemie und Technologie der Hefe*. Verlag Th. Steinkopff, Dresden, 1959.

A. S. Sergejewa, *Chemie des Holzes und der Cellulose*. Verlag Th. Steinkopff, Dresden, 1959.

# II. Sugar, starch, and glucose

Sweeteners used in antiquity and during the Middle Ages stemmed almost exclusively from insects. Manna mentioned in the Bible is still used by Bedouins. Tamarix manna consists of the excretions of cochineal, i.e. it is candied honeydew. The oldest sweetener used on a large scale is honey. In the old civilizations of Egypt, Greece, and India honey was of the greatest importance and almost the only sweetener available. It was taken from wild bees. The keeping of bees is more than 2,000 years old. In Germany bees were already kept in beehives as long ago as the early Middle Ages. Beekeepers formed their own guild. When cane sugar was discovered honey lost its importance and beekeeping continued on a much smaller scale. The beekeepers' guild was dissolved in the 18th century.

Almost the only sweeteners occuring in nature are glucose (dextrose) and fructose (levulose) in the sweet juices of fruit. The most important sugars produced industrially are beet sugar and cane sugar. Chemically these are disaccharides, saccharose, $C_{12}H_{22}O_{11}$, consisting of D-glucose (aldose) and L-fructose (ketose). Glucose is also produced industrially, by the hydrolysis of starch with diluted acids ('Dextropur'). Fructose is very difficult to crystallize. The high crystallizing ability of cane sugar was a favourable factor in the development of its industrial production. A disaccharide, cellobiose, has been mentioned earlier as a hydrolytic decomposition product of cellulose, consisting of two molecules of glucose. Cane sugar which rotates the plane of polarized light to the right (+), is composed analogously of one molecule glucose and one molecule fructose:

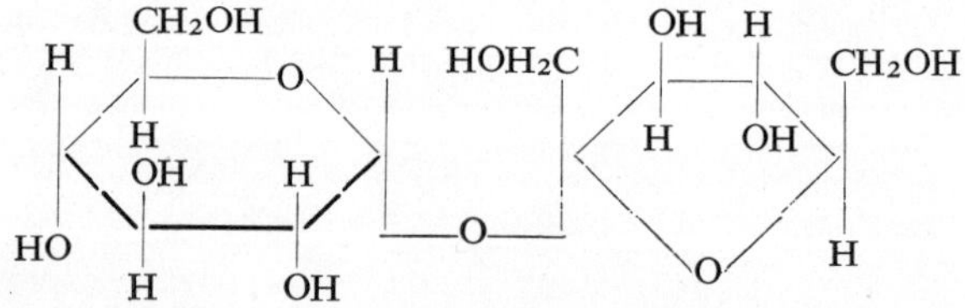

Cane sugar

Hydrolysis of cane sugar by diluted acids or enzymes (saccharase) reverses the rotation from right to left ('inversion'). D-glucose and L-fructose are formed of which L-fructose rotates more strongly to the left so that invert sugar is levorotatory. Inversion is carried out industrially for the production of artificial honey and confectionery. Invert sugar tends less to crystallization than saccharose; glucose contained in it may crystallize out (honey).

Technical polysaccharides, cellulose and starch, can be saccharified by hydrolysis, forming only D-glucose (wood sugar, starch sugar). Starch is hydrolysed industrially by acids or enzymes. In the germ of grain, particularly barley, the enzyme diastase which saccharifies starch is essential for the fermentation since only mono- and disaccharides are fermented by yeast. Diastase (amylase) is also found in some micro-organisms. Only the products of agriculture and forestry can, therefore, be raw materials for the fermentation industry proper. The sugar- and starch-containing products of agriculture cannot, because of their great volume and frequently high water content (sugar beet and potatoes), be transported over great distances for reasons of economy. Moreover, the by-products of the fermentation and sugar industries are re-used in agriculture. The extracted pulp from sugar beet manufacture, the protein-containing residual alcohol liquors from potato- and grain distilleries, and the pulp from starch factories are all important fodders. It is, therefore, important that the sugar, starch and fermentation industries and agriculture should be closely linked both physically and economically.

## 1. MANUFACTURE OF CANE AND BEET SUGAR

Sugar has been produced from sugar cane for 1,500 years, from sugar beet only since the beginning of the 19th century. In 1747 the Berlin chemist Markgraf discovered sugar in the native beet and found that it was chemically identical with the well-known cane sugar. His gifted and versatile pupil Achard developed an industrial process on the basis of this observation and in 1802 started the first sugar factory in Silesia. At that time sugar beet contained only 4–5 per cent sugar. In spite of his extraordinary energy, great organizational gifts, and support from the authorities, Achard had to overcome many technical and economic difficulties at first.

The beginnings of the German sugar beet industry were greatly favoured by the closure of the European ports imposed by Napoleon which led to a great reduction of cane sugar supplies. When the blockade was lifted beet sugar manufacture collapsed. Nevertheless, 50 small factories were in operation in France in 1830. In 1840, beet sugar manufacture was started again in Germany and great technological progress was made. Lime separation was improved, filtration of the juice through carbon was introduced, and evaporation was carried out under vacuum. Achard's enterprise was nevertheless looked upon with distrust by leading personalities. This can be gathered from a remark by Liebig, who called sugar manufacture 'a luxuriant hothouse plant which is being cultivated at the expense of the whole community and has no future'. The English cane sugar importers assessed the economic prospects of beet sugar manufacture much more favourably. They offered Achard large sums of money through intermediaries if he renounced his plans for the introduction of beet sugar manufacture, however, without any success. He continued to work on the development of his process and taught young men from many countries who came to work without salary in his factory in order to learn the manufacturing process.

In the last hundred years beet sugar production has increased enormously in almost all countries and the former luxury food has become a popular nutrient. Whilst in 1900 beet sugar production in the whole world amounted to 6 million tons, it had risen to 12·9 million tons by 1950. It is, however, still far behind cane sugar production which increased during the same period from 5·3 to 22·3 million tons. World sugar consumption per head of population varies greatly. In 1948, it was highest in Sweden and Denmark. In Germany it was then only half as high, probably due to the aftermash of the war.

World sugar production in 1951 amounted to 35–36 million tons, 13 million tons of which was from sugar beet, and 22·5 million tons from sugar cane; in 1960, 23 million tons came from sugar beet and 29·4 million tons from sugar cane.

Beet sugar manufacture

| | 1950/51 | 1960 | |
|---|---|---|---|
| U.S.S.R. | 2,150 | 6,363 | |
| U.S.A. | 1,814 | — | |
| Germany | 1,750 | 2,017 | (Federal Republic) |
| | — | 680 | (Democratic Republic) |
| France | 1,350 | 2,728 | |
| Poland | 925 | 1,380 | |
| Czechoslovakia | 775 | 881 | |
| Great Britain | 695 | 779 | |
| Cuba (1959) | — | 5,964 | (cane sugar) |

Sugar is taxed in almost all countries. This taxation originated probably at a time when sugar was considered a luxury, not a nutrient. The extraordinary importance of beet sugar manufacture for Germany is illustrated by the fact that during the period 1890 to 1914 she was second only to Cuba (sugar cane) in the production of sugar. In 1913 about 40 per cent of the production was exported. The highest German production was achieved in 1937/38 with 2·2 million tons for which 13·7 million tons sugar beet were processed. In 1950/51 production amounted to only 1·82 millions tons due to the consequences of the war.

Apart from the technological advances the success of the sugar beet industry owes most to the skill of the beet growers. Originally the *Beta vulgaris* had a sugar content of only 5 per cent. By improved cultivation this was increased to 16–20 per cent. The raw sugar yield from German beet rose as follows:

1840: 5·9 per cent
1870: 8·6 per cent
1890: 12·0 per cent
after 1910: 16·0 per cent

Germany used to be famous for its sugar beet seeds which were an important export product, particularly to the United States.

From one hectare of good agricultural land at least 25,000–30,000 kg sugar beets were harvested, corresponding to 4,500–5,000 kg sugar, i.e., calculated on the basis of nutritive value, one-and-a-half times the yield from potatoes and twice the yield from wheat. The cultivation of sugar beet is economic only in large areas, heavy soils, and with extensive mechanization of the working of the soil, cultivation and harvesting. The application of fertilizer requires much care and knowledge, for modern sugar beet is a highly bred individual which from a plant-physiological point of view is degenerate and reacts, therefore, very strongly to external influences. Good yields with high sugar contents can be achieved only with very intensive manuring. The beet contains 4–5 per cent marrow (cellular tissue), 91 per cent sap with a sugar content of 16–20 per cent, 1–1·5 per cent organic non-sugar substances, and 0·8 per cent ash (salts). The marrow consists mainly of cellulose, pectic substances, and the protoplasm of the cell. The sugar content of the sap is determined by measuring the rotation of polarized light in a polarimeter. The degree of rotation is proportional to the concentration of the solution and the length of the liquid layer. The sugar concentration can also be determined by testing with a hydrometer, i.e. density determination with the saccharimeter (Brix's scale), or by refractometric measurement, recently also by a rapid micro-method.

### Manufacture of beet sugar

The manufacture is divided into the extraction of the juice, its purification, evaporation, crystallization, and the separation of the crystals. Raw sugar was manufactured in Germany in 1937 in 211 sugar factories, in 1950/51 in 140. Before the war 20 refineries were in operation, now only 5. A large sugar factory processes daily 1,500–3,000 tons of sugar beet. Manufacture stretches over several months, the 'campaign'. During the remainder of the time the factories are closed. This is unavoidable since great losses of sugar are caused by the 'breathing' of the beet during storage.

The beet collected from the fields is first flushed by a water jet (1–2 atm) (Elfa flushing method) from the trucks into storage. From there it is transported by flushing with water into the factory according to the requirements of manufacture. The mixture of sugar beet and water has usually to be lifted which is done by air-lift pump.

The air-lift pump is a compressed-air water raising device. If compressed air is introduced into the bottom of a pipe which is filled with a liquid (this is best done by a distributor nozzle) an air-liquid mixture is formed in the pipe the specific weight of which is less than that of the liquid. This causes the liquid in the pipe to rise and it overflows when the weight of the air-liquid-column *a* is smaller than that of the liquid column *b* (see *Figure 115*). The lower the pipe dips into the liquid, or the higher the liquid level of a vessel with which the pipe communicates, the greater is the lifting capacity of the pump. In general, pumps are used for only small heights of lift and the comparatively cheap air produced by a

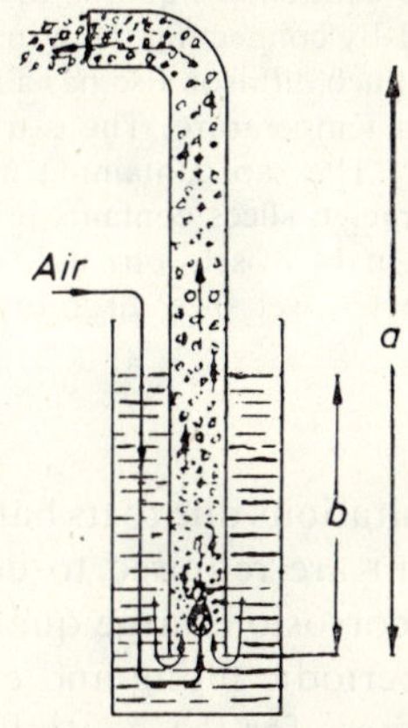


Figure 115. Principle of the air-lift pump

0·8 atm pressure fan is adequate (compressed air of more than 1 atm gauge pressure produced by compressors is considerably more expensive). Solid materials even of fairly large size, such as beet, potatoes, sand and gravel, and salts, floating in the liquid or which can be made to rise turbulently to the surface, overflow with the liquid. The air-lift pump is, therefore, suitable also for the transport of coarse materials. The same principle is also used for fine aeration and for the circulation of liquids with gases (see e.g. production of protein yeast, pp. 307–309).

The sugar beet is passed into the washing plant which consists of an open trough in which revolves a shaft with beaters attached. Rapid movement dislodges the sticky earth which often adheres very strongly to the roots. Stones, sand, and beet tops are removed by washing in preparation for the subsequent shredding of the sugar beet.

The cleaned beet is then carried by an elevator via the beet scales to the slicing machine which works on the principle of a cabbage shredder. Serrated knives are attached in the knife box to a horizontal revolving disc. The beets drop on to this disc from a cylinder above, their own weight producing the continuous sliding movement through the cylinder. They are sliced to 4 mm thickness so as to permit the sugar to pass easily from the cells into the water during the subsequent diffusion. The extraction of the slices is carried out in a diffusion battery.

*Extraction in the diffusion battery*

The sugar dissolved in the cell fluid can pass into the extraction liquid (water) only if the protoplasm and the semi-permeable plasm wall of the cell are denatured, or if the cells are opened mechanically. Diffusion at low temperatures ($<70°C$) would produce only little extraction (only the cut cells are leached); moreover, at such temperatures the desired denaturation of the protein constituents does not take place. However, at temperatures $>80°C$ the slices have a pasty character, the pectic substances swell ('scalding' of the slices); this must be prevented at all costs since the extraction vessel would become clogged. It follows that the extraction has to be carried out at temperatures between 70 and 80°C.

The diffusion battery consists of a number of vertical iron cylinders (8–16) with a capacity of 4–13 m$^3$ to which truncated cones are attached at both ends. The fresh slices are fed into the diffusors from the top through an aperture which can be shut, the extracted slices are emptied at the bottom through a hydraulically operated opening. In the lower conical part of each diffusor is fitted a sheet iron screen (diameter of the perforations 5–10 mm) through which the extraction liquid is discharged. The combination of the individual vessels into a battery is arranged by connecting them in series by means of a system of pipes (diagram of a diffusion battery see p. 287). Each diffusor also has a number of valves and a calorizor heated by steam which maintains the required temperature. The extraction is carried out with water which extracts the beet slices countercurrently. The sap containing most sugar is drained off the vessel with the fresh slices, the vessel with the extracted slices contains practically fresh water. In a diffusion battery of 16 vessels extraction is going on in 14 vessels, one diffusor is being filled and the diffusor longest in the cycle is being emptied. The extracted wet slices drop out at the bottom. The diffusion period per vessel is 50–90 minutes.

*Continuous extraction*

The diffusion battery described has a number of limitations due to its batchwise operation, such as: many operators and much manual labour are required to deal with the large number of valves, great possibilities of attack by corrosion, large quantities of diffusion waste water to be disposed of during emptying, periodic filling and emptying of slices. To overcome these difficulties a number of appliances for the continuous extraction of beet slices have been developed of which recently the 'tower process' has become the most important. It is a further development of Hildebrandt's extraction process described on p. 351.

The mode of operation of this process will be briefly described using the diagram of an extractor (Buckau-Wolf A. G.) consisting of two parts: the scalding trough and the tower part (Figure *116*).

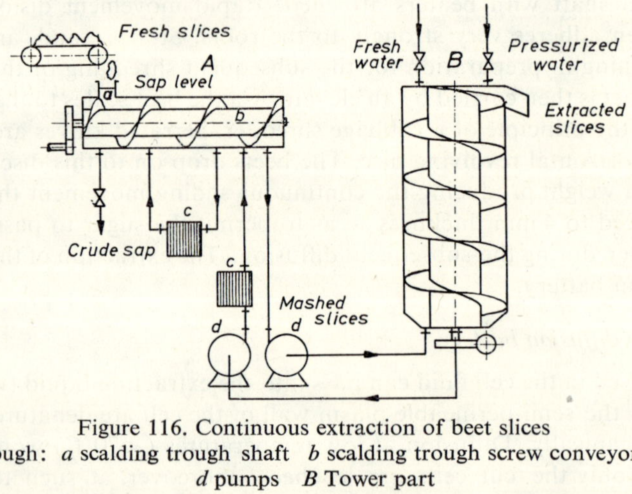


Figure 116. Continuous extraction of beet slices
*A* Scalding trough: *a* scalding trough shaft *b* scalding trough screw conveyor *c* calorizers *d* pumps *B* Tower part

The slices are first fed into the scalding trough (A) where they are picked up by a screw conveyor. During their passage through the trough they are scalded by revolving sap heated in calorizors (c) and finally they are pumped into the lower part of the extraction tower (B) proper. Here a slowly revolving shaft conveys them upwards. More and more sugar concentrates in the water flowing downwards in the opposite direction to the slices (tower juice). The extracted slices are discharged continuously from the upper part of the tower, where fresh water or pressurized water from the slice presses is introduced. After heating the tower juice is fed into the scalding trough whence it is extracted continuously as 'crude juice'.

The slices containing only about 6 per cent solids are crushed after extraction in presses, increasing their solids content to about 10 per cent. The continuously operating slice press consists of a conical shaft with augers revolving inside a cylinder cover. This principle which permits continuous operation is also used in the popular fruit presses in the kitchen. The wet slices are either carted off to neighbouring agricultural establishments, deposited in silos, used directly as cattle food, or dried. For this purpose large drum driers are used which are heated by furnace waste gases from the boiler house. Dry slices still contain about 10 per cent water.

The crude juice produced contains a number of non-sugar materials (proteins, pectins, aminoacids, invert sugar, inorganic substances) which have to be eliminated since they both hinder the crystallization of the saccharose (molasses-forming agent) and reduce the quality of the sugar. The crude juice is purified by the addition of lime ('liming'). Liming is subdivided into two stages. During the preliminary liming the crude juice is adjusted to a pH value of 10·8–11 by the addition of milk of lime. This causes proteins to coagulate, free acids are neutralized, insoluble calcium salts are precipitated, and the juice is sterilized. During the main liming an excess of CaO is added to the juice, producing decomposition of the invert sugar and saponification of the amides. The main liming is specially important because due to the excess of CaO a correspondingly large amount of $CaCO_3$ is formed during the first saturation thus rendering possible thorough filtration. The treatment with lime has three objects:

(1) It renders the slightly acid juice alkaline and protects it from inversion by natural vegetable acids, and from fermentation by yeasts and bacteria introduced during the further processing;
(2) Phosphoric acid, oxalic acid, and other organic acids, and iron and magnesium salts are precipitated and nitrogen compounds are decomposed;
(3) Cloudy colloidal substances preventing crystallization are enveloped and precipitated (purification with Wofatit see p. 327).

After liming the excess calcium has to be precipitated by carbon dioxide. This is done in the saturation plant. The saturation is carried out in a specially developed apparatus in two stages: first, carbon dioxide is introduced only up to pH 10·9. This is particularly important so as to prevent dissolution of the colloids precipitated during the preliminary liming. Then the juice is filtered. For the control of automatic dosage the alkalinity is determined by indicator paper, by titration, or electrometrically. During the second stage of saturation when the $Ca(OH)_2$ and KOH still present are converted to carbonates (optimal acidity), further precipitation takes place rendering more filtration necessary. A sugar factory requires, therefore, quicklime and carbon dioxide, necessitating the operation of a lime kiln.

The saturation juices are filtered in filter presses. Since the separated slurry has to be washed again these presses are equipped with washing devices. The design and mode of operation of a filter press is shown schematically in Figure 116. When the press is opened the frame contains the calcium carbonate slurry ('separation slurry'). In addition to calcium carbonate this slurry contains almost all the phosphoric acid taken from the soil, some nitrogen, potassium, and organic matter. It is an excellent fertilizer and is returned to the fields.

In multiple-effect evaporators the thin juice which has been treated with $SO_2$ in order to lighten the colour is evaporated to syrup. Multiple-effect evaporators require much less steam than single-body evaporators, since the heat of the vapour from the first stage of the evaporation is utilized for heating the next stage.

*Evaporation, concentration; general remarks on heat transfer*

The boiling temperature of a material at a given pressure is the temperature at which its vapour pressure reaches this pressure. Materials have characteristic vapour pressure curves which depending on their nature are either flat or steeply inclined. By dissolving a material which differs from the solvent the vapour pressure is lowered in proportion to the dissolved molecules, and the boiling point is raised. Almost all the solutions which in industry are concentrated by evaporation have boiling points which rise continually with increasing concentration. However, due to the weight of the liquid, the pressure at the heating surface of an evaporator is higher than in the vapour chamber where free evaporation is possible. The boiling temperature is correspondingly higher. This is particularly noticeable during evaporation under vacuum and must be taken into account in the design of equipment for concentration by evaporation.

Evaporation requires energy for heating and to overcome the force of attraction of the molecules, i.e. the 'heat of evaporation', which is a material constant. More energy is required to overcome the external pressure. The velocity of evaporation is proportional to the quantity of heat absorbed in a unit of time which in turn is proportional to the heating surface and its thermal conductivity and to the heat transfer. The size of the heating surface is mainly determined by the heat transfer coefficient. This coefficient can be calculated for the heat transfer from fuel → heating surface, but for the heat transfer from heating surface → boiling liquid only values derived from experience with the various designs for evaporators are available. Since evaporation is also affected by separation of salt, foaming, and incrustation, an evaporation test has to be carried out to determine the performance and dimensions of the heating surface. The coefficient of heat transmission, with the dimension kcal/m$^2$ h °C, usually forms the basis for the calculation of the heating surface. All designs for evaporators aim at the most favourable heat transfer to the boiling liquid, i.e. the thermal energy of the heating medium is to be transferred as rapidly as possible to the material to be evaporated.

Heat flow can be by convection, conduction, or radiation. Convection is the motion of a medium transferring the heat. In convection the heat transferred Q (in cal) is proportional to the surface $F$, the temperature difference $t_1 - t_2$ of the media, the time $z$ (in seconds), and the coefficient of heat transfer $\alpha$ which is equal to the quantity of heat transferred to the medium from 1 cm$^2$ of surface per second at a temperature difference of 1°, i.e. $Q = \alpha \cdot F \cdot z(t_1 - t_2)$. The heat transfer by heating surfaces is, however, also affected decisively by the internal heat conduction of the material. Heat propagates in bodies from warm places to cold ones by the thermal oscillations of the molecules. A characteristic of every material is the heat transfer coefficient, the specific thermal conductivity $\lambda$. It is equal to the quantity of heat transferred (in cal) through 1 cm$^2$ in 1 second, at a temperature difference of 1° between the two sides of a 1 cm thick plate of the material, i.e. $\lambda$ = cal °C$^{-1}$/cm$^{-1}$/sec$^{-1}$. In industry, calculations are based on kcal, m, and h. The heat transfer coefficient $\lambda$ for Ag at 20°C is 353; Cu 320; Fe 51; glass 0·65 (kcal/mh °C). The quantity of heat Q transferred by conduction is proportional to the heat transfer coefficient $\lambda$, surface $F$, time $z$, difference of temperature $t_1 - t_2$, and inversely proportional to the thickness $\delta$, i.e. $Q = \frac{\lambda}{\delta} \cdot F \cdot (t_1 - t_2) \cdot z$.

In evaporators, stills, coolers, reaction vessels with coils or heating jackets, preheaters, and heat exchangers, there are numerous points where heat transfer by convection and by conduction alternate. For instance, the simple heat exchange of two liquids through a flat wall gives two heat transfers and the heat resistance of the separating wall material. The process is composed of three stages:

(1) the heat transfer of the warm liquid to the wall,
(2) the conduction of the heat from the inner surface of the wall to the outer surface,
(3) the heat transfer from the outer surface to the cold liquid.

Instead of dividing the heat transfer into three parts it can also be related summarily to the difference of temperature between heat carrier and liquid when the fundamental quantity is the coefficient of heat transmission $k$.

$$k = \frac{1}{\frac{1}{\alpha_1} + \frac{\delta}{\lambda} + \frac{1}{\alpha_2}} \quad (1)$$

$$Q = k \cdot F \cdot (t_1 - t_2) \quad (2)$$

In equation 2 it was assumed that the temperatures $t_1$ and $t_2$ remain constant over the entire surface. However, both liquids flow along the surface and their temperatures change. The warm liquid cools, the cold liquid warms up. The final formula is, therefore,

$$Q = k \cdot F \cdot \Delta_m,$$

where

$$\Delta_m = \frac{\Delta_a - \Delta_e}{\ln \frac{\Delta_a}{\Delta_e}}$$

is the mean temperature difference. $\Delta_a$ is the initial temperature difference and $\Delta_e$ the final temperature difference of the exchanging liquids.

In heat radiation heat is transferred from one body to another by electromagnetic oscillations without the intervention of matter. A glowing body, e.g. heating gas or a heating surface, emits heat rays the wavelengths of which lie in the range from the infra-

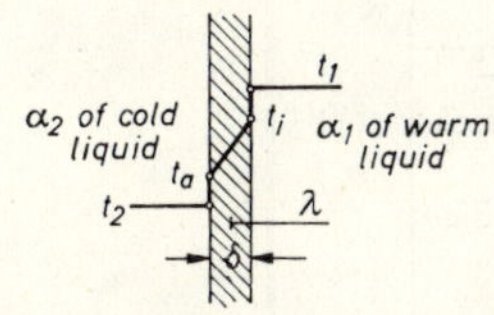


Figure 117

red to visible light. The heat rays are produced by the thermal motion of molecules and on impact impart thermal motion to other molecules. The quantity of heat radiated by a surface is proportional to the fourth power of the absolute temperature. Hot gases transfer heat mainly by radiation. In boiler firing, gas-heated furnaces, circulation gas firing, heat accumulators (regenerators), electric radiators and infra-red radiators most of the heat is transferred by radiation.

The heating elements used in evaporators, stills, and reaction vessels are:

(1) heating jackets (external heating) for small performance,
(2) coils and tubes the surface of which can be enlarged by fins,
(3) plates (e.g. in vacuum drying cabinets),
(4) lamina, or ribbed tubes (in calorifiers and car radiators),
(5) metal or ceramic rods, firebricks in heat accumulators (regenerators, Cowpers) operating on the reversing process principle.

In the direct evaporation methods the medium giving out heat (vapour, hot gas) is in direct contact with the material to be concentrated by evaporation. In some cases it is possible to let the flame burn directly in the material to be evaporated (submerged combustion). The liquid can be passed through spray columns into which hot gases, e.g. furnace waste gases, or steam, are introduced from below. Evaporation of foaming and crust-forming solutions can be carried out in the revolving tubular furnace into which the solution is injected, or in evaporator screws. Solutions of caustic soda are processed to caustic soda in evaporators with direct firing, or in open pans.

Evaporators are designed to produce a liquid layer of minimum thickness over a heating surface of maximum dimensions. The various types of evaporators are, therefore, characterized by the heating units and the liquid circulation.

Organic materials which are mostly sensitive to heat—sugar is one of these—are evaporated with minimum loss by indirect heating by steam under vacuum. In place of steam, heating bath liquids can also be used (see separate section on pp. 234, 235). Vacuum evaporation is the method causing least loss and offering most advantages from the heat economy point of view; it is one of the most widely used operations in chemical engineering. The diagram in Figure 118 shows a vacuum evaporator with barometric con-

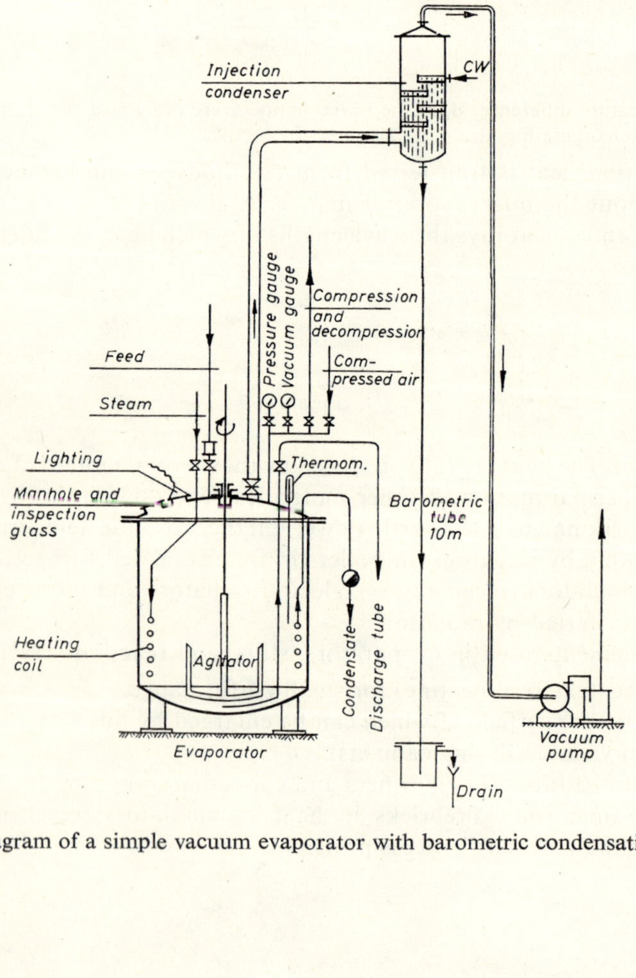


Figure 118. Diagram of a simple vacuum evaporator with barometric condensation discharge

densation discharge such as is used mostly when production is not very large. Instead of the barometric discharge the condensate can also be removed by an Elmo water ring vacuum pump (see p. 122, Figure 57).

The main problem in the evaporation of large quantities is thermal economy. In order to achieve the most favourable utilization of the heating vapours, multi-stage evaporation is used in which the waste vapours of the first stage are used to heat the next stage. In multiple-effect evaporators only the first body is heated by live steam, the others by

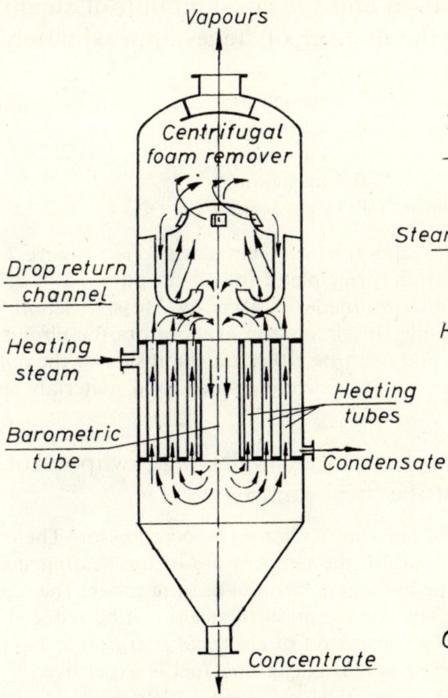


Robert evaporator

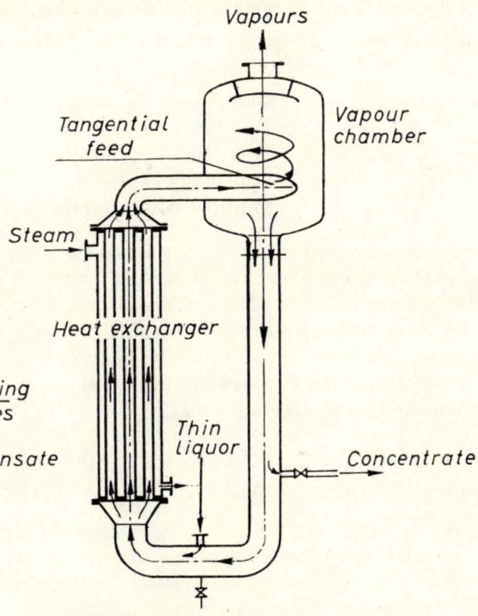


Circulation evaporator with external heater

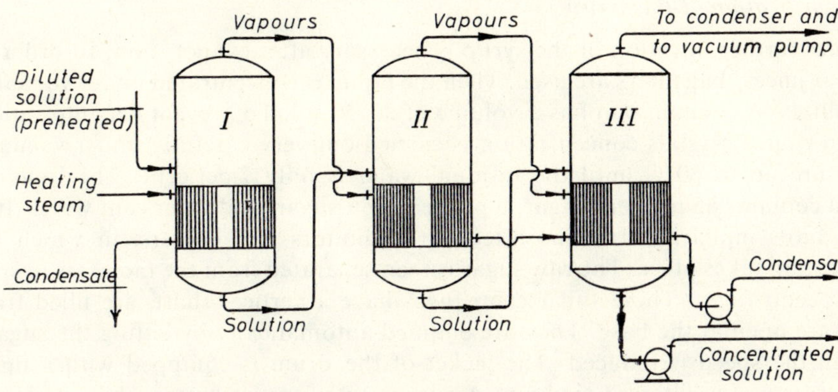


Figure 119. Arrangement of a three-stage evaporation plant

the vapours from the preceding ones. Boiling in the subsequent stages can be achieved only by reducing appropriately the pressure from one body to the next. This can be done by increasing the vacuum or by using a higher initial pressure. The most concentrated solution is thus exposed to the lowest temperature which is important for evaporation with low losses. Figure 119 shows the mode of operation of a multiple-effect evaporator such as is used in the sugar industry: Giant evaporators are used in cellulose factories where the solution has to be concentrated for combustion. The daily output is 500 to 1,000 m³. Up to six evaporator bodies are used and the consumption of steam per kg of evaporated water drops in practice with the number of stages approximately as follows:

Single-body evaporator 1·2 kg steam
Two-body evaporator 0·7 kg steam
Three-body evaporator 0·5 kg steam
Four-body evaporator 0·4 kg steam
Multiple-body evaporator 0·35 kg steam

The evaporator body may be, e.g., the Robert evaporator with circulation (see Figure 119). The liquid rises in the heating tubes by gas bubble formation (principle of the air-lift pump) and drops back through the barometric tube. The circulation evaporator made of glass by Messrs. Schott & Gen., Jena, in very large sizes also operates on this principle. In circulation evaporators the circulation can be augmented by impellers or pumps. The heating body can be either on the inside or the outside (see sketch). The design of the evaporators has to allow for easy cleaning since solid materials and crusts are often deposited on the boiling tubes.

Another possibility for utilizing the vapour heat in a single-stage evaporator, e.g. a circulation evaporator, consists in the use of the 'heat pump'.

Part of the vapours is condensed by increasing the pressure with a turbo-compressor. The condensation heat liberated raises the temperature of the vapours to the temperature of the heating steam. This newly produced heating steam is fed together with the live steam into the heating tubes. The liquid to be evaporated is preheated by the vapour condensate. The consumption of steam can be reduced down to 0·25 kg per kg of evaporated water. This is actually a conversion of electrical energy into heat but it is economical in plants where electricity (hydroelectric power) is cheap and fuel is expensive.

Evaporator plants with vapour compression are economical provided 1 kWh costs less than 1 kg bituminous coal of 6,500 thermal units.

*Second purification of the syrup*

Sometimes further filtration of the syrup is necessary after evaporation. In order to obtain 'clear juices' bag filters are used. Then the liquid is concentrated under vacuum until it crystallizes. A vacuum pan has a volume of 30–50 m³. To prevent decomposition and to obtain coarse crystals concentration is carried out very carefully under vacuum at a temperature up to 60°C until the residual water is only 7 per cent. The 'massecuite' obtained contains 85 per cent sugar, 8 per cent non-sugar, and 7 per cent water. It is discharged into 'minglers', troughs fitted with agitators and coolers, in which further crystallization takes place. The raw sugar is then separated from the molasses in specially designed centrifuges. These sugar centrifuges have a vertical shaft, are filled from the top and are open at the base. They are emptied automatically by letting the sugar drop out when the speed is reduced. The jacket of the drum is equipped with a fine sieve which serves as the filtering medium. A sugar centrifuge is shown schematically on p. 243.

The centrifuged crystals, with a grain size of 2–4 mm, are covered with syrup residues which give them a yellowish colour. They are designated as raw sugar I and contain 95–97 per cent saccharose. The run-off I from raw sugar I still contains 76–79 per cent sugar and is again concentrated under vaccum. The raw sugar II separating out crystallizes more slowly. After separation in a centrifuge molasses are obtained as run-off with a sugar content of about 50 per cent. Molasses with 80°Brix = 43°Baumé (Bé) contains in addition to sugar 20 per cent organic non-sugars, 10 per cent salts of which 5 per cent are K salts, and 20 per cent water. They contain many amino acids. They can be further desaccharified by the Strontian process. Because of its sugar content and the high contents of nitrogen and salts molasses is the best raw material for the manufacture of baking yeast and is a good animal feeding stuff. It is also fermented to alcohol in molasses alcohol plants.

From 100 kg sugar beet, with 17·5 per cent sugar, one obtains, calculated for 100 per cent, 13·65 kg raw sugar I, 1·78 kg raw sugar II, and 1·03 kg molasses. The loss is 1·04 kg.

### *Purification and refining of sugar*

The simplest method of purification is by washing (affination). It consists in simply covering the sugar in the centrifuge with syrup and water in the form of steam or a fine spray. The sugar is thus washed only superficially. This is adequate for most purposes since internally the crystals are pure by the nature of the crystallization process. Only light-coloured raw sugar with coarse crystals is suitable for the washing process.

Purification is carried out in affination centrifuges. The crystals are mixed with syrup, the so-called 'covering liquor', centrifuged, and topped up with a little water or steam. Mixing takes place in mixing ducts fitted above the centrifuge. Here, too, centrifuges with bottom discharge are used. After affination the crystals must be kept in motion until they are dry since they would otherwise cake together because of their somewhat sticky surface. Drying is carried out in cylinders with screw conveyors and warm air, or in barometric tubes through which preheated air flows upwards. Granulated sugar, sand or crystal sugar are obtained.

The refining of sugar is a re-crystallization process, i.e. the sugar is dissolved and then re-crystallized. The raw sugar covered with wash liquor in centrifuges is dissolved in hot water to a syrup with a sugar content of 60–70 per cent, the solution is adjusted to weak alkalinity, and after kieselguhr has been added to induce precipitation of the impurities it is filtered. Following this 'preliminary filtration' a second filtration is carried out through activated charcoal (boneblack, activated carbon, or charcoal). The clear syrup, the 'clearing liquor', is concentrated under vacuum down to a water content of about 10 per cent when crystallization takes place.

It can then be processed again to granulated sugar, or it can be poured into conical moulds with an open base, centrifuged in these and dried, thus obtaining sugar loaf. By pouring into suitable moulds, or by pressing, cube sugar is produced. Sugar candy is obtained by slow crystallization of the sugar solution in which are suspended strings on which large crystals form. Sugar candy made from sugar beet is white, that made from sugar cane, brown. The refining process is much simpler with cane sugar; whilst molasses from beet sugar has an unpleasant, irritating taste, the taste of cane sugar molasses is pleasantly aromatic.

Pure refined sugar contains 99·95 per cent saccharose and is thus one of the purest products of the chemical industry.

The amino acids contained in the raw sugar juices can be extracted by a new process developed at the Farbenfabrik Wolfen (by Griessbach and his colleagues) by means of 'Wofatit'. If this process were used on a large industrial scale it would be possible to obtain large quantities of glutamic acid which is important in pharmacology. Other nitrogen compounds present include: betain, aspartic acid, leucin, isoleucin, and tyrosine, which it might be possible to extract.

*Sugar from sugar cane*

By far the largest part (2/3 of world production) of the sugar produced comes from sugar cane which is grown in tropical countries, e.g. Cuba, Jamaica. In contrast to the processing of sugar beet, the sugar juice is obtained not in diffusion plants but by crushing the cane in cane crushing mills.

Cane crushing mills consist of three rolls, two bottom and one top roll, which are connected by gears. The top roll is held against the two bottom ones by hydraulic pressure, and the cane passes between them. Several cane mills form an assembly unit. Before they enter this unit hard cane stalks pass through a preliminary crusher.

The extracted sugar cane (bagasse) is used for fuel and is burnt wet in the boiler house. The juice extracted collects underneath the mills and after coarse impurities have been removed it is subjected to a similar chemical purification as beet juice.

New strains produced a great increase in the yield of sugar which led to temporary over-production. Cane sugar molasses is an important raw material for the fermentation and yeast industries.

LITERATURE

F. Brukner, W. Dörfeldt, *Fortschritte in der Zuckerindustrie*, Verlag Th. Steinkopff, Dresden und Leipzig, 1933.

*Technologie des Zuckers, Verein der Zuckerindustrie*, Verlag M. u. H. Schaper, Hannover, 1955.

Ost-Rassow, *Lehrbuch der chemischen Technologie*, 26th ed., J. A. Barth, Leipzig, 1955.

K. Heyns, in: Winnacker-Weingaertner: *Chem. Technologie*, Organische Technologie, II. Hanser Verlag, München, 1954.

P. Andres, Die Rübenzuckerindustrie, *Chem. Z.* 1952, 525, 591.

P. Herrmann, *Laboratoriumsbuch für die Zuckerfabrikation.* Knapp, Halle, 1949.

R. Griessbach, *Austauschadsorbentien in der Lebensmittelindustrie.* J. A. Barth, Leipzig, 1949.

## 2. MANUFACTURE OF STARCH

Starch $(C_6H_{10}O_5)x$ is the most widespread reserve material of the vegetable kingdom. It is formed from carbon dioxide and water during the assimilation of green plants in light. Starch can be detected in the chloroplasts of leaves. The carbohydrates in the plant are transported as water-soluble glucose and deposited in the seeds and roots as starch in spherical grains. Each type of starch has a very characteristic appearance and its origin can be identified under the microscope. In its non-isolated state in the form of agricultural products starch is the most important food for both man and animal. For many purposes it is isolated from these products.

Up to the end of the 1939–1945 war potatoes were the most important raw material for the manufacture of starch in Germany. 3–4 per cent of the total harvest, i.e. 1·5–2 million tons, were processed to starch up to 1944. After the war the production of starch in Western Germany was based almost exclusively on American maize whilst the potato remained the most important raw material in the DDR. In both parts of Germany wheat is also processed to starch. A cheap raw material for the manufacture of starch used to be scrap rice.

Starch is used in many branches of the food and numerous other industries. Pudding mixtures, noodles, and bakery products are made from it, it is used as a filler for pharmaceutical tablets, in the textile industry as a dressing and size. By partial hydrolytic decomposition vegetable glue, dextrin, and paste are obtained, and by complete decomposition glucose syrup, capillary syrup, and glucose.

Starch and starch products are produced in Germany in a large number of small plants which are closely associated with the agricultural industry since the waste products have to be used as animal fodder. In 1938, 157 plants were in existence most of which manufactured dry starch and wet starch, others glucose, dextrin, maize-, rice-, and wheat starch.

## Manufacture of starch from potatoes

The potato contains 16–22 per cent starch which is contained in the amniotic fluid in the shape of well-defined grains. It is important for the manufacture of starch that these grains should be as large as possible and well developed, which is the case only in ripe and healthy fruit. With sugar beet growers aim at a high sugar content, with potatoes at a high starch content and a good grain. Because of the high specific weight of starch the starch content of the potato can be determined approximately by determining its specific weight. Thus, for instance, a density of 1,080 corresponds to a starch content of 13·9 per cent and 1,140 to 26·7 per cent starch. The specific weight is determined by means of Reimann's potato scales.

Before processing for starch the potatoes have to be broken up. After careful clearing in agitator vats they are finely ground in a grating mill consisting of a revolving drum fitted with a large number of saw blades. During grating water is injected. Brush scrubbers then wash the exposed starch out of the grated material which is kept in continuous motion. The grating and brushing corresponds entirely to the action of the housewife in the kitchen when she makes raw potato dumplings. She too isolates the starch grains separating out of the amniotic fluid.

In the brush machines, horizontal arms fitted with brushes revolve on circular copper perforated bottoms. The starch grains pass through the screen as milk. The cell debris and hulls are ejected at the edge. They still contain some starch and have to be crushed once more in a wet grinding stage so as to release the last starch grains. This material is then brushed again. The residue is the 'pulp', an important animal fodder.

The raw starch is then refined. It is forced through fine revolving cylindrical screens to separate the last fibres. The large starch grains are separated from the small ones by 'flooding' in wide wooden or concrete channels in which runs a slow flow of water. This principle of wet preparation by grain size was discussed earlier in the section on the wet preparation of coal. First the large starch grains are deposited. They are marked as 'prime raw starch'. The separation of starch by hydrocyclones is particularly economical. The thin starch suspension is subjected to a gauge pressure of several atmospheres by means of a pump and the flow is introduced tangentially into many small cyclones arranged in series. As in a solid jacket centrifuge the starch grains are forced against the wall and are discharged as concentrate through the conical part at the base. The water overflows at the top. Concentration is carried out in several stages which can be combined with washing. The amniotic fluid together with the small grains runs into collecting pits where the last starch residues are deposited as 'secondary and slurry starch'. Starch can also be separated in separators.

The wet starch contains about 50 per cent water. Drying has to be carried out very carefully since already at 40°C swelling and agglutination take place which have to be avoided at all costs. Part of the water can be eliminated by means of filter centrifuges. The remainder of about 30–35 per cent is expelled by drying by various methods at 30°C. Drying installations used include chambers with drying plates, drying cylinders in which the starch is kept moving by means of a slatted drum, vacuum paddle driers, or Büttner driers. About 20 per cent water is left in the starch since it will always absorb this quantity again.

The 'pulp' which still contains one fifth of the potato substance and 10–20 per cent of starch sealed in the cell substance is reduced to a water content of about 70 per cent by dehydration in pits. In addition to its high cellulose content the pulp also contains proteins. When used as fodder it is mixed with oil cake or bran to improve it.

## Dehydrated potatoes

Dehydrated potatoes and dextrinized potato flour are important as feeding stuffs and for the production of foods. Boiled potatoes are mashed and dried on roller driers, or raw and peeled potatoes are cut into slices and dried in hot air in rotary driers.

## Starch from wheat

The wheat grain consists of the outer shell, the endosperm, and the germ. The endosperm forms the largest part of the material and contains the starch grains tightly packed. This shows in the shape of the grains which are circular and flattened resembling lentils. They are not much stratified and have diameters of 20–30 $\mu$; many are spherical with diameters of 2–8 $\mu$. The starch material is surrounded by a ring of angular grey aleurone cells. Wheat gluten consisting of protein materials is embedded in this aleurone layer but also in the rest of the endosperm. Wheat grains contain much less water than potatoes (13·5 per cent), but more nitrogen, and the starch is much more strongly embedded. The method of extraction of the starch is a consequence of this structure.

Formerly, wheat used to be processed to starch in many small plants by the 'acid process'. The wheat was crushed with water and was then left for lactic fermentation. The gluten became soluble and could be washed off the starch grains. However, the yield was unsatisfactory. The modern method is the 'sweet process' developed in Hungary. The wheat is not crushed but ground to flour, made into a dough with a little water, and thoroughly kneaded in Werner-Pfleiderer kneading machines, the raw starch becoming detached from the gluten-containing material. The viscous dough is then separated into raw starch milk and gluten dough in a trough by further mechanical treatment and with the addition of water. When this mixture is centrifuged in a solid jacket centrifuge, an outer ring forms of pure starch, an inner ring of impure starch, and in the middle, after the centrifuge has been stopped, is the viscous gluten liquid which is discharged. First the inner layer is peeled off, then separately the outer layer which contains pure wheat starch. Then the starch is dried, radial cracks form and the material crumbles into pieces. This is the origin of the name 'radial starch' which is a characteristic of the more valuable grain starch as compared with potato starch. 100 parts of wheat flour yield 60 per cent prime starch, 15 per cent gluten starch, 10 per cent gluten glue. The gluten glue is subjected to a sort of decay process at 50°C, rolled into sheets, and dried. It is used as shoemaker's glue. Glutamic acid is produced industrially by acid hydrolysis of wheat gluten.

## Starch from rice

Even more difficult is the production of starch from scrap rice because the granules are very small and firmly cemented with mucilage. Treatment with water to bring about disintegration is inadequate in this case. Soaking must be carried out with strongly diluted caustic soda of 0·3–0·6 per cent. The granules are so small that flooding is impossible and centrifuges have to be used for the separation.

## Starch from maize (corn)

The decomposition is carried out with sulphurous acid which reacts in a concentration of about 0·3 per cent $SO_2$ at 50°C for 48 hours.

Maize contains an exceptionally large amount of protein. It is composed of 60–65 per cent starch, 4·5 per cent fat, and 10–14 per cent protein. After decomposition with sulphurous acid the protein is contained in the discharged maize liquor, which is, therefore, after concentration a very high-grade animal fodder. Maize liquor is also an excellent culture medium for growing fungi and bacteria. It is, therefore, used as a nutrient fluid for the mould which is grown on a large scale for the production of penicillin. The germs yield the valuable corn oil.

The manufacture of starch from maize corresponds to that from wheat and rice. The production of corn starch increases continually because of the increasing popularity of corn flour for the manufacture of puddings and other foods, and because maize has the highest yield per acre. Very widely known is 'Mondamin' made by the Maizena Gesellschaft.

### Hydrolytic decomposition of starch

$$(C_6H_{10}O_5)\,x + x(H_2O) = x\,C_6H_{12}O_6 \quad \text{(D-glucose)}$$

Dextrin is obtained by partial hydrolysis of starch with diluted acids, e.g. hydrochloric acid or nitric acid, or by roasting in drums at 180–200°C. It is an amorphous intermediate the molecular weight of which lies between those of starch and sugar. It is used as a glue.

Malt sugar is obtained by hydrolytic decomposition of starch with the enzyme diastase, going as far as the disaccharide maltose. A solution of barley malt sugar is concentrated under vacuum and sterilized in airtight boxes under pressure (biomalt). The solution can also be dried to a non-adhesive fine powder in an atomizer drier by the Nubilosa- or Krause processes.

In the U.S.A. the hydrolysis of starch is no longer carried out with malt diastase but mainly with the fungus *Aspergillus oryceae.*

## 3. STARCH SUGAR

Starch sugar is starch decomposed down to the monosaccharide D-glucose. Saccharification is carried out with diluted hydrochloric acid. The solid starch sugar 'Dextropur' is obtained from the syrup of the starch saccharification by atomizing drying or by crystallization. Saccharification has to be carried out in a 15 per cent solution to ensure complete hydrolysis, at an acid concentration of 0·25 per cent HCl and a gauge pressure of 1·8–2·5 atm.

The small amount of acid is neutralized directly with soda; the traces of NaCl formed do not interfere. Solid starch sugar can also be obtained by concentrating the solution under vacuum and crystallizing it by cooling. A micro-crystalline mass of dextrose hydrate forms: $C_6H_{12}O_6 \cdot H_2O$. The pure starch syrup ('capillary syrup') is in many cases used directly, e.g. in the manufacture of sweets, fillings for chocolates, and the production of liqueurs. It is a cheap sugar of mild sweetness and with little tendency to crystallize.

Pure fructose can be obtained only from inulin, the reserve material of various tubers, by hydrolysis with diluted acids. Honey is a mixture of mainly fructose with glucose and a little saccharose. The crystals separating out of honey consist mainly of glucose. If the bees are fed not on fruit sugar (levulose) but on saccharose or starch sugar the honey will contain invert sugar or glucose since the bees possess an inverting enzyme.

# 4. PECTIC SUBSTANCES

Pectic substances the name of which is derived from the Greek pectos, 'solidified, gelatinized', consist of galacturonic acid chains which are partly esterified with methyl alcohol. The basic unit D-galacturonic acid is an aldehyde carboxylic acid:

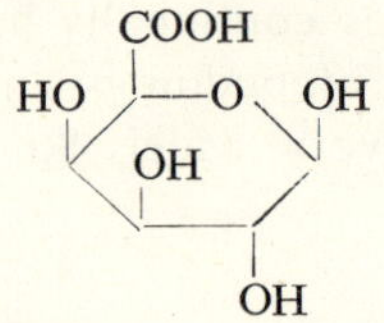


which yields D-galactose by reduction.

The molecules are bound glucosidically as with the polysaccharides. The constitution of pectin differs from that of starch (which will not be discussed here) by the fact that outside the six-membered ring are partly —COOH—, partly $—COOCH_3—$groups instead of the $—CH_2OH$ group, and that it is the D-galactose configuration, not the glucose configuration.

Pectin has a similar role in plants as lignin in lignified parts. It acts as a cement for the cellulose membranes. At the same time it has a regulating effect on the water economy because of its great capacity to swell. Pectin is found mainly in the fruits of many plants, apples, berries, particularly currants, lemons and oranges, but also in beets, in the stem of flax, and in many parts of plants.

In Germany it is produced industrially from the dried residues of cider- and grape juice manufacture. It should also be possible to obtain it from extracted beet chips. To isolate it from the plant parts these are extracted with diluted acids at an elevated temperature. The solutions are clarified and purified. In addition to pectin they also contain polyoses (araban). They are concentrated by evaporation under vacuum, or dried to pectin powder in an atomizing drier.

There are many uses for pectins in the kitchen and in the manufacture of foods as a gelatinizing agent (e.g. 'Opekta'). Pectin is also used in pharmacology, (e.g. as the haemostatic agent 'Sangostop').

LITERATURE

K. Heyns, in Winnacker-Weingaertner, *Organische Technologie, II*, p. 927. Hanser, München, 1954.
W. R. Aehnelt, *Stärke, Stärkesirup, Stärkezucker*: Steinkopff, Dresden, 1951.
Ullmann, *Encyklopädie*, 2nd ed., vol. 9. Urban & Schwarzenberg, 1932.

# III. The fermentation industry

A witty man once said that the essential difference between man and animal was the fact that man knew how to prepare alcoholic beverages and could, therfore, get drunk, and animals could not. (It has since been found that certain insects deliberately get intoxicated by fermented natural juices.) This explains why the manufacture of alcoholic beverages, i.e. the fermentation industry, is one of the oldest and most important trades. Since the earliest times in many different places men have invented quite independently of each other, the fermentation of sugary liquids. Fermentation is also the oldest catalytic process although only organic catalysts are used, enzymes produced by micro-organisms. The fermentation industry and the bio-synthetic processes occupy, therefore, an intermediate position between the chemical industry and agriculture.

With the exception of baker's yeast manufacture, modern fermentation processes operate mostly continuously (e.g. in the production of molasses and sulphite alcohol, and in the U.S.S.R. also potato- and grain-distillation). The great majority of all spirit distillation is operated continuously. Fermentation is a relatively slow process and requires, therefore, a great deal of space, particularly special processes such as citric acid fermentation. Unless very cheap ethylene is available, the production of spirits by fermentation is, nevertheless, the most economical method, much cheaper than e.g. from acetylene. Particularly in countries rich in carbohydrates or wood, alcohol is one of the cheapest chemical raw materials (alcohol from ethylene, see p. 174).

Apart from the fermentation of molasses from beet or cane sugar, starch products are used exclusively since they are cheaper than sugar. The starch has first to be hydrolysed with diastase to maltose. Yeasts possess an enzyme which splits maltose into glucose which is then converted into alcohol and carbon dioxide.

## General remarks on fermentation. Chemical basis

In spite of a great deal of research the chemical course of fermentation is still not completely known. In the presence of yeast fungi sugars decompose, with a yield of 94–95 per cent theoretical, according to the equation: $C_6H_{12}O_6 \rightarrow 2C_2H_5OH + 2CO_2$. This equation had been found as early as 1815 by Gay Lussac. 133 kcal of heat are liberated per kg glucose. Pasteur proved that this process is possible only in the living organism and that sterilization and denial of access to germs from the air prevents fermentation. It was assumed at first that fermentation was linked with the growth of the living yeast cells, the 'formed ferment', and belonged to their life process. In 1897 Eduard Buchner published an account of his sensational experiment in which he had succeeded in fermenting grape sugar (dextrose) with a cell-free yeast press juice obtained by triturating the yeast cells with sand and subsequent pressing. The 'cell-free fermentation' proved that fermentation was not linked to the cell but was caused by a substance which could be separated from it and which was given the name 'ferment' or 'enzyme'. This observation had a revolutionary significance for the ideas prevalent until then about enzymatic processes. Proof had been supplied that this was true catalysis with an organic catalyst. Buchner called the fermentation enzyme 'zymase'. It was discovered later that zymase was not a homogeneous enzyme but a complex of enzymes which affect catalytically the reaction during the individual stages of fermentation.

It is impossible to discuss here in detail the mechanism of fermentation which is very complicated and has been investigated by Neuberg, Harden and Young, Robison, Embden, Meyerhof, Nielsson, Warburg, *et al.* Reference should be made to text books of organic chemistry.

The first phase of fermentation is a phosphorylation (hexokinase) and the conversion of hexose diphosphoric acid (aldolase) into two triose phosphates: glycerine aldehyde and dihydroxyacetone which are in equilibrium. The glycerine aldehyde fraction is withdrawn from the equilibrium by converting it, by means of an enzyme the effective group of which is co-zymase, into phosphoglyceric acid. The two hydrogen atoms obtained in the dehydrogenation are transferred later in the reaction to acetaldehyde, hydrogenating it to ethyl alcohol.

The enzyme 'enolase' causes dehydration of the phosphoglyceric acid to phosphoenol pyruvic acid that supplies the phosphoric acid which is rich in energy and essential for the entire process and which is then saponified to pyruvic acid. C. Neuberg long ago recognized and isolated pyruvic acid as an intermediate. Another enzyme 'carboxylase' decomposes pyruvic acid into acetaldehyde and carbon dioxide: $CH_3—CO—COOH \rightarrow CH_3—CHO + CO_2$. As mentioned above the acetaldehyde is then hydrogenated to alcohol by the hydrogen absorbed by the co-zymase complex.

During fermentation small quantities of glycerine are always formed. Neuberg, Connstein and Lüdecke found that these quantities can be increased by intercepting the acetaldehyde as an intermediate by the addition of sulphite or bisulphite. The hydrogen of the co-zymase complex originally destined or the acetaldehyde is then transferred to the glycerine aldehyde- or dihydroxyacetonephosphoric acid products formed during the first stages, which are reduced to glycerine:

$$CH_2OH—CHOH—CHO + H_2 \rightarrow CH_2OH—CHOH—CH_2OH.$$

The industrial glycerine fermentation process (protol process) is based on this principle.

By-products of fermentation are acetaldehyde, acetals, glycerine, fusel oils, acetic acid, and succinic acid. The latter and the fusel oils (propyl-, butyl-, amyl- and higher alcohols) stem from the amino acids which always occur as admixtures of sugars from natural raw materials and which also originate from the protein metabolism of yeast. In addition to the fermentation metabolism, protein metabolism takes place in yeast as in all organisms and leads to the above-mentioned products.

Certain conditions have to be created to control the course of the fermentation: for its growth and propagation yeast requires potassium, phosphoric acid, magnesium, and nitrogen compounds as amino acids or ammonium salts. The sugar concentration must not be too high, with saccharomyces 12–15 per cent. Only certain wine yeasts tolerate higher concentrations. Yeast is inhibited in its autogenous alcohol when the concentration becomes too high. The optimum fermentation temperature is between 30 and 37°C. Below 5° and above 50°C fermentation is impossible. For special purposes the temperature is frequently kept very low, e.g. for the brewing of beer. Since the fermentation reaction is exothermic cooling must be provided unless adequate dissipation of the heat is possible. There is a very large number of fermentable yeast species. The choice depends on the nature of the substrate used.

In general monosaccharides are too expensive for the production of alcohol and cheap alcoholic beverages. Moreover, the taste of beet sugar fermented to a beverage is not attractive since aromatic substances are lacking. The position is different with crude cane sugar juice or cane sugar molasses from which the very tasty rum (Jamaica) is obtained by fermentation with special micro-organisms and subsequent distillation.

When yeasts grown on rice hydrolysate are used arrack can be obtained from cane sugar. Fruit juices are also fermented to beverages, frequently after sugar has been added, producing wines from berries and fruit. Wine, the noblest beverage, is made from a grape specially grown for this purpose. True brandy is produced by distillation of wine. The name 'Cognac' is reserved for the French product. In blended brandy part of the wine alcohol is replaced by fine spirit (e.g. from potatoes or corn). The replacement, in accordance with laws or regulations, of part of a high-grade beverage by a lower-grade one is called 'blending'; it represents a lawful adulteration to reduce the price but the product must be marked accordingly.

For the industrial mass production of alcohol and alcoholic beverages both sugar- and starch-containing raw materials are used; for alcohol potatoes, corn, and beet molasses (also sulphite spent liquors), for beer only barley. Since yeasts cannot ferment polysaccharides, starch has first to be hydrolysed by starch-cracking enzymes. Most suitable for this purpose is germinating barley which contains the enzyme diastase. This enzyme cracks starch into the disaccharide maltose which consists of two molecules glucose (grape sugar): $(C_6H_{10}O_5)x + \frac{x}{2}H_2O \rightarrow \frac{x}{2}C_{12}H_{22}O_{11}$.

The yeasts contain an enzyme 'maltase' which converts malt sugar into grape sugar: $C_{12}H_{22}O_{11} + H_2O \rightarrow 2C_6H_{12}O_6$.

Fermentation cannot begin until this reaction is complete.

There are also fermentations by bacteria which are carried out on an industrial scale, e.g. the manufacture of lactic acid from sugar by the rod-shaped lactic acid bacteria. To this category belong also the acetic fermentations by acetic bacteria which contain an enzyme system capable of oxidizing alcohol to acetic acid. Special cultures of bacteria for the production of certain solvents are grown and used. Butyl alcohol is produced by butyl bacteria from carbohydrates. As mentioned earlier, in the U.S.S.R., the U.S.A. and other countries an industry for the production of solvents, e.g. butanol, isopropanol, and acetone, from corn and maize is based on this principle. Butyric acid and citric acid fermentation by mould fungi are also carried out on an industrial scale. Glycerine ester fermentation was discussed earlier.

## 1. THE BREWING OF BEER

The production of beer from barley is divided into three stages:

(1) malting, the germination of barley during which diastase is formed, and kiln-drying by heating;
(2) production of the wort by mashing of malt, saccharification, boiling, filtration, and addition of hops;
(3) fermentation and storage.

### *Malting*

Malt factories operate as independent enterprises geographically separate from the breweries. The cleaned and graded barley is first steeped in water for 2–3 days in conical vessels, the steep tanks, at 15–20°C. The steeped barley is drained and transferred to the malt floors, well-aired spaces which have to be kept as cool as possible. Here the barley is spread in a layer of about 30 cm thickness. After only one day germination and spontaneous heating start. To obtain uniform growth the barley has to be turned frequently with a shovel. The temperature is 15–20°C.

The acrospire grows underneath the shell and emerges at the tip of the barley grain. At the opposite side the longer radicles are formed. Germination is complete after 7 to 9 days. Although by then the optimum of diastase formation has not been reached, the quantity produced is sufficient for the production of beer. During germination part of the starch is consumed, that remaining unconsumed is converted into sugar and alcohol during the following fermentation. For the distillation of alcohol germination is allowed to proceed longer in order to obtain the maximum possible diastase content. According to the time germination has taken, the malt is either 'short malt' or 'long malt'. The barley which after germination is called green malt contains about 50 per cent water.

In order to save the manual labour of turning the barley on the malt floor, malting is carried out in several plants in large drums which rotate slowly and through which passes moist air to cool the barley.

Then follows the kiln-drying of the malt. Its object is to interrupt the germination process, to dry the malt, and by slightly roasting it to introduce into the beer the aromatic substances and colouring matters. The type of kiln-drying used determines the character and taste of the beer.

Light beers are obtained from malt dried at a low temperature, dark ones from malt dried at higher temperatures. The dark colour is due to caramel formation during drying. Drying is carried out on roasting hurdles consisting of slotted metal sheets on which the malt is spread in layers of 10–30 cm thickness. Heat is supplied by hot air, the malt being turned frequently. During heating and turning the radicles which affect the taste unfavourably and contain little diastase drop off. They are screened and used as fodder. During drying the high temperature causes a weakening of the diastase enzyme which is a desirable feature. The degerminated cured malt is stored for several weeks; during this time the shells of the grains become more pliable which facilitates grinding in bruising mills for the mash and permits better extraction.

Malt can also be produced from rye or wheat; germinated oats contain too little diastase. The moisture and spent air escape from the drying ovens through a stack which is heated internally by a pipe conducting away the waste gases from the firing zone. A characteristic feature of malt factories are the many short wide stacks on the roofs of the drying kilns, fitted with covers to protect them from the rain and with rotary wind caps. The smell of the aromatic substances of the malt is usually noticeable in the vicinity.

*Production of the wort*

The wort is produced by 'mashing' (mixing) the crushed malt with water and by converting the starch into soluble materials. For this purpose two processes are required: the gelatinization of starch for which a temperature of at least 75°C is necessary, and the saccharification by diastase the optimum temperature for which is about 60–68°C. These two processes are essentially mutually exclusive. In order to fulfil both requirements the steps shown in Figure 120 are taken.

Part of the contents of the mash copper, e.g. one third, is removed several times, boiled in the mash pan and returned into the copper. Thus the favourable saccharification temperature in the mash copper is achieved whilst retaining an adequate amount of effective diastase. This process is carried out repeat-

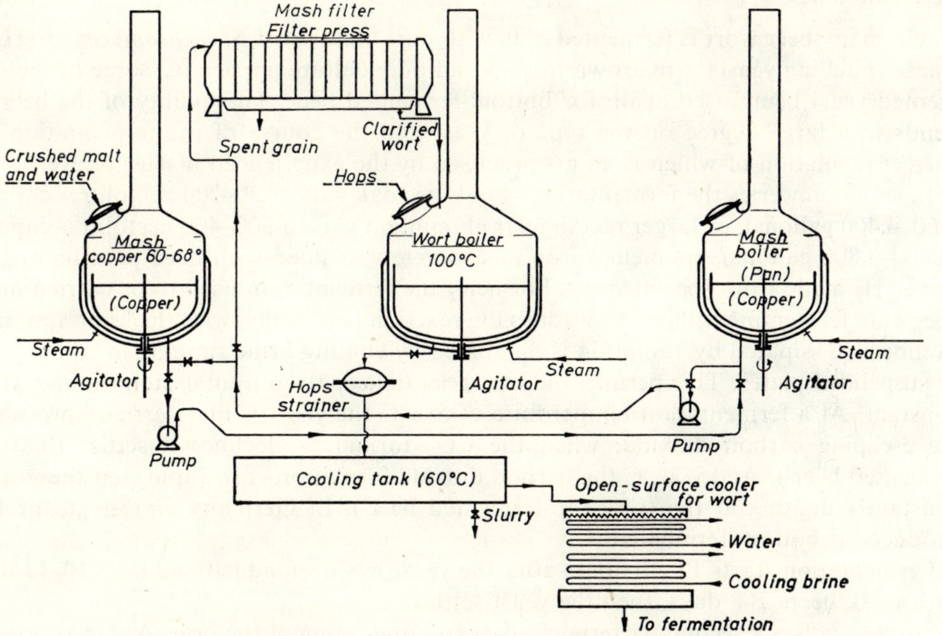


Figure 120. Diagram of the production of wort for beer

edly. The arrangement of the apparatus for the production of wort in the brew house is shown in Figure 120.

The equipment consists of a mash copper, a mash tub also made of copper, a straining vat, or a mash filter and a wort kettle. When the contents of the mash tub are completely saccharified by repeated treatment in the mash copper and by the action of the diastase, they are filtered through the mash filter. This may consist of a settling tub or a filter press. Here the insoluble components, the spent grain, are filtered off and washed out with hot water. They are used as cattle fodder. The wort passes into the wort boiler where it is mixed with hops (about 300 g per hectolitre) and boiled for 1·5 to 2 hours. The precipitation of proteins can be observed through an inspection glass. The wort then turns clear, it 'breaks', and simultaneously sterilization takes place and destruction of the enzymes which might later cause a runaway fermentation. The wort is conveyed via the hop strainer into the cooling tank where it is cooled to about 60°C and where the slurry settles. It is then passed into an open-surface cooler for it is necessary to cool it as rapidly as possible in order to avoid the dangerous temperature range between 20 and 50°C in which harmful schizomycetes develop quickly. In the open-surface cooler oxygen, required by the yeast for growth, is introduced into the wort. The upper part of the cooler is filled with cooling water, the lower part with ice water or cooling brine.

The concentration of the wort is measured by means of the saccharometer designed by Plato (formerly Balling). The content of extractive substances at equal density corresponds approximately to the content of a cane sugar solution. It is called 'original wort'. Light beers have an extract (original wort) content of 10 per cent, Bavarian ones 12–14 per cent, Bock and Salvator 16–18 per cent. Water has a marked effect on the character and quality of the beer. It has been found that in spite of the use of the same malt, the same equipment, mode of operation, and type of yeast it is never possible to produce the same beer at different places.

*Fermentation*

In Germany beer wort is fermented only with pure cultures of *Saccharomyces cerevisiae*. These 'pitching yeasts' are grown in a special pure culture apparatus, some of the sedimented yeast being used again for bottom-fermented beer. The quality of the beer depends to a large degree on the type of yeast and the course of the fermentation, the correct regulation of which is an art practised by the experienced master brewer.

In small breweries the fermentation vessels are oak vats of 30–200 hectolitres capacity (660–4,400 gallons), or larger rectangular aluminium vats of 300–400 hectolitres capacity (6,600–8,800 gallons), enamelled iron or concrete vats lined with hard paraffin or tiles, or vessels made from special steels. Formerly the fermentation used to be carried out in deep cold fermenting cellars. Nowadays the reaction heat which with the large apparatus cannot be dissipated by radiation is dissipated by cooling brine running in cooling coils or suspended plates. This permits the breweries to keep the output in summer and winter constant. At a fermentation temperature of 5–8°C the yeast is first carried upwards by the escaping carbon dioxide; when the $CO_2$ formation declines it settles ('bottom-fermented beer'). At 15–22°C the carbon dioxide formation is so rapid that the yeast is constantly floating at the top ('top-fermented beer'). In Germany almost all the beer produced is bottom-fermented.

Fermentation starts 12–20 hours after the yeast has been added and lasts 10–12 days, with weak beers 7–8 days, then the yeast settles.

Storage causes a secondary fermentation and upgrading of the beer. As far as possible it is stored in cooled cellars at 0–3°C. High-grade 'lager' beer is stored for 2–3 months, light draught beer 3–6 weeks.

Formerly, pitched storage casks were used generally. The large breweries store the beer now in large tanks and after storage it is decanted into smaller shipping casks. The beer is in this case lightly filtered once more without separating the colloids. Then the casks are sealed. To keep the beer fresh 0·3–0·4 per cent carbon dioxide is left in which is noticeable by the aroma when the beer is drawn directly from the casks. Ordinary beer has an alcohol content of 3·5–4 per cent, heavy beer such as 'Salvator' 4·8, 'porter' 6·7 per cent. Output of this truly popular beverage is enormous. In 1937/38, 776,958 tons cured malt, corresponding to 1 million tons barley, were consumed in Germany, and 43,676 breweries, mostly small ones, were in operation. Compared with former years the output and consumption of beer has dropped sharply. In Germany, output amounted to 1,369 million hl (30,118 million gallons) in 1912 and in 1937 to 43·6 million hl (960 million gallons), i.e. about 55 litres (97 pints) per head annually. In 1960, the total output of beer in both parts of Germany was 60·7 million hl (1,335 million gallons), i.e. 85 litres (150 pints) per head. Formerly, more beer used to be drunk since in the years 1898–1900, 118 litres (208 pints) were consumed per head of population. If it is considered that the share of women was smaller and that of children certainly negligible, this represents a remarkable quantity for the male population.

## 2. THE MANUFACTURE OF ALCOHOL

The ancients knew intoxicating beverages, but not alcohol. It was known that the 'intoxicating principle' evaporated during boiling but how to capture the volatile 'spirit' (sprit) remained a mystery. To produce aqueous alcohol became possible only when distillation was invented during the 12th century. It was given resounding names such as Aqua vitae or Spiritus vini and unlimited curative effects were ascribed to it. The residue was called 'phlegma'. The name 'alcohol' was coined by Paracelsus.

At the end of the Middle Ages, grain- and potable spirits were produced. Absolute alcohol was obtained for the first time in 1796 by the distillation of aqueous alcohol via potash and lime. During the years 1897–1902 the alcohol consumption in Germany per head of population amounted to 4·2 litres (7·4 pints), in 1935–1936: 0·98 litres (1·73 pints), calculated in litres of spirit of wine (1 W = 100 per cent). It will be seen that consumption has been declining even more strongly than that of beer.

Apart from the use of alcohol as a beverage it has also been employed as a fuel, formerly as an admixture to motor fuels, as a solvent in the pharmaceutical, cosmetic, and chemical industries, e.g. for re-crystallization, and as a chemical raw material for the manufacture of ether, acetic acid, ethyl acetate, acetaldehyde, and ethylene. In the U.S.S.R. and the U.S.A. it is also processed directly to butadiene. In 1933–1934 4·8 litres alcohol (8·4 pints) per head were used industrially.

Alcohol can be obtained synthetically from acetylene by the addition of water to acetaldehyde and reduction, and from ethylene by the addition of water. Alcohol from acetylene is relatively expensive and is produced by this method only in emergencies. On the other hand, alcohol is produced in the U.S.A. from ethylene so cheaply that the fermentation industry is economically threatened.

In Germany alcohol is produced almost exclusively by fermentation. In addition to the industrial production in molasses and spent sulphite liquor distilleries, baker's yeast factories and large potato and grain distilleries there are also numerous smaller agricultural distilleries where starchy raw materials are processed. The crude spirit produced is rectified in purification plants yielding various qualities of alcohol (finely-filtered alcohol, first-quality alcohol, second-quality alcohol, technical alcohol). When potatoes and grain are used distillation leaves a residue, the vinasse which has long been used as a valuable fodder (increase of milk and meat yields). In this way the raw materials used, with the exception of the fermented carbohydrates, return eventually to the soil as manure. The small agricultural distilleries are, therefore, of considerable importance to the national economy.

Grain spirits can be produced from crude grain alcohol after careful rectification which must be conducted so as to retain the aromatic substances from the grain in the fine alcohol. Fine grain alcohol is used also for various other spirits (gin, geneva). Whisky is a grain spirit which obtains its aroma and colour by long storage in slightly charred oak casks (America) or by the use of smoke-dried malt for the mash (Scotland). Original vodka is purest potato or grain spirit reduced to drinkable strength (U.S.S.R.). Fruits of all sorts are also fermented, often in very small plants (Southern Germany, Czechoslovakia, France). A separate branch is wine distillation.

Distillates produced from wine yield brandy; the name 'Cognac' may only be used for French products from the Charente wine-growing district.

Alcoholic beverages produced overseas include rum (Jamaica) and arrack (Java, Ceylon) fermented from cane sugar molasses either alone or with rice as raw material.

A general principle applying to spirits is that they are improved by storage, the taste and bouquet substances being 'rounded off'. French cognacs for instance are stored for 4–5 years, special top qualities for up to 25 years. Whisky is stored for 4–12 years.

The Exchequer has always been interested in alcohol. Whether in the form of import duties, excise duties, taxes, or other charges, a way was always found to obtain revenues. In Germany, the mashing room tax (1920), the brandy contingent (1887), the brandy monopoly and the distilling licence (1922) belong to this category. The latter two are still valid in West Germany, where all alcohol stills and collecting vessels are strictly supervised by the State and the removal of alcohol from the plants is rendered impossible by lead-sealing all screws and taps. Only for analysis purposes a controllable small quantity can be withdrawn by means of a special tap which is fitted with a counter. Alcohol taxes differ according to the use to which the alcohol is to be put. The same is true for denaturants for alcohol which is liable to a reduced rate of tax. Denaturing is usually carried out by excise officials with certain permitted denaturants on the consumer's premises.

Because of the tax position alcohol is not very popular as a raw material and solvent in chemical plants. All movements, decanting, additions, and withdrawals of untaxed alcohol are checked by excise officials. This applies particularly to chemical plants which have to use undenatured alcohol taxable at lower rates such as in the production of ether and ethylene. The German brandy monopoly law uses the term 'wine spirit' for 100 per cent ethyl alcohol as the unit for calculation, and the term 'brandy' for all alcohol/water mixtures. For the purpose of taxation the concentration is determined by measuring the density by means of the alcoholmeter, a very accurate hydrometer with a built-in thermometer. The wine spirit content can be read off from very accurate tables of density, taking into account the temperature. Only when the spirit has been denatured and rendered useless for drinking purposes is it free from state control. For industrial and chemical purposes various denaturants are used according to the intended use, e.g. benzene, toluene, petrol, acetic acid, chloroform, ether, oil of turpentine, etc. For fuel alcohol 2·5 per cent by volume of raw wood spirit and 1 per cent by volume pyridine bases are added. Germany's net revenue from the brandy tax in 1937/38 amounted to RM 285 mill.

In the German Democratic Republic the monopoly law and the brandy quota system have been abolished. The factories account to the Tax Authorities. The amount of alcohol to be produced, like that of other goods, is laid down as a production target within the framework of the Economic Plan.

Two technological examples are given below of the manufacture of alcohol: potato distillation and molasses distillation.

## (a) Potato distillation

As in brewing, the fermentation substrate in potato distillation is glucose. It is, therefore, necessary to convert the polysaccharide starch contained in the potato tubers first into a soluble form and then to decompose it by the action of malt amylase via several intermediate stages down to the disaccharide maltose which can be used for the yeast. The cracking of maltose to glucose is effected by the yeast itself by means of the α-glucosidase produced by it. In the fermentation phase the glucose is then converted into the end-products desired: alcohol and carbon dioxide. The alcohol produced is finally separated from the fermented mash by distillation and is obtained as crude spirit.

Potato spirit is produced almost exclusively in smaller plants which are closely linked with agriculture through their by-products vinasse and have a daily output of 6–15 hl (132–330 gallons) crude spirit. The process comprises five stages:

(1) malting,
(2) steaming of the potatoes,
(3) preparation of the mash and of the yeast,
(4) fermentation,
(5) distillation.

### *Malting*

In contrast to brewing, in distillation the important consideration is the saccharification of a large amount of starch stock by means of a little amylase from malt. Use is, therefore, made by preference of a long malt rich in enzymes, obtained from small-grain fodder barley with a high nitrogen content which is left to germinate for 12–15 days on the malting floor. To preserve the enzyme complex intact it is not kiln-dried but is used immediately as green malt. For 100 kg potatoes 1·5–2 kg barley = 2·0–3·0 kg wet malt is required. Of this, 0·3–0·4 kg is used as a nutrient addition to the yeast culture; the remainder is used for saccharification.

*Steaming of potatoes*

Since the potato starch is still enclosed in the plant cell complexes, heating the tubers to the gelatinization temperature of their starch (65°C) is not sufficient to render them susceptible to the action of the amylase. This is done more successfully when the raw material is heated to a higher temperature (95–100°C). Formerly large wooden vats were used in which the potatoes were boiled at atmospheric pressure by steam. The potatoes were then crushed in crushing cylinders and saccharified with malt. Nowadays conical-cylindrical pressure vessels made of steel sheet are used exclusively which are called after their inventor 'Henze' and are equipped with the necessary safety devices. In these steamers the potatoes which have first been thoroughly washed are disintegrated at steam pressure and are heated to nearly 143°C.

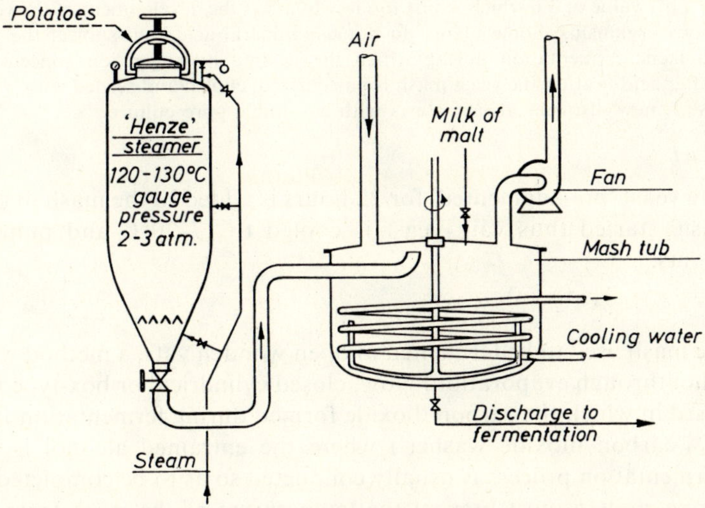


Figure 121. Steaming and washing of potatoes for the production of alcohol

Henze steamers (Figure 121) are built with capacities of between 2 and 6 $m^3$; they hold 1,300 to 4,000 kg potatoes. Steam can be introduced as required at various points. The filling hole, usually a manhole, for filling the potatoes is in the top plate of the 'Henze'. At the conical tip at the bottom is the sluice valve for emptying which is connected to the discharge pipe leading to the mash tub.

The steaming process takes about 60 minutes. When the potatoes are disintegrated the sluice valve is opened to admit them into the mash tub. As a result of the sudden pressure release the cell walls of the potatoes burst, the cell contents emerge and the amylase can begin to act on the dissolved starch.

*Preparation of the mash and yeast*

The mashing process in the distillery is carried out in a steel or copper tub fitted with a mechanically driven agitator and a cooling coil. The tubs are designed as flat cylinders with convex bottoms and capacities of 2,000–10,000 litres. First, some milk of malt is introduced into the mash tub. Then the contents of the Henze steamer are injected over a period of 30–45 minutes. The temperature is kept at between 52 and 56°C. Further quantities of malt are added in batches. Finally, the temperature is raised to 60–61 °C to pasteurize the contents. 20–30 minutes later cooling starts of the saccharified and highly fluid mash.

*The Delbrück method of preliminary augmentation of the yeast to mash*

To start fermentation, a top-fermented distillery yeast culture which has been strengthened in a separately treated yeast mash is used. In the preparation of the yeast Delbrück's principle of natural pure culture is applied so as to prevent the growth of contaminating organisms.

After completion of the saccharification about 5 per cent of the mash is removed from the tub and poured into small wooden tubs of 200–300 litres capacity or, in larger plants, into a yeast mash apparatus resembling the mash tub; malt is added to improve the nutrient content, and diluted sulphuric acid to acidify to a pH value of 3·6 which whilst too low to affect the development of the yeast is too high for many infectious organisms. Some plants do not use mineral acids but subject the yeast mash for twenty hours to lactic fermentation during which the desired hydrogen ion concentration appears automatically. After acidification the yeast mash is pasteurized, cooled and seeded with 'ripe yeast' from the previous day. A new distillery always starts with a suitable pure culture.

*Large-scale start*

During cooling yeast 'pre-augmented' for 22 hours is added to the mash in the mash tub. The 'new mash' started thus with yeast is cooled to 17–20°C and pumped into the fermentation vats.

*Fermentation*

Originally, the mash used to be fermented in open wooden vats, a method which entailed losses of alcohol through evaporation. Now, closed cylindrical or box-type iron fermenting vats are used in which the carbon dioxide formed during fermentation is led through a water bath ('carbon dioxide washer') where the entrained alcohol is collected for re-use. The fermentation process is usually conducted so as to be completed in 72 hours. During this time cooling must prevent the temperature of the mash from rising above 28–30°C.

Yield: Theoretically, 76·1 litres alcohol should be produced from 100 kg starch; the maximum yield is, however, 64 litres. 0·5–1 per cent starch is not disintegrated, 1–2 per cent not fermented, the remainder is consumed in other organic reactions.

*Distillation*

Crude alcohol is usually separated from the fermented mash in continuously operating single-column stills. In their lower, stripping, part these contain 10–12 bubble trays and in the upper, concentration part, 4–6. The apparatus is closed at the top by a dephlegmator which acts as a heat exchanger and in which the fermented mash is pre-heated and the issuing alcohol-steam mixture partly condensed and thus concentrated once more (see Figure 32).

The pre-heated mash passing over the mash floors through overflow pipes is completely stripped of alcohol on its way downwards by countercurrent steam and issues from the bottom tray as slops (vinasse). The vapours rich in alcohol leaving the dephlegmator are condensed in a coil or frame condenser and are collected as crude spirit consisting of about 85–88 per cent by weight alcohol, and some aldehyde, ester, fusel oil, and water.

From an apparatus equipped with a special concentrating column a product can be obtained from which most of the fusel oil has been extracted and which contains about 94 per cent by weight alcohol.

Fusel oil is the name of a group of by-products of alcohol fermentation which are contained in crude alcohol in quantities of 0·2–0·5 per cent. Fusel oil is not a homogeneous substance but a mixture the main components of which are the 'higher alcohols', in particular n-propyl alcohol, isobutyl alcohol, and amyl alcohol.

### (b) Distillation of molasses

In contrast to potato distilleries molasses distilleries are usually large works with a daily output of more than 100 hl (2,200 gallons) alcohol (in the Argentine one plant has a daily output of 2,500 hl (55,000 gallons)). In modern plants molasses is not used exclusively for the production of alcohol but all by-products are processed. Fermentation carbon dioxide is collected and either bottled in steel cylinders or made into solid 'dry ice'. The slops are utilized either biologically for the protein synthesis (manufacture of nutrient or fodder yeast), or chemically for the production of potash, ammonium sulphate, and crude potash for fertilizer, sodium cyanide for hardening steel, betain hydrochloride, succinic acid, etc.

This raw material does not require any treatment other than dilution of the alkaline beet molasses, neutralization, and acidification to the optimum pH values for the individual fermentation stages since the utilizable carbohydrates occur almost exclusively as saccharose and can, therefore, be converted directly by yeast.

In the past 50 years the fermentation methods have undergone various changes; the following systems are used:

(1) the filling method (process using a homogeneous mash),
(2) the part mash method (process using mashes of different concentrations),
(3) the feed method (in which concentrated mash is fed continuously into a mash with low sugar concentration until a fermentation vat is completely filled),
(4) the yeast recycle method.

This last process gives the highest alcohol yield since hardly any sugar is consumed for the formation of yeast cell substance. The method consists essentially of the separation of the yeast from the fermented molasses mash and immediate re-use to start fresh mash after exposure to an acid bath for four hours to rid it of infectious organisms. This 'Usines-de-Melle' process developed first in France is, however, not yet being used in the molasses distilleries of the German Democratic Republic because of the high capital costs of the separators required.

*Dehydration of alcohol*

For many industrial purposes and as an additive to motor fuel, alcohol has to be dehydrated since water would separate out during mixing with petrol and other solvents insoluble in water. On a large scale chemical dehydration with quicklime is too expensive and too complicated. In the 'Hiag' process dehydration is carried out with fused sodium-potassium-acetate which is regenerated. More advantageous and economical is azeotropic dehydration by distillation with an auxiliary liquid which gives a constant boiling mixture with alcohol and water.

This mixture is distilled in an efficient column. After condensation it separates in a separator (principle of the Florentine receiver) into an aqueous phase and a layer consisting of alcohol and the auxiliary liquid. The alcohol-solvent mixture is continuously returned into the still. This process can be operated both continuously and batchwise.

A suitable auxiliary liquid is, e.g., benzene which forms a ternary mixture of 18·3 per cent by weight alcohol, 7·7 per cent water, and 74 per cent benzene, with a boiling point of 64·86°C. With carbon tetrachloride a mixture forms of 10·3 per cent alcohol, 3·4 per cent water, and 86·3 per cent $CCl_4$, boiling point 61·8°C. The dehydration method of the Monopoly Administration uses carbon tetrachloride or trichloroethylene as an auxiliary liquid.

The former IG Farbenfabrik Wolfen (Wallis) developed a dehydration method which was used in a large plant at Leverkusen and in Sweden. It is based on the possibility of binding the water of the alcohol with carefully dehydrated gypsum. The alcohol was treated with gypsum in a continuously operating screw apparatus of very original design, the gypsum was then separated, the absorbed alcohol expelled by heating, and the gypsum then dehydrated in a special dehydration screw and recycled.

## LITERATURE

Ost-Rassow, *Lehrbuch der chemischen Technologie*, 26th ed., J. A. Barth, Leipzig, 1955.

Dimair, Koch and Aman, Die Gärungsindustrie, in Winnacker-Weingaertner, *Organische Technologie*, II. Hanser, München, 1954.

Foth-Drews, *Die Praxis des Brennereibetriebes*. Verlag Parey, Berlin, 1951.

H. Lüers, *Gärungsgetränke*, Verlag Hans Carl, Nürnberg, 1949.

H. Lüers, *Grundriß der Bierbrauerei*, Verlag Parey, Berlin, 1949.

E. Bergander, *Biochemie und Technologie der Hefe*, Verlag Th. Steinkopff, Dresden, 1959.

# IV. Technology of fats, oils, fatty acids, and soaps

## 1. GENERAL NOTES ON FATS

### (a) Formation of fats

Fats and fatty oils are products of the animal and vegetable kingdoms. Fatty acids can also be obtained by paraffin oxidation, oxosynthesis from olefins, and by polyenol-condensation from acetaldehyde, and then be esterified with glycerine to fats. Plants form fats by the assimilation process and store them together with the proteins and carbohydrates as reserves in the seeds. Fats constitute a reserve material for men and animals, too.

Body fat can be produced from food fats, carbohydrates and proteins by decomposition and reconstitution. It is deposited under the skin, on the inside wall of the stomach, and in the intestines.

### (b) Digestion of fats

Formerly it had been assumed that during digestion fats are decomposted completely into glycerine and free fatty acids. It is now known that a mixture forms in the intestines of free fatty acids, mono-, di-, and triglycerides in which process a special physiological part is played by the strongly surface-active monoglycerides as emulsifiers for the digestion of fats. Other emulsifiers for fats are bile acids. In spite of numerous investigations details of the mechanism of fat resorption are still not completely understood. The only fact known for certain is that in the cells of the mucosa of the intestines triglycerides are re-formed from the resorbed fatty mixture. Most of the glycerine required comes from the intermediary metabolism and only a small proportion from the food. After resorption the fatty acids appear always as triglycerides. If, e.g., fatty acid ethyl ester or fatty acid monoglyceride are used as fodder triglyceride is formed immediately in each case. It is even possible to use soap as fodder. Only a small part of the resorbed fatty acids is transported by the blood vessels. The greater part is transported in the chylus, the fluid of the lymphatic vessels of the digestive tract. If excessive quantities of fat are introduced into the organism foreign fat can be deposited directly in the fat storage tissues. Cell fat proper is, however, always of the same kind as the organism. The human organism requires certain essential fatty acids. Production of fat micro-organism see p. 308.

### (c) Chemical composition and properties of the fats

The main constituent is glyceride- or neutral fat; associated with it are the lipoids. These are usually present in much smaller quantities.

Most fats consist predominantly of mixed triglycerides of the acids $C_{16}$ and $C_{18}$. Some solid vegetable fats (cocoa butter, shea butter, Borneo tallow, etc.) contain mainly saturated fatty acids $C_{16}$ and $C_{18}$ and unsaturated $C_{18}$ acids. Others (palm-kernel oil, coconut oil) contain large quantities of short-chain saturated fatty acids ($C_6$—$C_{14}$), principally lauric acid. Liquid vegetable fats (sunflower oil, cottonseed oil, soya oil, etc.) consist mainly of unsaturated fatty acids of the $C_{18}$ series (oleic acid, linoleic acid). Rapeseed oil, mustard oil, and peanut oil contain many long-chain fatty acids, e.g. 40–50 per cent erucic acid ($\Delta_{13}$-docosenic acid). Animal fats contain in addition to small quantities of unsaturated fatty acids ($C_{18}$) mainly saturated ones with 16 and 18 C atoms. In milk fat (butter) on the other hand relatively large quantities of low-molecular fatty acid (from $C_4$ onwards) are present. Fish oils are characterized by a sometimes quite high content of polyenic acids, fatty acids with 20 and more C atoms being found occasionally.

Castor oil contains the unsaturated hydroxy fatty acid ricinoleic acid:

$$CH_3—(CH_2)_5—CHOH—CH_2—CH = CH—(CH_2)_7—COOH.$$

Drying oils, such as linseed oil, poppyseed oil, wood oil, oiticica oil, etc. contain fatty acids with several double bonds (e.g. linoleic acid):

$$CH_3—(CH_2)_4—CH = CH—CH_2—CH = CH—(CH_2)_7—COOH$$

and linolenic acid:

$$CH_3—CH_2—CH = CH—CH_2—CH = CH—CH_2—CH = CH—(CH_2)_7—COOH.$$

These oils are important in the manufacture of paints as bonding agent for pigments and for forming films.

Results of investigations in recent years have shown that contrary to earlier theories natural fats (with certain exceptions which will be discussed separately) contain also small quantities of odd-numbered, branched-chain, conjugated-unsaturated and transisomeric fatty acids.

Fats can be saponified most easily with alkalis:

$$C_3H_5(OCOC_{17}H_{35})_3 + 3NaOH—C_3H_5(OH)_3 + 3C_{17}H_{35}COONa.$$

Alkaline earths, lead oxide, and strong sulphuric and hydrochloric acids are other saponifying agents. Finally, saponification can also be carried out industrially with enzymes, the lipases, e.g. from castor beans. In this case, however, an equilibrium is established in which 5–10 per cent of unreacted fat is present. Saponification is, therefore, carried out in stages.

Conversely, lipases also build mono-, di-, and triglycerides from fatty acids and glycerine.

Fats can be distilled without decomposition only under a high vacuum, preferably using molecular distillation. If superheated steam is employed under vacuum partial saponification occurs. On the other hand, fatty acids can be distilled under vacuum without decomposition.

Pure saturated fats have neither taste nor smell, and keep well. Natural fats, however, are never free from impurities, particularly those that contain nitrogen. Impurities accelerate the tendency of fats to turn rancid under the influence of light and atmospheric oxygen.

The decomposition of fats is partly an autoxidation process in which the carbon chain disintegrates. Particularly liable to this process are the unsaturated fatty acids because they react very easily with oxygen. At first, the oxygen does not add to the double bonds but to a $CH_2$ group on one side of the double bond. Then follow rearrangement and cracking. If several double bonds are present the positions at which these occur change. The processes are very complicated and cannot be discussed here any further. Closely related to the process of fats turning rancid is the drying of oils in paint technology. In addition to oxidation, polymerization and poly-condensation also occur; these processes are important in the formation of films.

Fats can be protected against turning rancid by anti-oxidants. This is of practical importance for the preservation of edible fats and the method is used on a large scale in other countries. Natural fats are partly protected by natural anti-oxidants. These are removed during purification which accounts for the fact that natural fats often keep better than purified ones. However, the decay of fats is not only an autoxidation process during which foul-smelling decomposition products, ketones, aldehydes, and low fatty acids, are formed but enzymatic processes, saponification and decomposition by bacteria and fungi also take place, particularly with fats that contain water (butter, margarine). (Inhibitors see p. 395).

The decay of fats is accelerated appreciably by external influences, such as temperature, light, and air, and the presence of oxidizing agents. Apart from chlorophyll and haemin, even traces of heavy metals, particularly iron and copper, which may come from the equipment used for the production of the fats have a strong oxidizing effect. From this follow inevitably the counter-measures to be used in practice to improve the keeping properties of the fats, i.e. maximum exclusion of the effect of light and air, prevention of avoidable rises in temperature, and of contact with oxidizing compounds during production and storage.

### (d) Lipoids

A number of substances which in their chemical structure are only distantly related are collectively known as lipoids because they share several properties, e.g. solubility in organic solvents such as ether, acetone, petrol, benzene, carbon tetrachloride, etc. These compounds occur often in association with neutral fats. Lipoids include waxes, sterols, phosphatides, and carotenoids.

*Waxes*

Waxes are secreted as by-products by animal and vegetable organisms. Their function is frequently the protection of the surface and they are not reserve materials. They are fatty acid esters of the higher monovalent alcohols. Alcohols involved are aliphatic, e.g. myricyl alcohol in beeswax, $C_{30}H_{61}OH$, esterified with palmitic acid (about 75 per cent). Carnauba wax used for high-grade shoe creams and floor polishes to which it gives the gloss is derived from a Brazilian palm tree and is essentially a cerotic acid myricyl ester, $C_{25}H_{51}COOC_{30}H_{61}$. Grafting waxes contain resin alcohols (tannols).

Animal waxes contain cholesterol, $C_{27}H_{45}OH$, which is the main constituent of wool grease in the form of a fatty acid ester. 'Adeps lanae' widely used as a base for ointments ('Lanolin') is produced from sheep's wool grease. Vegetable waxes contain phytosterols which resemble cholesterol.

Sterols are contained in the unsaponifiable fraction of fats, e.g. ergosterol, $C_{28}H_{44}O$, from which vitamin D can be obtained by irradiation (e.g. of milk). Many fats contain vitamins soluble in fats. Cod liver oil is particularly rich in vitamins A and D.

*Phosphatides*

The phosphatides are very important constituents of many fats; they include the lecithines consisting of glycerine di-fatty acid esters in which the third alcohol group of the glycerine is esterified with the residue

$$\begin{array}{c} O \\ \| \\ HO{-}P{-}OCH_2{-}CH_2{-}\overset{(+)}{N}(CH_3)_3 \\ |^{(-)} \\ O \end{array}$$

i.e. with choline phosphoric acid. They are obtained from soya oil during refining and are used as emulsifiers for fats in water. They are also said to have a very favourable influence on the organism, particularly the nervous system, and are used as therapeutics.

## 2. PRODUCTION OF FATS

The following table shows the world production of the most important nutrient fats (just over 80 per cent) and fats used for industrial purposes, barely 20 per cent. Central Europe's relative fat requirements are 2–3 times larger than the world average and are met to only about half by home production.

### (a) Production of vegetable oils

Vegetable oils and fats can be isolated by two methods: mechanical pressing, or extraction with solvents.

Unlike sugar refineries oil mills operate continuously throughout the year, not seasonally. With few exceptions the seeds keep in general for at least six months. The important consideration is the provision of ample storage space at the factories to receive the tens of thousands of tons of oil seeds arriving together.

*Pressing*

The oil seeds are first crushed in roll crushing mills, any adhering impurities having been removed by screening and iron particles by magnets. Preliminary pressing is carried out in continuously operating screw presses of similar design to mincing machines or kitchen fruit presses. A tapering spindle revolves in a narrowing press chamber the walls of which are perforated to allow the oil to pass through. The yield obtained is incomplete, the oil content drops e.g. from 40 to 26 per cent. The diagram in Figure 122 shows the mode of operation of oil production by pressing in two stages.

Before being pressed the crushed oil fruits are heated in pans. After the first pressing the oil seeds are crushed again, first once more in a roll crushing mill, then in a five-roll frame; the two upper rolls are corrugated. After this very thorough crushing, pressing is carried out batchwise in hydraulic presses at a gauge pressure of 130–300 atm to remove

World fat production (in 1,000 tons)

| Type of fat | 1934–1938 (average) | 1948–1952 (average) | 1957 | 1965 (provisional) |
|---|---|---|---|---|
| *Food fats* | | | | |
| Cottonseed | 1,254 | 1,439 | 1,819 | 2,570 |
| Ground nut | 1,618 | 1,628 | 2,221 | 3,165 |
| Soya | 909 | 1,663 | 2,851 | 4,860 |
| Sunflower | 472 | 845 | 1,296 | 2.375 |
| Olive | 935 | 1,050 | 1,224 | 1,000 |
| Sesame | 558 | 606 | 591 | 610 |
| Maize | 73 | 118 | 133 | 235 |
| Total | 5,819 | 7,349 | 10,135 | 14,815 |
| *Palm oils* | | | | |
| Coconut | 1,483 | 1,696 | 2,138 | 2,225 |
| Palm kernel | 330 | 366 | 402 | 480 |
| Palm oil | 830 | 1,001 | 1,137 | 1,115 |
| Babassu kernel | 38 | 40 | 47 | 50 |
| Total | 2,681 | 3,103 | 3,724 | 3,870 |
| *Animal fats* | | | | |
| Butter and whey | 3,704 | 3,334 | 4,063 | 5.000 |
| Lard | 3,030 | 3,311 | 4,025 | 4,330 |
| Tallow | 1,483 | 2,172 | 3,186 | 4,155 |
| Total | 8,217 | 8,817 | 11,274 | 14,485 |
| *Sea animal oils* | | | | |
| Whale (less sperm oil) | 485 | 387 | 392 | 335 |
| Fish | 274 | 243 | 327 | 765 |
| Total | 759 | 630 | 719 | 1,100 |
| *Industrial oils* | | | | |
| Linseed (some used for food) | 1,040 | 1,240 | 1,040 | 1,075 |
| Castor | 145 | 185 | 221 | 250 |
| Rapeseed (some used for food) | 1,238 | 1,581 | 1,983 | (1958) 1,949 |
| Oiticica | 11 | 13 | 10 | (1958) 10 |
| Tung | 140 | 117 | 100 | 95 |
| Total | 2,574 | 3,136 | 3,354 | 3,379 |
| Grand total | 20,050 | 23,035 | 29,206 | 37,649 |

the remaining oil. Formerly, filter cloths of wool or camel hair into which the material was wrapped were used for this purpose. Since the consumption of spinning cloths is very high the use of screening presses is preferable.

The oil yield is not very satisfactory. Without preheating the press cake still frequently contains 7 per cent oil, if preheated, 4–6 per cent. From the economic point of view, the

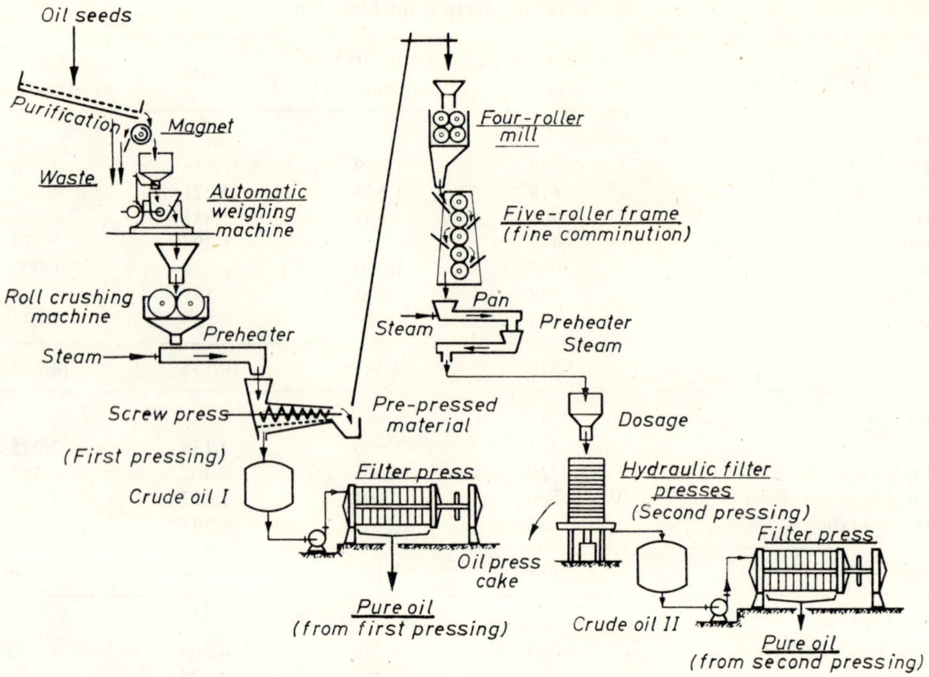


Figure 122. Extraction of oil from oil seeds in two pressing stages

extraction of the oil cake with solvents is, therefore, completely justified. Oil cakes are rich in proteins and carbohydrates and are a very valuable fodder material.

Modern oil mills use almost exclusively continuous high-performance screw presses which are designed on the mincing machine principle and have to be cooled.

### *Extraction*

The entire oil content can be extracted with organic solvents and even seeds or waste products such as germinating corn or maize can be utilized economically.

Oil extraction was still carried out in 1855 in France with the very dangerous carbon disulphide. Later, petrol, benzene, halogen hydrocarbons such as carbon tetrachloride or dichloroethylene were used. Very careful purification of the solvent is essential since small amounts of evil-smelling impurities spoil both the taste and the smell of the oil.

The method of extraction is identical to the 'Soxhlet' used in the laboratory. 'Cold extraction' is used for coconut and palm kernel fat, 'hot extraction' for soya beans and bones. For cold extraction an extraction battery of six iron cylinders is used through which flows liquid petrol. Operating on the counter-current principle explained earlier, the fresh charge is always brought into contact with the most saturated petrol. The petrol emerges with a fat content of 35 per cent, is distilled and the residues are finally removed from the fat by direct steam. The mode of operation of an extraction battery is shown on p. 287. Hot extraction is carried out as described in the Section on extraction of bones.

All fat extractions are rendered difficult by the fact that the perforated trays on which the material to be extracted is placed often become clogged.

Recently, continuous extraction plants have been introduced on a wider scale. They usually operate on the following principle: the crushed oil seeds are conveyed by worm gear through a U-shaped tube. In counterflow against the oil seeds flows the solvent which, saturated with fat, leaves through a sieve-type strainer on the seed-entry branch. The extracted material leaves at the top of the other branch of the U-tube. Movement through the horizontal part is slower causing damming up which is conducive to complete removal of the oil. The adsorbed solvent has to be expelled from the de-oiled seeds by means of steam. In other methods (Hansamühle) using the Paternoster principle the material to be extracted is sprayed with the solvent in perforated baskets.

The extracted oil is subjected to various purification operations (refining). These consist of filtration by filter presses, purification with sulphuric acid, and purification by adsorption on carbon, Florida earth, or other bleaching earths. Traces of fatty acids are removed with caustic soda. 'Perfume rancidity' which occurs with coconut oil and is due to ketones can be eliminated by injecting steam at a temperature of 150–170°C under vacuum. 'Palmin' is pure hydrated coconut fat. For every fat and every oil special purification methods have been developed which will not be discussed here in detail. It is possible to render fat containing much free fatty acid fit for consumption by treating it according to the 'Wecker' process under certain conditions with superheated steam and expelling the free fatty acids. The free fatty acid residue of 5–10 per cent in the refined fat has to be neutralized with alkali.

### (b) Production of animal fats

In Germany, the raw materials used for the production of animal fats are mainly the adipose tissues of pigs, cattle, and sheep. Whilst vegetable fat production is carried out almost exclusively by crushing and extraction, animal fats are usually rendered at as low a temperature as possible.

Fat from sea animals is obtained mainly from whales. The processing (rendering of the adipose tissues) of the harpooned animals is carried out nowadays on the mother ship of the whale fleet which is practically a floating oil factory.

Fish oil from herrings, menhaden, and sardines is obtained by boiling the fish with a little water and crushing them. The fat is then separated from the liquid phase.

#### *Production of bone fat*

For the removal of fat from bones horizontal boilers are used with perforated bottoms, a manhole at the top for feeding in the crushed bones and an aperture at the bottom for the discharge of the extracted bones. At least two appliances are used which are charged, extracted, and emptied alternately.

The extracting agent, e.g. petrol, distils from below through the bone charge in the still. Petrol and water are condensed in a cooler, the petrol is returned into the still. Water is separated in a separator. When the extraction is complete all petrol is distilled off and the bone is discharged from the still. The petrol has first been expelled from the degreased bones by steam.

About 6–9 per cent fat are obtained from bones. Another possibility is to centrifuge the fat from the crushed and heated dry bones. The ground fat-free bones were formerly used as phosphate fertilizers ('bone meal'). The protein substance is processed to bone glue. Boneblack is obtained by carbonization. Extracted crushed bones are also used as fodder, particularly for poultry.

A frequently used modern method for the extraction of bone fat is based on the 'Chayen impulse' principle which consists essentially in the decomposition of the crushed material with hot water by means of ultrasonic waves and separation of the emulsion produced by centrifuging at about 70°C.

## 3. HARDENING OF FATS

The possible uses of liquid fats are limited. People living in the moderate zone have become used to using semi-solid fats which spread easily. Fats of animal and vegetable origin possess a characteristic smell and taste which is not always popular; in particular fish oils are often rather unpalatable. In civilized countries the use of the butter substitute, margarine, is widely accepted and it is hard to imagine a household without it. For industrial fats, e.g. solid soaps and candles, too, hard fats or saturated fatty acids are desirable.

The fat industry achieved, therefore, a great success about 60 years ago (e.g. due to Normann) when hardening of fats by hydrogenation was introduced. In this process unsaturated fatty acids of oils are converted into saturated fatty acids by catalytic hydrogenation with hydrogen on a nickel catalyst. Hydrogenation is carried out in the liquid phase, and the nickel catalyst has, therefore, to be very finely dispersed.

The finely dispersed nickel catalyst is produced by precipitating nickel sulphate with a soda solution in the presence of kieselguhr as a carrier. The solution is then filtered, scrubbed, and reduced at as low a temperature as possible on plate- or drum roasters in a hydrogen stream. Since this treatment renders the catalyst highly active contact with air must be rigidly excluded. It drops directly from the roaster into the oil to be hydrogenated. The catalyst can also be produced in a hardening autoclave (hydrogenator) directly as finely dispersed nickel without a carrier. The nickel is then introduced into the oil in the form of nickel carbonate or nickel formate and heated in a hydrogen stream to between 230 and 260°C maximum.

When the catalyst is formed the temperature is reduced to the proper fat hardening temperature of 175–180°C. The hydrogenators used nowadays are long pressure cylinders fitted with agitators; the turbulent propulsion of the fat through the cylinders can also be carried out with gaseous hydrogen. The agitator pressure autoclaves are equipped with a heating jacket, internal heating coils, or Samesreuther tubes (see p. 234). The hydrogenation temperature of 175–180 °C can be obtained by heating with steam at a gauge pressure of 10–12 atm. For the formation of the nickel catalyst from formate arrangements are needed for heating to a maximum temperature of 260°C for short periods. This corresponds to a heating steam gauge pressure of about 60 atm. For this the central boiler plant of the hydrogenation works is usually inadequate and it would, moreover, be uneconomical to equip the boiler and steam piping for such high pressures if the required peak temperature in the pressure autoclave is needed only about once a week. The autoclave battery is, therefore, fitted with a special high-pressure steam generator.

Very high purity hydrogen is required. It is best produced electrolytically and numerous special electrolysers have been designed for the purpose. To harden 100 kg oleic acid about 8 $m^3$ hydrogen and 5 kWh are needed for the electrolytic production of hydrogen. Hydrogen from water gas can also be used after conversion (p. 71) but purification has to be very thorough since traces of $PH_3$, $AsH_3$ and $H_2S$ contaminate the catalyst. The samples taken to check the progress of the hydrogenation have ever-increasing melting points and smaller iodine numbers. Fats which are to be used in the food industry are hydrogenated to a melting point of 32–34°C. It is important for the heat of the hardened oil to be transferred in the heat exchanger to the un-hardened oil. When the hydrogenation is complete, oil and catalyst are separated in filter presses. The catalyst can be re-used. It is particularly easy to harden castor oil which on hardening can reach a melting point of up to 80°C, sesame, groundnut, rapeseed and cottonseed oil, and fish oil, particularly the cheap whale oils which are suitable for the manufacture of margarine. After hardening fats designed for the food industry are refined once more.

## 4. MANUFACTURE OF MARGARINE

The first margarine was produced in 1867 by the French chemist Mége-Mouriés who rendered beef tallow and allowed it to solidify again at a temperature between 23 and 25°C. The liquid residue (oleo-margarine) was emulsified in milk and the emulsion butterified. As early as 1870 the Dutch firm Jürgens acquired the process developed by Mége-Mouriés for the production of margarine. In modern margarine manufacture hydrogenated fats are used as raw materials together with solid and liquid vegetable fats.

Fats suitable for the manufacture of margarine are passed through a mixing kettle, with the addition of emulsifiers (e.g. lecithin), dye (carotenoids), aroma (diacetyl), vitamins (A and D), common salt, etc., and mixed in the churn with about 15 per cent water or acidified skimmed milk to a stable emulsion. The churn is a double-walled iron vessel lined with tin, with a capacity of 200 to 700 litres. A double agitator in the churn assures intimate mixing of the two phases. Emulsification takes about 20 minutes. The emerging emulsion (temperature 25–30°C) is solidified on an internally cooled rotary drum, then mixed with the quantity of starch prescribed for characterization (0·2 to 0·3 per cent), or with oil rich in essential fatty acids, and finally kneaded and wrapped.

In addition to this widely practised batchwise mode of operation continuous manufacture of margarine has been increasingly adopted in recent years. In the latter process all the mixing, churning, and cooling operations are carried out in a single machine.

## 5. MANUFACTURE OF SYNTHETIC FATS

### *Chemisynthesis*

Raw materials for the manufacture of synthetic fats are lignite paraffin, Fischer-Gatsch- and TTH (low temperature hydrogenation paraffin from which fatty acids are obtained by oxidation with the addition of suitable catalysts under defined conditions of reaction.

### *Biosynthesis*

Repeated attempts have been made to obtain fats from carbohydrates by means of micro-organisms. With surface fungi these methods are impracticable. Better prospects are offered by the use of fungi growing in water (e.g. *Fusarium*) and of growth yeasts of the type *Torula* or *Candida* grown in an N-deficient atmosphere (see p. 308). So far biological fat synthesis has not proved profitable. For some time autotrophic micro-organisms such as green algae have been used experimentally for fat synthesis. Because of their chlorophyll content these are able to synthesize fat directly from carbon dioxide and water.

## 6. FATTY ALCOHOLS

Fat is hydrogenated to fatty alcohol a ta hydrogenation temperature of 250°C and a gauge pressure of 200 atm, with copper, copper oxide-chromium oxide, or cobalt as catalyst. The hydrogenation of fatty acid esters of low alcohols is more convenient. The industrial production of fatty alcohols was carried out for the first time by Normann to whom the development of fat hardening owes a great deal, and by Schrauth and Bertsch in 1930 at the Deutsche Hydrierwerke at Rodleben.

$$R{-}COOR_1 + 2\,H_2 \rightarrow RCH_2OH + R_1OH$$
$$R{-}COOH + 2\,H_2 \rightarrow RCH_2OH + H_2O.$$

Hydrogenation is always carried out in the liquid phase, with equipment similar to that used in the hydrogenation of coal in the semi-solid phase. The catalyst is in suspension. Hydrogen gas is circulated. The equipment (a high-pressure tube) includes a preheater and a heat exchanger; for details see p. 126. Some ester is always left unhydrogenated to prevent the formation of too much hydrocarbon as a by-product.

Fat alcohols are also obtained by the 'Gerbet reaction' which consists in a doubling of the alcohol molecule, e.g. n-butanol → octanol via alcoholate.

Selective hydrogenation of unsaturated fatty acids to unsaturated alcohols is possible (according to Bouvault-Blanc with Na and low alcohols or catalytically with Cu-Cd).

Fat alcohols are very important for the manufacture of detergents suitable for hard water, alcohol sulphates $R{-}CH_2OSO_2{-}ONa$ obtained by sulphonation with sulphuric acid and oleum or chlorosulphonic acid (Gardinol, Texapon, Fewa and Onalkali, Cyclanol). By reacting fatty alcohols with ethylene oxide polyglycol ethers are obtained which are excellent emulsifiers (Emulphor). A mixture of fatty alcohol and fat alcohol sulphate is the ointment and emulsion base Lanette wax which is much used in medicine and cosmetics. Esters of phthalic acid and adipic acid with the fatty alcohols $C_6{-}C_8$ are very good plasticizers for plastics.

## 7. LIPOLYSIS, MANUFACTURE OF FATTY ACIDS AND GLYCERINE

Lipolysis can be carried out with alkali, in a neutral medium with water, or in strong acid. There are five different methods:

(a) autoclave decomposition,
(b) fermentation lipolysis,
(c) acidification,
(d) Twitchell decomposition,
(e) saponification with alkali.

Formerly, lipolysis was carried out mainly to produce a substitute for the increasingly expensive and rare beeswax for candles, i.e. for the production of stearic acid. For this purpose hard fats were saponified with caustic soda. In 1831 Milly replaced for the first time the expensive saponification with caustic soda by saponification with lime for which he still required the appreciable quantity of 12 per cent. The fat was boiled in open boilers with lime, lime soap was precipitated in insoluble form, decomposed with sulphuric acid, and finally the fatty acid was separated from the molten gypsum by pressing. In 1855 Milly found that caustic lime amounting to 2 per cent of the weight of the fat was sufficient if saponification was carried out in a pressure autoclave at 170–180°C and at a gauge pressure of 8–10 atm. This actually amounts to hydrolysis with water. The small quantities of mono- and diglycerides formed act as emulsifiers for the fat. The advantage of the method consisted in the fact that only small quantities of lime and sulphuric acid were required for the decomposition of the insignificant amount of calcium soap. It was, however, very difficult to separate the precipitated gypsum from the fatty acid. More recently, 1 per cent $Mg(OH)_2$ or 0·5 per cent ZnO has, therefore, been used in place of lime. The glycerine is then obtained in the aqueous phase in a concentration of 10–15 per cent.

### (a) Autoclave decomposition

Tall vertical cylinders of copper or special steel, with a capacity of 8–25 $m^3$, are used for modern decomposition autoclaves. Fatty acids corrode iron, of course, but not copper, high-grade steel, or high-purity aluminium. Heat is supplied by steam through immersion tubes which after completion of the decomposition are used to empty the autoclaves by pressure. 90 per cent of the fat is saponified in four hours at 170–180°C. Further decomposition takes place very slowly since a hydrolysis equilibrium between fat, fatty acid, glycerine and water is established. In order to complete the decomposition the aqueous glycerine has to be removed and decomposed again. A very dilute aqueous glycerine forms which is used with the charge for the next decomposition cycle. Decomposition without metallic hydroxides is also possible provided it is carried out at a somewhat higher temperature. Autoclave decomposition can also be carried out continuously, at 240°C and a gauge pressure of 45 atm; this is done, e.g., in the USA.

To produce pure white fatty acids, particularly pure white stearins for candles, the fatty acids have to be distilled under vacuum with steam. This method is often used very successfully for the purification of sensitive materials, particularly those that are sensitive to air. It combines the advantages of steam distillation, i.e. lowering of the boiling point due to aggregation of the partial pressures and easier boiling, and of vacuum distillation, i.e. lowering of the boiling point, rapid transport of the vapours to condensation, and protection from atmospheric oxygen, giving an effect similar to an inert gas.

Glycerine is obtained from the aqueous glycerine by concentration and distillation under vacuum (see later).

### (b) Fermentation lipolysis with lipases

In practice only castor-oil (ricinoleic acid) lipase can be used for lipolysis. Oil fruits contain lipases to mobilize their fat reserves during germination. Castor oil seeds are particularly rich in lipases. They are crushed, mixed with water, and the cell residues are separated by centrifuging. The mixture is fermented to remove the carbohydrates. The lipases separate out at the top of the liquid as cream.

As usual in fermentation reactions the optimum temperature is about 35–40°C. Lipolysis is, therefore, a process in which the materials are treated very carefully. The equipment used consists of a boiler which is cone-shaped at the bottom, is fitted with a heating coil, and contains an annular perforated tube for agitation with air. Water and enzymes equal to 6 per cent of the weight of the oil are added. Agitation with air produces an emulsion. When the decomposition is complete the mixture is allowed to settle at 80°C with the addition of a small amount of sulphuric acid, and after 24 hours separation is carried out into fatty acid, aqueous glycerine, and an emulsified interlayer which is used for the next charge. This process is rather expensive and is used comparatively rarely.

### (c) Acidification

The process of acidification with strong sulphuric acid can be used only with unsaturated fats. The sulphuric acid is added to the double bonds and a sulphuric acid ester forms, e.g. from oleic acid the ester

$$C_{17}H_{34}\left\langle\begin{matrix}COOH\\OSO_2OH\end{matrix}\right.$$

These esters are very effective dispersing agents and form emulsifiers and catalysts for the subsequent hydrolysis with water. The process is cheap, rapid, and the yield of

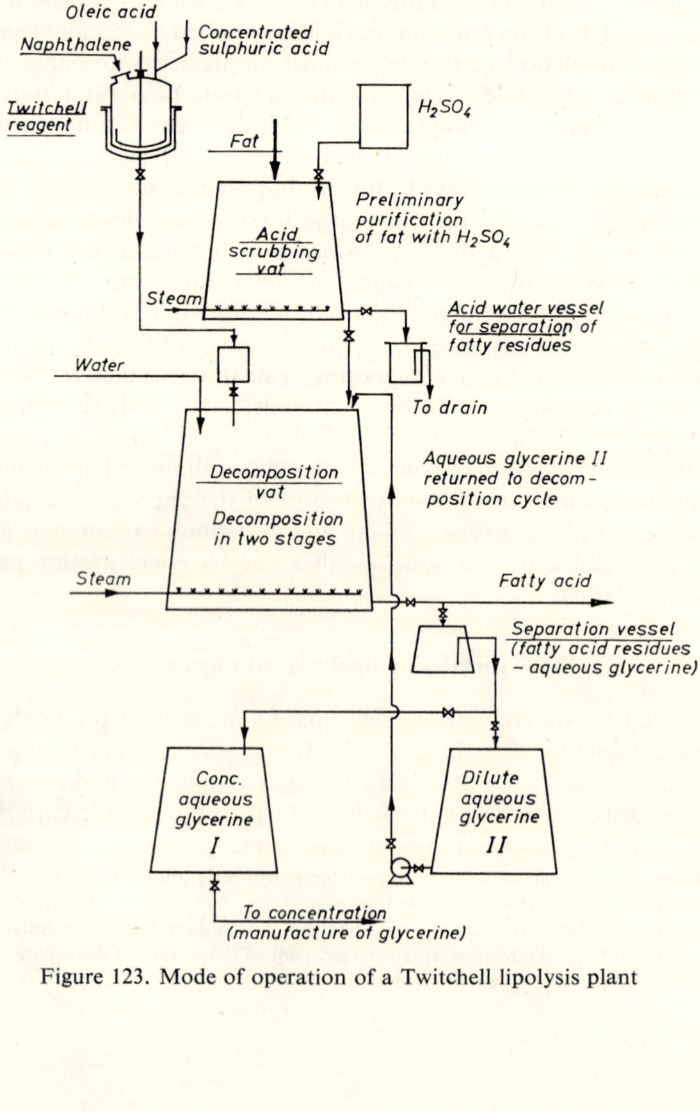


Figure 123. Mode of operation of a Twitchell lipolysis plant

fatty acids is good. A disadvantage is the dark colouring and the partial decomposition of the fat and of the glycerine by the strong sulphuric acid.

The dried fat is processed with sulphuric acid (2 per cent) in cast iron boilers, agitation being carried out at 110°C. The formation of the sulphuric acid ester takes place very rapidly and is complete within ten minutes. Then decomposition is carried out in boilers with direct steam at atmospheric pressure. In addition to the saponification the sulphuric acid residue is also split off again and hydroxy acid forms which is found as a small admixture in the fatty acid mixture. During distillation of the fatty acids, however, the hydroxy acid is dehydrated again and the double bond re-forms, partial isomerization taking place, e.g. iso-oleic acid forming from the oleic acid.

### (d) Twitchell decomposition

The process derives its name from its inventor Twitchell and consists in preparing the emulsifying catalyst separately (Twitchell reagent, Figure 123). This prevents the partial decomposition of the fat by sulphuric acid which has such a deleterious effect in the acidification process. The reagent is prepared from oleic acid, naphthalene, and concentrated sulphuric acid in a special boiler. The double bond of the oleic acid condenses with the naphthalene which is sulphonated simultaneously. The compound resembles the foaming, wetting and dispersing agent 'Nekal' which is butylated and sulponated naphthalene.

Before decomposition the fat undergoes preliminary purification in an agitator vessel with sulphuric acid, mucoid substances which inhibit catalysis being removed.

After discharge of the acid which is precipitated at the bottom, the pre-purified fat is run into the decomposition vat underneath. There it is mixed with 55 per cent water and 1–2 per cent Twitchell reagent and decomposed by injection of direct steam. In this process, too, a hydrolysis equilibrium is established so that at first only 90 per cent of the fat is saponified. The aqueous glycerine is, therefore, discharged into vat I, the fat mixture again mixed with water and reagent and the decomposition completed. The dilute aqueous glycerine II obtained is stored separately and used with the next decomposition charge. After decomposition the fatty acid can be used immediately for the manufacture of soap. The aqueous glycerine is collected in a small separation vessel in front of the vats, in which entrained fatty acid fractions collect. The concentrated aqueous glycerine I is freed from sulphuric acid and sulphonic acid with barium carbonate and can then be processed to glycerine.

*Purification of fatty acids by distillation*

Under high vacuum fatty acids distil without decomposition. Formerly, they used to be distilled at 15–17 mm in a pump vacuum, now they are distilled at 2–5 mm in the vacuum produced by a steam ejector. Fatty acids are distilled particularly easily and without harmful effects by the injection of steam. Distillation is carried out in the usual manner mostly batchwise. Continuous operation is also possible. Lurgi have developed special equipment for this purpose in which the fatty acids are heated in a still and distilled continuously under vacuum by the injection of steam. They condense with water on a very large tube condenser. The water is precipitated in other condensers. A vacuum is produced by three-stage steam ejectors with an injection condenser. To save steam an Elmo water ring vacuum pump is fitted in front of these. The condensed water runs off through a barometer tube and the precipitated fatty acid is extracted continuously by pumps.

Distillation yields almost without loss pure white odourless fatty acids used for pure soaps. They could also be re-esterified with glycerine so that pure edible fats can be produced from impure fats. Vacuum steam distillation permits also the separation of fatty acids from the fats without harmful effects so that the fat residue can be used as edible fat.

A special process, the 'Wecker process', which deacidifies the oil has been developed by the Veredlungsgesellschaft für Fette und Öle, Frankfurt, for the purification of fats which have turned rancid. The fat is processed continuously under a high vacuum in a horizontal boiler with steam supersaturated with water and, therefore, containing a water mist. The volatile components are distilled and are condensed in a cooler. The neutral oil runs off continuously through a cooler into the neutral oil collector. The Wecker process can also be used for the purification of fatty acids by distillation, the residual neutral oil being cracked again in each succeeding cycle. The final residue is stearin pitch which in its hard and high-melting form is a valuable product.

## 8. SAPONIFICATION WITH ALKALI, SOAPS

Saponification with alkali is a process carried out nowadays only for the purposes of soap manufacture. For saponification the fat is boiled with caustic soda until all the glycerine is cracked off. After addition of NaCl the boiling lye separates into two layers, the upper one consisting of soap which is liquid when hot, and the lower one, the 'bottom lye', being a solution of NaCl and glycerine. The bottom lye contains only 4–6 per cent glycerine and 8–15 per cent sodium chloride, also free alkali and soap. Processing is, therefore, difficult. Since glycerine is a much-sought after-product, a more convenient raw material for the manufacture of soap are free fatty acids obtained by one of the cracking processes described earlier. Fatty acids can, moreover, be purified much more easily than fat.

Various methods are used for the production of soap from fatty acids. In carbonate saponification the calculated amount of fatty acid is added to the boiling soda solution, causing very violent foaming. The bicarbonate formed is decomposed by the injection of steam, the $CO_2$ being expelled. If there is a residue of neutral fat, often present in fatty acids, further saponification with caustic soda can be carried out.

If pure fatty acids are used an anhydrous reaction directly with one molecule of calcined soda is possible. The mixture of soap and bicarbonate of soda obtained can be worked up directly into pieces ('milling'). In another process neutral fats can be saponified directly under pressure with a solution of soda.

When soap is produced in a liquid medium, the semi-liquid 'curd' is poured into iron moulds which can be taken apart and in which the soap solidifies in a few days; the bottom lye or water has first to be drained off. The blocks of soap removed from the opened mould are cut with steel wires. Faster operation is achieved with a cooling press such as has been designed by e.g. 'Miag'. It resembles a filter press and consists of moulding frames placed between cooling plates through which water flows. Cooling takes only 1–2 hours. After opening the press the slabs of soap obtained are cut. Soaps produced from semi-liquid soap curd are called curd soaps.

The best soaps are produced from highest-grade fats or purified fatty acids. First, a basic soap (pure curd soap) is produced by the process described above, this is then shredded and dried either by placing on conveyor driers in drying kilns, or by pouring the molten soap on to heated drums. The water content must not exceed 8–15 per cent. Modern soap manufacture up to the wrapped product is often carried out continuously.

Perfumes, dyes, and fillers, e.g. Tylose, are added to the soap which is then extruded, milled, cut into pieces, and shaped in hand presses. This gives toilet soaps their attractive appearance. Drying is necessary for the sake of both the pressing and the prevention of shrinking which occurs in ordinary curd soap. Because of recent dermatological findings fatty acids with chain lengths above $C_{16}$ are now preferred for toilet soaps; the lower ones cause irritation of the skin.

### *Various types of soap*

Filled soaps are solidified soap paste produced from fats with mainly low fatty acids, e.g. coconut fat. They contain water, glycerine, salts, and the excess caustic potash solution and their cleansing effect is very strong.

Soft soaps for domestic use and for use in the textile industry are produced from cheap vegetable oils with caustic potash solution to which is added 15 per cent potash to improve the consistency. They are also a kind of paste soap and contain some free alkali. Potash soaps are, of course, softer than soda soaps.

Soap powders are produced by spraying and drying pure soap in an atomizer drier. Washing powders are atomized with soda and salts.

Formerly, the bleaching washing powder 'Persil' for instance, consisted of 35 per cent pure soap, 12 per cent soda, 13 per cent pyrophosphate, 6 per cent sodium bicarbonate, 2·5 per cent magnesium silicate and 8 per cent sodium perborate. Synthetic detergents, e.g. Mersolat, are now often used in the production of washing powders.

Solvent soaps contain hydrocarbons, halogenated hydrocarbons, turpentine or tetralin, i.e. solvents with boiling points between 150 and 250°C. The solvents have to be very finely dispersed in the soap so that they remain emulsified when the soap is dissolved in water. Olein and castor oil potash soaps are best suited for the production of solvent soaps. In order to promote fine dispersion the solvent is best added during saponification. To produce such soaps the fatty acids can also be mixed with the solvent mixture and then with soda; this method yields solid solvent soaps. Solvent soaps are used mainly in the textile industry and for special industrial cleansing purposes; although they are very suitable for domestic use they are, unfortunately, very little used for this purpose in Germany.

## 9. MANUFACTURE AND PURIFICATION OF GLYCERINE

Glycerine is produced in a very simple manner from the aqueous spent liquors of fat hydrolysis by the autoclave or Twitchell process. The liquors are purified with milk of lime or barium hydroxide and ammonium oxalate, filtered, and evaporated under vacuum. The crude glycerine residue of about 80 per cent contains many organic impurities. Production of glycerine from the bottom lyes obtained during saponification with alkali is more difficult. In many cases these bottom lyes are, therefore, not processed and the valuable glycerine is lost. These glycerine liquors have recently often been freed from salt with ion exchange resins to render them suitable for use. The bottom lyes contain 4–10 per cent glycerine, 8–15 per cent NaCl, and alkali and fatty acid salts. First, neutralization is carried out, the fatty acids are precipitated with aluminium sulphate, and the liquors are purified with milk of lime. Then evaporation is carried out in several stages in equipment so designed as to permit the periodic removal of the precipitated salt.

Very high purity is required of glycerine intended for pharmaceutical purposes, for cosmetics, and particularly for explosives glycerine. Special distillation processes have been developed for the purification. Crude glycerine is heated under vacuum in a still fitted with a steam coil. The same steam pre-heats expanded steam in a superheater. This superheated steam is introduced directly into the glycerine which evaporates together with the water. Fractional condensation in several collecting vessels follows. In the first collecting vessel pure glycerine condenses, in the others aqueous glycerine which is recycled. Finally, the water is precipitated.

The manufacture of glycerine from propene has already been discussed (p. 177).

## LITERATURE


H. Ost, and B. Rassow, *Lehrbuch der chemischen Technologie*, 26th edn, Leipzig, J. A. Barth, 1955.

W. Diemair, I. Koch, H. Aman, in: Winnacker und Weingaertner, *Chem. Technologie*, vol. 4, Org. Technologie II, p. 901, München, Hanser Verlag, 1954.

H. Schönfeld, *Chemie und Gewinnung der Fette*. Wien, Springer Verlag, 1936/37.

R. Lüde, Die Gewinnung von Fetten und Ölen. *Technische Fortschrittsberichte*, vol. 47, Dresden, Verlag Th. Steinkopff, 1948.

E. Kadmer, *Fettöle, Wachse, Harze, und ihre Verarbeitungsprodukte*. Hannover, Wissenschaftl. Verlagsanstalt, 1949.

J. Marcusson, F. Püschel und A. Dierichs, *Die Untersuchung der Fette und Öle*. 4th edn, Halle, Knapp, 1952.

R. Lüde, Die Raffination von Fetten und fetten Ölen. *Technische Fortschrittsberichte*, 2nd edn, vol. 58 Dresden und Leipzig, Verlag Th. Steinkopff, 1962.

H. I. Watermann, *Hydrogenation of Fatty Oils*. New York, Elsevier Publishing Co., 1951.

H. P. Kaufmann und J. G. Thieme, *Neuzeitliche Technologie der Fette und Fettprodukte*. I. Lieferung: Die Rohstoffe, ihre Gewinnung, ihr Transport und ihre Lagerung, Münster, Aschendorffsche Verlagsbuchhandlung, 1956.

H. P. Kaufmann, *Analyse der Fette und Fettprodukte*. Vols. 1 and 2. Berlin/Göttingen/Heidelberg, Springer, 1958.

# V. Tanning and the manufacture of leather. Tannins

## General remarks

The manufacture of leather, which is older than the use of textiles, has always aimed at protecting animal hides from hardening and decay. At first, leather was used only for clothing, later its use as a raw material for a variety of essential goods also increased enormously. It is still mainly used in clothing, e.g. shoes, gloves, and other leather apparel. Many useful and ornamental objects are made of leather, e.g. saddles and bridles, handbags, briefcases, suitcases, wallets, purses, wrappings for a great variety of articles, bookbindings, belts, and carrying straps. Leather used to be the only raw material for many industrial goods, such as drive belts, seals, valve components, car components, e.g. car body suspensions, bellows, and elastic parts of machinery, but nowadays it has been mostly replaced by other materials such as rubber, plastics, or combinations of these with textile fibres. Plastics have in many cases supplanted leather because they are cheaper and for some special purposes are even more suitable. Nothing, however, can compare with leather for general usefulness and beauty.

The production and treatment of leather is one of the oldest crafts and one of the earliest arts practised by man, as is shown by finds from the pre-Christian era. In the Middle Ages the tanners formed their own guild and many old street names point to the practice of this craft, almost always near a river. When the steam engine was introduced leather factories equipped with machines increasingly replaced the old craftsmen's shops. Centralized collection of hides, the international hide trade and its customs have greatly contributed to the industrial development of leather manufacture.

In the German Federal Republic the value of the production of leather in 1950 amounted to 675 million DM. The proportion of employees in the leather industry in 1950 was 5·5 per cent of total employees which shows the great importance of this branch of industry.

## Animal hides as raw materials

The raw material for the production of leather is the animal hide. Its nature mainly determines the character of the leather produced. Raw hides are obtained from all types of cattle, such as bulls, oxen, cows, and calves, from sheep, goats, pigs, horses, and for special purposes from game, water animals and reptiles. To this must be added the furry animals proper, both wild and domesticated, the hides of which also have to be tanned and processed.

Whilst Nature has created in the natural fibres, e.g. cotton, wool, and silk, unidimensional macromolecules that are models in respect of their chemical and micro-structure to which their strength and heat insulation properties are due, the animal hide is an ideal example of a good raw material that can be put to use directly.

The skin of mammals consists histologically of three parts: epidermis, cutis or corium (the true skin), and subcutis (the subcutaneous connective tissue). With furs care is taken to preserve the hair formed by the epidermis, and with reptiles and sea animals the scales. In all other cases the epidermis and the subcutaneous connective tissue are removed for the manufacture of leather.

The main constituents of the corium are the scleroproteins of which collagen is the most important for the formation of leather. It constitutes 98 per cent of the dry substance of the corium. Other components are elastin which also contributes to the formation of leather during tanning, and keratin, the main constituent of wool and hair, which is removed. Collagen is, by its chemical structure, a protein with a special micro-

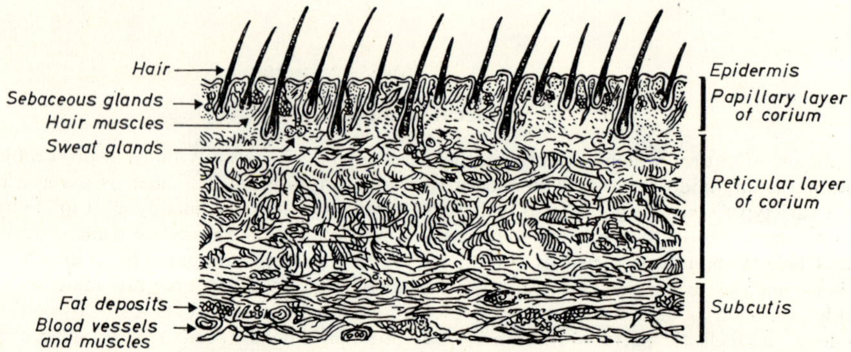


Figure 124. Diagram of the histological construction of mammal skin (from F. Stather, *Leder und Kunstleder*, 3rd edn, Berlin, Akademie-Verlag, 1959)

structure and arrangement of its amino acids. By using the methods of X-ray structural analysis and electron microscopy it has proved possible to obtain an insight into the microstructure of the collagen fibre crystals. Each amino acid recurs periodically according to its frequency in the collagen molecule. The dicarboxylic acids are distributed in the peptide chain in such a way that they face a basic amino acid of the neighbouring

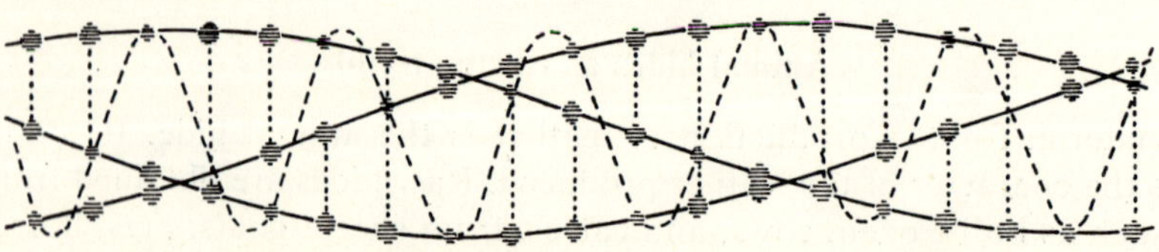

Figure 125. Helical construction of the collagen fibre

polypeptide chain, so that the peptide chains attract each other electrostatically. As a result, collagen swells in acid and alkaline media, with simultaneous expansion of the lattice, first reversibly with longitudinal shrinkage of the fibres, then, particularly in heat, irreversibly, and finally it hydrolyses ('cementation').

Collagen contains as amino acids mainly glycine, a large amount of proline, also hydroxyproline, tyrosine, lysine, arginine, aspartic acid and glutamic acid, but no sulphur-containing amino acids such as cystine which by forming S–S bonds is partly responsible for the chemical and mechanical strength of wool and hair. The collagen fibre does not need 'cross-linking' by primary valence bonds. Its great tensile strength and elasticity, due to the peptide chains being prevented from slipping past each other, is produced by their intertwining helical arrangement, in opposing rotational senses, a further strengthening being obtained by the saline bonds and hydrogen bridges, as schematically shown in Figure 125 (O. Bayer, *Angew. Chemie*, 1959, **71**, 147).

# STAGES IN THE MANUFACTURE OF LEATHER

The fresh hide with its water content of 50–70 per cent is a good nutrient medium for putrefactive bacteria. Growth of the bacteria is slowed down if the moisture drops below 30 per cent. Drying to a water content of 12–15 per cent, or dehydration by curing with salt, represents, therefore, an adequate preserving treatment.

## 1. Tannery wet-work in the beamhouse

The first stage of leather manufacture comprises the tannery wet-work operations. In these, epidermis and subcutaneous connective tissue are removed and the corium proper is prepared for tanning. This stage consists of the following operations:

(1) Soaking (accelerated by the addition of a wetting agent).
(2) Liming in paddle vats, to loosen the hair, using a solution of lime and sodium sulphide, or painting the skins with a paste consisting of lime and sodium sulphide.
(3) Deliming, to reduce the alkalinity and shrinkage (using organic acids and ammonium salts).
(4) Bating, to improve the grain by proteolytic action (using tryptic enzymes).
(5) Pickling, to arrest the bating process for mineral tannage (treatment with acids and salts).
(6) Mechanical operations (unhairing, fleshing, scraping).

## 2. Tanning

The next operation is the tanning process proper in which, according to the tannins and hides used, different equipment and tanning times are employed and, as a result, leathers which differ completely in their properties are obtained.

### (a) Polyaromatic tanning — Tanning with vegetable and aromatic synthetic tannins

Vegetable tanning for the production of 'heavy' leathers such as sole, inner sole, transmission belt, and harness leather, is carried out in vats. The hides are either suspended (rocker yard) or laid horizontally (layer yard). The process is very protracted and lasts at least 70 days, in the old pit tanning process up to two years. The shorter time required for the modern tanning process as compared with the old pit tanning and the original bark tanning is due to the use of easily soluble tanning extracts and auxiliary tannins in place of tanning barks, and mainly to a kind of 'counter-current extraction' of the tanning liquor through the hides.

Modern tanning is carried out in rectangular concrete or wooden vats which are arranged in a continuous series in such a way that the tanning liquor runs off the surface of one vat into the bottom of the next when fresh tanning liquor is added at one end of the row. The hides to be tanned are moved periodically from vat to vat in the direction of the fresh tanning liquor, i.e. of the highest concentration. Thus the fresh hides come into contact with the largely extracted liquor, the tanned ones with the fresh liquor (counter-current principle). It is very important to maintain an optimum pH value (acid).

Uniform superficial tanning and the diffusion can be accelerated by agitating the hides. A very great acceleration of the tanning process is achieved by "drum tanning", in which the process is carried out in revolving drums. Paddle tanning and drum tanning are also used in combination, e.g. when paddle tanning or drum tanning are followed by deep tanning in the drum with a strongly penetrating "filling" tannin, i.e. one which imparts weight or stability. The shortest tanning period in drum tanning is four days, in the Soviet 3-phase process.

Leather for saddles, bags, and shoe uppers must not contain an excess of tannin. It is usually tanned in drums with quick-acting tannins which give a light colour. Heavy types of leather are sold by weight, and tannins which increase the weight (rendement tannins) are, therefore, particularly desirable for these.

### *Tannins*

The vegetable tannins used are mainly quebracho, chestnut, oak, mimosa, and in the German Democratic Republic, pine bark. These contain polyphenols (pyrocatechol, resorcinol, pyrogallol, and 1, 3, 5-trihydroxybenzene) in condensed form as carboxylic acid derivatives, or in pyran ring groups. The basic types are gallo-tannins, ellagic tannins, catechins, brasilin, and haemathydroxylin. They are chemically classified into hydrolysable and condensed tannins. Pine bark tannin belongs to the hydroxystilbene group; its solubility in water and its acid nature are due to an accumulation of phenolic groups. Vegetable tannins can be used in the form of crushed tannin-containing barks or woods (old pit tanning process), usually, however, extracts are prepared (see chapter on native bark extracts on p. 313).

All tanning woods or extracts except oak and pine have to be imported.

### *Syntans*

To achieve independence from imports, the production of synthetic tannins was developed in Germany. In 1937, 66,060 tons pure tannins were consumed in Germany, of which 53,810 tons were imported. By 1943, the production of synthetic tannins had already reached 19,700 tons. The 'pure tannin content' of a solution is determined by measuring the quantity of tannin absorbed during filtration through a standard hide powder. The result is misleading in that it shows too high a content since only part of the absorbed material has leather-forming properties. The real tannin content is indicated better by the 'tan value' and the 'combining value'. (See F. Stather, *Gerbereichemie*.)

Synthetic aromatic tannins are divided into true synthetic tannins ('exchange syntans') any quantity of which can be used in combination with vegetable tannins or alone and which yield a high-grade leather, and auxiliary syntans which have long been known and are used to improve the solubility of natural tannins (phlobaphene solution) or to accelerate the tanning process.

The main problem in the manufacture of synthetic tannins is the problem of cost since natural tanning extracts are very cheap. Only very cheap raw materials, such as simple phenols, crude pyrocatechol, naphthalene, naphthols, or lignin sulphonic acid, can be used. This severely limits the scope for the production and use of synthetic tannins.

The synthesis of the various syntans is fundamentally very similar. Phenols, phenol sulphonic acids, naphthalene or naphthol sulphonic acids are linked by nucleus-linking agents, such as formaldehyde, acetone (e.g. in the first stage diphenyl-methane- or 'Dian' formation) or by a sulphonic acid residue by means of sulphuric acid and then condensed to higher-molecular compounds which, however, still have to be water-soluble. If sulphonic acid groups are absent they can be introduced with formaldehyde and bisulphite (sulphitation). These tannins yield mainly very light-coloured or even white leathers.

Another important principle of the synthesis is the production of a resin component by the principle described above, and subsequent condensation with lignin sulphonic acid to enlarge the molecule, and solubilization using formaldehyde as a linking agent.

Lignin sulphonic acid which as the ammonium salt in purified form (free of iron and calcium) can be used only as a tannin diluent (see p. 310), forms, together with a resin component, the base for a whole series of exchange tannins.

Exchange tannins were marketed for the first time by the former IG as Tanigan-extra brands. They are now known under the names of Tanigans (Bayer), Basyntans (BASF), Irgatans (Geigy), Wofagans (VEB Farbenfabrik Wolfen), Boemotans (VEB Fettchemie- und Fewawerk) and Orotan TV (Rohm and Haas, U.S.A.). When used in combination with natural tannins they prevent sludge formation, accelerate tanning right through (i.e. no dead zone forms in the centre of the cross-section of the leather), and cause the leather to be light-coloured.

The first and most important tannin of this kind was Tanigan Extra A (Leverkusen) which is also produced as Boemotan US by the Fettchemie- und Fewawerk.

Dihydroxydipenylsulphonic acid is produced from phenol and fuming sulphuric acid, alkaline sulphitization is carried out with formalin and bisulphite by the introduction of $—CH_2—SO_3Na$ groups, and condensation with lignin sulphonic acid. In the manufacture of Tanigan Extra A the separation of the tannin from the brine is very difficult. It is achieved by salting out with ammonium sulphate. The tannin separates out in resin form and has to be separated from the brine as a resin in a specially designed kneading press.

These difficulties were avoided in the Farbenfabrik Wolfen process (Rieche, Rudolph, Müller) for the production of lignin tannins which is based on yeastified beechwood sulphite spent liquor. The basic calcium salt of lignin sulphonic acid, which does not dissolve easily, is precipitated with lime, converted with sulphuric acid into the easily soluble neutral calcium salt of lignin sulphonic acid, and separated from the gypsum. Then follows condensation with a cresol-formaldehyde-, dihydroxydiphenylsulphonic acid, or empyreumatic oil-formaldehyde resin (from crude pyrocatechol) to form tannin. The Ca ions are precipitated with sulphuric acid and the solution is adjusted with ammonium hydroxide to a certain pH value. The tannin solution at this stage is already rather concentrated and can be evaporated even further. Complete deliming and freeing from iron are then carried out for the first time with exchange resins.

The diagram in Figure 126 shows the manufacture of Tanigan Extra A, Boemotan US (From Stather, *Gerbereichemie und Gerbereitechnologie*).

The following diagrammatic formula shows the structure of a typical lignin sulphonic acid tannin such as Tanigan Extra A:

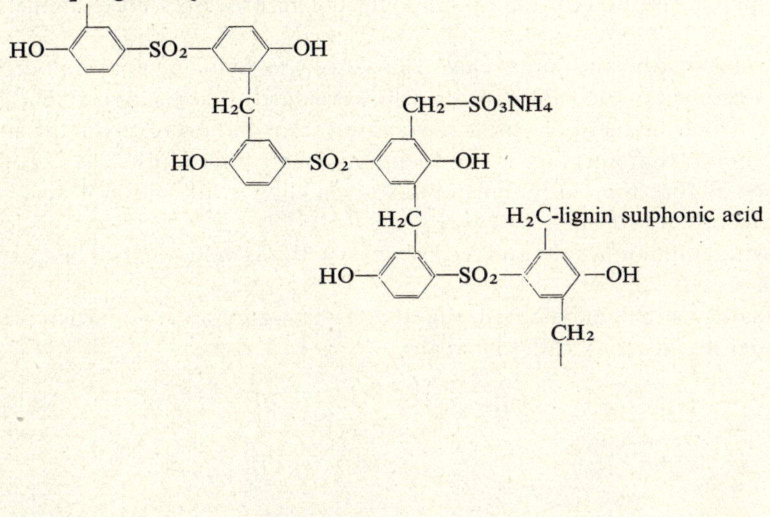

Another route to the conversion of lignin sulphonic acid into a high-grade tannin resembling quebracho leads to decomposition and demethylation via an alkali fusion (Mauthe, Leverkusen). F. Stather, Freiberg, further improved the tanning properties of these 'lignophenols', i.e. phenols from lignin, by

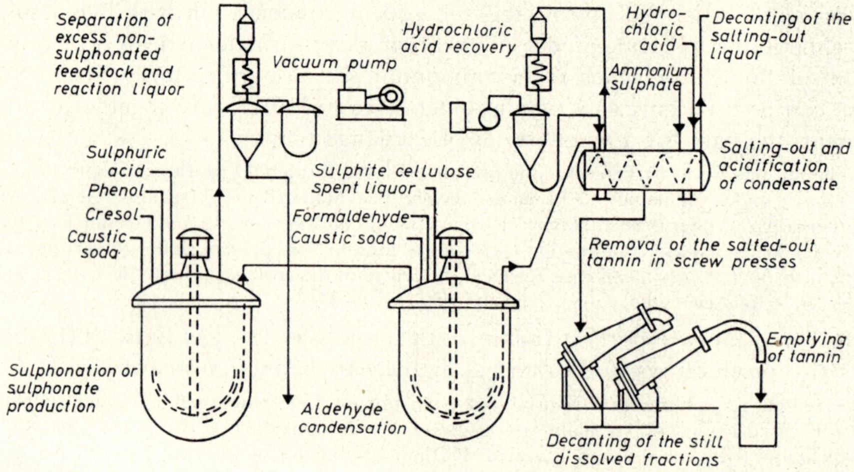


Figure 126. Diagram of the manufacture of Tanigan Extra A

partial condensation of lignin sulphonic acid, or by sulphitization with formaldehyde and bisulphite in the presence of phenols. Although this is certainly the best route to the complete replacement of natural tannins by lignin tannins these methods are not yet being applied in industry.

### (b) Mineral tanning

Mineral tanning is carried out exclusively in drums. The duration of the tanning cycle is 8–24 hours. The salts of trivalent chromium are used. These are formed by reduction of sodium or potassium dichromate with glucose or molasses in acid solution. The acid has to be partially neutralized. In an aqueous solution the chromium salts hydrolyse. Hydroxy salts form, yielding polynuclear 'ol-compounds' with a higher polymer structure. With increasing basicity the chromium salt solutions become colloidal, a condition which is essential for the tanning effect. The processes are extremely complicated.

The basic chromium salt liquors can be produced by reducing agents in the tanneries themselves. Chrome tannins ('Basochrom'), however, are usually supplied ready-made by the industry which first utilizes the dichromate for oxidation (e.g. for the production of anthraquinone from anthracene) and suitably treats and adjusts the chromium sulphate liquors. Before the chromium tannage, 'pickling' with salt and acid (described earlier) has to be carried out.

Tanning with aluminium salts and recently also with zirconium salts is being carried out on a modest scale.

Chrome leathers are light coloured, slightly greenish-grey, and are particularly suitable for shoe upper leather (box calf, chevreau).

### (c) Other tanning methods

A certain amount of chamois tanning with fats and fish oils is also being carried out. A considerable improvement of this method is 'Immergan' (BASF) or 'Leunagan' tanning with paraffin sulphonic chlorides ('Mersol'). The unhaired hide is treated with sulphonic chloride in the presence of soda. HCl cracks off with basic groups of the collagen, and paraffin sulphonamide groups form. When this method is used the 'fat residue' is chemically bound and cannot be washed out, whilst stuffing with fats and fish oils only impregnates the hide. Tanning can also be carried out with aldehydes, e.g. formaldehyde, alone.

Finally, resin tannins, i.e. condensation products of urea or melamine with formaldehyde, are also used for tanning.

### (d) Chemistry of tanning

Only the general principles of the chemistry of tanning are understood. The bonding of the vegetable tannins and of the synthetic polyphenol tannins is based on the formation of hydrogen bridges to the collagen of the skin. Sulphonic acid groups may effect an additional ionic bond to the basic points.

Mineral tannins are bonded by incorporation of the carboxylic acid groups of the collagen into the appropriate metal complexes. Tanning with sulphonic chlorides and with aldehydes, mentioned earlier, are typical examples of the reactions of principal valencies with the amino groups of the collagen.

In spite of these different chemical bonds a feature common to all tannins is that they react at least bi-functionally with the collagen fibrils and thus cause cross-linking of the collagen structure and thereby an increase in its strength and elasticity, and a reduction in its swelling capacity.

## 3. DRESSING OF LEATHER

The last stage of tanning is the dressing which comprises dyeing, lubrication, and surface treatment. The individual operations vary according to the method of tannage used and the properties required in the leather. Lubrication may be either simple oiling, or greasing manually or in the drum, or baking molten fats into the dry leather. These operations are mentioned in ascending order of the amount of fat introduced and consequently of the impermeability and the stability of the leather. Insoluble fats are used for the lubrication. Chromium leather is lubricated by 'fatliquoring' which is done almost exclusively in drums with fatty materials rendered water-soluble by emulsifiers.

Leather is dyed nowadays only with aniline dyes, applied either with a brush or in the drum.

There is no essential difference between leather dyes and textile dyes. In leather dyeing, too, metal complex dyes are used for specially fast dyes (glove leather).

The most widely used modern method of drying shoe upper leather is adhesive drying in which the wet leather is pasted on to sheets of glass and thus dried under tension. The leather is mechanically softened by 'staking'. Leather, particularly shoe upper leather, leather for fancy goods, etc., is dyed either by the classical method with casein coating dyes which contain casein as a binding agent, pigments as colouring matter, and sulphonated oils as plasticizers, or, as in the case of the quantitatively very important leathers with smoothed grain (Kona leather), by using aqueous Perbunan solutions as additional binders. The coating is applied with either brushes or spray guns. Between the individual coatings the leather is dried and it may also be smoothed by hydraulic presses. At this stage an artificial grain may be embossed in the leather. Leathers dressed exclusively by the classical method, using protein coating dyes, are buffed instead of smoothed.

The finished leathers are sold by size except for heavy leathers which are sold by weight. The measurements for these purposes are carried out after completion of the process of leather manufacture.

## 4. MAIN DEVELOPMENT PROBLEMS IN TANNING

Intensive efforts are being made everywhere to shorten the duration of the tanning of sole leather with vegetable tannins whilst preserving the favourable properties of the old tanning process and as far as possible improving wear resistance, reducing the loss from washing, and maintaining impermeability.

A further problem is the development of machinery. The quality of the surface is largely the result of the dressing and of the machines used. Many leathers, particularly shoe uppers, leave a lot to be desired in this respect. Dressing and other processes used in the leather industry still show strong traces of their origin in a craft. The trend is towards further mechanisation.

Yet another problem is the development of new tannins and auxiliary substances, the principal aim being the production of high-grade synthetic tannins to replace vegetable tannins, the creation of improved coating dyes and of special auxiliary substances, such as wetting agents and water-repellent agents.

### LITERATURE

F. Stather, *Gerbereichemie und Gerbereitechnologie*, 2nd edn, Berlin, Akademie-Verlag, 1951.

G. Otto, Die Lederherstellung, in Winnacker-Weingaertner, *Chem. Technologie*, 1st edn, Organ. Technologie II, München, Hanser Verlag, 1954.

H. Hersfeld, *Grundlagen der Lederherstellung*, Dresden and Leipzig, Verlag Th. Steinkopff, 1950.

F. Stather, *Leder und Kunstleder*, Berlin, Akademie Verlag, 1959.

## *C. The manufacture of commodities by chemical processing of natural and synthetic raw materials*

# I. Synthetic fibres

Total world production of fibre materials at the present time amounts to about 20 million tons annually. These are mainly natural fibres for the production of which arable and grazing land is required. Since the area available for this purpose is more or less constant a very large increase in the production of natural fibres cannot be expected. In the last 50 years the production of wool and cotton has increased, therefore, by only 30 per cent. In 1936, 12,360,000 tons of textile fibre materials, including natural silk, were produced for clothing.

However, the world population increases annually by about 20 million for whose feeding the area under cultivation has to be constantly enlarged. Hundreds of millions of people live almost without clothing or make do with a loin cloth. They will not remain satisfied for ever with this state of affairs. It can be assumed that in the undeveloped areas about 800 million people have a wholly inadequate supply of clothing. The raising of living standards in other parts of the world involves an increasing demand for textiles that cannot be met by natural fibres. This creates great possibilities for development with their attendant problems—for the synthetic fibre industry.

Fibre materials are, however, used not only for clothing; the requirements for decorative materials, carpets, and above all for industrial textiles, particularly for filters, are increasing constantly. At present the production of tissues for fibre bases (car tyres, conveyor belts, etc.) is almost as large as that of filter materials which are made mainly of synthetic fibres.

In part, the synthetic fibres are produced from natural raw materials, such as cellulose and proteins, the remainder synthetically.

They have been created by the genius of the chemist in collaboration with physicists, engineers, and textile experts. Synthetic fibres produced by the conversion of natural raw materials are also called 'semi-synthetic fibres', those from chemical raw materials produced by the chemist himself 'fully synthetic fibres'. This then is a first basic classification of synthetic fibres.

1966 World production (in 1,000 tons)

| Cotton | Wool | Synth. fibres | Cellulose fibres | Natural silk |
|---|---|---|---|---|
| 10,307·6 | 1,485·8 | 2,476·4 | 3,321·6 | 32·6 |

All figures taken from *Textile Organon*, June, 1967

The proportion of synthetic fibres in world production is still surprisingly small. However, the enormous increase in the world production of rayon and staple fibre in the years 1900, 1926, and 1951, from 100 tons to 96,000 tons to 1·8 million tons shows clearly the great prospects which the future holds for synthetic fibres. The raw material basis of the semi-synthetic fibres will always remain limited; that of the fully synthetic ones, however, is for the time being unlimited, being based on coal, petroleum, and natural gas. Compared with the production of rayon and staple fibres that of fully synthetic fibres is at present still rather small: in 1957 for instance 2,871,600 tons synthetic fibres were produced, of which 409,600 tons were fully synthetic fibres, i.e. about 14 per cent.

Of the 2 million tons of synthetic fibres produced in 1952 in the whole world, i.e. about 10 per cent of the total production of fibres, the proportion of the fully synthetic fibres was only about 7·5 per cent, 78 per cent were viscose and cuprammonium rayon, 14·5 per cent acetyl-cellulose fibres.

Cotton is, therefore, the most important fibre, then comes sheep's wool, followed by the other natural cellulose fibres, linen, hemp, China grass, etc., and natural silk.

| 1961 | Rayon + staple | Fully synthetic fibres | Total |
|---|---|---|---|
| BRD | 230,600 tons | 65,100 tons | 295,700 tons |
| DDR | 190,200 tons | 11,000 tons | 201,200 tons |
| | 420,800 tons | 76,100 tons | 496,900 tons |

In 1951, Germany produced 254,000 tons rayon and staple fibres and was thus in second place in the world production of semi-synthetic fibres, and led in the production of staple fibre (1951: 184,000 tons).

Fully synthetic fibres have been developed only in the last twenty-five years, industrial production began about 1940. World capacity of fully synthetic fibres is estimated at 1,800,000 tons per annum (1963: production 1,331,000 tons). In the U.S.A. the production of synthetic fibres had increased to 638,000 tons by 1964.

World production of synthetic fibres in 1964 was 1,688,300 tons of which

903,600 tons were polyamides,
337,900 tons were polyester,
300,700 tons were polyacryls, and
146,100 tons were other fibres.

The first fully synthetic fibre (PC-fibre) was produced about 25 years ago at Agfa-Wolfen (Rein). The re-chlorinated polyvinylchloride which is more soluble than the non-chlorinated PCU, was successfully spun to yield the PC fibre which is chemically inert and rot-proof. Only then were the possibilities of the fully synthetic fibre realized. Perlon (Dederon) was developed industrially by the former IG Farbenindustrie and was first produced by Aceta, Berlin-Lichtenberg (Schlack). Nylon was invented by the American Carothers and was put on the market by Du Pont. The polyacrylonitrile fibre (PAN fibre) was first spun by Agfa Wolfen (R. Rein), after dimethylformamide, $(CH_3)_2NCHO$, had been identified as a suitable solvent and swelling agent. Its use became economic by the discovery of a new process for the production of acrylonitrile from acetylene and hydrocyanic acid in 1939 by O. Bayer and P. Kurtz in the former IG plant at Leverkusen. The same product was marketed by Du Pont, U.S.A., under the name of Orlon. (Petrochemical production of acrylonitrile see p. 178).

H. Rein and W. Zerweck developed the manufacture of this synthetic fibre at Casella, Offenbach. Other synthetic fibres are the vinyl fibres 'Saran' and 'Vinyon' (U.S.A.), and 'Rhovyl' and 'Thermovyl' (Rhodiaceta). About 80 synthetic fibres are now in production.

### *Fibre structure*

All fibre-forming materials are high polymers and consist of large linear chain molecules ('linear polymers') produced by condensation or polymerization. As far as their internal structure is concerned they are, therefore, mainly unidimensional, whilst most film- and foil-forming synthetic substances are laminar or two-dimensional, and heavy materials usually three-dimensional. The basic units of linear polymers have a certain orientation in the fibre. They may agglomerate to molecular aggregates of colloidal size, and to submicroscopic crystalline structures. The colloidal sizes are between 10 and $10^4$ Å, the aggregates between 50 and 200 Å. The natural fibres cotton, wool, silk, and hair also have this structure.

Figure 127 (From Houwink, *Grundriß der Technologie der synthet. Hochmolekularen.* Leipzig, Akad. Verlagsges. Geest u. Portig K.-G., 1952) shows their microstructures greatly simplified.

Amorphous Crystalline micelles (crystallites) Orientated crystallites

Figure 127. Diagram showing the structure of fibre forming macromolecular substances

The orientation is recognizable, for instance, by the interference in the x-ray spectrum. X-ray structural analysis is, therefore, of great importance in fibre research since it gives information about the degree of orientation. The proportion of less and more highly orientated molecules varies greatly in the different fibres, in cotton about 70 per cent, in regenerated cellulose much less. The molecular groups are arranged in even higher formations called fibrils which are visible under the electron microscope. The properties and strength of the fibres are influenced by the degree of orientation, but particularly by the secondary valence-type cross-linkages of the filamentary molecules, i.e. by hydrogen bridges and ionic bonds and, in wool and hair, also by cystine bridges. All mild treatment of the fibres such as dyeing and finishing, acts preferentially in the less orientated regions and in the cavities.

Fibre-forming polymers can be divided basically into three main groups:

(1) Those with a paraffin structure with and without substituents. Polyethylene (polythene) can be considered as the basic substance although the chain is branched. Possible substituents are: chlorine, fluorine, the carboxylic ester and nitrile groups, the —OH and ether groups. Being true chain molecules they are thermoplastic.

(2) High polymers in which the molecules are linked by oxygen atoms. These include the polyesters, particularly the important Terylene fibre from terephthalic acid and glycol, as well as cellulose itself (acetal bonds) and its derivatives.

(3) Fibres with acid amide bonds, i.e. wool, silk, synthetic fibres from proteins, and the polyamide fibres of the Perlon, Dederon (German Democratic Republic) and Nylon types. Similar to these are the high polymers from isocyanates and glycols, i.e. the polyurethanes (Perlon U, formerly Igamid, now Ultramid, by O. Bayer and colleagues).

$$\text{OCN—R—NCO} + \text{HO—R'—OH} \longrightarrow (\text{—OC—NH—R—NH—COO—R'—O—})_n$$

*Methods of formation, spinning methods*

High polymers to be used in the textile industry have, then, to form filamentary molecules and must be capable of being spun, in the sense of the activity of the spider and the silkworm, not that of the girl at the spinning wheel who only makes yarn from ready-made fibres and does not actually spin. Spinning comprises two processes: formation in the liquid or plastic stage, and strengthening of the filaments. To this has to be added orientation, i.e. longitudinal arrangement of the molecules to produce certain mechanical properties, which is achieved by 'stretching'.

(1) Formation in the wet spinning process is by
   (a) precipitation of the filament from the solvent, e.g. with water (PC fibre, cuprammonium rayon);
   (b) precipitation by chemical conversion of the solution, e.g. with acid or salt solution (viscose rayon, protein fibres, Wolpryla, Prelana).

(2) Formation in the dry spinning process by solution of the polymer and evaporation of the solvent (acetate silk, Orlon).

(3) Formation can also be effected by melting, forcing out a fine jet from a nozzle (spinneret), and cooling (fusion spinning process, e.g. Nylon, Perlon, Dederon, and Lanon).

The type of spinning machinery used depends on which of these three methods is operated.

In every case the solution or melt has to be adjusted to a given viscosity and has to be free of any trace of gas bubbles and solid particles before it enters the spinning nozzles. The supply to the nozzle is measured exactly by a geared pump (spinning pump). The spinning nozzles are removable discs made from various materials, e.g. Au/Pt and Ta are mostly used for viscose, special steel for polyamides and polyester; Ni is used only for acetate silk, together with, more recently, special steel.

Diameter of bore: 0·06–0·08 mm for viscose,
0·25–·035 mm for polycondensate fibres,
0·1 mm for cuprammonium rayon only,
1 mm for bristles only.

After leaving the nozzle the liquid filament enters a setting (spinning) bath or a drying stack (see Figure 128) where coagulation takes place. If the filament leaves the nozzle as a melt it is led through a cooling stack.

The liquid jet or melt issues from the nozzle having the same thickness as the bore, i.e. it is rather thick; as a result of the spinning 'pull' the filament formed is, however, very fine since the 'drawing off' takes place with considerable speed.

| Drawing-off speeds: | | |
|---|---|---|
| | viscose | 60–100 m/min, |
| | acetate | 300–600 m/min, |
| | polycondensates | 800–1,200 m/min. |

The denier of the filament depends on the drawing-off speed (see p. 378). The yarn is taken off usually by one of two methods:

by winding on to bobbins or reels, or
by coiling into centrifugal pots (pot spinning process).

Before the yarn is wound on to a bobbin or reel it may also be led over revolving rollers which then determine the drawing-off speed. Figure 128 shows the arrangement in the bobbin and pot spinning processes. As it enters the rotor (pot) of the centrifuge the filament is twisted. It is wound parallel on to the bobbins. To distribute the windings uniformly a filament guide imparts to the filament a shuttling motion. The twisting of the filament into yarn follows as a separate operation.

A special spinning process has been developed for cuprammonium rayon (Figure 128a). The spinning nozzle (spinneret) is fitted closely above a glass funnel; water entering the jacket surrounding the funnel overflows into it. Water and filament issue at the bottom of the funnel in concurrent flow. The filament

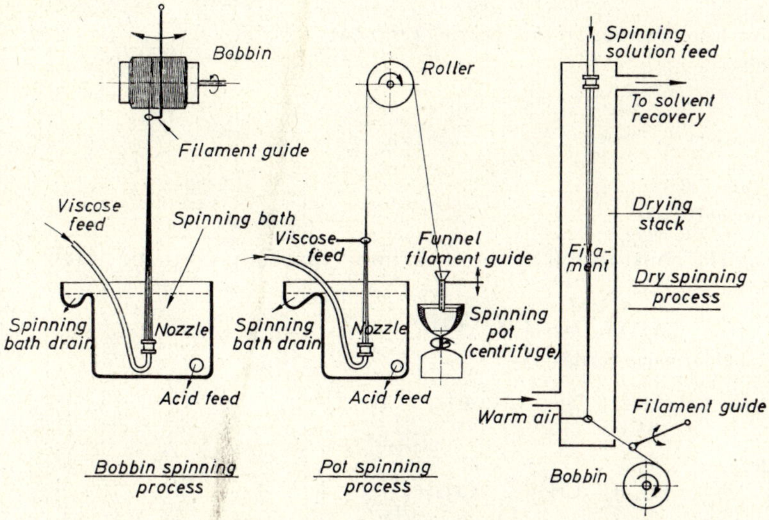


Figure 128. Spinning processes

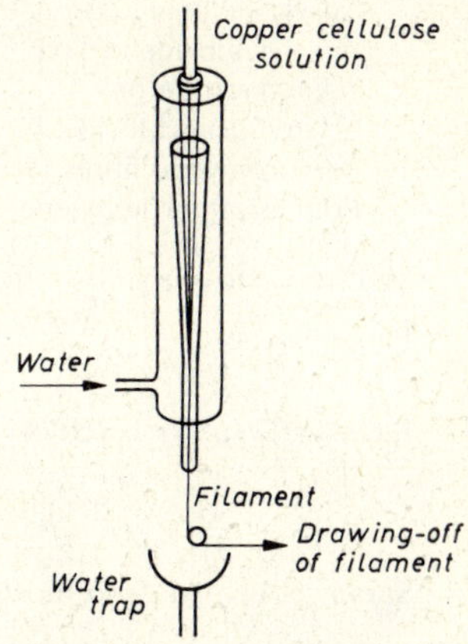


Figure 128a

passes through a bath, then over the roller to the reel. In the wet spinning process extensive washing operations follow the spinning.

If the filaments after precipitation are twisted and spooled they are called filament rayon yarn. If, on the other hand, the filaments are cut or shredded into flakes so that they resemble wool and cotton they are called rayon staple fibre and the resulting textile yarn 'spun rayon yarn'.

They are then processed, either alone or in combination with other fibres, into yarns and are particularly suitable for the production of mixed fabrics.

The following Table gives the most important synthetic fibres.

## I. Fibres from natural polymers

(a) Fibres from regenerated cellulose and cellulose derivatives:
Viscose rayon and staple fibre,
Cuprammonium rayon and staple fibre,
Acetate silk and staple fibre.

(b) Protein fibres:
Casein (Lanital, Fibrolane, Aralac),
Groundnut protein (Ardil),
Soya protein (Silkool),
Maize zein (Vicara).

(c) Sugar derivatives:
Alginate fibres (from alginic acid obtained from algae) (Tubular foils are produced from alginic acid which are used as digestible synthetic sausage casing (Wolff & Co., Bomlitz)).

## II. Fibres from synthetic polymers

Polyamide silk and fibres (Dederon, Perlon, Nylon, Rilsan, Kapron, Silon, Steelon, Grilon),
Polyurethane fibres (Perlon U),
Polyvinylchloride fibres (PeCe, Rhovyl, Chlorin),
Polyvinylidenechloride fibres (Saran),
Polyacrylonitrile fibres (Wolpryla, Prelana, Orlon, Dralon, Acrilan, Redon, Creslan),
Polyvinyl alcohol fibres (Kuralon, Vinylon),

Polyethylene fibres (Courlene),
Polyester fibres (Lanon, Terylene, Dacron, Trevira, Diolen, Terlenba).
Copolymerizate fibres from:
Acrylonitrile with
vinylchloride or
vinylacetate or } (Vinyon N, Dynel)
vinylidenechloride
Polypropylene fibres (Recoon),
Polytetrafluoroethylene fibres (Teflon).

The following list shows the constitution of the most important fully-synthetic fibres:

Polyethylene

$$-CH_2-CH_2-CH_2-CH_2-$$

Igelit PCU (polyvinylchloride), chlorinated further PeCe

$$-CH_2-\underset{Cl}{CH}-CH_2-\underset{Cl}{CH}-CH_2-\underset{Cl}{CH}-$$

Saran (polyvinylidenechloride)

$$-CH_2-\underset{Cl_2}{C}-CH_2-\underset{Cl_2}{C}-CH_2-\underset{Cl_2}{C}-$$

Perlon, Dederon (poly-ε-aminocaproic acid)

$$-HN-CH_2-(CH_2)_4-CO-NH-CH_2-(CH_2)_4-CO-NH-CH_2-(CH_2)_4-CO-$$

Nylon (condensation product from hexamethylenediamine and adipic acid)

$$-HN-(CH_2)_6-NH-CO-(CH_2)_4-CO-NH-(CH_2)_6-NH-CO-(CH_2)_4-CO-$$

Perlon U (from glycols and di-isocyanate) polyurethane

$$-O-(CH_2)_n-O-CO-NH-(CH_2)_m-NH-CO-O-(CH_2)_n-O-$$

Orlon, Wolpryla, Prelana, Dralon (polyacrylonitrile)

$$-CH_2-\underset{CN}{\underset{|}{CH}}-CH_2-\underset{CN}{\underset{|}{CH}}-CH_2-\underset{CN}{\underset{|}{CH}}-$$

Terylene, Lanon, Trevira, Diolen (terephthalic acid polyester)

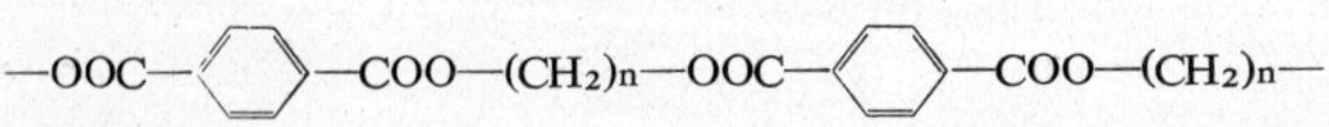


*Physical properties*

The nature and origin of the fibre can be identified from the fibre cross-section under the microscope; thus, e.g. the fibre solidified from a melt has an almost circular cross-section and a smooth surface. Cuprammonium rayon also is almost round and very fine since it is precipitated directly as hydrated cellulose. Fibres produced by the wet spinning process (salt and acid precipitation bath) have a very uneven surface because the salt precipitates them first as a 40 per cent gel which then shrinks in the acid bath. It has recently proved possible to produce structured polycondensate fibres, the "profile fibres" (Böhringer and Bolland).

The prototype of the macromolecular structure of a natural fibre molecule is cellulose. Natural and synthetic fibres have molecular weights of upwards of 10,000. The number of basic molecules forming a chain molecule gives the degree of polymerization, which varies greatly. Thus for instance native cellulose contains 2,000–3,000 glucose molecules, whereas regenerated cellulose (rayon) owing to depolymerization contains only 250 to

500 molecules. The determination of the degree of polymerization by molecular gravimetry (e.g. by end group determination, osmotic methods, or viscosity determination) yields a statistical mean value since the substances are polymer homologues.

Single crystals examined crystallographically by von Laue's method give x-ray spot diagrams. Irregular powders of crystalline substances yield Debye-Scherrer concentric rings (ring patterns). In well ordered natural fibres (e.g. China grass, and also cotton), owing to the partial orientation of the ordered zones with respect to the fibre axis, Debye-Scherrer pictures show a change of the ring patterns to a sickle shape (fibre patterns). With increasing orientation the fibre x-ray pattern becomes increasingly well defined.

It is possible to determine by x-ray analysis:

(1) whether the material is amorphous or crystalline,
(2) whether in an amorphous substance the dimensions of the structural unit are altered by polymerization,
(3) whether the substance is only partly or completely crystalline,
(4) whether the crystals are orientated (fibre pattern),
(5) the dimensions of the crystallites, and lattice expansion owing to swelling.

Without x-radiography our knowledge of the microstructure of high polymers would be very incomplete.

### *Stretching*

In the wet spinning process stretching takes place immediately after the filament leaves the nozzle whilst it is still in the gel phase. The orientation and increase in strength achieved by it is not as large as that achieved by the subsequent stretching of the finished filament. With viscose filaments stretching is possible only in hot water and this is done particularly in the case of tyre cord rayons (hot stretching). For cellulose fibres a stretching ratio of 1:1·8 is normal. Fully synthetic fibres have to be stretched much more strongly to attain optimum mechanical properties. Polycondensate fibres are stretched 1:3 to 1:5 cold and hot, polymerizate fibres 1:6 to 1:10 only hot.

However, true cold stretching does not exist, for the filament warms up during stretching just like a mild steel wire does during drawing. The carbon chains of the synthetic fibres spun from the melt are much more highly zig-zagged than the relatively rigid cellulose molecules which preserve a certain order in the dissolved and swollen states. The former have therefore to be stretched more strongly so as to obtain an orderly arrangement of the molecules.

This order is, however, not to be compared with the crystalline state. There are only more highly and less highly ordered zones in the fibres. Stretching achieves only the orientation of existing ordered zones in the cellulose fibres and not the creation of new ones, but the molecules do not 'forget' this order when the material is dissolved.

### *Bonds, orientation and fibre strength*

The cohesion and, therefore, the strength of the fibre is based on two types of bonds:

(1) 'Primary bonds' which include heteropolar and ionic bonds as well as homopolar or atomic bonds, and
(2) 'Secondary bonds' which are based on the attraction of dipoles and on exchange (dispersion) forces.

The latter are a consequence of electron movements in the atoms which produce constantly changing dipoles and continuously induce new dipoles in neighbouring atoms. Whilst in a low-molecular carbon chain the secondary bonds are weak, e.g. in paraffins, with increasing molecular weight they add up and may eventually even surpass the primary bonds. In such a case the melting point is higher than the decomposition temperature; this happens e.g. with polyacrylonitrile.

As mentioned earlier the orientation of chain molecules of adequate size leads during stretching to the formation of crystalline zones. This is the maximum effect of the formation of secondary bonds. During the formation of crystal lattices the side groups of the chain molecules fit themselves into the potential holes of the neighbouring molecules. A sort of gear meshing is formed and in polar compounds as the molecules approach each other secondary bonds become effective, e.g. hydrogen bridges. It is then difficult to loosen these by pulling. With increasing tension the molecules at first keep springing back into this gear meshing until they finally slide past each other. A partially crystalline structure has less ductility with greater strength; such behaviour is normal for a true fibre. Rubber must have great elasticity and great strength; its strength is proportional to the degree of its crystallinity in the extended state. A fibre has its maximum strength in one preferred direction, the direction of orientation. When the molecules are orientated in a parallel direction each is held by the neighbouring molecule along its entire contact surface. Unorientated tangled molecules cannot possess high strength since each molecule is in contact with other molecules at a few points only.

This has been shown very elegantly by experiments made with natural silk, a strongly orientated fibre with great tensile strength. If the silk is dissolved and spun without orientation its strength falls to almost one hundredth of its former value. If the fibre is spun with simultaneous strong stretching its strength increases even beyond that of the original fibre. It also follows from these considerations that the strength of a fibre increases with increasing size of the molecule. For similar reasons a yarn is the stronger the longer are the individual fibres from which it was spun. For worsted yarn combed long fibres are used.

The stress-strain diagram is used to determine the mechanical strength of solid materials, particularly also that of fibres.

In ideally elastic materials (e.g. steel) stress and strain up to a certain limit are proportional. If this 'flow limit' is exceeded with increasing strain the material begins to flow and eventually breaks. The elastic part of the tension, i.e. the zone in which the stretched material springs back into its original position when the tension is relaxed, is of special

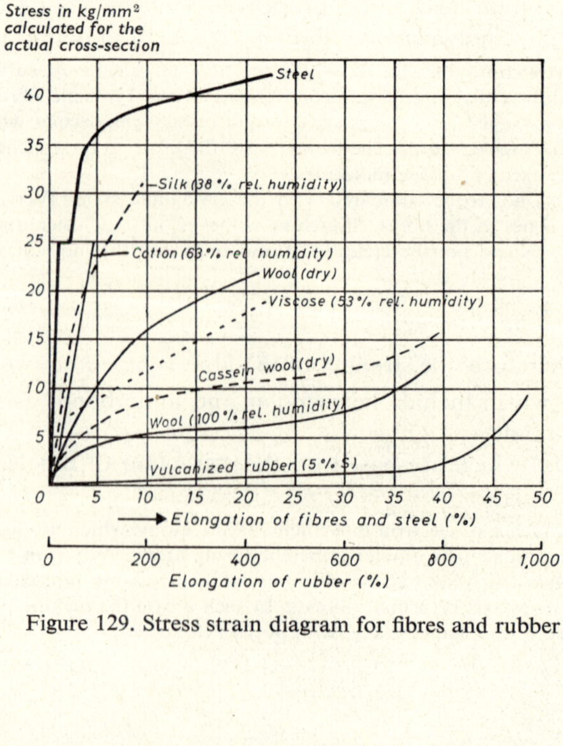


Figure 129. Stress strain diagram for fibres and rubber

importance for the assessment of fibres (see Figure 129) (From Houwink, *Grundriß der Technologie der synthetischen Hochmolekularen*).

*Characteristics of yarns*

The weaver is interested in the yarn number, the 'denier'. There are three indices of strength:

Denier: 1 = weight of 9,000 m in g,
Grex (U.S.A.): 1 = weight of 10,000 m in g,
Metric No. = number of m per 1 g of filament
(450 Denier = 500 Grex = 20 Nm).

A titre of 120 den can consist e.g. of 24 single fibres with titres of 5 den each. The usual titres are about 12-1,650 den (Perlon Monofil for stockings to viscose cord silk). The thick ribbon in cellulose manufacture has titres from 500,000 to 1 million den, but only up to the stage when it is shredded into flakes.

Strength is defined as g load/den at which rupture occurs. Elongation (%) indicates the percentage by which the fibre stretches before breaking. Breaking kilometer (Rkm) is the length of the fibre at which is breaks as a result of its own weight. An objective comparison of the different types of fibres is possible, however, only in terms of material strength in kg/mm² ('absolute strength'). If two fibres with the same titre have the same breaking strength (e.g. g/den or Rkm) their material strength is the same only if they have the same specific weight. Otherwise their fibre cross sections are inversely proportional to their specific weights. For an exact comparison the calculation must be based on the formula Rkm $\cdot \gamma(\gamma =$ spec. weight). Synthetic fibres have widely differing elongations and breaking strengths, e.g.:

| | Elongation % | Breaking kilometer, Rkm |
|---|---|---|
| Nylon, Perlon, Lanon | 16–30 | 35–70 |
| Wool | 20–40 | 9–18 |
| Viscose rayon (normal) | 15–22 | 12–20 |
| Cotton | 3–10 | 17–40 |

Elongation may be elastic elongation which is reversible, or permanent elongation. Atmospheric humidity and humidity of the fibre greatly affect these values. The hydrophobic fibres, e.g. acetate silk and PC fibres, are relatively insensitive, regenerated celluloses are very sensitive. It is remarkable that the strength of cotton increases when it absorbs moisture, a fact that may be explained by its internal microcrystalline structure. Wet wool will stretch, the elongation is, however, irreversible.

The chemical strength of fibres is of great interest, particularly for their industrial uses (filter material). Cellulose fibres are attacked by acids (hydrolysis) and oxidants, and are fairly resistant to alkalis. (If even the smallest amount of acid is left in the cellulose fibre from the spinning process or other processes and dries in it, the fibre will become brittle). On the other hand, acetate silk is saponified by alkali. PC fibre, Orlon, and polythene have the highest chemical strengths because of their paraffin chain structure (e. g. PC fibre is resistant to acids, even HCl and $HNO_3$, and alkalis). But they are sensitive to organic solvents (swelling). The most resistant in every respect is the polytetrafluoroethylene fibre (Teflon). It is unattacked by hot aqua regia.

Nylon, Perlon, polyesters have vulnerable spots and are less resistant. Whilst they have great chemical strength they present difficulties in dyeing. The fibres absorb only special dyes and cannot be delustered by subsequent treatment.

This limitation can be overcome by dyeing and delustering during spinning, i. e. incorporation into the fibre of dyes or finely dispersed pigments which are soluble in organic media. With such pigments, e.g. the phthalocyanines, colour fastness of the highest degree can be achieved.

## 1. MANUFACTURE OF SEMI-SYNTHETIC FIBRES

### (a) Cellulose fibres

In 1880 the French Count Chardonnet developed the industrial production of nitrocellulose fibre from collodion; this fibre is impressive because of its magnificent lustre; it is, however, inflammable (Chardonnet silk). Mixed with PC fibre it becomes non-inflammable.

Between 1890 and 1895 the Germans Fremery and Urban developed the manufacture of rayon; they based their process on the observation that cellulose is soluble in ammoniacal copper oxide and can be precipitated with water. At about the same time the Englishmen Cross and Bevan discovered the formation of viscose from alkali cellulose and carbon disulphide (cellulose xanthate) and the fact that regenerated cellulose can be re-precipitated with acids. The first rayon was cellulose acetate (Schützenberger, 1869) the full industrial development of which did not begin until after 1920 although as early as the turn of the century small plants operated in the U.S.A. and in England. After 1930 came the great growth of the staple fibres as a substitute for cotton and to some extent for wool. The first purely synthetic fibre was the PC fibre (1935), then followed in 1938 Nylon, Dederon, and Perlon and finally Perlon U. During the war the polyacrylonitrile fibre was created (PAN, Orlon), later Dralon and Wolpryla, and in 1950 Terylene (polyester fibre).

*Cuprammonium rayon*

The cuprammonium process is based on the dissolution of cellulose in 25 per cent ammonia and a 40 per cent solution of copper sulphate. Specially pure cellulose with a content of 96–98 per cent $\alpha$-cellulose, and cotton linters which have to be thoroughly prepurified are used as raw material. Agitating for two hours produces a paste with a cellulose content of 13 per cent to which is added 8 per cent caustic soda. The basic copper sulphate is converted into copper hydroxide and a clear spinning solution is formed with a cellulose content of 7·5 per cent. The formation of the solution (24°C) is promoted by the injection of limited amounts of air which reduces the viscosity. If the viscosity drops too rapidly the oxidation process can be stopped by the addition of sodium sulphite.

In order to filter and homogenize the solution it is pressed through a fine-mesh nickel wire net in filter presses; the pressure is then released under vacuum and a fairly large amount of ammonia escapes. The degree of polymerization and therefore the viscosity of the solution do not change during long storage with exclusion of air and light, unlike the viscose solution (cellulose xanthate) which continues to mature.

Then follows spinning using the stretch spinning process described earlier, by precipitating with water at 35°C. The coagulation of the filament takes place as a result of the absorption of copper and ammonia by the water. In order to complete the coagulation the fibres are sprayed with dilute sulphuric acid and then washed thoroughly. The individual filaments of cuprammonium rayon agglutinate easily and yarn is, therefore, not always produced. J. P. Bemberg developed a fully continuous cellulose-upgrading, -dissolution, and -spinning process for cuprammonium rayon.

Cuprammonium rayon had a hard struggle in competing against viscose rayon. The economics of its production depend on the complete recovery of the copper and as far as possible of the ammonia from the dilute acid waste liquors. Not until this became possible by the development of the Wofatit-Haltmeier process (Dormagen) could the cuprammonium rayon process compete against the cheaper viscose process. The adsorbent for copper used is a Wofatit resin (cation exchanger) formed from phenol-formaldehyde with $SO_3H$ groups. The copper is dissolved out of the resin after saturation with sulphuric acid formed in the spinning process (regeneration of the resin) and is precipitated with soda as the basic salt $CuSO_4 \cdot 3Cu(OH)_2aq$. The copper slurry is concentrated in Dorr thickeners by decanting and is recovered in rotary filters. The ammonia is separated from the spinning waste liquors using steam under vaccum, the Haltmeier process making use of the waste heat in heat exchangers. Those parts of the equipment which come into contact with acid liquids are gummed or made of V4A steel. Cuprammonium rayon gives very delicate smooth tissues with a mat sheen and at one time was widely used, therefore, for ladies' stocking and underwear. World production of cuprammonium rayon is shrinking. It has been completely displaced by polyamide fibres for the manufacture of stockings, but is still used for many special applications as a weaving and knitting yarn.

## *Viscose rayon*

It was given its name after the spinning solution which is highly viscous (viscose solution). The viscose process consists of 8 main operations (see Figure 130) (From F. Endress, Winnacker-Weingaertner, *Chem. Technologie*, Org. Techn. I, Hanser-Verlag).

(1) Formation of the alkali cellulose and primary maturing ('alkalization') (a–e).

(2) Reaction of the alkali cellulose with carbon disulphide, simultaneous formation of xanthate (f)-('Sulphidation'). Xanthate is a soluble ester salt of dithiocarbonic acid.

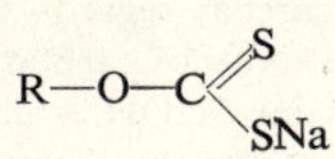

(3) Dissolution of the cellulose-xanthate to yield the viscose solution. Pressure release. Secondary maturing (partial hydrolysis of the ester) until the desired degree of viscosity is achieved ('solution') (g–l).

(4) Spinning by forcing the viscose filament into Na- and Zn-sulphate containing sulphuric acid, causing the xanthate to decompose into cellulose + $CS_2$ ('spinning') (m–t).

(5) Draining to remove every trace of acid by washing ('deacidification').

(6) Desulphurization of the fibre to remove colloidal sulphur with sodium sulphide or sodium sulphite, scrubbing and subsequent bleaching with NaOCl or $H_2O_2$, or recently also with chlorine dioxide ($ClO_2$) which does not decompose the cellulose, removal of adsorbed traces of metal by scrubbing with diluted acid, washing with water and finishing with textile improving agents ('aftertreatment').

(7) Drying which must be carried out slowly and carefully, and

(8) Textile processing by twisting and reeling.

Alkali cellulose is produced by soaking cellulose pulp for 2 1/2 hours in 18–19 per cent caustic soda at 16°C. Some of the hemicelluloses are leached out.

The discharged liquor is processed by dialysis. After the liquor has been removed by pressing in hydraulic presses the alkali cellulose is shredded and 'matured' in steel vessels with excess of air at 20–24°C for 35–40 hours; this "maturing" consists of a partial decomposition of the cellulose, i.e. a reduction of the molecules. Recently, methods have been developed for continuous alkalization and 'maturing'. The cellulose in the form of flakes can be soaked in alkali and pressed in screw presses.

Cellulose xanthate ('xanthate') is produced by reacting cellulose with carbon disulphide for 2–3 hours at $< 30$°C in water-cooled mixing drums ('barattes'). The degree of conversion is rather low. A glucose molecule has 3 OH-groups of which in the fresh xanthate only half a molecule is esterified, i.e. for two glucose molecules only one. The xanthate is nevertheless soluble in dilute NaOH. In place of the xanthate mixing drums large kneaders of the Werner-Pfleiderer type are also used. 'Sulphidation' must

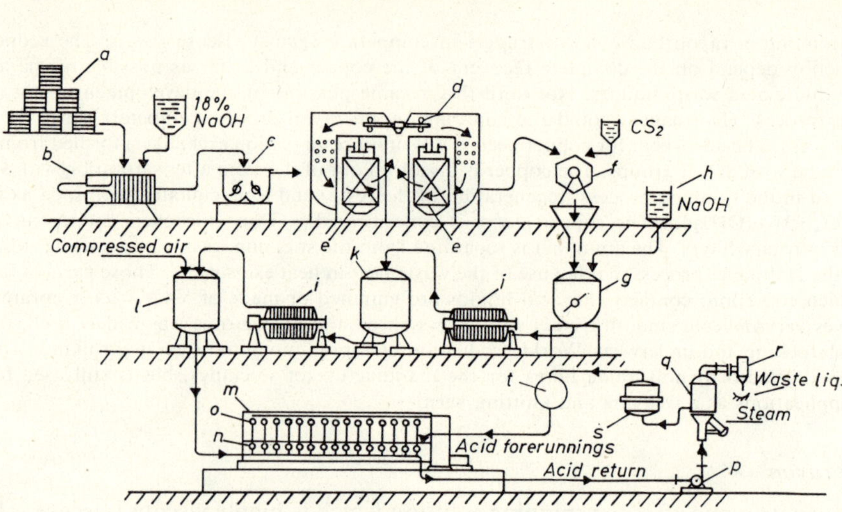


Figure 130. Diagram of the manufacture of viscose rayon
*a* cellulose tank *b* steeping presses *c* shredder *d* aging *e* aging cars *f* baratte and measuring vessel *g* dissolver *h* dissolving liquor *i* filter presses *k* secondary maturing tank *l* spinning tank *m* spinning machine *n* viscose pumps *o* spinning bobbins *p* acid pump *q* evaporator *r* condenser *s* pebble filter *t* bath tank

be carried out with the greatest possible care as regards both fire prevention and industrial hygiene because of the great toxicity and high explosivity of carbon disulphide.

Dissolution of the xanthate with 7 per cent NaOH is carried out in dissolvers, horizontal vessels with horizontal agitators and internal fittings for the disintegration; the material is homogenized and broken up by pumping it through a 'pulverizer'. After filtration in frame filter presses using cotton filters the solution which now has the consistence of honey is left for the 'secondary maturing' which lasts 8–12 hours at 13–15°C. Part of the ester saponifies to hydrated cellulose and $CS_2$. The latter is, however, not free but forms sodium di- or trithiocarbonate with the excess caustic soda which gradually decompose into $Na_2S + Na_2CO_3$. (This is the reason why during spinning $H_2S$ occurs with $H_2SO_4$). Viscosity increases. Once more filtration is carried out and the pressure is released during a whole day in tall vertical vessels under high vacuum. From these 'spinning vessels' the viscose runs into the spinning machines.

'Modifiers' (e.g. cyclohexylamine) are very important for the spinning process, particularly for high-strength tyre cord silks.

The spinning process has been described earlier. Spinning nozzles are made of noble metal alloys or tantalum. For viscose rayon according to the denier 15–120 mesh, for viscose cord silk 600–750 mesh, for rayon staple 3,000–20,000 mesh are used. Excess alkali is first neutralized with sulphuric acid. Free cellulose dithiocarbonic acid decomposes into cellulose and $CS_2$:

$$=\text{CH—O—CS—SNa} \xrightarrow{H_2SO_4} =\text{CH—O—CS—SH} \longrightarrow =\text{CH—OH} + CS_2$$

In addition to sulphuric acid about double the quantity of sodium sulphate containing $ZnSO_4$ and $MgSO_4$ is used (saturated solution). In salt-free baths decomposition takes place too rapidly, i.e. in practice no intermediary formation of cellulose xanthate gel occurs which because of its high swelling value is particularly easily deformable (orientation) whilst cellulose hydrate (decomposition product) is deformable only with great loss of elasticity (brittle fibres!). The first phase of spinning is, therefore, merely a shift of concentration and only a limited amount of cellulose hydrate is produced. The filament is at first only a plastic and deformable gel tube with a circular cross section, enveloped in an epidermis membrane of zinc cellulose xanthate (Klare and Gröbe). Cellulose hydrate does not form until the acid bath stage during which the filament shrinks and its surface assumes its characteristic fissured appearance.

The behaviour of the xanthate has been utilized to produce the crinkled 'XT fibre' by first making xanthate fibres then cutting them and treating them with acid. During the formation of cellulose hydrate irregular shrinking takes place which produces the crinkling of the fibre.

Regenerated cellulose is obtained after heating to 90°C in water or during drying. The individual phases can be seen by x-ray spectrography. First, the x-ray spectrum of alkali cellulose is obtained, then that of cellulose hydrate which resembles it. The diagram of native cellulose does not, therefore, reappear.

Further shrinking takes place on the bobbin. Sulphuric acid and sodium sulphate solution are recycled in the spinning machine and are gradually replaced by fresh solutions.

During spinning $CS_2$ and, due to the $Na_2S$ content, $H_2S$ are liberated and very thorough suction has, therefore, to be applied to the spinning machine. A spinning machine with 100 spinning points produces about 110 kg rayon of 120 den per day. In order to prevent traces of acid remaining in the filament after drying which could cause a reduction in strength by hydrolysis, a very careful deacidification must be carried out. This is done by washing the filament on the bobbins thoroughly with water of 40°C by the vacuum washing method, but nowadays the modern method of pressure washing is mostly used. The fibre contains colloidal sulphur which is dissolved out (desulphurization), then follow, as stated earlier, washing, bleaching, acidification, washing with wash-active agents (e.g. soap) and softened water, and finally drying with hot air in drying chambers.

Continuous operation has been developed furthest in the manufacture of rayon staple fibre (first used by Agfa-Wolfen about 30 years ago) where the band spinning method is used. In the spinning machines the individual filaments are gathered into ribbons which are subjected to the treatment described above. The last operation is shredding on roller frames. The rayon staple flakes are pressed into bales and dispatched.

Important problems for the viscose rayon industry are the large amounts of sulphuric acid required, and disposal of the sodium sulphate and spinning waste liquors produced. 1,100–1,200 kg sulphuric acid are consumed per 1,000 kg fibre. The method depends, therefore, on the production of sulphuric acid. So far the usual practice has been to crystallize Glauber's salt and to calcine it. Recovery of the sulphuric acid is a difficult problem. One of the technical difficulties of $Na_2SO_4$ electrolysis is caused by the high consumption of electricity which is about double that of NaCl electrolysis. Methods are being developed to produce sulphuric acid from $Na_2SO_4$ with HCl.

Carbon disulphide is recovered by adsorption methods or by a continuous condensation method.

### *Acetate rayon*

Since cellulose has 3 free OH groups per hexose molecule it can be triacetylated with acetic anhydride. Cellulose triacetate is, however, soluble only in chloroform, trichloroethylene and similar solvents which are unsuitable for spinning. The triacetate obtained is, therefore, partially saponified with water until an intermediate phase between triacetate and diacetate is reached which is soluble in acetone in which it forms spinnable solutions. In recent times triacetate fibre has also been produced, e.g. 'Tricel' by British Celanese and Courtaulds Ltd. by dry spinning from trichloroethylene or methylene chloride.

Esterification is carried out in the presence of sulphuric acid which causes a partial decomposition of the cellulose. The operation has, therefore, to be carried out rapidly and at a low temperature. The degree of polymerization of the cellulose drops by about half. Various viscosities can be obtained by controlled acid hydrolysis (depolymerization), similarly as in the primary maturing in the viscose process. The method is divided into three stages:

Acetylation and partial saponification for the production of 'Cellit', preparation of the spinning solution, and the spinning process.

### *Production of Cellit K*

For the acetylation, high-grade cellulose in the form of crepe paper or usually cotton linters is shredded (in a ball mill) and pre-treated in a bronze vat with 30–40 per cent acetic acid. After 2–3 hours of kneading at 50°C a friable mass is obtained which is conveyed via a silo into a bronze acetylizer fitted with cooling jacket and a horizontal agitator. There acetic anhydride, a very small quantity of sulphuric acid and methylene chloride are added to dissipate the heat and the mass is kneaded for 5–6 hours. The evaporating methylene chloride prevents by cooling undesirable decomposition by acetolysis and is constantly recycled via a reflux condenser. The degree of esterification can be controlled by measuring the viscosity with Höppler's dropping ball viscometer.

Then, saponification is carried out in an agitator vessel with the addition of 0·5 parts of water and a little sulphuric acid at a temperature which rises spontaneously to 80°C. In 6–8 hours a degree of esterification of 2·25–2·5 is reached. The methylene chloride is distilled off. When the desired degree of saponification is reached the solution is buffered using sodium acetate and the process stops. A 30 per cent cellulose acetate solution in 75 per cent acetic acid is obtained which is precipitated with 8–10 per cent acetic acid. The acetate separates as a lump which is broken up in a crushing machine, separated from the mother liquor on a screen cloth, crushed again, and washed in a washing machine (spraying on a perforated plate conveyor) operating on the countercurrent principle, to obtain 18 per cent acetic acid. (The concentration is important for the economical recovery of the acetic acid, which is further concentrated directly to 99 per cent acid by azeotropic dehydration with ethyl acetate). The acetate is then centrifuged and, travelling via a charging valve, dried in a vacuum paddle drier. It is then discharged through another valve, to be filled in sacks ('Cellit K').

### *Preparation of the spinning solution*

The spinning solution is obtained by dissolving the Cellit in acetone or methylene chloride (20–25 per cent solution) in copper or VA (corrosion-resistant) steel agitator vessels since Cellit always reacts slightly acid by separating acetic acid. After filtration of the very highly viscous solution through cotton filters and the usual pressure release the solution is spun, using the dry spinning process (see above), through nozzles with 0·5 mm diameter of bore at 60°C in a drying stack of 4 m length through which passes hot air. Since acetyl cellulose is an excellent non-conductor the filament is passed through a finishing agent which renders it conductive so as to prevent electrostatic charges. Then follows the twisting of the filament (dry spinning process, see Figure 128).

The cost price of acetate rayon raw materials used to be 5–7 times that of viscose. Moreover, the equipment has to be of noble material because of the strongly corrosive properties of acetic acid. The maximum possible recovery of the acetic acid and of the solvents (acetone or methylene chloride) is of the greatest importance for the economics of the process. Acetone is recovered almost completely from the waste gas by activated carbon. As a result of almost 100 per cent recovery of the acetic acid (acetylation) and of the solvents, and due to the rise in cellulose prices, acetate rayon is now cheaper or no more expensive than viscose rayon!

After spinning the fibre is finished and does not require any further wet treatment or drying. It gives very fine and attractive tissues with an elegant sheen. The production and use of acetate rayon will, however, always remain limited.

Dyeing has to be done with special dyes, such as Cellit, Celliton, Cellitazol, Setacal, or Celanthrene dyes; since acetyl cellulose is not hydrophilic and does not swell in water ordinary dyes do not impregnate the fibre.

Acetate rayon's share of world production has increased from about 7–8 per cent of viscose production in 1937, to about 10 per cent of viscose production in 1957.

### (b) Protein fibres

Protein fibres were created in response to the demand for a fibre resembling as closely as possible the natural protein fibres, wool and silk. One has, however, to bear in mind that the fibre-forming proteins, wool keratin and silk fibroin, which possess an x-ray pattern and contain crystallites, have a different microstructure from the food proteins whose structure is almost amorphous. It is fundamentally unnatural for such food proteins to be used for the formation of fibres. The strength of these protein fibres, particularly when wet, is, therefore, low and they can only be processed in conjunction with true fibres, e.g. wool.

The spinning process of protein fibres is nearly always based on the preparation of a colloidal solution of maximum concentration from protein (e.g. casein) and alkali, and spinning into dilute sulphuric acid after deaeration and maturing. The addition of formaldehyde and aluminium salts or zinc sulphate to the spinning bath initiates a sort of tanning process in which the fibre hardens. It is also possible to apply after-treatment with formaldehyde. The filament is then cut into staple fibres. As regards feel, dye absorption, and heat retention protein fibres resemble wool but they cannot be milled by themselves since their surface has not the scaly structure of that of wool. It is a well-known fact that the felting of wool during milling is due to this scaly structure.

A table of the most important protein fibres has been given earlier. Of these, the Italian Lanital from casein is being produced in the largest quantities. The proteins required for protein fibres, casein, soya protein, groundnut and corn meal protein, are high-grade foods and feeding stuffs which are lost to human and animal consumption if they are used for the manufacture of fibres. Protein produced by microbiological synthesis does not appear to have been used as yet for the manufacture of fibres.

Quite apart from their not particularly favourable mechanical properties it is, therefore, not likely that the production of protein fibres will develop to any large extent. Some day they will be replaced by fully synthetic fibres with a wool-type structure. It is, however, also possible that these fully synthetic fibres will give a new momentum to the production of protein fibres since mixtures of these two types often have very good properties, e.g. Vicara and Terylene.

## 2. FULLY SYNTHETIC FIBRES

Fully synthetic fibres are divided into two groups: polymerization and polycondensation products. These two types are represented by the formulae on p. 376. As with plastics, the trend of development of synthetic fibres will be to apply to polymerizates a condensation aftertreatment with certain groups in order to alter their properties, or subsequently to polymerize polycondensates containing vinyl groups.

Fully synthetic fibres are produced from organic solvents by wet or dry spinning or by fusion spinning. (Glass and silicic acid fibres spun from the melt must also be considered as 'fully synthetic inorganic fibres', and asbestos as a 'natural inorganic fibre').

Stretching, in some cases to 1,000 per cent, is of great importance for the strength of all synthetic fibres.

### (a) Polymerizates as fibres

*PC fibre*

The first fully synthetic fibre was the PC fibre (Agfa-Wolfen, Rein). Polyvinylchloride (PVC) which is one of the most important plastics hardly dissolves in almost any solvent.

It is possible, by chlorination in tetrachloroethane, to introduce 1 chlorine atom per vinyl residue up to a chlorine content of 55–63 per cent Cl.

The perchlorinated polyvinylchloride (63 per cent Cl) is soluble in acetone and can be spun with warm water by the wet spinning process or in the dry spinning stack. This material has a low softening point and shrinks even in boiling water. Highly chlorinated PC with 67 per cent Cl has much more favourable properties but can be spun only from dimethylformamide. Since perchlorinated polyvinylchloride is rather expensive its solubility has also been raised by polymerizing into it other vinyl compounds thus making it possible to spin it. Mixed polymerizates are produced e.g. with asymmetric dichloroethylene (PC 120, Saran, Velon), with acrylonitrile (Vinyon N, Dynel-fibre) or with vinyl acetate (Vinyon = 90 per cent vinyl chloride + 10 per cent vinyl acetate). It is remarkable that more than half of the PC fibre consists of chlorine; it is used mainly as a filter material for industrial purposes. It is also used for underclothing with anti-rheumatic properties which easily accumulates electric charges. It is chemically stable and rotproof but is sensitive to heat and organic solvents.

### *Polyacrylonitrile fibre*

It seemed at first as if of all the polymer fibres the polyacrylonitrile fibre might one day become the most important ('PAN', 'Dralon' Bayer, 'Orlon' Du Pont, 'Wolcrylon' in the German Democratic Republic). The Terylene fibre has, however, better prospects because of its greater abrasion resistance and lesser absorption of dirt, in spite of its higher price.

Acrylonitrile used to be produced from ethylene oxide and hydrocyanic acid. From the $\beta$-oxypropionitrile (ethylene cyanhydrin) obtained in the first stage water was cracked off. The starting material was, therefore, ethylene. This process gave poor yields and was expensive. In 1940 Bayer-Leverkusen succeeded in adding hydrocyanic acid to acetylene in the liquid phase with aqueous acid cuprous chloride solution as the catalyst.

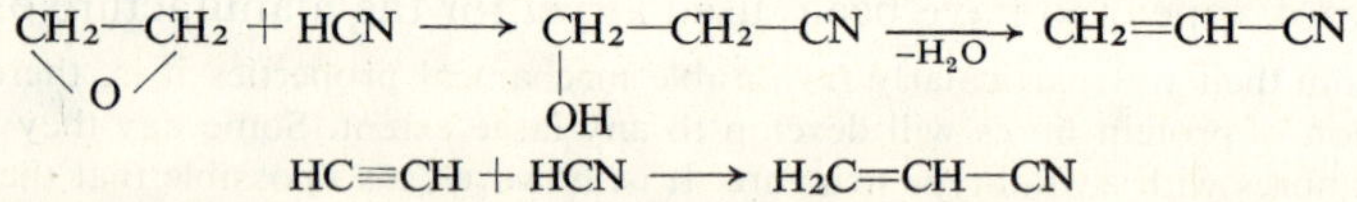

This brought the price of acrylonitrile down by one half.

Another route of industrial importance leads to acrylonitrile via allylchloride produced from propene by high-temperature chlorination (see p. 177). With ammonia allylamine is obtained which can be dehydrogenated in air on a silver catalyst to acrylonitrile. In the same way methacrylonitrile is obtained from methallyl chloride. Yet another path to acrylonitrile goes via lactic acid nitrile from acetaldehyde and hydrocyanic acid (K. Sennewald, *Erdöl u. Kohle*, 1959, **12**, 364), and via the ammoxidation of propene.

Polyacrylonitrile cannot be used by itself as a plastic since its decomposition point lies below its melting point. It is, however, widely used in mixed polymerizates, e.g. in the synthetic rubber Buna N.

Because of its thermal instability polyacrylonitrile, with a molecular weight of about 100,000, could not be spun from the melt and for a long time no suitable solvent for it was known. Not until the former IG (Rein) found dimethylformamide, $(CH_3)_2N{-}CHO$, to be a solvent did it become possible to spin polyacrylonitrile fibre. Solutions are obtained with a polyacrylonitrile content of 20–25 per cent which are spun by the wet spinning process in water containing polyglycols or dimethylformamide, or by the dry spinning process with evaporation of dimethylformamide in heated stacks (e.g. Dralon, Bayer). After hot stretching by a factor of six a fibre with a wool-type character is obtained. Of the fibres so far known the polyacrylonitrile fibre has the greatest resistance to light and weather. Other properties correspond approximately to those of Nylon and Perlon. Tensile strength is about 4/5 of that of Perlon and Nylon, but resistance to abrasion is considerably lower.

### (b) Polycondensates as fibres

*Polyamides*

By condensation of dicarboxylic acids with diamines or of aminocarboxylic acids with each other and by simultaneous elimination of water polycondensates are obtained which, like the protein molecule, are connected by carboxylic acid amide groups R—NH—CO—R. The principle was found long before its practical application by Carothers, U.S.A. By prolonged heating these polyamides become very high-molecular compounds; the formulae for their structures were given earlier. The first usable polyamide was Nylon (Du Pont) which is obtained from adipic acid $HOOC—(CH_2)_4—COOH$ and hexamethylenediamine $H_2N—(CH_2)_6—NH_2$ in the mole ratio 1:1 ('Polyamide 66'). Other dicarboxylic acids with more than six C-atoms, and diamines are also suitable for the synthesis of polyamides, e.g. sebacic acid (octane dicarboxylic acid) which can by obtained by alkaline cracking of castor oil, or by oxidation of undecylenic acid from the cracking of castor oil. Pimelic acid ($C_7$) and ketopimelic acid have also been used successfully for condensation. Recently, the polyamide of 11-aminoundecylic acid ('Rilsan', France) which is an excellent plastic has become very important (M. Genas, *Angew. Chem.*, 1962, **74,** 535). The 11-aminoundecylic acid, too, is produced from undecylenic acid via 11-bromoundecylic acid with ammonia. Finally, azelaic acid obtained by ozonization of oleic acid is being used increasingly for polyamides.

Adipic acid is produced industrially be oxidation of cyclohexanone with nitric acid. Another route to adipic acid is that devised by Reppe via 1,6-hexandiol (see under 'Syntheses with acetylene'). Of the many other possibilities only the oxidation of hexahydropyrocatechol will be mentioned here which can be obtained easily by hydrogenation of pyrocatechol (from phenosolvan extracts). Hexamethylenediamine can be produced from adipic acid by reaction with ammonia on a dehydrating catalyst (adipinodinitrile) and subsequent hydrogenation in an ammonia atmosphere. Amines can be produced also in a single process directly from acids, ammonia, and hydrogen if a catalyst is used which has both a dehydrating and a hydrogenating action (Leuna).

Nowadays, hexamethylenediamine is mainly produced from tetrahydrofuran (Reppe) or from furfurol which is made cheaply by Messrs. Quaker Oats from oats husks (Du Pont). In recent years the U.S.S.R. has built large plants for the production of furfurol from agricultural waste products.

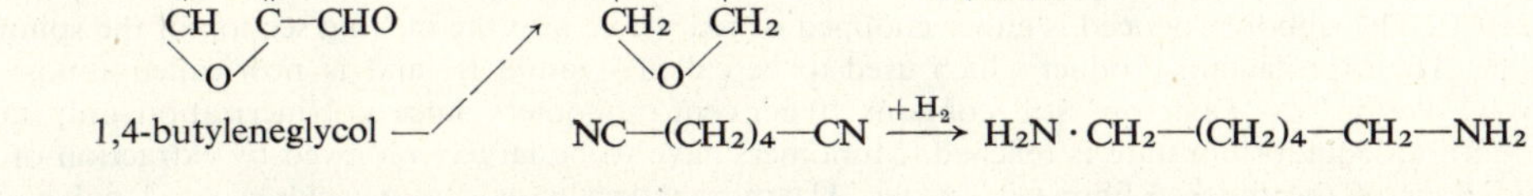


Carothers realized also the possibility of the polycondensation of aminocarboxylic acids and attempted to polycondense ε-aminocaproic acid-lactam. After several unsuccessful experiments he became convinced that caprolactam could not be polymerized. The credit for the solution of the problem of polymerization of caprolactam must go to P. Schlack (Aceta, Berlin-Lichtenberg) who added water so that aminocaproic acid is constantly being formed which then condenses. Thus the Perlon fibre was created in Germany. Lactam polymerizes without splitting off water.

Caprolactam is produced from cyclohexanone oxime in a continuous process by a Beckmann rearrangement in strong sulphuric acid.

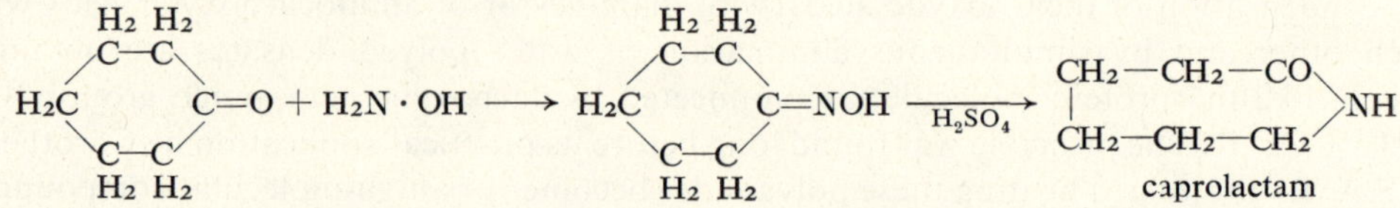


From cyclohexanol the route to cyclohexanone is by catalytic oxidation with atmospheric oxygen on a zinc-iron or Ag catalyst. Cyclohexanone can also be obtained from benzene (see K. Smeykal and H. I. Naumann, *Chem. Techn.*, 1961, **13**, 132).

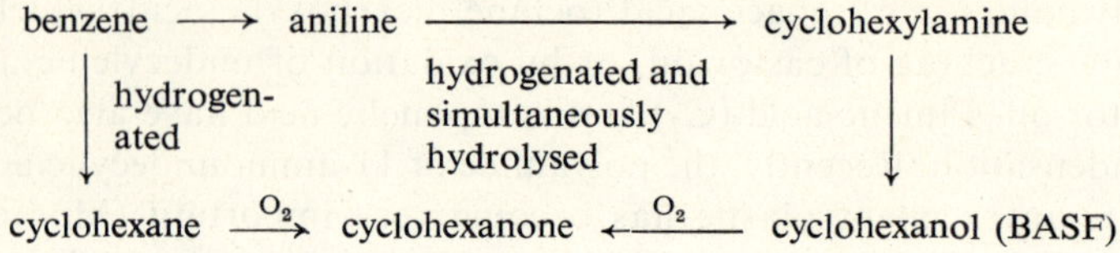


Polymerization takes place in two stages up to a molecular weight of at least 12,000. Two methods are used: polymerization in autoclaves under pressure, and continuous polymerization at atmospheric pressure in a VK tube (a VK tube is a heated tube, developed by Dr. H. Ludewig, in which caprolactam is polycondensed continuously (with complete exclusion of air) and the molten product is spun directly. The spinning nozzles are at the bottom).

In the autoclave method the polyamide-forming material, salts of diamines and dicarboxylic acids, ω-aminocarboxylic acids or lactam, is heated with 8–40 per cent water in an autoclave of special steel to a temperature 20–40°C above the melting point of the polyamide formed (see H. Klare, Die grosstechnischen Verfahren zur Herstellung von Perlonseide; *Faserforschung und Textiltechnik*, 1951, **2**, 1). Pressure is maintained at 10–25 atm gauge pressure by blowing off steam. When the polymerization (polycondensation) is complete the pressure is released, the autoclave is exhausted or flushed with nitrogen which it completely free of oxygen, and the melt is extruded by nitrogen through slot-shaped nozzles so that it solidifies as a ribbon. This method gives good results only with diamines and dicarboxylic acids. Aminocarboxylic acids are condensed more elegantly by the continuous process at atmospheric pressure (H. Ludewig, Kontin. Herstellung u. Verarbeitung von Perlon nach dem VK Verfahren. *Faserforschung und Textiltechnik*, 1951, **2**, 341; *Chem. Techn.*, 1952, **4**, 523); it is, of course, also possible to polycondense lactam in autoclaves.

The condensation process takes place in a tube of 8 m length with diphenyl heating for 6 hours at 250–260°C. The ribbon produced is either chopped or fed whole into the melting section of the spinning machine. The caprolactam product which used to be called 'Igamid B' and is now called almost exclusively 'Perlon' or 'Dederon' still contains 10 per cent monomers since polymerization only takes place until an equilibrium state is reached. Monomers have to be largely removed by extraction of the Igamid B or of the finished fibre with water. 11-aminoundecylic acid also yields a good polyamide. Polyamides are spun from the melt (see Figure 131) (after F. Endress, in Winnacker-Weingaertner, *Chem. Techn.*, Hanser-Verlag, München).

Chips, formerly a ribbon, are melted with complete exclusion of oxygen, and forced through nozzles; melting temperature 215°C, spinning temperature above 250°C. The spinning temperature has to be maintained very accurately. The most widely used modern spinning method is the direct spinning method from VK tubes through distributor ducts heated with diphenyl. The drawing-off of the 0·3 mm thick filament is carried out at the very rapid rate of 350–1,200 m/min. The bundle of fibres solidifies in a cooling stack. The filament is stretched only after it has been cold finished with slight twisting between two rollers (ratio 1:3·5 to 1:4·5). Cold stretching requires very strong forces, the bobbins bear a high tensile strain. High atmospheric humidity in the workrooms and wetting with conducting finishing agents prevent electrostatic charges on the filament.

Bristles as well as fibres are produced from polyamides.

Finally, mention must be made of Perlon U developed by Bayer-Leverkusen (O. Bayer *et al., Angew. Chem.*, 1947, **59,** 257; H. Rinke, *Angew. Chem.*, 1962, **74,** 612) which is obtained from 1,6-hexane di-isocyanate and 1,4-butylene glycol (polyurethane formation. For formula see above). Hexane di-isocyanate is produced from hexamethylene-

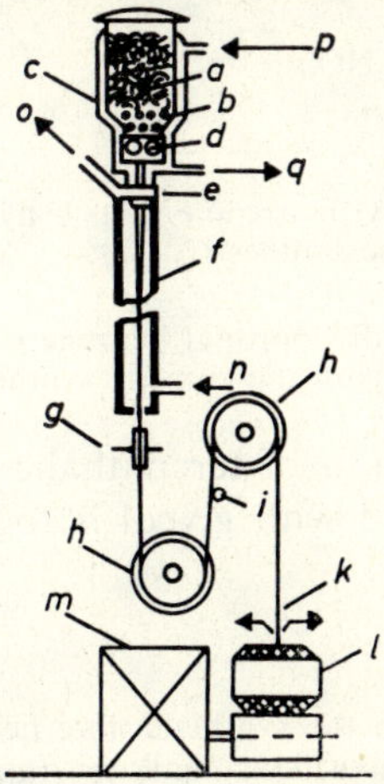


Figure 131. Melt-flow spinning process for polyamides
*a* melting chamber *b* melting grate *c* steam jacket *d* heated pump *e* heated nozzle *f* spinning shaft *g* roller *h* stretching rollers *i* stretching point *k* twisting *l* spinning reel *m* motor *n* cooling air *o* waste air *p* steam inlet *q* condensate outlet

diamine and phosgene, 1,4-butylene glycol from 1,4-butindiol (see acetylene syntheses). Because of its rigidity Perlon U is particularly suitable for the production of synthetic bristles. Polyurethanes are mainly important for plastics for which aromatic di- and tri-isocyanates with polyvalent alcohols are also used.

*Polyesters: Terylene fibre, Lanon, Trevira, Diolen, Dacron*

The Terylene fibre (J. Goodman, *Angew. Chem.*, 1962, **74,** 606) (from England) is a polyester of terephthalic acid with glycol (formula p. 376). The fibre has excellent elasticity and strength and is particularly suitable for clothing because it does not irritate the skin. Considerable development may still be expected from it.

The method of spinning Terylene (Lanon fibre) is analogous to that of Perlon or Dederon; the spinning temperature is somewhat higher corresponding to the higher melting point of 253°C (Perlon: 215°C). However, stretching has to be carried out hot (above 80°C); a heated stretching post is usually employed.

The main problem is to produce terephthalic acid cheaply enough. In this respect the petroleum-processing countries are in a favourable position economically since they obtain a xylene mixture with a high p-xylene content by the aromatizing polymerization of butane from natural gases. Isobutylene can be dimerized very easily and dehydrogenated to p-xylene.

A difficult problem is the separation of m- and p-xylene; it is achieved by low-temperature cooling and crystallization. Bituminous coal xylene contains only a little p-xylene. Oxidation of p-xylene is carried out catalytically with air in the liquid phase in two stages via toluic and toluic acid acid ester (Chemische Werke, Witten, früher Innhausen, Katzschmann).

Great efforts were made in Germany to find a synthetic method for producing terephthalic acid, such as e.g. reaction of benzene or toluene with formaldehyde and hydrochloric acid, followed by complete chlorination of the side chains and saponification. This method has been abandoned. Of interest is the rearrangement of the K-salt of o-phthalic acid in the presence of a Zn or Cd catalyst under pressure with $CO_2$ (Racke at Henkel's, Chem. Werke, Buna). This rearrangement is technically difficult because it has to be carried out in the solid state at a high temperature and a plastic phase intervenes (equipment: reaction band (a reaction band is an endless pan-shaped wide steel strip passing through, for instance, a furnace) or fluidized bed furnace). Other promising methods are:

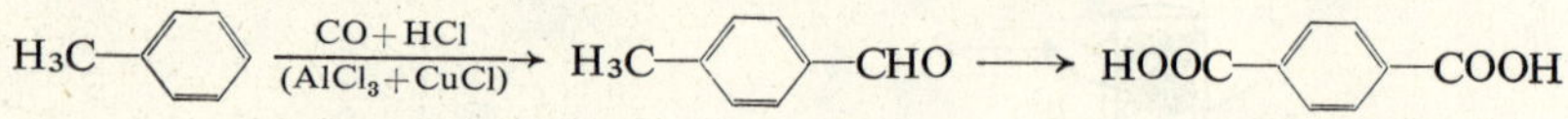

Gattermann-Koch synthesis

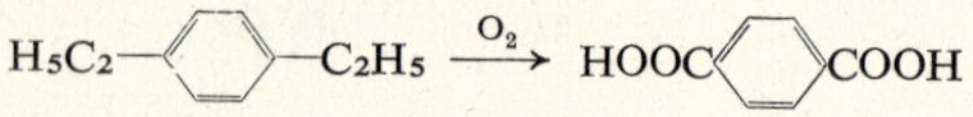

(diethyl benzene e.g. as a by-product from the styrene synthesis)

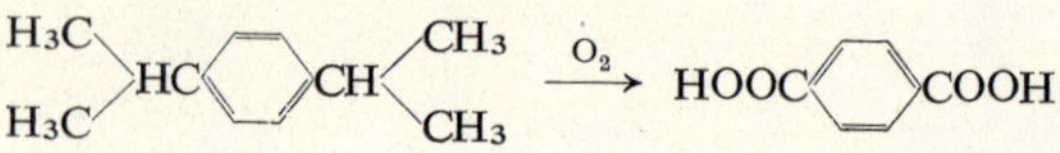

(di-isopropyl benzene e.g. also as a by-product from the cumene synthesis, e.g. at BASF)

The polyester is obtained by trans-esterification of terephthalic acid dimethyl ester with glycol since direct esterification of the acid with glycol is too difficult and gives rise to decomposition.

### *Prospects*

One of the most important problems yet to be solved in the synthetic fibre field is the production of filaments with the properties of wool. Many attempts have been made to attain this goal, mainly by producing mixed fibres, and the possibilities are by no means exhausted. The chemist has developed so many new chemical types of fibres that the synthetic fibre and textile experts should now be in a position to produce the filament best fitted for any given use by suitably mixing the fibres. An indication of this is provided by the fact that mixed fibres of polyacrylonitrile and nylon possess much better mechanical properties than either of these filaments alone.

In order to improve the properties of fibres single molecules containing 'active' groups are being experimentally introduced into the polymerizates, e.g. —OH or —COOH groups into vinyl polymers. These make possible a reaction with other molecules, e.g. ethylene oxide, amino acids, and isocyanates, which impart specific properties to the polymers, e.g. adhesion to surfaces, ability to accept dyes, etc. Another route is the synthesis of defined high polymers from smaller fractions, e.g. by coupling with di-isocyanate. Routes to such filaments with specific properties are via 'block' or 'graft' polymerizates (see p. 398, Plastics).

It has recently been shown (Ziegler, Natta) that 'isotactic' isomers of ethylene and propylene, due to their symmetrical structure, also have properties suitable for filaments (G. Bier, *Kunststoffe*, 1958, **48,** 354; *Skinners's Silk and Rayon Record*, 1957, **32,** 623; G. Natta, *Angew. Chem.* 1956, **68,** 393; *Melliand Textilber.*, 1957, **38,** 239).

## LITERATURE

F. Endress, 'Die künstlichen Fasern', in: Winnacker-Weingaertner, *Chem. Technologie*, Organische Technologie I. München, Hanser-Verlag.

F. Endress, *Chem. Ing. Techn.* 1951, **23,** 265.

H. Rein, *Melliand Textilber.*, 1949, **20,** 243, 299.

Houwink, *Grundriß der Technologie der synthetischen Hochmolekularen*. Leipzig, Akadem. Verlagsges. Geest & Portig K.-G., 1952.

H. Klare, *Technologie und Chemie der synthetischen Fasern aus Polyamiden*. Berlin, Verlag Technik, 1954.

*Chemische Textilfasern, Filme und Folien, Grundlagen und Technologie*. R. Pummerer (Ed.), Stuttgart, Enke, 1953–1954.

Z. A. Rogowin, *Chemiefasern, Grundlagen der Chemie und Technologie*. Leipzig, Fachbuchverlag, 1960.

P. Schlack, Stand der Technik auf dem Gebiet der Chemiefasern, Tendenzen und Ziele in Forschung und Entwicklung. *Melliand Textilber.*, 1962, **43,** 543, 681, 802.

O. Bayer, Zur Entwicklung und Problematik des organischen Makromoleküls. *Angew. Chem.*, 1959, **71,** 145.

# II. Raw materials for lacquers, plastics, elastomers, exchange resins, silicones

### General remarks on the chemistry of macromolecular substances, their formation and their properties

Plastics are those modern organic materials and bonding agents which, like the natural materials cellulose, wool, silk, rubber, natural resins, starch, and gelatine, are high-molecular organic compounds. Owing to the great variety of highly developed types available they have long lost the character of substitute materials so that the term 'plastics' in almost all languages appears to be preferred to 'artificial materials'. The dividing line between plastics and elastic materials, i.e. between natural and synthetic rubbers, tends to become blurred by the most recent developments. Resins proper are the still viscous or easily-melting raw materials for plastics, lacquers, and adhesives, with a medium molecular weight, which like the drying oils, do not attain the high-molecular state until the drying and film-forming stage, or by an additional hardening process. These materials are invariably macromolecular compounds.

The term 'macromolecule' was coined by H. Staudinger (Comprehensive account: O. Bayer, Zur Entwicklung und Problematik des organischen Makromoleküls. *Angew. Chem.*, 1959, **71,** 145–152). It is not always possible to define unambiguously when a molecule is a macromolecule but it is generally so called when the molecules contain more than 1,500 atoms and the molecular weight exceeds 10,000. However, the macromolecular substances used as plastics have much higher molecular weights. For regenerated cellulose these are between 75,000 and 100,000, for natural rubber about 150,000 to 200,000, for polystyrene up to 500,000, for the polycondensate from caprolactam, Dederon or Perlon (and also 'Nylon') up to 32,000. From the molecular weight it is possible to calculate the length of the macromolecule which should be of the order of magnitude of several 10,000 Å. Measurements have shown the size to be much smaller which suggests that the macromolecules form tangled skeins.

The size of the molecule can also be defined by the degree of polymerization, i.e. the number of single molecules combined into a macromolecule by polymerization or polycondensation. These two parameters of macromolecular substances are never absolute values but indicate the mean molecular weight or the mean degree of polymerization. Plastics are mixtures, necessarily formed during the syntheses of 'polymer homologues', a name given to molecules of different degree of polymerization or condensation. It is worth noting that some natural polymerizates and condensates have a somewhat more uniform molecular size and degree of polydispersion than synthetic ones, from which a different manner of formation in nature may be deduced (e.g. natural rubber).

The synthesis of high-molecular plastics and elastic materials from low-molecular basic substances can take place by polymerization, polycondensation, or poly-addition, or by a combination of these processes, which differ fundamentally from each other in the way the reaction goes.

## *Polymerization*

The polymerization reaction takes place in three main stages. It is a radical or ion chain reaction which once initiated is self-propagating. A large number of investigations has been carried out into the kinetics of these reactions (Review articles e.g.: W. Kern, *Chem. Z.*, 1952, **76**, 667–672; P. D. Bartlett, *Angew. Chem.*, 1955, **67**, 45–52). Only the fundamental aspects can be discussed here. Radical polymerization takes the following course:

(1) Initiating reaction. It consists in the formation of a reactive nucleus, e.g. a radical, which is created by the activation of a monomer molecule. The activation energy required is of the order of magnitude of 10–25 kcal/mole. Single molecules can be activated by light absorption, by the formation of a peroxide from the olefin with atmospheric oxygen and its subsequent decomposition, by cleavage of active molecules such as peroxides, by a decomposition reaction of unstable substances during which radicals are formed, and finally by heat. In the simplest case the double bond of the monomer opens out, i.e. single molecules have 'lone' electrons to which other monomers add forming a new radical. This leads to the second stage, the growth of the chain.

(2) Chain growth. The chain grows by the addition of further molecules to the continuoulsy forming radicals. Thus a macro-radical is created. This process is the polymerization proper:

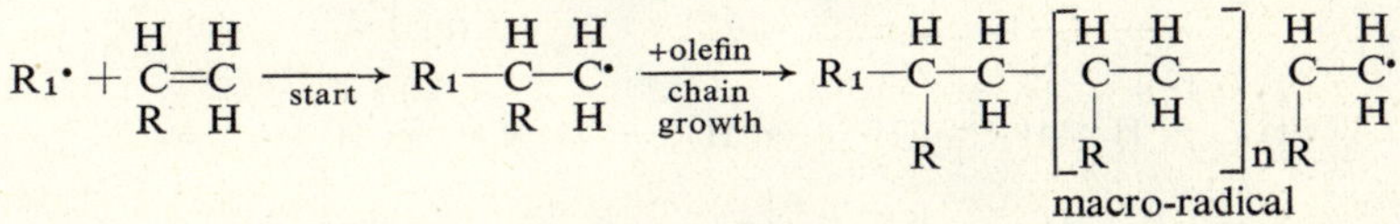


The chain grows very rapidly, almost instantaneously, once an active nucleus is present so that the speed of polymerization depends essentially on the number of nuclei formed. The chain would continue growing indefinitely if the macro-radical did not react also with other substances present in the reaction mixture, e.g. with a hydrogen atom of its own or of another molecule. The radical site can become inaccessible because of chain tangling.

(3) Termination reaction. The chain reaction can stop either by the migration of a hydrogen atom or by an external reaction or by cyclization. In the first case the newly formed macromolecule terminates in a double bond, in the second in the group which caused the termination by reaction with the macro-radical, e.g. water, alcohol, carboxylic acid, or an inhibitor molecule. It has been proved that where peroxides are used as initiators in the polymerization of styrene and other monomers the peroxide catalyst is incorporated into the chain.

The role of atmospheric oxygen in termination reactions is particularly important. It may initiate, by nucleation, the formation of a chain (preferably in the dark), or cause the breaking off of a chain by reaction with the radical (the chain carrier) at a very early stage of the growth reaction, e.g. in polymerization initiated by ultra-violet light. It may well be that molecular oxygen by itself always prevents polymerization. Where it initiates it the action is probably not direct but leads first to the formation of an unstable peroxide and its decomposition. Finally, oxygen may also cause cross-linking.

A special type of termination reaction is the 'chain transfer' (radical transfer). In this case the energy of the macro-radical is transferred to a monomer which thereby becomes the starter of a new chain. The newly formed radical can then react with other radicals. In many more cases, however, a side-chain grows on the new radical by the addition of olefin. The side-chains in high-pressure polyethylene are formed in this way. Intra-molecular radical transfer also occurs. Radical transfer can be utilized in industry in graft-polymerization:

Radical transfer:

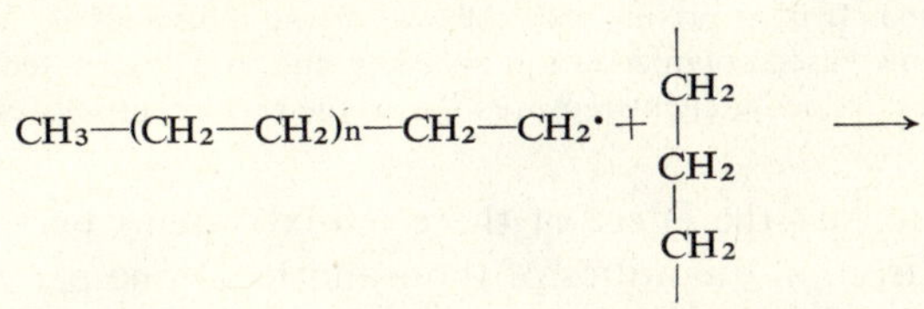

Graft polymerization:

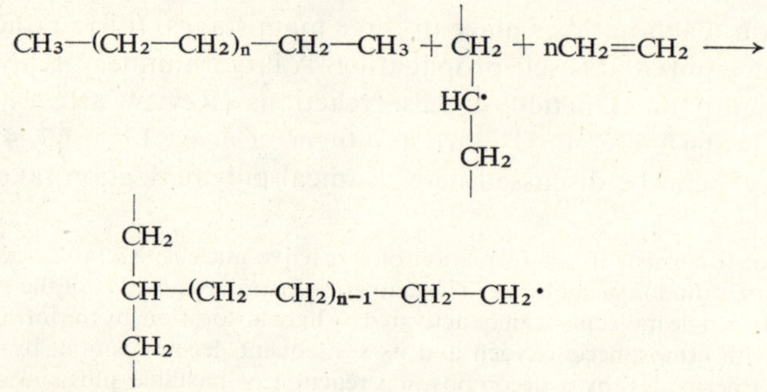

Such 'radical points' in macro-molecules can be formed by radical transfer or by energetic radiation. Ionic polymerization is caused by inorganic catalysts and takes the following course:

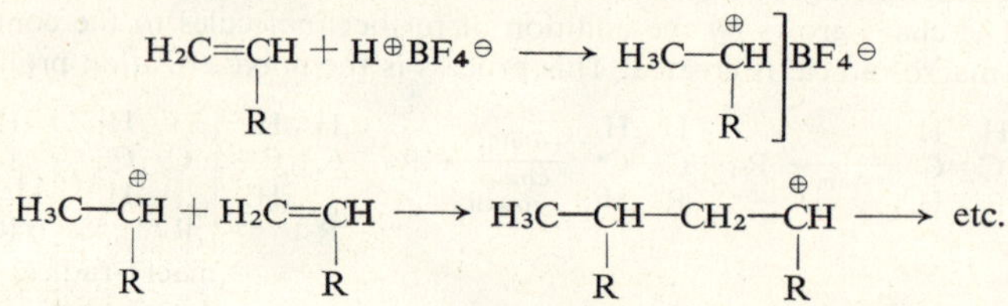

The speed of polymerization and the nature of the polymerizate obtained depends on the structure of the monomer, the ease with which it polymerizes, and the three factors mentioned earlier, start, growth, and break-off of chain. Temperature, solvent, and concentration naturally affect it as well. The speed is, however, mainly a function of the initiating and of the terminating reactions.

If many reaction chains start simultaneously the probability of chain termination is greatest, i.e. a low-molecular polymerizate is obtained. Generally it is found in practice that rapid polymerization at elevated temperatures and using a large quantity of catalyst, or an active catalyst, yields low polymers.

The individual intermediate processes can be affected catalytically by catalysts and inhibitors (reaction retarding agents). Examples of polymerization catalysts are peroxides which initiate the starting reaction by decomposition into radicals, or other radical formers, inorganic catalysts with an incomplete octet such as $BF_3$ and $AlCl_3$ which by addition to the olefin polarize it to such an extent that it is capable of adding further monomers alkali metals act as catalysts in the polymerization of butadienes. They transfer electrons to the olefin so that in the 1- or 1,2-position are 'lone' electron pairs which cause the gradual addition of further butadiene molecules.

The observation made at the former IG Farbenindustrie that reducing agents in an alkaline medium in the presence of peroxides cause a particularly large increase in the polymerization speed was very important for the technology of polymerization. The initiating factor here is the reaction between an oxidizing agent and a reducing agent ('redox catalysis').

This effect can be visualized as an acceleration of the decomposition of the peroxide by reducing agents. An increased decomposition of peroxide entails the formation of an increasing number of nuclei, i.e. more chains are started. It is surprising that the size of the molecules is not reduced, contrary to what happens when the increased polymerization speed is due to a higher temperature. This may be caused by the fact that the reducing agent removes the oxygen which promotes the termination of the chains.

It is possible to intensify the effect of these redox systems by the addition of some metallic salts, e.g. of iron, in quantities of thousandths of one per cent.

The following substances can be used for polymerization in an aqueous emulsion: $H_2O_2$ and persulphates, together with triethanolamine, rongalite, sulphite, sulphinic acid salts, dihydroxyacetone, and dextrose, as well as metallic salts e.g. Na-ferropyrophosphate.

Organic peroxides are used as catalysts for block polymerization, polymerization in solvents, and suspension polymerization, e.g. dibenzoyl peroxide, lauryl peroxide, cyclohexanone-, tertiary butyl-, and cumene hydroperoxide. At present about 30 organic peroxides are manufactured for use in polymerization processes. Suitable reducing agents are those materials mentioned above that are soluble in organic media, and metallic salts e.g. lead-, iron-, and cobalt naphthenate.

### *The effect of inhibitors*

The termination of polymerization can be brought about by inhibitors. These are also added to prevent the premature polymerization of monomers by atmospheric oxygen, e.g. during distillation and storage ('stabilization'), the autoxidation of oxidizable substances, and the 'peroxide effect' in addition and substitution reactions. This ability of inhibitors to forestall in general consequent reactions of the peroxide effect points to the similarity of the mechanism of their protective action. It also supports the correctness of the opinion expressed earlier that molecular oxygen initiates polymerization not directly but via the formation of peroxides and their decomposition. Active centres form, therefore, by peroxide decomposition. Their number can never be large, and small quantities of inhibitors are, therefore, adequate for the interception of these nuclei by reaction. When inhibitors (chain breakers) are active it is always a sign of a radical chain reaction.

It is sufficient to carry away the activation energy of these few active centres to prevent the onset of oxidation, polymerization, or of the 'peroxide effect', e.g. during bromination. Age resistors of rubber also belong in this group of inhibitors.

An inhibitor has to (1) cause the termination of chains, (2) destroy peroxides, and (3) bind metal traces (form chelates).

A large number of inhibitors are known and are in use. Most of them belong to the following groups: phenols (hydroquinone), amines, aminophenols (benzylaminophenols), quinones, and mercapto-compounds. The latter are particularly suitable for the promotion of 1,4-additions and the prevention of 1,2-additions.

### *Cross-linking*

If during polymerization only two reactive points form on a molecule it is called 2-reactive and can form only unidimensional chain molecules, such as the monovinyl compounds, dicarboxylic acids, and dioles. These high polymers are thermoplastic. Higher reactive substances, e.g. phenol, 8- and higher valence alcohols, are able to form two-and three-dimensional molecules. These can be the result of bridge formation, sometimes delayed, between the chain molecules. A net-like structure develops and the process is called 'cross-linking'.

Cross-linking occurs during polymerization of butadienes and also in all radical polymerizations by intramolecular radical transfer. It is also characteristic of the condensation of phenoplastics and aminoplastics. Cross-linking can also be caused during polymerization of vinyl compounds by the addition of divinyl compounds, such as divinyl benzene for styrene, or subsequently by vulcanization with simultaneous formation of sulphur bridges. Di-isocyanates or dicarboxylic acids make possible the cross-linking of linear molecules containing OH-groups.

In the United States a method based on the treatment with energetic radiation, e.g. $\gamma$-rays, has been developed for cross-linking in finished articles such as polythene bottles, thereby raising the softening points.

In hair keratin the polypeptide chains are linked by cystine bridges (—S—S—). If these are detached by reducing agents (e.g. thioglycollic acid, 'cold wave liquid') the hair becomes plastic and can be 'set'. The S—S bridge and with it the former rigidity of the keratin molecule are then restored by oxidation.

The degree of cross-linking determines, therefore, the strength and the mechanical properties of macro-molecular materials. Phenolformaldehyde condensates are thermoplastic as long as there is not much cross-linking. They become hard and brittle when cross-linking methylene bridges form during 'hardening' under pressure and at high temperature.

In addition to cross-linking by main valencies, the properties of natural and synthetic high polymers are strongly affected by secondary bonds between the chain molecules. Thus, in hair in addition to the —SS— main valency bonds there are ionic bonds between the acidic amide groups which are dissolved even by water or weak alkali. The nature of secondary bonds was discussed earlier in the section on fibres.

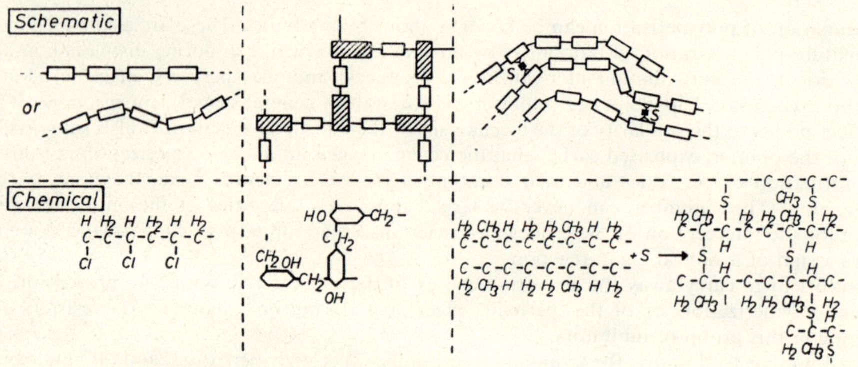


Figure 132. The three main types of macromolecule formation

Figure 132 (from Houwink, *Grundriß der Technologie der synthetischen Hochmolekularen.* Leipzig, Akad. Verlagsges. Geest & Portig K.-G., 1952) shows very clearly the three main principles of the formation of macro-molecules. With these schematic representations one must always bear in mind that the structures of macro-molecules from asymmetrical monomers are not as regular as shown in the illustration. In trivial terms, the molecule has a 'head' and a 'tail' which makes possible many different groupings, according to whether head-tail, or head-head, or tail-tail links form and in what sequence. The sequence of the monomers is a function of the polymerization conditions. At low temperatures the alternating structure (head-tail link) is prevalent in vinyl derivatives. This is a consequence of polarity factors and resonance stabilization. Flory showed that polyvinyl acetate produced at 35°C has a ratio of vicinal to alternating structures of 1:110; the vicinal proportion increases at higher temperatures. The chemical reactions of polymers, particularly their stability, depend on their structure: on the whole, vicinals are more reactive and less stable.

Using the catalysts found by Ziegler and Natta (K. Ziegler, Das Mülheimer Niederdruckpolyäthylen-Verfahren, *Kunststoffe*, 1955, **45**, 506; G. Natta, Stereospezifische Katalyse und isotaktische Polymere, *Angew. Chemie*, 1956, **68**, 393; Fünf Jahre Forschung an der stereospezifischen Polymerisation, *SPE-Journal*, 1959, **15/5**, 373–382) it is possible to produce polymers of very regular structure. These isotactic polymers are distinguished from the normal ones by greater density, hardness, and tendency to crystallize, and some of them also by higher softening temperatures. This entails, however, lower adhesion and welding capacities. The polymerization mechanism is anionic with alkyl halides of the intermediate shell elements, as e.g. by reaction products of aluminium alkyls with titanium tetrachloride, orientated adsorption on the surface of the catalyst leading to stereospecific polymerization.

## *Steric configuration of polymers*

As long as 15 years ago it was recognized that high polymers with asymmetrical C-atoms, e.g. polystyrene, had different physical properties although their molecular weights were identical. (Huggins explains this phenomenon by steric symmetry or asymmetry.)

Natta devoted a great deal of effort to the production of sterically defined polyolefins, and proved that they had stereochemically different structures. He coined the term 'tactic' for polymers with sterically ordered configuration, those without order 'atactic', and those with uniform d- and l-configuration 'isotactic'. The latter have a high degree of crystallinity, low solubility, high melting points, and higher strength. Isotactics can be drawn to form strong threads.

*Isoprene syntheses*

Ziegler catalysts open up new possibilities of producing isoprene from olefins. Of the reactions shown below the most promising is the dimerization of propylene with subsequent cracking.

Mixed dimerization of propylene and ethylene:

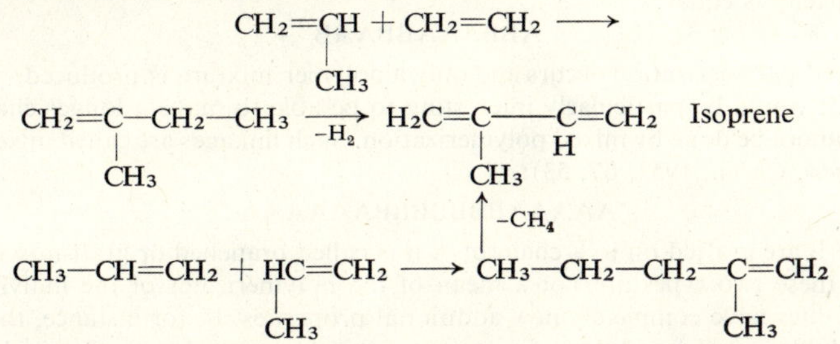

*High polymers produced by condensation*

Whilst the polymerization described above always runs its full course, in the condensation of monomers, e.g. dehydration by heat, as in all esterifications an equilibrium establishes itself. By removing the reaction product formed (water) the equilibrium can be shifted towards the polycondensate in accordance with the law of mass action. This, however, gives rise to technological difficulties. As the reaction progresses the rate of growth of the molecule decreases. With polycondensates, e.g. polyamides from diamines and dicarboxylic acids or from caprolactam, one is content with smaller molecular sizes than are acceptable with polymerizates. The most important types of condensation resins were discussed in the section on synthetic fibres. This group also includes the pheno- and amino-plastics.

*Mixed polymerizates and mixed condensates*

Homogeneous natural polymerizates are rubber and gutta-percha, a homogeneous condensate is cellulose, whilst wool keratin and natural silk are very complex natural mixed condensates.

The possibility of producing mixed polymerizates (first observed by Klatte in 1914) created the first opportunity of using the polymers of many vinyl compounds. The properties of mixed polymerizates as far as solubility, softening point, etc. are concerned, are in general halfway between the homogeneous components. These properties are, however, only rarely obtained by after-mixing of the polymer components. As the most important example of the creation of new properties may be mentioned Buna N (Perbunan), a mixed polymerizate of butadiene with acrylonitrile, and Buna S, butadiene with styrene. It was this mixed polymerization of butadiene which first made possible the use of synthetic rubber. Since vinyl compounds differ greatly in their readiness to polymerize the production of a mixed polymerizate requires very many preliminary experiments and the technological execution has to be carried out very carefully. Polymerization is usually carried out in emulsion, recently in many cases in suspension also; the mutual adjustment of catalysts, reducing agents, and regulators, and of the physical conditions has to be very exact. In some industrially important cases it is still necessary to work largely empirically.

Some monomers which alone do not polymerize, e.g. maleic acid anhydride, can also take part in mixed polymerization.

If two monomers A and B present in equal quantities were mixed ideally during polymerization the result would be:

ABABABABAB

Statistically, however, polymerization takes place in approximately the following way, provided their polymerization capacity is equal:

ABBAAABBAAB

Often, no true mixed polymerization occurs and only a polymer mixture is produced. The processes are very complicated. It would be particularly interesting to be able to make a longer chain of A followed by one of B. This cannot be done by mixed polymerization. Such linkages are called mixed block polymers (see H. Mark, *Angew. Chem.*, 1955, **67**, 53):

AAAAABBBBBBAAAAAA

If side chains of B are grafted on to a chain of A it is called branched or graft-mixed polymerization. The properties of these two types are not a mean of the polymerizates of the individual components but these polymerizates have completely new additional properties. If, for instance, the polymerizate of component A is soluble in water and that of component B in organic solvents, the block or branch mixed polymer is soluble in both media. Branch mixed polymerizates can be obtained by distributing along the main chain reactive groups on to which side chains can be condensed or polymerized. Such groups are the —OH, —$NH_2$, or the —COOH groups. Or an OOH group is produced by autoxidation and, by its decomposition on the polymer molecule, a radical point is produced. This can also be produced by energetic radiation. At this point polymerization of vinyl compounds then takes place (formation of branches). There is another possibility of constructing block mixed polymerizates in such a way that one may speak of the building block principle, or of the assembly of finished components, or of high polymers made to measure. The first block polymers and mixed block polymers (Vulkollane) were produced by O. Bayer and his colleagues at Leverkusen (*Angew. Chem.*, 1948, **59**, 263; 1950, **62**, 57). The method is based on the production of different polymers; at first comparatively low-molecular weight polymers are produced by condensation of dicarboxylic acid esters with glycols, which are then linked with bi-functional reactive molecules, e.g. di-isocyanate, thereby increasing the molecular weight by a factor of about ten.

It is a well-known fact that the properties of plastics are extensively altered by the addition of 'softeners' (see later), high boiling substances compatible with the plastic, and of soft resins ('external plasticizing'). Using the principle described above it is also possible to incorporate into them by chemical means substances with a plasticizing effect ('internal plasticizing'). The polymerization or condensation of single molecules with side chains which impart certain properties to the complete molecule, e.g. improve its adhesiveness to foils or metal surfaces, affect its dye absorption, etc. is also important.

### *Polymerization techniques*

Nature, too, uses the principle of radical formation for the construction of macromolecules, e.g. the enzymatic dehydropolymerization of coniferyl alcohol to lignin (Freudenberg). In industry only the following four methods are available:

(1) Polymerization of the undiluted liquid monomer: 'block polymerization'.

(2) Monomer and polymerizate are in solution: 'solution polymerization'.

(3) Most important in industry is the 'emulsion polymerization' (dispersion polymerization) in which the more or less soluble monomer is emulsified in water by an emulsifier and polymerized with water-soluble catalysts with simultaneous stirring and heating. A suspension of very fine particles of the polymer forms (latex) which is coagulated.

(4) In suspension- or bead polymerization, as distinct from emulsion polymerization, in which drops of the monomer dispersed in water polymerize, i.e. they are produced in small blocks as granules or beads. The monomers can also be polymerized with a catalyst atomized in air or a gas. The use of organophilic catalysts (peroxides dissolved in monomers) is important. Suspension polymerization is increasingly displacing emulsion polymerization; it combines the advantages of emulsion polymerization with those of block polymerization. The heat of reaction can be dissipated conveniently in the aqueous dispersion, and the polymerizates produced are transparent and exceptionally pure, i.e. free of emulsifiers and salts, with a minimum of catalyst incorporated. The molecular weights and degree of branching are usually higher so that processing is somewhat more difficult; this is the case particularly with PVC. Because of their greater purity and transparency the proportion of suspension polymerizates is increasing.

Block polymerization is used only when polymerizates with the required properties can be obtained in no other way. The factors causing polymerization are heat, light, and catalysts, e.g. peroxides and $BF_3$. It is very difficult to dissipate the heat of reaction; material whose viscosity increases continuously loses heat only very slowly. Special apparatus had to be developed for the different products. The most important examples of block polymerization are: polyvinylalkyl ether, polystyrene, and the polymerizates of methacrylic acid (Piacryl, Plexiglas, Resartglas).

The technique for the polymerization of styrene is as follows (Wulff and Dorrer, BASF): First, a preliminary polymerization is carried out for two to three days in an agitator vessel with the addition of 0·3 to 0·5 per cent dibenzoylperoxide at 70–80°C. The still liquid mass is conveyed into an aluminium tower consisting of several sections in which from top to bottom the temperature rises from 100 to 220°C. A feeding screw at the bottom of the tower squeezes the plastic polymerizate through a slot; the polystyrene emerges in the shape of a strip which cools on an endless conveyor belt and is then chopped into pieces (figure 133) (after Thinius and Bier in Winnacker-Weingaertner, *Chem. Technologie*, Organ. Technologie II, Plastics).

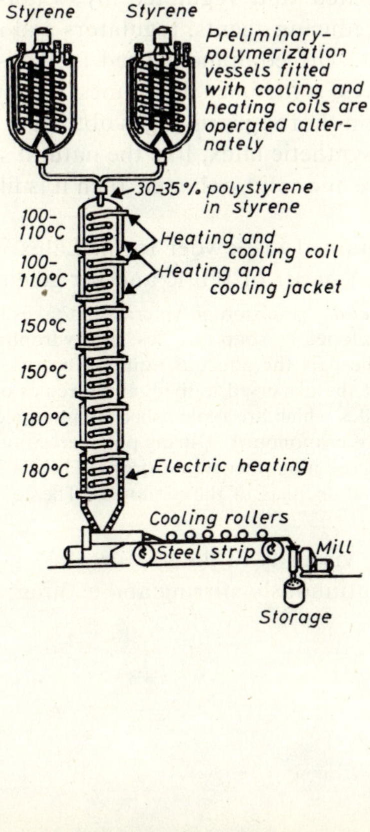


Figure 133. Diagram of continuous strip polymerization of styrene

Methacrylic acid esters are polymerized in ovens with the addition of benzoylperoxide at temperatures between 50 and 90°C. Since the product is often used as 'organic glass' the aim is to conduct the polymerization slowly and uniformly by accurately controlling dissipation of the heat. The degree of polymerization in the interior of a thick body is, nevertheless, always somewhat lower than on the outside since polymerization takes place more rapidly inside which results in more frequent breaking of the chains.

Solution polymerization is used when the object is to obtain the polymerizate directly in solution. As far as possible solvents are employed which dissolve both monomers and polymers. It is, however, also possible to begin the polymerization in a solution of the monomer and to finish with the onset of insolubility, i.e. the precipitation of the polymer. Very high polymers can be produced in this way.

The catalysts should be soluble in organic solvents, e.g. organic peroxides. The degree of polymerization depends, apart from physical factors, on the concentrations of the monomer and catalyst and on the nature of the solvent, e.g. its polarity. The molecular weights are lower than with block and emulsion polymerization. These differences show plainly in vinyl chloride. Solution polymerization leads to polyvinyl chlorides which are easily soluble in most solvents. Emulsion polymerization leads to products which are difficult to dissolve. Methacrylic acid ester can also be polymerized in solvents.

Solution polymerization is used mainly for the production of mixed polymerizates of maleic acid anhydride with vinyl compounds, e.g. with vinyl acetate and vinyl ethers. The solvent process is the most costly and subsequent removal of the solvent from the polymerizate often presents technical difficulties.

The most economical and most widely used processes are emulsion polymerization and bead- or suspension polymerization.

Many monomers are slightly soluble in water. Fine dispersion (20–40 per cent monomer) is assisted by soaps or synthetic emulsifiers and maintained by a protective colloid (emulsification). The reaction is initiated and regulated by water-soluble catalysts (e.g. $H_2O_2$, persulphates, perborate), reducing agents, regulators and inhibitors, adjustment to a certain pH value, and heat. The equipment used includes agitator vessels, usually a set (cascade), rotary autoclaves, sometimes flow tubes, in other words, mainly devices in which by agitation and motion thorough mixing is obtained, and which can be cooled. At the end of the reaction the synthetic latex, like the natural sap of hevea (rubber), is precipitated by salts to produce the solid polymer, then it is filtered, or concentrated in an atomizing drier.

After drying the polymerizate is obtained in powder form. Latex is also often used directly for the impregnation of webs of paper or fabric and for glueing.

According to Harkins, Hohenstein, Mark, *et al.*, emulsion polymerization takes the following course: Groups of molecules of the monomer are enveloped by soap micelles, the hydrophobic part of the soap molecule facing the monomer, the hydrophilic part the aqueous liquid. Nucleation takes place at the monomer partly in the micelles, and partly at the dispersed individual molecules of the monomer. The nuclei initiate polymerization inside the micelles which are replenished by monomer from drops of the emulsion. The drops act, therefore, as a reserve of monomer. During polymerization it is possible to observe shrinking or complete disappearance of the monomer liquid drops. The diameter of the polymer drops is much smaller than that of the original droplets of the monomer. The course of the process is influenced by the presence of the polymer.

Since in emulsion polymerization the viscosity of the mixture does not increase during the reaction there is no difficulty in continuously stirring and cooling; this is impossible in block- and solution polymerization.

If emulsion polymerization is carried out continuously care must be taken to prevent the fresh monomer from mixing with material that is already largely polymerized. This is achieved by agitator vessels arranged in series (see Buna, Figure 143) or by a flow pipe. With groups of agitator vessels it is particularly easy to observe the progress of the process by sampling from the individual vessels, e.g. by determining the density, and to heat or to cool, add reagents, etc. as required. The emulsion is then precipitated in the last vessel by the addition of common salt. It is then drawn off by suction on rotary filters and washed. In the manufacture of Buna a paper machine is used for the isolation of the butadiene polymerizate in the form of a strip. The following are polymerized in emulsion: certain types of styrene (with molecular weights of about 1 million), vinyl chloride and various mixed polymerizates, vinyl acetate, acrylic acid ester, acrylonitrile, and butadiene. Polyvinyl chloride and methacrylic ester are often polymerized in suspension.

### Physical properties, mechanical behaviour, and testing of plastics

The problem of mechanical strength as a function of micro-structure, of primary and secondary bonds, of orientation, crystallite formation, and cross-linking, has already been discussed in the section on synthetic fibres. There the specific importance of the stress-strain diagram was shown. In plastics cross-linking leads to a considerable increase in the strength, e.g. in rubber by vulcanization (cross-linking by S-bridges) up to fifteen-fold. Heat resistance also increases as a consequence of cross-linking, i.e. the softening point is higher. However, cross-linking has the most striking effect on the hardness of a plastic, in rubber it increases 30-fold. The highly cross-linked phenolformaldehyde resins are even harder.

A special position among plastics is occupied by the so-called 'high-elastic' materials. These behave like an ideal gas even though the physical basis is different (for details see Houwink, *Z. Phys. Chem.*, (A), 1938, **183**, 209).

The tension in a stretched natural rubber in the tensile range of 0–400 per cent increases in proportion to the absolute temperature. It was shown by calculation (K. H. Meyer) that elongation and contraction take place without any change in internal energy. This finding made it possible for Kuhn to define his two conditions for high elasticity. In compounds consisting of large molecules single groups of atoms in the molecule can join firmly together ('freeze') so that they are in a condition of complete rest, the whole molecules can no longer oscillate, i.e. the Brownian molecular motion is arrested at this point. However, other atom groups in the molecule can still oscillate. According to Kuhn, whilst the Brownian macro-motion is arrested, Brownian micro-movement continues. In highly elastic materials Brownian micro-motion should be as free as possible, and macro-motion should be prevented. The reason is that if the former is inhibited the molecules adhere to each other firmly along their whole length and elongation would lead to a change in the internal energy. On the other hand with uninhibited Brownian macro-motion the molecules would slide past each other during elongation and there would be no resilience. Since Brownian micro- and macro-motion are functions of temperature, high-elasticity is limited to certain temperature ranges. The high-elastic state is a sort of intermediate between the liquid and solid states. The principal factors influencing elasticity are the secondary linkage forces.

The deformability of plastics is greatly affected by plasticity which increases during heating (thermoplasticity). It becomes apparent only after the flow limit has been exceeded, i.e. above the elastic part of elongation, and can be characterized by the stress-strain diagram. In the ideal case ('Newtonian flow') the flow velocity is proportional to the shear stress and indirectly proportional to the viscosity coefficient. This applies, however, only to low-molecular materials. In high polymers the flow velocity does not increase uniformly in proportion to the stress, but first increases slowly and later more rapidly. In most plastics viscosity is so high at room temperature that they cannot be deformed mechanically. The viscosity coefficient decreases for each 10°C rise by approximately an order of ten eventually making deformation possible. The softening point of plastics which can be processed as thermoplastics has to be below their decomposition temperature so that they can be sprayed or pressed.

Polymers which are largely crystalline, such as polyamides, low-pressure polyethylene, and isotactic polymers, have a narrow softening range and when used in injection moulding require extremely accurate temperature adjustment of the heating system and special spray orifices capable of being closed.

Crystalline substances in which the intermolecular ordering forces are equal and collapse simultaneously at a certain temperature have a sharply defined melting point. In polymers molecules of different sizes are present. Their arrangement is maintained by various forces which during heating are gradually released. They have, therefore, not a melting but a softening range. This property renders them easily deformable at slightly elevated temperatures. Materials with a sharply defined melting point (e.g. metals) can be shaped only by casting.

Finally, it should be mentioned that the elastic properties of a plastic decrease continuously as the temperature drops since due to the inhibition of Brownian micro-movement the molecules eventually set and the material becomes brittle. There is an embrittlement temperature during cooling which corresponds to the softening temperature during heating. A practical example is the cracking of polyvinyl chloride gutters and fall pipes easily brought about by impact at low winter temperatures. The strength of plastics is greatly affected by any fillers used. Substances with long fibres increase their strength much more than those with short fibres. Phenolformaldehyde resins have greater strength when moulded with fillers, e.g. wood meal, cellulose- or textile fibres, than without them.

*Testing methods*

Certain testing methods have been developed to characterize the physical properties of plastics. The figures obtained are determined by conventional methods. They differ in the various countries and are, therefore, not comparable. In Germany particularly phenoplastics were standardized at an early stage and certain minimum values were fixed. The determination and control of the standards are carried out by the national material testing stations. The finished articles are stamped with a control mark for standardized plastics. Standardization of aminoplastics and thermoplastics is at present being carried out in Germany and on an international scale. The properties are measured on specimens (rods, cubes, plates, cylinders, some with tapered centre e.g. for tensile tests, rings, etc.) with specified dimensions to which the testing apparatus is adjusted.

To characterize the mechanical properties, tensile strength, modulus of elasticity, elongation at fracture, and permanent elongation are determined on automatically recording apparatus (e.g. from VEB Werkstoffprüfmaschinen Leipzig, and from Messrs. Schopper). Bending strength is determined with a standard test rod supported at the ends which is loaded in the centre with a specified load. Impact bending strength is determined by an impact of specified force. When the notched rod is fraturated this value is called the notch impact strength. The strength is usually much reduced by a notch. Hardness can be measured by impressing a steel ball and measuring the depth of the indentation produced under load (ball-thrust hardness). Tearing strength and ultimate strength are important for rubber. Fatigue tests are also carried out with moving mechanical devices which bend, buckle, crumple, twist spirally, etc. A counter records the number of movements required to fracture the specimen.

Stability of shape during heating (e.g. thermal stability according to Martens) is determined by measuring the deformation with or without load. Incandescent strength is a measure of the flammability on contact with a glowing rod heated to 950°C. For materials used in electrical engineering the electrical properties are also determined, e.g. resistance, dielectric strength, surface resistance, dielectric loss factor, and the dielectric constant. It is important to remember that a humid atmosphere or storage in water affect these values. Other data required are those bearing on processing, such as temperature during pressing, drawing, spraying, and the degree of cutting and non-cutting workability. Of interest for thermoplastics and hardenable resins is the ability to flow for which DIN Standard 53478 has been developed on the basis of the flow test. The closing time of the mould is determined according to DIN Standard 53465. The beaker produced during moulding is used for the determination of mould stiffness, i.e. the strength of the beaker in deformation. Flow behaviour and mould stiffness give the hardening time required by hardenable resins. The most interesting chemical effects are water absorption and capability of swelling in solvents, which are determined by soaking for a given period at constant temperature, and subsequent weighing. Resistance to ageing in oxygen is particularly important for rubber.

The test results are listed in a table and compared with the corresponding minimum values for the type. Taking as an example a phenolformaldehyde resin containing 40 per cent phenol resin and the remainder wood meal and other additions (Type 31, formerly Type S) we obtain:

| Test results for phenoplastics Type 31 | Measured | DIN minima |
|---|---|---|
| Bending strength, kg/cm$^2$ | 800 | 700 |
| Impact bending strength, cm-kg/cm$^2$ | 7·0 | 6·0 |
| Notch stength, cm-kg/cm$^2$ | 1·8 | 1·5 |
| Non-deformability according to Martens, °C | 150 | 125 |
| Incandescent strength (grades 0–5) | 3 | 3 |
| Water absorption, mg/dm$^2$ surface in 4 days | 150 | max. 180 |
| Surface resistance, ohm | $10^8$ | $10^8$ |

### *Solubility, swelling, plasticizing*

The first stage of dissolution is swelling. Whether the high polymer will merely swell to a limited degree or whether it will undergo unlimited swelling depends on its nature and on the nature of the solvent, particularly its polarity.

Swelling and solution are the results of various energetic reactions between the polymer and the solvent. Expenditure of energy is necessary for the separation of the molecules of the solvent as well as of the polymer whilst energy is released in the reaction of the solvent with the polymer. Swelling occurs only when the latter process predominates, i.e. when heat is released. In the other case the energy required can also be supplied by heating. In the swelling and solution of macro-molecular materials the solvent molecules are linked by van der Waals forces ('solvation'). A solvate layer or solvate sheath forms. The degree of its dissolving power depends on how many groups of the macromolecules are 'solvated' by the solvent.

Swelling of a polymer is always found when at some points molecules are still held together by dipolar forces (secondary bonds) or by primary bonds (e.g. cross-linking) whilst the other parts of the molecule separate with simultaneous interstratification by solvent molecules. If the molecules are held together by dipoles they can be more easily separated by polar solvents than when they are linked by principal valency bonds (cross-linking). A very few cross-linking points are often sufficient to prevent complete dissolution whilst permitting an extraordinarily strong swelling. In general the solubility of polymers decreases with increasing molecular size since the secondary linkage forces increase simultaneously.

Polymers can also be dissolved by high-boiling solvents whereupon they acquire new mechanical properties. In many cases they become capable of being processed only by swelling. They soften, become plastic, usually also elastic. Tensile strength is reduced, notch strength increases.

This process is called 'external plasticizing'. The swelling agents or solvents used are called 'plasticizers'. The concept was taken from painting technology, where a plasticizer is added to prevent embrittlement of the coat of paint. A sort of intermediate position is occupied by 'thinning agents' which are usually slowly evaporating solvents, or agents for keeping the paint in a 'thin' condition, which prevent too rapid drying of the coat of paint and promote its even spread. A plasticizer has to have a low vapour pressure so that it does not evaporate either at room temperature, at the temperature of usage, or in the temperature range in which the polymer is processed, otherwise embrittlement of the material takes place. The volatility of a plasticizer is tested by heating a test mixture. During cooling the plasticizer must not solidify or crystallize in the plastic material. The plasticizer also raises the cold resistance. This is tested by cooling and sudden buckling. Finally, the plasticizer has to be compatible with the solvents, the polymer, and the pigments, it must not precipitate, nor exude from the film. This compatibility may need to be restricted only to given quantities within a narrow range of temperature. Another requirement is that plasticizers have to be odourless and physiologically harmless. This is particularly important in view of the fact that plasticized plastics are used increasingly for the packaging of foodstuffs.

'Internal plasticizing' is obtained by the incorporation of elastic parts of polymer chains into rigid ones. This can be achieved by mixed polymerization or by the linking of rigid 'blocks' by elastic links, as discussed earlier. An example of a material plasticized by internal plasticizing is the mixed polymerizate of styrene with butadiene.

Plasticizers are added to the polymers cold or, when necessary, hot. For this purpose, heatable kneading machines are used, e.g. of the Werner-Pfleiderer type, or calenders, sets of mixing rollers, and kneading pumps. The plasticized plastics can be rolled directly on calenders into thin foils, and shaped continuously by screw presses (extruders) into sections and tubes.

If the swelling is unlimited, dissolution takes place. In good solvents the macromolecules occur singly, in poorer ones, particularly at higher concentrations, they are combined in molecular aggregates. The molecules can have either an elongated shape or be coiled. Aggregation of any type causes an increase in the viscosity of the solution. The molecular weight of high polymers can be calculated from the viscosity by Staudinger's viscosity law (Staudinger and Husemann, *Liebigs Ann.*, 1937, **527**, 195).

This law applies only to linear molecules. The molecular weight can also be determined by means of the ultracentrifuge, the use of semi-permeable walls (osmotic method), the measurement of the diffusion velocity, the measurement of light diffusion using the diffusion photometer with transparent solutions, the determination of precipitability by solvents (G. V. Schulz), and finally by the determination of end groups.

Solubility and swelling are important for the practical use of high polymers. Whilst raw materials for lacquers should be easily soluble in the usual solvents, it is desirable that they should not swell in as many others as possible. Plastics and rubber should swell as little as possible in benzene and oil. In paint technology high concentration of the solution together with low viscosity is often desirable. Not all soluble, film-forming high polymers and lacquer raw materials combine these properties and they can, therefore, be used only for special purposes despite their other favourable properties. An expedient used is the emulsification of the film-forming substance, e.g. in water. These emulsions (artifical latex) are highly fluid and can be applied with a brush or sprayed. Emulsions of polyacryl esters (Acronal, BASF;

Plexigum, Röhm and Haas; Acrylit D, Buna) are particularly suitable for use as bonding agents for leather coating colours. PC mixed polymerizates or per-chlorinated PC are also used for latex colours.

*Solvents for plastics*

In the lacquer and plastics industries solvents are used to render materials suitable for processing by dissolution. During processing they are mostly lost by evaporation, or they can be recovered in special plants using compression, condensation, or adsorption methods.

It is of great practical importance that some solvents mixed with non-solvents are capable of dissolving a material. This property is essential for the dilution value of lacquer solutions with non-solvents for the purposes of dilution. A lacquer mixture contains, therefore, solvents and diluents. The rheological properties (flow behaviour) of high polymers in mixtures with solvents, diluents, plasticizers, and pigments are important for their use in practice. If the mixture is appropriately composed and concentrated a range is obtained in which the mixture is a gel, then becomes thixotropic and finally liquid. A certain degree of thixotropism is desirable for paints to prevent running. (Thixotropism is an intermediate phase between solid and liquid. A thixotropic substance at rest forms a gel which is liquefied by stirring).

In addition to their chemical properties, certain physical ones are of special importance for the practical use of solvents: apart from their dissolving power and ability to mix with other solvents or diluents, these are their boiling ranges, their vaporizability and vapour pressure, to which are related ease of evaporation and volatility, viscosity, flammability (flash point), explosion limits in mixtures with air, and their physiological effects. The determination of these physical values is described in the appropriate text books and will not be discussed here. Mention will only be made of the fact that the speed of evaporation of lacquer solvents is more important for their practical use than the boiling point since a solvent with a higher boiling point can be more volatile than a solvent with a lower one. The evaporation indices of solvents are determined by means of various methods, e.g. the filter paper method, and are usually based on the evaporation time of ether = 1 (see e.g. DIN 53170).

The evaporation indices of important solvents are as follows:

| | | | |
|---|---|---|---|
| ethyl ether | 1 | cyclohexane | 40 |
| ethyl alcohol | 8·3 | decalin | 94 |
| ethyl acetate | 2·9 | methanol | 6·3 |
| ethylene glycol | 43 | methylene chloride | 1·8 |
| acetone | 2 | carbon tetrachloride | 4 |
| petrol according to boiling range | 3·5–40 | tetralin | 190 |
| butanol | 33 | trichloroethylene | 3·8 |
| n-butyl acetate | 11·8 | toluene | 6·1 |
| chlorobenzene | 12·5 | xylene | 13·5 |

In practice, solvents are divided into three groups: low boiling (b. p. < 100°C), medium boiling (100–150°C), and high boiling (above 150°C).

The solvents can be grouped according to their principal uses or on chemical considerations. In classification I many solvents appear in several groups.

I. Classification by use

Solvents for:

oils, fats, waxes
collodion cotton
cellulose ester
cellulose ether and chlorinated rubber
natural resins and rubber
synthetic resins, polymerizates

II. Classification by chemical class

1. aliphatic hydrocarbons, mineral oils
2. aromatic hydrocarbons
3. hydroaromatic hydrocarbons
4. halogenated hydrocarbons
5. terpenes and derivatives
6. alcohols
7. glycols and derivatives
8. esters
9. ketones
10. ethers
11. carbon disulphide

There are many special solvents which are marketed under various names and consist mainly of mixtures. Their polar character is of importance for their dissolving power. Non-polar solvents include the aliphatic hydrocarbons. Alcohols, ketones, and many esters are strongly polar, halogenated hydrocarbons and most aromatic and hydroaromatic hydrocarbons are weakly polar.

The manufacture and properties of most solvents have been discussed in various other sections of this book.

*Plasticizers*

The nature of the plasticizing effect and the main requirements have already been discussed. Here follows only a short survey of the materials themselves and their most important uses.

The versatility required of plasticizers has greatly increased with the growth of plastics technology. Their molecular weight can be as high as 1,000.

Whilst the purpose of plasticizers was at first merely to maintain the lacquer film in a sufficiently elastic state they now have to make possible the thermoplastic processing of many plastics, or the removal of brittleness and hardness from high polymers and an increase in their plasticity and, above all, their elasticity. In addition to these requirements, very high water resistance and electrical insulating properties are often expected. Frequently they are also required to reduce the flammability of fils. The effect of plasticizers can be assessed only after lengthy testing since when solvents are used in processing (lacquers, photographic film, imitation leather) residual quantities are often retained in the coating layers blurring the appearance. Discoloration, embrittlement, sweating, incompatibility, reactions with pigments, and other disturbances often only occur after long periods.

No system of plasticizers can be drawn up since they are derived from nearly all of the many branches of organic chemistry and, according to the resin or plastic used, have to satisfy the most varied requirements. Only a few plasticizer types used on a large scale will be mentioned here.

The most important are phosphoric acid esters and phthalic acid esters. Phosphoric acid esters are produced from alcohols or phenols and phosphorus oxychloride.

*Phosphoric acid ester*

| | Boiling range at 20 mm |
|---|---|
| Trichloroethyl phosphate (Cetamoll Q) | 210–220 °C |
| Isobutyl phosphate | 150–160 °C |
| Tricresyl phosphate | 275–280 °C |
| Triphenyl phosphate (m. p. 45 °C) | 260 °C |

Tricresyl phosphate is one of the most versatile plasticizers. Since it contains the toxic phosphate of o-cresol certain conditions have been laid down for the maximum content of o-cresyl phosphate on delivery. It is unsuitable for the packaging of foodstuffs.

*Phthalic acid ester*

Diethylphthalate (Palatinol A) is exceptionally compatible with acetyl cellulose but it is also suitable for nitrocellulose and many resins and rubber. Less volatile is di-octylphthalate (Palatinol AH) which is specially suitable for polyvinylchloride and has excellent resistance to cold. Di-butylphthalate (Palatinol C, Vestinol C) is one of the most widely used plasticizers for nitrocellulose lacquers and adhesives. The Palatinols F, HS, K, L, O, V, and others are phthalic acid esters of a variety of alcohols.

*Other plasticizers*

Other good plasticizers are acetic acid esters of higher alcohols, and esters of the higher fatty acids.

A new group are the plasticizers obtained in the forerunnings during the oxidation of paraffins to fatty acids. These are esterified with higher and polyvalent alcohols.

Glycols, their esters and ethers, and other polyvalent alcohols, such as glycerine and the by-product of 1,3-butylene glycol manufacture, hexantriol, are plasticizers or constituents of plasticizers on a basis of esters or ethers. Among ketones, camphor is the most important plasticizer for nitrocellulose (celluloid).

Toluene sulphamides are used under the name of 'Plastol'. 'Clophen' is a chlorinated diphenyl which imparts specially strong weather and water resistance to lacquers. Diphenyl is produced by 'pyrogenic synthesis' from benzene. Pure 'Clophenes' have excellent electrical insulating properties and small electric losses.

'Lutonals' are polyvinyl ethers. 'Mesamoll' is the cresyl ester of mepasin sulphochloride (Mersol) and is used a great deal, e.g. for polyvinylchloride.

A large number of other plasticizers with fancy names and various letter and number symbols exists which cannot be discussed here.

## The most important plastics (synthetic substances and lacquer raw materials)

Until the discovery of celluloid from nitrocellulose and camphor in 1865 (Parkes) only natural products were available for paints and industrial materials. Celluloid is a semi-synthetic plastic; for like acetyl cellulose invented in 1894 it is based on natural cellulose. The first fully synthetic plastics were Röhm's polyacrylic acid esters for which at first, however, no practical use could be found. Another observation made at the turn of the century by Spitteler was that casein could be hardened with formaldehyde (Galalith). Then followed Bakeland's phenolformaldehyde resins (1907). One of the most important discoveries in the plastics field was the polymerization of vinyl compounds (vinyl chloride, vinyl acetate) by Klatte (Chem. Fabrik Griesheim), 1912–1914, which, however, was not used fully on an industrial scale until 20 years later. The first synthetic rubber was developed during the first world war by Hofmann.

The alkyd resins, products of the esterification of phthalic anhydride with glycerine, condensed with fatty acids, resin acids, or oils, were produced by Kienle in 1927. Today they are still the most important synthetic resins for lacquers. The main development in the field of plastics have taken place in the last 30 years, from 1940 onwards at a very rapid pace. One of the most important technological advances in the field of synthetic high polymers was the development of emulsion polymerization and of redox catalysis, the nature of which has already been discussed. A new era began with the synthesis of the polyamides and polyesters (Carothers), and the discovery of the usefulness of di-isocyanates and of butadiene rubber (former IG).

In recent years a further very strong impetus has been given to the development of synthetic high polymers by the following new materials and processes: the use of epoxy resins which are hardenable by poly-addition, the introduction of glass-fibre-reinforced plastics using mainly unsaturated polyester resins and epoxy resins, the processes for the manufacture of isotactic polymers, and some new linear polyesters and polyethers.

The following table (Franzke, *Chem. Ind.*, 1952, **4,** 205) shows that world production of plastics increased during 1933–1951 almost fifteen fold, and yet several countries are only at the threshold of their development in this field.

World production of plastics

| | |
|---|---|
| 1900 | 20,000 tons |
| 1913 | 50,000 tons |
| 1929 | 85,000 tons |
| 1933 | 110,000 tons |
| 1939 | 350,000 tons |
| 1947 | 870,000 tons |
| 1951 | 1,500,000 tons |

In recent years world production of plastics has developed as follows:

| | |
|---|---|
| 1952 | 1,850,000 tons |
| 1954 | 2,650,000 tons |
| 1956 | 3,800,000 tons |
| 1957 | 4,400,000 tons |
| 1958 | 4,800,000 tons |
| 1961 | 7,741,000 tons[1] |
| 1962 | 8,860,000 tons |
| 1963 | 10,255,000 tons |
| 1964 | 11,950,000 tons |

[1] In 1960 the German Federal Republic produced more than 1 million tons.

These figures do not include synthetic rubber: 1957 (world) about 1,770,000 tons, 1960 (DDR) 86,800 tons, (BRD) 81,100 tons, nor natural rubber: 1960 (world) 2,030,000 tons, and cellulose hydrate foil: 1957 (U.S.A.) 250,000 tons.

Of the total world production the share of the various types of plastics is as follows:

| *Type* | 1957, in % | 1959/1960 in % |
|---|---|---|
| PVC and vinyl resins | 21 | 23·3 |
| Phenoplastics | 17 | 10·9 |
| Aminoplastics | 10 | 9·6 |
| Alkyd resins, also modified | 16 | 10·4 |
| Polystyrene | 12 | 13·7 |
| Polyethylene | 10 | 15·2 |
| Cellulose derivatives (esters and ethers) | 7 | 4·6 |
| Polymethacrylic acid- and polyacrylic acid esters | 3 | 2·0 |
| Polyamides | 1 } | 10·0 (Polyamides and other plastics together) |
| Other plastics including silicones | 3 } | |

Of these, polyethylene and the isotactic polyolefins, particularly polypropylene, show the highest rate of growth because of their excellent properties and the low cost price of monomers produced by the petrochemical industry.

Plastic materials which two decades ago were used mainly in the electrical engineering industry are now used in almost all branches of daily life and industry including nuclear and rocket technology.

# 1. LACQUER RAW MATERIALS AND SYNTHETICS FROM NATURAL PRODUCTS

## (a) Natural lacquer resins

The manufacture of rosin which is one of the balsamic resins, and its properties have already been discussed on p. 314. The disadvantages of rosin are its low softening point, its high acid number (abietic acid) and, as a consequence, low resistance to water and reactivity with basic pigments. It is, therefore, neutralized with $Ca(OH)_2$ or ZnO which reduces its acid number and greatly improves its properties. Valuable modified products for lacquers and adhesives are obtained by hydrogenating rosin.

Lacquer resins are produced from rosin also by esterification. This must be carried out at a high temperature e.g. with glycerine or diglycol. More than half of all rosin is used for the sizing of paper (p. 292). An Al-salt is produced on the paper.

Much more important lacquers are produced from the copals which belong to the 'recent fossil' resins (amber is a fossil resin). They are very hard but are insoluble in fat oils and lacquer solvents in their natural state. They are, therefore, subjected to a cracking process by heating them up to 350°C in fusion kettles and letting them react with linseed oil after cooling to about 200°C. Much skill and experience is required in carrying out a copal fusion in such a way that a clear and good lacquer resin is obtained.

Shellac is a refined product of stick-lac which in tropical countries is secreted on certain trees by lac insects. The processing of stick-lac to shellac requires much manual labour and is very laborious. Shellac is readily soluble in spirit and is used mainly as spirit varnish. Its most important use is in French polish. When it is still moistened by spirit it is plastic and can be worked up to a beautiful high gloss polish by rubbing with a spirit-moistened cloth ball and putting on successive new layers (polishing). The correct polishing of a concert grand piano takes many weeks. There are also rapid polishes consisting of synthetic resins usually with collodion cotton, and which are increasingly replacing shellac polishes.

Amber is a fossil tree resin (oldest tertiary). It is deposited by the sea on the shore or mined (coast of Samland). Pieces unsuitable for the manufacture of jewellery are, like copal, subjected to a fusion process rendering them oil-soluble. Amber lacquer is distinguished by its exceptional hardness and beautiful gloss.

Japan and China lacquers are balsamic resins originating from Asiatic lacquer trees. Oxidative drying of the films of these lacquers, which produces their great hardnes, strength, and high gloss, is carried out with the participation of an enzyme which is contained in the natural resin. The beautiful Japanese and Chinese lacquer finish has great mechanical and chemical strength.

## (b) Drying oils

Almost all the drying oils used as binders (film-formers) for paints come from the vegetable kingdom. Drying in air produces a film. This film formation is based on polymerization of unsaturated fatty acids with simultaneous cross-linking. Fatty acids capable of such cross-linking are often unsaturated.

The polymerization is started by an autoxidation process in which peroxides occur as intermediate products. Drying is, therefore, accelerated by atmospheric oxygen, light, and oxidation catalysts, the drying substances or siccatives. Particularly effective are Co-, Pb-, and Mn-salts of oil- and resin acids, and the corresponding naphthenates. Drying oils are used either alone or in conjunction with lacquer resins.

### *Chemistry of oil drying*

It used to be assumed that atmospheric oxygen adds first to a double bond of the fatty acid molecule (I). It was found, however, that here, too, as in all autoxidations, oxygen inserts itself into an activated CH-bond which is next to a double bond (II, III) (See A. Rieche, *Z. Angew. Chem.*, 1937, **50,** 520; 1938, **51,** 707; *Wiss. Ann. d. DAdW*, 1953, **2,** 735). This corresponds exactly to an autoxidation process which proceeds as a radical chain reaction (see pp. 206, 447):

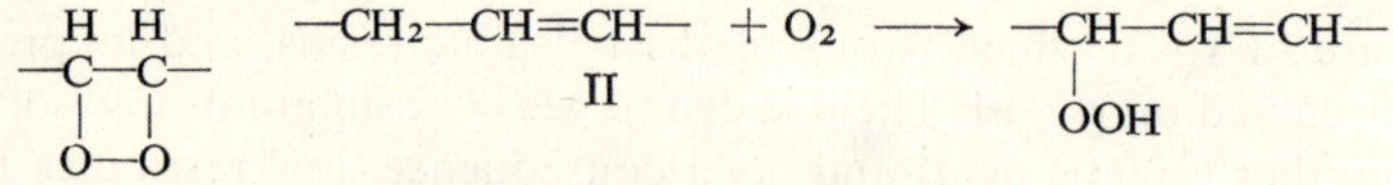

I III

The formation of a hydro-peroxide has been confirmed by numerous workers (A. Rieche, *Z. Angew. Chem.*, 1937, **50**, 520; 1938, **51**, 707; *Wiss. Ann. d. DAdW*, 1953, **2**, 735; Review articles: W. Treibs, *Fette und Seifen*, 1950, **52,** 549; K. Meier, *Chem. Z.*, 1952, **76,** 218. See also Kern and Willersinn, *Angew. Chem.*, 1955, **67,** 573). The double bonds of multiple unsaturated fatty acids are shifted from the isolated into the conjugated arrangement when hydro-peroxides are formed (Treibs, Farmer, *et al.*). This is a consequence of the occurrence of mesomeric marginal forms of the C-radical ('Farmer mechanism').

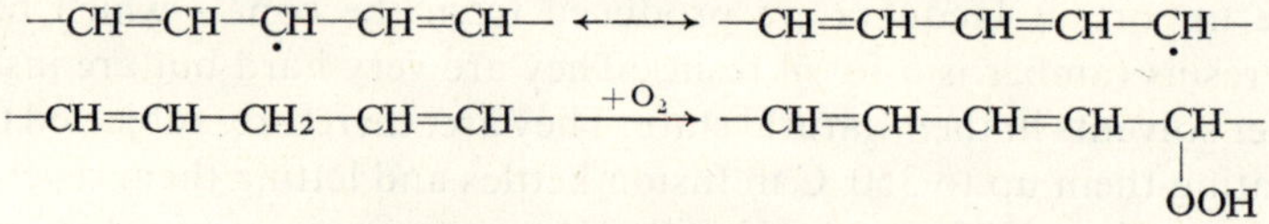

When several double bonds are present, those hydro-peroxides in which the double bonds can become conjugated are formed preferentially. The formation of the primarily formed radical and with it the tendency to autoxidation are dependent on the activation of the CH-bond (A. Rieche *et al.*, *Fette, Seifen, Anstrichmittel*, 1962, **64,** 198).

According to several authors, hydro-peroxide formation predominates at ordinary temperatures, whereas at higher temperatures attack on the double bond predominates. The linking of the fatty acid molecules can be effected by peroxide groups which can transform into ether bonds, as in 'Linoxyn'. However, of greater importance for film formation is the polymerization of double bonds initiated by the peroxide catalyst. Since the butadiene configuration has also to be taken into account, cross-linking also occurs, e.g. as in rubber. As in graft polymerization (see p. 398), side chains and cross-linking points can form on the fatty acid residues bearing the HOO groups by polymerization of olefins.

Up to the beginning of film formation an induction period occurs caused usually by the presence in the oils of natural antioxidants. Following autoxidation, fractions of fatty acid molecules also occur such as low fatty acids, aldehydes, and ketones, as happens when natural oils and fats turn rancid.

The first binder for oil paints was poppyseed oil, the most important modern drying oil is linseed oil. Wood oil and Brazilian oiticica oil also form good films during drying. Sunflower, cottonseed, and soya oil by themselves do not form films, they have to be specially prepared for use in paints, or used in combination with drying oils.

Castor oil, too, can only be used in paints after it has been chemically transformed.

The technology of oil extraction from seeds has been described earlier (p. 348).

Converting linseed oil into linseed oil varnish increases the drying speed eight-fold.

In the manufacture of varnish linseed oil is first dehydrated by heating to 130°C, causing it to foam violently. Then metal oxides (Mn or Pb) ground with linseed oil or prepared siccatives are added as oxidation catalysts, and the mixture is heated to 220–250°C. If air is injected, heating to 150°C for a short period is sufficient. However, paint films produced in this way are not adequately weatherproof. 'Stand oil' or 'thick oil' for outside paintwork is produced by hot polymerization of linseed oil at 285–300°C with exclusion of air.

This reduces both the number of double bonds and the drying speed and increases the resistance of the films to weather.

China wood oil (or Tung oil) has been known in China as a varnish for 3,000 years. It is extracted from the seeds of euphorbiacaeas, the fruit of which contain about 50 per cent oil. The main constituent of the oil is the triglyceride of eleostearic acid, an octodecatrienic acid occurring in two stereomeric forms. This is responsible for the strong tendency of the oil to polymerize. Owing to the triple conjugated character of the double bonds they easily undergo cross-linking or react with other resins which are capable of adding to conjugated systems. Wood oil is unsuitable for foodstuffs since it is toxic. The wood oil film dries particularly quickly under the influence of light on its surface and then presents a frosted appearance. This can be prevented by a preliminary treatment similar to the boiling of stand oil.

Castor oil from the bean of the castor oil plant consists mainly of the triglyceride of ricinoleic acid. It has no drying properties but is a widely used plasticizer for nitro-lacquers and imitation leathers on a collodion cotton basis, and as a raw material for synthetic lacquer resins, such as alkyd resins.

*Oil cloth, linoleum*

Oil cloth is a very old product. A woven material is used as an underlayer, e.g. calico or fustian. After preliminary treatment with an aqueous size (e.g. from starch) an oil film is applied in several layers with intervals for drying. The material used contains linseed thick oil as a binder, siccatives, some petroleum as diluent, and pigments. The coats, of which there may be 3 to 5 in number, have slightly different compositions. They are applied with a coating machine. The final operations are rotary printing and application of a protective lacquer (nitrolacquer, resin ester lacquer, Albertol).

Imitation leather is a product similar to oil cloth in which synthetic binders, such as vinyl resins, are nowadays used, in addition to nitrocellulose.

Linoleum has practically all the good properties one could wish for from a floor covering. Its thermal insulation and durability are unsurpassed, although the abrasion resistance of a rubber floor is greater.

Its manufacture is rather complicated and requires a great deal of experience. The method described in 1863 by the Englishman Walton is still in use today.

Linoleum contains about 40 per cent of linoleum cement, a material consisting of linoxyn, rosin, and a natural or synthetic copal, 35–40 per cent cork powder or wood meal or a mixture of these, and 20 per cent pigment dye.

Linoxyn is produced from linseed oil by a lengthy oxidation process. It is then treated at 150°C with colophony and copal until it forms a gel.

The finely ground fillers and pigment dyes are combined with the linoleum cement in mixing machines and the material is rolled on to jute and polished by heated 4-roller calenders. The linoleum is finally subjected in the drying house to a secondary oxidation (maturing) at an elevated temperature, lasting several weeks. It is obvious that this process determines the quality of the product. Recently, floor coverings made of synthetic materials or of coloured rubber have supplanted linoleum in many cases.

### (c) Semi-synthetic lacquer raw materials

Combinations of drying oils with vinyl polymerizates have been produced by polymerizing oils partly oxidized by atmospheric oxygen with vinyl monomers, such as styrene, methyl styrene, cyclopentadiene, or maleic anhydride.

Alkyd resins are extremely important because of their resistance to weather and their elasticity. They are constructed from a dicarboxylic (usually phthalic acid, but also maleic acid, adipic acid, or sebacic acid), a polyvalent alcohol (glycerine, also trimethylol propane or -ethane, pentaerythritol, hexantriol), and a natural oil mixed so as to contain 40–50 per cent fatty acid.

The most important oil for alkyd resins is linseed oil, followed by castor oil. Tall oil and the forerunnings from the fatty acids $C_5$ to $C_9$ produced by paraffin oxidation are also suitable. For the production of alkyd resins the components are heated to a high temperature, e.g. 250°C. The resins are marketed under a variety of trade names.

The same group includes the KM resins (BASF) which are obtained by condensation of colophony with maleic acid anhydride and subsequent esterification with polyvalent alcohols.

The oldest semi-synthetic lacquer raw materials are the natural resin or oil modified phenol resins of L. Berend's which were first marketed under the name of 'Albertoles' by Messrs. Albert. They are produced by esterification of resin or fatty acid with phenol formaldehyde resins. The organic acid introduced in excess is esterified with a polyvalent alcohol.

To produce alkyl phenol resins, higher alkylated or arylated phenols are condensed with alkalies using excess formaldehyde.

'Dian' (diphenylol propane) can also be used for this purpose; it is obtained by condensation of phenol with acetone in the presence of sulphuric acid:

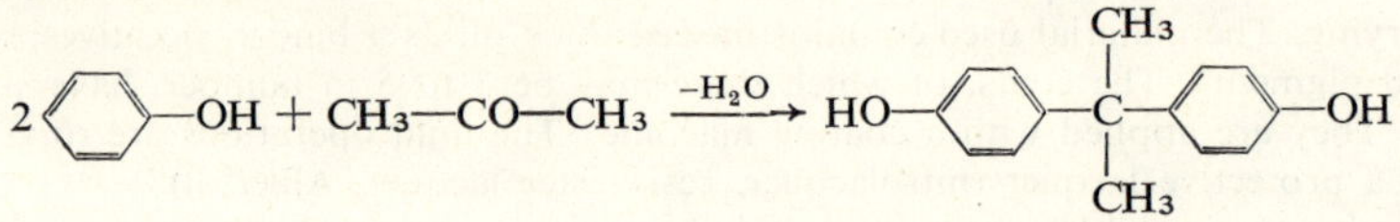

The quadruple-functional nature of 'Dian' is not utilized because some of the OH-groups etherify before the formaldehyde condensation with alcohols. This reduces its hydrophilic properties, and better compatibility with hydrocarbons and oils is achieved.

Typical examples of alkyl phenol resins are the Superbeckazites of Reichhold-Chemie, and the Alresenes of Chemische Werke Albert. In recent times Dian has become one of the most important synthetic intermediates, namely for epoxy resins and the polycarbonates ('Makrolon').

*Cellulose derivatives*

It became possible to use nitrocellulose as a raw material for varnishes only after a polymerhomologue series of collodion cotton with graduated specific viscosity had been produced by partial decomposition of the cellulose molecule. This led to the production of lacquers of higher concentration and lower viscosity. Among cellulose derivatives nitrocellulose is still the most important raw material for lacquers.

For nitration, cellulose has to be purified and bleached so as to have a purity of 99 per cent. Nitration is carried out with a mixed acid consisting of 60 per cent $H_2SO_4$, 24 per cent $HNO_3$, and 16 per cent $H_2O$. The degree of nitration is then 10–12 per cent N (collodion cotton) so that solubility is good. 'A-cottons' with 10·5 to 11 per cent N are soluble in alcohol, 'E-cottons' with upwards of 12 per cent N are soluble in esters. Nitrocelluloses with higher N-contents are difficult to dissolve and are used as gun cotton. As with all nitrations, here, too, the degree of nitration depends on the water content of the residual acid. Temperature and duration of nitration affect solubility and viscosity. With longer duration both the molecular weight and the viscosity decrease. A nitration centrifuge (Troisdorf) can be used as a nitration vessel. The acid is then centrifuged off, the residues are driven out with water so as to obtain a spent acid with the highest possible concentration, and finally scrubbing is carried out with large quantities of water. Some plants prefer a beating engine for the comminution during scrubbing. Boiling with water for many hours removes residual acid and unstable ester fractions to such a degree that the collodion cotton becomes stable ('stabilization'). The water is then largely removed in centrifuges or presses.

Nitrocelluloses are dissolved in ketones (acetone), esters (acetic acid ester, butyl acetate), and glycol ethers. The films require plasticizing (see section on 'Plasticizers'). Due to its low viscosity and the rapid evaporation of the solvents used, lacquers can be applied by spraying ('spraying lacquers'). Combinations of nitrocellulose and lacquer resins, e.g. with shellac (nitrolacquers), can be highly polished. Nitro-combination lacquers with alkyd resins give highly weatherproof glossy coats e.g. for car lacquers. They are shipped moistened with alcohol.

Other cellulose esters that can be used in the production of lacquers are: cellulose acetate (esterification degree between di- and tri-acetate), propionate, and cellulose acetobutyrate. Ethers with important practical applications are ethyl cellulose and benzyl cellulose. These are produced from alkali cellulose with ethyl chloride or benzyl chloride.

Methyl cellulose (Tylose), hydroxyethyl cellulose (Ethoxose) and cellulose glycol acid are water-soluble binders and thickening agents. They are also used as finishing agents in the textile industry.

'Cellophane' (Kalle), 'Transparit' (Wolff & Co.), 'Wilaphan' (VEB Zellstoff- und Zellwollewerke, Wittenberge) are foils made of cellulose hydrate. Their production is basically identical with that of viscose and cuprammonium rayon (q.v.), the only difference being that the solution emerges from a slot in a trough instead of from a spinning nozzle and flows on to a cylinder rotating in the acid coagulating bath. The foil is pulled off this cylinder by rollers. Cellulose hydrate foils are sensitive to water. They can be made 'weatherproof' by varnishing with nitrocellulose lacquers. Cellophane is one of the synthetic materials made of natural products to be discussed below.

### (d) Synthetic materials made from natural products

*Artificial horn*

Proteins like casein, blood albumen and gelatins are treated with a weak alkali, pressed and then hardened with formaldehyde. This process yields artificial horn (Galalith). Webs of paper swelled by soaking in a zinc chloride solution are pressed together, washed, dried, rolled on a calender and pressed in moulds, giving true 'Vulcan fibre', a material of high mechanical strength which is used for suitcases and car components and in the electrical industries.

The oldest thermoplastic, celluloid, is produced by working nitrocellulose with camphor in an alcoholic solution in a kneader, filtering it under high pressure, adding admixtures in a mixer, homogenizing it, freeing it from alcohol, and pressing it into blocks at 100°C. Celluloid is still widely used because it is easily processed and because of its good mechanical properties.

A particularly important use of collodion cotton is that for films in photography and cinematography. A disadvantage of nitrocellulose films is their high flammability. 'Safety film' (acetyl cellulose, cellulose triacetate) has not been able to displace nitrocellulose film since it lacks the great strength of the latter. Polycarbonates, e.g. that of Dian, behave much more favourably and may one day supplant celluloid film since they have high strength and maintain their dimensions even under multiple stresses (Makrolon).

### Technical methods of processing synthetic materials

*Films and foils*

Films are produced from solutions in film casting machines. The solvent is reclaimed. There are dipping roller, cylinder, and band casting machines. The method of working of the latter two is shown in Figure 134 (After K. Thinius and G. Bier, 'Die Kunststoffe' in Winnacker-Weingaertner, *Chem. Technologie*, Organische Technologie II, München, Hanser-Verlag, 1954).

When several layers of plasticized synthetic films and pigments are applied to webs of fabric one obtains imitation leather. A heated calender is used for this process (Figure 135). A pattern can be imprinted on the surface by means of an embossing machine.

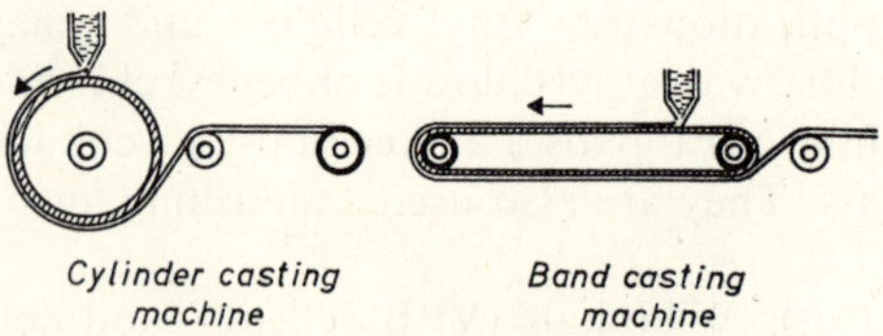


Figure 134. Casting of films

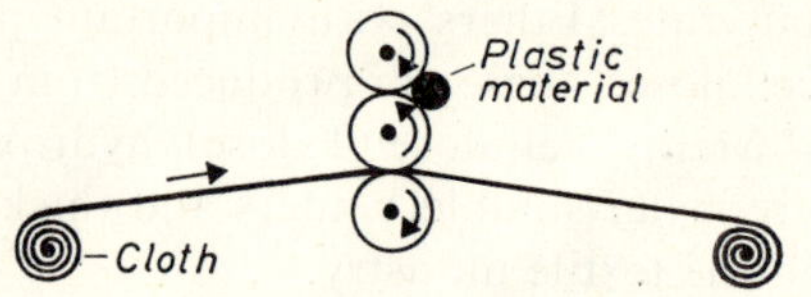


Figure 135. Manufacture of imitation leather on a calender

*Processing of thermoplastics*

Thermoplastics can be rolled into foils on heated calenders. Pipes, rods, and sections are pressed in a screw press with a heated mouthpiece to which is connected the mould (see Figure 136) (from Houwink, *Grundriß der Technologie der synthetischen Hochmolekularen.* Leipzig, Akad. Verlagsges. Geest und Portig K. G., 1952). By means of

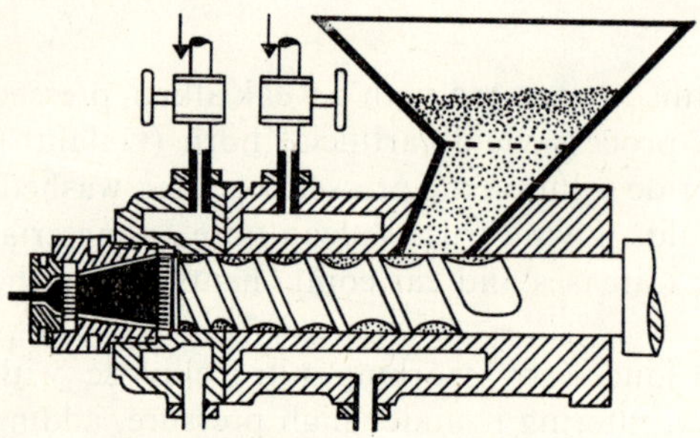

Figure 136. Screw press

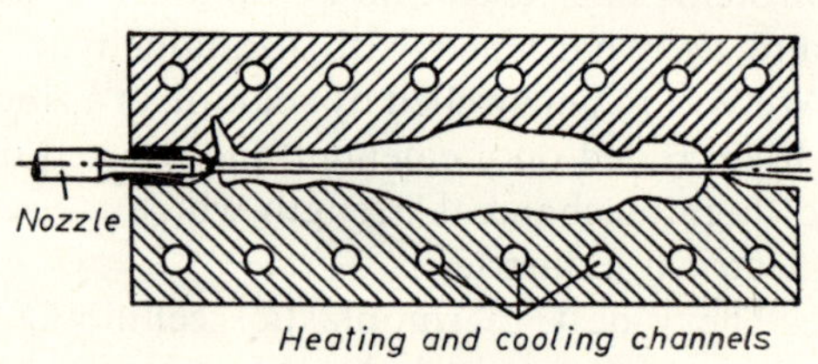


Figure 137. Blowing of hollow bodies from two plates in a heated mould

screw spraying machines (extruders), wires and pipes are coated with plastics; this is the cheapest method of manufacturing insulated wires. For hollow objects such as toys and bottles two plates are inflated whilst hot in a mould and then cooled, as shown in Figure 137 (after K. Thinius/G. Bier, 'Die Kunststoffe' in Winnacker-Weingaertner, *Chem. Technologie*, Organische Technologie II, München, Hanser-Verlag, 1954). The blowing of hollow bodies as shown in Figure 138 is similar. In the 'Holofol process' invented by W. Opavski, single foils and plates of plastic and elastic materials are split down the middle by means of a preliminary treatment, and inflated into hollow bodies, e.g. pipes and toy animals.

Sprayed die castings are produced in the injection-moulding machine (Figure 139) (After Thinius/G. Bier, 'Die Kunststoffe' in Winnacker-Weingaertner, *Chem. Technologie*, Organische Technologie II, München, Hanser-Verlag, 1954). It consists of a heated cylinder and die into which the powdered material or pellets are fed through a feed funnel. The liquefied material is sprayed through a spraying duct into a cooled mould where it solidifies. The mould is opened and the casting removed. Thermoplastics, too, can now be moulded in an injection press (jet-moulding process). The oldest process is the treatment of thermoplastics in dies (Eichengrün, 1919). The powder which may be pre-heated is fed into a hot die, pressed in a hydraulic press, cooled under pressure, and the finished object is removed from the mould. A more suitable process involves pre-forming the material into blanks. This is the method which is used almost exclusively for hardenable resins (e.g. the phenol-formaldehyde rapid pressing materials). Since these harden inside the die it is possible to mould them hot. This has the advantage that the dies need not be cooled.

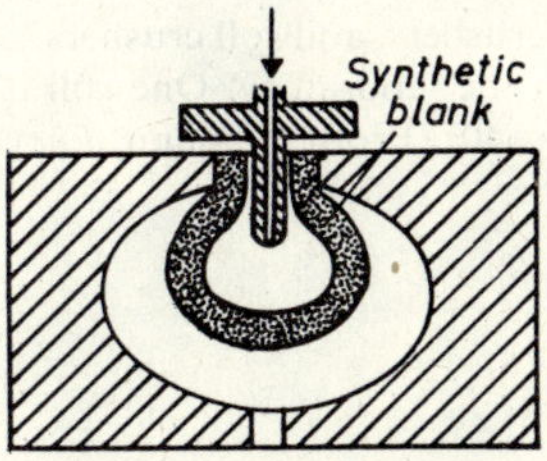


Figure 138. Blowing of hollow bodies

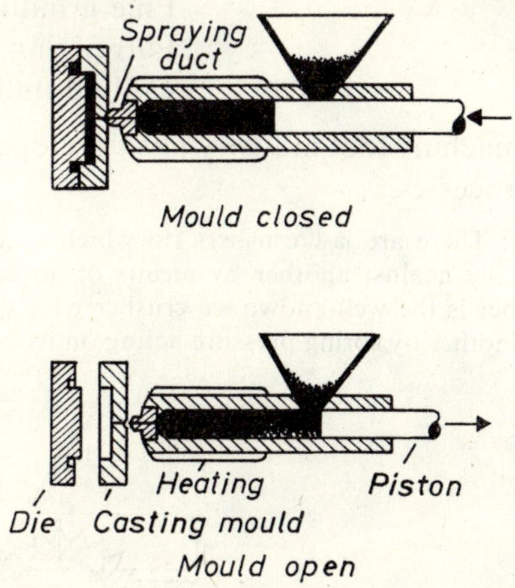


Figure 139. Injection moulding machine

Cellulose derivatives and certain plasticized vinyl compounds can be processed also by the hammer press method. The material, pre-formed as a strip or rod, is heated up to the melting point and then a piece is pressed by one blow through a die into a cold mould. The blank solidifies within a few seconds and is ejected.

*Comminution, grinding, mixing*

Mills belong to the oldest technical appliances of mankind. In chemical engineering wherever solid materials are processed they almost always have to be comminuted. Plastics are ground to render them more easily soluble, and fillers and pigments are pulverized to a fine powder. In mixing devices liquid binders or plastics are processed with solid ones, e.g. plasticizers, fillers, and pigments are incorporated into plastics.

The possibilities of comminution and the devices used are so numerous that they can be discussed here in outline only.

The 'degree of comminution' is the ratio of the particle size before and after comminution. The degraded particles have very different particle sizes. The proportion of the various sizes is characterized by the particle size distribution curve. If the material is to be separated by particle size it is necessary to classify it during or after grinding. For this purpose grinding equipment is often equipped with sieves or air elutriators.

Comminution involves a considerable expenditure of energy which is expressed in kWh per kg of material to be ground, or by its reciprocal value, i.e. the specific throughput kg/kWh. The finer the material is ground the greater is the energy expenditure. The mechanical efficiency is only about 2 per cent and a large amount of heat is generated during comminution. For this reason wet grinding is preferred for heat-sensitive or easily ignitable materials. Any numerical values given apply always to one type of machine only.

According to the type of comminution the processes are classified into:

Coarse and fine crushing,
Granulation and gritting,
Fine grinding,
Pulverizing,
Colloid milling.

The machine tool industry has developed a very large number of types of machines for these processes.

Crushers: There are jaw crushers (in which a grooved plate which swivels in one direction is pressed rhythmically against another by means of an eccentric), gyratory crushers, and roll crushers. A simple roll crusher is the well-known ice crusher with spiked rolls, used in the laboratory. One roll is pressed against another by spring pressure acting on its bearings (see Figure 140) (From Ullmann, *Encyklopädie*

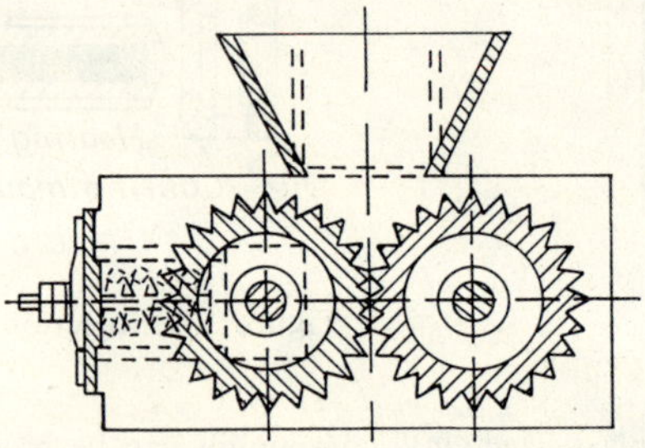

Figure 140. Spiked roll crusher

*d. techn. Chemie*, 3rd edn, vol. 1, München-Berlin, Verlag Urban & Schwarzenberg, 1951). There are various other designs for crushers such as the cam crusher in which revolving cams engage with a crusher grid.

The mode of operation of crushing rolls corresponds in principle to the pug mill in which heavy movable suspended rollers run on a grinding face. The difference is that in roll crushers the rolls and the grinding face are enclosed and the rolls are pressed against the support by centrifugal force or by springs.

In ball and tube mills grinding is carried out in a drum containing the grinding element, usually balls or rods. These mills are extensively used in the laboratory and in chemical engineering not only for grinding but also for mixing because of their great versatility and the possibility of manufacturing them in any size. Except for strongly caking materials; substances of any consistency can be crushed. If the material has a tendency to bake on to the walls of the drum this is sometimes prevented by the use of tetrahedrons instead of spheres. Ball mills can also be equipped with a screening jacket and a discharge device. The Concentra mill (see Figure 61) (Krupp-Gruson) has already been discussed in the section on Coal Hydrogenation.

The reciprocating ball mill (Siebtechnik G. m. b. H.) works like a ball mill with grinding elements which, however, are contained in a pan-shaped trough suspended by springs. Grinding efficiency depends on frequency and amplitude of the oscillations. The oscillations are produced by the rotation of an eccentric weight.

Beater mills and stud mills are suitable for continuous working and fine grinding. They contain rapidly rotating grinding elements which throw the material violently against breaker plates where it disintegrates as a result of the impact. In these mills semi-soft and soft materials can be comminuted particularly easily, with hard materials the wear is very great. A special type of beater mills are the hammer mills in which movable hammers revolve on a rotor and pass close by a screen and the housing. The perplex mill has a rotating disc fitted with bosses or studs which rotates close to a similarly studded grinding ring. Stud mills with one rotating and one fixed stud disc operate on a similar principle. In disintegrators a studded disc rotates in the opposite direction to studded rings. Figure 141 illustrates a stud mill (Kolloplex mill, Alpine design) in cross section (from Ullmann, *Enzyklopädie d. techn. Chemie*, 3rd edn, vol. 1, München-Berlin, Verlag Urban & Schwarzenberg).

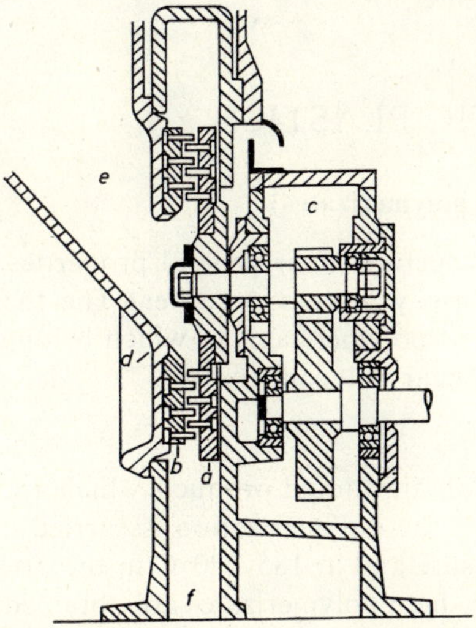


Figure 141. Kolloplex mill
*a* rotary disc *b* fixed disc with round studs
*c* drive *d* door *e* feed inlet *f* discharge outlet

The beater illustrated in Figure 107 is used for the comminution of solid materials (mainly by disruption) in liquids. The wood grinding machine used in the paper industry has also been discussed earlier (e.g. Figure 106). Pneumatic crushing mills which are still in the development stage and are particularly suitable for fillers and pigments used in the plastics and varnish industries, are very efficient and versatile. They are a variation of the colloid mill.

The predecessor of the roller frame used for ginding is the age-old corn grinding mill with millstones between which the corn is ground. Multiple roller frames are used for the comminution and mixing of plastic materials. These were discussed earlier in the chapter on the extraction of fats and oils in the section on devices for the comminution of oil seeds (see Figure 122). Roller frames are of prime importance for the manufacture of ointments and pastes, for the mixing of pigments and binders in the manufacture of paints and printing inks, and for the incorporation of plasticizers, fillers, and pigments, or the mixing of hardenable moulded plastics. The most robust type of mixing is carried out with machines of this type. Steam-heated double rolls of a particularly sturdy design have been developed for many purposes, e.g. the homogenizing of solid phenolformaldehyde resins with wood meal, hardeners, pigments, and other additives. The effect of the roller frame is reinforced by a slight difference in the number of revolutions of the rollers (friction). The temperature of the rollers is measured by placing on them a thermocouple with a measuring instrument.

Mixing (not comminution) is carried out in kneaders (e.g. Werner-Pfleiderer design) with S-shaped kneading arms moving in opposite directions. They are widely used in the plastics industry in sizes up to 500 kg capacity, usually equipped with heating devices. The trough can be turned upside down and the arms move so that the material is ejected.

The mixing of powder material is carried out in mixing drums of various designs which are usually fitted with internal devices for raising and agitating the material. In a new design by Gebr. Lödige, the plough share mixer, plough share-shaped members revolve on a rapidly rotating shaft inside a slowly rotating drum.

The Buss-Ko-Kneter permits simultaneous continuous mixing, homogenizing, and plasticizing in the plastic phase. Its particularly intensive dispersing and plasticizing effect is due to the fact that the mixing worm, in addition to its rotary motion, carries out a reciprocating movement in the direction of its axis, and that the profile of the worm and of the external wall are designed in such a way that there is a minimum of dead space and a maximum of shearing effect.

# 2. FULLY SYNTHETIC PLASTICS

## (a) Thermoplastics (vinyl polymerizates)

The manufacture of thermoplastics, their microstructure, their general properties, and methods for their testing and processing were briefly discussed earlier. The thermoplastics can be divided into two main groups: vinyl polymerizates to which belong also the polymerizates of the olefins and dienes, and linear polycondensates.

*Polyethylene*

By polymerizing ethylene with $AlCl_3$ or $BF_3$ one obtains liquid products which are used as lubricating oils (e.g. 'Buna oil', 'Leuna oil'). If the polymerization is carried out at pressures of 1,000–1,500 atmospheres gauge pressure and at 185–220°C in the presence of some $O_2$ as a catalyst (up to 0·01 per cent), very high polymerizates are obtained with a molecular weight of 18,000–26,000, even up to 35,000. The process was developed by Fawcett and Gibson. Polyethylene has been manufactured in England since 1936 (Polythene, ICI), in the USA since 1942 (Alkathene, Parowax), by the former IG Ludwigshafen since 1942, and is now manufactured by the BASF (Lupolen H).

Polymerization is carried out continously in a slightly inclined high pressure tube made of V2A steel into which is injected very carefully purified ethylene with small amounts of oxygen as a catalyst at a gauge pressure of 500 atmospheres. The polymerization pressure is produced during heating. The onset of polymerization becomes apparent by a drop in pressure. Only 18 per cent of the material reacts, excess ethylene is returned into the polymerization with gradual release of pressure after the polymerizate has been separated.

High pressure polyethylene does not possess the ideal unbranched chain of a paraffin (I) but has methyl group branches (II):

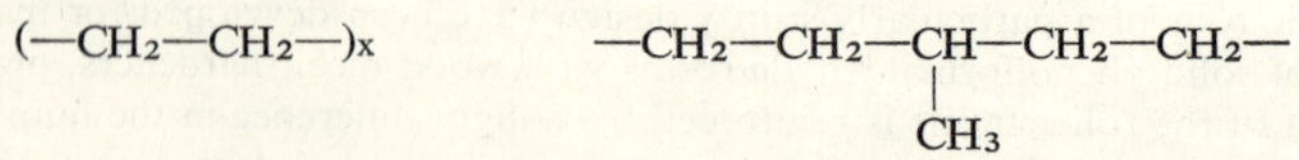

I II

From what has been said earlier about the dependence of the solubility of high polymers on their microstructure it is obvious that polyethylene will be the more difficult to dissolve the fewer methyl side chains there are present. According to H. Hopff, the oxygen used for the initiation of polymerization is present in the form of OH groups; polymerization catalysts or their transformation products are generally incorporated into the polymer chain.

Polymerization of ethylene can be carried out also in solvents (e.g. methanol) at 110–120°C and at gauge pressures of up to 300 atmospheres. The wax-like product obtained has a molecular weight of 2,000–3,000 (Lupolen N, former IG Wachs A). Recently K. Ziegler developed a method (*Angew. Chem.*, 1955, **67**, 541) for the polymerization without pressure using metal alkyls as catalysts. The high polymers obtained are particularly suitable for plastics (e.g. Hostalen, Hoechst). The catalyst consists of $Al(C_2H_5)_3 + TiCl_4$ and is very sensitive to air and humidity. The removal of the catalyst from the polyethylene entails a laborious procedure.

Technical polyethylene melts at 100–120°C and is partly crystalline. It is a very light plastic, pliable to hard, mechanically strong, very resistant to cold and water, impermeable to water, chemically very stable, and has good electrical insulating properties. However, it begins to soften at 70–80°C. By means of Ziegler catalysts other $\alpha$-olefins can also be transformed into high polymer solid plastic-type products. These have the following softening points:

| | |
|---|---|
| Polyethylene | 135° |
| Polypropylene | 165° |
| Polybutene-(l) | 128° |
| Polypentene-(l) | 80° |
| Polyhexene-(l) | 55° |
| Polydodecene-(l) | 45° |

Using the Philippson process, olefins can be polymerized at a gauge pressure of up to 35 atmospheres also on $Al_2O_3$-chromate or $SiO_2$–$Al_2O_3$-chromate catalysts.

As mentioned earlier, it is possible to raise the softening temperature by irradiation of finished polyethylene goods, so that no deformation occurs at working temperatures above 120°C (e.g. sterilization with hot steam).

Polyethylene has found wide use as a cheap and very high-grade plastic on account of its versatility for use as foils, tubes, pipes, vessels such as bottles, and as a lining for metal containers.

As discribed on p. 396, polypropylene (Hostalen PPH and PPN, Hoechst, and Moplen, Montecatini) is produced by isotactic polymerization by the process developed by G. Natta, and has a softening point of about 156°C. In this, and also partly in other properties it surpasses polyethylene and can, therefore, be considered as one of the most important thermoplastic materials of the future.

The diagram on page 420 shows the continuous olefin polymerization process of the Farbwerke Hoechst (S. Sommer, S. Wagener, and H. Ebner, *Kunststoffe*, 1959, **49**, 500).

### *Polystyrene*

Polystyrene is also derived from ethylene. In the presence of acid catalysts, e.g. $AlCl_3$, ethylene is added to benzene, and the ethylbenzene formed is dehydrogenated catalytically in a furnace to styrene. As described earlier, it can be polymerized continuously into a strip by the principle of block polymerization (see Figure 133) (after K. Thinius and G. Bier, in Winnacker-Weingaertner, *Chem. Technologie*), or in emulsion. Its molecular weight is very high (200,000 to 500,000). Since it flows easily and solidifies rapidly on cooling it is particularly suitable for injection moulding.

It possesses particularly outstanding electrical properties (specific resistance $10^{13}$–$10^{14}$ ohm/cm, loss factor tang $\delta$ 0·001–0·003 at 50 c/s) and is, therefore, used widely for radio and television components and other articles in electrical engineering. Many goods for everyday use are manufactured from this material, e.g. drinking vessels, zips, boxes, combs, and fountain pens. Its strength is improved by stretching (Styroflex, BASF). Other trade names are: Trolitul, Styrophan, Styron, Vestron. Polystyrene is also used as a varnishing resin (Ronilla L). Mixed polymerizates are also marketed, the most important of these being Buna S. Small additions of divinyl benzene during polymerization cause strong crosslinking. "Styropor" is an excellent foam insulation material based on polystyrene.

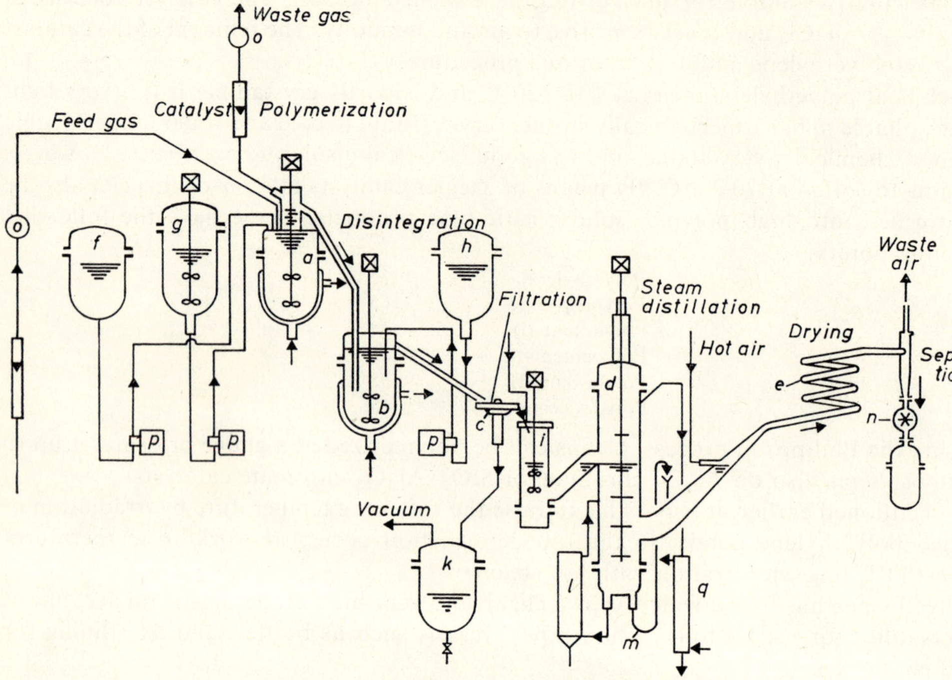


Figure 142. Diagram showing continuous olefin polymerization
*a* polymerization vessel *b* agitator vessel *c* plane filter *d* steam distillation *e* dryer *f* tank containing dispersing agent *g* tank containing catalyst suspension *h* alcohol tank *i* stripper *k* vacuum tank *l* continuous evaporator *m* discharge pipe *n* starwheel feeder *o* gas meter *p* pump *q* condenser

*Polyvinylchloride*

One of the most important and most versatile thermoplastics is polyvinylchloride of which 403,000 tons were produced in the USA in 1957, including mixed polymerizates. (For the share of PC production in the world production of plastics see p. 408).

The production of the monomer, the polymerization of which takes place continuously in emulsion or in suspension, and the mixed polymerization with other vinyl compounds have all been discussed earlier. The emulsion polymerizate is processed into a powder by atomized drying or on roller dryers.

PVC has good strength, small water absorption, good electrical properties, high resistance to chemicals, can be easily worked, glued, and welded, and has a low softening point (75°C approximately). Without added plasticizers it is brittle at low temperatures. PC is widely used as an exchange material for metal pipes, and for linings, valves, and armatures. It can easily be rolled into foils, pressed into plates, and worked into profiles and pipes by extrusion. With plasticizers (tricresyl phosphate, 'Mesamoll', dicarboxylic acid esters) it yields elastic foils. It is marketed under many names, e.g. PVC, Ekadur, Ekalit, Decelith, Hostalit C, Vestolit, Vinylite. Mixed polymerizates are Mipolam and Hostalit. The suspension polymerizate which is the purest product is stabilized against HCl cleavage by organo-tin compounds.

Since pure polymerizates of the asymmetrical dichloroethylene (vinylidene chloride) are difficult to work, mixed polymerizates with vinyl chloride or acrylonitrile (Hopff) are produced which are marketed as Diorid, Diophan (BASF), Vestan (Hüls), and Saran (Dow).

*Polytrifluorochloroethylene and polytetrafluoroethylene*

After Schloffer and Scherer (Hoechst, 1934) had found that trifluorochloroethylene $ClFC=CF_2$ could be polymerized although it no longer has a hydrogen atom, Du Pont found that tetrafluoroethylene also forms a polymer. Monomeric tetrafluoroethylene can be obtained by a remarkable reaction via difluorocarbene from 2 mole difluorochloromethane (obtained from chloroform and HF) at 600–800°C in platinized tubes:

$$2\,CHClF_2 \rightarrow CF_2 = CF_2 + 2\,HCl.$$

Tetrafluoroethylene can undergo spontaneous explosive decomposition. Polymerization, which is also carried out in the aqueous phase, yields from both trifluorochloroethylene and tetrafluoroethylene plastics with very high softening points and unsurpassed resistance against even the most corrosive chemicals; they are, therefore, particularly suitable as seals (e.g. Hostaflon and Hostaflon TF, Hoechst).

Decomposition temperature >400°C, thermally stable below 280°C.

*Polyvinyl acetate, polyvinyl alcohol, and polyvinyl acetals*

Polyvinyl acetate has a very low softening point and cannot, therefore, be used as a raw material for plastics but it is suitable as a raw material for varnishes, mainly as a dispersing agent, for paints, as a basis for primers and adhesives, e.g. for glueing plywood and veneers (Mowilith, Mowicoll, Vinapas). Polyvinylacetate is the ester of the enol form of polymeric acetaldehyde and can be saponified to polyvinyl alcohol. Technically the esterification is carried out with methanol:

$$(-CH_2-\overset{|}{C}H-O-\overset{\overset{O}{\|}}{C}-CH_3)_x + xCH_3OH \rightarrow (-CH_2-\overset{|}{C}HOH)_x + xCH_3-COOCH_3$$

Polyvinyl alcohol (Vinarol, Mowiol, Polyviol) is resistant to fuels and is, therefore, used for lining tanks and pipes. It is soluble in water and such a solution is very suitable as a thickening agent, protective colloid, and size for textiles.

Polyvinyl alcohol can be acetalized with aldehydes yielding polyvinyl acetals which are used mainly as raw materials for varnishes and as cold-resistant and light-fast interlayers for laminated glass (Mowital), in addition to their use as raw materials for plastics.

Ethers of polyvinyl alcohol are the polyvinyl ethers mentioned earlier which are obtained from acetylene and alcoholate with subsequent polymerization with $BF_3$ (Reppe). There is a variety of uses for these materials as insulating materials, raw materials for varnishes, and adhesives for laminated glass.

*Polyacrylic esters*

Of the polyacrylic esters and methacrylic esters the polymeric methyl ester of methacrylic acid,

$$H_2C=\overset{\overset{CH_3}{|}}{C}-COOCH_3$$

is the best known and most important. Because of its crystal clear transparency it is used as organic glass ('Piacryl', VEB Stickstoffwerke Piesteritz, 'Plexiglas', Röhm and Haas). As compared with silicate glass, the polymerizate has great advantages; it is much lighter, its impact bending strength is better by a factor of seven, it does not shatter, it transmits more light rays (from 330 m$\mu$ upwards), can be worked mechanically extremely easily, and can be cemented. However, its surface hardness is considerably lower. The manufacture of monomeric acrylic ester was discussed on pages 172 and 212. Methacrylic ester is obtained by the addition of HCN to acetone and reaction with $CH_3OH$ and $H_2SO_4$.

Polyacrylic ester is used alone and in mixed polymerizates mainly in the form of dispersions as a binder (Acrylit D, VEB Chemische Werke Buna, Acronal, BASF, Plexigum, Röhm and Haas).

The various methods of polymerization were described on p. 398.

Polyacrylic acid salts are used as thickeners and dispersing agents (Rohagit, Röhm und Haas, Viskol, BASF) and as soil improvers (Krilium, Monsanto).

*Other polymerizates*

Polyisobutylene, the manufacture of which was mentioned earlier, is used as Oppanol B (BASF) either alone or combined with rubber for acid-resistant linings of chemical apparatus, or as sealing and insulating material in the building industry. Oppanol C (BASF) is polyvinyl isobutyl ester. Mixed polymerizates of isobutylene with some dienes are manufactured as butyl rubber. It can be vulcanized to a certain extent. Polyvinyl carbazole (Luvikan, BASF, and Trolitul EH, Dynamit Nobel AG) is used as a very high-melting plastic.

### (b) Linear, non hardenable polycondensates

Since linear polycondensates form threads and are used by preference for synthetic fibres, their manufacture and processing were discussed in detail in the section on synthetic fibres. Linear polycondensates are, of course, very important also as plastics.

Here it should be added that mixed condensates can also be produced. Particularly suitable for this purpose is caprolactam, and so is its cleavage product, $\varepsilon$-aminocaproic acid. An important mixed polyamide is Igamid 6 A, consisting of 40 per cent caprolactam and 60 per cent AH salt, i.e. hexamethylenediamine adipate. Having a larger plastic range it is more thermoplastic and more soluble than the condensates of its components. Other substances used for mixed condensation are ketopimelic acid and p,p'-diamino-dicyclohexyl methane. Mixed condensation is a process with many possible variations.

At first, processing of polycondensates by plasticizing them with solvents or plasticizers proved difficult since those used up to that time for vinyl polymers were found to be unsuitable.

The manufacture of foils is very important because they are very strong (e.g. "Perfol", "Lyafol"). They can, if required, be produced from solutions containing plasticizers with coagulating organic solvents in strip casting machines (two-dimensional deformation!).

Another processing method is based on preliminary soaking of the polyamide chips in alcohol as a result of which they melt at considerably lower temperatures than the dry amide. They can then be shaped, pressed into plates, or rolled into foils like normal thermoplastics.

For the dry processing of polyamides pressure and high temperatures (about 275°C) are required.

Polyamide can be dyed either during manufacture (batch dyeing) or subsequently. If batch dyeing is used, only specially heat-resistant dyes such as phthalocyanin pigments or inorganic pigments can be used because of the high injection temperature. For subsequent dyeing many dyes are suitable, e.g. acetate silk dyes (Celliton fast dyes), acid absorbing wool dyes, vat dyes, etc.

Polyamides have very high mechanical strength, e.g. an impact bending strength of 7,150 cm kg/cm$^2$; the comparable figure for a phenolformaldehyde resin with wood meal (type 31 to DIN 7708) is 7, for the same with shredded fibre (type 71 to DIN 7708) it is 12. On the other hand they are very hydrophilic. Their main disadvantages are their low resistance to weather, their low fastness to light, and their tendency to discolour. The polyamides of caprolactam or of the AH salt absorb water relatively freely, the polycondensate of sebacic acid with hexamethylene diamine (Ultramid S, BASF) less so, and least of all the polyamino undecylic acid ('Rilsan', France). Linear polyurethanes also absorb water less than the first named polyamides. Very important among the linear polyesters (Carothers) is polyterephthalic acid glycol ester (Whinfield and Dickson), mentioned already in the section on fibrous substances, for the production of transparent foils with high strength and heat resistance ('Hostaphan').

A newcomer to the film-forming condensates is the polycarbonate of diphenylol propane (Dian) ('Makrolon', Bayer (W. Hechelhammer and G. Peilstöcker, 'Makrolon, ein thermoplastischer Kunststoff aus der Gruppe der Polykarbonate', *Kunststoffe*, 1959, **49**, 3–8, and 95–98), 'Lexan', USA). It is produced by treating dian in solution in methylene chloride with phosgene at about 30°C, or from dian and a carbonate, particularly diphenyl carbonate, by interchange of ester radicals at high temperatures, with subsequent distillation of phenol under vacuum.

The outstanding properties of Makrolon which is as transparent as glass are its high softening temperature (softening starts at 140°C), wide thermoplastic range, and exceptionally high cold impact strength. Even at the temperature of liquid air it is not brittle like glass. It excels polyamide in almost every respect. It is suitable for the production of foils ('Makrofol', Bayer (W. Hechelhammer and G. Peilstöcker, 'Makrolon, ein thermoplastischer Kunststoff aus der Gruppe der Polykarbonate', *Kunststoffe*, 1959, **49**, 3–8, and 95–98)), for injection moulding, continuous extrusion casting of pipes, sections, etc. and, in the form of plates and foils, for deep drawing. So far it is the only film material capable of replacing nitrocellulose for photographic and cinematographic film. For polyformaldehyde (Delrin) see p. 151.

An advance in polycondensation methods constitutes contact polycondensation in which solutions of two reactive components, e.g. alcohol or amines and acyl chloride, are brought together in two different immiscible solvents. The polycondensate formed at the very large interface is extracted continuously; polycondensates have much higher molecular weights than can be achieved by the usual condensation methods.

### (c) Hardenable polycondensates

By careful condensation of formaldehyde with phenols, urea, melamine, aniline, or other materials, thermoplastic preliminary condensates are formed which when heated and, if required, with the addition of more formaldehyde, condense further into three-dimensional structures; this renders them very hard and completely insoluble. They are, therefore, called hardenable resins, the most important of which are the phenolformaldehyde resins.

*Phenolformaldehyde resins*

Suitable phenols are phenol itself, cresols in mixtures, of which m-cresol is the most reactive, and, for special resins, resorcinol. The phenol mixture may contain a certain amount of xylenol.

Formaldehyde is condensed in alkaline or acid media, yielding resins with fundamentally different properties. Novolacquers which are fusible, soluble, and stable, are formed in an acid medium, using less than 1 mole formaldehyde per mole phenol. Novolacquers are thermoplastics. In this form they do not harden any more, but they react when formaldehyde and ammonia, in the form of hexamethylenetetramine, are added and they harden in the process.

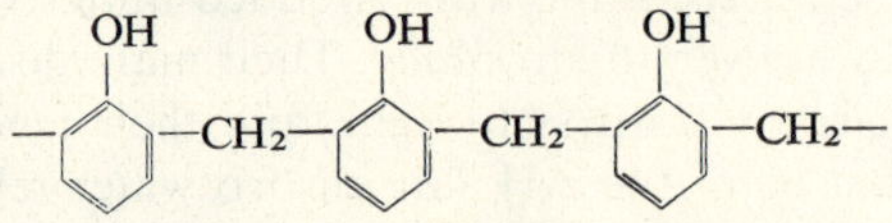

*Novolacquer* (schematically)

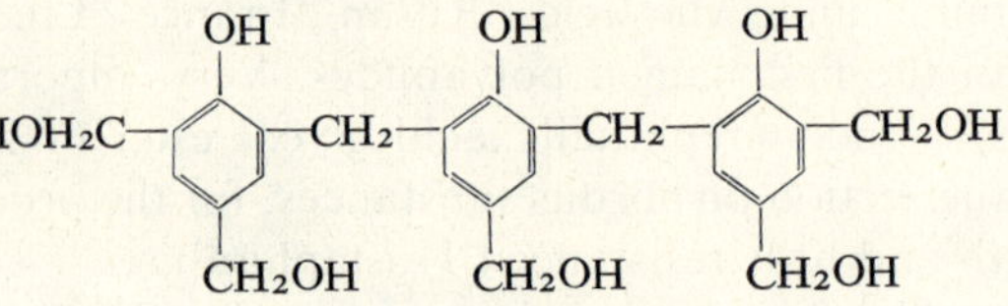

*Resol* (schematically)

The hardening is caused by the linking of phenol molecules by methylene bridges or methyl ether groups $—C_6H_4—CH_2—O—CH_2—C_6H_4$. Only methylol groups which are favourably placed in space in relation to each other can react with each other or with a phenol group, the others cannot. They form 'loose spots' in the macromolecule.

By condensing in an alkaline medium (with about 1·5 mole formaldehyde) one obtains resoles (A state) which at first are still fusible but become progressively harder during heating since they contain condensable methylol groups. They pass through the B state as resitol when they are still plastic and elastic, and finally become resite (C state) which is infusible. Hardening can also be carried out by means of acids without raising the temperature (cold-hardening resins).

Phenolformaldehyde resins are used as casting resins and varnishing resins (e.g. for stoving lacquers for the linings of reaction vessels, drying chambers, etc.). Asplit is a cold-hardening phenolformaldehyde resin with quartz sand, 'Haveg' the same with asbestos fibres.

The main use of phenolformaldehyde resins is in the field of moulded plastics containing fillers. Novolacquers are first mixed wit hfillers, mainly wood meal, magnesium oxide, hardener (hexamethylenetetramine), and stearic acid as lubricant, homogenized in a set of mixing rollers at 110°C, and moulded above 150°C at high pressure (400 to 500 kg/cm²).

It is advisable to pre-heat the material in a stove to 80–100°C before moulding. Depending on the thickness of the wall of the moulded article, the hardening time (standing time of the press) is usually between 2 and 10 minutes. After opening of the press the moulded article is removed hot.

Paper soaked in resol is marketed under the name of Tegofilm-Goldschmidt and is used for glueing plywood. Wood or textile laminated materials have great strength so that they can be used for bearing boxes and gear wheels. Laminated glass fibre materials have the highest tensile and compressive strength. The most important materials for their production are epoxy resins and unsaturated polyesters. The glass fibres must not contain alkali. Laminated glass fibre materials are used for building boats, light structural components, etc.

Self-hardening resins on a resol basis are also very important, e.g. for the production of plastic bodies for small cars.

If textile clippings, paper shreds or wood chips are used as fillers, the material is impregnated in a kneading machine with an alcoholic resol solution and a hardener, then it is dried and moulded. The products obtained have extraordinarily high mechanical strength. Hard board is also made of wood shavings and phenolformaldehyde resin (see p. 291).

Some of the phenol can be replaced by lignin if the latter is condensed before the formaldehyde treatment with phenol in the presence of acid catalysts (Rieche and Gnüchtel, *Plaste und Kautschuk*, 1954, **1**, 36). Phenolformaldehyde resins and lignin formaldehyde resins are used as binders for foundry moulding sand. The pre-formed mould covers are hardened in a stove and can be used for specially valuable and complicated castings.

For the manufacture of large objects special resins have recently been developed which can be moulded at low pressures (1–10 kg/cm$^2$).

*Urea-formaldehyde condensate* (carbamide plastics, aminoplastics)

Urea, obtained from ammonia and carbon dioxide under pressure by a method developed by BASF, readily forms methylol ureas with formaldehyde in the presence of acid catalysts:

$$H_2N{-}CO{-}NH{-}CH_2OH; \quad HOH_2C{-}NH{-}CO{-}NH{-}CH_2OH.$$

By splitting off water, these transform easily into methylene urea,

$$H_2C{=}N{-}CO{-}NH_2,$$

the intermediate stage in the formation of urea-formaldehyde resin. In the production of the preliminary stages of the resin the aim is the formation of water-soluble methylol-methylene-ureas (I):

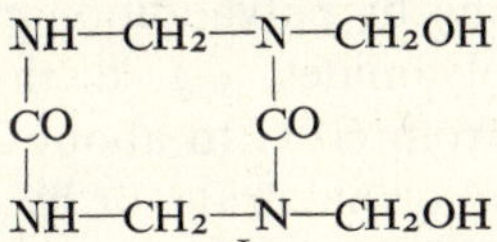
I

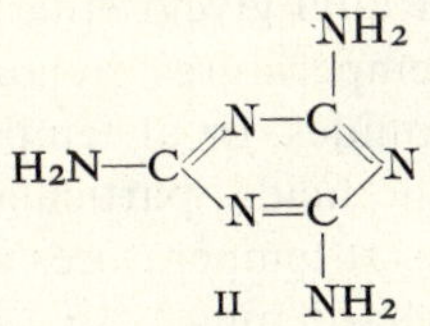
II

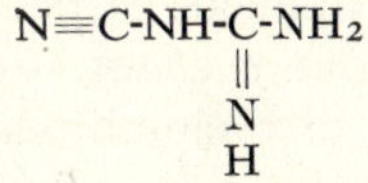
III

The reactive methylol- and imino groups react during hardening into a cross-linked three-dimensional macromolecule with simultaneous splitting off of water.

Thio-urea $H_2N{-}CS{-}NH_2$, dicyandiamide (III), and melamine (II) react in a similar way. In the latter all the hydrogen atoms can react with formaldehyde, thus forming a multiple cross-linked structure.

The filler used most widely is cellulose since clear transparent materials can be produced with it. Impregnation with the aqueous resin solution is carried out in a kneading machine, followed by drying under vacuum. Then lubricants, accelerator, and, if required, pigments are added and the mixture is finely ground and moulded hot.

Aminoplastics are very important for the glueing of plywood and veneers (Kaurit, BASF, K-glue, Leuna, Didi wood glue, Piesteritz, urea formaldehyde resin). For these corn flour can be used as a filler. Like the phenolplastics they are very suitable for laminated plastics. For the production of laminated plastics, tissues, paper, or thin wood veneers are soaked with the hardening glass, resol, or aminoplastic, either in the liquid phase or in solution, and are pressed in a plate press, hardening taking place in the process.

Hardening is produced by heating, or in the cold by the addition of ammonium chloride solution. Melamine resins are most resistant to water; they are used to impart crease resistance to textiles and for the production of paper possessing great wet strength. Aminoplastics can also be used for the production of foam plastics with a surprisingly low weight and excellent heat and sound insulation properties (Iporka, BASF, Piatherm, Piesteritz).

Aminoplastics are marketed under the names of Ultrapas, Maprenal, Uformit, Plastural, Meladur, etc.

*Polycondensates produced by polyaddition or polymerization*

In addition to phenolplastics and aminoplastics the more recent hardenable polycondensates have gained an important share of the market; they harden without splitting off volatile reaction products, i.e. by polyaddition or polymerization. This group consists of the polyurethanes, the epoxy resins, and the unsaturated polyester resins.

A brief description has been given on p. 389 of the polyaddition method with isocyanates developed at the beginning of the last war (O. Bayer). It can be used in many fields. Linear polyurethanes from hexamethylenediamine and Diolen are suitable for injection moulding and the production of threads, bristles, and foils. Polyurethane rubber which has very great abrasive strength is produced by the incorporation of trivalent alcohols and a small amount of cross-linking. If di- or tri-isocyanates (desmodures) are added to the polyester types with free OH groups (desmophenes), lacquer films, tough materials (Vulkollan), or rubber-elastic substances are produced according to the type and mixture ratio used. Foam materials (Moltopren) are produced by the splitting off of $CO_2$ during the addition of water and, if required, dioles, urea, etc.

Commercial epoxy resins usually consist of condensation products of diphenylolpropane with epichlorhydrin in the production of which alkali was used (R. Wegler, Chemie der Polyepoxyde, *Angew. Chem.*, 1955, **67,** 582). These resins which are liquid, viscous, or solid, depending on their molecular weight, are hardened by the addition of reactive thinners, in particular of allyl glycid ether. Hardening by polyaddition can be carried out even at ordinary temperatures by organic polyamides, e.g. diethylenetriamine or cyclo-aliphatic polyamides, or at temperatures from 60°C to about 220°C by the anhydrides of dicarboxylic acids, particularly phthalic anhydride, or by direct etherification of the epoxy groups at temperatures above 200°C.

Formation of epoxy resin and hardening:

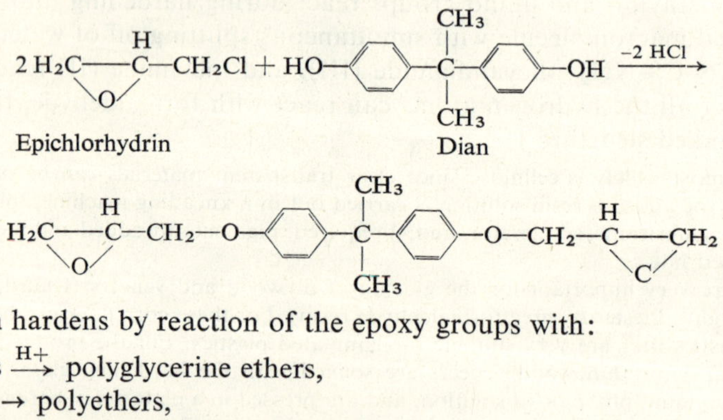


Epoxy resin hardens by reaction of the epoxy groups with:

(1) themselves $\xrightarrow{H^+}$ polyglycerine ethers,
(2) diphenols → polyethers,
(3) dicarboxylic acid anhydrides → polyesters,
(4) diamines → polyimines,
(5) the OH groups released from the epoxy ring of allyl glycid ethers or dicarboxylic acid anhydrides → cross-linking.

Unsaturated polyester resins are produced from a low-molecular condensate (from unsaturated dicarboxylic acid and allyl alcohol with glycols) by the addition of monomers, particularly styrene, and polymerization catalysts (peroxide, organic amine and cobalt naphthenate, or similar substances). In contrast to epoxy resins, the polymerization velocity can be regulated to a much larger degree by the catalysts used. Hardening can be achieved in a few seconds at low temperatures which is impossible with epoxy resins and polyurethanes.

The great practical importance of unsaturated polyester resins (as well as of epoxy resins) is due to their ease of hardening without splitting off of volatile components, and the possibility of using them in the form of castings, as lacquers or glues, particularly for metal glueing, and in combination with glass fibres.

Glass-fibre reinforced laminates utilize the extraordinary high strength of spun glass which far exceeds that of the best steels, making possible the production of structural components with high strength and low weight. Their range of application extends from wash basins via bath tubs to motor and lifeboats of 15 m length and stressed structures in aircraft construction. A bath tub made of glass fibre reinforced polyester, for instance, weighs only one tenth of an enamelled cast iron tub, a fact of special importance in the equipping of vehicles, ships, etc. Owing to the fact that a polyester boat is practically a single polymerized unit it is considerably stronger than a wooden boat and moreover practically unsinkable because of the cavities or foam particles polymerized into it.

## 3. ELASTOMERS

*Synthetic rubber, Buna*

Synopsis of the constitution of various types of rubber

Natural rubber

$$(-CH_2-\overset{CH_3}{\overset{|}{C}}=\overset{H}{\overset{|}{C}}-CH_2-)_n \quad \text{(cis-form configuration)}$$

Polychloroprene, Neoprene

$$(-CH_2-\overset{Cl}{\overset{|}{C}}=\overset{H}{\overset{|}{C}}-CH_2-)_n$$

Buna S

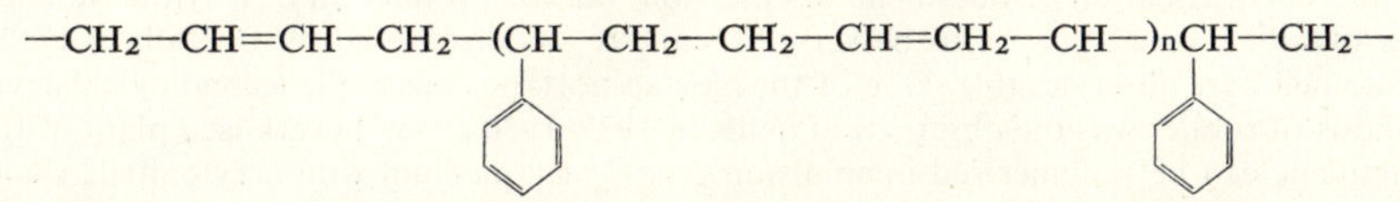

Apart from 1,4-addition to butadiene there is always 1,2-addition. See following pages.

(Number) buna

$$(-CH_2-\overset{H}{C}=\overset{H}{C}-CH_2-)_n$$

(Pure butadiene polymerized with Na; the number refers to the degree of polymerization)

Methyl rubber

$$(-CH_2-\overset{CH_3}{\overset{|}{C}}=\overset{CH_3}{\overset{|}{C}}-CH_2)_n$$

Natural rubber is a high-polymeric isoprene $(C_5H_8)_n$. In 1902 Harries tried in vain to transform isoprene into rubber by prolonged heating (300°C). He proved by ozone decomposition that isoprene is present in natural rubber as a building element. In 1909 F. Hofmann and K. Coutelle (C. Duisberg, of Bayer A. G., with great foresight supported the development of synthetic rubber) succeeded in obtaining a rubber-type i.e. an elastic and vulcanizable product, by heating isoprene for eight days at 120–200°C (vulcanization, i.e. the conversion of the thermoplastic caoutchouc into the elastic material rubber by sulphur, was discovered in 1844 by the Englishman Ch. Goodyear).

Subsequently Hofmann discovered the polymerizability of butadiene and dimethylbutadiene.

During the 1914–1918 war, Hofmann developed at Bayer's a technique for the production of methyl rubber W and H (soft and hard) from dimethylbutadiene. The dimethylbutadiene was obtained as follows:

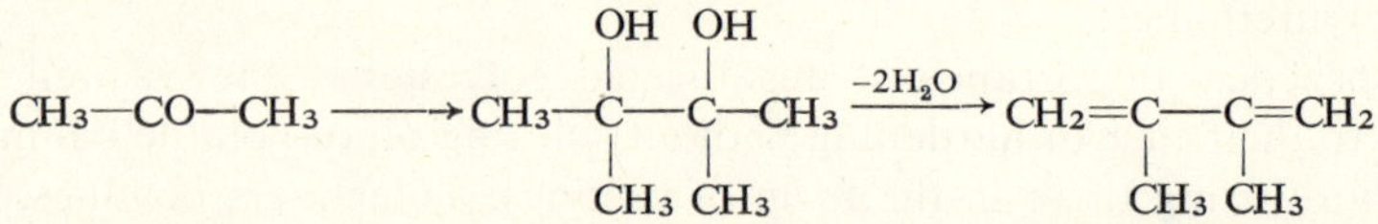

acetone pinacol 2,3-dimethylbutadiene-(1,3)

W. Reppe improved the method for the production of isoprene. He obtained isoprene from acetone and acetylene via 2-methylbutinol $(CH_3)_2COH—C{\equiv}CH$ and methylbutenol $(CH_3)_2COH—CH{=}CH_2$. Isoprene can also be obtained by dehydrogenation of the isopentane fraction of natural gas and from propene (see p. 397).

Since new possibilities for its polymerization were discovered (Ziegler-Natta), isoprene has become important as a substitute for natural rubber. This 'synthetic natural rubber' has better properties than butadiene rubber. The essential point is that there is a poly-cis-1,4-polymerization as in natural rubber. This has been a major development in the 50 years' history of rubber synthesis. Isoprene rubber is produced principally in the U.S.A. and the U.S.S.R.

Pure 1,2-polymerization of butadiene can also be achieved with $(C_2H_5)_3Al + TiCl_4$, yielding a rubber which carries all the olefin residues in the side chain (see the right-hand side of the formula below).

This method was greatly improved by the use of sodium for the polymerization of isoprene (discovered originally by Harries and the Englishmen Matthews and Strange). Later, butadiene was used almost exclusively for rubber synthesis, mainly because it was easier to produce industrially. The various methods for the industrial production of butadiene have been discussed in detail earlier (see e.g. pp. 179 and 120).

'Number Buna' is produced continuously in a horizontal autoclave with cooled or heated walls and a cooled screw conveyor. First, butadiene is injected by a dosage pump. Two more pumps add measured quantities of dioxane as a regulator and potassium or sodium in powder form as pulp. The polymerizate emerges in the form of a strand at the end of the screw conveyor. 'Number Buna' is, therefore, a block polymerizate.

At first, the polymerization of butadiene in emulsion, i.e. the production of a synthetic latex similar to the latex of rubber trees, proved to be very difficult. It was discovered by Gottlob in 1912, but the products obtained were unserviceable. One of the most important steps in the technological development of the synthesis of rubber was the observation made in 1929–1930 at the Leverkusen plant of the former IG that butadiene can be polymerized in emulsion in a warm medium with acrylonitrile (Konrad and Kleiner) and with styrene (Bock and Tschunkur) (Buna N, Buna S).

During the 1939–1945 war, cold polymerization (−10°C) of butadiene with styrene by the redox method was developed at Leverkusen; this process has been used industrially since 1945 in the U.S.A. ('cold rubber') and its importance continues to grow. Polymerization of low-temperature rubber is carried out in emulsion at −5°C with specially active organic peroxides and reducing agents. Low-temperature rubber (TTK) is stereospecifically more uniform than normal Buna and will probably some time in the future replace it almost completely. A resin soap from transformed collophonium (a mixture of dehydro- and dihydro abietic acid) is of prime importance for the polymerization.

Chlorobutadiene rubber developed in the U.S.A. (see p. 322 for the production of chlorobutadiene) can be processed more easily than Buna S and for some time it seemed that chlorobutadiene rubber would overtake Buna. The success achieved in making Buna more plastic and, therefore, more easily workable by oxidative thermal decomposition tilted the decision in favour of butadiene rubber. The cause of the difficult workability of synthetic rubber as compared with natural rubber is the stronger cross-linking of the former. Addition of the butadiene molecule takes place at the 1,4 and 1,2 positions:

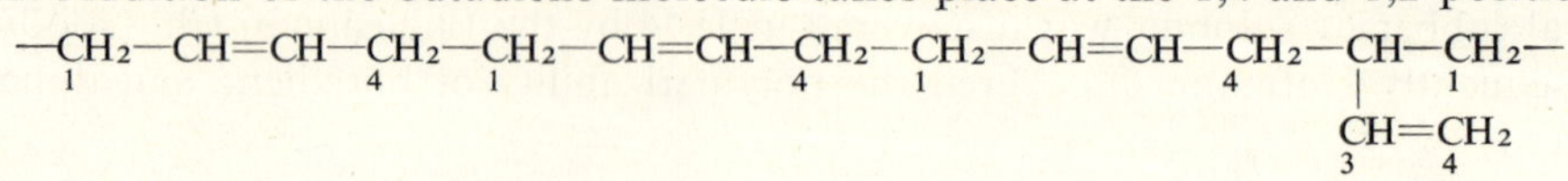

The latter configuration easily leads to cross-linking of the molecules and thereby to a reduction in the plasticity. Natural rubber on the other hand consists mainly of linear molecules. Buna therefore has a three-dimensional structure.

Only since 1935 has the modern Buna synthesis been used on a large industrial scale.

Polymerization of Buna N and S is carried out in a system of 5 to 12 cooled V2A or lead-lined pressure vessels with agitators in series (each with about 10 $m^3$ capacity) (see Figure 143) (Weinbrenner and Konrad in Winnacker-Weingaertner, *Chem. Technologie*, Organische Technologie, Hanser-Verlag, München).

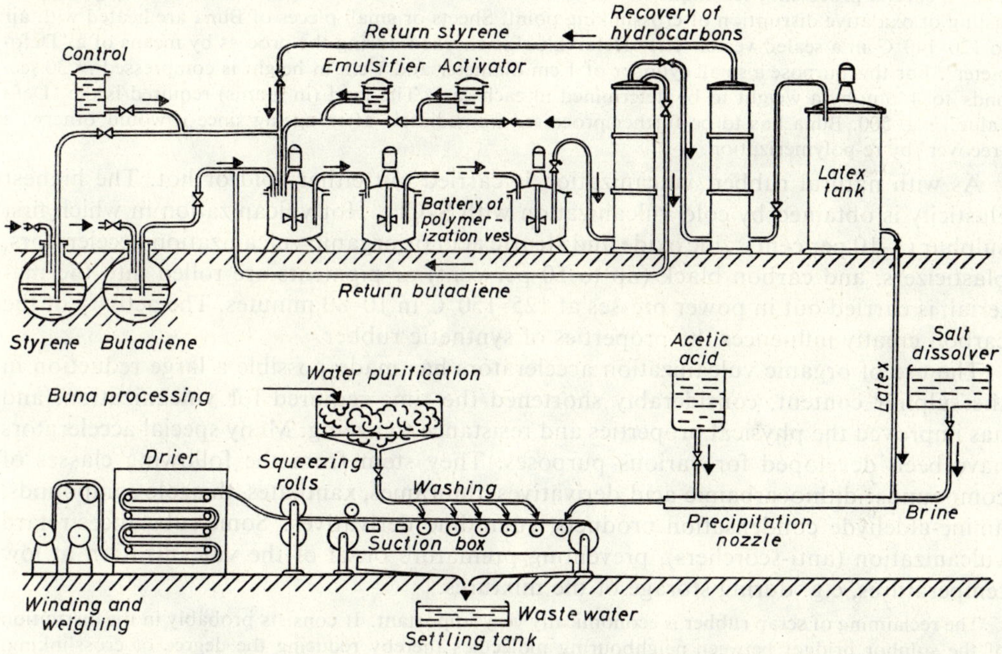


Figure 143. Diagram of the manufacture of butadiene rubber (emulsion polymerization)

The first vessel contains butadiene and styrene in the aqueous phase. To these are added via dosage pumps polymerization accelerators (catalysts, such as sodium persulphate), emulsifiers and wetting agents (e.g. soaps, paraffin sulphonic acids such as mersolate, Nekal, sodium oleate; these are important since with a larger reaction as a result of the 'sponge' effect of the polymerizate on the monomer the polymerization takes a different course in the monomer than in the soap micelle), protective colloids (e.g. gelatines, albumens), regulators (a solution of di-isopropylxanthogen-disulphide 'Diproxid' $(CH_3)_2$-CH—O—CS—S—S—CS—O—$CH(CH_3)_2$ in styrene), and buffers, such as sodiumpyrophosphate.

Checking and regulation of the process is carried out by measuring the density, viscosity, and pH of the contents of the vessels. Polymerization takes place at about 40°C and is complete in about 30 hours. In the first vessel the gauge pressure is 8 atmospheres, decreasing to 6 atmospheres in the last vessel. After the reaction is about 60 per cent completed and after the latex has emerged from the last vessel, the process is interrupted by adding a suspension of the age resistor phenyl-$\beta$-naphthylamine, and sodium sulphite is also added to promote decomposition of the persulphate residues.

Any unreacted butadiene and styrene are expelled from the latex in columns using steam and are recycled. The latex is then processed into sheets in a manner very similar to that employed with natural latex (Buna-S-band). The latex, diluted with a sodium chloride solution, is precipitated in a flow pipe in two stages using a solution of calcium chloride and dilute acetic acid. Dehydration is carried out on a wire cloth, then the material is washed carefully, vacuum-desiccated, and is formed on a paper machine combined with drying oven into a strip which eventually is cut into sheets.

The Buna-S types contain 10–45 per cent styrene, e.g. S 1 for car tyres contains 26 per cent, and Buna SS 45 per cent styrene. The more styrene it contains the higher is its plasticity. 'Perbunan' (Buna N) contains 27 per cent acrylonitrile, 'Perbunan extra' 35 per cent. It is not affected by petrol or oil.

The advantages and limitations of Buna as compared with natural rubber are due to its stronger crosslinking. Buna is more compact, less easily soluble, more resistant to abrasion, ageing, and heat, but is less elastic and plastic and has a lower tearing strength. The double bonds in the side chains cause 'recovery phenomena', i.e. re-polymerization of Buna which may have become depolymerized by heat. Thus a special processing technique has to be used. The molecule is subjected to artificial ageing consisting of oxidative disruption of crosslinking point. Sheets or small pieces of Buna are heated with air to 120–140°C in a sealed vessel. The 'Defo value' is checked during the process by means of a 'Defometer'. For this purpose a small cylinder of 1 cm diameter and 1 cm in height is compressed in 30 seconds to 4 mm by a weight to be determined in each case. The load (in grams) required is the 'Defo value', e.g. 500. Buna has to be further processed immediately after ageing since it would otherwise 'recover' by re-polymerization.

As with natural rubber, vulcanization is carried out either cold or hot. The highest elasticity is obtained by cold vulcanization with $S_2Cl_2$. Hot vulcanization in which first sulphur (1–10 per cent), zinc oxide and stearic acid or organic vulcanization accelerators, plasticizers, and carbon black (up to 50 per cent) or pigments are rolled into the material, is carried out in power presses at 125–150°C in 10–80 minutes. The activity of the carbon greatly influences the properties of synthetic rubber.

The use of organic vulcanization accelerators has made possible a large reduction in the sulphur content, considerably shortened the time required for vulcanization, and has improved the physical properties and resistance to ageing. Many special accelerators have been developed for various purposes. They stem from the following classes of compounds: dithiocarbamic acid derivatives, thiurames, xanthates, thiazole compounds, amine-aldehyde condensation products, guanidine derivatives. Some substances retard vulcanization (anti-scorchers), preventing premature onset of the vulcanization at low temperatures, e.g. during storage of the mixtures.

The reclaiming of scrap rubber is economically very important. It consists probably in the disruption of the sulphur bridges between neighbouring molecules thereby reducing the degree of crosslinking. Reclaiming is usually carried out by heating to 100°C with diluted sulphuric acid, to 175–200°C with alkali under pressure, or by steam with the addition of plasticizers (mineral oils, fatty acids). The addition of special reclaiming auxiliaries (such as thiophenols, mercaptans, aromatic nitrosulphides, etc.) is particularly useful for the reclaiming of synthetic rubber.

Polysulphide rubbers or thioplastics are polycondensation products of alkene chlorides with alkali polysulphides, i.e. alkene polysulphides. They are very cheap, and resistant to oil, but have an unpleasant smell. One possible method is to condense e.g. ethylene chloride (Thiokol) or dichlorodiethylether (Perduren G, Bayer) or dichloroethylformaldehyde (Perduran H) with potassium polysulphide. The products contain at least 80 per cent sulphur and can be vulcanized with ZnO without adding any more sulphur.

The Vulkollanes (O. Bayer) have excellent rubber-type properties; they can be synthesized from aromatic di-isocyanates, e.g. naphthalene-1,5-di-isocyanate, and a polyester (urethane formation). The action of water or glycols on terminal isocyanate groups causes crosslinking. Vulkollanes have very great tearing strength (molecular weight 300,000 to 400,000).

If high-molecular polymer methyl siloxanes (molecular weight 300,000 to 400,000), e.g.

$$C_2H_5O(CH_3)_2SiO—(Si(CH_3)_2O)_n—Si(CH_3)_2OC_2H_5,$$

are crosslinked by peroxides in the presence of fillers, (e.g. finely dispersed silicic acid and kieselguhr), the product is silicone rubber (other types of rubber can also be 'vulcanized' with peroxides), which, like all silicones, has a very great resistance to heat (e.g. thermally stable up to 200°C). On the other hand, the mechanical properties at normal temperatures are inferior to those of natural and synthetic rubber. This applies also to their resistance to water. Their use is, therefore, limited mainly to sealing materials and waterproofing agents for very low and very high temperatures, also in combination with Clophen (perchlorodiphenyl) for the electrical engineering industry.

## 4. ION-EXCHANGE RESINS

Ion-exchange resins are solid substances having acidic or basic groups on the surface, which must be porous and capable of swelling but must not contain resin fractions that can be washed out.

Acid resins are cation exchangers. With acids they become charged with H ions (regeneration) which they can exchange for cations. When treated with alkali salts, they become charged with alkali ions which can be displaced by polyvalent ions, e.g. Ca- or Fe-ions (regeneration with NaCl). Resins with basic groups, e.g. amino- or quaternary ammonium groups, are anion exchangers. By combining cation and anion exchangers it is possible, for instance, to desalinate water without distillation (water for high pressure boilers and also for chemical and analytical purposes).

The various types of exchangers are divided into natural, partially synthetic, and fully synthetic exchangers:

(1) Natural mineral cation exchangers: greensands, zeolites.
(2) Synthetic aluminium silicates and aluminium oxides.
(3) Natural organic exchangers, e.g. humic substances (in the soil), starch, cotton, paper (see e.g. chromatographic separation methods).
(4) Introduction of acid groups into carbon, e.g. by sulphonation or COOH groups by oxidation.
(5) Synthetic exchange resins, e.g. Wofatit, Levatit, which can be formulated for any desired purpose.

Cation exchangers are resins from polyphenols or phenolcarboxylic acids, condensed with nucleus-linking agents. Their 'acid character' is reinforced by the introduction of sulphonic acid groups. Polystyrene, crosslinked by divinyl benzene, is sulphonated and then becomes a highly efficient cation exchanger (e.g. Natrantit).

Anion exchangers are obtained by condensation of aromatic bases such as aniline and phenylanisidine with formaldehyde. To increase their basicity the amino groups can be quarternerized by alkylation.

Another way of producing anion exchangers consists in introducing $-CH_2Cl$ groups into polystyrene resin with chlorodimethyl ether $ClCH_2-O-CH_3$ (from methanol, hydrochloric acid, and formaldehyde) and letting the chloromethyl groups react in the swelled state of the resin with amines.

Some exchangers and redox exchangers also form complexes. Exchanger resins find many and varied uses in industry and science, e.g. for the purification of solutions (thin sugar juice), for the separation of ions of all types, and for the recovery of metals from lean waste solutions (Cu in the cuprammonium rayon process). They act either as acid or as basic catalysts; for instance, esterification can be carried out in the presence of water with cation exchangers. It is important when using and recovering exchange resins to operate in a column using the counter-flow principle discussed earlier in the section on extraction methods.

*Prospects*

The development of the newer types of plastics is based not only on the fact that the monomer raw materials are available in large quantities at favourable prices but also to a great extent on the ever-increasing knowledge of the relationship between structure and properties of the high-polymer compounds. Isotactic polymerization of olefins and dienes leads to a large number of very valuable polymers, the technological exploitation of which has only just started with polypropylene (A. Schley and F. Schülde, 'Hostalen PPH, low-pressure polypropylene for injection moulding', *Kunststoffe*, 1959, **49**, 102) and the isotactic polyisoprene. With short-chain branched olefin hydrocarbons it is possible to produce impact-resistant thermoplastics with softening temperatures up to about 250°C whilst olefins with long unbranched side chains yield rubber-type materials with new improved properties, e.g. greater resistance to ageing and oxygen.

One of the more recent polymerizates is high-molecular polyhydroxymethylene. The high polymer of formaldehyde is obtained from a highly purified raw material with exclusion of water by an anionic polymerization mechanism. Due to the large intermolecular forces, manifested as high crystallinity and density, the material is insoluble at normal temperatures in all solvents and, being a polymer ether, is resistant to many media (Delrin, Dupont, USA).

Of great practical importance are the various recent copolymerizates in which the negative properties of the component polymers are neutralized. Thus, copolymerization of styrene with small additions of butadiene or acrylonitrile, or both (Novodur, Bayer), yields impact resistant deep-drawing types which are widely used for the manufacture of furniture, refrigerators, cabinets, etc. Plexidur (Röhm and Haas), an acrylonitrile-containing copolymerizate of methacrylic acid methyl ester, has exceptional resistance to impact. It is in fact an almost unbreakable organic glass. Finally, mention must be made of foam polystyrene (Styropor, BASF) which is marketed in granular form and can be foamed in situ by heating. Other possibilities for the manufacture of new products are offered by graft polymerization, i.e. the action of monomers on polymerizates in the presence of polymerization catalysts or strong radiation. Research into graft polymerization is still going on.

## 5. SILICONES

Silicone chemistry has been developed only in the last twenty five years and has grown very quickly in this short period. The technology of these compounds is of quite recent date (Cf. R. Müller, Über den augenblicklichen Stand und die voraussichtliche zukünftige Entwicklung der Siliconchemie. *J. Prakt. Chem.*, 1960, **10,** (4), 184–203. W. Noll, Zur Chemie und Technologie der Silicone. *Angew. Chem.*, 1954, **66,** 41).

The origins of the processing of carbon silicide compounds go back to the sixties of the last century (Friedel-Crafts, later Ladenburg). Silicone chemistry received a strong impetus by the application (simultaneously by Dilthey and the American Kipping) of the Grignard reaction for the production of Si-organic compounds; this is still the most important method used for their preparation. The development of silicone chemistry is due mainly to Kipping (1900–1944) whose original purpose was to study the problem of the two tetravalent elements carbon and silicon and the behaviour of their analogous compounds, and who forecast no great future for the practical use of these compounds. Very soon, however, two American firms (General Electric and Dow Corning Glass Work) recognized the great importance of the organo-silicon compounds. In Germany, work in this field was done by R. Müller, Radebeul, and, under licence from American firms, by Bayer, Wacker, and Goldschmidt.

Silicones are derived from the hypothetical monomer compounds $R_2Si = 0$ the structure of which is analogous to that of ketones $R_2C = O$; to this they also owe their name: silicone = ketone, which, however, is no longer quite correct. The scientific name for this class of compounds is siloxanes.

However, unlike the C = O double bond, the Si = O 'double bond' is not viable at low temperatures, and the ease of polymerization is due to the formation of simple Si–O compounds:

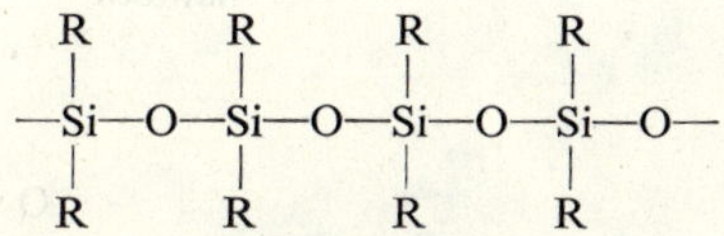

(polysiloxane)

According to the number of alkyl- or aryl residues linked to the silicon atom, a residue of oxygen atoms remains which determines the 'functionality' of the hypothetical monomer siloxanes. The four basic groups of structural elements are:

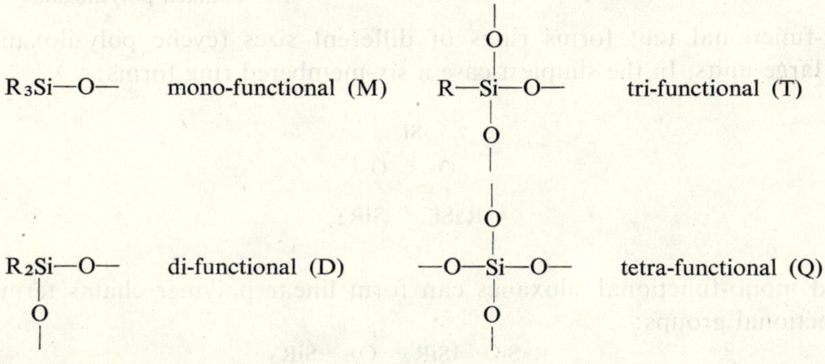


The letters MDTQ have become standard designations in silicone chemistry and are abbreviations signifying functionality.

The combination of these structural elements into dimer, annular, linearly-polymer, two- or three-dimensionally crosslinked structures can be arranged in a great variety of ways. Changing the organic residue R makes possible a large number of further variations. Depending on structure, polymer siloxanes are liquid, waxy, firmly thermoplastic, elastic like rubber, or hard to brittle like glass.

The two main methods used for the manufacture are the Grignard reaction and the Rochow-Müller synthesis from Si and alkyl halide in the presence of copper as a catalyst. When silicon tetrachloride reacts with alkyl- or aryl magnesium chloride (Grignard reaction), organic silicone chlorides form (alkyl- or aryl halogen silanes). These hydrolyse very easily with water to alkyl silanols. The silanoles are very unstable, the Si-OH groups transform into Si–O–Si bonds, i.e. siloxane- or polysiloxane groups. The more OH groups there are present in the silanol, the more unstable it is. The nature of the polymers obtained depends, therefore, in the final analysis on the chlorosilane used and on the conditions of hydrolysis. The hydrolysis is carried out by adding water and, if necessary, a solvent to the chlorosilanes with simultaneous reduction in the temperature.

The reaction of alkyl- or aryl magnesium chloride with silicon tetrachloride produces three series of compounds:

(1) 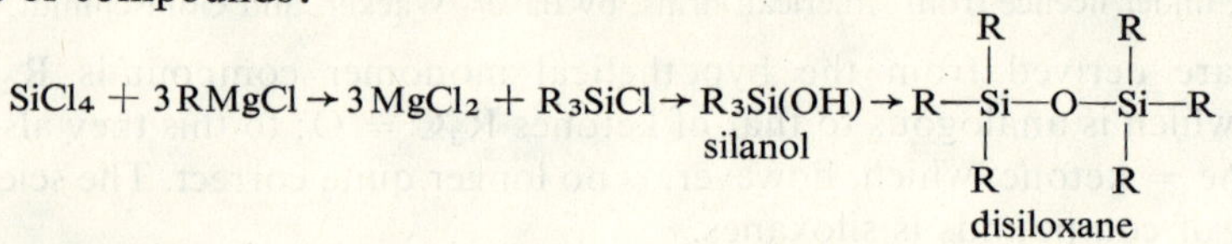


(2) 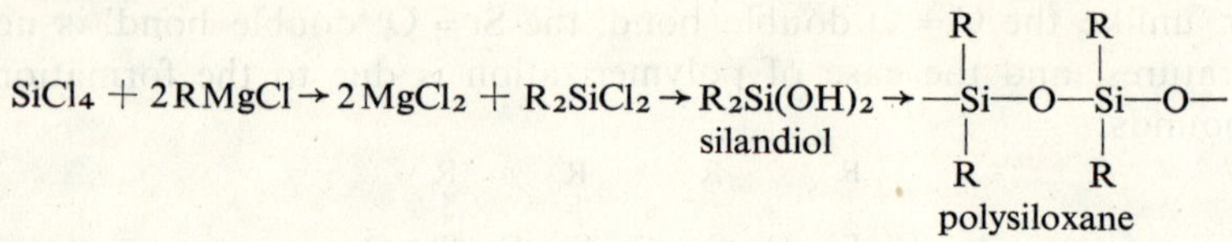


(3) 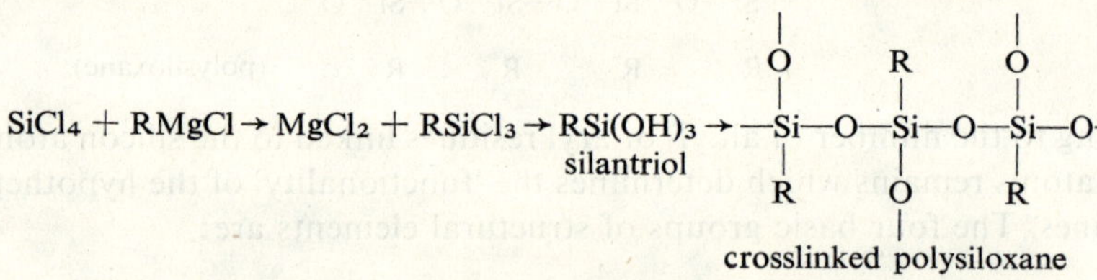


The di-functional unit forms rings of different sizes (cyclic polysiloxanes) up to infinitely large units. In the simplest case a six-membered ring forms:

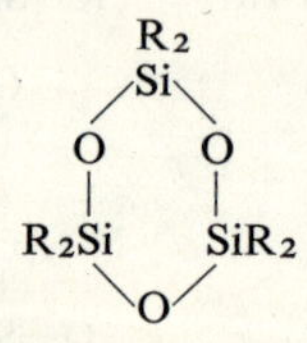


Di- and mono-functional siloxanes can form linear polymer chains terminating in mono-functional groups:

$$R_3SiO—(SiR_2—O)_x—SiR_3$$

Tri-functional siloxane units yield crosslinked molecules. The combination of di-functional and tri-functional units produces chains with three dimensional cross-linkages. This shows the extraordinarily large number of possible siloxane macromolecule structures which is further increased by the addition of tetrafunctional groups.

Methyl silanols can also be produced via silicic acid ester with Mg, and alkyl halides via methyl alkoxysilanes:

$$2\,Mg + Si(OR)_4 + 2\,CH_3Cl \rightarrow Si(CH_3)_2(OR)_2 + 2\,Mg(OR)Cl.$$

The Rochow-Müller synthesis is represented by the following summary relation:

$$Si + 2\,CH_3Cl \xrightarrow[Cu]{} Si(CH_3)_2Cl_2$$

The actual reaction is, however, much more complicated. Copper with methyl chloride forms copper(I)-chloride and Cu-methyl. Since the latter has a very short life, the silicon and the catalyst have to be very finely dispersed and well mixed for the reaction.

The most important raw materials for polysiloxanes are now methyl chlorosilanes and phenyl chlorosilanes.

Linear polysiloxanes are marketed as silicone oils with different viscosities, boiling points, pour points, and with mean molecular weights of 3,000 to 30,000. They are very resistant to air and humidity and have a very high electrical insulating capacity. Silicone oils have a great advantage over hydrocarbon oils in that their viscosity is less dependent on temperature.

Solid silicones are used in solution as varnishes. They have to be thickened by heating during which treatment a very complicated condensation polymerization takes place (P. J. Flory, *Principles of Polymer Chemistry*, Cornell University Press, Ithaca, New York, 1953, p. 37).

Atmospheric oxygen plays a similar role in this process as in oil drying. Silicones are very resistant to heat and are water-repellent. Textiles, ceramics, and paper can be provided with silicone protective films which render them water-repellent. Liquid silicones are physiologically harmless, break down foam very efficiently, and can also be used as heating fluids. Silicone rubber has already been mentioned previously. Notable is the resistance to prolonged heating and the non-dependence of their low elasticity on temperature. Silicones make possible the solution of a variety of problems, but their high price inhibits their use on a large scale.

## LITERATURE

K. Thinius, *Hochpolymere, Herstellung, Eigenschaften und Anwendung als Kunststoffe*. Fachbuchverlag, Leipzig, 1952.

R. Houwink, *Grundriß der Technologie der synthetischen Hochmolekularen*. Akad. Verlagsges. Geest & Portig K.-G., Leipzig, 1952.

W. Schaefer, *Einführung in das Kunststoffgebiet*. Akad. Verlagsges. Geest & Portig K.-G., Leipzig, 1951.

K. Thinius and G. Bier in Winnacker-Weingaertner, *Chem. Techn.*, Org. Techn. II, Hanser-Verlag, München, 1954; 2nd edn, Winnacker-Küchler II, vol. 4, 1959.

F. Runge, *Einführung in die Chemie und Technologie der Kunststoffe*. 2nd edn, Akademie-Verlag, Berlin, 1959.

H. Gnamm, *Die Lösungsmittel und Weichmachungsmittel*. Wissenschaftl. Verlagsges. m. b. H., Stuttgart 1950.

W. Breuers and H. Luttropp, *Buna*. VEB Verlag Technik, Berlin, 1954.

E. G. Rochow, *Einführung in die Chemie der Silicone*. Verlag Chemie, Weinheim, 1952.

A. Hunyar, *Chemie der Silicone*. Verlag Technik, Berlin, 1957.

W. Noll, *Chemie und Technologie der Silicone*. Verlag Chemie, Weinheim, 1960.

# III. Textile auxiliaries and washing raw materials

The group of surface active materials, textile auxiliaries and washing raw materials, includes capillary-active and capillary-inactive materials.

The capillary-active materials form four classes of compounds:

(1) Anion-active compounds
(e.g. soaps, fatty alcohol sulphates, sulphonic acids, $R{-}COO^{\ominus}Na^{\oplus}$, $R{-}SO_3{}^{\ominus}Na^{\oplus}$, carboxymethyl cellulose).

(2) Cation-active compounds
(e.g. alkyl ammonium salts, condensation products of fatty acids with amines as salts).

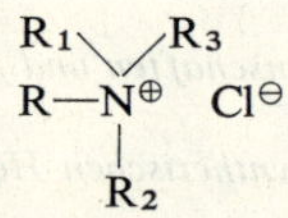

(R = long-chain alkyl radical)

(3) Amphoteric compounds (ampholytes)
(e.g. betaines).

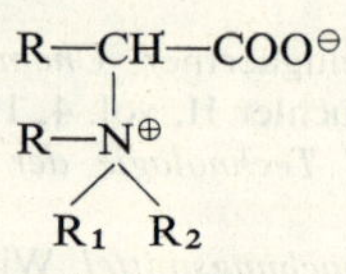

(4) Non-ionic compounds
(e.g. polyethylene oxide products, monoesters of polyhydric alcohols).

Capillary-inactive materials are:

(1) Natural colloids
(e.g. starch, mucilage, glue, gelatins, alginic acids).

(2) Modified natural substances
(e.g. cellulose derivatives, degradation products of protein).

(3) Polycondensates and polymerizates
(e.g. phenoplastics, aminoplastics, polyamides, silicones, vinylpolymers).

(4) Other special auxiliaries
(enzyme preparations, optical brighteners, water softening agents, and complex-forming substances).

It will be seen that a large number of products from the most varied classes of compounds have been developed for the industrial processing or the improvement of the quality of textiles (fibres and tissues) and for their cleaning.

At first, soap was the only auxiliary, then came its derivatives and those of natural oils. These products are still the cheapest. In spite of their higher prices, synthetic agents have, developed greatly in the last thirty years because they had special effects far exceeding those of soap and of the simple derivatives of oils. This development was promoted by the increased requirements made on natural fibres and the many new semi-synthetic and fully synthetic fibres which had to be processed, dyed, and finished.

These agents are important not only for textiles but also for the manufacture and upgrading of leather (tanning auxiliaries). By value and volume washing and cleaning agents still occupy the first place; apart from the textile industry they are also used in the household, in laundries, and for personal cleansing. The production and sale of synthetic washing agents is increasing rapidly especially in the U.S.A. where the production of soap continues to decrease. The production of synthetic washing agents, of semi-synthetic and synthetic fibres and of plastics has developed approximately parallel as regards volume; in 1953, it amounted to about 7–10 million tons per year each, which shows the great importance of this branch of industry. This is not surprising if one considers that plastics have a very long life but fibres only a limited one and that washing agents, like food, are used by everybody and that even the textile auxiliaries used in industry are never recovered.

A very serious effluent problem has been created by the steadily increasing consumption in laundries and households of synthetic washing agents, detergents, particularly of alkyl- and alkaryl sulphonic acids (F. Sierp and H. Thiele, *Vom Wasser*. 1954, **21,** 197–246; W. Winter, *Wasserwirtschaft-Wassertechnik*, 1959, **9,** 15).

Whilst soaps and alcohol sulphates are biologically decomposed in the sewage works, this is not the case with other washing agents. Strong formation of foam and flotation of dirt occurs in rivers, particularly at dams and sluices. The respiration of aquatic animals breathing through gills is impeded by higher concentrations of surface active materials, and water birds are endangered when the film of fat is stripped off their feathers.

### (a) Capillary-active materials

*General remarks*

Capillary-active materials have a typical structure: they always consist of a hydrophobic residue and a hydrophilic group.

 (symbols).

The hydrophobic residues can either interrupted by, or carry, hydrophilic groups.

They concentrate at the surface and cause a change in the surface tension. The hydrophobic (lipophilic) residue (usually a long-chain aliphatic or araliphatic hydrocarbon residue) has an affinity for the non-aqueous phase, the hydrophilic one for the aqueous phase. This pronounced symmetrical structure is the cause of the surface (capillary) active effect. The long chain hydrocarbon residues of the capillary-active substances are not 'filamentary'. As was shown in the section on plastics here, too, the length of the molecules is considerably shorter than would be calculated from a filamentary arrangement. They can be considered as compact small rods.

In an aliphatic hydrocarbon the hydrophobic residue has a length of 10–18 carbon atoms, corresponding to the natural fatty acids. Other raw materials for the 'fatty residue' are petroleum hydrocarbons, synthetic hydrocarbons, and cracking products, which can also be branched. The aliphatic residue can be linked with an aromatic residue (alkaryls) which in that case carries the hydrophilic group. The aliphatic residue can then be shorter than in purely aliphatic capillary-active products without impairing the effect.

The hydrophilic groups can be salt-forming or non-salt-forming. The solubility of these substances possessing a salt-forming group, in particular with the $—O—SO_3H$ or $—SO_3H$ group, is usually better than that of non-salt-forming ones.

Hydrophilic groups in capillary-active materials:

salt-forming:

$—COOH$, $—OSO_3H$, $—SO_3H$ (acid)
$—NH_2$, $=HN$, $\equiv N$, $\equiv N^{\oplus}$ (basic)
(good water solubility)

non-salt-forming:

$—COO—$, $—CO—NH—$, $—O—$, $—\underset{\displaystyle H}{\overset{\displaystyle OH}{C}}—$
(poor water solubility)

$—CHOH—CHOH—CH_2OH$
$—CH_2—CH_2—O—CH_2—CH_2—O—CH_2—CH_2—O—$
(very good water solubility)

The properties of the capillary-active material depend on the length, structure, and type of the hydrophobic residue and the nature and position of the hydrophilic residues. In different classes of compounds they are strongly affected by the pH value.

The hydrophobic chain can be interrupted by a hydrophilic residue, or the hydrophobic part of the molecule can incorporate one or several hydrophilic groups. In the latter case the solubility is increased but the polarity, which is important for the surface activity, is reduced. The incorporation into, or addition to the chain of various residues affects the affinity for the fibre. Thus, by varying these structural principles the effect of a detergent or a textile auxiliary can be varied.

*Orientation of capillary-active substances and detergent effect*

A consequence of the polar structure of the capillary-active substance is a definite orientation at the interface of aqueous and non-aqueous layers with air and at the interfaces of liquid and solid materials with each other. If a fatty acid is spread on a water surface in a very thin, if possible monomolecular layer, the molecules orient themselves with the carboxylic groups towards the water; the reverse occurs when hexane or oil form an interface with air. At the interface of hexane or oil and water the 'fatty tails' are in the hexane or oil layer, the carboxylic groups in the water layer (Langmuir and Harkins):

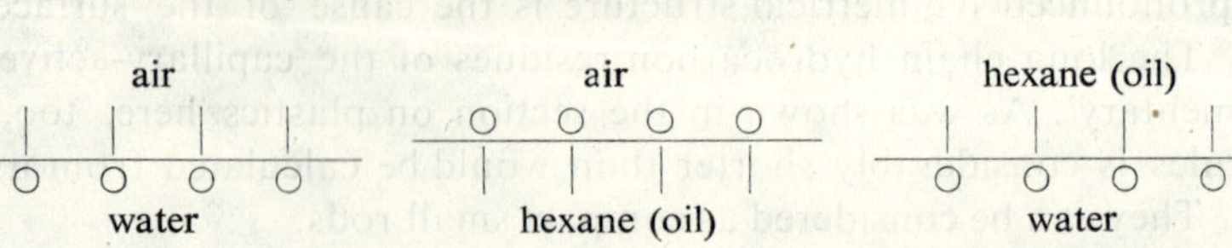


If the thickness of the layer is greater only the molecules lying directly at the surface orient themselves in the way shown above. Those lying above orient themselves randomly (purely statistically). The molecular layer covering the water surface reduces its high surface tension. The larger the water surface, of course, the higher the expenditure of energy because of the strong dipolar forces of the water that have to be overcome. This work is much less when one has to separate only the fatty acid residues covering the water surface.

These concepts apply similarly to solid-liquid systems, where the surface-active substances also orient themselves at the interface. According to whether the hydrophilic groups face the water layer or not an increase of wettability (wetting effect, wetting agent) or a decrease (rendering the material water-repellent, hydrophobic) occurs.

*Principles of the washing process*

With increasing concentration the equivalent conductivity of alkali salts of longer-chain fatty acids, unlike that of short-chain ones, decreases sharply and then increases slightly. It is concluded from this that in solution an agglomeration of dissociated and non-dissociated molecules occurs, the degree of which depends on the concentration. Soap molecule aggregates called 'micelles' form. In the interior of the micelle are the fatty residues, on the outer shell the polar groups which are partly dissociated. Thus one obtains a relatively high 'solubility' of the soap which, however, is not a true one because

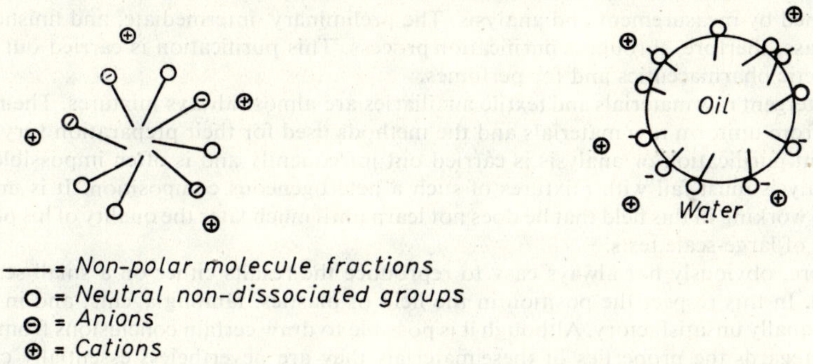


Figure 144. Left: Distribution of the ions at the surface of a soap micelle. Right: Oil droplets emulsified in water with the aid of a capillary-active substance

it is not based on complete dissociation of the molecules in the solvent. Into such a micelle other substances can be incorporated (see p. 398, Emulsion polymerization) which are thus 'dissolved' in water. This fact is utilized for the suspension of oil or solid materials in water. Oil droplets suspended in water in the presence of a capillary-active material are prevented from re-coalescing since they are enveloped by the polar substance and have thus the character of ions. The hydrophobic part faces the oil, the hydrophilic one the water. The following illustrations, taken from Winnacker-Weingaertner, demonstrate this process very clearly (Figure 144).

Capillary-active materials also envelop solid particles in the manner described so that the hydrophobic residue faces the surface of the particle and the hydrophilic one the water. This renders the particle wettable by water. The particles are 'dispersed'.

If liquids are dispersed an 'emulsion' is obtained; with solid particles a 'suspension'.

The detergent action of a capillary-active material is based on this effect. If for example a fibre is soiled by soot and oil the detergent displaces the dirt from the fibre and suspends it in the washing liquid. The particle has to be kept in suspension and must not flocculate ('dirt-suspending capacity' of washing liquid).

In addition to this capillary-active effect an important role is played in the washing process by electrostatic processes. The electrostatic repulsion due to a charge of equal polarity in the dirt particles and fibre is increased by soap type materials so that separation occurs. An increase in the pH value has a similar effect (reinforcement of the washing action by an increase in the alkalinity of the washing liquid). This is true only for anion-active detergents.

The processes described are fully effective only when the contact surfaces between the materials are constantly renewed i.e. when they are kept in motion mechanically. Mechanical motion by shaking, stirring, or flow improves, therefore, the detergent effect (washing machine).

## Properties of detergents and textile auxiliaries and their testing in practice

In the synthesis of most organic materials the aim is almost always to obtain clearly defined products of technical purity that are uniform chemical units with specific physical and chemical properties which can be determined by measurement and analysis. The preliminary, intermediate, and finished products, solvents, etc. pass, therfore, through a purification process. This purification is carried out more rigorously for synthetic pharmaceutics and for perfumes.

However, detergent raw materials and textile auxiliaries are almost always mixtures. Their production does not start from uniform raw materials and the methods used for their preparation very rarely coincide. Subsequent purification or analysis is carried out infrequently and is often impossible. The usual methods of analysis must fail with mixtures of such a heterogeneous composition. It is unsatisfactory for the chemist working in this field that he does not learn until much later the quality of his product, after the completion of large-scale tests.

It is, therefore, obviously not always easy to reproduce the results either on a small scale or on an industrial scale. In this respect the position in the field of plastics, tanning agents, and in part also in that of dyes is equally unsatisfactory. Although it is possible to draw certain conclusions from the chemical structure as regards the properties of these materials they are nevertheless essentially characterized by their effects. Much research work has been carried out in order to define the dependence of various effects on the composition and some interesting laws have been found, although mainly for pure materials and mixtures of definite composition.

To assess detergents and textile auxiliaries used in practice the only way is almost always the practical test which is often laborious. Certain empirical test methods have been developed for this purpose which as far as possible are adapted to practical usage. These tests are carried out at the manufacturer's in applications laboratories which also supervise production, since the commercial products are also checked by methods adjusted to their applications. It is by no means sufficient to test only the washing effect of a detergent but many other requirements have to be satisfied before it can be considered a serviceable detergent.

As in the assessment of dyes, a reference specimen, called a manufacturer's standard, is used for comparison. The individual manufactured batches may deviate from the properties of this standard only within predetermined limits. An important part is played by the mixing of batches with different properties which in many cases makes possible an adjustment so that the commercial product obtained conforms to the standard.

The most important properties of detergents and textile auxiliaries which determine their serviceability will be discussed briefly below. They are:

Solubility,
Response to hard water,
Resistance to acids and alkali,
Wetting capacity,
Dispersive capacity,
Emulsifying capacity,
Foaming capacity,
Detergent capacity,
Specific effects.

In capillary-active compounds solubility in water decreases with an increasing number of C atoms in the hydrophobic residue whilst the capillary activity increases. The most important influence on solubility is the nature of the hydrophilic residue and of the cation belonging to it. Of these the sulphonic acid and sulphonic ester groups are the most effective in anion-active compounds, less so carboxylic acid groups. The most easily soluble are triethanolamine- and ammonium salts, followed by potassium- and sodium salts; alkaline earth salts (e.g. calcium soap) and metal salts are more difficult and very difficult respectively. In cation-active substances solubility depends largely on the anion, chlorides being more easily soluble than sulphates. They can be precipitated by high-molecular anions. From this follows that a quaternary ammonium salt, a cation-active material, is precipitated with a sulphate- or carboxylic anion. Anion- and cation-active materials must not, therefore, be mixed in use.

Non-ionic compounds, of the polyglycol ether type, are more soluble the larger the number of glycol residues they contain. There is a fundamental difference between ionic and non-ionic compounds as regards their solubility as a function of temperature. Whilst the former, such as soap, alcohol sulphates and alkylsulphonic salts, become more soluble with increasing temperature, the solubility of polyglycol ethers decreases with rising temperature so that flocculation occurs. This points to a fundamental difference in the nature of the solution. Non-ionic compounds dissolve by surrounding themselves with water films by means of secondary valencies and form a kind of hydrate. During heating these hydrates cleave and the materials become difficultly soluble. Additions of alkali and salts have a similar effect; on the other hand, organic solubilizing intermediates such as alcohols improve solubility.

Many capillary-active materials precipitate with calcium and magnesium salts and, therefore, also with hard water; they are too easily affected by hard water. Higher fatty acids, i.e. capillary-active substances with carboxylic acid groups, are particularly sensitive and precipitate calcium soap. Calcium soap has no detergent effect; on the contrary, it may precipitate in the fibre and clog it. The same applies to other metal salts, such as iron and aluminium. The hardness of water is determined analytically by means of a soap solution. Not all salts of sulphonic acids and sulphonic esters are completely unaffected by hard water; this applies particularly to low-sulphonated unsaturated oils. However, the precipitates formed disperse easily and do not become fixed on the fibre. Non-ionic materials on the other hand are completely resistant to hard water.

All higher fatty acids and products obtained by mild sulphonation of oils precipitate with acids because they still contain non-sulphonated fractions; they are not resistant to acids. Mineral acids saponify sulphonic acid esters to fatty alcohols. On the other hand, true sulphonic acids and substances containing polyglycol residues or derivatives of polyhydric alcohols with a larger number of OH groups are resistant. Only acid-resistant materials may, therefore, be used in acid liquors such as are used in the finishing of wool.

Even more important is resistance to alkali since many upgrading processes, particularly of cellulose fibres, are carried out in alkaline media or even in a strong solution of caustic soda e.g. the mercerizing of cotton.

All compounds with carbonic ester groups are incompletely resistant to alkali, but sulphonic esters are resistant, and so are compounds with carboxylic and sulphonic acid groups and polyglycol ether residues.

The capillary-active effect is made apparent most clearly by the wetting capacity, i.e. by the degree to which the material reduces the surface tension between water and the textile fibre. There are various test methods based on measuring how rapidly a piece of cloth sinks when it is immersed below the surface of the solution.

The dependence of wettability on the length of the hydrophobic residue, or on the number of glycol residues in polyglycol ethers, as shown in appropriate tests, is interesting. For fatty alcohol sulphonates the optimum is at 25°C and a chain length of $C_{14}$.

The wetting ability of long-chain compounds with terminal hydrophilic groups, i.e. those that are particularly suitable as detergents, is comparatively low at room temperature but improves with increasing temperature. They are 'hot wetting agents'. Compounds with a hydrophilic group near the middle of the chain and relatively short alkyl- or alkaryl residues on the other hand have good wetting ability in the cold ('cold wetting agents'). Dispersions of solid materials are used in many different fields, e.g. dyes, synthetic resins, delustering agents for synthetic fibres, sprays for plant protection, etc. In order to prevent flocculation (agglomeration), dispersing agents are added. In many cases grinding of the finely dispersed solid material with the dispersing agent is particularly effective. Disintegration is improved and during 'dissolution' dispersion occurs immediately (peptization). Of special importance is the dispersion of calcium soap which at first is held in solution at high soap concentrations and then precipitates during dilution ('rinsing'). The addition of calcium soap dispersing agents to the rinsing water prevents this. For the same reason every detergent should contain as good a calcium soap dispersing agent as possible.

Compounds of very different classes of materials have a dispersing and protective-colloid effect. Substances with a carbon amide group, such as fatty acid methyl tauride (Igepon T, Hostapon T), and fatty acid-protein condensation products are particularly effective.

When oils or difficultly soluble organic liquids are dispersed in water, the droplets re-coalesce after initial dispersion, due to the great surface tension of the water, and finally complete separation into two layers takes place.

In the presence of capillary-active materials possessing emulsifying capacity, dispersion of either the oil droplets in water (oil-in-water emulsion O/W) or the water droplets in oil (water-in-oil-emulsion W/O) is dependent on their structure. The dispersed fraction is called the 'inner phase', the surrounding medium the 'outer phase'. Most emulsions used in practice are oil-in-water emulsions (e.g. milk, rubber latex).

An oil-in-water emulsion is recognized quite simply by the fact that a drop of it placed on water disperses immediately. If it is a water-in-oil emulsion, the drop floats undispersed (there are oil-soluble and oil-insoluble emulsifiers).

It depends on the emulsifier as to which of the two types of emulsion forms. Emulsifiers with strongly hydrophilic residues, such as alkali salts of fatty acids, alcohol sulphates, sulphonic acids, highly hydroxyethylated products, usually give O/W emulsions. The molecules of the emulsifier are oriented at the interface of the oil droplets in such a way that the 'fatty tails' face the oil, the hydrophilic groups the water, they ionize, and repel each other. Emulsifiers with a slightly hydrophilic character, such as low-hydroxyethylated fatty alcohols, form W/O emulsions.

An oil-in-water emulsion can be transformed into a water-in-oil emulsion (phase reversal) by adding $CaCl_2$ to an O/W emulsion, stabilized with sodium oleate, thus altering the emulsifier.

Emulsions are cloudy like milk because the droplets are fairly large (1–50 $\mu$m). They are visible under the microscope. By strong mechanical dispersion in emulsifying and homogenizing machines and using suitable emulsifiers, dispersion can be continued so far (particle size 0·1 $\mu$m) that the liquid appears clear and can be recognized as an emulsion only by means of the Tyndall effect. Recently liquids have been successfully dispersed in each other by ultrasound, but emulsions can also be flocculated by ultrasound.

The greater the degree of dispersion, the less tendency the droplets have to coalesce. Electrolytes have a very strong effect on emulsions because they change the state of charge of the particles. Thus, many emulsions flocculate when salt is added, a fact utilized frequently in practice. Emulsions are least stable at the isoelectric point, i.e. the point where the potential difference between the dispersed particles and the surrounding medium is zero.

Many types of emulsions are used for the finishing and water-proofing of textiles. They can be liquid, pasty, or in the form of a salve, and have to be flocculated on the fibre.

If gas bubbles are conducted through a liquid they normally rise and escape. Surface tension opposes the formation of foam layers. Capillary-active materials make possible the development of foam, the main factor in this, but not directly dependent on it, being the reduction of the surface tension. The foaming capacity of soaps and alkyl sulphates is directly related to the chain length. It has, however, no significance for the washing capacity and other properties of practical importance. Excellent foamers can, therefore, be very bad washing, emulsifying, or wetting agents. Since the opinion that foaming and washing are related is, however, deeply rooted among consumers, this factor cannot be altogether disregarded in the development of detergents.

Low-foaming agents are now being offered for washing machines. Electrolytes and salts causing hardness of the water reduce the foaming capacity. To determine foaming capacity one measures foam volume, stability of the foam, and the amount of liquid supported.

*Washing capacity*

In practice the washing capacity of an agent does not depend only on the agent itself but also, on the one hand, on the chemical nature of the yarn or tissue to be washed and the nature of its surface, and on the other hand on the type of dirt, e.g. its state of charge. Other important factors are the state of the washing liquor and the washing conditions, since the effect of the capillary-active material is strongly influenced by other substances in the washing liquor.

The washing process comprises separate complex chemical and physical processes.

First, the detergent has to wet the fibre and dirt well. The fibre swells and this alone already loosens the dirt particles which are not only between individual fibres but have penetrated into the submicroscopic cavities. In washing textiles a prolonged period of soaking is, therefore, advantageous; for efficient soaking needs time. The longer the dirt acts on the textiles the more strongly it adheres. Fully synthetic fibres hardly swell at all.

Water soluble dirty substances dissolve in the washing liquor, oily and fatty particles are emulsified, solid ones are dispersed. In an alkaline liquor fats and oils saponify at higher temperatures.

A great part in cleaning is played by changing the wetting capacity and the electrical charge. The molecules of the capillary-active material concentrate at the surface of oil-soiled fibres. This loosens and emulsifies the oil droplets. Textile fibre and dirt are both negatively charged by the attachment of OH ions; this is reinforced by the presence of alkali and further by anion-active detergents to such a degree that the dirt is electrostatically repelled by the fibre. It will be seen that cation-active agents, by giving the dirt a positive charge, increase the adhesion of the dirt by electrostatic attraction. A cleaning effect is achieved only when the charge on the fibre is reversed by operating in an acid liquor and by using a large quantity of detergent.

The importance of the dirt-carrying capacity of the washing liquor has already been mentioned. It can be reinforced by protective colloids such as water-soluble cellulose derivatives (e.g. carboxymethyl cellulose).

In testing detergents the concentration has to be considered. With increasing concentration the washing effect increases up to a maximum, and then decreases. This is dug to the quantity of monomolecular detergent ions present which are carriers of the washine effect.

Detergents are used not only for cleaning soiled textiles but in the textile and dye industries also for the removal of additives and softeners applied to facilitate the processing of the fibre. Since these are well-defined 'dirts' the effect of a detergent on them can be assessed easily.

To test the washing effect on dirts such as occur in the household and in industry, a large number of artifical dirts has been developed which are applied to standardized test tissues. They usually contain a fat or oil constituent, and also staining substances and various pigments such as Sudan dyes (fat-soluble dyes), soot, or 'artificial road dust' made up of clay, iron oxide and soot. These mixtures have always to be applied in the same manner. Washing is carried out under strictly defined conditions, either manually or in washing machines.

To assess the washing effect, the degree of degreasing and the white content, i.e. the brightening, are determined by visual comparison or by photometer. An important part is played in washing by changes in wettability.

In these tests certain laws governing the dependence of the washing effect on constitution have been formulated:

With aliphatic compounds an adequate washing effect occurs only when the hydrophobic residue has at least 12 C atoms. With increasing chain length solubility decreases which has an unfavourable effect on the washing capacity. Aliphatic-aromatic residues are a kind of substitute for aliphatic C atoms. (An aliphatic chain of 8 C atoms is then sufficient.) As far as possible the hydrophilic group should be a terminal one, or near the end of the chain. Several hydrophilic groups disturb the polar character and reduce the washing effect, and so do those nearer the middle of the chain.

With non-ionic compounds of the polyglycol ether type the washing effect depends on the degree of hydroxyethylation and is strongly dependent on temperature; the degreasing and brightening effects do not run parallel.

### *Specific effects of textile auxiliaries*

In textile dyeing an important part is played by levelling agents, i.e. materials which slow down the absorption of dyes on the fibre and thereby produce a more even dyeing (levelling). There are fibre-affinitive and dye-affinitive levelling agents:

(1) The capillary-active material itself has an affinity for the fibre, is absorbed, blocks the dye-affinitive positions and releases them only slowly for the dye. Fibre-affinitive levelling agents are sulphonates, sulphonic acids, and fatty acid condensates.

(2) The levelling agent forms loose addition compounds with the dissolved dye from which the dye is released gradually so that it can be absorbed. These dye-affinitive levelling agents include polyethylenehydroxide ethers of fatty alcohols and alkyl phenols.

The improvement of the appearance and of the 'feel' by the application of certain preparations (brightening, softening) is very important for the finish of textiles. This is based on the deposition of fatty materials on the surface of the fibre. Cation-active compounds with a fatty residue, because of their positive charge, are particularly strongly absorbed by the negatively charged fibre. The various agents impart to the fibre a characteristic surface which becomes most obvious when the fibre is handled with a firm grip. Assessment requires much practical experience. The 'feel' is called 'full', 'soft', 'rustling', 'smooth', etc.

Preparations are also used to fix other materials on the fibre, e.g. white pigments to reduce the unpleasant shine by delustering, or alumina or zirconium salts to render the fibres and tissues water-repellent. Synthetic fibres have to be delustered during spinning.

## The production of capillary-active materials

The industrial production of fats, fatty acids, and soaps has been dicussed earlier. Natural fats and oils are still an important raw material basis for detergents and textile auxiliaries. The chart below shows the raw materials, the preliminary and intermediate products, and some important detergents and textile auxiliaries and their trade names.

As the chart in Figure 145 shows, the fatty alcohols obtained by catalytic hydrogenation of fatty acids or of their esters, or by the Bouveault-Blanc method with sodium and low monovalent alcohols, are of great importance. In the latter method the double bonds of unsaturated fatty acids are not hydrogenated. Raw materials for fatty alcohols are spermaceti and sperm oil (from the sperm whale), and vegetable fatty acids, e.g. coconut fatty acid ($C_{12}$ and $C_{14}$). The products obtained are 'Lorol', cetyl-, stearyl-, and oleyl alcohol. The fatty acids obtained as by-products of paraffin oxidation are, however, nowadays also used mainly for the production of fatty alcohols since it is less advantageous to use them directly for the production of soaps because of the unpleasant smell. Since paraffin oxidation is a particularly interesting industrial process it will be discussed in greater detail below.

### *Paraffin oxidation*

The first attempts to oxidize paraffins to fatty acids were made over a hundred years ago. Experiments were started at the BASF, Oppau, in 1921, and in 1926 the first experimental plant was built there in which lignite paraffin was used as a raw material. In 1931 the former IG and the Standard Oil Co. built

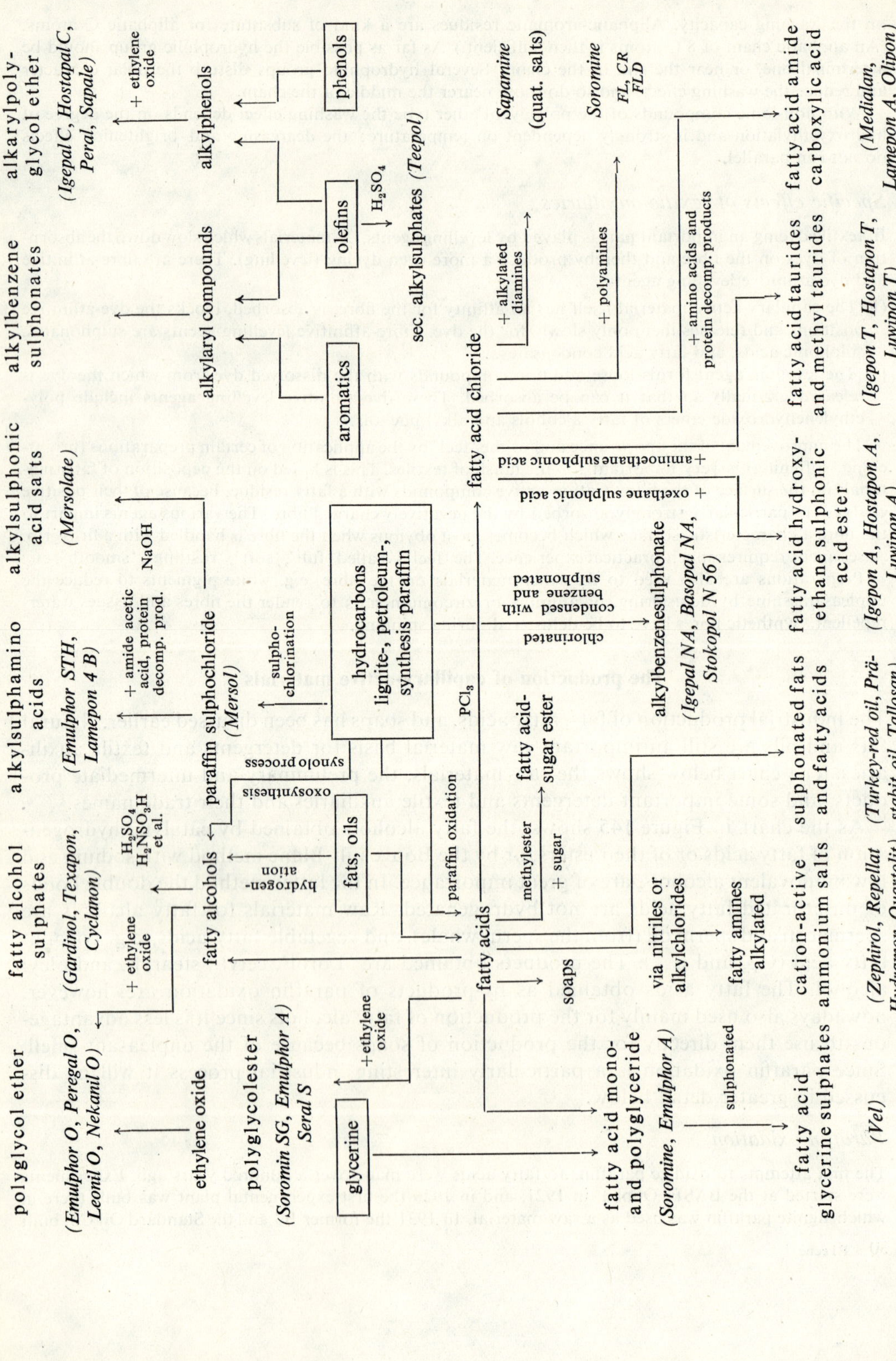


Figure 145. Raw materials, preliminary and intermediate products and some detergents and textile auxiliaries

a plant at Baton Rouge, U.S.A., with a capacity of 90 tons per month, based on petroleum paraffin. The first large-scale German plant was built in 1936/1937 by the Deutsche Fettsäurewerke at Witten (Ruhr) (Henkel & Co. and Imhausen & Co.) for an annual throughput of 40,000 tons paraffin, then followed Ludwigshafen-Oppau with 20,000 tons per annum. A plant for pressure oxidation (gauge pressure of 25 atm) was built during the war by Hubbe and Fahrenholtz, at Magdeburg, which, however, never became operational.

In 1948 the Deutsche Hydrierwerke built a large-scale plant at Rodleben which now supplies the major part of the fatty acids requirements for fatty alcohol production in the German Democratic Republic.

Natural paraffins from lignite and petroleum can be used as raw materials but synthetic paraffins from the Fischer-Tropsch plant ('Fischer-Gatsch'), with chain lengths of $C_{18}$–$C_{30}$, are used nowadays almost exclusively. From these are obtained fatty acids with chain lengths of $C_{10}$–$C_{20}$ which are most suitable for detergents, and which should have as few branches as possible (this depends on the nature of the paraffin; TTH paraffin (low-temperature-hydrogenation paraffin) is particularly suitable).

The oxidation is carried out using atmospheric oxygen in the presence of 0·3 per cent $KMnO_4$ which forms paraffin-soluble manganese and potassium soaps, at 100–130°C, in oxidation furnaces made of special steel or high-purity aluminium, of 30–60 m³ capacity. 40–60 m³ finely dispersed air per hour and per ton of paraffin are required, the reaction taking 21–23 hours. It is interrupted when an acid number of 70–75 and a saponification number of 120–127 is reached. Only about one third of the paraffin is, therefore, oxidized since otherwise the decomposition is too strong and too large a quantity of low fatty acids is produced. The unoxidized paraffin is continuously recycled. It contains reaction nuclei so that the induction period which is rather long when fresh paraffin is used, is appreciably shortened. The reaction is strongly exothermic and efficient cooling is, therefore, necessary.

The chemistry of paraffin oxidation is not yet fully understood. Like other autoxidations it is a radical chain reaction in which hydro-peroxides form first (A. Rieche, see diagram on p. 206). The primary attack of oxygen is statistical so that every C atom of the paraffin chain is in an equally favourable position. In this respect the process resembles the chlorination of paraffins (Asinger) and the sulphochlorination of paraffins with $SO_2$ and $Cl_2$ in the presence of light (Mersol-process) which will be discussed later. Paraffin oxidation is probably similar to the mechanism involved in ketone oxidation (Langenbeck and Pritzkow):

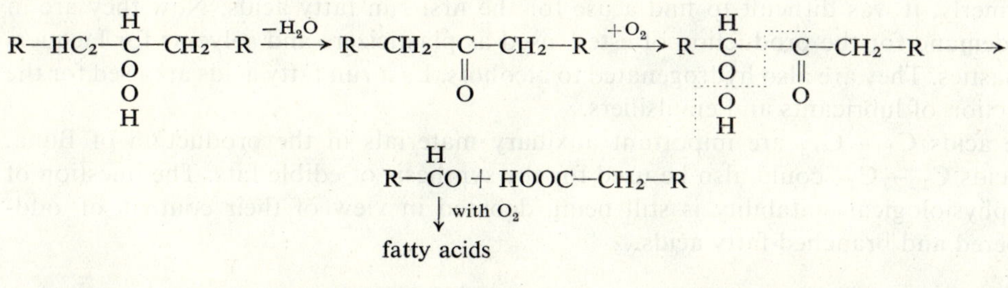

Many secondary reactions also take place. It has not yet proved possible to influence the industrial oxidation process so that the oxygen attacks preferentially at the terminal groups thus forming longer-chain acids.

The crude oxidation product is a mixture of all even and odd-numbered saturated fatty acids from $C_1$—$C_{22}$, and also branched chain and olefin-, hydroxy-, aldehyde-, keto- and dicarboxylic acids. In addition, alcohols, aldehydes, ketones, esters, lactones, and estolides also form; the mixture is, therefore, very heterogeneous.

The unused air passes through the cooler, where the 'condenser water' and the 'condenser oil' condense. The former is a 25–30 per cent aqueous solution of low fatty acids and can be used for the production of esters. Condenser oil contains volatile alcohols, esters, aldehydes, lactones, and also fatty acids.

During the final treatment the crude product is first washed in columns with water, low-molecular fatty acids, their Mn- and K salts, and dicarboxylic acids being washed out. Then follows 'saponification', i.e. the oxidation mixture is treated at 95–100°C with soda and a solution of caustic soda at atmospheric pressure, or at 185°C under pressure. If the mixture was saponified at atmospheric pressure a subsequent treatment at 185°C and a gauge pressure of 25 atm has to be carried out. During the first (atmospheric) saponification the fatty acids present dissolve as soaps, with simultaneous precipitation of part (one third) of the unoxidized paraffin.

During the pressure treatment any esters formed are saponified and a further third of the paraffin precipitates.

Then follows vigorous steam treatment of the crude soap by passing it very quickly together with superheated steam through long steam-heated pipes at 330–370°C, or by forcing the crude product, using the Oppau principle (Harder furnace), through a pressure pipe at 320°C in 20 minutes. When the pressure is released the steam carries off most of the residual paraffin and a refined soap is obtained. This method is very expensive and involves loss of material. Attempts are being made to replace it by a more economical one.

The dry soap which now contains very little paraffin is made into a paste with water to give a 23–30 per cent soap. The 'soap paste' is decomposed with $H_2SO_4$ or HCl in acid-resistant digesters. The free fatty acid is washed and decomposed by distillation into fractions under a high vacuum. The maximum yield is 75–80 per cent of crude fatty acid:

| | |
|---|---|
| First run fatty acids | 12–15 per cent $C_4$—$C_9$ |
| Main run fatty acids | 60 per cent $C_9$—$C_{20}$ |
| Last run fatty acids | 4–8 per cent $C_{21}$—$C_{23}$ |
| Residual fatty acids | 10–15 per cent |

Formerly, it was difficult to find a use for the first run fatty acids. Now they are in great demand for the production of esters used as plasticizers and solvents for lacquers and plastics. They are also hydrogenated to alcohols. Last run fatty acids are used for the production of lubricants and emulsifiers.

The acids $C_{11}$—$C_{15}$ are important auxiliary materials in the production of Buna. The acids $C_{12}$—$C_{18}$ could also be used for the synthesis of edible fats. The question of their physiological suitability is still being debated in view of their content of odd-numbered and branched fatty acids.

### *Various textile auxiliaries and detergents*

Some of the oldest textile auxiliaries are sulphonated unsaturated vegetable oils and fatty acids. Sulphonation can be carried out under mild conditions so that the fats do not saponify. The products, therefore, still have the character of oils but are to a certain extent water-soluble. One of these is Turkey-red oil, another is Tallosan. Products with higher solubility because of their higher degree of sulphonation are obtained when the sulphonation is carried out under rigorous conditions, or when the non-sulphonated fraction is extracted before neutralization by means of a solvent (Monopol brilliant oil, Prästabit oil V).

Unsaturated fatty acids yield sulphuric acid esters with addition of $HOSO_3H$ to the double bond and esterification of hydroxy groups. Under more rigorous conditions condensation of the carboxylic group with double bonds occurs and the formation of estolid takes place:

$$R{-}CH{-}CH_2({-}CH_2)_x{-}COOH$$
$$|$$
$$O$$
$$|$$
$$R{-}CH{-}CH_2{-}(CH_2)_x{-}C{=}O \qquad \text{Estolid}$$
$$|$$
$$OSO_3H$$

*Fatty acid derivatives*

Important fatty acid derivatives are obtained by condensation of fatty acid chlorides produced from fatty acids and $PCl_3$, by condensation with hydroxyethane sulphonic acid, aminoalkyl sulphonic acids such as taurin, and with amino acids or products of protein decomposition:

$R{-}CO{-}O{-}CH_2{-}CH_2{-}SO_3Na$ *Igepon A, Hostapon A, Luwipon A*

$R{-}CO{-}N(CH_3){-}CH_2{-}CH_2{-}SO_3Na$ *Igapon T, Hostapon T, Luwipon T*

$R{-}CO{-}N(CH_3){-}CH_2{-}COONa$ *Medialan A*

$R{-}CO{-}NH{-}CH(R_1){-}CO{-}NH{-}(CH(R_2){-}CO{-}NH)_x{-}CH(R_3){-}COONa$ *Lamepon A* *Olipon*

These fatty acid derivatives are completely unaffected by hard water and, except for Medialan, have excellent washing properties, particularly for delicate materials (wool, silk).

*Fatty alcohol sulphonates*

$R{-}CH_2{-}O{-}SO_3Na$ fatty alcohol sulphate

An equally valuable group of detergents and textile auxiliaries are the fatty alcohol sulphonates obtained from fatty alcohols with excess sulphuric acid (production of fatty alcohols, see p. 354). H. Bertsch was the first to recognize the great industrial importance of higher molecular weight primary alkyl sulphates, produced first by Dumas in 1836, as detergents and textile auxiliaries (DRP 640 997, 659 277). During sulphonation an equilibrium always establishes itself, i.e. a certain quantity of fatty alcohol is always left over but is held in emulsion by the sulphate during dissolution in water. Another important method is the sulphonation of saturated fatty alcohols with chlorosulphonic acid. No excess of chlorosulphonic acid is required since no water forms during the reaction. The reaction terminates here.

$$R{-}CH_2OH + H_2SO_4 \rightarrow R{-}CH_2{-}O{-}SO_2{-}OH + H_2O$$
$$R{-}CH_2OH + Cl{-}SO_3H \rightarrow R{-}CH_2{-}O{-}SO_2{-}OH + HCl$$

One uses as an emulsion basis for pharmaceutics and cosmetics a mixture of sulphonate with a large quantity of fatty alcohol which can be mixed with any desired amount of water (e.g. 'Lanette wax') and yields an oil-in-water emulsion. Detergents based on alcohol sulphonates are e.g. Gardinol WO ($C_{12}$—$C_{16}$), CA ($C_{16}$—$C_{18}$) and Texapon ($C_{12}$), Fewa, Onalkali, and Cyclanon.

Unsaturated alcohols are sulphonated with sulphuric acid at the double bond also, forming disulphonic acid esters. This reduces the detergent effect but increases the emulsifying effect. Methods exist also for the sulphonation of unsaturated fatty alcohols where the double bond is kept intact, e.g. with the sodium salt of chlorosulphonic acid, amidosulphonic acid, $SO_3$ + pyridine, *et al.* With these methods esterification is more c omplete than with sulphuric acid.

*Sulphochlorination of saturated hydrocarbons, Mersol, Mersolate*

Higher aliphatic saturated hydrocarbons began to be used for syntheses comparatively late although they occur in both petroleum and lignite paraffin and can be produced cheaply by the Fischer-Tropsch synthesis. They are too unreactive and too non-uniform. Of the utmost importance not only for the whole detergent and textile auxiliaries industry but also for the supply of detergents in Germany was the process worked out shortly before the 1939–1945 war by P. Herold and his colleagues at Leuna for the sulphochlorination of hydrocarbons with $SO_2$ and chlorine in ultra-violet light. The process was developed very rapidly for large-scale industrial use. It was also the first chemical industrial process in which ultra-violet light was used as an activator. The reaction was first described by the American Reed in 1936 but was not exploited industrially. The Mersol process is used in the DDR at the VEB Leuna-Werke 'Walter Ulbricht' and at the VEB Farbenfabrik Wolfen.

When $SO_2$ and $Cl_2$ react with saturated hydrocarbons the products formed under certain conditions are mostly chlorides of true sulphonic acids. In addition small quantities of halogenated hydrocarbons form. In regulating this process one distinguishes between 'saponifiable chlorine' of the sulphochloride group and 'chain chlorine'.

The reaction is carried out in the liquid phase, the most favourable temperature being 40–50°C.

Sulphochlorination is a chain reaction. The raw material has to be fully saturated and free of O-compounds. Olefins have a very disturbing influence on the process, and so have aromatics. Impurities cause the chain to break. The best raw material is Kogasin II with boiling points between 240 and 340°C corresponding to a chain length of $C_{12}$—$C_{18}$. The crude product is, however, unsuitable and has to be purified. At first it was subjected to $SO_2$ extraction by Edeleanu's method on an experimental scale. Later it was found that high pressure hydrogenation suffices. The pure hydrocarbon obtained is called 'Mepasin', the sulphochloride 'Mersol' and the saponification products 'Mersolates'.

A quantum of light with a wavelength of 3,000–5,000 Å causes a chlorine molecule to dissociate into chlorine atoms (dissociation energy 57 kcal/mole). These start a chain reaction. The quantum yield for industrial batches is 2,000–3,000 so that comparatively small quantities of light are sufficient:

$$Cl_2 \xrightarrow{h\cdot\nu} 2Cl^{\bullet} \qquad R^{\bullet} \text{ (from peroxide} + SO_2 \longrightarrow \text{decomposition)}$$

$$RH + Cl^{\bullet} \longrightarrow R^{\bullet} + HCl$$

$$R^{\bullet} + SO_2 \longrightarrow R\text{—}\dot{S}O_2$$

$$R\text{—}\dot{S}O_2 + Cl_2 \longrightarrow R\text{—}SO_2Cl + Cl^{\bullet}$$

At the beginning of the reaction only one hydrogen atom of the hydrocarbon is substituted. The attack takes place mainly at the secondary and tertiary H atoms; it is purely statistical, i.e. every C atom is equally vulnerable so that a mixture of sulphochlorides is obtained (Herold, Asinger). When, however, a large part of the hydrocarbon is consumed further sulphochloride groups enter the carbon chain of the sulphochlorides already formed. Di- and polysulphochlorides form.

For reasons discussed earlier the washing effect of the higher sulphonated hydrocarbons is smaller. Mersol D, sulphochlorinated up to 80 per cent, contains 50 per cent monosulphochlorides but also 30 per cent di- and polysulphochlorides, in addition to 20 per cent raw material. Mersolate D obtained from it by saponification with a solution of caustic soda has a smaller washing effect than Mersolate H (50 per cent sulphochlorinated) or Mersolate 30 (30 per cent sulphochlorinated). After saponification the non-reacted Mepasin settles ('unreacted oil') and can be put back into the mixture or used for other purposes. The separation of unreacted oil is more complete when steam distillation or extraction with methanol is used, resulting in a great improvement in the detergent properties. For this purpose Mersol can e.g. be heated in a pressure coil with NaOH followed by pressure release.

The technical equipment for sulphochlorination consists of a column, with acid-resistant lining, which contains numerous wide glass tubes arranged tangentially. Inside these are mercury pressure lamps or, more recently, suitable fluorescent tubes. Since the process is strongly exothermic the mixture has to be cooled by pumping it through a stoneware cooler. Chlorine and $SO_2$ are introduced in the appropriate proportions through a perforated coil in the lower part of the column. The process is interrupted when the required quantitiy of saponifiable chlorine is reached. It is also possible to operate continuously by using the flow process. In this case vertical tubes, lined with Havek or Phenytal, and provided with windows for illumination, are cascaded in series. The necessary quantity of $SO_2 + Cl_2$ can be introduced into each tower in such amounts as are required from time to time by the degree of reaction of the mixture. With the same degree of reaction better products are thus obtained than with batch operation since accurate regulation of the process is possible. Phenytal is a phenolformaldehyde resin which is applied in liquid form and then hardened by heating. It is resistant to acid and HCl.

Paraffin sulphochlorides take part in many different reactions. With amido acetic acid one obtains alkyl-sulphamido acetic acid which is an excellent emulsifier (e.g. for cutting oils in the metal-working industries). With ammonia one obtains sulphamides and disulphamides. The emulsifier STH contains these in admixture with alkylsulphamido acetic acid and unreacted oil.

Materials of the Lamepon type are obtained with the decomposition products of protein. With phenols one obtains alkylsulphonic acid phenyl esters, e.g. 'Mesamoll', a plasticizer for polyvinylchloride.

Mersolate is used in detergents for coarse and delicate work, for the production of washing powders for 'whites', and for numerous textile auxiliaries. It is, however, somewhat more affected by hard water than alcohol sulphonates and Igepones. It is also less suitable for personal hygiene, although satisfactory from the physiological point of view, because of its strongly degreasing effect on the skin.

An interesting use for alkylsulphochlorides was found in leather manufacture. They also react with the amino groups of the collagen of the animal skin, and with sulphochloride and soda a kind of oil dressing is achieved which corresponds approximately to the tanning of chamois leather ('Immergan tanning', BASF).

### *Sulphoxidation*

True alkylsulphonic acids are obtained directly by a unique reaction called sulphoxidation which was discovered at Hoechst (Platz, Orthner, Graf, Schimmelschmidt *et al.*) and is used there. It consists in the reaction of $SO_2$ and $O_2$ with a paraffin, e.g. Kogasin.

To initiate sulphoxidation free radicals must first be formed, e.g. ultraviolet light, peracids, a little ozone, or a short sharp change in temperature.

### *Aralkylsulphonates*

In 1916 Günther (BASF) was the first to realize that sulphonated alkylated aromatic hydrocarbons are surface-active substances. These were the first textile auxiliaries to be produced without using fats. Nekal which was not marketed by BASF until 1924 and is a short-chain alkylated and sulphonated naphthalene, has hardly any detergent property but excellent wetting, dispersing, and foaming properties. (Nekal BX is di-isobutylnaphthalene sulphonic acid). Alkylation is carried out with alcohols and sulphuric acid, with subsequent or simultaneous sulphonation, or with an alkylhalide (Friedel-Crafts).

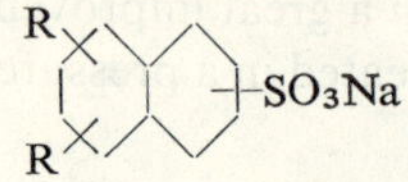


Nekal type (Bevekale)
(wetting effect)

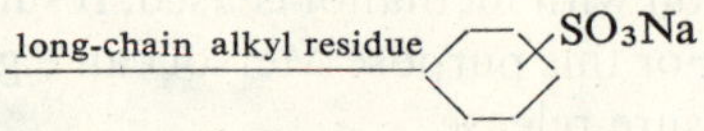


Igepal NA type
(detergent effect)

The products acquire a detergent property when benzene is used as an aromatic and longer-chain alkyl residues are substituted. It is possible, for instance, to chlorinate a hydrocarbon of $C_{12}$—$C_{14}$ from the Fischer-Tropsch synthesis, condense the chloralkane by the Friedel-Crafts method with benzene, and then sulphonate (first product Igepal NA, BASF, now Basopal NA).

Alkylbenzene sulphonates first attained their greatest importance in the U.S.A. and in volume excel other detergents. 65 per cent of all detergents produced in the Western world are based on alkylbenzene sulphonic acid. The paraffin raw materials are petroleum paraffins or polymerizates of natural gas or cracking olefins, e.g. tetrapropylene or tri-isobutylene (dodecene). These products are now quite generally used as detergents in Western Europe and Federal Germany.

### *Hydroxyalkylation products of fatty acids, fatty alcohols, and alkyl phenols*

Under pressure and in the presence of acid catalysts the action of ethylene oxide on fatty acids produces fatty acid polyglycol esters, and on fatty alcohols and alkyl phenols it produces polyglycol ether:

$$R—C(=O)—O(CH_2—CH_2O)_nH$$ *e.g. Emulphor A, Soromin SG, Seral S*

$$R—CH_2O(CH_2—CH_2O)_nH$$ *Emulphor O, Nekanil O, Peregal O*

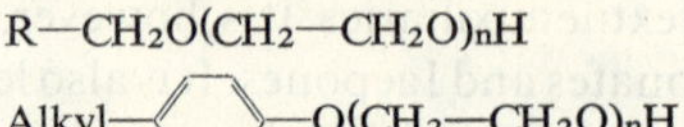


(also naphthalene) *Igepal C = Hostapal C, Sapal, Peral*

By changing the residue carrying the alkyl group and the degree of hydroxyethylation, many variations are possible which allow modification of the properties in the desired direction resulting in better wetting, emulsifying, or washing ability. A very large number of special products is produced for the most varied purposes. They are non-ionic and their solubility is based on the addition of water molecules to the hydrophilic residue. When heated they become more difficult to dissolve in water.

*Cation-active textile auxiliaries*

For reasons discussed earlier these have no detergent effect on textiles, they are, however, very important as textile finishing agents, since they are absorbed very easily because of their charge of opposite polarity to that of the fibre. With various dyes they form difficultly-soluble materials and can, therefore, be used to increase colour-fastness.

The least complex members of this group are the salts of primary, secondary, and tertiary amines; owing to their low solubility and stability they are, however, unimportant. On the other hand, the quaternary ammonium compounds are strong bases, form easily soluble salts and are widely used as textile auxiliaries. They are obtained from tertiary amines and alkylhalides or alkylsulphates, e.g.:

$$C_{16}H_{33}—Br + N(CH_3)_3 \longrightarrow [C_{16}H_{33}N(CH_3)_3]^+Br^-$$

The most important members, however, are those that are obtained from alkylhalides and pyridine:

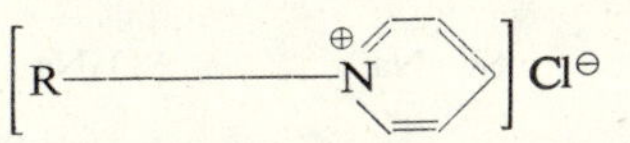

(e.g. Repellat)

## (b) Capillary-inactive compounds

In the textile industry, both natural colloids and modified natural substances and plastics are used as additives ('size'), as thickeners for printing colours, and for the finishing of fibres and textiles. Capillary-inactive materials are also used in conjunction with capillary-active ones for stretching and to improve the properties. One uses as fibre-glueing materials (size) for the preparation of yarns for weaving and knitting natural colloids such as starch or glue, or semi-synthetic ones, e.g. the etherification products of starch and cellulose, and also plastics, such as polyvinyl alcohols and polyacrylic acid salts. Oils, e.g. linseed oil or mineral oil, come from solvents or from an emulsion (softeners). These agents increase the tearing strength and smoothness. After finishing the materials have to be desized. Starch size can be removed best with enzyme preparations (amylases from malt). Other materials used for finishing are plastics, e.g. phenolformaldehyde resins in their first stage, or aminoplastics. They reduce the swelling ability of cellulose. The principal materials used for creaseproofing are urea formaldehyde resins or melamine resins.

Fabrics are made water-repellent and waterproof by impregnation with rubber- or plastic dispersions, or by hot-rolling them with plastic plasticizer mixtures on calenders ('backing'). Water-repellent fabrics for coats, tents, etc., should preserve their porosity. This is achieved by saturating them with a soap solution, and subsequent treatment with alumina solutions. Aluminium soap forms on the fibre. However, a more effective treatment is one using aluminous paraffin dispersions or wax dispersions (Ramasit, Imprägnol), or reaction of the fibre with reactive compounds, e.g. with long-chain alkylhydroxymethylpyridinium salts or substituted ethylene ureas. Many other methods have been developed for rendering fabrics water-repellent. Recently, silicones have been used with great success for this purpose.

'Optical brighteners' have become important as additives to detergents and for optical bleaching'. They have replaced the housewife's laundry blue. They are based on the transformation of incident short-wave light into long-wave blue light by fluorescence which makes the yellow tinge of the fibre appear white. The neutralization of the yellow tinge is additive, i.e. no loss of light occurs as happens when fabrics are dyed blue with dyes but, on the contrary, additional radiation is emitted.

The same materials can also be used as protection against the destructive action of ultraviolet rays (Ultrasan). For instance, stilbene derivatives have such properties. Blankophor B (Bayer) is obtained by condensation of two moles cyanuric chloride with diaminostilbene disulphonic acid and two moles aniline.

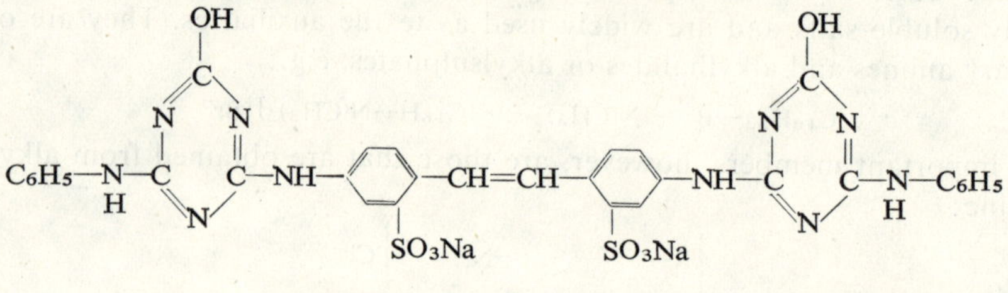

## LITERATURE


A. Chwala, *Textilhilfsmittel, ihre Chemie, Kolloidchemie und Anwendung*, Wien, 1939.

Winnacker-Weingaertner, *Chem. Technologie*, Org. Techn. II. 2nd ed., Winnacker-Küchler, Org. Techn. II, **4**, 153. Hanser-Verlag, München, 1954.

H. Walland, *Die Chemie der textilen Hilfsstoffe und ihre Anwendung*. Arbeitsgemeinschaft der Fachbuch- u. Fachzeitschr.-Verleger im A. Ziemsen-Verlag, Wittenberg, 1948.

A. Lottermoser, *Kurze Einführung in die Kolloidchemie*. Verlag Theodor Steinkopff, Dresden and Leipzig, 1948.

H. Frotscher, *Chemie und physikalische Chemie der Textilhilfsmittel*. VEB Verlag Technik, Berlin. Vol. 1, 1954, vol. 2, 1955.

K. Lindner, *Textilhilfsmittel und Waschrohstoffe*. Wiss. Verlagsges., Stuttgart, 1954.

H. Stüpel, *Synthetische und Reinigungsmittel*. Konradin-Verlag, R. Kohlhammer, Stuttgart, 1954.

A. M. Schwartz, and J. W. Perry, *Surface Active Agents*. Interscience Publishers, New York, vol. 1, 1949, vol. 2, 1958.

G. Gawalek, *Waschmittel und Netzmittel*. WTB Akademie-Verlag, Berlin, 1962.

# IV. Pharmaceutical products

## Introduction

The field of pharmaceutical products is so wide, the number of compounds so large, their properties and methods of production so varied that it is impossible to offer even a shortened comprehensive survey. The isolation of pharmaceutical products from natural products, the processes for the manufacture of synthetic remedies and the equipment used in the production are similar to many methods used in other industries and discussed in earlier sections. A plant for the production of synthetic medicines is very similar to that used for the production of dye intermediates, and the intermediates for pharmaceutical products and plant protective agents are in many cases identical with these. There were very good reasons for the large dye works to be the first to start the production of synthetic medicines; for here all the necessary conditions already existed. Even today preliminary products for both dyes and pharmaceutical products are frequently manufactured with the same equipment for reasons of economy. Only when the synthesis proper of the medicine begins—and particularly when the product is to be given its final form by the most careful purification—is special apparatus used, the material and condition of which should both be exemplary as far as tidiness and cleanliness are concerned. The same applies, of course, to conditions in the manufacturing, storage, dispensing and packing rooms.

Chemical research in the field of pharmaceutical products, too, was undertaken in the factories by intermediate-product and dye chemists since they had the necessary experience with synthetics. 'These dye chaps will make anything for you' said Paul Ehrlich, the grand old man of chemotherapy, appreciatively on one occasion since the dye chemists were able to supply him with any desired combination of groups for pharmacological experiments. Later the chemical plants installed their own special laboratories for the synthesis of pharmaceutical products but outside laboratories still frequently supply materials for pharmacological investigation. Quite recently many instances have occurred where the collaboration of all the laboratories of a firm with departments of chemotherapy has been most beneficial in the creation of new synthetic medicines.

In the present book only a brief description of the development of the pharmaceutical industry can be given. This will be followed by a survey of the most important groups of pharmaceutical products and of their significance. Finally, the technical aspects of the manufacture of a few typical representatives of these groups will be discussed. The reader can use the various specialized books available in large numbers for any further information in this field he may require.

### *History, Development of the manufacture of pharmaceutical products*

Right up to the eighties of the last century the few medicines then available were produced from natural materials of mineral, vegetable, or animal origin. Their effect was sometimes doubtful. Only relatively few materials with a precise therapeutic effect were known. Three admirable discoveries made in earlier times were: the use of cinchona bark against malaria, of *radix ipecacuanhae* against amoebic dysentery, and of mercury against syphilis.

The active substances had been isolated earlier from various drugs. Thus, Sertürner had succeeded as early as 1805 in producing morphine from opium, in 1820 Pelletier and Caventou isolated quinine from cinchona bark, and in 1860 Niemann isolated cocaine from coca leaves. But it took another 50–100 years for the constitution of these medicines to become known. However, only the knowledge of the chemical structure of these compounds made it possible to start thinking about their synthesis. The investigations of the constitution and attempts at syntheses frequently led to the discovery of new materials and of synthetic pharmaceutical products, even if they sometimes did not reach their original objective. It is a well known fact, for instance, that W. H. Perkins, who was assistant to A. W. von Hofmann, obtained in 1856, mauvein, the first synthetic dye during the course of an experiment designed to produce quinine from aniline.

Chloroform, which had been known since the days of J. von Liebig, was used as an anaesthetic for the first time by Simpson in 1847; ether has been used for the same purpose since 1848.

Some of the first synthetic pharmaceutical products obtained industrially were chloroform (anaesthetic), chloral hydrate (remedy against insomnia), salicylic acid (a remedy against rheumatism, the first large-scale synthesis of a medicine, in production since 1847 at Chem. Fabrik von Heyden, Radebeul), and iodoform (a bactericidal agent).

At first the manufacture of medicines was entirely in the hands of pharmaceutical chemists, and the first pharmaceutical factories originated from these pharmacies (e.g. Merck, Riedel, Schering, Boehringer, Gehe).

The synthesis of pharmaceutical products proper dates from about 1880. At that time the large German dye works also began to manufacture pharmaceutical products. Since they already possessed plants for the production of intermediates the conditions for the industrial synthesis of pharmaceutical products were in their case particularly favourable.

In 1884 Knorr first produced antipyrine which the pharmacologist Filehne recognized as an excellent antipyretic (fever-suppressing agent). This was the year of birth of the large-scale industrial synthesis of pharmaceutical products and also the start of the collaboration between chemist and medical man which was to prove so fruitful in the future. The Farbwerke Hoechst began to manufacture antipyrine. Four years later Duisberg and Hinsberg, working at Bayer, Elberfeld, prepared phenacetin which was the first of many pharmaceutical products developed there. A little earlier Kalle had produced antifebrin. Both antipyrine and phenacetin are still used widely, particularly in combination preparations.

In 1893, Stolz (Hoechst) developed pyramidone from antipyrine, which was not marketed, however, until it had been subjected to careful clinical tests over several years. In spite of its advantages its general acceptance was very slow because of the preference for the well-tested antipyrine. In 1899 Bayer's aspirin was brought on the market which is considerably less toxic than salicylic acid. In 1903 E. Fischer and Mehring discovered the sedative properties of barbituric acid derivatives.

The first synthetic local anaesthetic was Anaesthesin (p-aminobenzoic acid ethyl ester). Soon after, Einhorn (1905) discovered Novocain (p-aminobenzoic acid ester of N-diethylaminoethyl alcohol) which is much more effective and can be used for a greater variety of purposes. There followed a long series of local anaesthetics for special purposes; these are still being developed.

Paul Ehrlich opened a new chapter in the development of chemotherapy (pharmacological action brought to bear on live pathogenic agents in the body) when he discovered dyes which selectively stained and killed pathogenic agents without harming the tissues. The first were Trypan Blue, Trypaflavine, both effective against trypanosoma, and Rivanol. Of greater importance were the systematic investigations carried out by Ehrlich, Uhlenhut and others into organic arsenic compounds and their effectiveness against *Spirochaeta pallida*, the germ that causes syphilis; this led eventually to Salvarsan and the chemotherapy of infectious diseases. This work was applied industrially at the Farbwerke Hoechst where Salvarsan has been produced from 1910 onwards. The new feature of Ehrlich's working method was to change by systematic chemical modification materials which had shown themselves to be useful only under certain conditions; this was done in such a way that they became as far as possible harmless to the body whilst maintaining or even improving their efficiency. The great successes that followed were due to this method of working. At the suggestion of Ehrlich's pupil, Roehl, the drug for combating sleeping sickness 'Bayer 205' (Germanin) was developed at Bayer's in 1916. When the development cycle of the malaria parasite was discovered and the canary test was evolved it became possible to synthesize anti-malarial agents. In 1924, at the end of a long series of experiments carried out at Bayer's, Schulemann, Schönhöfer and Wingler produced Plasmochin (6-methoxy-8-diethylamino-isopentylamino-quinoline). Further progress constituted Atebrin (2-methoxy-6-chloro-9-diethyl-amino-isopentyl-amino-acridine), produced in 1934 by Mietzsch, Mauss, and Kikuth. As a result of this Resochin (Andersag) followed eventually in 1937. With these preparations and particularly with further developments from them (e.g. Paludrin), the malaria-producing germ can now be attacked at almost all stages of its development. A new era in the fight against bacterial infection was introduced by the discovery of the chemotherapeutic effect of the azo dye Prontosil by Domagk, Klarer, and Mietzsch (1932) at Bayer's. A large number of sulphonamides followed, most of which are still of importance despite the development of antibiotics. We shall mention here only sulphanilamide, sulphathiazole, sulphapyrimidine, sulphaguanidine, sulphathiourea and, as a representative of the more recently developed long-acting sulphonamides, Lederkyn. In the last few years certain sulphonamides have grown in importance because of their diuretic action, others because they reduce blood sugar.

The latest phase in chemotherapy was initiated by the discovery of antibiotics, particularly for the treatment of bacterial infections. As early as 1929 Fleming recognized that penicillin which can be isolated from certain types of moulds had bactericidal properties of unheard-of efficiency. Manufacture was started in the U.S.A. and England in 1939/1940, but in Germany not until after the 1939/1945 war. Penicillin was followed by more and more new substances with antibiotic effects such as streptomycin, aureomycin, terramycin, and chloramphenicol some of which have very specific effects against certain pathogenic agents. Streptomycin in particular is a highly efficient agent against tuberculosis bacilli, and other tuberculostatics were discovered soon afterwards, e.g. the thiosemicarbazone of acetylaminobenzaldehyde (Conteben, Domagk), p-amino-salicylic acid (PAS, Lehmann), and isonicotinic acid hydrazide (INH) which was discovered at about the same time by Bayer and by Hoffmann de la Roche.

Since 1937, further groups of pharmaceutical compounds have been developed, e.g.

the antihistamines. Many diseases, particularly allergic ones, are produced by the proteinogenic amine histamine which results from the decarboxylation of histidine. The physiological effects of histamine are neutralized by the antihistamines. Other groups to be mentioned here are ganglion-blocking, muscle-relaxing, spasmolytic, sympatholytic, and many others.

Only a little later than the synthesis of pharmaceutical products came the use of vaccines (1890) and serum therapy which are indelibly linked with the names of Robert Koch and E. v. Behring, Robert Koch's pupil. In collaboration with Robert Koch the Farbwerke Hoechst as early as 1892 had started the industrial production of 'Alt-Tuberkulin' which was obtained from tubercle bacilli and later proved suitable only for diagnostic purposes ('tuberculin reaction').

In principle, there are two possible ways of protecting human beings and animals by antibodies indroduced into the blood stream: (1) passive immunization when antibody-containing serum is injected which originates from a diseased organism already immune against the disease (serum therapy); and (2) active immunization when a vaccine is introduced into the organism causing it to form antibodies against the antigen. The vaccinated person, therefore, becomes ill with the infectious disease although in a very mild form. Vaccines are preparations of attenuated living or dead pathogenic agents, i.e. germs, viruses, or their metabolic products (e.g. toxoids). Active immunization is effective for very long periods, sometimes for life (e.g. vaccination against smallpox).

After costly experiments extending over many years had been carried out the first serum found which proved effective was the diphtheria serum.

Vaccines against smallpox were followed by numerous further vaccines, e.g. against typhus, cholera, gas gangrene, tetanus, and recently poliomyelitis and tuberculosis (Calmette vaccination). These sera were produced first at Hoechst, later at the Behring Werke, Marburg, a subsidiary of the Farbwerke Hoechst. In the DDR sera and vaccines are manufactured mainly by the Forschungsinstitut für Impfstoffe Dessau and by the VEB Serumwerk Bernburg.

Following their discovery and production in pure form since about 1930 the therapeutic importance of vitamins has steadily increased. Vitamin A, C, $D_2$ and $D_3$, and $B_1$ are produced synthetically or semi-synthetically. Vitamin $B_{12}$ has recently been obtained as a by-product in the manufacture of streptomycin.

The therapeutic use of hormones is in the midst of an extremely rapid development. The first hormone to be isolated was adrenaline (methylaminoethanolpyrocatechol) which is produced synthetically (Stolz, 1904). From it are derived other cardiovascular substances, e.g. Sympatol. The next hormone to be produced was the thyroid gland hormone the iodine-containing component of which is thyroxin (Kendall, 1914; Harington, 1926).

However, the most important therapeutic discovery of all was that of the hormone of the islands of Langerhans in the pancreas, insulin, which regulates sugar metabolism. V. Mehring and Minkowski showed in 1889 that diabetes is due to the lack of a hormone of the pancreas, and in 1922 Banting and Best succeeded in isolating it. The first producers of insulin in Germany were the Farbwerke Hoechst (since 1936, crystalline insulin).

Steroid hormones (suprarenal cortical and sexual hormones) are obtained by partial synthesis; research on these did not start until 1929.

Of the hormones which have become important therapeutically in recent years we shall only mention the suprarenal cortical hormones. These include gluco-corticoids which affect the sugar metabolism, and mineralo-corticoids regulating the mineral metabolism. The third group are the androgens (sex hormones). Of these, cortisone, prednisone, and prednisolone have proved effective pharmological substances.

Of the hypophysis anterior lobe hormones the adrenocorticotropic hormone (ACTH) is used very widely, particularly in the U.S.A. It causes increased production by the suprarenal cortex of corticosteroids which, like cortisone, have a favourable effect on rheumatic and allergy diseases. Recently, the opinion has been expressed that the anti-rheumatic action of salicylic acid preparations also takes place via the suprarenal cortex.

*Economic aspects*

The production of pharmaceutical products is quantitatively small in comparison with many other products of the chemical industry but is nevertheless very valuable. The development costs of a new preparation are frequently very high. It is often necessary to spend large amounts on animal experiments and klinical tests before a preparation can be put on the market and begins to show a profit. Usually many compounds have to be synthesized and tested pharmacologically first before a suitable one is found, and in many cases the whole expenditure is completely wasted. When Mietzsch gave his first comprehensive report on sulphonamides 1,700 substances had already been produced and tested.

It is, therefore, not surprising that the prices of newly developed medicines are high initially and remain high in the case of those which are not produced in large quantities, which is true of most of them. One has to take into consideration the fact that the production of some pharmaceutical products is unprofitable and is continued by the firms only for reasons of prestige. Other quite substantial items, particularly for pharmaceutical products, are: production in marketable form, packaging, and expert information given to the doctors (propaganda). Suitable presentation is at least as important for pharmaceutical products as for foodstuffs. The psychological effect of presentation and name is important in this field and must not be underrated. The use of unattractive cheap makeshift packaging often has very unfavourable effects. One may look on this as 'imagination' on the part of the patient but it is a fact that such inessential factors may have an effect on the therapeutic results.

Before the 1939–1945 war the economic centre of gravity of the market were synthetic pharmaceutical products. In both production and exports Germany led all other countries. 40 per cent of world trade was in German hands and German pharmaceutical products were everywhere accepted as the best available. As a result of the war most of the export markets were lost. Many of the importing countries became self-sufficient or imported from the U.S.A., Switzerland, or England. Up to the end of the war the most important pharmaceutical factories were in Western Germany so that the DDR was faced with supply difficulties. These were solved by the erection of numerous factories and nowadays some pharmaceutical products are even exported. During the same period the share of the U.S.A. in world trade rose from 14 to 45 per cent, that of Switzerland from 9 to 14 per cent, and the turnover in the U.S.A. rose from \$ 300 million to about \$ 1,500 million. These changes were, however, caused also by a growing shift in the production of pharmaceutical products from synthetic products to antibiotics, vitamins, and hormones, of which antibiotics are the largest by value. From this scientific and technological development which started during the war Germany was completely excluded until the end of the war. In the U.S.A., approximately one quarter to one third of the total trade in pharmaceutical products is in antibiotics, the share of vitamins and hormones together is almost the same. One may assume that more than half the total U.S.A. production of pharmaceutical products by value is obtained by biochemical methods. This development is starting only very slowly in Germany.

## Classification of pharmaceutical products into groups

According to their origin, pharmaceutical products could be divided into three large groups:

(1) Vegetable materials, either concentrated or isolated active substances such as alkaloids, volatile oils, and carbohydrates.

(2) Vitamins, hormones, enzymes, and proteins, produced partly from animal organs and plants or synthetically. This group also includes sera and vaccines.

(3) Synthetic pharmaceutical products.

All three groups include pharmaceutical products of the most varied chemical classes and effects, as well as auxiliary materials for the preparation of pharmaceutical products, so that this classification is purely formal and unsatisfactory.

Pharmaceutical products can be classified more logically according to their pharmacological effects and field of application. The doctor is more familiar with such a classification and he understands it better. If, therefore, one wishes to classify pharmaceutical products by pharmacological principles one has to start from their effect on the functions of the body and its organs. One then arrives at the following scheme into which all pharmaceutical products readily fit (see also the classification in H. P. Kaufmann, *Arzneimittel-Synthese*, Springer, Berlin, Göttingen, Heidelberg, 1953, and Ehrhart in Winnacker-Weingaertner, *Chem. Tech.*, vol. IV, 'Die Arzneimittel', Hanser-Verlag, München, 1954):

I. Agents for the prophylaxis and therapy of infectious diseases
- (a) Disinfectants
- (b) Agents used in chemotherapy
  - (1) Against bacterial infections,
  - (2) Against protozoic infections.

II. Biocatalysts
Vitamins, hormones, enzymes, sera, vaccines.

III. Pharmaco-therapeutical agents
Compounds acting on
- (a) The central nervous system
  (narcotics, hypnotics, antipyretics, analgesics, spasmolytics, anti-arthritics)
- (b) The peripheral nervous system
  (local anaesthetics)
- (c) The autonomic nervous system
  (compounds acting on sympatheticus and parasympatheticus)
- (d) Cardio-vascular system
- (e) Digestion, diuresis.

# 1. AGENTS FOR THE PROPHYLAXIS AND THERAPY OF INFECTIOUS DISEASES

## (a) Disinfectants

Disinfectants are used for the prevention of infectious diseases. They have been known since antiquity. However, full scientific knowledge of the manner of their application and effect was obtained only by Semmelweis, Pasteur, Robert Koch, and Lister. Since then the systematic fight against human and animal epidemics, and asepsis in obstetrics and surgery have become possible. Disinfectants either act as protoplasma toxins or they interfere with the enzyme system of the germs.

The number of disinfectants is extremely high in accordance with their many and varied uses. Here only a few important organic compounds will be mentioned.

Apart from certain alcohols, the most widely used disinfectants are still the phenols, e.g. cresols, often emulsified and reinforced by soap or synthetic detergents. The aim is to render phenols odourless which with cresols and xylenols has been partially achieved by halogenation (e.g. Raschit = chloro-m-cresol; Raluben).

A new method for the production of odourless phenol derivatives consists in the linking of phenol residues by methylene bridges, e.g. hexachlorodihydroxydiphenylmethane ('Hexachlorophene'), a component of many germicidal ointments and bactericidal or deodorant soaps.

Of the polyphenols, resorcinol, used widely in dermatology, is distinguished by its non-toxicity and negligible corrosive action.

According to recent findings, salicylic acid anilide and substituted salicylic acid amides (Multifungin, Jadid, Actol) are very effective germicides and fungicides.

In addition to phenols a number of nitrogen-containing compounds are also important disinfectants. Thus, 8-hydroxyquinoline sulphate is the active substance of 'Chinosol' preparations; halogenated 8-hydroxyquinolines are also recommended for the disinfection of wounds (Yatren, Vioform). Alkyl-dimethyl-benzylammonium salts with a long alkyl residue (e.g. Zephirol) are both surface active and germicidal. For large-scale disinfection, compounds with active chlorine atoms are used with the grouping $R{-}SO_2{-}\underset{\substack{|\\ Na}}{N}Cl$, e.g. the sodium salt of p-toluenesulphamide chloride (Chloramine). For the disinfection of the skin and mucosa the yellow pigment trypaflavine (Panflavin) is on the market:

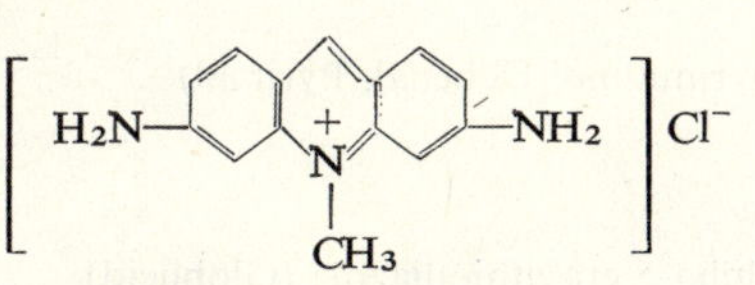

Trypaflavine

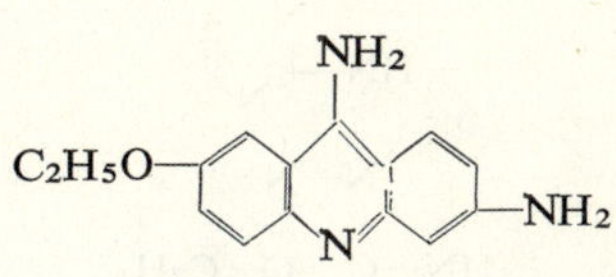

Rivanol

The intestinal disinfectant Rivanol is also an acridine derivate and belongs to the disinfectants which have a localized effect in the body (e.g. in the intestine).

The synthesis of griseofulvin, 7-chloro-2',4,6-trimethoxy-6'-methyl-gris-2'-en-3,4'-dione

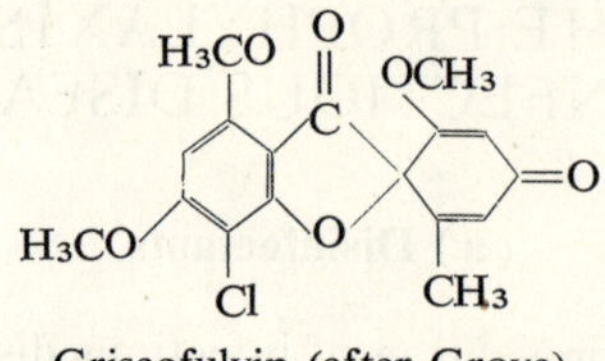

Griseofulvin (after Grove)

(e.g. Likuden, Hoechst) made possible for the first time oral treatment of diseases caused by skin fungi (mycosis). It increases the resistance of the epidermis, nails and hair to pathogenic skin fungi.

The germicidal effect of tetrahydrothiadiazine thiones (Afungin, Fungiplex) which are effective on a wide front against germs, yeasts, and fungi, is due to the cleavage of mustard oil.

### (b) Products used in chemotherapy

#### (1) Against bacterial infections

*Sulphonamides*

The first preparation which permitted an attack on bacterial infections in the body to be carried out in the blood was the red azo dye Prontosil rubrum. It is produced by diazotization of p-aminobenzenesulphamide followed by coupling with m-phenylenediamine.

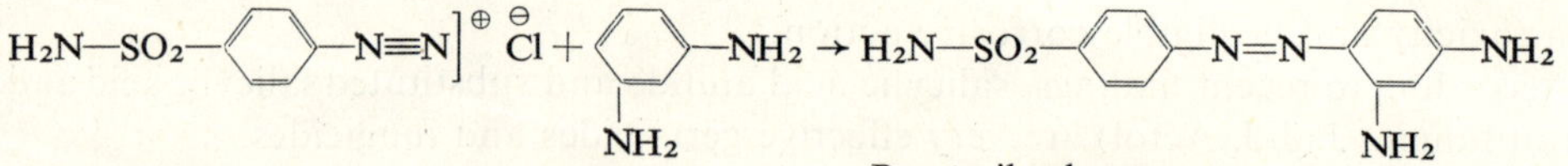

Prontosil rubrum

The realization that the effect of Prontosil was essentially linked to the sulphanilamide grouping led to the synthesis of a large number of sulphonamides; the chemotherapeutical cone of dispersion widened considerably as the sulphonamide group was substituted by heterocyclic residues, e.g.:

(general formula)

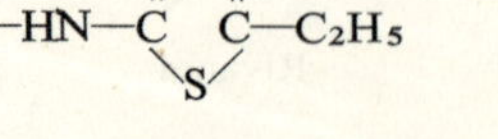

sulphathiazole (Eleudron, Cibazol)

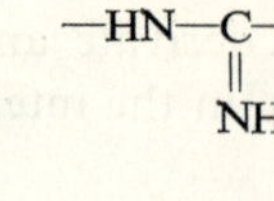

sulphapyrimidine (Debenal, Pyrimal)



sulpha-5-ethylthiodiazole (Globucid)



sulphaguanidine (Resulfon)

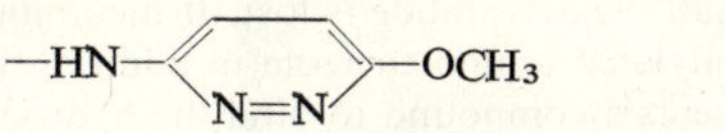

Lederkyn (long-acting sulphonamide) (Depovernile)

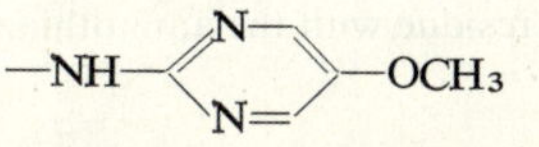

sulphomethoxypyrimidine, (Durenat) (long-acting sulphonamide)

Whilst Prontosil is effective only against streptococci and staphylococci, sulphapyridine is also effective against gonococci, meningococci, and pneumococci, and sulphathiazole additionally against bacterial dysentery. The syntheses were designed to widen progressively the spectrum of effectiveness against bacteria and this has now been achieved to the highest degree by mixtures of various sulphonamides. Euvernil (sulphurea) is used particularly for the disinfection of the urinary passages.

### *Production of sulphathiazole*

The production of sulphathiazole which is still one of the most important members of the sulphonamide group will be discussed below as a typical case.

The raw material for many sulphonamides is N-acetylsulphanilyl chloride. It is obtained by introducing acetanilide into excess chlorosulphonic acid. During this reaction some 4,4'-diacetaminodiphenylsulphone is also produced. Ice is added to the reaction mixture and the product is aqueous N-acetylsulphanilyl chloride. Since at first it was used for the production of Prontosil it was given the technical name 'Prontylchloride'. It decomposes rather easily in a moist condition, and has to be processed immediately.

The second component, 2-aminothiazole, is produced from thiourea and chloroacetaldehyde or $\alpha$, $\beta$-dichloroethylether, which reacts like chloroacetaldehyde and is obtained by photo-chlorination of diethylether. 1,2-dichloroethylacetate (obtained by chlorination of vinylacetate) can also be used. 2-aminothiazole is reacted with prontylchloride in an aqueous suspension by the addition of soda. During the reaction, however, 2 prontyl residues enter the aminothiazole. A residue is cleaved off by saponification with a weak solution of caustic soda, the acetamide group being saponified with a stronger one, and sulphathiazole is obtained as a soluble Na salt which is precipitated as free sulphathiazole by neutralization with acid. The synthesis is illustrated by the following schematic representation of the quations:

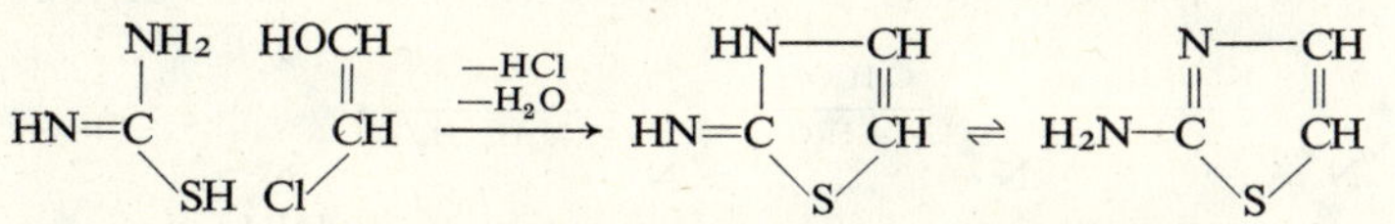

thiourea chloroacetaldehyde 2-aminothiazole (imino-form) (amino-form)

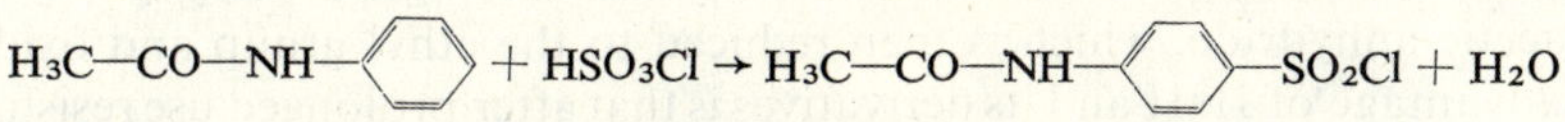

p-acetaminobenzene sulphochloride ('Prontylchloride')

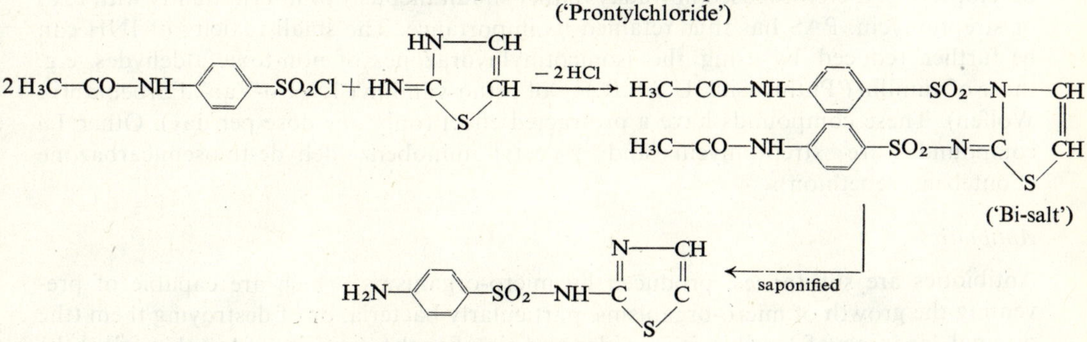

('Bi-salt')

sulphathiazole (*Eleudron*)

Sulphanilic acid is also produced so that in this synthesis half the acetanilide is lost. If the amount of prontylchloride is reduced one obtains nevertheless disulphonylated amino thiazole in addition to unreacted aminothiazole. Using an organic solvent and pyridine as a compound to bind the hydrochloric acid it is possible to achieve the reaction of only one prontyl residue with the aminothiazole.

*Tuberculostatics (TB compounds)*

It has at last become possible to fight tuberculosis in the host organism with bacteriostatic agents (tuberculostatics). An important compound is p-aminosalicylic acid (PAS), evolved by Lehmann, which is used in very large quantities in the form of its Na- or the more compatible Ca salt. It is produced by carboxylation of m-aminophenol with $CO_2$ under pressure. The addition of the COOH group takes place so easily that it is possible to work in aqueous solution. Fusion of metanilic acid with alkali produces m-aminophenol.

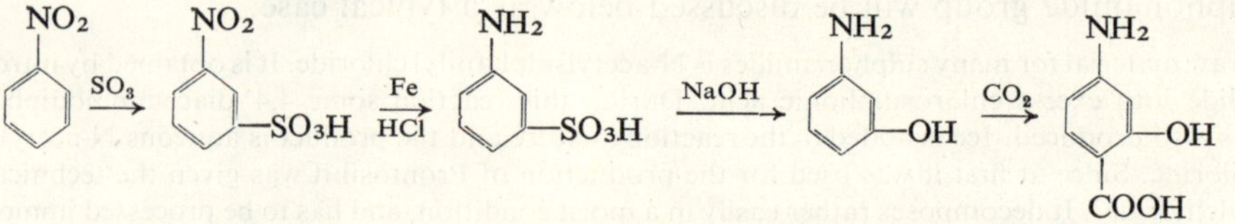


PAS

Recently much more effective tuberculostatics than PAS have been found which are derived from isonicotinic acid. Neoteben (Bayer) and Rimifon (Geigy) are isonicotinic acid hydrazide (INH). It can be obtained from $\gamma$-picoline (from tar) by oxidation with potassium permanganate, subsequent esterification with ethanol, and reaction with hydrazine hydrate.

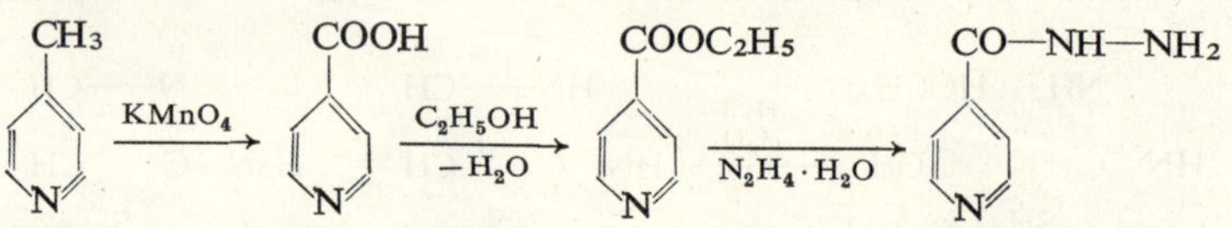


INH

One can also use pyridine as a raw material by introducing a p-acetyl group using zinc dust and acetic anhydride, which is then reduced to the ethyl group and oxidized.

The disadvantage of INH and its derivatives is that after prolonged use resistant strains develops. It is, therefore, used nowadays either simultaneously or intermittently with PAS or streptomycin. PAS has thus retained its importance. The small toxicity of INH can be further reduced by using the isonicotinylhydrazones of non-toxic aldehydes, e.g. those of vanillin ('Phthivasid', U.S.S.R.) or of ortho-vanillin ('INH-o-Van', Farbenfabrik Wolfen). These compounds have a protracted effect (only one dose per day). Other Tb compounds are streptomycin and p-acetyl-aminobenzaldehyde-thiosemicarbazone (Conteben, Tebethion).

*Antibiotics*

Antibiotics are substances, produced by micro-organisms, which are capable of preventing the growth of micro-organisms, particularly bacteria, or of destroying them (the original meaning of 'antibiosis' is wider and signifies the opposite of 'symbiosis'). Like sulphonamides many antibiotics have a selective effect on certain micro-organisms. This is called 'specific antimicrobial spectrum'. Other antibiotics have a very wide effective spectrum stretching from bacteria via the rickettsia to the large viruses.

In order to achieve specificity for certain groups of pathogenic agents it is often necessary to use several antibiotics, or an antibiotic simultaneously with sulphonamides. From the therapeutic point of view it is not always advisable to use single substances or mixtures with the widest possible effectiveness to fight bacterial infections in the body because they also largely destroy its natural bacterial flora. The latest investigations have shown that this may lead to too great an increase of fungi in various organs and consequently to serious damage. The most perfect form of chemotherapy when using sulphonamides and antibiotics is, therefore, the concentrated attack on the pathogenic germs with large doses (shock therapy). The large number of preparations is, therefore, fully justified.

This multiplicity is also desirable for the following reason: A new problem in the treatment of bacterial infections arises from the fact that when bacteriostatic agents are used repeatedly, gradually only varieties of the bacterial strain resistant to the agent develop. As a result the effect of the agent is weakened, or completely eliminated. Success can be achieved by changing the agent, e.g. from sulphonamides to an antibiotic, or by the combined use of various preparations. For it is rather unlikely that in one strain of bacteria mutations should occur accidentally which are resistant to two bacteriostatic substances. A condition is, however, that they attack at different stages in the course of the vital functions of the organism.

The first antibiotic used therapeutically was penicillin. Fleming discovered it when he observed that certain types of germs did not develop in the neighbourhood of colonies of the mould *Penicillium notatum*.

Since then systematic investigations have begun in many countries into all available types of fungi, bacteria, and many lower plants to discover whether they contained antibiotic substances; many such substances have in fact been found. A large number of these is already being produced industrially and used in therapy.

The search for micro-organisms which produce active substances, and the breeding of such varieties which give the highest yield have become routine methods in many research laboratories.

As raw material they use soil samples 1 g of which often contains several milliards of germs. The soil to be investigated is 'planted' on agar plates inoculated with test germs. After incubation one can then see the antibiotic effect of individual soil germs from the retardation of the growth of the test germ in their vicinity. These germs are isolated and cultured in pure form in a lengthy selection process. Many micro-organisms produce not one but frequently several active substances.

A number of different types of penicillin were found in this way and are available commercially. They all have the same basic structure and are called penicillin I–IV in England and F, G, X, and K in the U.S.A.

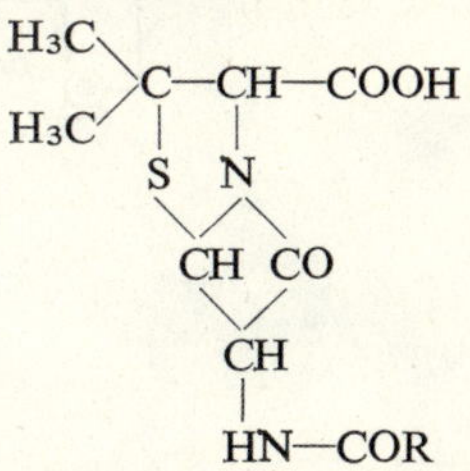

Basic structure of penicillin

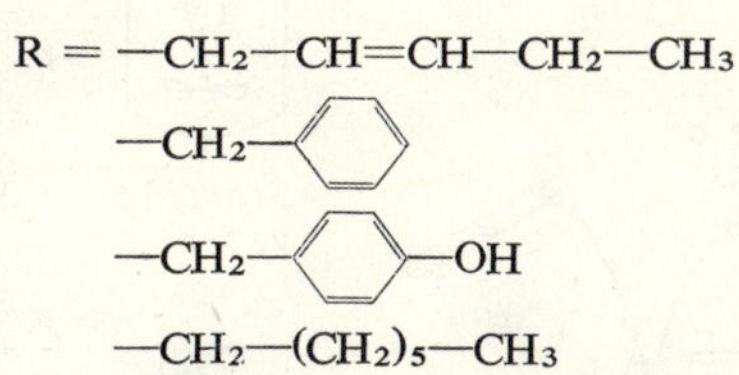

(I, F)
(II, G)
(III, X)
(IV, K)

Penicillin I-IV or F, G, X, K

The most important of the penicillins is benzylpenicillin (II, G).

Penicillin is effective against most pathogenic bacteria which respond also to sulphonamides. It retards the growth of gonococci, Gram-positive bacteria and Gram-negative cocci even in dilutions of 1:50,000,000 though not that of Gram-negative bacteria. Staining by Gram's method allows the classification of bacteria in two groups: 'Gram-positive' germs bind aniline dyes, with subsequent iodine treatment, so firmly that they cannot be washed out with alcohol, 'Gram-negative' ones release the dye. Gram-positive types are e.g. staphylococci, streptococci, diphtheria bacilli, and actinomycetes, Gram-negative types are gonococci, meningococci, and coli, dysentery and typhus germs. Unlike sulphonamides it is also effective against syphilis, relapsing fever, Weil's disease, etc. It is ineffective against virus infections for which so far no effective chemotherapeutic compound exists. Aureomycin and chloramphenicol are reputed to have some effect on viruses.

## *Production of penicillin*

The production of penicillin includes the following manufacturing stages (see Figure 146, after G. Ehrhart, in Winnacker-Weingaertner, *Chem. Technologie*). Other antibiotics are produced in a similar manner.

### Pure culture

In the pure culture laboratory the strain of fungi is bred continuously in pure culture and maintained at its highest efficiency. In the course of many inoculations of this strain the tendency to form antibiotics decreases as a result of mutations. The most effective mutants have, therefore, to be continuously 'worked up' by choosing suitable conditions

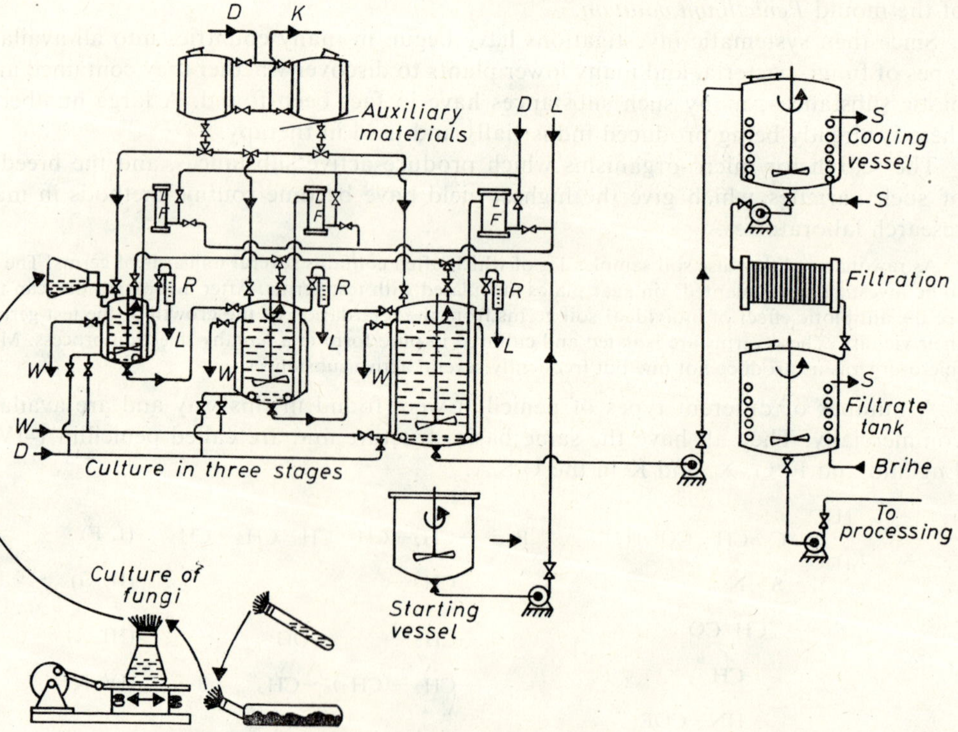


Figure 146. Culture installation for the production of penicillin
*D* steam *K* condensate *L* air *LF* air filter *R* rotameter for measuring air flow *S* cooling brine *W* cooling water

for the culture. The nature of the culture medium is of great importance. It is possible, for instance, to increase the yield of penicillin by several hundred per cent by using maize liquor, a by-product in the manufacture of corn flour, as a culture medium. Similarly, the addition of phenyl acetic acid during the fermentation process causes an increase in the yield of benzyl penicillin (G) since phenyl acetic acid is one of the units in the penicillin G molecule.

The pure cultures of the fungi are best stored in permanent form (spores), mixed with sterile soil. For breeding the vegetative form they are inoculated into an agar medium and incubated at 24°C. After 1–2 days the mycelium forms and a few days later develops into a green culture bearing conidia. The young spores are then transferred into Erlenmeyer flasks containing a nutrient solution and are processed further under continuous agitation at 24°C in sterile air. During this treatment more spores start forming briskly. The contents of the flask are a preliminary culture for the fermentation.

Fermentation, culture in three stages

Formerly the cultures were grown as surface cultures in many flasks with flat bottoms standing in heated chambers. This method involved a great deal of manual work and has now been replaced entirely by submerged cultures in aerated tanks. When the culture is agitated and strongly aerated the mycelium also grows below the surface and is dispersed throughout the whole liquid. Technical development, therefore, took a course similar to the microbiological manufacture of protein (see p. 302), except that the latter is carried out continuously as a flow process without using preliminary cultures. The manufacture of antibiotics has not as yet progressed as far as this. (Even in industrial bacterial fermentations fully continuous methods are not yet used; the process is always started afresh in several stages from pure cultures, see pp. 308, 335). The aeration problems are the same as in the yeast culture processes which are already being applied on a very large scale but complete sterilization with exclusion of external air is necessary. Figure 147 shows a schematic sketch of a fermenter (after Ehrhart in Winnacker-Weingaertner) three sizes of which are used in the culture process as preliminary, intermediate, and production fermenters respectively; the quantity of culture liquor is multiplied by about ten from stage to stage.

The simple arrangement using an agitator and perforated jet pipe has been replaced in many modern plants by special designs by Stich, Lurgi, MAN, Waldhof, and others. All important accessories of the fermenter are shown in Figure 147.

The nutrient solution contains sugar and maize steeping liquor which supplies the carbohydrates, nitrogen, and salts; it also contains nutrient salts such as potassium phosphate, nitrate, magnesium sulphate, zinc sulphate and Ca salts. It is mixed in special feed vessels and sterilized there or in the fermenter itself at 120°C under pressure. All inlets and outlets have to be kept sterile by steam or electric heating. All external infection must be avoided and the ventilation air from the fans has, therefore, also to be rendered germ-free by heating, filtration, or disinfectants.

To control foaming, anti-foam agents such as fats and oils, silicones or synthetic special antifoaming agents are used. Since these may, however, have a deleterious effect it is better to destroy the foam mechanically in the fermenter itself. 30–60 hours after the introduction of the preliminary culture a mycelium paste forms in the preliminary

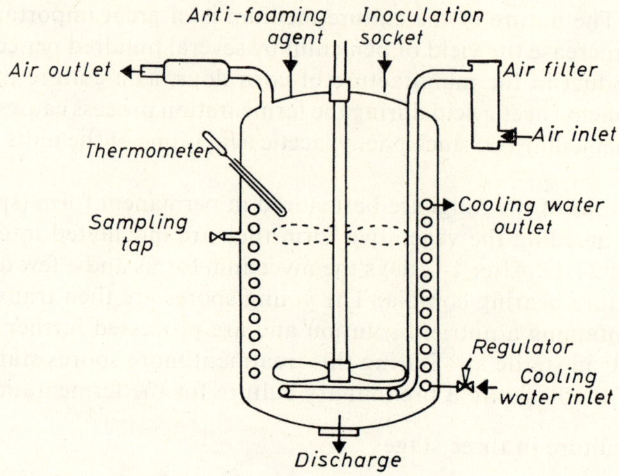


Figure 147. Sketch of a fermenter

fermenter at 25°C which is then forced through a pipe system, sterilized by means of steam, into the intermediary fermenter containing the nutrient solution. After 14–24 hours this intermediate culture is ready for the inoculation of the production fermenter.

Isolation and purification of penicillin

Penicillin is contained in the nutrient fluid in a concentration of thousandths of one per cent. The contents of the production fermenter pass into a cooling tank where they are cooled to +2°C; the process is then stopped. Then the fungi paste is filtered on filter presses, pressure filters, or rotary filters. The filtrate is acidified with hydrochloric acid to pH 2 to 2·5 so as to liberate the salt-bound penicillin and thus render it extractable. Possible extracting agents are acetic ester, butyl acetate, chloroform, and other solvents. Extraction is carried out on the counterflow principle using one of the continuous methods described on p. 59. The penicillin is converted back into the liquid phase from the organic solvent as a Na-salt by means of a buffer solution (approximately a fifty-fold concentration as compared with the culture filtrate). From the aqueous solution the penicillin can either be separated in crystalline form as a salt using suitable bases, or be extracted again with organic solvents and then precipitated with bases. The final product is produced from these stable penicillin salts.

Like all solutions used for injections, penicillin solution, too, has to be free of live or dead germs and organic impurities to prevent undesirable side effects in the patient. These happened frequently with the first penicillin preparations that came on to the market. Sterile solutions are obtained by filtration through bacterial filters and processing in rooms sterilized with ultra-violet rays or by aerosols and ventilated using germ-free air.

The activity of antibiotics is checked by biological testing methods which are used in addition to chemical determination methods. The active substance in a certain concentration and for a fixed number of germs has to have a constant antibiotic effect when compared with a test organism.

*Streptomycin and other antibiotics*

After penicillin, streptomycin is the most interesting and most thoroughly investigated antibiotic; it was isolated from *Streptomyces griseus* in 1943 by Waksman, Bugie and

Schatz. Unlike penicillin it is effective also against Gram-negative and acid-resistant bacteria and notably against tubercle bacteria.

Chemically, streptomycin is a trisaccharide, the constitution of which was established after very difficult investigations carried out jointly by several research institutes. Its industrial production is particularly difficult because of the very toxic impurities adhering to it. In the DDR, streptomycin is produced by the VEB Jenapharm. To isolate streptomycin from the nutrient fluid the mycelium is filtered and adsorbed on suspended carbon. The carbon adsorbate is filtered, washed with alcohol (to remove impurities) and then the streptomycin is eluted with hydrochloric acid and alcohol. Ion-exchange resins have also been used successfully for isolation purposes.

Other important antibiotics used in therapeutics are hydroxytetracycline (Terramycin), chlortetracycline (Aureomycin) which have very broad effective spectra, and chloromycetin which is marketed also under the names of Chloramphenicol and Chloronitrin. The latter is effective also against typhus bacteria and is now produced synthetically. Griseofulvin, which can be obtained from several types of penicillium, has already been mentioned on p. 462 (peroral agent against skin mycoses). Effective polypeptides isolated from bacteria are Bacitracin, Thyrotricin, and Gramicidin.

With antibiotics we are only at the threshold of great developments. As happened in the case of the discovery of the alkaloids new antibiotics are constantly being discovered. Bacteria, fungi, lower plants, lichens, mosses, and higher plants are used as raw materials. The presence of antibiotics is first recognized in a preliminary test, they are then isolated and produced in pure form. Then follows the pharmacological investigation and elucidation of the structure. Synthesis, or at least partial synthesis, of many antibiotics should become possible one day.

### (2) Chemotherapy of infections caused by spirocheetes and protozoa

*Arsenic preparations, Salvarsan*

Chemotherapy began with the introduction of Salvarsan. The first precursor of the group of effective arsenic compounds was Atoxyl which had a curative effect on experimental infections by trypanosoma and spirochaetes (e.g. syphilis). Robert Koch had also treated sleeping sickness with Atoxyl, but this frequently led to blindness. The 'chemotherapeutical index', the ratio of curative dose to toxic dose, was too unfavourable. By way of example, in rabbit lues the index for mercury is 1:1 to 1:2, for Salvarsan 1:20 to 1:30, i.e. the 'therapeutic width' of Salvarsan is about 20 times larger.

Acetylatoxyl is much less toxic. After Ehrlich had succeeded in elucidating the structure of Atoxyl which originally had been assigned a wrong formula he eventually arrived via other arsenic compounds at Salvarsan, dihydroxydiaminoarsenobenzene.

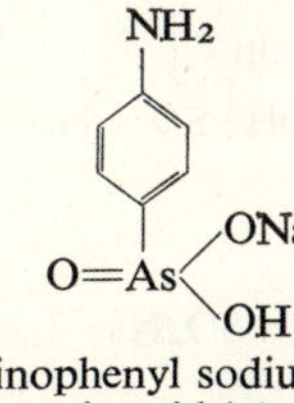


p-aminophenyl sodium salt of arsonic acid '*Atoxyl*'

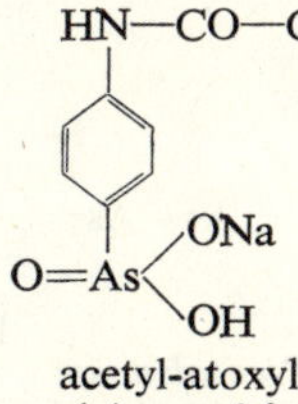


acetyl-atoxyl '*Arsacetin*'

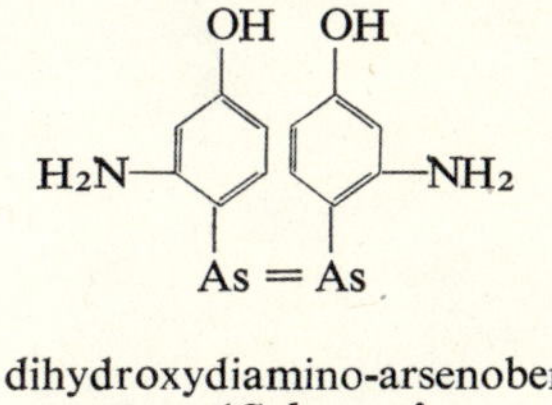


dihydroxydiamino-arsenobenzene '*Salvarsan*'

Since Salvarsan sometimes created problems in use because it was difficult to dissolve and oxidized easily, Ehrlich introduced into one of the two amino groups a —$CH_2$—$SO_2Na$-group via formaldehyde sulphoxylate (Neosalvarsan). In Neosalvarsan there is, moreover, a second Rongalit ($H_2C \cdot OH \cdot SO_2Na$) molecule loosely bound. Salvarsan was followed by a whole series of arsonic acids, arsine oxides, and arsenobenzenes of which only Spirocid (Stovarsol), Tryparsamide, Mapharsen, Solusalvarsan, and Spirotrypan shall be mentioned.

*Production of Salvarsan*

The raw material for Salvarsan is 3-amino-4-hydroxybenzenearsonic acid which is obtained by the nitration of 4-hydroxybenzenearsonic acid and its subsequent reduction. 4-hydroxybenzenearsonic acid can be produced from phenol by heating with arsenic acid to 150°C, or from arsanilic acid (p-aminophenylarsonic acid) by diazotization and subsequent boiling. For the nitration, 4-hydroxyphenylarsonic acid is mixed with concentrated sulphuric acid and the calculated quantity of nitrating acid is added slowly with constant stirring at a temperature below 0°C. The mixture is left standing for several hours below 10°C and is then poured into water from which 3-nitro-4-hydroxybenzenearsonic acid gradually precipitates.

By using a mild reducing agent, e.g. sodium amalgam, only the nitro group is reduced, and one obtains 3-amino-4-hydroxybenzenearsonic acid. Various strong reducing agents are suitable for the reduction of the arsonic acid group to the arsenical compound. Arsine oxide occurs as an intermediate stage which under certain conditions it is possible to fix (Mapharsen). The reduction is carried out by rapid heating, with direct steam to 67°C, of a mixture of 3-amino-4-hydroxybenzenearsonic acid, sodium dithionite ($Na_2S_2O_4$), bisulphite, and caustic soda solution. The Salvarsan base is drawn off by suction, dissolved again in methanol and, after addition of alcoholic hydrochloric acid until the Congo acid reaction occurs, is precipitated as the dihydrochloride with a large quantity of ether, this is drawn off by suction, and the yellow powder obtained is then dried in a vacuum at 35°C. Being an o-aminophenol derivative, Salvarsan is very sensitive to oxidation.

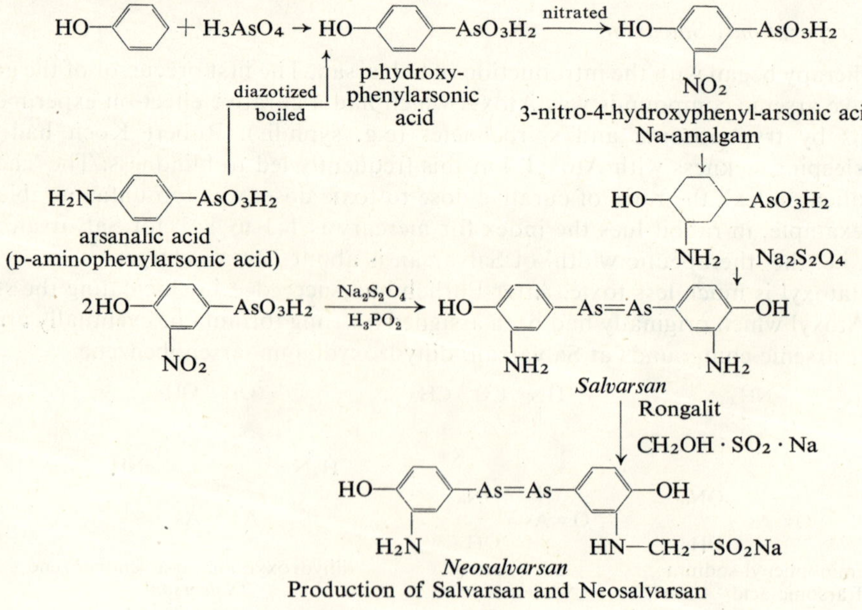


Production of Salvarsan and Neosalvarsan

It is marketed in solution as the dihydrochloride in ampoules filled with inert gas. Before the injection is given the hydrochloric acid has to be neutralized by the appropriate quantity of caustic soda solution.

Salvarsan is rendered less easily oxidizable by the introduction of reducing groups into the molecule. To produce Neosalvarsan a solution of Salvarsan in methanol and water is mixed with Rongalit in silver equipment and the precipitate formed is dissolved in soda.

Neosalvarsan is precipitated as a yellow powder with an alcohol-ether mixture. After suction, filtration and drying a further precipitation from water is carried out. Neosalvarsan can be dissolved in sterilized water and injected directly.

*Antimony compounds*

The therapeutic use of antimony was disputed for centuries because of its serious side effects. However, great success was achieved by H. Schmidt and his colleagues at Bayer's when they produced only slightly toxic benzene-derivatives of antimony. Neostibosan and Solustibosan are high-molecular complex compounds with a very complicated structure containing several p-aminophenylstibinic acid residues

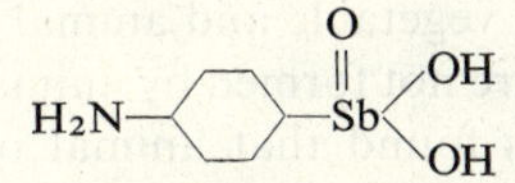


which proved very effective against tropic protozoa infections (Kala-azar, leishmaniasis).

*Antimalaria agents*

This group of chemotherapeutical compounds includes agents effective against malaria infection, the most widespread protozoic disease on earth (almost one third of mankind was at one time infected). It is transmitted by the female anopheles fly which injects the parasites into the human blood stream where they can propagate asexually (sporogony). Pharmacological research has succeeded in designing synthetic agents, in addition to the long-known quinine, which attack the development cycle of the malaria parasite at different points. Of the three main development phases, sporozoites, schizonts, and gametes, only the two latter phases have so far proved susceptible. Quinine, cinchona alkaloids, and Atebrin are effective against schizonts, Plasmochin, Certuna, and Resochin against gametes. Atebrin (Mietzsch, Maus, and Kikuth) and Plasmochin (Schulemann, Schönhöfer, Wingler, and Roehl) are amongst the most important contributions of German pharmaceutical research to the health and welfare of mankind. They are worthy successors to Germanin, Bayer 205, the sleeping sickness remedy (Heymann, 1921).

Germanin was developed by systematic changes of the structure of azo dyes effective against trypanosoma infections. Atebrin is a logical development of acridine dyes utilizing the experiences gained in the synthesis of quinine and Plasmochin. For the synthesis of Plasmochin, starting from quinine, the quinoline nucleus was used as a base. It could be considered as a synthetic alkaloid.

Atebrin is obtained from 2,4-dichlorobenzoic acid by reaction with p-anisidine in the presence of alkali, subsequent cyclization to the acridine derivative with $POCl_3$, and finally reaction with 1-diethylamino-4-amino-pentane.

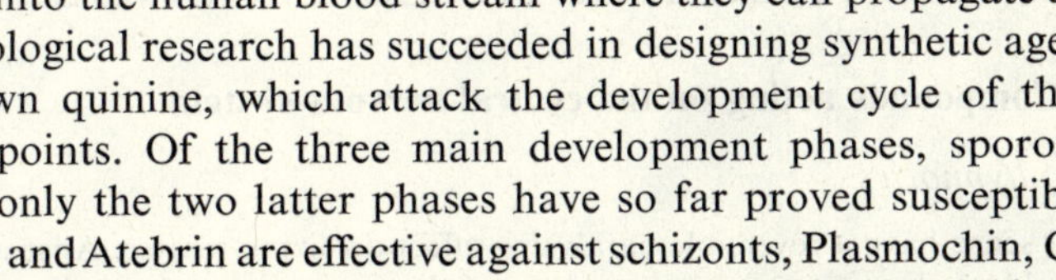


Germanin, Bayer 205

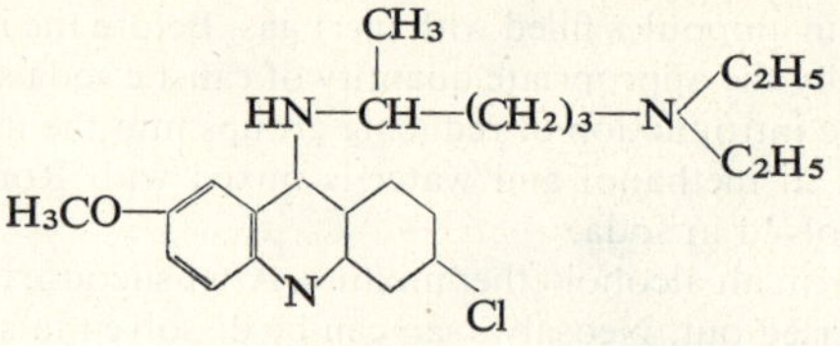

Atebrin

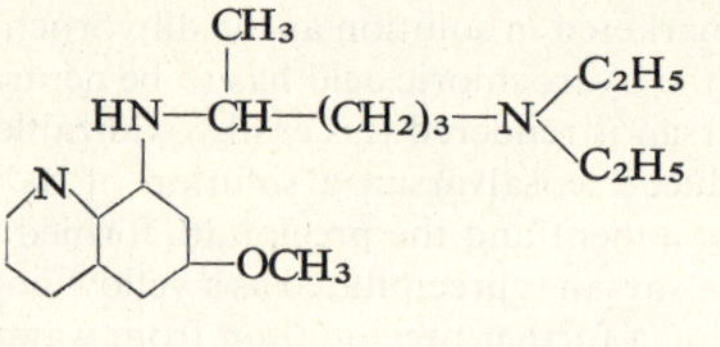

Plasmochin

## 2. BIOCATALYSTS

The infinitely wide field of vitamins, hormones, enzymes, sera, and vaccines belonging to this groups cannot be discussed here.

Only some of these biocatalysts are produced by synthesis, most of these are isolated from natural materials of vegetable and animal origin. Formerly it was assumed that, unlike hormones, vitamins are not formed by animal organisms. However, this has proved to be incorrect since it was found that animal organisms can form Vitamin A and Vitamin D from preliminary stages, and that many animals can themselves build Vitamin C. To produce vitamins one often uses complicated synthetic methods, and also industrialmicrobiological ones; biosynthetical processes are used in some of the stages in the manufacture of hormones.

## 3. PHARMACOTHERAPEUTIC COMPOUNDS

### (a) Compounds acting on the central nervous system

*Narcotics, anaesthetics, hypnotics*

Anaesthetics in the general sense have a reversible paralysing effect on living matter. Anaesthetics that can be used in therapy should produce a temporary paralysis of the sense organs, loss of consciousness, loss of sensitivity and of the reflexes of the central nervous system. The attack takes place preferentially via the lipoid-containing organs, and the period during which an anaesthetic is effective is related to the dispersion coefficient in lipoids and water. Chloroform, because of its poor solubility in water, has, therefore, a long-lasting effect.

Anaesthetics are divided into inhalation anaesthetics such as ether, chloroform, ethyl chloride, and trichloroethylene, gas anaesthetics such as nitrous oxide, acetylene, and cyclopropane, and finally, non-inhalation anaesthetics which are administered per os, rectally, or intravenously. Among the latter are compounds used for basal anaesthesia, i.e. the anaesthesia applied before full anaesthesia to quieten the patient and to save anaesthetics. Avertin (tribromoethylalcohol, $CBr_3 \cdot CH_2OH$) is used for this purpose; it is produced by Meerwein's method from tribromoacetaldehyde by reduction with aluminium alcoholate. Formerly, it was produced by the Willstätter process from bromal by biological reduction using growing yeast.

Evipan-sodium which actually belongs to the group of hypnotics is used for short-duration anaesthesia and basal anaesthesia by injection.

Since sleep is an early symptom of anaesthesia it can also be induced by small quantities of anaesthetics but their effect is too short and they are too toxic. Hypnotics should have a mild but longer-lasting effect on the cerebral cortex. They are divided into sleep-inducing, medium-term, and long-acting hypnotics. They belong to a variety of organic compounds.

The oldest and mildest hypnotic is alcohol taken in moderate doses. The first synthetic hypnotic was chloral hydrate $Cl_3C—CH(OH)_2$. A harmless hypnotic is paraldehyde. The hypnotic effect is usually reinforced by the introduction of halogen into alcohols or aldehydes. Apart from some carbamic acid derivatives (Urethan, Voluntal) which are no longer in general use, most hypnotics are derived from urea, barbituric acid, and hydantoin. These include e.g. bromisovaleryl urea (Bromural)

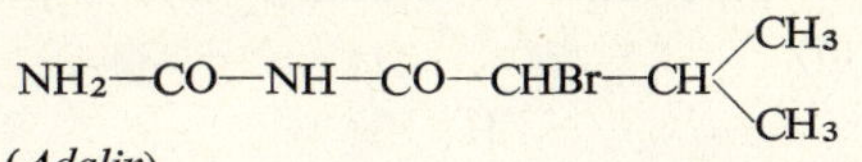

and bromodiethylacetyl urea (*Adalin*)

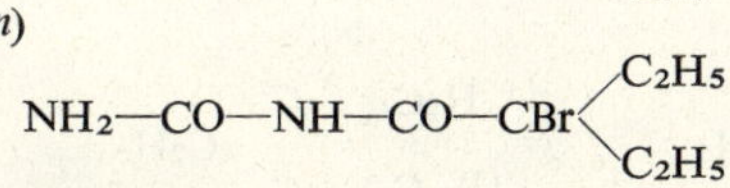

Adalin can be produced from diethylmalonic acid by decarboxylation to diethylacetic acid which is brominated with bromine or phosphorus tribromide to diethyl-bromoacetylbromide. The material is then reacted with urea:

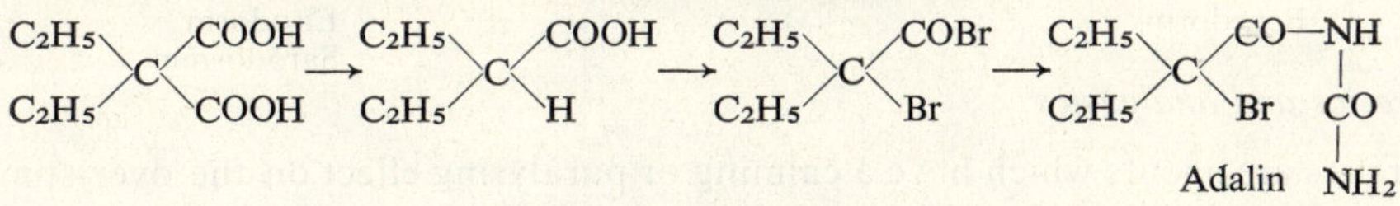

The most widely used hypnotics contain the malonyl urea skeleton (barbituric acid). Only barbituric acids substituted in the 5 position are effective; it will be seen that their structure resembles that of Adalin (trisubstituted acetylurea) described above. The oldest preparation of this kind is Veronal (E. Fischer). The method devised by E. Fischer is still the most important and the simplest for the industrial manufacture of the 5-substituted barbituric acids. Intermediate products are the appropriate malonic esters which can be obtained from the sodium compound of malonic ester with aklyl halide. The route to many barbiturates is as follows:

chloroacetic acid $\xrightarrow{+KCN}$ cyanoacetic acid → malonic acid → Na-malonic ester → alkyl malonic ester $\xrightarrow{+urea}$ alkyl barbiturate

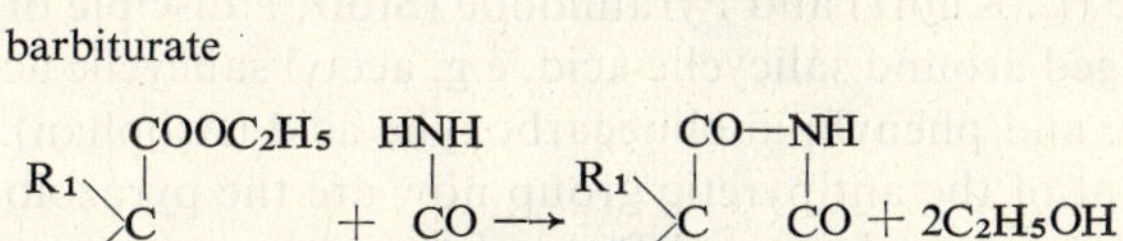

Malonic acid diamides, esteramides and chlorides or dichlorides can also be condensed with urea as intermediate products.

By varying the radical $R_1$ and $R_2$ hypnotics with very diverse pharmocological effects are obtained. In the DDR barbituric acid derivatives are produced by the VEB Arzneimittelwerk Dresden, in the Federal Republic, e.g. Bayer, Leverkusen, and other firms.

$R_1$ and $R_2$ = diethyl: Veronal, Ethylbarbital,
ethyl, phenyl: Luminal, Lepinal, Phenylbarbital,
cyclohexenyl, ethyl: Phanodorm,

cyclohexenyl, methyl, N-methyl: Evipan, Hexabarbital,
ethyl, crotyl-: Kalypnon,
bromallyl, isopropyl: Noctal.

Since hypnotics containing barbituric acid often have undesirable side effects and other limitations, sedatives and hypnotics are now being produced which belong neither to the ureides nor to the barbiturates, are very effective and have little toxicity. These include e.g. 3,3-diethyl-2,4-dihydroxy-tetrahydropyridine (Persedon, Benedorm), 3,3-diethyl-2,4-dihydroxy-5-methylpiperidine (Noludar), and α-phenyl-α-ethylglutaric acid amide (Doriden, Elrodorm).

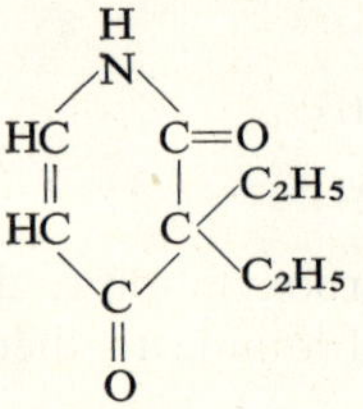

Persedon
Benedorm

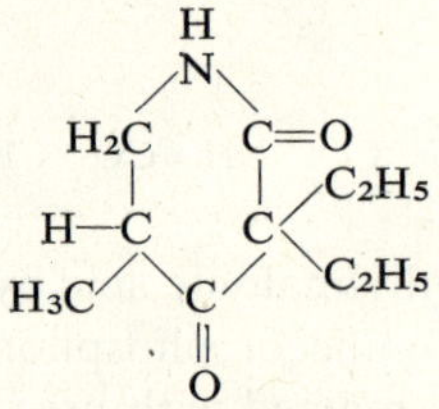

Noludar

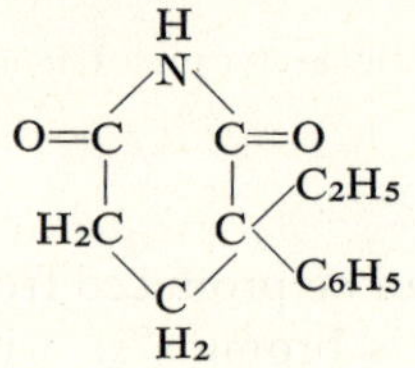

Doriden
Elrodorm
Sarodormin

*Antipyretics and analgesics*

Antipyretics are agents which have a calming or paralysing effect on the over-stimulated temperature centre in the brain. Since their effect is not strictly specific, a narcotic effect is exercised also on the areas adjacent to the temperature centre which explains why they act both as antipyretics and as analgesics.

As body temperature is a resultant of heat formation and heat loss, lowering of the temperature can be initiated either by antipyretics reducing the formation of heat (quinine), by restricting metabolism, or by increasing the heat loss by increasing evaporation through the skin and respiration (e.g. salicylic acid, Antipyrine). Formerly, the therapeutic significance of the antipyretic effect was greatly overrated. Now, antipyretics have retained their importance mainly because they are also analgesics.

Compounds from very different classes of materials have an analgesic effect. Development proceeded via Acetanilide (now without much importance), Phenacetin and Lactophenin to Antipyrine (L. Knorr) and Pyramidone (Stolz, a disciple of Knorr's, Filehne). Another series is ranged around salicyclic acid, e.g. acetyl salicyclic acid (Aspirin), acetyl salicyclic acid amide, and phenylquinolinecarboxylic acid (Atophan).

The most important of the antipyretic group now are the pyrazolone derivatives, e.g. Antipyrine, Pyramidone, Novalgin, and Butazolidin.

*Manufacture of Antipyrine and Pyramidone (aminophenazone)*

1-phenyl-3-methylpyrazolone is an important intermediate for the manufacture of both dyes (e.g. pyrazolone dyes as pigments; pyrazolones in the 4 position can be coupled with diazotized bases) and Antipyrine and Pyramidone.

1-phenyl-3-methylpyrazolone is produced from phenylhydrazine and acetoacetic ester. Formerly, acetoacetic ester was obtained by the usual synthesis from ethyl acetate, sodium, and alcohol. The sodium alcoholate required was later produced from an alcoholic solution of caustic soda by azeotropic dehydration with benzene, using equipment such as is used for azeotropic esterification (former I. G. Hoechst). More recently, diketene, which can be obtained easily from acetic acid or from acetone, or ts derivative acetoacetic amide are used in place of acetoacetic ester for the pyrazolone synthesis.

Phenylhydrazine is reacted with acetoacetic ester containing about 10 per cent of 85 per cent alcohol, first in the cold with agitation, followed by boiling. After cooling 1-phenyl-3-methylpyrazolone crystallizes out, is centrifuged off and washed. Methylation in the 2 position can be carried out with dimethyl sulphate in an aqueous methanol solution, in a xylene solution, or with methyl chloride in the presence of methanol at 100°C and at a gauge pressure of 10 atm. The crude antipyrine is dissolved and recrystallized.

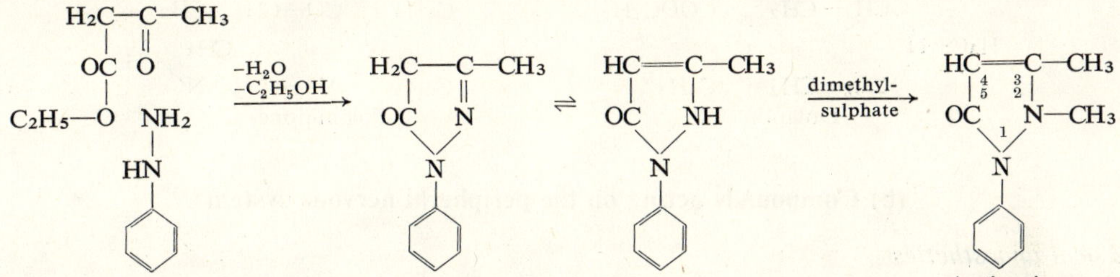

Antipyrine

To convert Antipyrine into aminophenazone [1-phenyl-2,3-dimethyl-4-dimethylaminopyrazolone-(5)] first an amino group has to be introduced into the 4 position. For this purpose, Antipyrine is reacted in sulphuric acid solution with sodium nitrite to 4-nitrosoantipyrine which is then reduced with disulphite to 4-aminoantipyrine. 4-sulphaminoantipyrine formed as an intermediate is cleaved by boiling with sulphuric acid. Methylation of the amino group is carried out by boiling for several hours with formaldehyde and formic acid in aqueous solution.

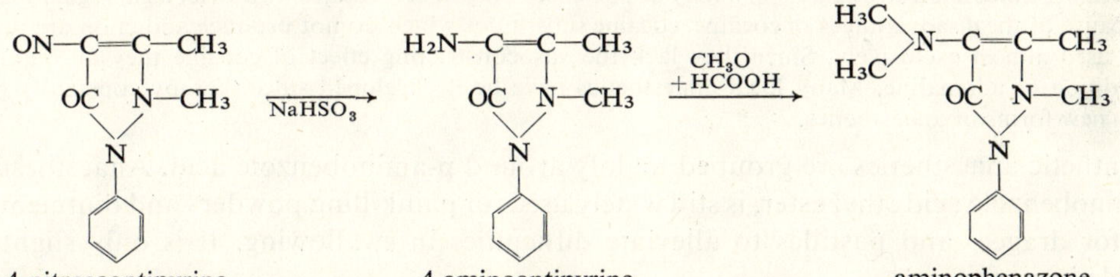

4-nitrosoantipyrine 4-aminoantipyrine aminophenazone Pyramidone

4-aminoantipyrine is the starting material for other antipyretics, e.g. by letting concentrated aqueous formaldehyde solution and sodium bisulphite solution react with 4-methylaminoantipyrine one obtains the corresponding 4-methylaminomethanesulphonic acid, Novalgin.

When hydrazobenzene is condensed in the presence of Na-alcoholate with n-butylmalonic ester one obtains the sodium salt of 1,2-diphenyl-3,5-dioxo-4-butyl-parazolidin (Butazolidin) which by itself and with Pyramidone is an excellent antirheumatic agent (Irgapyrin-Geigy, Wofapyrin).

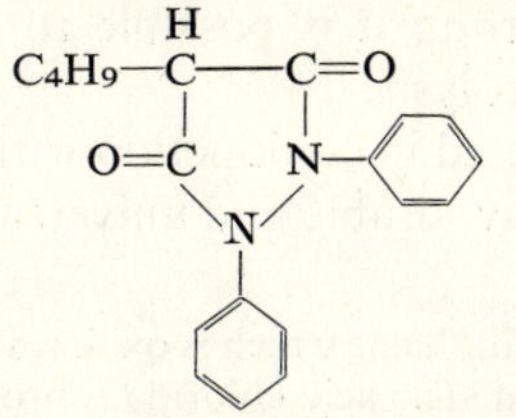

Butazolidin
Butalidone

The effective analgesic component of Pyramidone is particularly in evidence in a very large number of mixed powders, frequently containing an addition of caffeine which produces an enhanced effect exceeding the purely additive effect of the individual components.

Analgesics with the most potent effect belong to the opium group. They are alkaloids, e.g. morphine obtained from *Papaver somniferum*. Heroin, Dilaudid, Dicodid are some of the semi-synthetic morphine alkaloids. The disadvantage of all these compounds is the danger of producing addiction. 1-methyl-4-phenylpiperidine-4-carboxylic acid ethyl ester (Hoechst) and Polamidone, 2-dimethylamino-4,4-diphenylheptanon-(5) (Hoechst) are synthetic compounds with an effect similar to, but even exceeding hat of morphine.

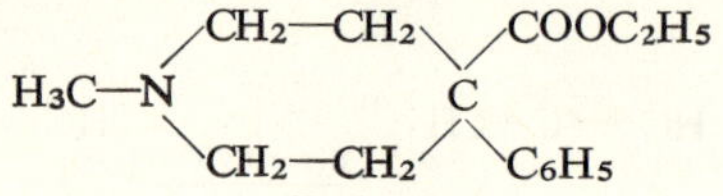

Dolantin

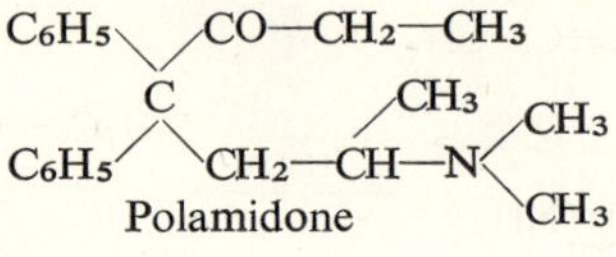

Polamidone

### (b) Compounds acting on the peripheral nervous system

*Local anaesthetics*

Local anaesthetics act by paralysing the sensitive nerve ends. The oldest natural local anaesthetic is cocaine, obtained from the leaves of the coca bush in the Andes, which was used by the Incas as an intoxicant. Chewing coca leaves is supposed to remove any feeling of hunger, exhaustion, and worry. Physical and mental decay follows. This addiction is partly a consequence of bad social conditions.

Cocaine was produced by Wöhler in 1860. Koller discovered it again in 1884 and used it in the treatment of eye diseases. Schleich and Bier were the first to introduce it into surgery. It is an excellent mucosal anaesthetic, particularly important in ophthalmology. Cocaine is unstable which is a disadvantage for sterilization. Since it is a strong poison it may be dispensed only in accordance with strict legal regulations.

Because of the disadvantages of cocaine, cocaine substitutes which do not produce addiction are now being used almost exclusively. Since they lack the vasoconstricting effect of cocaine they all require an addition of adrenaline. Many local anaesthetics have been developed since therapy constantly requires new forms of anaesthetics.

Synthetic anaesthetics are grouped mainly around p-aminobenzoic acid. Anaesthesin, p-aminobenzoic acid ethyl ester, is still widely used for painkilling powders and ointments, and for dragées and pastilles to alleviate difficulties in swallowing. It is only slightly soluble in water.

The greatest progress constituted the introduction of Novocaine (Procain, in the DDR Jenacain) by Einhorn (1905) in which p-aminobenzoic acid is esterified with diethylaminoethanol

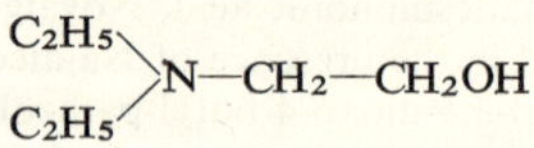

By introducing this strongly basic group it is possible to obtain salts that dissolve easily in water and are resistant to hydrolysis.

Since it has a vasodilatory effect it is used in conjunction with Suprarenin (Adrenaline) which cancels out this effect. So far its favourable and universal properties have not been excelled.

The starting material for Novocaine is p-nitrotoluene which is oxidized in an aqueous suspension with potassium permanganate to p-nitrobenzoic acid. The acid chloride is produced with phosphorus pentachloride and converted with ethylenechlorohydrin into 4-nitrobenzoic acid chloroethylester. By heating with diethylamine under pressure one obtains diethylaminoethylester which is reduced by Béchamps's method with iron to 4-aminobenzoic acid diethylaminoethylester (Novocaine). One can also produce diethylaminoethylalcohol from ethylene oxide and diethylamine and subsequently esterify it with p-nitrobenzoic acid or p-aminobenzoic acid.

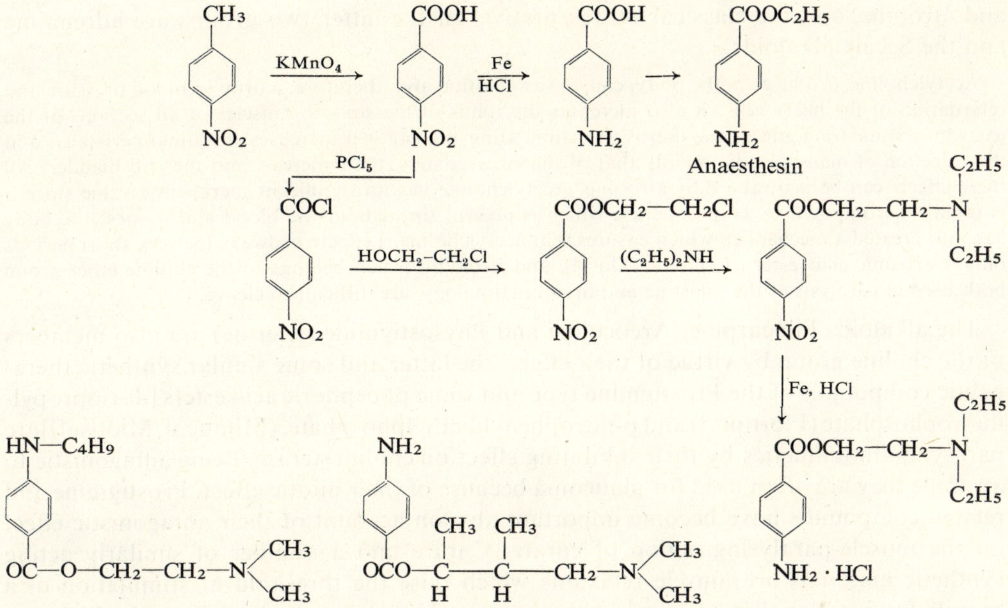


Pantocaine Tutocaine Novocaine

Tutocaine contains a 1,2-dimethyl-3-dimethylaminopropanol residue. Pantocaine and Larocaine are also very important surface anaesthetics of this type. Many other compounds have been developed for special purposes which contain a p-aminobenzoic acid- or a p-aminosalicylic acid residue (and derive, therefore, from PAS) (e.g. Oxyprocaine, Salicaine). The old Orthoform contains the 4-amino-3-hydroxy-benzoic acid residue. In addition to these there are also other very important local anaesthetics which belong to other classes of chemical substances. Thus, Percaine, Xylocaine, and Hostacaine are acid amides, and Falicaine (by Profft) is a keto base (4-propoxyphenyl-piperidinoethylketone-hydrochloride).

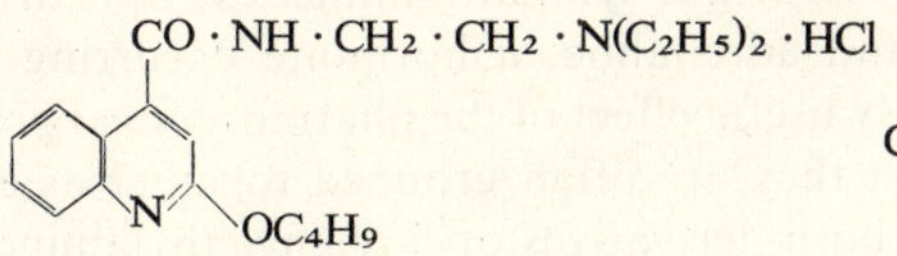


Percaine

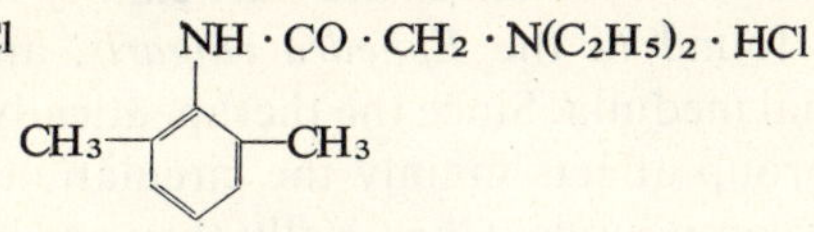


Xylocaine

### (c) Compounds acting on the autonomic nervous system

The autonomic nervous system regulates the functions of the smooth muscles—which can be influenced consciously only to a small degree—such as the activity of the heart and of the glandular apparaturs.

Since the autonomic (= vegetative) nervous system can be divided morphologically and functionally into two nerve fibre systems (sympatheticus and parasympatheticus = vagus), the pharmacotherapeutics affecting it are divided into those stimulating the vagus (parasympathomimetics) or paralysing it (parasympatholytics), and those stimulating the sympatheticus (sympathomimetics) or paralysing it (sympatholytics). The most important representatives of the first two groups of therapeutic agents are acetylcholine and atropine, and the classical representatives of the latter two groups are adrenaline and the Secale alkaloids.

Acetylcholine, produced in the body, causes vasodilation and, therefore, a drop in blood pressure and retardation of the heart beat; it also increases the tonus of the smooth muscles of all sections of the gastro-intestinal tract and of the uterus. By stimulating the vagus it increases intestinal peristalsis and the secretion of many glands, mainly that of the sweat glands, the pancreas, and the gall-bladder. All these effects can be neutralized by atropine. Acetylcholine has no significant therapeutic value since it is rapidly decomposed by cholinesterase which is present throughout the blood and tissues; the body has thus created a mechanism which ensures that acetylcholine is effective always for very short periods only. Carbamic acid ester of choline (Doryl), and Esmodil, which belongs to the choline ether group both used in paralysis of the intestine and in ophthalmology, are difficult to cleave.

The alkaloids Pilocarpine, Areocoline and Physostigmine (Eserine) are also members of the choline group by virtue of their effect; the latter and some similar synthetic therapeutic compounds of the Prostigmine type and some phosphoric acid esters [di-isopropylfluorophosphate (Floropryl) and p-nitrophenyldiethylphosphate, (Mintacol, Miotisal)] are parasympathomimetics by their inhibiting effect on cholinesterase. Being antagonistic to atropine they are often used for glaucoma because of their miotic effect. Prostigmine and related compounds have become important also on account of their antagonistic effect on the muscle-paralysing action of curare. Curare and a number of similarly acting synthetic materials are muscle relaxants which raise the threshold of stimulation of a muscle for acetylcholine released at the end plates of the motor nerves, and which are, therefore, very important in modern anaesthetic techniques.

The prototype of parasympatholytics is atropine which is very important in ophthalmology because of its mydriatic effect. Dilation of the pupils is often required for diagnostic or therapeutic purposes. It is also used as a spasmolytic since, being antagonistic to acetylcholine, it relaxes parasympathetic contraction of smooth-muscle organs, e.g. the intestinal tract and the bronchi. Papaverine is used for the same purpose; it inhibits the muscle directly as opposed to the more neurotropic effect of atropine.

Two natural substances are particularly important sympathomimetics: Ephedrine, an alkaloid found in the *Ephedra vulgaris*, and adrenaline, a hormone occurring in the suprarenal medulla. Since the therapeutically useful effect of the pharmaceutical products of this group affects mainly the circulation they are often grouped together as cardiovascular compounds. Chemically they are both derivatives of $\beta$-phenylethylamine.

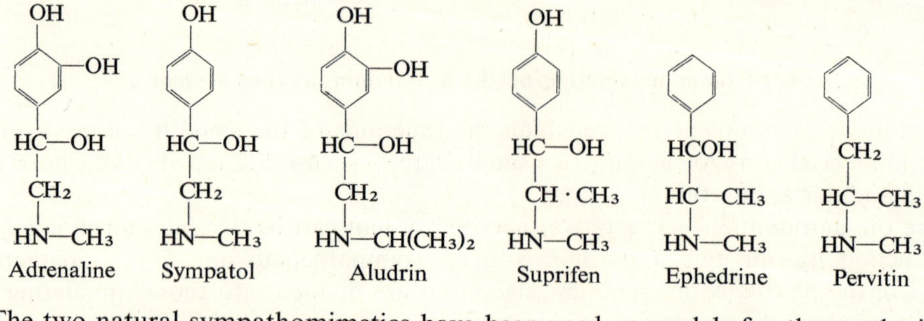

Adrenaline Sympatol Aludrin Suprifen Ephedrine Pervitin

The two natural sympathomimetics have been used as models for the synthesis of a large number of products of similar structure and similar pharmaceutical action. The small differences between the various remedies of this group can be divided into sympathomimetics with an adrenaline-type effect, and sympathomimetics with an ephedrine-type effect. Ephedrine has a moderate but longer-lasting contracting effect on the blood vessels and thus causes a correspondingly lasting increase in blood pressure, to which is added a central stimulation. Adrenaline has a very strong effect on the blood vessels which, however, is of only short duration, and central stimulation is completely absent. Adrenaline-type synthetic sympathomimetics include Sympatol, Corbasil, and Aludrin, the latter being remarkable in that it has hardly any effect on blood pressure but stimulates the heart beat and, above all, has a much stronger spasmolytic effect particularly on the bronchial muscles so that it is used especially against asthma.

The synthesis of Sympatol starts from p-benzoylhydroxy-$\omega$-chloroacetophenone which is reacted with N-methylbenzylamine to p-benzoylhydroxy-$\omega$-(N-methyl-benzyl-amino)-acetophenone. Sympatol is produced from this by hydrolysis with hydrochloric acid and by catalytic reduction.

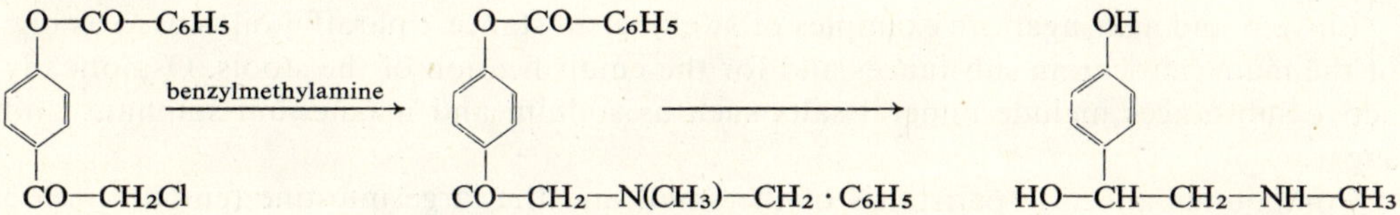

Sympatol

Ephedrine-type synthetic compounds are Veritol and Suprifen. This group of sympathomimetics also includes the 'Weckamines' in which the central stimulation becomes the main effect. 'Weckamines' are basic compounds which produce a strong stimulation of the central nervous system so that they are used to resuscitate people [German: *weck* = to wake] poisoned by sleeping drugs for instance. The name applies also to substances like 'Pervitin' because of the 'wide awake' and 'lively' condition they produce in man. In man this central stimulation causes mainly an increase in psychic activity so that these preparations, such as Pervitin = Weckamine, are frequently used in psychiatry.

### (d) Compounds acting on the heart and the blood vessels

Besides the glycosides with cardiac effect (Digitalis, Strophantine) and the general spasmolytics, nitrites and nitrates (nitroglycerine), caffeine also is a widely used cardiac remedy. It is produced by extraction of coffee (manufacture of caffeine-free coffee) and of cocoa husks, as well as synthetically (e.g. at VEB Arzneimittelwerk Dresden). This group also includes Theophylline and Theobromine which are at the same time diuretics.

One of the most potent preparations stimulating circulation (analeptics) is the convulsive poison strychnine. Like other convulsion-inducing preparations of this group it is used as an arousing agent e.g. after poisoning by sleeping pills. In 1925, having started from work on the centrally-stimulating effect of camphor, K. F. Schmidt and Hildebrandt arrived finally at Cardiazole (cyclopentamethylenetetrazole) via Hexatone. Unlike camphor it is soluble in water and when injected subcutaneously it has an excellent arousing effect in cases of poisoning and especially an outstandingly stimulating effect on even very sluggish circulation. It is produced by reaction of excess hydrazoic acid with cyclohexanone in benzene or ether as solvent and $P_2O_5$ or $ZnCl_2$ as dehydrating agent:

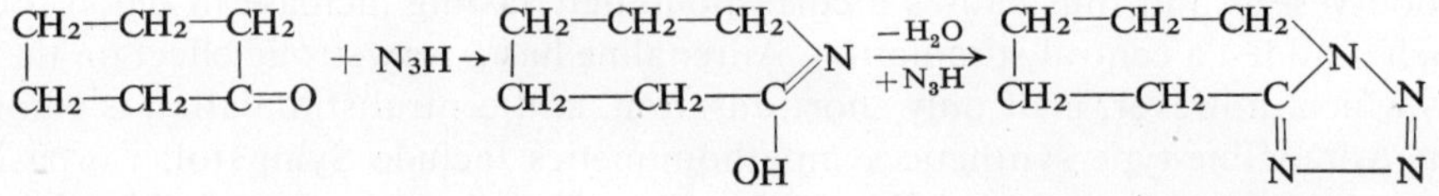

cyclopentamethylenetetrazole
*Cardiazole*

## (e) Compounds acting on digestion and diuresis

The large number of laxatives (laxantia) can be divided according to their pharmacological and chemical aspects into swelling substances, lubricating substances, osmotically acting substances, and such compounds which stimulate intestinal peristalsis.

Linseed and agar-agar are examples of swelling substances: paraffin oil is used as one of the main lubricating substances and for the emulsification of the stools. Osmotically active substances include mineral salts such as sodium and magnesium sulphate, and sugar.

Compounds affecting peristalsis of the small and the large intestine (emodines) are castor oil and constituents of another series of plants and drugs the active substances of which are anthraquinone compounds and some of which are also produced synthetically. Istizin (1.8-dihydroxyanthraquinone) is very effective and non-irritating.

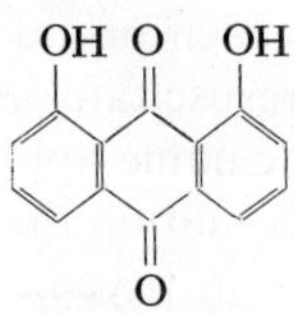

Istizin

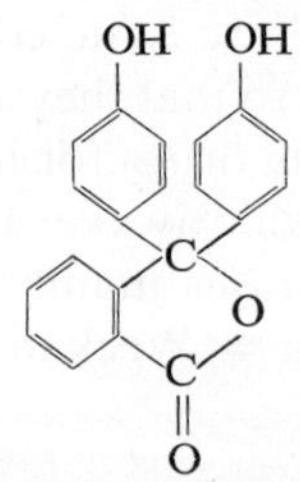

Phenolphthalein

The effective substance in very many laxatives is phenolphthalein which is produced by the condensation of phthalic anhydride with phenol, using a condensing agent (e.g. concentrated sulphuric acid).

Tannin preparations and swelling substances (pectins) are used as constipating agents (obstipantia).

Very important progress has been achieved recently in the field of diuretics. Whilst formerly purine derivatives (Theophylline, Theobromine) and organic mercury compounds (these promote the excretion of $Cl^{\ominus}$) were used which, however, had to be injected, now peroral diuretics are available commercially (Diamox, Nephramid, Chlorothiazide, Dihydrochlorothiazide) the effect of which is based partly on the inhibition of carbonic anhydrase, i.e. they cause in the main excretion of $Na^{\oplus}$ and $K^{\oplus}$. Rodiuran-Boehringer, 6-trifluoromethyl-7-sulphaminyl-3,4-dihydro-1,2,4-benzothiodiazin-1,1-dioxide, is a diuretic which promotes mainly the excretion of salt. Orpidan also is a diuretic applied perorally but not an inhibitor of carbonic anhydrase.

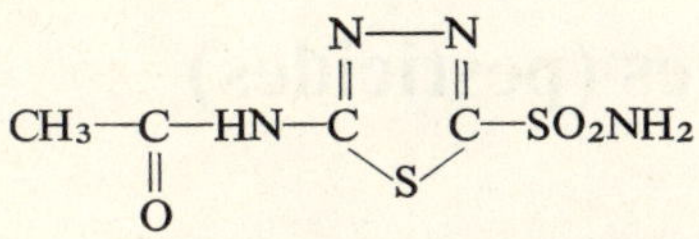

*Diamox, Nephramid*

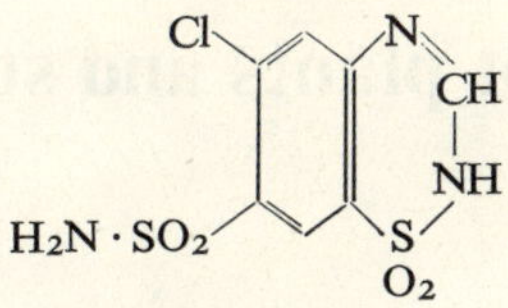

*Chlorothiazid*

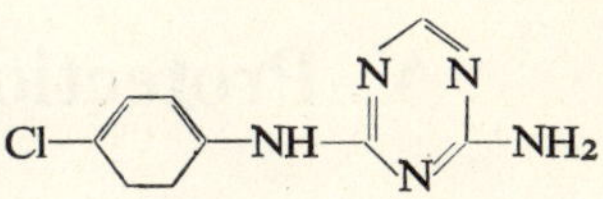

*Orpidan, Triazurol*

LITERATURE

H. P. Kaufmann, *Arzneimittel-Synthese.* Springer-Verlag, Berlin/Göttingen/Heidelberg, 1953.

Ullmann, *Encyklopädie.* 3rd., vol. 5. Chapter Chemotherapy. Urban & Schwarzenberg, München/Berlin, 1954.

G. Ehrhardt in Winnacker-Weingaertner, *Chem. Technologie*, vol. 4. Carl Hanser Verlag, München, 1955.

G. Ehrhardt in Winnacker-Küchler, 2nd ed. II, vol. 4, 958, 1960.

J. Klosa, *Entwicklung und Chemie der Heilmittel.* vols. I–III. Verlag Technik, Berlin, 1952.

F. Eichholtz, *Lehrbuch der Pharmakologie.* 9th ed. Springer-Verlag Berlin/Göttingen/Heidelberg, 1957.

Poulsson, *Lehrbuch der Pharmakologie*, revised by G. Liljestrand. 16th ed. S. Hirzel Verlag, Leipzig, 1949.

F. Hauschild, *Pharmakologie und Grundlagen der Toxikologie.* VEB Georg Thieme, Leipzig, 1956.

# V. Protection of plants and stores (pesticides)

To this day, agriculture harvests only what is left by pests of the plant foods originally sown. About 20 per cent of the world's harvest is lost year after year. Thus about 33 million tons bread grains and rice are destroyed annually which could feed an additional 150 million people. Plant protection is often faced with unexpected tasks by the importation of pests. The time it takes to bring food, animal fodder, and seeds from distant countries is continually growing shorter. As a consequence the survival rate of pests is very high. Owing to the lack of both defence by the plants themselves and natural enemies the imported pests are able to multiply very rapidly.

For pesticides the problem of costs is particularly important. Only cheap preparations can, therefore, be employed, or in the case of expensive ones only those of high efficiency, i.e. those which need to be applied only in minimum concentrations. Another property that is required of modern plant protection agents is that as far as possible they must be harmless to man and to useful animals. In many cases a selective effect against certain insects is desired, for not all insects are harmful.

The mode of application of the plant protection agent has the greatest influence on efficiency and this depends on the way it is fashioned into a commercial product. The effect can be reinforced by the presence of surface-active substances, e.g. wetting agents. Such materials are often added in order to ensure that during dilution of the concentrates, which may be solids, pastes, or liquids, the active substances remain in solution or in emulsified form (for spraying). Commercial preparations in powder form (for dusting) contain active substances in very finely ground form diluted with solid materials such as slate dust or kaolin.

The most important pests and plant diseases are listed below:

Animal pests:

(1) Rodents: field mouse, vole, house mouse, hamster, grey rat, brown rat.

(2) Insects and worms: plant louse, shield louse, ground flea, bark beetle, corn weevil, cockchafer, colorado beetle, rape blossom beetle.
Winther moth, Cochylis roserana, Black Arches, fruit worm.
Cabbage fly, turnip fly, onion fly.
Mites (Red Spider); nematodes.

Plant pests:

(1) Plant diseases caused by fungi:
fruit scab (*fusicladium*), monilia or brown rot of fruit, vine- and hop leaf scorch, club root of cabbage, potato blight (*phytophtora*), wheat bunt, loose smuts of oats, ear blight of corn (*fusarium*), mildew of vine (*oidium*).

(2) Plant diseases caused by bacteria:
soft rot of cabbage, black leg of potato, wilt of tomatoes.

(3) Plant diseases caused by viruses (viroses):
top necrosis and tuber blotch of potato, leaf roll diseases, mosaic diseases.

(4) Weeds.

# 1. INSECTICIDES

## (a) Inorganic and plant insecticides

Up to about 1940, that is to say, until a large number of highly efficient synthetic 'contact insecticides' and 'systemic insecticides' was developed, the principal materials available for the fight against harmful insects, apart from some arsenic compounds, were a number of substances derived from plants.

Arsenic compounds (Paris or Schweinfurt Green, lead arsenate, and calcium arsenate) are typical stomach poisons. Their importance has decreased in the last twenty years and the use of arsenic compounds, particularly in viticulture, is now prohibited.

One of the substances which derive from plants and are used as insecticides is tobacco nicotine which played an important part for a long time; it has to be used with extreme care because of its high toxicity to man.

Anabasine which occurs in some Asiatic plants is very similar to nicotine. The Soviet Union produces and exports it on a large scale.

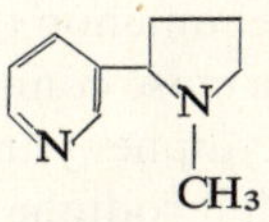

Nicotine

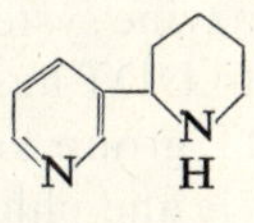

Anabasine

Besides rotenone (obtained from various types of Derris), the active pyrethrum substances are very important vegetable insecticides. They are obtained from the dried blossoms of some types of chrysanthemum and are distinguished particularly by the fact that they are completely harmless to warm-blooded animals. After the total synthesis of these complicated plant constituents had been successfully accomplished, pyrethrine-type substances were also synthesized of which allethrine is widely used, particularly in the U.S.A.

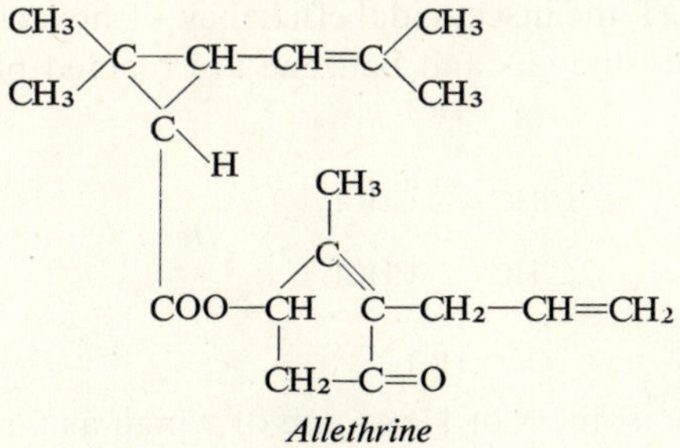

*Allethrine*

## (b) Synthetic insecticides

The synthetic insecticides which have been developed since about 1940 can be divided into three groups:

The DDT group,
hexachlorocyclohexanes and their derivatives,
the organophosphorus compounds.

*The DDT group*

The basic substance is *d*ichloro*d*iphenyl*t*richloroethane (DDT) which can be produced relatively easily by condensation of 2 moles chlorobenzene with chloral.

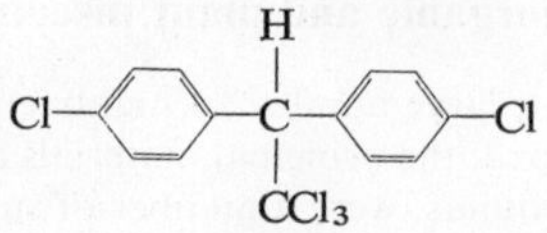

*DDT*

The excellent insecticidal properties of DDT were discovered in 1939 by the Swiss Dr. Paul Müller in Basle. Since it is almost harmless to man it has become outstandingly important for the control of disease-carrying vermin. It is used with great success in typhus epidemics (carried by lice), pest epidemics (carried by rat fleas) and malaria infections (carried by anopheles).

DDT is very highly efficient and has a rapid, lasting, and weather-resistant effect on plant pests, without harming the plant. It is a typical contact insecticide which acts by the contact of the insects with the active substance. The poison passes through the legs or the lipoid substances of the top layer of the skin of the insect into the nervous system and there blocks the vital enzyme systems. The phenomenon that insects (e.g. flies) after a time may become resistant to DDT has obviously a close connection with this mechanism. Other members of the DDT group are dimethoxydiphenyltrichloromethane (Methoxychlor) produced from anisole and chloral, p,p'-difluorodiphenyltrichloroethane which is known under the name of Gix and used for the control of flies and midges, and dichlorodiphenyldichloroethane (DDD).

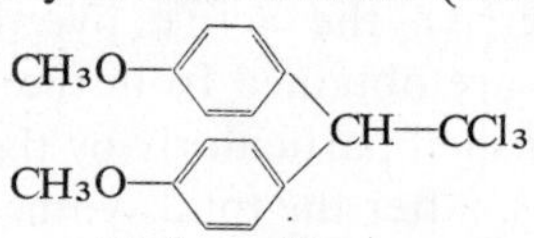

*Methoxychlor*

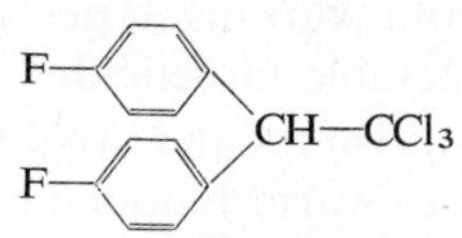

*Gix*

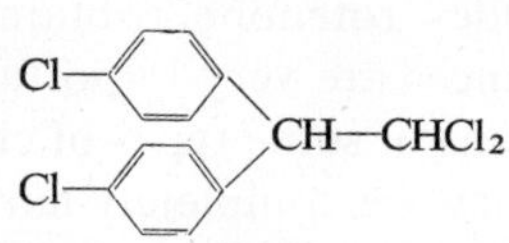

*DDD*

*Hexachlorocyclohexane (Hexa) and its derivatives*

Only a few years after DDT the insecticidal efficiency of hexachlorocyclohexane became known. It forms when chlorine gas and benzene are reacted photochemically.

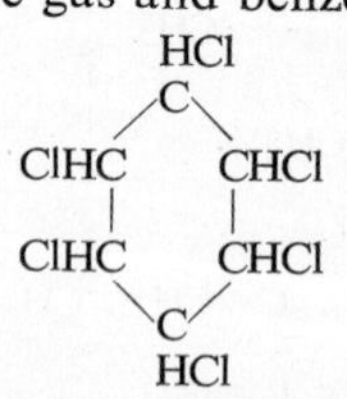

*Hexa*

There are eight possible isomers of Hexa one of which as a racemate must be capable of being split up into its optical antipodes. The industrially produced 'Rohhexa' (crude Hexa) contains about 60–70 per cent $\alpha$-isomer, 5–12 per cent $\beta$-isomer, 10–15 per cent $\gamma$-isomer, 6–10 per cent $\delta$-isomer, and 3–6 per cent $\varepsilon$-isomer. Of these isomers only the $\gamma$-isomer has an excellent insecticidal effect (Gammexan or Lindan).

Gammexan is a contact poison, as well as a stomach- and a respiratory poison, and thus its range of effectiveness is exceptionally wide. It loses its effect relatively rapidly, however, owing to its volatility. Non-purified technical Hexa has an unfavourable effect on the smell and taste of fruits and plants.

The manufacture of pure Gammexan is comparatively expensive since crude Hexa contains at the most only 15 per cent of it. In order to render its production more profitable many attempts have been made to find a use for the 85 per cent by-products. These can be converted into 1,2,4,-trichlorobenzene from which a weed killer, trichlorophenoxyacetic acid, can be produced. It can also be processed into pentachlorophenol which is used as a wood preservative.

An insecticide resembling Hexa is Chlordan, developed in the U.S.A. Its effect is similar to that of Hexa but it is a little less volatile. Chlordan is produced by the addition of hexachlorocyclopentadiene to cyclopentadiene followed by chlorination.

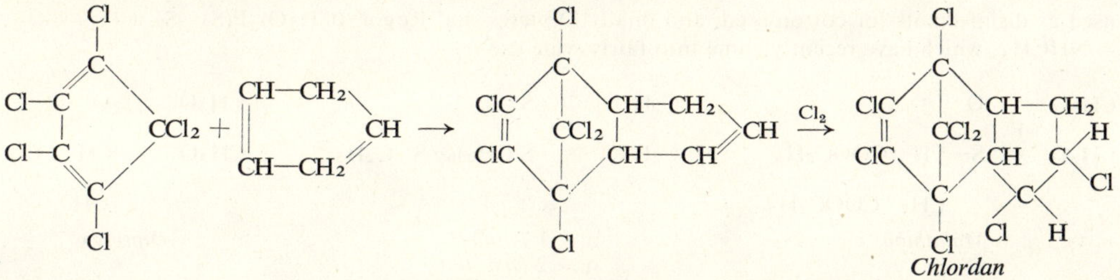

*Chlordan*

Like Chlordan, the insecticides Aldrin and Dieldrin are also produced on the principle of the Diels-Alder diene synthesis.

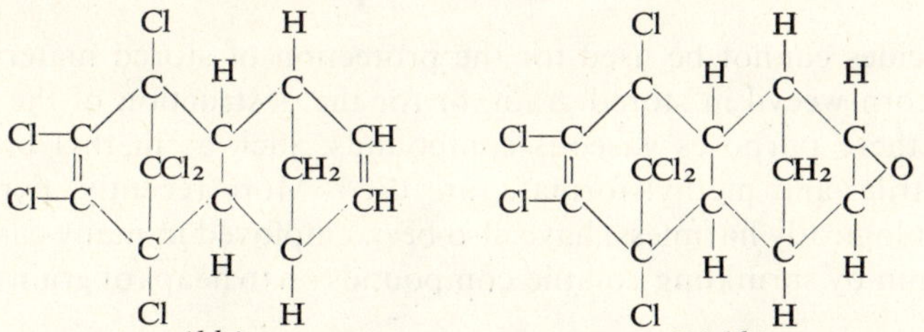

*Aldrin* *Dieldrin*

Aldrin has proved particularly valuable in emergencies involving pests (locusts, cotton pest) because it has an excellent instantaneous effect. An outstanding feature of both Dieldrin and Toxaphen is their low volatility. The latter compound is obtained by chlorination of $\alpha$-pinene, an intermediate of the production of camphor; its constitution is not yet known. Since it is completely harmless to bees it can be used even during the flowering period of the rape plant for the control of the rape blossom beetle. In addition to its effect as a stomach and contact poison it presumably also has a repellent effect against approaching insects.

*Organophosphorus compounds*

The third large group of synthetic insecticides, the organophosphorus compounds, was discovered and developed by G. Schrader in the plant protection laboratory of Farbenfabriken Bayer. The first preparation from this group that has become very widely used is diethyl- (or dimethyl-)p-nitrophenyl-thiophosphate which is used for spraying or dusting under the names E 605, Parathion, or Wofatox.

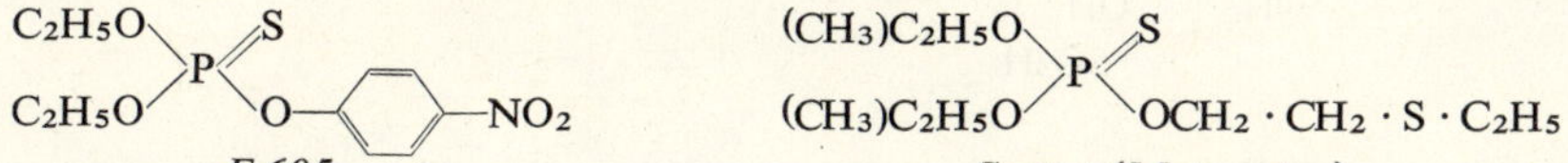

*E 605* *Systox (Metasystox)*

In Systox and Metasystox the nitrophenol residue is exchanged for an ethylthioethanol residue. The two latter thiophosphates belong to the group of systemic insecticides which are absorbed by the roots and the leaves of the plant and are thus able to protect it for several weeks against sucking insects.

Since the thiophosphates mentioned so far are highly toxic to warm-blooded animals a search has been made for new effective members of this group that would be less toxic to man and animal. Of the many new organic phosphorus preparations we shall mention here only the contact insecticide Malathion which is 200 times less toxic than E 605, and the systemic insecticides Thimet and Disyston which are used as disinfectants for cotton seed, and finally Dipterex and Rogor, $(CH_3O)_2P(S)—S—CH_2—CO—$ $—NHCH_3$, which have recently come into fairly wide use.

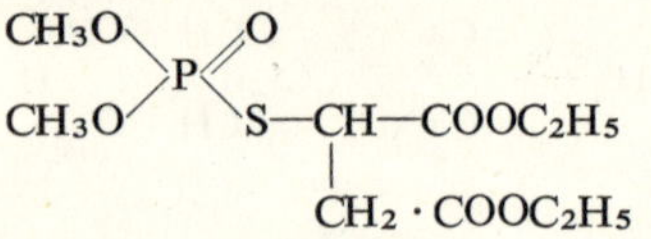

*Malathion*

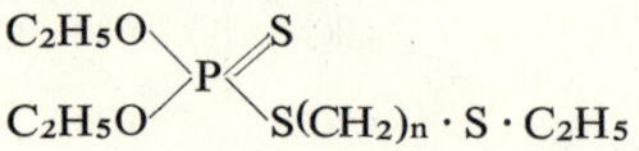

n = 1 *Thimet*
n = 2 *Disyston*

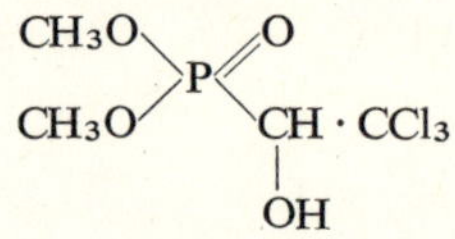

*Dipterex*

### (c) Insecticides with specific uses

Contact insecticides cannot be used for the protection of stored materials, e.g. for the control of the corn weevil in stored grain, or for the destruction of the eggs and larvae of insects; for these purposes gaseous compounds, such as methyl bromide, ethylene oxide, acrylonitrile, and methyl formate, are used. More recently, pyrethrum extracts which are toxicologically harmless, have also been employed in many cases. Grain can be freed from vermin by sprinkling volatile compounds on to heaps of grain or by fumigating silos.

Wood preserving agents have a fungicidal effect in addition to their insecticidal one. Apart from fluorides, dichromates, and arsenates, the agents employed for this purpose are particularly tar oils, pentachlorophenol, dinitrophenol, and chloronaphthalene.

The vapours of chemicals such as naphthalene, camphor, dichlorobenzene, and hexachloroethane which have a repellent effect on pests have been used for a long time to protect wool, furs, and feathers against attack by moths. Recently, woollen fabrics have been successfully mothproofed by impregnating the fibres with substances which deter or even kill the insects. These mothproofing agents include the Eulanes. Eulan-Neu is a condensation product of benzaldehyde-2-sulphonic acid with two moles of 2,4-dichlorophenol; Eulan-NK is formed as a quaternary salt of triphenylphosphine with dichlorobenzylchloride.

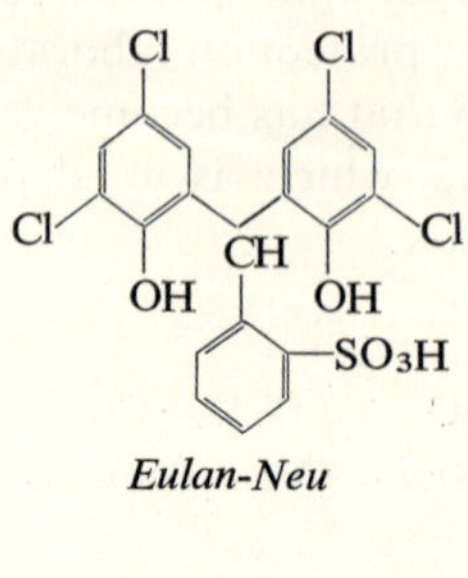

*Eulan-Neu*

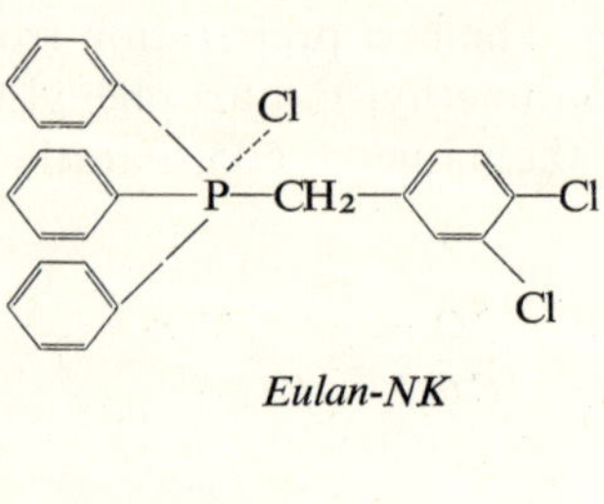

*Eulan-NK*

## 2. OVICIDES AND ACARICIDES

Ovicides and acaricides are agents used specially against the eggs of insects and for the control of mites and their eggs (red spider). Mites which cause much damage in fruit- and vegetable gardening cannot be controlled by modern contact insecticides. For spraying fruit trees in winter one uses 2,4-dinitro-o-cresol and carbolic oil; during the summer months these cannot be used because of their strongly phytotoxic properties. Of the large number of newly developed agents mention will be made here only of Ovotran (p-chlorophenyl-p-chlorobenzene sulphonate), Dimite, developed in England, (bis-p-chlorophenylmethyl carbinol), and Chlorbenzilat by Geigy (4,4′-dichlorobenzilic acid ethyl ester); the latter is particularly effective against bee mites.

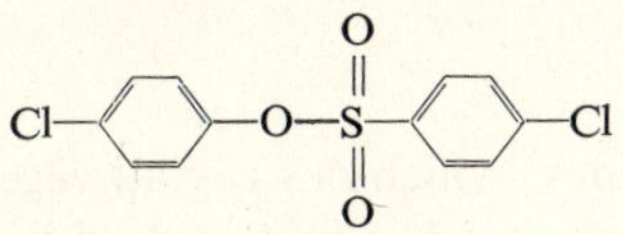

*Ovotran*

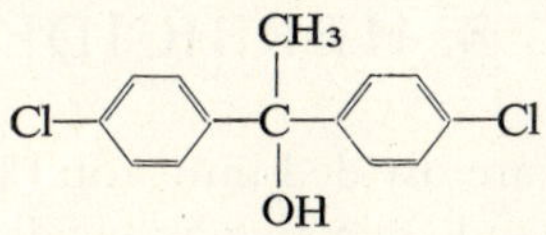

*Dimite*

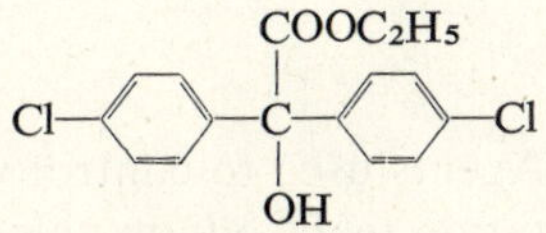

*Chlorbenzilat*

## 3. FUNGICIDES

Compounds with a fungicidal effect are used for the control of fungal-parasitic plant diseases. Mercurial compounds such as cresolmercurycyanide, ethylmercurychloride, methoxyethylmercuryacetylide, phenylmercuryacetate, etc. are employed especially for the disinfection of seed.

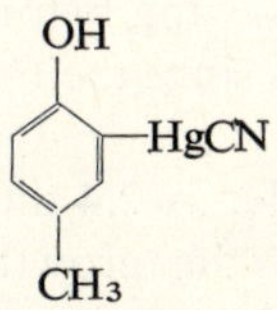

cresolmercurycyanide
*Germisan*

Whilst seed dressings containing mercury are equally effective against the four most important grain diseases: bunt of wheat, loose smuts of oats, ear blight of corn, and leaf stripe of barley, it has not yet proved possible to find compounds not containing mercury with such universal efficiency.

To control fungi occurring in viticulture, fruit growing, hop growing, horticulture, and potato growing, compounds containing copper and sulphur are used, i.e. agents based on copper sulphate, copper oxychloride, compounds with monovalent copper, lime-sulphur mixtures, and calcium polysulphides. In those cases where inorganic copper- and sulphur-containing compounds have proved to be too strongly phytotoxic (and also to save copper) certain organic sulphur compounds can be used. Thus, tetramethylthiuramdisulphide (Thiuram) and some dithiocarbamates (Fuklasin = zinc-dimethyldithiocarbamate) are valuable agents used to control fruit scab.

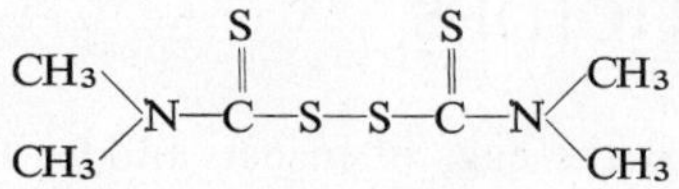

*Thiuram*

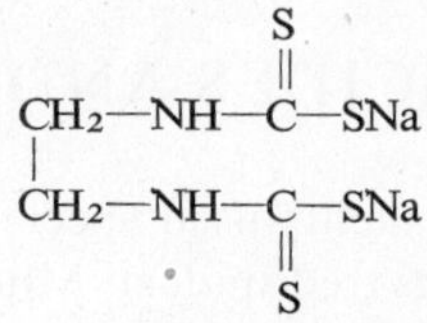

*Nabam*

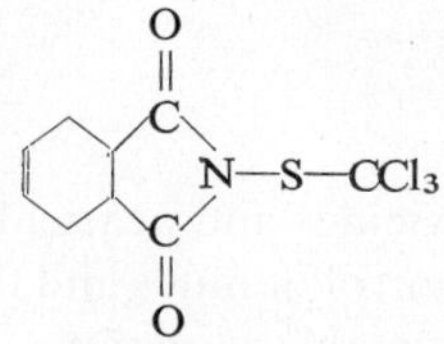

*Captan*

Nabam is used very successfully against phytophtora, and Captan which can be obtained by the reaction of phthalimide with perchloromethylmercaptan has proved particularly useful for the control of fusicladium and peronospora. Organotin compounds are also used (Brestan).

## 4. HERBICIDES

Agents used to control weeds are divided into 'total herbicides' which destroy all vegetation (e.g. sodium chlorate), and 'selective herbicides' which attack only the weed but not the useful plant. Of the latter, calcium cyanamide (lime-nitrogen) acts also as a fertilizer. Other selective herbicides are dinitro-o-cresol (Hedolit), 4,6-dinitro-2-sec.-butylphenol (Dinoseb), pentachlorophenol, and isopropyl-N-phenylcarbamate (IPC).

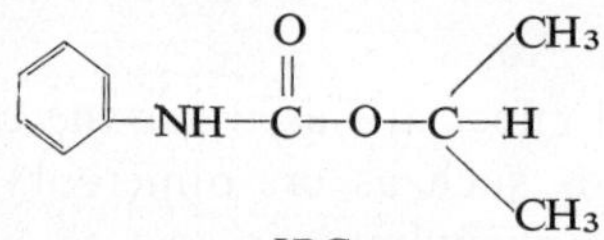

*IPC*

They are applied preferably after sowing before the first shoots come up. They have to be used in carefully measured doses since their mordant effect might otherwise damage the useful plant, too.

A particularly important weed killer is 2,4-dichlorophenoxyacetic acid (2,4-D). It is a growth substance which increases the growth of plants beyond the physiological measure so that they eventually wilt due to exhaustion. 2,4-D does not affect grasses and corn provided certain concentrations are not exceeded, and is, therefore, used for the control of dicotyledonous weeds. It is produced from dichlorophenol and monochloroacetic acid in an alkaline medium and is applied in the form of a sodium salt or triethylamine salt in aqueous solution.

Trichlorophenoxyacetic acid is used for the same purposes as 2,4-D; it can destroy some lignified plants which are resistant to 2,4,-D.

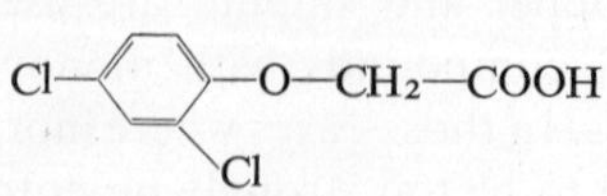

*2,4-D*

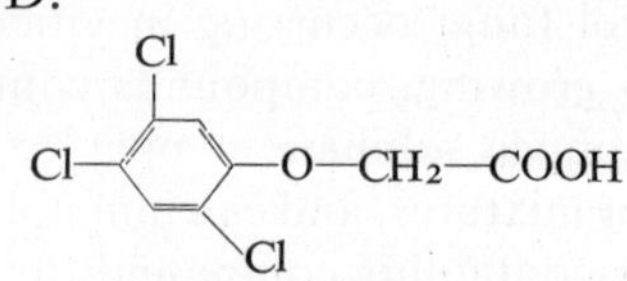

*trichlorophenoxyacetic acid*

Tetrachlorobenzene and trichloroacetic acid are used as herbicides particularly to control monocotyledonous weeds, e.g. grasses.

# 5. RODENTICIDES

To combat rodents a number of strong poisons have been used for a long time, such as strychnine, zinc phosphide, and thallium sulphate, also preparations produced from squill, a plant occurring in Mediterranean countries and belonging to the Liliaceae. Its active ingredient is the steroid 'scillieroside'.

Since 1945, $\alpha$-naphthylthiourea has been used as a rodenticide under the name Antu. It is a specific poison against brown rats and can be obtained from $\alpha$-naphthylamine and ammonium rhodanide.

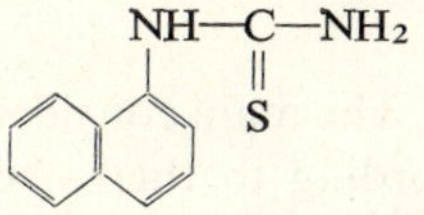

*Antu*

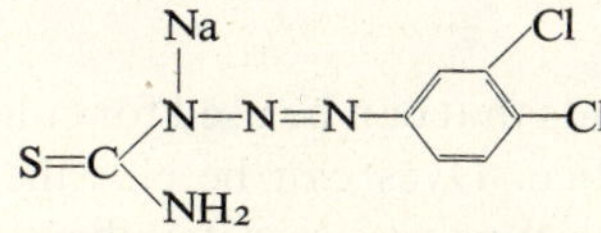

*Promurit*

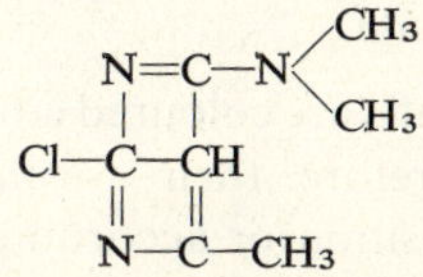

*Castrix*

Since various setbacks have been suffered with the use of this compound it is now less important than it used to be. Another thiourea-containing agent is Promurit, a diazoamine compound produced from diazotized dichloroaniline and sodium cyanamide to which $H_2S$ is subsequently added. Poison grains containing the active substance Castrix which is considerably more poisonous than thallium compounds are particularly suitable for the control of fieldmice and voles.

Chemically, Castrix is a chloro-dimethylamino-methylpyrimidine; it is employed as a chlorohydrate.

Rodenticides with an entirely different poisoning action have been found among the derivatives of 4-hydroxycoumarin. They act by inhibiting the coagulation of the blood. Their advantage is that they are equally effective against all rodents and do not produce defensive reactions in them so that the rodents gradually become chronically poisoned. The best-known preparations of this type are Warfarin and Cumachlor of which the first one is easily obtainable by a Michael addition of 4-hydroxycoumarin to benzalacetone.

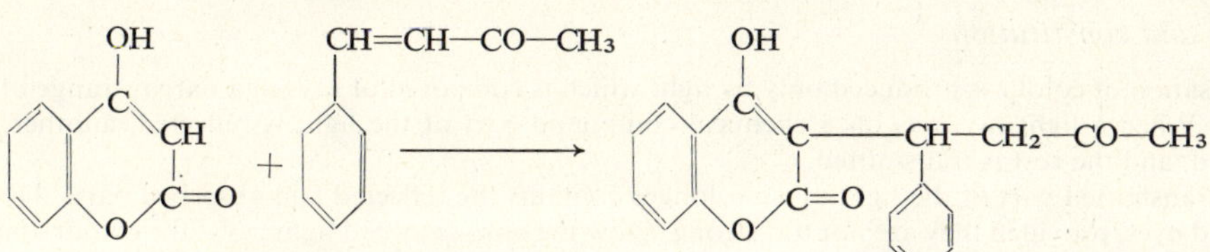

4-hydroxycoumarin *Warfarin*

## LITERATURE

H. Frensch in Winnacker-Weingaertner, *Chemische Technologie* vol. 4. Carl Hanser-Verlag, München 1954, 2nd ed Winnacker-Küchler II, vol. 4, 1960.

W. Perkow, *Die Insektizide*. Dr. Alfred Hüthig Verlag, Heidelberg, 1956.

H. Fürst, *Chemie und Pflanzenschutz*. VEB Verlag Technik, Berlin, 1959.

# VI. Synthetic organic dyestuffs

*Definition*

Dyestuffs are coloured compounds that can be used for colouring. The main consideration is, therefore, their use in practice. Dyes can be classified according to their chemical composition or according to the type of use. Classification according to use yields the following groups:

Pigments, insoluble dyes applied with binding agents in decorating, painting, and printing.

Dyes soluble in organic materials, for the transparent dyeing of lacquers and plastics.

Textile dyes:

(1) soluble in water and absorbed from the solution.
(2) dissolved temporarily and deposited in insoluble form on the fibre.
   (a) vat dyes (sulphur dyes),
   (b) mordant dyes.
(3) dyes developed on the fibre (from soluble components),
(4) insoluble dyes absorbed from the suspension.
(5) insoluble dyes (pigments) fixed with binding agents.

Although the most serviceable classification of dyes would be that according to their use (for instance see A. Schaeffer, *Chemie der Farbstoffe und deren Verwendung*) we shall employ a different basis because groups based on use include representatives of completely different chemical character produced by different technological methods. We shall, therefore, deal with dyes classified according to their characteristic colouring groups as is done generally in textbooks.

*Colour and constitution*

The sensation of colour is produced only by light which is composed of rays in a narrow range of wavelengths. When a light ray falls on a chemical compound part of the light is reflected, another part is absorbed, and the rest is transmitted.

The transmitted part of the light rays no longer contains the reflected and absorbed parts. Dye solutions and dyes, provided they are not too strong, show the non-absorbed light rays; the colour appearing is, therefore, complementary to the absorbed one.

All organic molecules absorb light, mostly, however, in the ultra-violet, i.e. in wave length ranges $<4{,}000$ Å. They appear, therefore, colourless. Light absorption is a consequence of oscillations of electrons. The more firmly bound electrons respond to light of short wave length. The more mobile the electrons are, the more easily can they be excited by light of longer wavelengths. The molecules then absorb in the visible region and appear coloured. Mobility of electrons is increased by accumulation of double bonds in the molecule. Dyes almost always contain aromatic systems which in the ultro-violet have a characteristic absorption. This absorption is shifted into the visible region by the introduction of suitable groups. The absorption bands shift from violet via green and yellow to red, the colour of the compound moves from yellow via red and violet to green. When the absorption bands reach red, absorption bands in the ultra-violet can shift into the visible region (first into violet) by the introduction of further suitable groups. The compound then appears again yellow ('second order yellow'). These absorption processes are, however, never quite clear and neat, just as the dependence of the colour on the constitution never is quite straightforward. Certain predictions can only be made within a group of dyes with basically similar constitutions.

According to Schütze's somewhat obsolete terminology the migration of absorption towards the longer wavelengths (yellow → green) is called deepening of colour produced by 'bathochrome' groups, that towards the shorter wavelengths (green → yellow) heightening of colour produced by 'hypsochrome' groups.

Absorption of the visible light

| Absorbed light | | Visible colour |
|---|---|---|
| Wavelentgh Å | Colour | produced |
| 4,250 | indigo-blue | yellow |
| 4,500 | blue | orange |
| 4,900 | bluish-green | red |
| 5,100 | green | purple |
| 5,500 | yellow | indigo-blue |
| 5,900 | orange | blue |
| 6,400 | red | bluish-green |

A concept published in 1876 by O. N. Witt has proved very useful when considering the dependence of colour on constitution: the chromophore theory, based on C. Graebe's and C. Liebermann's observations and ideas (1868).

'Chromophores '(Greek chroma = colour, phoros = carrier) are unsaturated groups one or more of which, when introduced into a molecule, colour it, e.g.

benzene → nitrosobenzene (yellowish green)
benzene → azobenzene (orange)
acetone → diacetyl (yellowish green)

The 'depth' of colour increases with the accumulation of chromophores in a molecule, particularly when they are conjugated.

The ethylene bond is a relatively weak chromophore. In conjugation with aromatic double bonds colour (a weak yellow) first occurs with diphenylpolyene which contains three ethylene bonds, $C_6H_5$-$(CH = CH)_3$-$C_6H_5$. With consecutive double bonds, on the other hand, e.g. in tetraphenylbutatriene

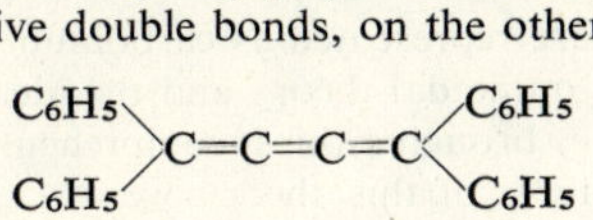


(R. Kuhn, 1938) a deep yellow colour occurs in benzene solution.

In the absence of certain groups in the aromatic nuclei the depth of colour increases (colour-intensifying substitutes) e.g. $—NR_2 > NHR > NH_2 > OH > OCH_3$. Following another suggestion by Witt, these groups are called 'auxochrome'. Dilthey, and later Wizinger, showed that the effect of auxochrome groups is often a function of their ability to form salts. There are, however, many dyes that are incapable of forming salts. According to R. Kuhn these have to be considered as inner salts or hybrid ions. This explains for instance the colour and the high melting point of quinoline yellow which is obtained from quinaldine and phthalic anhydride:

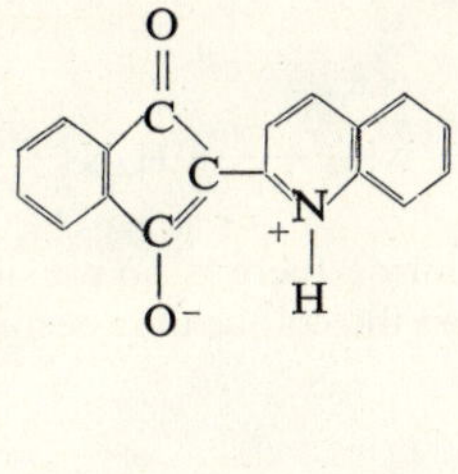

The deep colour of indigo and the exclusive occurence of the trans-configuration support this theory:

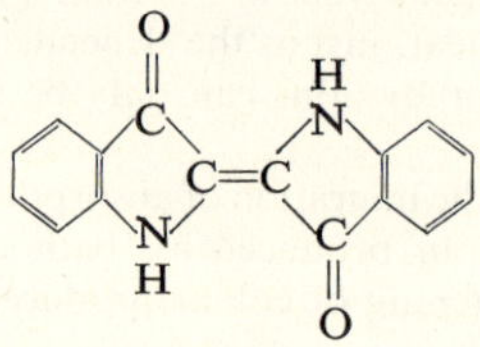

Indigo

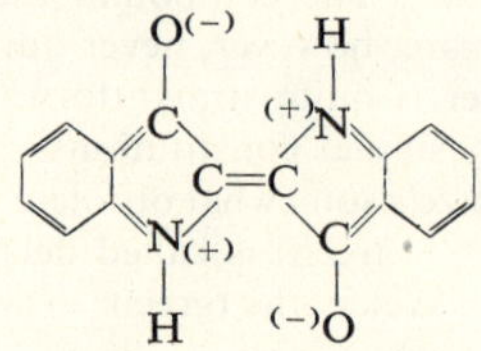

Indigo as an inner salt
(according to R. Kuhn,
tetrapolar, ionic structure)

In accordance with the coloured picrates of hydrocarbons which are dipolar complexes, R. Kuhn explained the deep colouring of indigo and the double molecular weight of blue solutions in polar solvents (in the vapour phase and in non-polar solvents indigo is red) by the presence of such dipolar complexes, two molecules being attracted to each other by their four charges. All these concepts are important for the whole of dye chemistry.

Thorough systematic investigations have been carried out in the field of triarylmethane dyes in order to determine the causes of colour. Almost all processes could be adequately explained by the quinoidal formulae introduced first by Nietzky. They were, and still are, of first rate importance for dye chemists. Accordingly, triarylmethane dyes with the $NH_2$ group in the p-position are best regarded as imonium salts, e.g.:

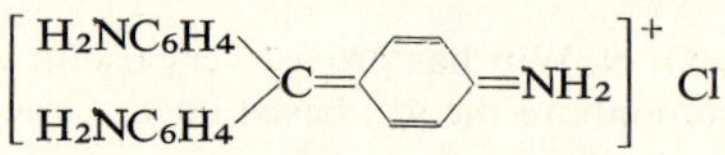

Salts of triarylcarbinols, e.g. those of trimethoxy-or trichlorotriphenylmethylcarbinol are, however, also coloured (A. v. Baeyer). These are carbonium salts. Dilthey and Wizinger preferred, therefore, the carbonium formula for triarylmethane dyes:

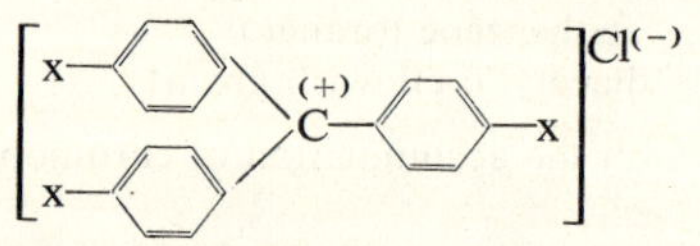

Three types of representation have been in use since then:

In Winnacker-Weingaertner they are represented as carbonium salts, in textbooks, e.g. in 'Fieser', mostly quinoidal. In addition to the quinoidal theory and the ideas about salt formation as a colour-intensifying factor the mesomer theory brough a third, comprehensive explanation for the colour of the carbonium salts, too. The basic features of this theory were vaguely recognized by A. v. Baeyer as early as 1907 and he used it to explain the colour of Döbner's violet.

Dyes that are salts or hybrid ions are resonant systems. A deepening of the colour by a shift of the absorption towards the longer wavelengths is to be expected with resonance hybrids because of the particularly great mobility of the electrons of their chromophore systems. Thus, the great depth of colour of relatively simple dye molecules, such as the triarylmethane series, the acridine series, methylene blue, and the indigo- and thioindigo series, can be explained by salt formation and the presence of two or more resonance hybrids. An example is Döbner's violet the ion of which (mono-HCl salt) occurs in two structures:

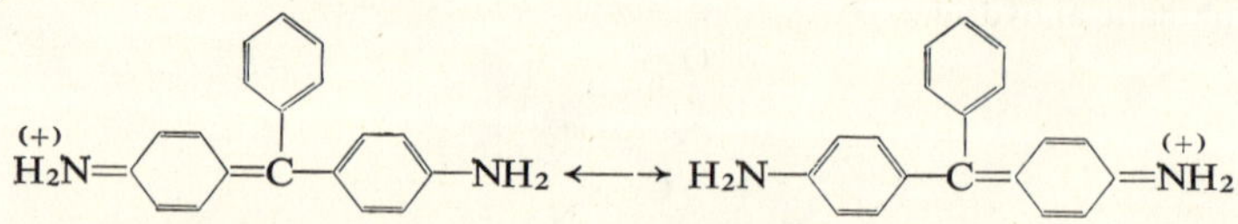

On the other hand, with bis-hydrochloride there is no possibility of resonance and it is, therefore, only yellow. By adding HCl (A. v. Baeyer) the colour thus becomes lighter, i.e. tending to yellow.

A third possible mesomeric form could be the carbonium salt:

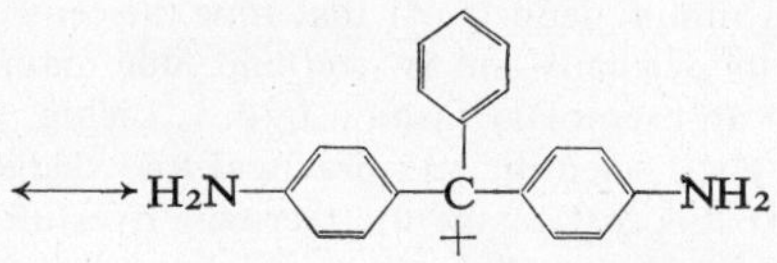

The resonance theory is not yet at a stage where it can explain satisfactorily all the phenomena. The quinoidal form and the carbonium form both occur in the equilibrium of resonance hybrids. We cannot, however, merely to be rational, write them always in both forms and we are, therefore, justified to write either one or the other, since they are both correct.

## *History*

For many thousands of years man has been tempted by the pleasure derived from colour and the wish to imitate colourful Nature to produce dyes. At first, he used mineral and vegetable dyes to colour objects and also his own body. Textiles were dyed with mineral dyes and the juices of wild plants. How much inventiveness has been necessary to discover and develop dyeing processes, such as the purple of the purple shell, the most valuable dye of ancient times, indigo, madder, logwood, yellowwood, catechu, or cochineal (from the cochineal insect)! It is easy to understand the high price of purple when one considers that Schunk in 1879 isolated 7 mg from 400 purple shells during his investigation of the dye, and later Friedländer (1909–1911) 1·4 g of the dye from 12,000 shells. Friedländer elucidated its constitution and showed that the purple dye was 6,6'-dibromoindigo.

Until modern times, the cultivation of dye plants played an important part and only stopped when it became possible to produce synthetic dyes much more cheaply. Logwood is still important, e.g. for silk dyeing and for the dyeing of Persian lamb furs (very dark blue), catechu for the dyeing of silk. Incidentally, the fastness of these natural dyestuffs is far exceeded by that of modern synthetic dyes.

The first synthetic dye used for dyeing silk was pioric acid (1849). As early as 1834 Runge had obtained yellow aurin from phenol which, however, could not be used for dyeing purposes, and he had also produced from phenol and aniline triphenylmethane dyes of unknown constitution. The first practical dye was mauveine, an accidental discovery of W. H. Perkin's who intended to make quinine from impure aniline and oxidized it with potassium dichromate. During the reaction—as Runge had also observed—mainly aniline black formed together with a small quantity of a violet dye. Great credit is due to W. H. Perkin for having developed an industrial process from this observation. Perkin called the dye 'aniline purple', later it got the name mauveine in France (mallow in French = mauve). Its structure was not elucidated until much later:

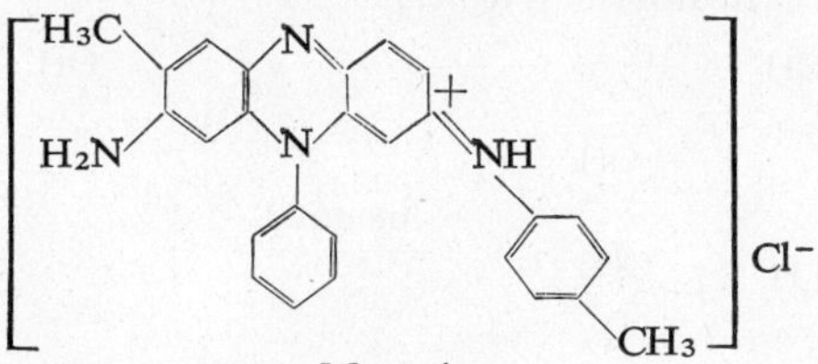

Mauveine

As will be seen from the formula, Perkin's aniline contained o- and p-toluidine. Perkin's success gave rise to many similar experiments, i.e. oxidation of crude aniline. Independently of each other, Nathanson, A. W. v. Hofmann, and Verguin obtained fuchsin which at first was produced only by the French firm Renard & France, Lyon, for silk dyeing. All these experiments led purely empirically to further triarylmethane dyes but their structures were still unknown.

A. W. v. Hofmann, whilst working in London, laid the foundations of scientific research in dyestuffs by establishing the chemical composition of the bases used and the dyes obtainable from them. Research proper, however, did not begin until 1865 after A. Kekulé developed his valency theory and the now generally accepted benzene formula (the six-membered ring with double bonds). The first great success based on these concepts was the establishment of the structure of alizarine, a dye from the madder root, by Graebe and Liebermann (1868). The synthesis of alizarine by Farbwerke Hoechst, the BASF Ludwigshafen, and Perkin followed almost simultaneously. At that time the centre of gravity in dyestuff research moved from England via France to Germany and Switzerland, due mainly to the fact that owing to the training of chemists organized in an exemplary fashion by J. v. Liebig, Germany had available a large number of young chemists with good scientific and practical knowledge who had learnt to carry out independent research. Already at this early stage the German dyestuff plants were establishing their own large research laboratories. In about 1865 when the new era of the aniline dyes began, the principal substances available to textile printers and dyers, in addition to the 'classical' dyes mentioned earlier, were the following:

*Aniline black* played an important part as oxidation black for dyeing and printing. The following operations were carried out:

(1) Impregnation of the fibre or imprinting of a concentrated mixture of aniline salt, copper sulphate, sodium chlorate, ammonium chloride, aluminium acetate,
(2) suspension in a chamber at 30–35°C or steaming,
(3) development by chroming in a dye bath.

The establishment of the structure of aniline black proved very difficult. R. Willstätter worked out the formula for indamine. This preliminary stage is obtainable only in mild conditions. Eight aniline molecules form 'pernigraniline'.

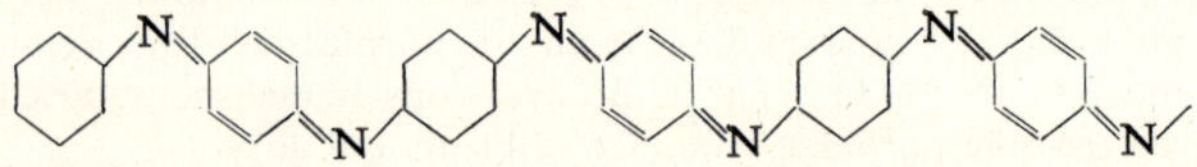

Green evolved the following formula:

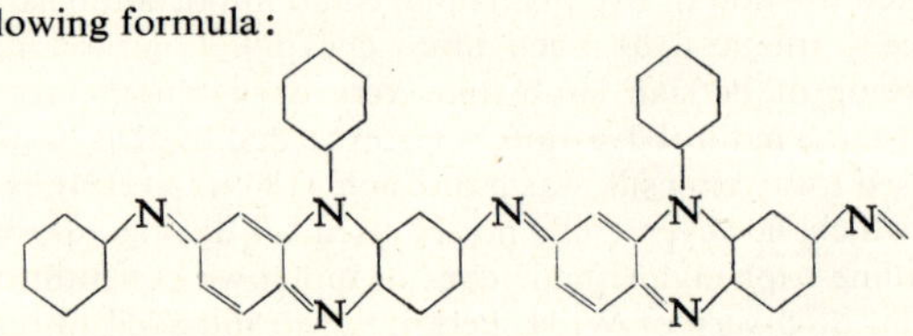

During further oxidation azines form due to the entry of more aniline molecules.

*Logwood* is a mordant dye of high fastness which is used as an oxidation dye. According to the type of mordant used different shades are obtained; with alum mordant, grey, with chromium and iron, dark blue to black (for Persian lamb furs). Logwood was used for silk, wool, cotton printing, and furs. The main pigment in logwood is haematoxylin (Pfeiffer):

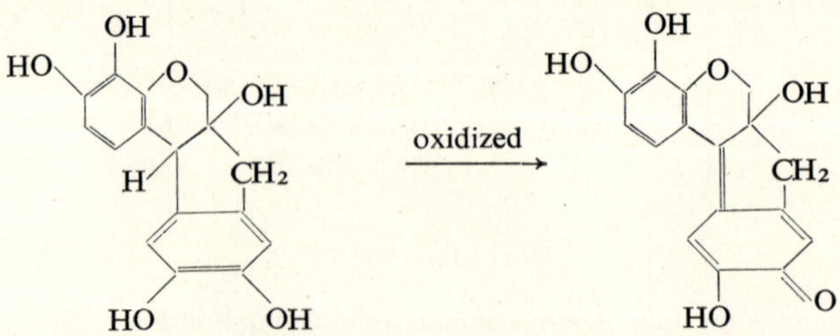


Haematoxylin Haematein

*Catechu* is one of the anthoxanthidines, i.e. the flavone dyes. The basic structure is flavone.

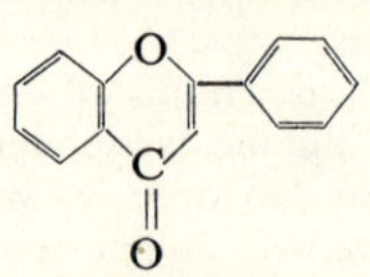


Flavone

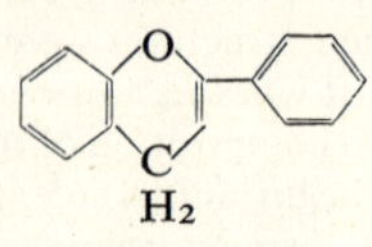


Flavan

Both benzene rings contain OH groups. Catechu is a mordant dye used in cotton printing. This type resembles the natural tannins. Very similar to it is *quercetin* found in the bark of the North American dyer's oak (*Quercus tinctoria*):

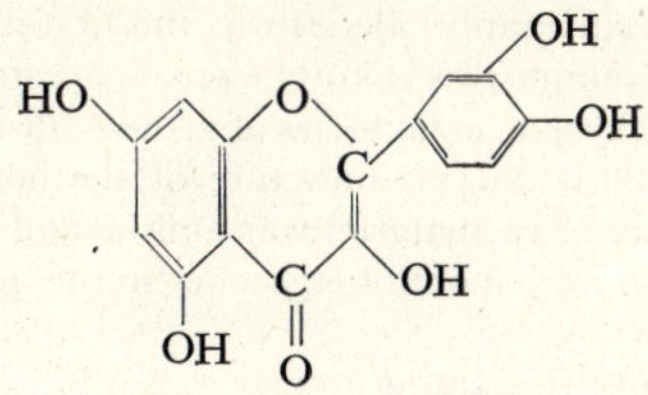

Quercetin

*Rhamnetin* (from berries of purging buckthorn and buckthorn) has a lateral pyrocatechol residue instead of a resorcinol. On mordants it dyes a brownish orange.

Catechu, an extract from Indian trees and bushes, contains catechol. The structure of 1-epi-catechol was established by Freudenberg:

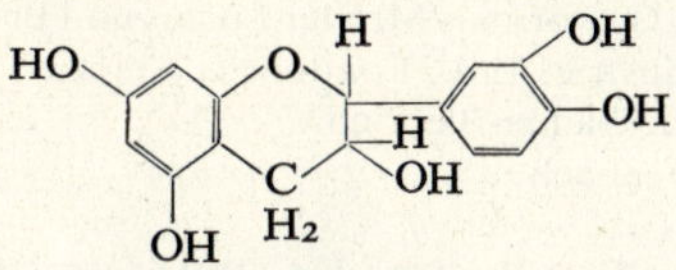

Many inorganic pigments were produced in the fibre for printing and dyeing, e.g. lead chromate and iron oxide. 'Mineral khaki', produced by precipitating iron oxide or chromium oxide on cotton, is still used e.g. for tent cloth and shirting for the tropics. Such dyes are largely rot and insect-proof.

However, all these dyes could not compete in luminosity and beauty with the new synthetic dyestuffs which were developed one after the other in the period that followed.

After the discovery of mauveine and fuchsin which were at first only silk dyes and were, therefore, first produced in Lyons, the silk centre of Europe, research and development were devoted almost exclusively to derivatives of rosaniline which had luminous clear hues but very limited fastness. The basic substance is rosaniline, i.e. 'fuchsin without the methyl group'.

From it are derived the following dyes: Hofmann's violet = 'triethyl fuchsin', aniline blue = 'diphenyl fuchsin', methyl violet 6 B = hexamethylrosaniline.

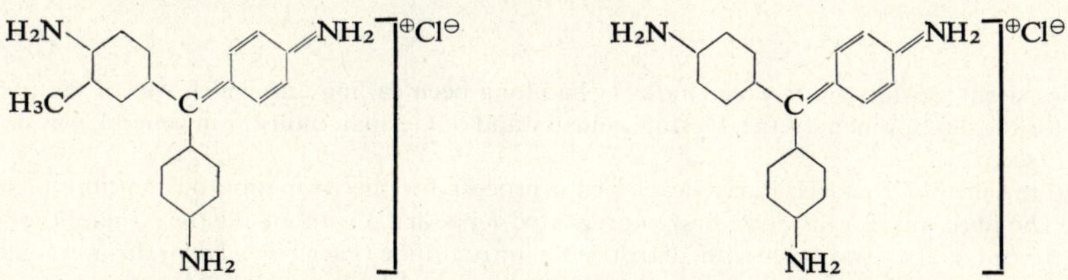

Fuchsin Rosaniline

All these were only basic dyes. It was a great step forward when in 1862 Nicholson heated aniline blue with concentrated $H_2SO_4$ and obtained water-soluble mono-, di-, and trisulphonic acids of the triarylmethane dyes. According to the conditions, the products were alkali blue, methyl blue, water blue or Hoechst new blue.

This represented a fundamentally new method for the transformation of dyes that could be dissolved only with difficulty or not at all into water-soluble and acid dyes with affinity to wool. This principle was later widely used, for instance for azo dyes.

One of the most remarkable observations of the greatest significance for dyestuff chemistry was the discovery of the diazonium compounds by Peter Griess (1858) which was shortly followed by the production of aminoazobenzene, the first azo compond. At almost the same time Martius obtained the first serviceable azo dye by the diazotization of m-phenylenediamine. In 1863 Lightfoot isolated aniline black which Runge, Perkin, and Hofmann had produced earlier.

Martius yellow (1-hydroxy-2,4-dinitronaphthalene) was the first nitro dye (1864).

As a consequence of the rapid developments taking place in dyestuffs technology in the years 1859 to 1867, many dye factories were established or factories that had till then been manufacturing vegetable and mineral dyestuffs were converted to the manufacture of synthetic organic dyes. These were called aniline dyes because the first important raw material was aniline and its derivatives. The most important firms and their dates of establishment were mentioned earlier in the introduction to the chapter Aromatic intermediates.

The abbreviations of the names of firms stand for:

| | |
|---|---|
| BASF | Badische Anilin- und Sodafabrik A. G., Ludwigshafen |
| Bayer | Farbenfabriken Bayer A. G., Leverkusen |
| Cassella | Leopold Cassella A. G., now: Cassella Farbwerke Mainkur A. G., Frankfurt (Main) |
| Ciba | Ciba A. G., Basle |
| Du Pont | E. I. du Pont De Nemours & Co., Wilmington, Delaware |
| Geigy | I. R. Geigy A. G., Basle |
| Hoechst | Farbwerke Hoechst A. G., vormals Meister Lucius und Brüning, Frankfurt (Main)—Hoechst |
| ICI | Imperial Chemical Industries Ltd., London |
| Kalle | Kalle & Co. A. G., Wiesbaden-Biebrich |
| Offenbach | Naphthol-Chemie, Offenbach |
| Sandoz | Sandoz A. G., Basle |
| Wolfen | VEB Farbenfabrik Wolfen, Wolfen, Krs. Bitterfeld (früher Agfa-Aktiengesellschaft für Anilinfabrikation, Berlin-Lichtenberg) |

A. v. Baeyer (1871) discovered the phthalein dyes phenolphthalein, fluorescein, eosin, phloxine, erythrosin, and Rose Bengale. These were first marketed by the BASF—shortly after determination and publication of the constitution by A. W. v. Hofmann—followed by other firms since at that time no patent protection existed as yet in Germany. Other dyes shared the same fate, such as the azo dye orange II from sulphanilic acid → β-naphthol which at the time was the most important dye in use. With azo dyes, the arrow → signifies that a diazotized amine, here e.g. sulphanilic acid, was coupled with a phenol, here β-naphthol.

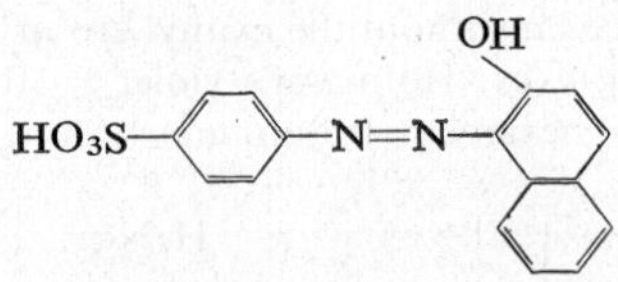


German patent legislation for which industry had long been calling and which was of the utmost importance for the development of the dyestuff industry and of German industry in general, was not passed until 1877 (see p. 3).

In 1880 the firm of Read Holliday developed a process for the formation on the fibre of insoluble azo dyes. The fibre was, for instance, first impregnated ('padded') with an alkaline β-naphthol solution and then treated in a second bath with diazotized p-nitroaniline (Para red). From this were later developed the naphthol-AS-dyes [AS = alkali salts] which contain 2,3-hydroxynaphthoic acid arylides as coupling components. The alkali salts of hydroxynaphthoic acid arylides have a special affinity for the fibre.

Other important dates in the history of dyestuff chemistry are:

1868: establishment of the constitution of indigo by A. v. Baeyer
1880: first synthesis of indigo from o-nitrocinnamic acid

This was followed by a whole series of indigo syntheses with the result that synthetic indigo became cheaper and eventually displaced natural indigo.

After 1880, a large number of dyestuffs was developed from a variety of classes of materials; discussion of their historical development would entail dealing with the whole of dyestuff chemistry.

Only a few groups can be mentioned here; viz:

Pyrazolone dyes, benzidine dyes (directly absorbed substantive dyes), sulphur dyes, indanthren blue, thioindigoes, Naphthol AS, up to the phthalocyanines and the latest process, their formation on the fibre in printing.

Individual dyes (excluding mixtures) marketed at present number about 3,200. Azo dyes (about 1,600) are still the largest group; about 250 vat dyes are now available commercially.

In 1936/1937 world production amounted to 244,000 tons per annum, of which Germany's share was 27 per cent. The average price of world production was RM 3·40 per kg, in Germany RM 5·50 per kg, so that Germany's share by value was 40 per cent (RM 830 millions).

In 1949–1951 average world production amounted to 260,000 tons (value £ 270 million). Germany's share has dropped appreciably. At the end of both world wars German dyestuff processes and patents became known in the victorious countries and available to their industries.

## 1. DIARYL- AND TRIARYLMETHANE DYES

The most important dyestuffs of the diarylmethane series are the auramines, e.g. Auramin O (BASF), produced first by Caro and Kern from p'-tetramethyldiaminodiphenylketone (Michler's ketone), later by Sandmeyer from p,p'-tetramethyldiaminodiphenylmethane with sulphur in the presence of NaCl and $NH_4Cl$ in a stream of ammonia at 200°C:

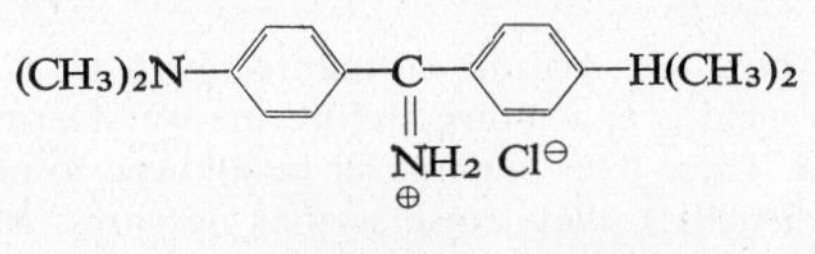


Lacquers which are very fast to light and which are made from auramines with phosphotungstic acid and phosphotungstic-molybdic acid are now used for colouring paper.

Whilst mono-p-amino derivatives of triarylmethane in the form of their salts are coloured compounds but not dyes, those with amino groups in the 2- and 3-p-positions yield servicable dyes. Diaminotriarylmethane dyes include O. Fischer's malachite green which is obtained from 1 mole benzaldehyde and 2 moles dimethylaniline with zinc chloride first as a leuco base which is then transformed with oxidants into the dye:

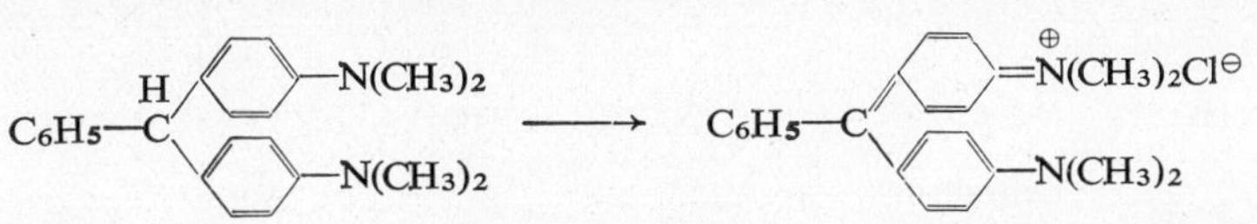


Benzaldehyde or its derivatives, e.g. chlorobenzaldehyde, can be used. The dyes obtained are green and, in some cases, blue. The pigments are dyed on silk or on cotton mordanted with tannin or catanin. Pigments that are fast to light can be produced with the phosphotungstic acids mentioned earlier. Acid absorbing wool dyes are obtained by the reaction of benzaldehyde carrying one or two sulphonic acid groups with dialkylarylamines, e.g. Xylylen Blue VS (Patent Blue VF). Others carry a benzylsulphonic acid radical instead of an ethyl group on the nitrogen (Patent Blue AE, Guinea Green B, Light Green SF). The two brands of green are now produced by Bayer under the names of Acilan Green B and Acilan Green SFG.

The first of the triaminotriphenylmethane dyes was fuchsin which is now obtained from a mixture of aniline and o- and p-toluidine by oxidation with nitrobenzene using Fe or FeCl as a catalyst (formula on p. 495). According to the raw material used, fuchsins with different numbers of methyl groups in the o-position to the amino groups are formed.

Fuchsin is now mainly an intermediate for aryl fuchsins which are obtained by treating aniline at 150–180°C, cleaving off $NH_3$, in a similar way to the production of diphenylamine from aniline (aniline blue brands). Water-soluble blue pigments are obtained by sulphonation; according to the degree of sulphonation, these are called alkali blue, water blue, and cotton blue.

### *N-alkylated fuchsins*

In 1861 Lauth obtained methyl violet by oxidation of dimethylaniline with air and $Cu^{II}$-salts (according to A. W. Hofmann a fuchsin methylated on nitrogen; formation and constitution were established by E. and O. Fischer in 1878). The methane carbon atom is formed from an aldehyde group derived from a $CH_3$ group. A pure hexamethyl fuchsin is crystal violet which is obtained from Michler's ketone (produced from dimethylaniline + phosgene + $ZnCl_2$) with dimethylaniline and $POCl_3$ via the keto-chloride

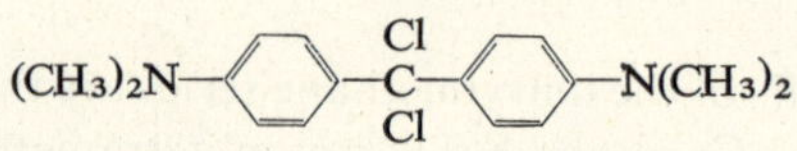


which condenses with dimethylaniline to form the pigment directly. This reaction, which is still used on a considerable scale, can also be carried out in one step (dimethylaniline + phosgene).

Crystal violet is used for dyeing silk, cotton mordanted with tannin, typewriter ribbons, copying pencils and inks, etc., since its yield is enormous. Before the war Germany produced 400 tons methyl violet and 90 tons crystal violet. These dyes must never be allowed to penetrate under the skin (danger of injuries caused by copying pencils!), they cause serious necroses. Michler's ketone and alkyl- and arylnaphthylamines yield diarylnaphthylmethane dyes, e.g. with ethyl-α-naphthylamine Viktoria Blue R, with phenyl-α-naphthylamine Viktoria Blue B (BASF).

### *Hydroxytriarylmethane dyes*

The basic material for this group of dyes is benzaurin:

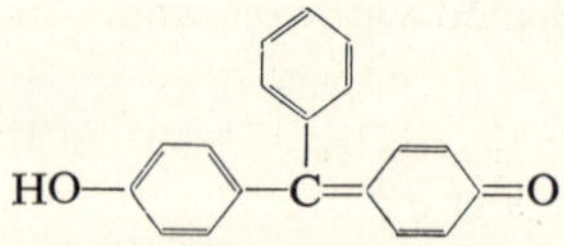


Benzaurin

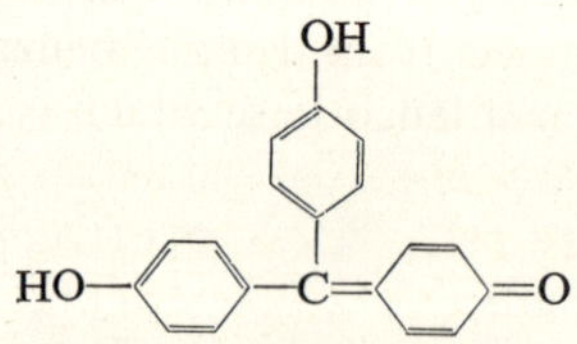


Aurin

The oldest and best-known dye, phenolphthalein (A. v. Baeyer, 1871) is formed from 1 mole phthalic anhydride and 2 moles phenol. It is useless as a dye, but this group of dyes is important because of its indicating properties, e.g. thymol phthalein (reversible change from colourless to greenish-blue). Phenolphthalein is colourless. The red disodium salt is formed with NaOH via carbinol, addition of further NaOH giving the colourless trisodium salt.

The problem of the changes in the constitution of the phthalein system produced during these reactions is dealt with in textbooks on organic chemistry. Here is shown only the way in which the formula for the red form as a carbonium ion is written:

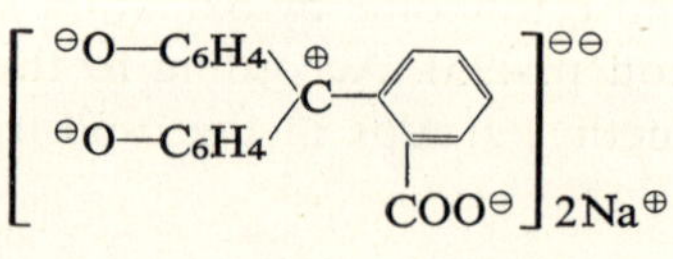

When 3 moles salicylic acid are reacted with 1 mole formaldehyde in the presence of nitrosylsulphuric acid (Sandmeyer), one obtains aurintricarboxylic acid (Chrome ruby). The salicylic acid groups produce chromability. Messrs. Geigy have developed a whole group of chromable triphenylmethane dyes for dyeing wool and for the chrome printing of cotton (Eriochrome and Chromoxan dyes) which contain two o-cresotinic acid

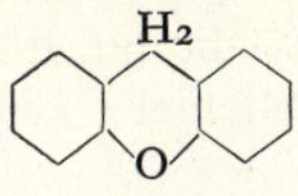

Xanthene

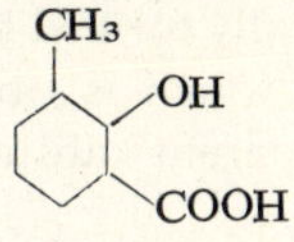

o-Cresotinic acid

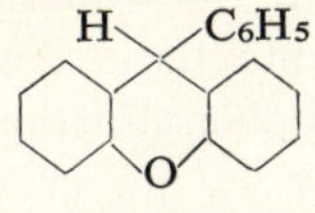

Phenylxanthene

residues in addition to a third aromatic ring. The most important dye of this series is Eriochromazurol (Geigy) produced from 2,6-dichlorobenzaldehyde and o-cresotinic acid. Xanthene dyes are obtained by linking two aromatic radicals in triarylmethane dyes by an oxygen bridge.

Pyronines are derived from xanthene, rosamines from phenylxanthene.

Sulphorhodamine is formed from 1 mole benzaldehyde-2,4-disulphonic acid and 2 moles diethyl-m-aminophenol. The most important rhodamines are obtained from phthalic anhydride and m-aminophenol:

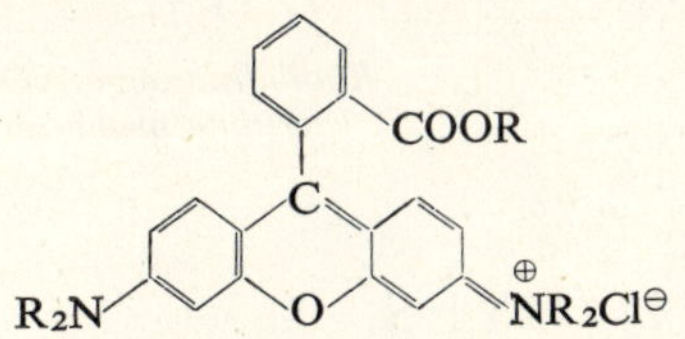


$R = C_2H_5$ : *Rhodamine B extra*

Fluorescein is a phenylxanthene dye with phenolic groups; it has been known since 1871 (Baeyer). It is produced by the condensation of 1 mole phthalic anhydride and 2 moles resorcinol. It is yellow and exhibits a green fluorescence. The tetrabromine compound is eosin (Caro) which dyes a luminous pink with yellowish-red fluorescence. Into the phenylxanthene dyes, too, a salicylic acid configuration was introduced, i.e. OH in the ortho position to the COOH group, e.g. by using hydroxytrimellitic acid anhydride, yielding chrome dyes (Hoechst), e.g. Chromoxanbrilliant-red BL (Hoechst) from hydroxytrimellitic acid and 3-diethylaminophenol.

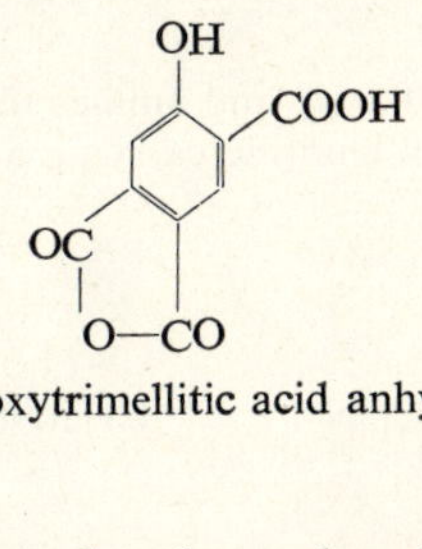

Hydroxytrimellitic acid anhydride

## *Acridine dyes*

The linking of two radicals in diphenylmethane dyes by a nitrogen atom which may also be alkylated or arylated, yields acridine dyes. Only one dye from this group will be mentioned here, Trypaflavine which has a strong bactericidal effect and is, therefore, used therapeutically (e.g. in 'Panflavin', Cassella).

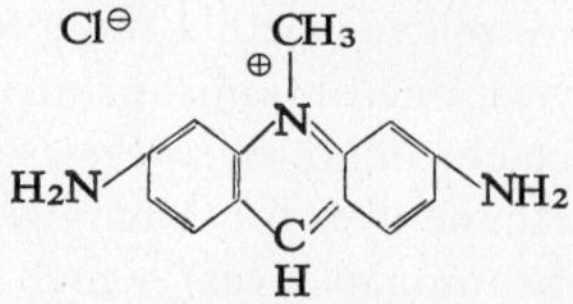

In these dyes the acridine ring is usually formed by the splitting off of $NH_3$ from o,o′-diaminodiphenylmethane derivatives which is obtained by the nitration of p,p′-tetramethyldiaminodiphenylmethane (the parent substance of Michler's ketone), mentioned repeatedly before:

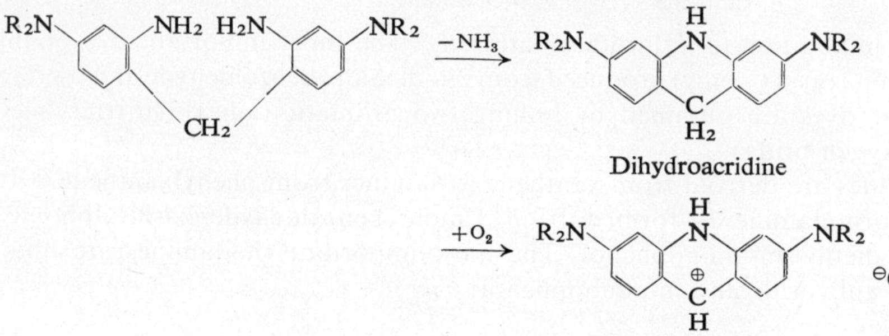

*Rhodulinorange NO*
(*acridine orange*)

## 2. AZINE DYES

We have met earlier the most important azine dye, aniline black, the constitution of which was established by Green. The parent substance of the dyes still in use (safranines, indulines) is phenazine or phenylphenazonium:

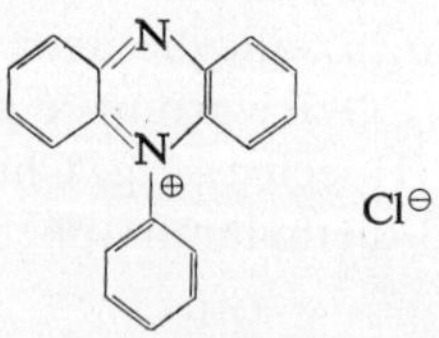

Phenylphenazonium chlorides can add $NH_3$ and amines forming a leuco compound which is easily oxidized to the coloured phenylazonium chloride carrying an amino or arylamino group (amination reaction).

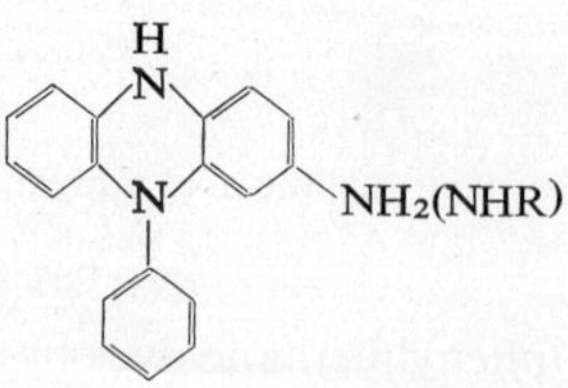

This group includes the basic red Safranin T with two methyl groups in the benzene radicals of acridine (Williams, 1859), obtained from 1 mole each of p-toluylenediamine, o-toluidine, and aniline by oxidation, and Supranol Blue BL which contains sulphonic acid groups.

### *Oxazines and thiazines*

Oxazines and thiazines are based on the following ring systems:

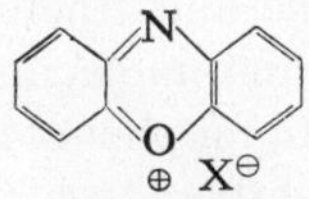

Oxazine
(phenazoxonium salt)

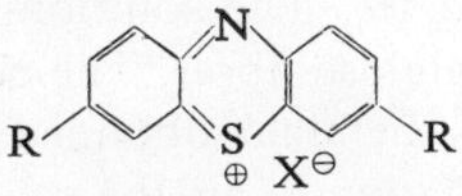

Thiazine
(phenazthionium salt)
R = $NH_2$ *Lauth's violet*
R = —$N(CH_3)_2$ *methylene blue*

The first thiazine dye was Lauth's violet (1876) obtained by the action of oxidizing agents on a solution of o-phenylenediamine saturated with $H_2S$. In 1877, Caro produced methylene blue, the tetramethyl derivative of Lauth's violet, in an analogous way from p-aminodimethylaniline. Bernstein established its constitution in 1883. Pure methylene blue is used in therapy. It is reduced biologically to the colourless leucomethylene blue, and oxidized back to a dye by active oxygen (detection reaction). It is still used for dyeing cotton (mordanted with tannin and catanol) and silk which assume a vivid blue colour, and for colour lakes which with tannins have only moderate fastness to light but with green earth their fastness to light is good.

### *Organic pigments* (*General remarks*)

Colour lakes (pigments), i.e. precipitates which can be dissolved only with difficulty or not at all, formed from water-soluble organic dyes, have already been mentioned on several occasions. They are employed in a variety of applications:

(1) together with binders, for painting, printing, wall coverings applied as a thin film, recently also for textiles;
(2) for coloured pencils, coloured crayons, Indian inks, and oil paints for artists;
(3) for colouring synthetic materials: linoleum, nitrocellulose, rubber, hardenable moulded resins (no transparent colouring).

The forms in which they are used are: acid pigments as Ca, Ba, or Zn salts, basic pigments as tannates with tannin or synthetic tanning substitutes, precipitated with tamol or catanol, or precipitated on mineral carriers ('substrates'). Such substrates may be:

(a) synthetic carriers: aluminium hydroxide, barium sulphate, titanium oxide, or zinc oxide;
(b) natural carriers: kaolin, gypsum, chalk, green earth. The carriers affect depth of colour, shade, body, and glaze.

Mechanical mixing of insoluble pigments with carriers is also carried out. Very thorough grinding and mixing with the binder, preferably on roller frames, is essential.

Glazing colours are transparent in oily or aqueous binders, e.g. water colours through which the paper must be visible. Glazing substrates, e.g. aluminium hydroxide, are important for colour printing.

Body colours reflect light, and reflecting substrates such as white lead or titanium oxide have, therefore, to be used for them.

The most transparent dyes are those dissolved in aqueous binders or in lacquers and oils.

## 3. SULPHUR DYES

An observation of very great importance was made in 1893 by Vidal, namely that organic compounds yield dyes when they are heated to 100–200°C in a solvent with sodium polysulphide and sulphur. These dyes are insoluble in water but form leuco compounds with sodium sulphide which have a great affinity for the cellulose fibre. In this, these 'sulphur dyes' resemble vat dyes but are cheaper and more easily reduced than anthraquinone and indigoes. The solution is strongly alkaline and, therefore, suitable only for cotton and cellulose fibres. The compounds are of high molecular weight. The formula below shows the black obtained by Vidal from p-aminophenol or o- and p-aminophenol. After oxidation in air the shade is at first greyish-blue to black, after treatment with iron chloride or chromates it turns jet black. The constitution of many sulphur dyes was established by Weinberg and Zerweck.

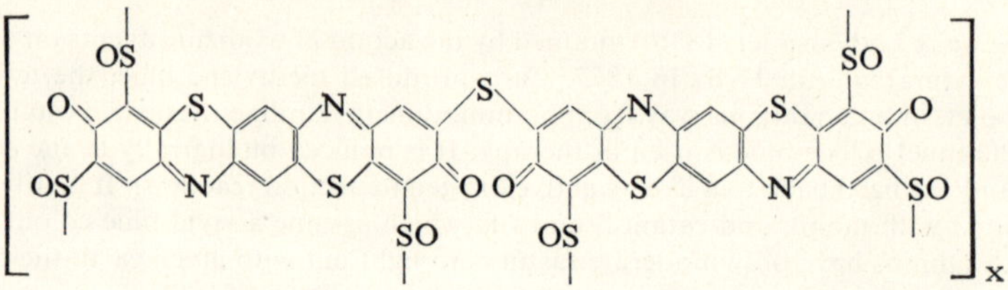

*Vidal black*

The advantages of sulphur dyes are: the comparatively low price, good fastness to water, easy colouring methods; the disadvantage is that their fastness to bleach, chlorine, and light is often not very good.

The first black that could be used without oxidative after-treatment came from Kalischer, Cassella (1897) and was obtained by polysulphide fusion of 4-hydroxy-2',4'-dinitrodiphenylamine

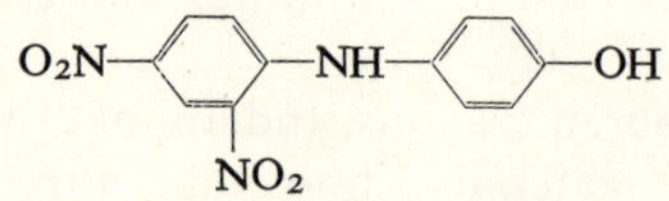

which can be produced easily from 2,4-dinitrochlorobenzene and p-aminophenol (Immedial black V).

Sulphurization of indoaniline (from p-nitrosophenol and carbazol or N-ethyl carbazol)

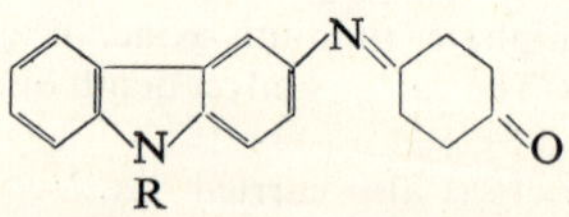

Hydron Blue R (R = H)
Hydron Blue G (R = $C_2H_5$)

produces the beautiful hydron blues (Haas and Herz, 1908, Cassella). They are vat dyes and are dyed from the hydrosulphite vat.

Indocarbon CL (Agfa, marketed by Cassella in 1927) which is fast to light and chlorine and constituted a great step forward as compared with the former sulphur black is obtained by sulphurization of 4-hydroxyphenyl-$\beta$-naphthylamine.

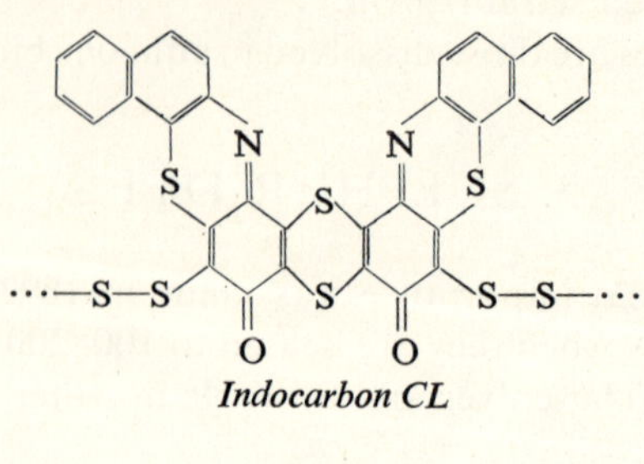

*Indocarbon CL*

This dye does not rot the fibre even during prolonged storage.

The production of sulphur dyes requires very careful methods of working which have to be developed empirically since these dyes are not homogeneous chemical individuals, but mixtures. The shade depends to a great extent also on the conditions of sulphurization, in particular on the ratio Na : S of the polysulphide.

Sulphurization can be carried out in agitator vessels with aqueous or alcoholic polysulphide, or by melting with sulphur in tanks, on iron, in baking apparatus, or in heated drums; hydrogen sulphide escapes. Another treatment is with sulphur chloride and sulphur. From the filtrates of the dyes sodium thiosulphate (fixing salt for photography) is obtained as a by-product. The group names for sulphur dyes produced by the German dyestuff factories are 'Immedial' (Cassella) and 'Schwefel' (Wolfen).

Sulphur dyes are mixtures of high-molecular condensation products which, according to their colour, contain benzene rings with thiazol rings and thiazine rings. The benzene nuclei are linked by —S—S— bridges which are transformed during vatting into —S—Na groups, the dyes becoming soluble in the process. Leuco sulphur dyes are now available commercially also in the form of soluble thiosulphonic acids (Hydrosol, Thionol) which dissociate in an alkaline medium and are adsorbed by the fibre.

The latest development in the field of sulphur dyes is the use for sulphurization of higher condensed aromatic ring systems, such as dinaphthocarbazole or polynitrodecacyclene. The dyes obtained are especially fast to light and washing.

## General observations on the fastness of textile dyes

The problem of 'fastness' of textile dyes has been mentioned repeatedly. The essential aspects of this problem will, therefore, be summarized below.

One has to distinguish between:

(1) fastness in manufacture, i.e. fastness of the dye in the various processes to which the dyed fibre is subjected until the goods are ready for scale. These include: fastness to washing, bleaching (chlorine), sulphur, milling, and hot pressing.

(2) fastness in use, i.e. fastness of the dyed textiles in use, mainly to light, washing, perspiration, ironing, and rubbing.

The degree of fastness required has to be correlated to the uses and to the textiles employed. Exaggerated requirements in this respect lead to unnecessary increases in cost since in general the 'faster' all-round a dye is the higher is its price.

The Committee on Fastness of the Standard Commission for the Testing of Materials (Echtheitskommission im Fachnormenausschuß für Materialprüfung) has prepared agreed testing methods which are based on the requirements in practice and are carried out on specimen colourings. The fastness is expressed in numbers which indicate the resistance of a colouring to the effects of manufacture and use. Reference dyes are used for these measurements. The manufacturers of the dyestuffs state the fastness number in their catalogues and manuals. The light-fastness scale has a range of 1–8; simultaneously certain reference dyes (blue scale) which are dyed on the same fibre are exposed to light since the nature of the fibre also affects fastness to light. Part of the coloured material is shielded from light so that fading shows plainly. 'Eight' signifies the highest fastness to light and indicates that the dye tested fades to the same degree as the reference dye 8. Testing is carried out in sunlight under glass, since not all artifical sources of light correspond to conditions in practice. Other fastness scales have ranges from 1 to 5.

# 4. INDIGO AND THIOINDIGO

Before the BASF Ludwigshafen started manufacturing synthetic indigo in 1897, the indigo required was obtained exclusively from plants grown in Java and in Bengal (Indigofera).

After A. v. Baeyer established its constitution in 1868 he himself carried out the first syntheses (1880). Altogether 30 different indigo syntheses were discovered, but only two synthetic processes, developed by Heumann, were used industrially. In 1897, 17 years after Baeyer had established the constitution, BASF succeeded in carrying out the first but very costly industrial synthesis. Four years later (1901) an improved synthesis was carried out at Farbwerke Hoechst. The difficulties were caused mainly by the fact that in addition to the dyestuff synthesis the large-scale industrial manufacture of the preliminary products and intermediates had first to be developed and set up.

Heumann's first synthesis was based on phenylglycine-o-carboxylic acid (discovered in 1890).

Naphthalene was oxidized to phthalic acid with $SO_3$ and $HgSO_4$. The catalytic oxidation process in air (see p. 270) was at that time still unknown. Phthalic acid was transformed into phthalimide with $NH_3$. Then followed decomposition with NaOCl by Hofmann's method to anthranilic acid. From this, phenylglycine-o-carboxylic acid was produced with chloroacetic acid:

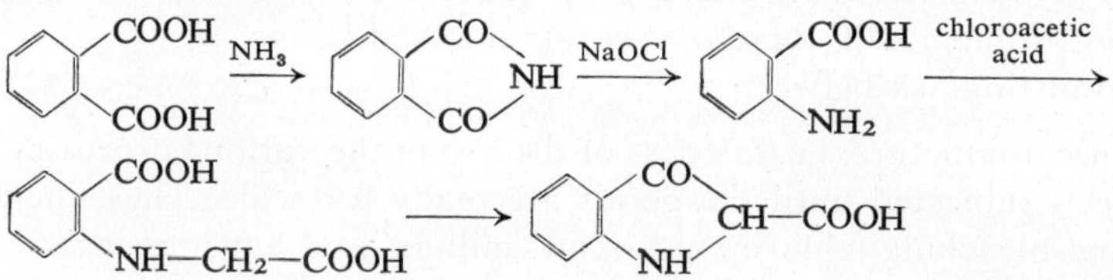


Alkali fusion yielded indoxylic acid which was oxidized in air to indigo. This synthesis was carried out at BASF on a very large scale for 30 years. The second method developed by Heumann was from phenylglycine via alkaline hydroxide fusion to indoxyl:

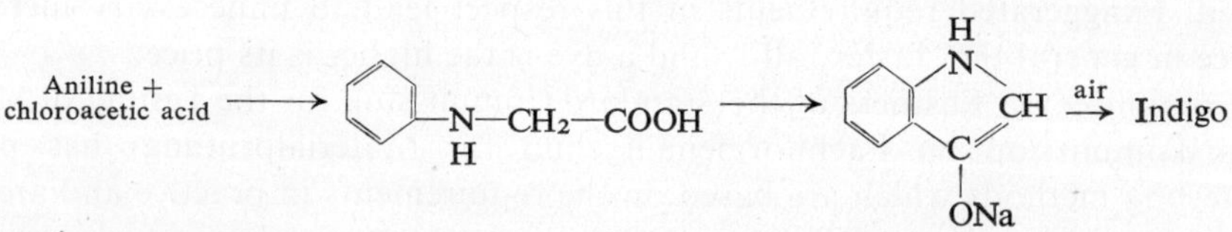


With this synthesis the yield was very low so that BASF did not use it.

Then in 1901 Pfleger, of the Deutsche Gold- und Silberscheideanstalt Frankfurt (Main), who was looking for a use of sodium amide as a condensing agent, discovered that in Heumann's second synthesis an increase in yield was achieved. This process was developed at Hoechst and eventually replaced the former method.

Phenylglycine was first produced from aniline + chloroacetic acid, later from 2 moles aniline and 1 mole formaldehyde with 1 mole HCN (from NaCN + $CO_2$):

$$C_6H_5{-}NH{-}CH_2{-}NH{-}C_6H_5 + HCN \rightarrow C_6H_5{-}NH{-}CH_2CN + C_6H_5NH_2$$

Phenylglycine is obtained by saponification of N cyano methyl aniline.

Cyclization to indoxyl is carried out with sodium amide at 190–210°C in iron agitator vessels with very close control of temperature. The yield calculated for aniline is 94 per cent of the theoretical! The sodium salt of indoxyl is dissolved in large vats and oxidized to indigo by introducing air. Schützenberger and De Lalande in 1872 were the first to use sodium dithionite ('hydrosulphite') for the reduction: The dye is available commercially as the free acid of leuco-indigo or as the Na salt in the form of a paste. Leuco-indigo is produced by hydrogenation using an Ni catalyst. Indigo is no longer of any importance.

Indigo can be halogenized when water is excluded. The most important of the halogen indigoes is indigo tetrabromide (Ciba Blue BB and Brilliant Indigo 4 B). These dyes are more luminous and have a redder tinge. The purple dye of the purple snail is an indigo bromide. Indigo bromides are still important.

*Thioindigo*

As indigo is produced in Heumann's synthesis from anthranilic acid via phenylglycine-o-carboxylic acid, so thioindigo can be obtained in a similar manner from thiophenol-o-carboxylic acid (thiosalicylic acid) (Friedländer):

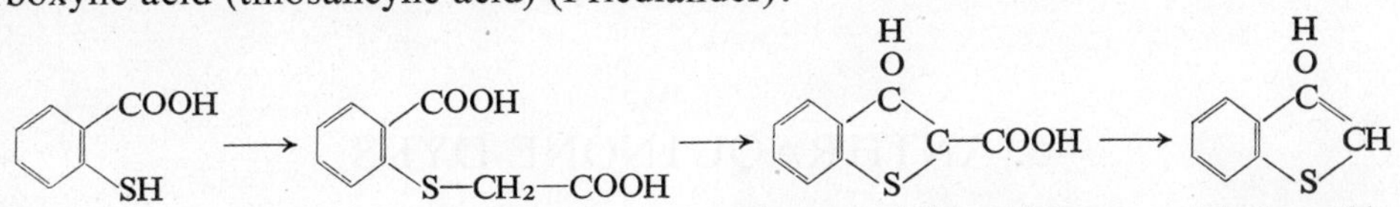

3-hydroxy-1-thionaphthene-2-carboxylic acid    3-hydroxythionaphthene

Thiosalicylic acid can be obtained from diazotized anthranilic acid and sodium polysulphide by reduction with zinc dust.

Since 1914 a general method discovered by Herz (Cassella) has been in use for the production of thiophenols as raw materials for thioindigoes. Primary aromatic amines are reacted with $S_2Cl_2$ to thiazthionium chlorides (Herz's intermediate compound), chlorination in the p-position to the amino group usually taking place simultaneously:

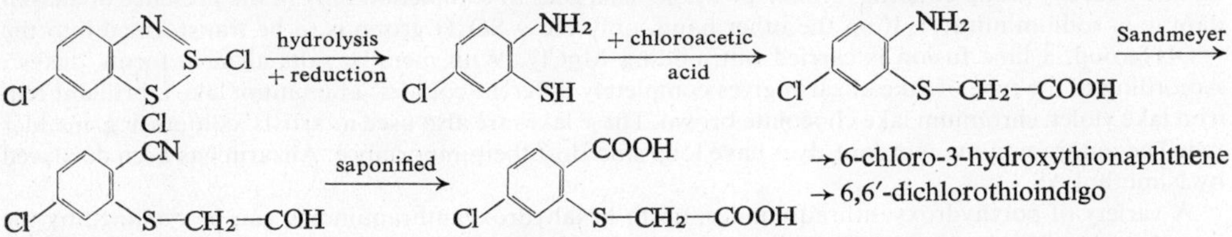

The following is a general method for the production of asymmetric thioindigoes:

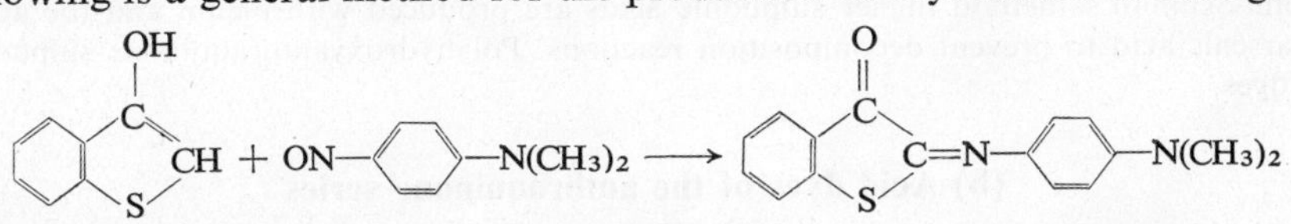

In the alkaline condensation of hydroxythionaphthene with p-nitrosodimethylaniline it reacts in its keto form to the anil of diketohydrothionaphthene. This anil can then be condensed with any hydroxythionaphthene to asymmetric thioindigoes (Indanthrene brilliant pink 3 B, Indanthrene print pink FFB).

Condensation can also be carried out in this way with o-diketones such as isatin or acenaphthenequinone, e.g.

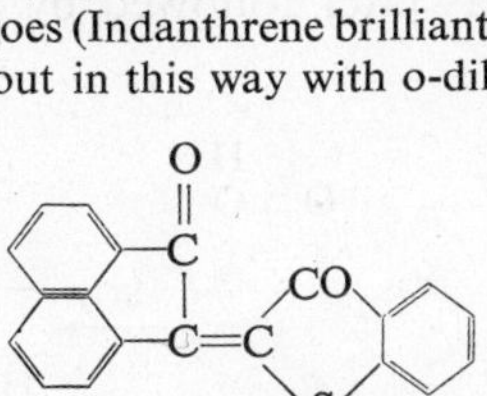

Algol Scarlet GG (Hoechst) (Ciba Scarlet G-extra)

Acenaphthenequinone is obtained from acenaphthene with ethyl nitrite. The thioindigoes still retain their importance.

*Leuco vat dye esters (Indigosoles (D. u. H.), Anthrasoles (Hoechst))*

In 1921 Bader and Sunder (Durand and Huguenin, Basle) found that the sulphuric acid esters of leuco dyes of the indigo- and indanthrene groups were stable compounds and could be used in dyeing. They are produced by reacting the leuco compounds with chlorosulphonic acid and pyridine at 50°C. The pyridine salt which forms first is transformed with caustic soda into the Na salt. To develop the dyed or printed leuco ester salt on the fibre the material is treated with a sodium nitrite solution, pressed, and the ester is saponified in a second bath with sulphuric acid; the nitrous acid liberated during this process oxidizes the leuco compound to a dye.

## 5. ANTHRAQUINONE DYES

### (a) Anthraquinone mordant dyes

One of the oldest dyes is alizarin which has been used for thousands of years and is obtained from the madder root (a member of the Rubicaea family). It was also the first anthraquinone dye to be produced synthetically after its constitution had been established by Graebe and Liebermann (1871). The way to the synthesis was discovered almost simultaneously by Graebe, Liebermann, and Caro, as well as by Perkin and by Rieser (Hoechst). Anthraquinone is sulphonated with oleum and the salt of the anthraquinone-2-sulphonic acid ('silver salt') is heated under pressure with a 20 per cent NaOH solution, a further hydroxy group entering in the 1-position. This goes to completion only in the presence of an oxidant (e.g. sodium nitrate). If on the other hand, only the —$SO_3H$ group is to be transformed into the —OH group, a lime fusion is carried out, adding $MgCl_2$. With metallic salts alizarin forms 'lakes'. According to the type of lake alizarin gives completely different colours: aluminium lake a brilliant red, iron lake violet, chromium lake chocolate brown. These lakes are also used as artists' colours, e.g. madder red. The anthraquinone mordant dyes have long since lost their importance. Alizarin has been displaced by Naphthol AS.

A variety of polyhydroxyanthraquinones, up to hexahydroxyanthraquinone, can be obtained by the introduction of further sulphonic acid groups into anthraquinone or hydroxyanthraquinone, and fusing. Using Bohn-Schmidt's method higher sulphonic acids are produced with oleum and the addition of boric or arsenic acid to prevent decomposition reactions. Polyhydroxyanthraquinone sulphonic acids are wool dyes.

### (b) Acid dyes of the anthraquinone series

The first acid dye of the anthraquinone series for dyeing wool was R. E. Schmidt's Alizarinsaphirol B in 1897 which was later followed by a whole series of acid dyes with excellent fastness to light.

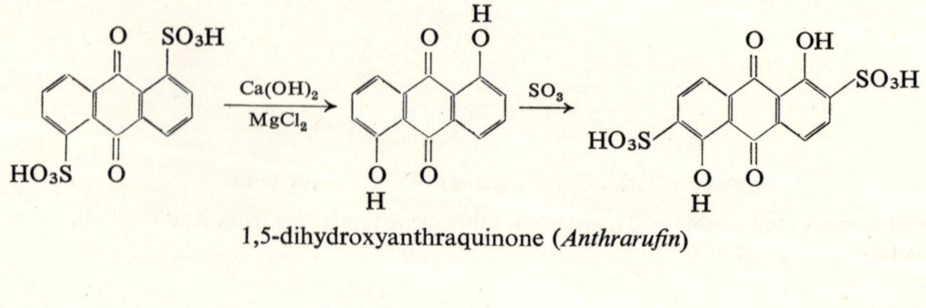


1,5-dihydroxyanthraquinone (*Anthrarufin*)

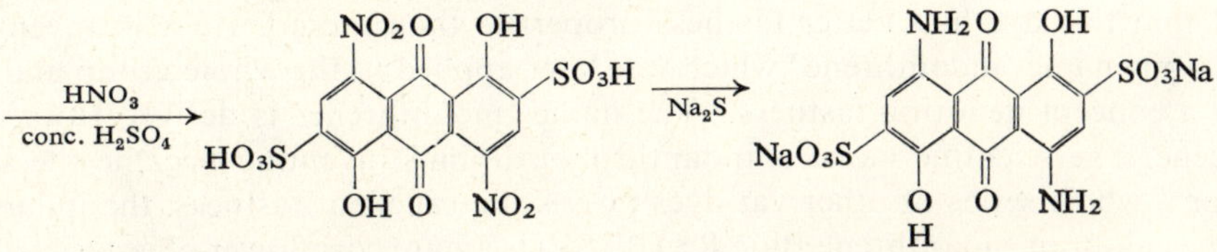

*Alizarinsaphirol B*
(= *Acilansaphirol B, Bayer*)

By splitting off an $HSO_3$ group with $H_2SO_4$ in the presence of boric acid at 125°C one obtains Alizarinsaphirol SE (Acilansaphirol SE, Bayer). These two dyes give wool a beautiful blue colour with a reddish tinge. Alumina lakes are true blue pigments.

Other acid alizarin dyes contain one or two arylamino groups carrying sulphonic acid residues. They can be introduced by exchanging with the bromine in aminobromoanthraquinone sulphonic acid in the presence of finely dispersed copper, or by substituting OH groups by heating with primary aromatic amines in the presence of boric acid.

From 1-bromo-4-aminoanthraquinone-3-sulphonic acid and aniline one obtains:

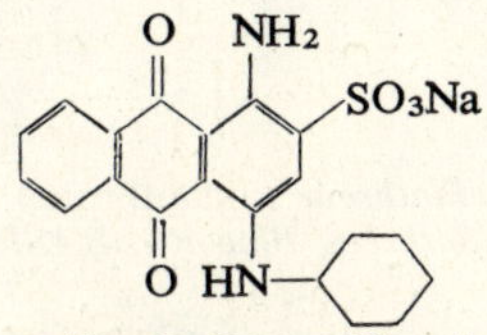

*Alizarin Direct Blue A*
(*AcilanDirectBlueA,Bayer*)

from quinizarin and p-toluidine, sulphonated:

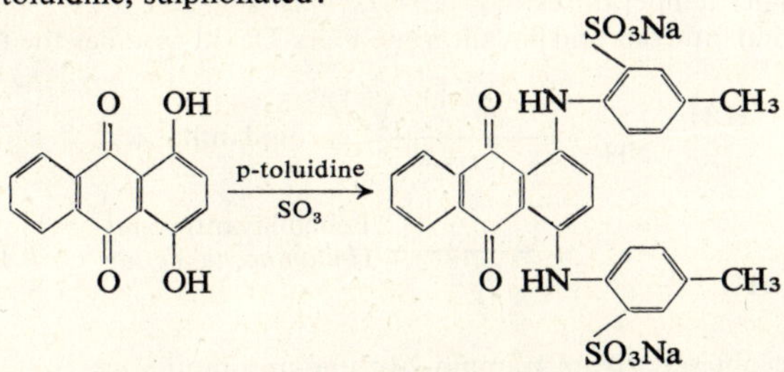

*Alizarincyanine Green G-extra* (*Bayer*)

A classic example of the use of Skraup's quinoline synthesis is the production of alizarin blue:

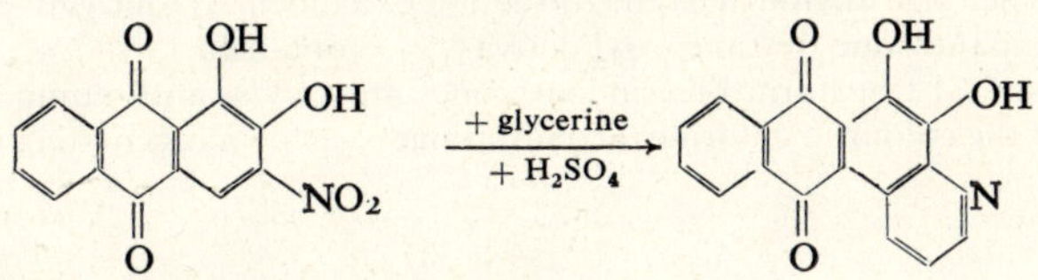

All pigments of this series are good vivid wool dyes.

## (c) Anthraquinone vat dyes

This is the largest and most important series of anthraquinone dyes. The first dye was Indanthrene Blue RS, obtained by René Bohn in 1901 by alkaline fusion of 2-aminoanthraquinone at 200–300°C. Bohn realized the full significance of this discovery and showed that this dye had better fastness properties than indigo. He also invented the rather happy name 'indanthrene' which was later applied to the whole group and finally became a concept denoting fastness. (The name 'indanthrene' is derived from indigo-anthracene. The scientific name is indanthrone). Bohn's invention became the starting point for a whole series of other vat dyes possessing excellent fastness, the indanthrene dyes. The colour of Indanthrene Blue RS (BASF) is a pure cornflower blue, its fastness to light is 8, its fastness to water, washing, acid, sulphur, rubbing, and perspiration is 5, the highest degree, but its fastness to chlorine is only 3. The constitution of indanthrene was established by R. Scholl:

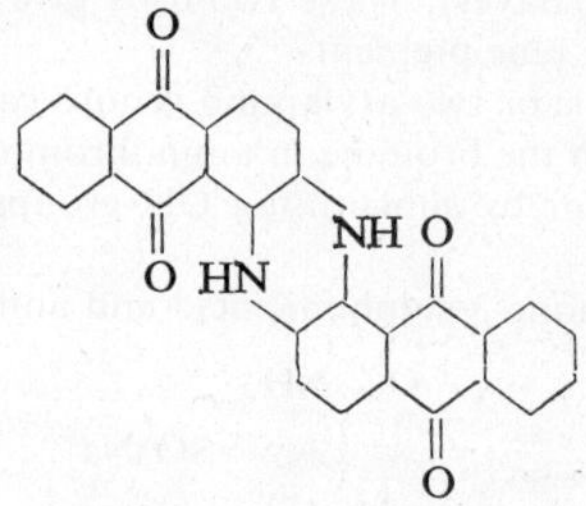


*Indanthrene* (indanthrone)
(*Indanthrene Blue RS, BASF*)

Technically, the alkaline fusion is carried out with a KOH-NaOH mixture with the addition of sodium nitrate. In spite of the efforts made over many years it has proved impossible to increase the yield beyond 50 per cent. A by-product of the fusion is alizarin. The yellow dye flavanthrone is also produced, larger quantities forming at higher temperatures.

For the formation of indanthrone and flavanthrone Fierz-David assumes the following intermediate:

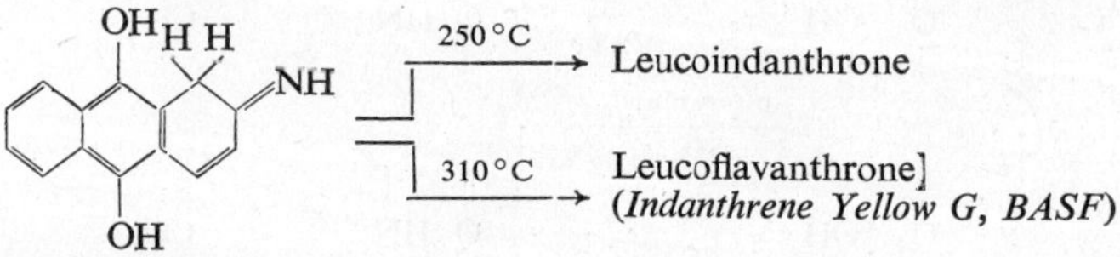


Indanthrone can also be obtained from 1-amino-2-bromo-anthraquinone. Whilst indanthrone cannot be produced by this process, it can be used for producing its derivatives (Tomaschewski).

With sodium dithionite, $Na_2S_2O_4$, one obtains as the first stage in the reduction (vatting) the blue dihydroindanthrone which was identified by the formation of a dibenzoyl derivative. By further reduction the brown tetrahydroindanthrone (tetrabenzoyl derivative) is obtained.

Oxidants such as NaOCl transform blue indanthrone into yellow anthraquinoneazine. To this fact is due the sensitivity of the colour to chlorine (greenish tinge). Reducing agents (e.g. dithionite) regenerate indanthrone.

### *Flavanthrone*

When the alkali fusion is carried out at temperatures in excess of 300°C, flavanthrone (Indanthrene Yellow G) instead of indanthrone is formed. Flavanthrone was also discovered by R. Bohn who called the yellow dye 'Flavanthrene'. Its constitution was established by R. Scholl (1907). The yields from fusions of 2-aminoanthraquinone with $AlCl_3$ or boiling with antimony pentachloride or titanium tetrachloride in nitrobenzene are better than those obtained in the alkali fusion. This dye still retains its importance.

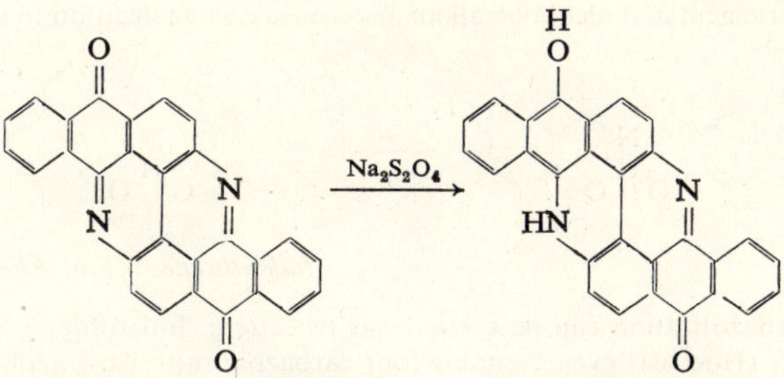

Flavanthrone (*Flavanthrene*) (*Indanthrene Yellow B*)

Leucoflavanthrone (*dihydroflavanthrone*)

Flavanthrone is based on dinaphthylenedi-imine or dehydrodinaphthylenedi-imine as a ring system which was produced in 1940 by Rieche and Rudolph, and simultaneously by Zeh, Leverkusen, by a different route:

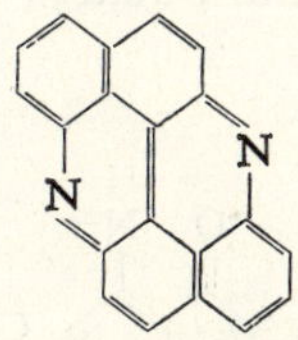

Dehydrodinaphthylenedi-imine

The discovery of the two anthraquinone derivatives indanthrone and flavanthrone started off systematic work on anthraquinone chemistry in all dye factories.

### *Acylaminoanthraquinones and anthrimides*

Anthraquinone is yellow and its vat is red. It is not yet a dye, but comparatively simply constructed anthraquinone derivatives are vat dyes, e.g. benzoyl-1-aminoanthraquinone which, however, is no longer of any importance.

Acylation is carried out in high-boiling solvents with acid chlorides, with the addition of HCl-binding substances.

The dibenzoyl compound of 1,4-diaminoanthraquinone is Indanthrene Red 5 GK (Bayer),

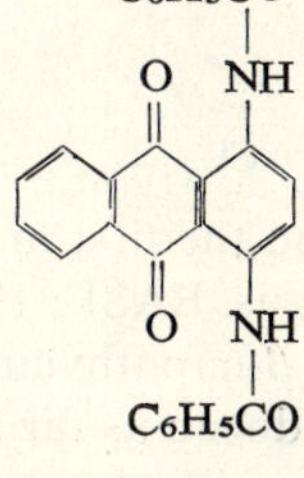

that of 1,5-diaminoanthraquinone is Indanthrene Yellow GK (Bayer).

The linkage of two anthraquinone radicals via an imino group (anthrimides) is achieved by reacting chloroanthraquinones with aminoanthraquinones (orange dyes).

1,1′-anthrimides can be transformed into carbazoles with aluminium chloride ('carbazolization reaction'). For this purpose, as in vat dye chemistry generally, an aluminium chloride—sodium chloride melt is frequently used (85 parts $AlCl_3$ and 15 parts NaCl, melting point 100°C, G. Kränzlein, Hoechst). The advantage of this mixture is that it melts at a lower temperature and is more highly fluid than $AlCl_3$ alone, and it sublimates less.

Concentrated sulphuric acid and alcoholic alkali also cause carbazolization in some cases.

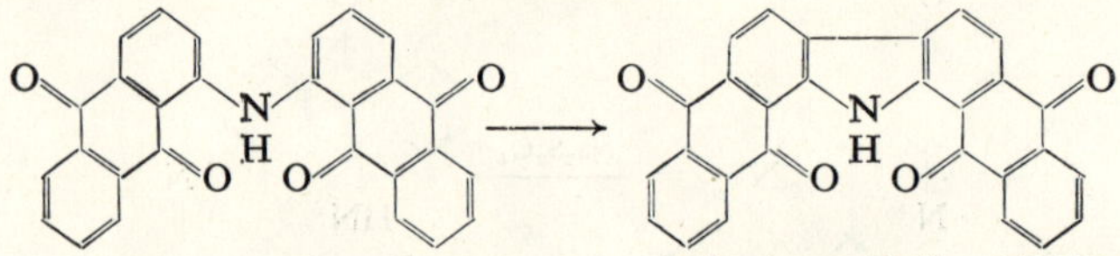

*Indanthrene Yellow FFRK*

With dianthrimides carbazolization can be carried out twice (e.g. Indanthrene Yellow 3 R, Hoechst). Indanthrene Khaki GG (Hoechst) even contains four carbazole radicals. Finally, anthraquinones can also be linked by pyridine rings.

*Anthraquinoneacridones*

Anthraquinoneacridones can be obtained by cyclization from the o-carboxylic acids of aryl-1-aminoanthraquinones with sulphuric acid or phosphorus pentoxide (via the chlorides of the acids):

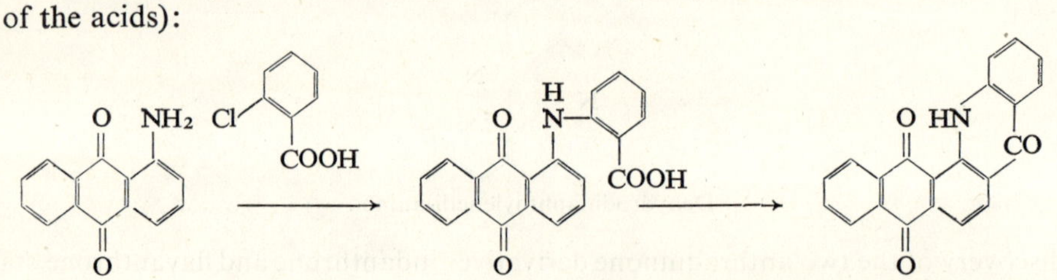

Anthraquinone-2-benzacridone

This dye is also formed from chlorobenzene and 1-aminoanthraquinone-2-carboxylic acid. Anthraquinoneacridones chlorinated to various degrees in the benzene fraction (Indanthrene Red Violet RRK, Indanthrene Brilliant Pink BBL (BASF)) are of some importance.

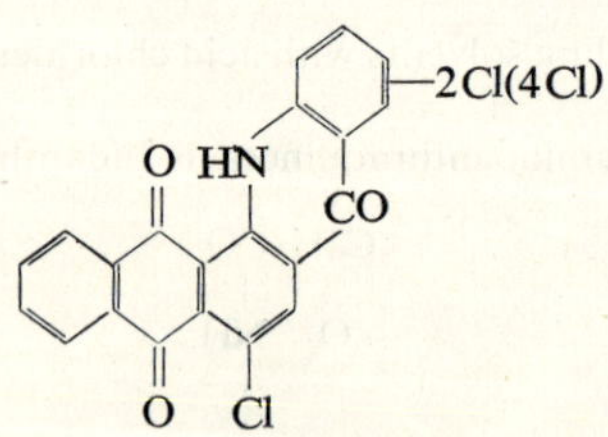

The first anthraquinoneacridone and the first indanthrene dye to give a clear red was Indanthrene Red RK by A. Lüttringhaus, BASF, 1910. It is obtained e.g. from 1-chloroanthraquinone-2-carboxylic acid and $\beta$-naphthylamine.

Anthraquinone derivatives are produced as far as possible from anthraquinone itself by the introduction of substituents. In many cases, however, synthesis is either the only possible way of producing the substituted anthraquinone, or it is more advantageous to use this method. For this purpose, phthalic anhydride or substituted phthalic anhydrides are used which with aromatics yield o-benzoylbenzoic acids when reacted with aluminium chloride under mild conditions (about 50°C). Cyclization to anthraquinone is carried out with concentrated sulphuric acid. An example is the synthesis of 2-methylanthraquinone to yield Indanthrene Red RK:

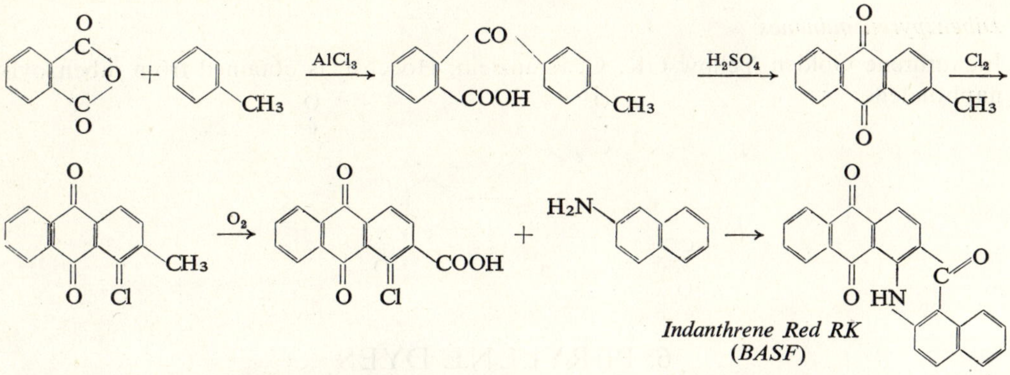

*Indanthrene Red RK (BASF)*

Indanthrene Violet FFBN (by Ullmann, 1909) contains two acridone rings.

*Pyranthrone and anthanthrone*

If CH groups are substituted for the N atoms in flavanthrone one obtains the ring system of pyranthrone which was produced by R. Schulz from 1-chloro-2-methylanthraquinone with Cu powder (Ullmann's reaction) via 1,1′,2,2′-dimethyldianthraquinonyl.

This is transformed into pyranthrone by melting with zinc chloride or potassium butoxide:

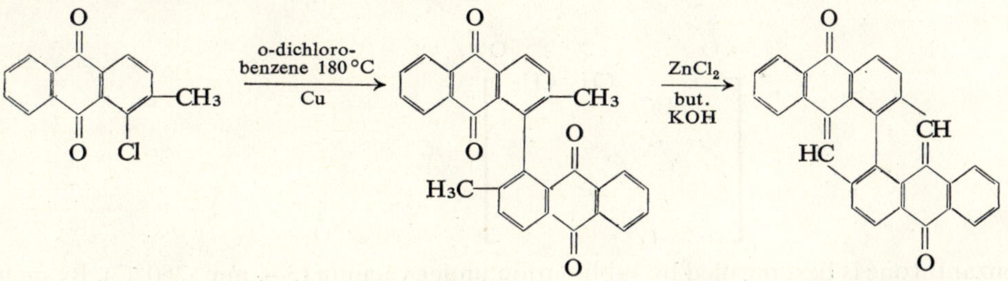

Pyranthrone
(*Indanthrene Golden Orange BASF*)

Bromo-pyranthrones are used in industry.

Similarly, only the halogen derivatives of anthanthrone (Kalb, 1914) are used.

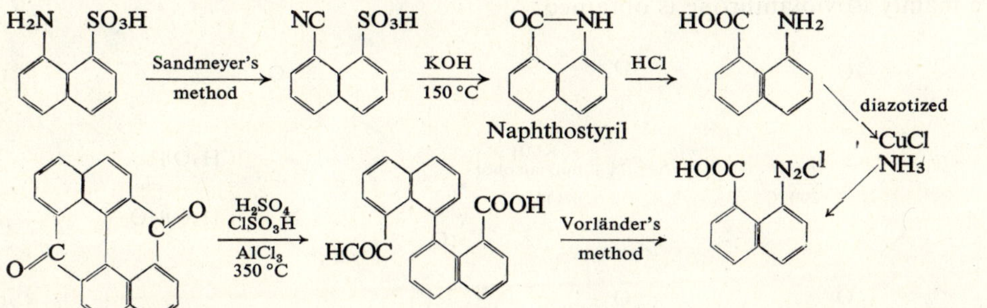

Anthanthrone

The most important representative of this group is dibromoanthanthrone (Indanthrene Brilliant Orange RK, Cassella).

*Dibenzpyrenequinones*

Indanthrene Golden Yellow GK, G. Kränzlein, Hoechst, is obtained from dibenzoyl-naphthalene:

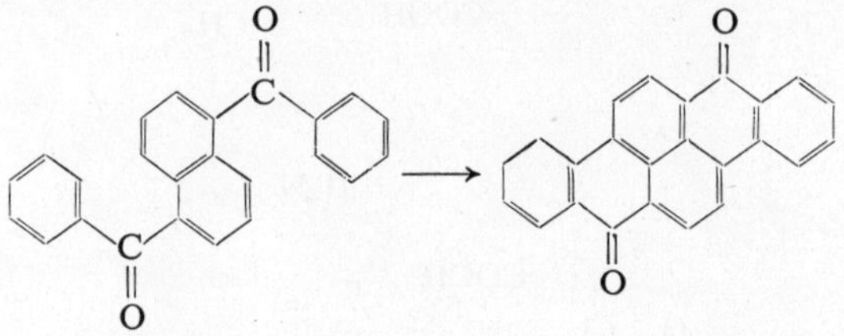

## 6. PERYLENE DYES

### (a) Violanthrones and isoviolanthrones

The starting material for violanthrones and isoviolanthrones is benzanthrone which is produced from anthraquinone and glycerol with an 80 per cent sulphuric acid solution (acrolein forming from the glycerine), in the presence of finely divided iron.

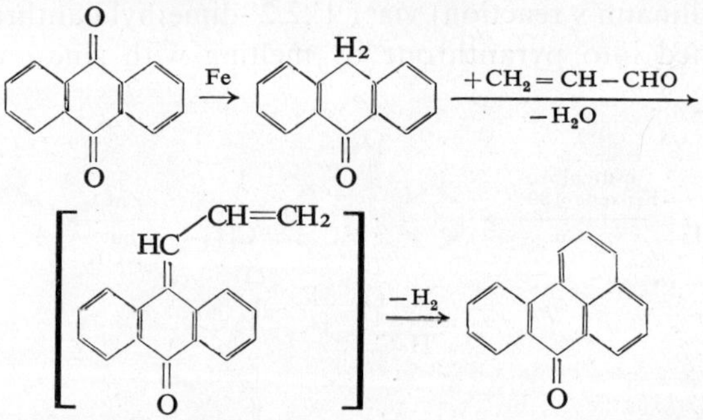


Benzanthrone is best purified by sublimation under vacuum (3–4 mm, 250°C). By melting with KOH at 180–225°C violanthrone (Bally, 1905), a perylene ring system formed by the splitting off of 4 H atoms, is produced. According to the conditions, using a KOH-NaOH mixture with the addition of sodium acetate, violanthrone, using alcoholic potash lye mainly isoviolanthrone is obtained:

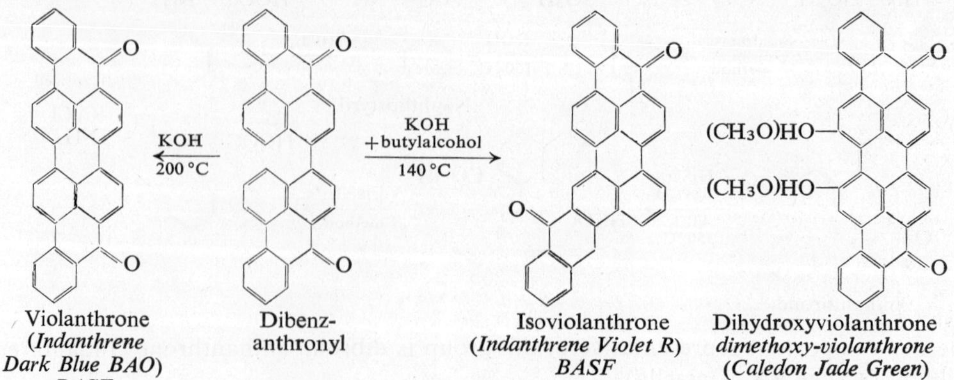


Violanthrone (*Indanthrene Dark Blue BAO*) *BASF* — Dibenz-anthronyl — Isoviolanthrone (*Indanthrene Violet R*) *BASF* — Dihydroxyviolanthrone *dimethoxy-violanthrone* (*Caledon Jade Green*)

These dyes can be separated as leuco compounds. Violanthrone can be transformed by Isler and Bally's method (1912), i.e. in sulphuric acid with $MnO_2$ into dihydroxyviolanthrone which was useless as a vat dye on account of the hydroxy groups. In 1920, Davis, Thomson, and Thomas carried out the methylation and obtained an extremely bright and very fast green dye which they called 'Caledon Jade Green' I.C.I. (Indanthrene Brilliant Green B and FFB, BASF). It is more convenient to obtain it from dibenzanthronyl with subsequent oxidizing condensation.

Other dyes with perylene groups are produced by melting 1,8-naphthalene-dicarboxylic acid imide with KOH. As in violanthrone production a perylene forms, perylene-3,4,9,10-tetracarboxylic acid di-imide. The corresponding methyl imide is Indanthrene Red GG, BASF, the anisidide Indanthrene Scarlet R, Bayer. The latter is produced from perylene tetracarboxylic acid and p-anisidine:

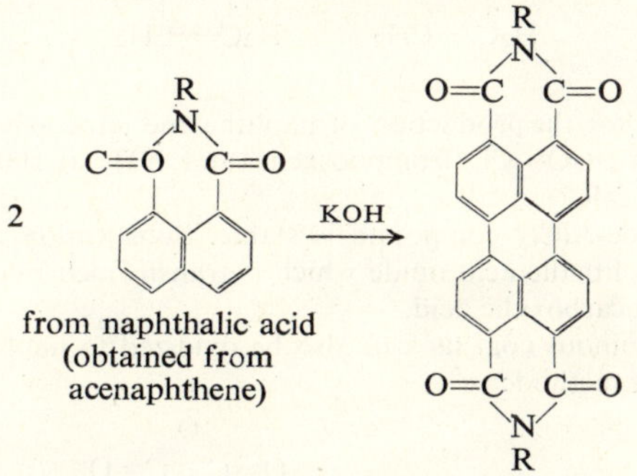


## (b) Naphthalene tetracarboxylic acid dyes

An interesting pair of dyes is derived from naphthalene-1,4,5,8-tetracarboxylic acid. These are probably the most expensive vat dyes on the market but their depth of colour, brightness, and fastness surpass all others.

If naphthalene carboxylic acid is condensed with o-phenylene-diamine a mixture of two imidazol dyes forms which dyes a brilliant scarlet (Indanthrene Scarlet GG, Hoechst). The relationship of these two dyes is that of the cis- and trans-forms and they can be separated via the alkali salts of the leuco compounds, variously soluble in alcohol, with alcoholic potash (Hoechst):

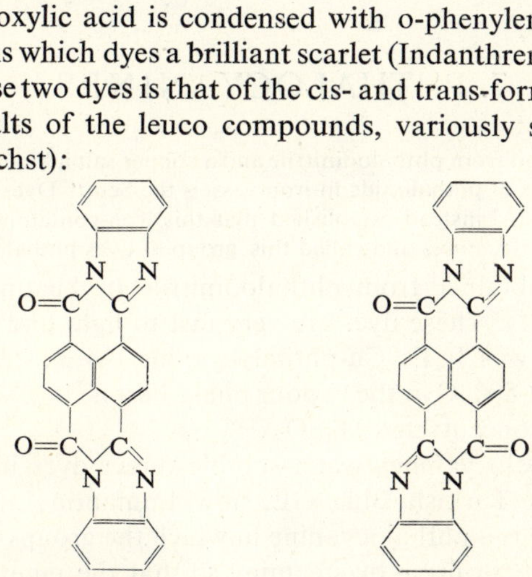

Indanthrene Bordeaux HRR (Hoechst)

Indanthrene Brilliant Orange GR (Hoechst)

Substituted o-diamines are also reactive. Thus it is possible to obtain a brown dye as a cis-trans mixture, one of the isomers of which is a black dye.

Naphthalene tetracarboxylic acid is produced from acenaphthene and malodinitrile:

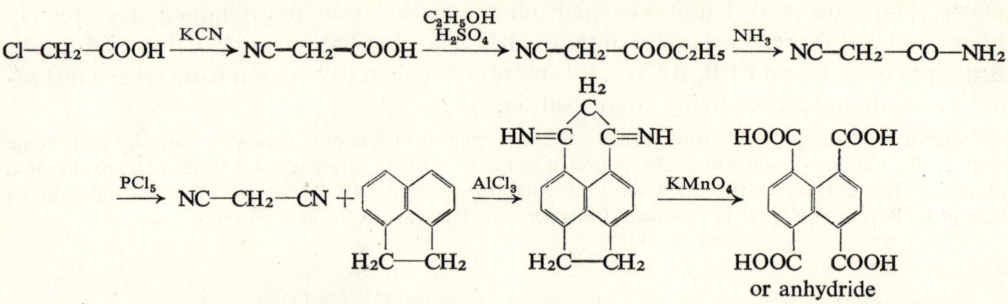

An interesting alternative for the production of naphthalene tetracarboxylic acid is the reaction of carbamic acid chloride $H_2N—CO—Cl$ (fromphosgene and $NH_3$ at 500°C, Hopff, BASF) with acenaphthene in the presence of $AlCl_3$.

The carbamic acid chloride-$AlCl_3$ compound is stable. Substitution takes place exclusively in the 1,8-position and yields acenaphthalic acid imide which is transformed by saponification and oxidation into naphthalene-1,4,5,8-tetracarboxylic acid.

Finally, pyrene from bituminous coal tar can also be oxidized to naphthalene tetracarboxylic acid. This is probably the cheapest method:

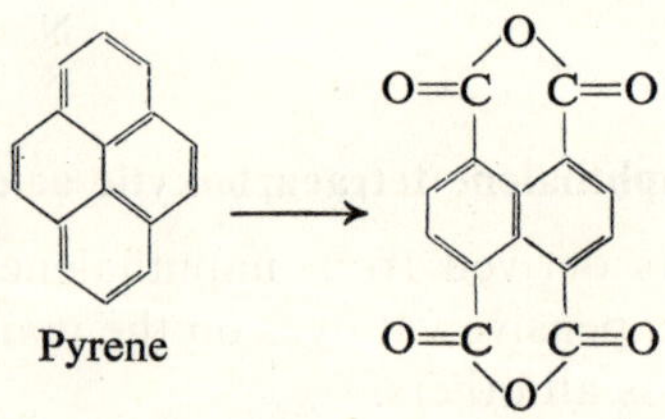

## 7. PHTHALOCYANINES

In 1927, De Diesbach obtained from phthalodinitrile and a copper salt a blue compound which was very stable. During the production of phthalimide in iron vessels the Scotts Dyes Company obtained a blue dye as a by-product. In 1932, Linstead established that this iron-containing dye and De Diesbach's Cu compound had identical structures and called this group of dyes phthalocyanines.

Phthalocyanines are obtained from phthalodinitrile by heating with metals, particularly copper and its salts. These dyes are very fast to light and have great luminosity. The first representative was ICI's Cu-phthalocyanine (Monastral Blue, Drescher and Wyler). It is stable e.g. at 500°C in the vapour phase despite its 28 per cent metal content. It is also stable in cold concentrated $H_2SO_4$.

By sulphonating phthalocyanines water-soluble direct dyes in vivid shades are obtained, e.g. Sirius Light Turkish Blue GL, now Lumantin Light Turkish Blue GL, BASF. ICI's Alcian Blue is a phthalocyanine in which the groups which render it soluble in water are split off after printing by steaming so that the pigment forms on the fibre. This is achieved by the introduction of quaternary ammonium groups which split off during steaming.

Finally, in 1951, Bayer succeeded in producing in Phthalogen-Brilliant Blue IF 3 G a mixture which after printing forms the dye from the elements during steaming. It contains aminoiminoisoindoline which is obtained by the following synthesis (Rösch and Wolf) (*Angew. Chem.* 1956, **68**, 133).

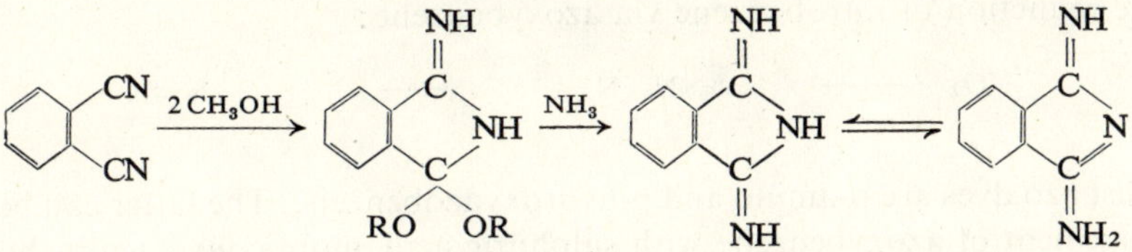

In addition, the discovery of these preliminary products provided an entirely new explanation for the formation of phthalocyanines. Aminoiminoisoindoline is a very reactive substance which, by self-condensation or with its dihydro compound, can form chains in the manner of an indamine condensation with elimination of $NH_3$; in the presence of copper salts these chains easily form a ring to yield Cu-phthalocyanine:

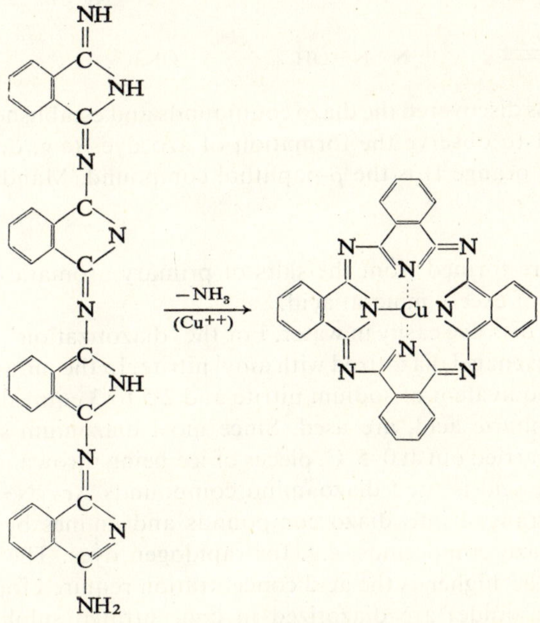

The constitution of the phthalocyanines resembles that of haemine and chlorophyll. In these the iron or magnesium is linked by 4 pyrrole nuclei.

Another industrial process for the production of phthalocyanines is the reaction of 4 moles phthalic anhydride and 1 mole copper(II)-chloride with urea in the presence of boric acid with the addition of catalysts such as ammonium vanadate at 150–190°C and in the presence of dichlorobenzene as a thinner. Co- and Ni-salts are also used. Under pressure in autoclaves 15–16 atoms of chlorine can be introduced, lower degrees of chlorination are obtained without water in a neutral medium. Chlorinated phthalocyanines are green dyes, e.g. Heliogen Green G (BASF) = Monastral Green S (ICI). This field is still being actively developed, particularly so the 'phthalogen principle', i.e. the formation of the phthalocyanine on the fibre, and the development of new shades.

# 8. AZO DYES

The parent substance of azo dyes is the orange-coloured azobenzene which is obtained by alkaline reduction of nitrobenzene via azoxybenzene:

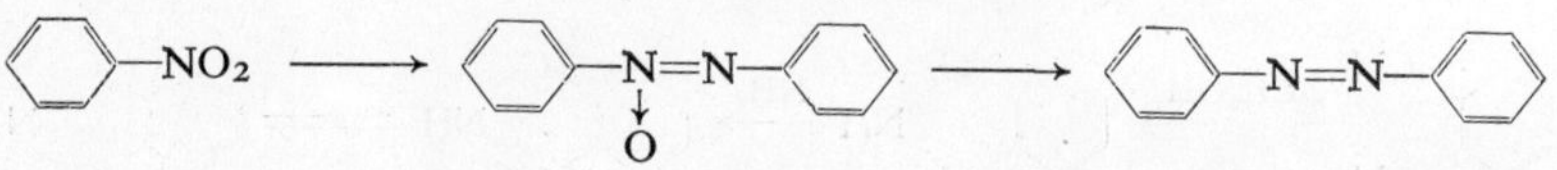

The simplest azo dyes are p-amino and p-hydroxyazobenzene. The latter can be obtained by rearrangement of azoxybenzene with sulphuric acid, more conveniently, however, by reaction of diazotized aniline with phenol. This reaction is called coupling. The diazo compound is the 'active component', the compound capable of coupling (e.g. a phenol or aromatic amine) the 'passive component' (diazotized aniline + phenol = p-hydroxyazobenzene; expressed in abbreviated form by: aniline → phenol).

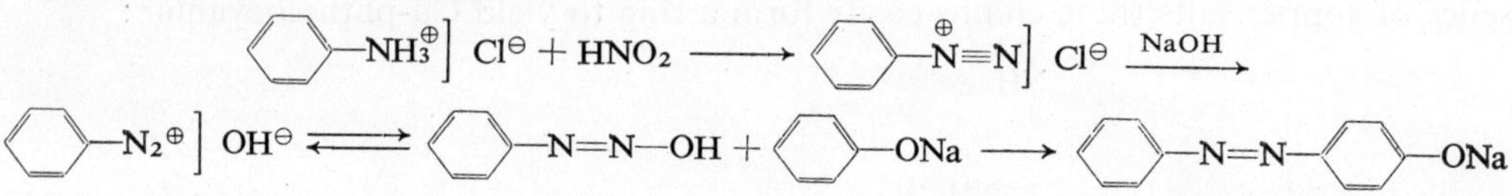

In 1858 Peter Griess discovered the diazo compounds and established their constitution. He was also the first to observe the formation of azo dyes (e.g. orange I = sulphanilic acid → α-naphthol; orange II is the β-naphthol compound, Mandarin G).

*Diazotization*

Diazo compounds are formed from the salts of primary aromatic amines with nitrous acid in the presence of excess mineral acid.

Almost all of them dissolve easily in water. For the 'diazotization', Griess used vapours of nitrous acid, Knoevenagel diazotized with amyl nitrite. In the industrial process usually a little more than 1 equivalent of sodium nitrite and 2·5 to 3 equivalents of hydrochloric acid, sometimes sulphuric acid, are used. Since most diazonium salts are sensitive to heat the reaction is carried out at 0–5°C, pieces of ice being thrown into the diazotization solution. If too little acid is used diazoamino compounds Ar—N=N—NH—Ar often form which are rearranged into diazo compounds and amines by acids. They can be used as stabilized diazo compounds, e.g. for rapidogen dyes. The less basic an amine (e.g. nitraniline) is, the higher is the acid concentration required for diazotization. Very weak bases (e.g. picramide) are diazotized in concentrated sulphuric acid or glacial acetic acid with nitrosyl sulphuric acid.

The conditions of diazotization, particularly the degree of acidity and the temperature, vary strongly according to the chemical nature of the amine and the direction of coupling desired. They have to be determined by a series of tests.

The cooled diazonium salt solutions are used directly for the production of the dyestuffs. Solid diazonium salts are unstable and are very explosive. They can, however, be isolated as complex salts, e.g. zinc chloride double salts, or as naphthalene sulphonates; these are stable and are used in dyestuff technology as active components ('fast salts'), e.g. for rapid fast dyes in naphthol AS printing.

*Coupling*

When diazonium salts are treated carefully with a solution of caustic soda, the diazonium cation first absorbs a hydroxyl ion; diazonium hydroxide forms as an unstable intermediates which reacts with another hydroxyl ion forming the diazotate ion.

$$[Ar-\overset{\oplus}{N}\equiv N \rightleftharpoons Ar-\bar{N}=\bar{N}^{\oplus}] \xrightarrow{+OH^{\ominus}} (Ar-\bar{N}=\bar{N}-OH) \xrightarrow[-H_2O]{+OH^{\ominus}} [Ar-\bar{N}=\bar{N}-\underline{\bar{O}}|]^{-}$$

First normal, easily coupling diazotates form which rearrange when heated with strong alkalilye into isomeric isodiazotates which couple with difficulty.

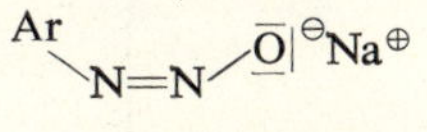

Normal diazotate
(couples)

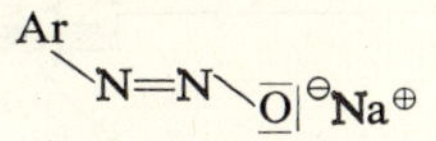

Isodiazotate
(couples with difficulty)

For the production of dyestuffs the phenolate or amine is introduced into a more or less strongly alkaline or acetic acid solution, then the diazonium solution is added with constant stirring, or alternatively it is the diazonium solution that is introduced into the alkali or acid solution. The position of coupling depends mainly on the nature of the diazo compound and, in compounds which have two positions capable of coupling, such as α-naphthol and α-naphthylamine, also on the pH. In the 1-naphthol and 1-naphthylamine series the steering influence of sulphonic acid groups is very marked. When the preferred position is occupied coupling occurs at another position or when for instance in β-naphthol the 1-position, being the only point capable of coupling, is occupied the substituent is eliminated.

The important aminonaphthol sulphonic acids, discussed earlier in the chapter on 'Intermediates', such as gamma acid, I-acid, and H-acid, have two positions capable of coupling. Thus, for instance, it is possible first to couple in an alkali medium with a diazo compound in the position next to the OH group, then in an acid medium with another diazo compound in the position next to the $NH_2$ group:

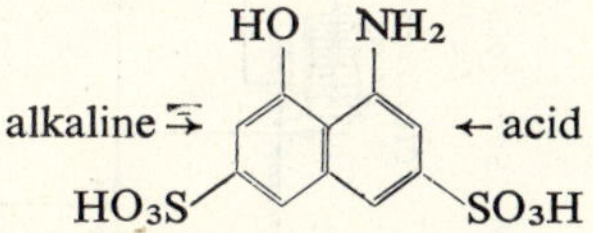


Coupling of H-acid

This is one of the routes to disazo dyes. Polyvalent phenols also have multiple coupling positions.

Another route is via the 'tetrazotization' of diamines (e.g. benzidine) and coupling with two passive components to yield disazo dyes. If azo dyes in their turn can be diazotized simply or multiply, the products may be trisazo dyes or tetrakisazo dyes.

The number of compounds that can be transformed into azo dyes is very large. There is an almost infinite number of possible combinations. Azo dyes can have the most extreme properties. Their production is comparatively simple and can be carried out in an aqueous medium. For these reasons azo dyes are still the most important and most versatile class of dyestuffs. At present, probably about 1,600 azo dyes are on the market. A list of the more important components used in diazotization, together with those most commonly used for coupling, can be found in Ullmann, *Enzyklopädie*, 3rd ed., vol. 4, 78–81.

*Equipment*

An azo dye factory consists of several storeys or platforms (usually 4) and a basement for the pressure vessels. It is linked with a large refrigeration plant since very large quantities of ice in lumps are required

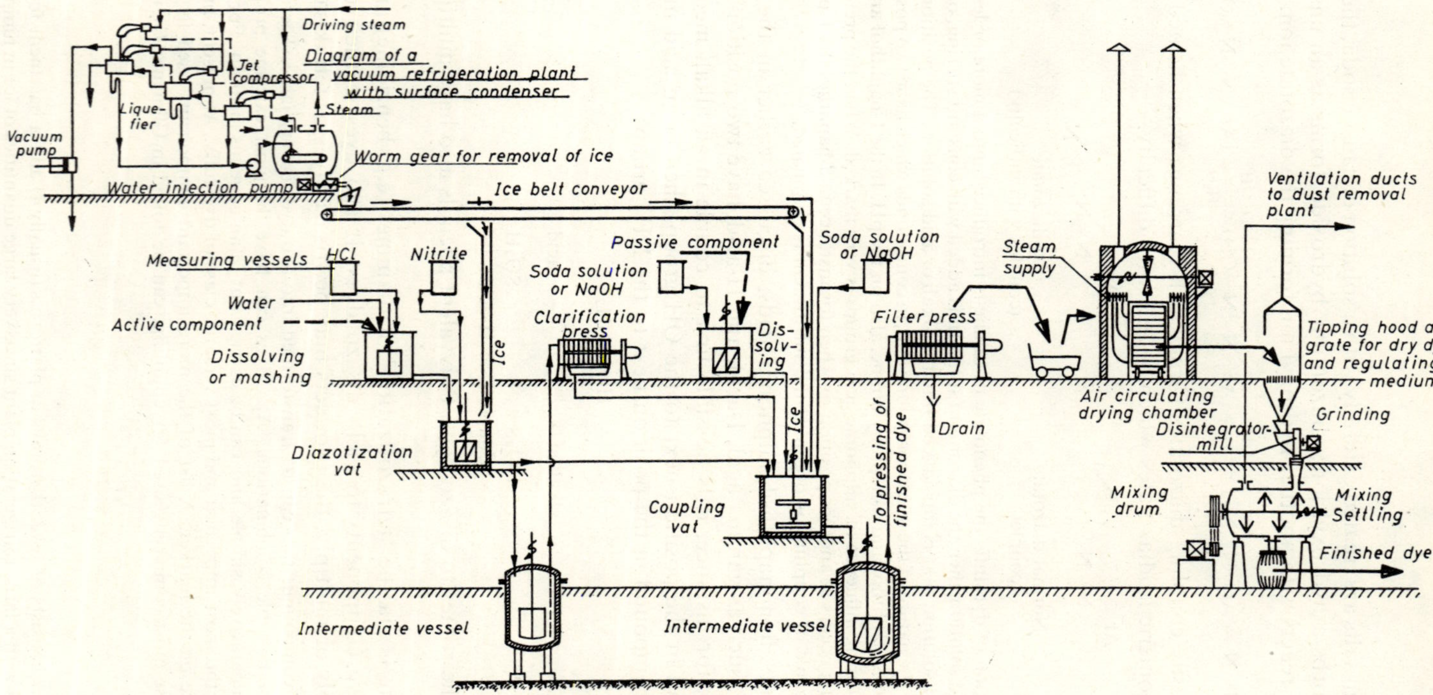


Figure 148. Diagram of azo dyes production

(Kältetechnik und Eiserzeugung: Ullmann, *Enzyklopädie*, 1, 310). In older systems ice bars were produced in cooling chambers which were placed in a brine vessel cooled by a refrigerating machine; they were then broken up in crushers and conveyed in wagons to the point of consumption. Modern plants operate on the cooled roller principle. A roller through which flows a coolant is dipped into a basin filled with water. The layer of ice formed is scraped off and is transported on conveyor belts directly into the diazotization vats. Alternatively, a vacuum ice production plant incorporating a steam jet refrigerator can be used (see Figure 148).

Equipment used in the production of azo dyes consists of storage vessels, agitating vats, filter presses, pressure vessels for pressing, drying installation, etc. customarily used in intermediate chemical engineering. Only in rare cases, e.g. when pyridine or glacial acetic acid are used, is it necessary to employ closed vessels. The vats used for dissolving, diazotizing, coupling, and saponification are, therefore, usually open vessels. Since many azo dyes are sensitive to metals the equipment should not contain iron or copper components. It is made of wood (pitch pine), or of iron lined with brick, rubber, or plastic. Pipes are mostly lined with rubber or made of plastics, the chambers and frames of the filter presses are usually made of wood.

The diagram in Figure 148 shows the arrangement of the equipment in an azo dye factory.

## Some azo dyes

In view of the enormous number of industrially important azo dyes it is impossible within the framework of the present book to give even a short survey of this field. Only a few important groups and types are, therefore, described below.

### (a) Azo dyes for wool

Azo dyes for wool can be divided into four groups:

(1) acid dyes,
(2) chrome-developed and mordant dyes,
(3) one-dip chrome dyes, and,
(4) acid chrome complex dyes.

*Mono azo dyes*

The first azo dyes were acid dyes for which naphthol or naphthol sulphonic acids were used as coupling components, e.g. orange I and II (1876). The latter is produced by Farbenfabrik Wolfen under the name Mandarin G.

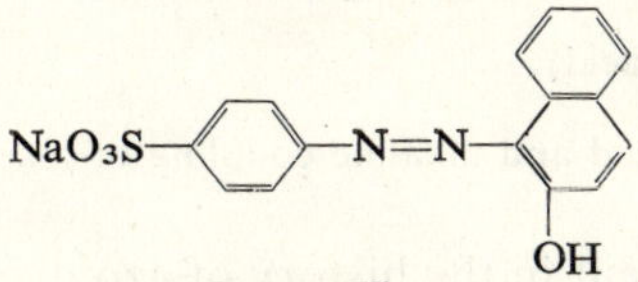


Orange II

The first red azo dye, Fast Red AV, now Basolan Red AV (BASF) or New Red Wo (Wolfen) (naphthionic acid → $\beta$-naphthol) was discovered by Caro in 1877. R- and G-salts were first produced in 1878 at Hoechst by Baum and the Ponceau dyes (e.g. xylidines → R-salt) were discovered in the same year. During this work the significance of the position of the sulphonic acid group for tint and dyeing properties was first realized. This group includes the red and Bordeaux dyes Naphthol Red SE, now Acilan Red Bayer (naphthionic acid → R-salt), Bordeaux R (1-naphthylamine → R-salt), Fast

Red E (naphthionic acid → Schäffer salt). The beautiful Supramin Red dyes, e.g. GG, B (Bayer), etc., which have high levelling capacity and better fastness to light and washing contain Nevile-Winther acid (naphtholsulphonic acid-1,4) as a coupling component. Coupling with chromotropic acid (1,8-dihydroxy-naphthalene-3,6-disulphonic acid) results in the blue and violet series, e.g. Viktoria Violet 4 BS now Acilan Violet 4 BS Bayer (4-nitraniline → chromotropic acid) and Azo Acid Blue B (4-amino-1-N-methyl-acetanilide → chromotropic acid). (See p. 269 for the production of chromotropic acid and H-acid.)

Metanil Yellow (Wolfen) (aniline-3-sulphonic acid → diphenylamine, Hepp, 1880) was the first acid yellow of the azo series and also the first wool dye with an amine as a coupling component.

The coupling of amines with diazonium solutions is acid coupling (e.g. acetic acid). Usually the diazo solution is prepared and the amine is added as such or in the form of a salt solution. When the amine is soluble only with difficulty, e.g. diphenylamine, it can be finely divided by means of an emulsifier (soft soap); this is done in the production of Metanil Yellow. Coupling, particularly that of amines, is often rather slow (sometimes 1–2 days). Frequently, the dyes are separated by salting out, then follows pressing in the filter press, washing with a sodium chloride solution, and drying.

Monoazo dyes are an important group which uses aminonaphthol sulphonic acids as coupling components. Gamma acid ($\gamma$-acid = aminonaphthol sulphonic acid-2,8,6), phenyl-gamma-acid (with a 2-phenylamino group), and H-acid are employed in their production (see pp. 268, 269). One obtains red, brown, blue, violet to black dyes. Gamma acid is usually coupled in acid media e.g. Anthralan Red B (p-aminodiphenylsulphon → $\gamma$-acid) and Anthralan Red G (Hoechst) (acet.-p-phenylenediamine-sulphonic acid → $\gamma$-acid) the coupling taking place in the 1-position next to the amino group. In the production of Supramin Brown R (Bayer) diazotized 4-nitro-2-anisidine is coupled in alkaline media with phenyl-$\gamma$-acid, the azo group taking the 7-position next to the OH group:

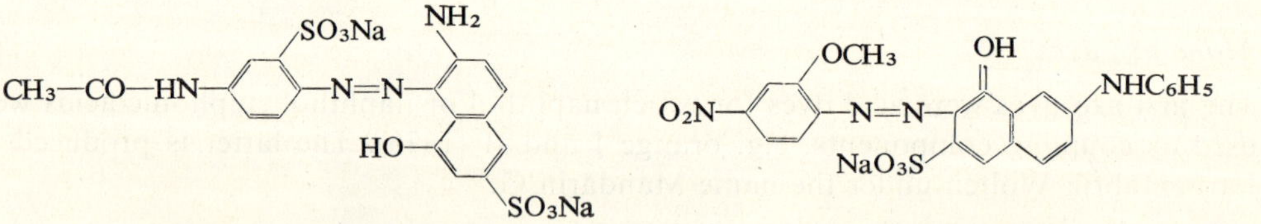


Anthralan Red G (acid) Supramin Brown R (alkaline)

Acid and alkaline coupling of $\gamma$-acid

A particularly important event in the history of azo dyes was the discovery of pyrazolone dyes. The first of these was tartrazine (Ziegler, 1884). Dyes of the pyrazolone series have retained their great importance right up to the present. The coupling component is 1-phenyl-pyrazolone with a variety of substituents in the 3-position of the pyrazolone radical and in the 2,6-position of the phenyl nucleus. Pyrazolones are usually produced from acetoacetic ester or oxalacetic ester and phenylhydrazines. The shades of this dye range from yellow to reddish orange. Tartrazine O (Hoechst) is now produced from 1-(4'-sulphophenyl)-3-carboxy-5-pyrazolone by coupling with diazotized aniline-4-sulphonic acid (sulphanilic acid).

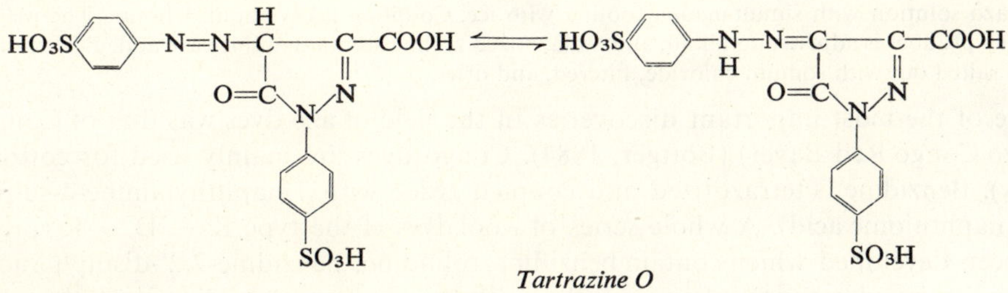

*Tartrazine O*

Fast Light Yellow 3 G (Hoechst) is produced in an analogous manner from aniline-2-sulphonic acid and the 2'-sulphonic acid of the above pyrazolone.

*Disazo dyes for wool*

The first disazo dye was put on the market in 1879 by Nietzki (Kalle) under the name Biebricher Scarlet (p-aminoazobenzene-2,4′-disulphonic acid → β-naphthol). In 1883 followed Brilliant Crocein M (MMO) by M. Hoffmann, Bayer, now Acilan Crocein M (MMO) which is still important:

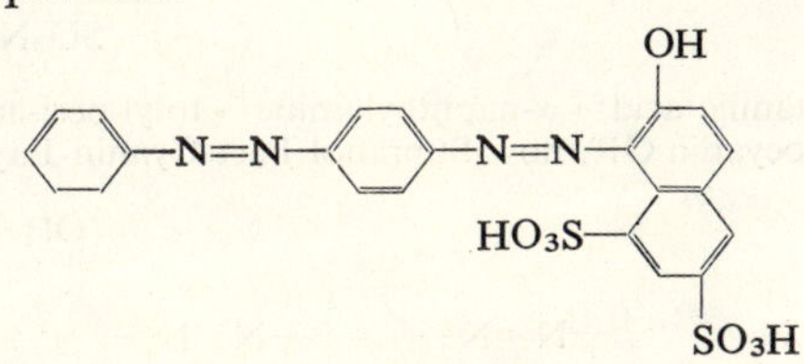

Brilliant Crocein M = aminoazobenzene → G-acid

The large group of primary disazo dyes $\begin{smallmatrix}D_1\searrow \\ D_2\nearrow\end{smallmatrix}K$ (D = diazotization component, K = coupling component) includes important wool black dyes. ('Primary' disazo dyes are formed when a phenol is coupled twice with 2 diazonium salt molecules, or a diamine is diazotized twice and coupled with 2 phenol molecules ($D_1 \rightarrow K \leftarrow D_2$, $K \leftarrow D \rightarrow K$). 'Secondary' disazo dyes are formed by coupling a diazotization component with an amine capable of coupling (M = middle component), further diazotization and coupling (D → M → K). The first of these was Naphthol Black B (Cassella) by A. v. Weinberg in 1885 (2-naphthylamine-6,8-disulphonic acid → R-salt ← α-naphthylamine). Naphthol Blue-Black discovered in 1891 at Mainkur (Amido Black 10 B, now Acilan Black 10 B Bayer) is still important (p-nitraniline $\xrightarrow{\text{acid}}$ H-acid $\xleftarrow{\text{alkal.}}$ aniline).

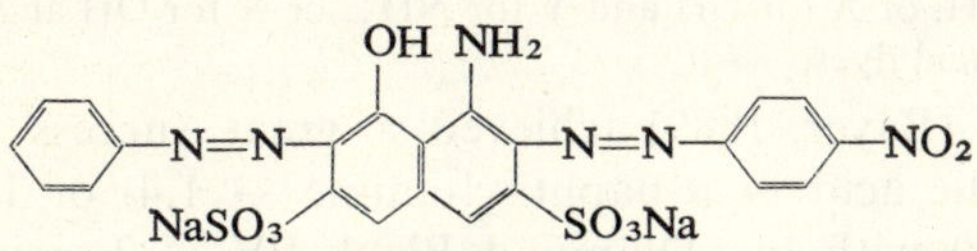

Amido Black 10B

Production of a diazo dye (Amido Black 10 B)

**p-Nitraniline is diazotized by adding a large excess of very dilute HCl to an aqueous suspension of the amine in a sodium nitrite solution which has been cooled to <10°C. H-acid solution is introduced into**

this diazo solution with simultaneous cooling with ice. Coupling takes about 3 hours. The pH of the coupling product is adjusted to alkali, and a diazotized aniline solution and ice are added. Then the mixture is salted out with sodium chloride, filtered, and dried.

One of the most important discoveries in the field of azo dyes was that of Congo red (Benzo Congo Red Bayer) (Böttger, 1884). Congo dyes are mainly used for cotton (see below). Benzidine is tetrazotized and coupled twice with 1-naphthylamine-4-sulphonic acid (naphthionic acid). A whole series of wool dyes of the type K ← D → K (primary) has been developed which contain benzidine, toluidine, benzidine-2,2'-disulphonic acid, diaminodiphenylmethane, diaminostilbenedisulphonic acid, or diaminodiphenylcyclohexane as a bifunctional amine, and are reacted with two equal or different coupling components (G-salt, phenol, pyrazolone, β-naphthol, Schäffer salt, etc.). Commercial dyes are Polar red, Polar yellow, Polar orange, sulphone yellow, and red Supranol dyes.

Secondary disazo dyes (D → M → K, M = middle component) include sulphocyanine dyes by Ph. Ott (Bayer) and Herzberg's famous Neroles (Agfa), e.g. Nerol VL, still produced by Wolfen.

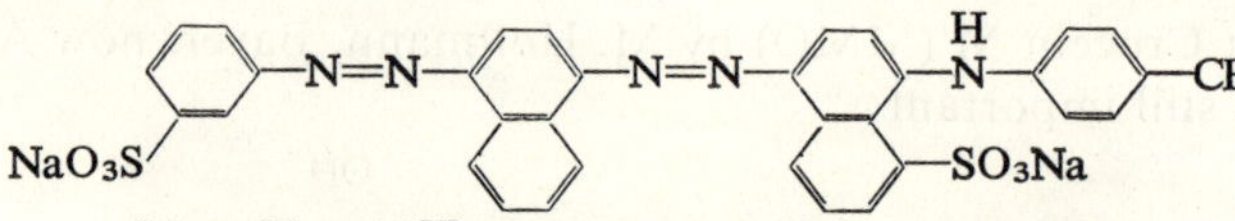

Metanilic acid → α-naphthylamine → tolyl-peri-acid
Sulfocyanin GR, now Supranol Fast Cyanin Bayer

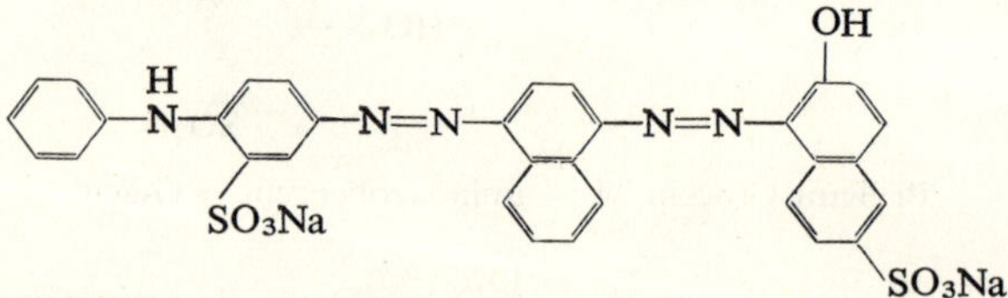

p-aminodiphenyl-aminosulphonic acid → α-naphthyl-amine → Schäffer acid
Nerol VL

### *Chrome dyes*

The first chrome dye was discovered by Nietzki in 1887 (Alizarin Yellow GG, Bayer, = metanitraniline → salicylic acid). From the diazo compounds of a wide variety of naphthylaminosulphonic acids, salicylic acid dyes are obtained which produce true mordant dyes on wool by after-treatment with chromate or $Cr^{III}$-salts. Another important finding was Erdmann's and Borgmann's discovery in 1893 that azo compounds with structural formulae of the type

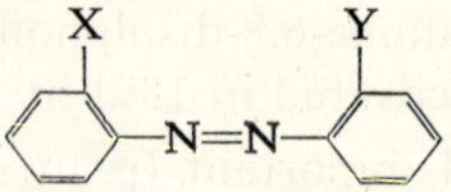

where X and Y stand for OH, or X for OH and Y for $NH_2$, or X for OH and Y for COOH, are faster chrome dyes than salicylic acid dyes.

Lauch and Krekeler (Bayer, 1889) achieved a great success with their Diamond Black F = aminosalicylic acid → α-naphthyl-amine → 1,4- or 1,5-naphtholsulphonic acid, and Kahn (Bayer) with his Diamond Black PV = 2-aminophenol-4-sulphonic acid → 1,5-dihydroxynaphthalene. This group also includes Eriochrome Black T (Hagenbach, Geigy) = 6-nitro-1-amino-2-naphthol-4-sulphonic acid → α-naphthol. These chrome black dyes have replaced natural logwood black in wool dyeing; logwood black is now used only for dyeing furs (Persian lamb black).

Afterchroming dyes were obtained in all shades, e.g. Chrome Yellow A and R, now Diamond Chrome Yellow A, R, (Bayer), Acid Alizarin Flavin R, and Acid Alizarin Red G (Hoechst), Fast Mordant Blue B, now Fast Diamond Blue BL (Bayer), Eriochrome Green H (Geigy).

In 1900, Agfa was granted patent rights for a dyeing process using one-dip chrome dyes. Certain dyes, metachrome dyes, can be used in the presence of chromate and ammonium sulphate (metachrome mordanting). This greatly simplifies the process and means much less harsh treatment of the wool.

Metachrome dyes are marketed in a variety of shades by various dye works. As diazotization components they use halogen- or nitro-substituted o-aminophenols and a pyrazolone, naphthol-sulphonic acids or chloro-m-phenylenediamine, and p-acetaminophenol as a coupling component.

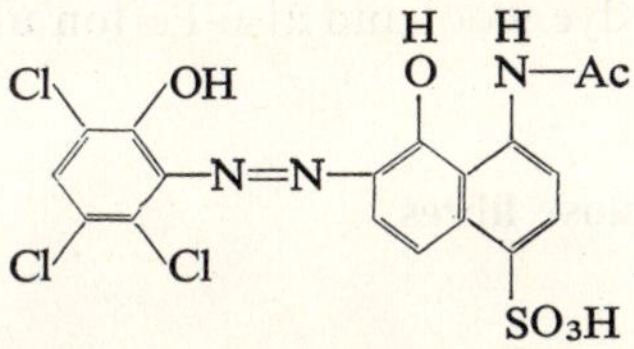

Metachromic Brilliant Blue BL (Wolfen)

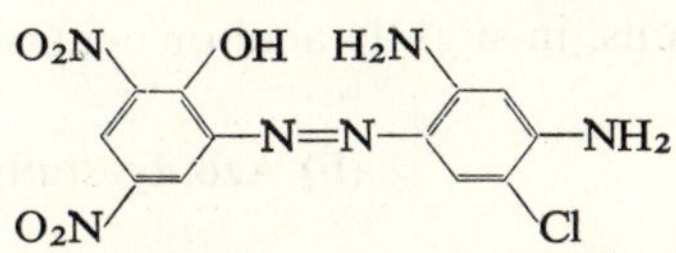

Metachrome Brown BL (Wolfen)

During chroming complex chrome compounds form which contain 1 or 2 moles of dye per atom of chrome.

Such metal complexes can also be produced in situ by heating the dyes with chrome$^{III}$-formate to 105–125°C (R. Bohn, BASF, 1912). In this way the important Palatine fast dyes, BASF, were produced (Neolan dyes by Ciba, Ultralan dyes by ICI). The main diazotization components used are o-aminophenols, e.g. amino-nitrophenol sulphonic acids and aminonaphthol sulphonic acids. Coupling components are acetanilide (for yellow), pyrazolones (orange and pink), naphthols and naphthylaminosulphonic acids.

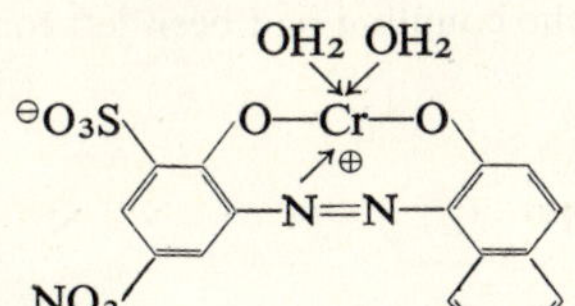

Palatine Fast Brown BRRN (1:1 complex)

These dyes are adsorbed on the fibre only in a strongly acid solution. Recently, this limitation was eliminated by introducing in place of the solubilizing sulpho groups other hydrophilic radicals, e.g. the following groups:

$NH_2$—$SO_2$—('Violan fast dyes', BASF)

$CH_3$—$SO_2$—('Irgalans', Geigy)

—O—$CH_2$—$SO_2$—, $H_3C$—NH—$SO_2$—('Isolans'. Bayer).

Other groups belonging into this category are the 'Cibalanes' (Ciba), the 'Ortolanes' (BASF), and the 'Wolfalanes' (Wolfen).

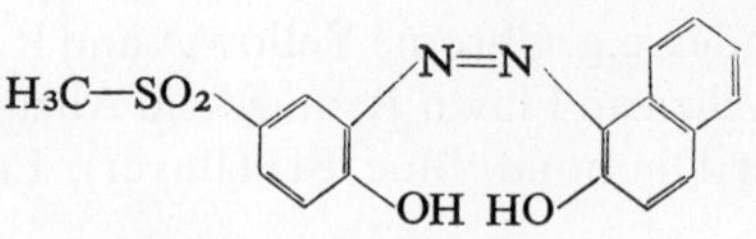


Irgalan Brown Violet DL (Geigy)

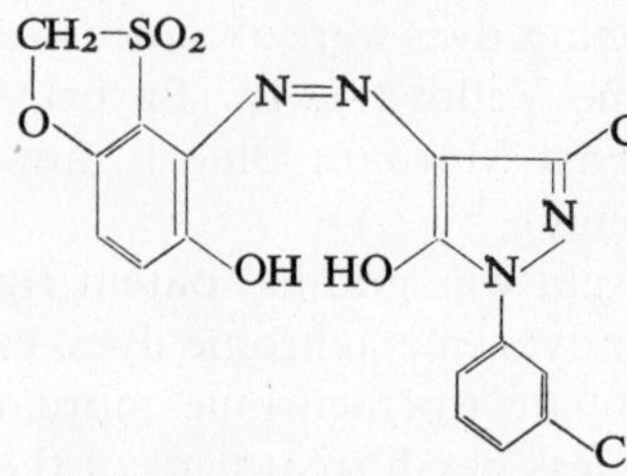


Isolan Bordeaux (Bayer)

Chrome complexes without sulphonic acid groups contain 2 dyestuff molecules per chrome molecule (1:2 complexes').

The new classes of dyestuffs make it possible to dye wool and also Perlon under very mild conditions, in slightly acid or neutral media.

### (b) Azo dyestuffs for cellulose fibres

*Substantive or direct dyes*

Cellulose fibres (cotton, linen, rayon, staple fibre, etc.) can be dyed only in neutral or alkaline media but not in acid media since this would lead to hydrolysis. Substantive or direct dyes must be absorbed in these fibres and produce a fast colouring, The nature of 'substantivity' has not yet been definitely established but a relationship between direct affinity and constitution has been found which is utilized in the formulation of new dyestuffs. A straight chain of 8 or 9 conjugated double bonds is essential for substantivity. For this reason, 1,4-linkages with benzene nuclei, and 1,4-;1,5-; and 2,6-linkages in the naphthalene series, e.g. by azo groups, result in substantivity.

I-acid (aminonaphtholsulphonic acid-2,5,7), benzidine and its 3,3'-substitution products, and 4,4' diaminostilbene are particularly suitable intermediates. Congo red mentioned earlier was the first cotton dye of this type although it was not fast to acid and not very fast to light, but it constituted the first stage in the development of cotton dyes. Duisberg (Bayer) discovered Benzopurpurin 4 B (o-toluidine → 2 mole naphthionic acid) in 1885, and Benzopurpurin 10 B is obtained in an analogous manner with dianisidine. Coupling of tetrazotized diamines often takes a very long time. Duisberg discovered his dye 'by accident' because a beaker with the coupling had been left for several days at his working place.

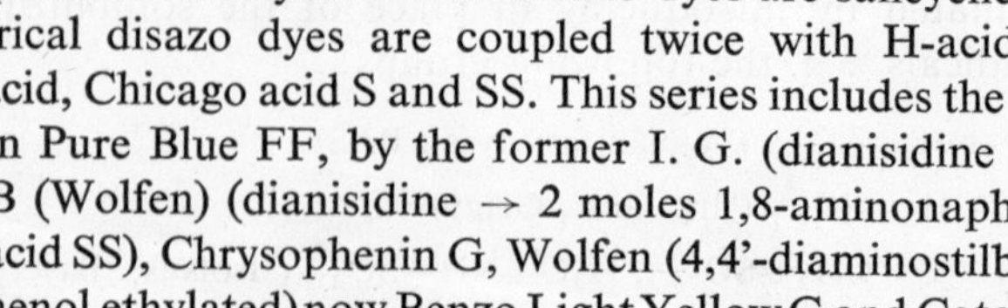


Congo Red

Coupling components for asymmetrical disazo dyes are salicyclic acid, $\gamma$-acid, I-acid, G-acid. Symmetrical disazo dyes are coupled twice with H-acid, naphthionic acid, Nevile-Winther acid, Chicago acid S and SS. This series includes the following important dyestuffs: Diamin Pure Blue FF, by the former I. G. (dianisidine → 2 moles H-acid), Chicago Blue 6 B (Wolfen) (dianisidine → 2 moles 1,8-aminonaphthol-2,4-disulphonic acid = Chicago acid SS), Chrysophenin G, Wolfen (4,4'-diaminostilbene-2,2'-disulphonic acid → 2 moles phenol ethylated) now Benzo Light Yellow G and Cotolan Yellow G, Bayer.

Finally, in addition to the diphenyl derivatives, diaryl ureas or imides are also used for the synthesis of direct dyes, e.g. those of I-acid:

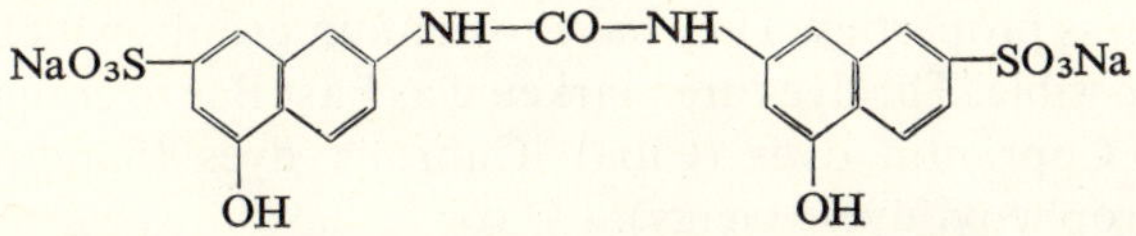


Ureid of I-acid

By coupling with diazotized aniline and substituted anilines one obtains from this the benzo fast scarlets (Benzo Fast Scarlet 4 BS, Columbia Fast Scarlet 4 BS, Wolfen = aniline → ureid of I-acid ← acet-p-phenylene-diamine). These are fast dyes, fast even to acids.

A series of direct dyes has been produced which etch white when reduced (for etch printing), e.g. Diamin Black BH (Cassella), Zambesi Dark Blue BH, Wolfen

= benzidine ↗ H-acid, ↘ γ-acid.

Finally, the important group of the Sirius- and Sirius Light dyes has to be mentioned. The latter are as fast to light as indanthrenes. They are trisazo- and tetrakisazo dyes. A particularly beautiful dye is Sirius Light Blue 6 G (Bayer) = aniline-2,6-disulphonic acid → α-naphthylamine → 2-ethoxy-Cleve-acid-6 → phenyl-I-acid.

Among the fastest direct dyes (Sirius light dyes or Chlorantin light dyes) of the orange and brown series are the condensation dyes obtained from dinitrostilbene disulphonic acid and aminoazo dyes. In view of the excess of oxygen in the nitro groups smooth formation of the azo compounds is not to be expected. A reducing agent ($Na_2S$, glucose) is, therefore, used for after-treatment. The products should be mainly tetrakisazo dyes:

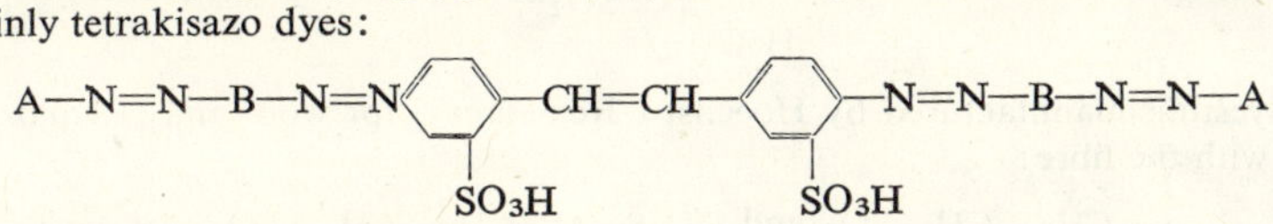


To produce e.g. Sirius Light Orange GGL, 4-aminoazobenzenesulphonic acid is condensed in an alkaline medium with the stilbene derivative.

Amino azo compounds can also be linked by phosgene. Sirius- and Sirius Light dyes (Bayer), Solamin- and Solamin Light dyes (Wolfen), and Chlorantin- and Chlorantin Light dyes (Ciba) can be obtained by this process, too.

Large dye molecules can also be synthesized from aminoazo compounds by reacting these with cyanuric chloride; this process yields disazo dyes with triazine rings (Ciba).

Tetrakisazo dyes obtained by diazotization and coupling lead to the black series Cotonerol A Wolfen

= diaminodiphenylaminesulphonic acid ↗ γ-acid → 1,3-phenylenediamine, ↘ γ-acid → 1,3-phenylenediamine.

Columbia Fast Black G Wolfen is produced by coupling p-nitraniline twice with H-acid, reducing both nitro groups, tetrazotizing, and coupling twice with 1,3-phenylenediamine.

The Sirius- and Chlorantin dyes are also marketed as copper complexes. This improves considerably their fastness to light and washing. Such dyes were obtained for the first time in 1915 by Ciba from o,o'-dihydroxyazo compounds with copper sulphate. The effect of aftercoppering has long been known and is particularly impressive with Chicago Blue 6 B, Wolfen (dianisidine → 2 moles 1-amino-8-naphthol-2,4-disulphonic acid).

Substantive dis-, tris-, and also tetrakisazo dyes of very varied constitutions can be converted into very stable complexes by aftertreatment with copper salts; this greatly improves their fastness properties. They have to contain groups which render the formation of complexes possible. The dyes are marketed as Fast Benzo copper dyes, now Benzocupuroles (Bayer), Coprantin dyes (Ciba), Cuprofix dyes (Sandoz), Cuproxon dyes (Wolfen), and Cuprophenyl dyes (Geigy).

Further important progress in the production of metal complex dyes was achieved by Pfitzner (BASF 1948) when he succeeded in introducing into o-hydroxyazo dyes a second OH group which is necessary for the formation of complexes, using $H_2O_2$ and copper acetate, the copper complex being formed simultaneously:

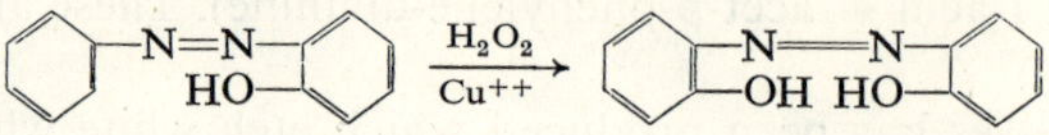


(Cu complex)

## *Dyes reacting with the cellulose fibre*

Recently, ICI introduced a new series of dyes, called 'Procion'. Highly water-soluble aminoazo dyes with two sulphonic acid groups are condensed with cyanuric chloride (intermediates for the Chlorantin Light dyes mentioned earlier). These compounds in molecular dispersion react with the positions of the cyanuric acid ring which are still occupied by Cl; in an alkaline medium they react with the OH groups of the cellulose. Procion dyes of the phthalocyanine series are also produced. Ciba has put on the market similar dyes in which two chlorine atoms of the cyanuric chloride are exchanged for amino groups of a dye (Cibacron dyes).

Earlier, BASF had introduced into dyestuffs the ethyleneimine radical as the group reacting with the fibre

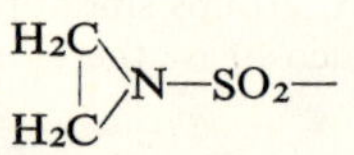


A group of dyestuffs manufactured by Hoechst ('Remalanes' for wool and 'Remazoles') contains groups reacting with the fibre:

$$CH_2{=}CH{-}SO_2 \text{ and } HO_3S{-}O{-}CH_2{-}CH_2{-}SO_2.$$

## *Diazotization dyestuffs*

These dyes also belong to those having one or more diazotizable amino groups which make possible diazotization on the fibre and coupling with a passive component ('developer'). During developing the shade changes appreciably, of course, for a new dye forms on the fibre with usually improved fastness to wetting. Such dyes are marketed under the name 'diazo-', e.g. Diazo Light Yellow GG, 2 GLL, Diazo Brown 3 R, now Benzamin Light Yellow GGS, and Benzamin Brown 3 R, Bayer. This group of dyestuffs also includes Zambesi Red 4 B and Zambesi Black D and V, Wolfen.

## *Insoluble dyestuffs produced on the fibre* (*Naphthol AS dyes*)

The principle of the production of insoluble azo dyes (pigments) on the fibre was enunciated by T. and R. Holiday in 1880. Cotton was soaked in β-naphthol, pressed, and dried ('bottoming'). The impregnated fibre was then treated with a solution of diazotized p-nitraniline, and Para Red formed:

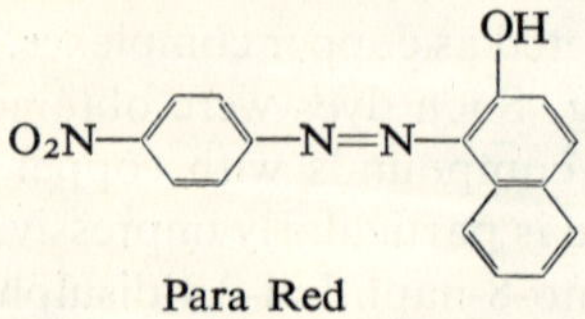


Para Red

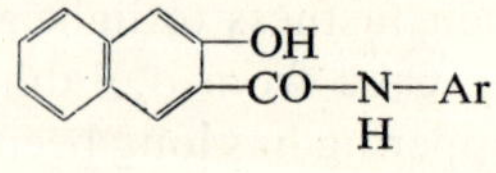


Naphthol AS

By varying the diazotization component it was possible to produce a great variety of shades. Since during the manufacture of these dyes cooling with ice was carried out they were called 'ice colours'. The colouring methods were developed in particular by Ullrich and Gallois (Hoechst). Following the work carried out by Griesheim-Elektron, $\beta$-hydroxynaphthoic acid arylides were first used in 1912 for bottoming which represented great progress as compared with $\beta$-naphthol. Cotton dyeing and printing entered a new era as the result of work by Winther, Laska, and Zitscher (Offenbach) in the naphthol AS field; para-red and alizarin red became obsolete. Naphthol AS is considerably more substantive than $\beta$-naphthol, the solutions are more permanent and the colourings and prints have much better fastness properties.

Naphthols and bases exist now for any shade desired. It is surprising that one obtains all shades from greenish yellow to deep blue with naphthol AS dyes which are monoazo dyes. Blue bases (diazotization components) are e.g. Variamin Blue B (4-methoxy-4'-aminodiphenylamine) and Fast Blue base BB (1-amino-4-benzoylamino-2,5-diethoxy-benzene).

The acid amide radical can be varied to a great extent. Naphthol AS products are mostly derived from 2,3-hydroxy-naphthoic acid by using various amines for formation of the acid amide (substituted anilines and naphthylamines).

Condensation of hydroxynaphthoic acid with arylamine is carried out with $PCl_3$ in a neutral solvent.

Arylides of other aromatic hydroxycarboxylic acids are also used, such as hydroxy-diphenyleneoxide carboxylic acid (AS-BT), hydroxycarbazole carboxylic acid (AS-LB), both for brown to black shades, or hydroxyanthracene carboxylic acid (AS-GR) for green, and finally the arylides of acetoacetic ester (AS-G, LG, L 3 G, L 4 G) for yellow shades. Active components for coupling are the diazotized naphthol AS bases, also in the form of hydrochlorides or dye salts, the diazotized bases in stabilized form. They are naphthalene sulphonic acid salts, zinc chloride double compounds, or hydrofluoric acid salts.

Bases and colouring salts available commercially which can be combined with any naphthol AS are called e.g. Fast Blue B Base (dianisidine), Fast Yellow GC Base (HCl salt of 2-chloraniline), Fast Blue Salt B (diazotized dianisidine-zinc chloride double salt), Fast Yellow Salt GC, etc. There are about 50 fast bases and fast salts on the market. Combined with about 20–25 naphthols this gives an infinite variety of possibilities for colouring and printing. Naphthol AS products are sold in Germany mainly by Naphtholchemie Offenbach which now forms part of the Farbwerke Hoechst.

Since the German patents expired many non-German dyestuff works have been active in this field and are offering naphthol AS assortments under different names (Ciba e.g. 'Cibanaphthol', Geigy 'Irganaphthol', ICI 'Brenthol').

A large proportion of the naphthol AS dyes (about 65 combinations) have the fastness properties of indanthrene dyes.

Although the introduction of colouring salts greatly simplified printing and colouring, two basic operations, viz. bottoming and development, were still required. For printing the whole area is bottomed and after coupling the naphthol has to be removed from the un-dyed places. This is avoided by using new groups of dyestuffs which contain a mixture of the naphthol and the diazonium compound in stabilized form and which are reacted on the fibre. They are marketed under the names Rapid Fast Dyes (Offenbach), Rapidogenes (Bayer), and Rapidazoles (Offenbach).

For Rapid Fast Dyes the diazonium compound is converted into an anti-(iso-)diazotate which is stable in alkali and yields the coupling diazo compound only when treated with acids. For Rapidogenes the diazo compound and the amine are converted into the insoluble diazoamino compound which is sufficiently stable. It is decomposed by organic acids and one obtains again the original coupling diazo compound.

The Rapidogen principle:

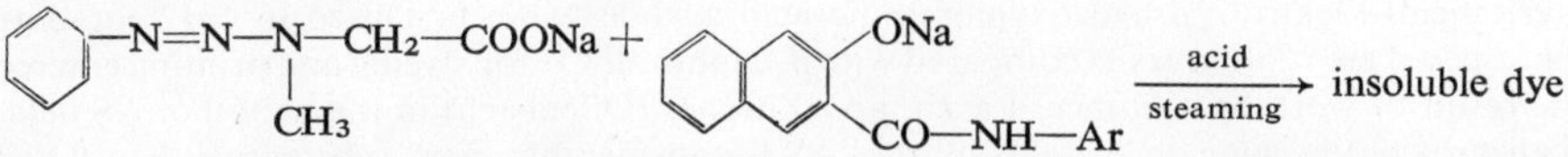

These two classes of dyestuffs could, however, not be supplied in all shades since only some bases form stable isodiazotates and diazoamino compounds. The Rapidazol dyes which are mixtures of diazosulphonates, e.g. of Variamin Blue bases, with naphthols, formed, therefore, an important supplement to this range of dyestuffs. They are developed into pigments by mere steaming in the presence of alkali.

Recently, a Rapidogen-N series has been put on to the market (Bayer, 'Neutrogen', Francolor) which can be developed by steaming without acids. In this case, phenylglycine-o-carboxylic acid is used to stabilize the diazo compound. This makes possible their use in combination with leuco compounds of the vat dyes.

### (c) Azo dyes for synthetic fibres and leather

The problem of dyeing acetate rayon, the first synthetic fibre, was solved by the introduction of Celliton-, Celliton fast-, and Celliton mordant dyes. Acetate fibre dyed with direct dyes remains white. The fibre was dyed with water-insoluble finely dispersed dyes which are soluble in cellulose acetate and are absorbed directly. Mainly highly dispersed paste dyes are, therefore, marketed. Solubility is improved by the presence of hydroxyethyl- and dihydroxypropyl radicals (former I. G. Ludwigshafen, and Wolfen), e.g. in Celliton mordant scarlet B:

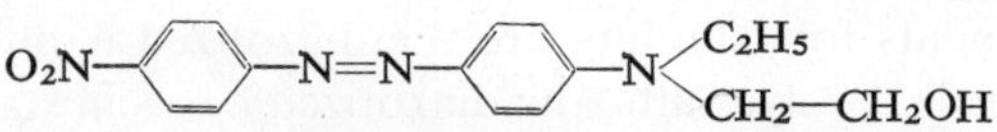

Eventually water-soluble acetate dyes were also developed. Cellit- and Cellit fast dyes (Bayer) contain one or two nitro groups, a secondary or tertiary amino group, and a sulphonic acid group. In the Solacet range (British Dyestuffs Co. and ICI) the sulphonic acid group is contained in the form of an alkylene-sulphuric acid ester:

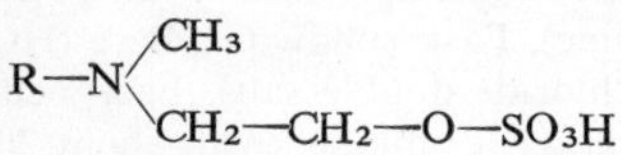

Acetate fibre also absorbs amines from solution or suspension which are diazotized on the fibre and then coupled (Cellitazoles, Cibacetdiazo dyes).

Some of the dyes used for acetate rayon dyeing can also be used for dyeing polyamide fibres ('Perlon', 'Dederon', 'Nylon'). Chrome and cobalt complex monoazo dyes as employed for colouring varnishes and plastics (Zapon fast dyes) are used for the same purpose. Perliton dyes (BASF) are insoluble in water. They can be coloured in dispersion. Perlitazoles correspond approximately to Cellitazoles. The synthetic fibres mentioned above can also be dyed with some naphthol AS dyes, the Ofnacet dyes (Offenbach).

For dyeing polyamide fibres, BASF put on the market the 'Vialon fast dyes' range which are chrome complex dyes containing —$SO_2$—$NH_2$ groups to make them soluble in water.

Great difficulties were at first encountered in dyeing polyacrylonitrile fibres ('Orlon', 'Dralon', 'Wolcrylon') for which all known classes of dyestuffs failed. The first special dyes for these fibres were the Astrazon dyes (Bayer) which form part of the polymethine series and contain an ammonium salt group. This observation stimulated the introduction of basic radicals into dyestuff molecules of various types (azo dyes, anthraquinone dyes).

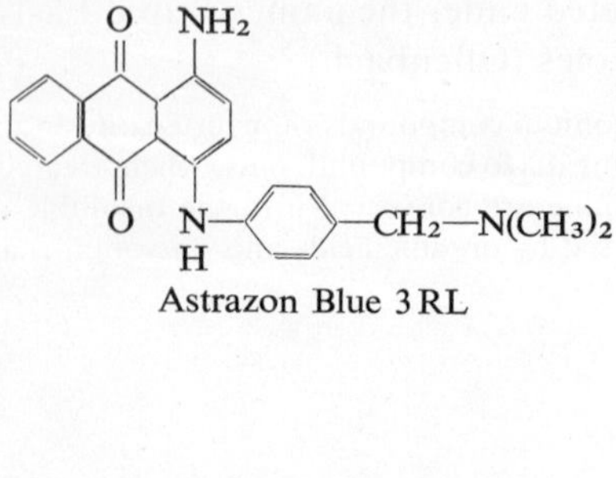

Astrazon Blue 3RL

The dyeing of terephthalic acid glycol ester fibre (Terylene, Trevira, Lanon, Diolen) proved difficult. This fibre is very hydrophobic and the chains of molecules are packed so tightly that they absorb dyestuff molecules only with difficulty. To start the dyeing process 'carriers' are used (dichlorobenzenes, terephthalic acid dimethyl ester) which swell the fibre and concentrate the dye at the surface of the fibre. The colouring can then be carried out with suspension dyes, similarly as with acetate rayon. Bayer developed the 'Resolin' dyes for this purpose. However, since the spinning temperature of the Terylene fibre is very high (300°C) dyes which are sufficiently stable to be used during the spinning operation are not yet available.

With synthetic fibres spin dyeing plays an important part. Finely dispersed fast pigment dyes are incorporated into the spinning solution before it is spun. This is the type of dyeing which produces the best fastness properties, and the expensive dyeing operation is eliminated. Spin dyeing is used mainly for the standard shades black, brown, and blue.

Leather can be dyed with many basic, acid, and chrome developing dyes. The most important shades are brown and black. Special ranges (azo dyes) have, however, also been developed for dyeing leather which cover the surface well (pigments), those that cover the surface but dye the cut, only to a lesser extent, and those that penetrate the leather. These were marketed formerly under the name of Igenal dyes, but currently as Wogenat, Baygenal dyes, acid leather browns, and Erganil dyes.

### (d) Dyes for foodstuffs, dyes soluble in alcohol, esters, and fats

Dimethylaminoazobenzene ('butter yellow') was permitted for a long time but its use is now prohibited because of its carcinogenic nature. A whole series of water-soluble azo dyes has proved harmless in extended tests. In 1952 a list was drawn up in the Federal Republic containing 18 dyes which may be used. It includes well known azo dyes such as Fast Yellow, Tartrazin, Azo Ruby, Fast Red E., Orange GGN, etc.

Azo dyes soluble in spirit and other polar solvents such as are used for varnishes, and in cellulose esters and natural and synthetic resins are marketed under the names Zaponecht, Irisol, Orasol, and Grasol dyes. They produce transparent colourings so that an aluminium foil covered with a yellow varnish appears golden.

For colouring waxes, paraffins, oils and fats, lacquers for wax crayons, lipsticks, stamping inks and printing inks, oil-soluble dyes have been developed which were originally produced only by the Farbenfabrik Wolfen under the name 'Sudan dyes' but are now sold under the same name by BASF and the Farbwerke Hoechst, by Bayer under the name Ceres dyes, by Cassella as Ceresin dyes.

The first Sudan dye was 4-aminoazobenzene (Sudan yellow R). Sudan orange G (aniline → resorcinol) has been known since 1875, butter yellow = Sudan Yellow GG (aniline → dimethylaniline) since 1877. They are azo dyes without any solubilizing groups or with groups solubilizing in fats. As a coupling component cyclohexylphenol has proved to be specially suitable for this purpose and in several cases amines with cyclohexyl residues have similarly proved to be suitable diazotization components (Farbenfabrik Wolfen). The fastness to sublimation of the Sudan dyes is particularly important for the graphic trades.

### (e) Pigments

For the colours yellow, orange, and red, azo dyes are used by preference as pigments. One distinguishes between pigment dyes proper, i.e. insoluble azo dyes, and colour lakes which are obtained from soluble dyes by precipitation with metallic salts.

Coloured pigments are well-known old azo dyes, such as para red (4-nitraniline → $\beta$-naphthol) and coupling products of many arylamines with $\beta$-naphthol which formerly were used as ice colours. Particularly important were acetoacetic arylides (Hansa Yellow range, Hoechst) and 2,3-hydroxynaphthoic acid arylides (naphthol AS products) for permanent red, bordeaux, violet, and brown. Disazo dyes are also used. Vulcan fast dyes are insoluble azo dyes for the rubber industry; they are resistant to the vulcanization process and are derived from benzidine.

For the production of colour lake pigment dyes existing acid azo dyes, such as Orange II, Thiazol Yellow G, Beizen Yellow GL, Bordeaux R, and many others were at first precipitated with barium and calcium salts, less frequently with other metallic salts.

Soon, however, special dyes were developed which were particularly suitable for the laking process. The most important coupling component is still $\beta$-naphthol which is reacted with the diazonium compounds of various substitution products of aniline sulphonic acid and with that of 2-naphthylamine-1-sulphonic acid. An important part is also played by R-salt and phenylpyrazolones with sulphonic acid groups in the phenyl nucleus.

These dyes are marketed under names such as Lake Red, Pigment Lake Red, Lithol Red, Helioorange, Permanent Red, Saturn Yellow, etc. The best pigment dyes with the best fastness properties, the phthalocyanines, have been discussed earlier.

The importance of pigment dyes continues to increase. Processes have been developed in which pigments are fixed on the fibre with high-molecular binding agents so that they are fast to washing and rubbing; shirt materials can thus be printed as in newspaper rotary printing and three-colour printing can be carried out without aftertreatment (e.g. 'Acramine', Bayer, 'Aridyes', Int. Chem. Corp.).

### *Prospects and problems*

After the war the necessity to save manpower and wages became very urgent in the mass production of dyed textiles. For this reason the price of the dyestuff is now not as important as the simplicity of the colouring process. 'Finishing' is being perfected more and more, and so is the final preparation of the dyestuffs, the production of finely dispersed pastes and finely pulverized pulp which give colloidal solutions in water and react suddenly during vatting. It has proved possible to use pigment dyes increasingly for textile dyeing, particularly for printing, by the use of very fine dispersions and of suitable binding agents. Since no after-treatment is required with these processes they are likely to become more and more widely used for the cheap mass production of printed cotton goods, particularly since excellent fastness properties are achieved.

An important development in dyestuff technology aims at an improvement in the fastness properties of existing water-soluble dyestuffs on the fibre by after-treatment with textile auxiliaries. As a result of the great increase in the production of plastics (e.g. polyethylene, polypropylene) pigment dyes and soluble dyes for transparent colouring are required in increasing quantities. New dyeing problems are constantly being created by the production of new textile fibres. Usually it is necessary to produce special dyestuff ranges for these new developments.

# 9. THE USE OF TEXTILE DYES FOR DYEING AND PRINTING

## (a) Acid dyes

Acid dyes are water-soluble dyes with acid solubilizing groups which are coloured from the solution. Their main field of application is wool dyeing.

According to their absorptive capacity they are dyed from an acid bath with the addition of sulphuric acid at an elevated temperature. Often Glauber's salt which has a certain 'stripping' capacity is added. This is important for good levelling. Too rapidly absorbing dyes give uneven colouring. A levelling agent is, therefore, added to slow down absorption. To utilize all of the dye liquor, acetic acid or sulphuric acid are added towards the end of the colouring. On the basis of their levelling capacities acid dyes can be divided into various groups.

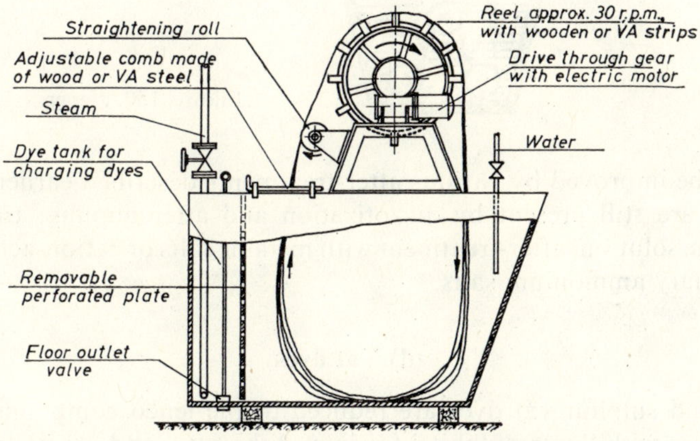


Figure 149. Reel vat for dye works

Yarns ('strands') are dyed on the reel vat (Figure 149). After colouring the material is pressed, rinsed, and soaped at boiling temperature; only then the colouring becomes clear. It is important that the dye liquor should be completely exhausted.

## (b) Chrome and mordant dyes

Dyeing with mordant dyes is rather cumbersome and is hardly practised any more. First, bottom mordanting is required, for cotton by repeated treatment with aluminium salts, for wool once with potassium dichromate and an organic acid. Then follows acetic dyeing or printing. After printing the material is steamed for fixing.

The acid after-chroming dyes for wool are first coloured like acid wool dyes, then potassium dichromate and acid are added to the bath; a chrome lake forms which has a different colour but improved fastness properties.

The simplest process is dyeing with single-bath chrome dyes (e.g. meta chrome dyes). They are used like acid wool dyes with the addition of potassium chromate and ammonium sulphate. All types of loose wool and finished textiles are dyed in this way and excellent fastness properties are obtained.

### (c) Direct dyes

Direct dyes (substantive dyes) are water-soluble and are used for dyeing cellulose fibres, mainly cotton. The dyes are absorbed from a bath containing Glauber's salt without any other addition. A 'jigger' is used for dyeing webs of woven cloth (Figure 150) (from Winnacker-Weingaertner). The web of cloth is pulled through the dyeing vat containing the dyeing liquor with the aid of rollers, the cloth being wound from one roller on to the other. Repeated winding backwards and forwards makes it possible for the cloth to pass several times through the bath. The fastness to wetting which with direct dyes often is not very good can be improved by various after-treatments described earlier, e.g. when free amino groups are still present by diazotization and aftercoupling, treatment with a diazotized amine solution, after-treatment with metallic salts or cation-active compounds, such as quaternary ammonium salts.

Figure 150. Jigger

### (d) Vat dyes

Sulphur dyes and sulphur vat dyes are reduced to the leuco compounds with sodium sulphide, vat dyes with 'hydrosulphite' (sodium dithionite) and caustic soda. The leuco compounds usually have a deep colour and are absorbed in the cotton from a strongly alkaline medium. Oxidation to the dye takes place during hanging in air or by treatment with oxidants. The colouring is then rinsed and soaped at boiling temperature when the dye develops its full clarity and brightness.

Special processes for continuous operation in vat dyeing have been developed. In continuous dyeing of fabrics the material is first padded through the Foulard padding machine (1) in Figure 151. Preliminary treatment, e.g. impregnation with $\beta$-naphthol or naphthol AS in development dyeing or with finely divided vat pigments, can be carried out with it (pigment padding process). After pressing the web of

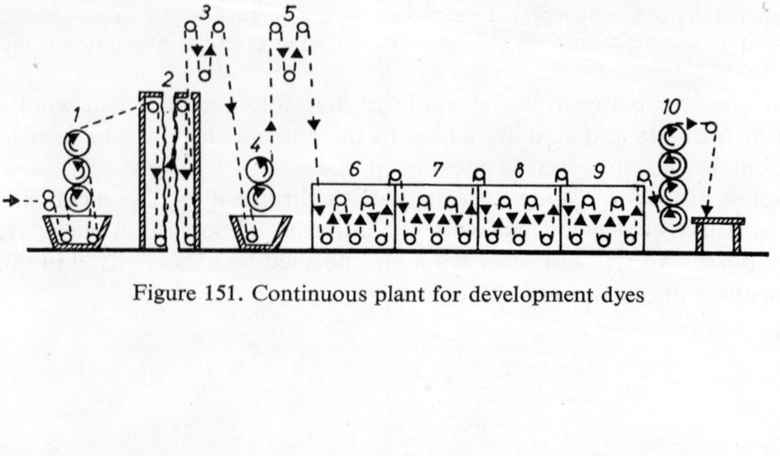


Figure 151. Continuous plant for development dyes

cloth is passed through a second bath with sodium hydrosulphite and caustic soda where vatting and absorption of the leuco dye takes place. Then the dye is developed in air or in an oxidation bath.

The continuous padding-steaming process is very similar. After padding, the cloth passes through a drying room, then through the hydrosulphite solution, followed by the processes of steaming, oxidizing in a series of baths, soaping, rinsing, and finally drying on drying rollers (Figure 151) (from Winnacker-Weingaertner).

Cloth is printed in a very similar way. First, the very finely divided vat dye is printed on to the material together with Rongalit C (hydroxymethanesulphinic acid Na), potash, glycerine and a sugar concentrate, starch, and tragacanth. Then the cloth passes through a steam vessel where reduction and fixation of the leuco dye take place. Further processing is as described above.

Because of the large amount of cloth printing being carried out with vat dyes, particularly fine and easily reducible dye pastes (Suprafix brands) have been put on the market.

For dyeing wool special wool vat dyes which need little alkali ($NH_3$) and are absorbed in wool at low temperatures have been developed (Helindron dyes, Hoechst).

Dyeing and printing with water-soluble leuco vat dyes (Indigosol, Soledon and Anthrasol dyes) is particularly easy. The dyeing, padding, or printing of the ester with the calculated quantity of nitrite is followed by a sulphuric acid bath in which the ester is saponified. The liberated nitrous acid oxidizes the leuco compound to a dye. The material is then neutralized, soaped, etc.

### (e) Developing dyes

First the fibre is treated with an alkaline solution of a naphthol ($\beta$-naphthol, naphthol AS) (bottoming). This is done on a Foulard, Figure 151 (1). After pressing the bottomed material is dried (2) and passes through an air duct (3) into the developing bath (4) which contains a cooled diazonium salt solution. There coupling takes place. Then, as seen in the illustration, further treatment is applied (rinsing bath 6, soap baths 7 and 8, rinsing bath 9, drying roller 10). The operations are greatly simplified when Rapidogen or Rapid fast dyes are used in the cloth printing process. In that case the material is placed in an acid bath, is then steamed and treated further as usual. Multi-coloured printing can also be produced in combination with Indigosoles since these are developed in the same way as Rapid fast and Rapidogen dyes.

### LITERATURE


Fritz Mayer, *Chemie der organischen Farbstoffe*, Springer, Berlin, 1924.

H. E. Fierz-David, and L. Blangey, *Farbenchemie*, Springer, Wien, 1952.

L. F. Fieser, and M. Fieser, *Lehrbuch der organischen Chemie*, Verlag Chemie, Weinheim, 1954.

K. Venkataraman, *The Chemistry of Synthetic Dyes*, 2 vols. Academic Press Inc., New York, 1952.

In Winnacker-Weingaertner, *Chem. Techn.*, Org. Techn., II. K. Holzach, and W. Haage, *Die organischen Farbstoffe*, Hanser-Verlag, München, 1954; 2nd ed., II, vol. 4, p. 1.

Ullmann's *Encyklopädie*, 3rd ed., vol. 4, p. 76–163, Azofarbstoffe, Verlag Urban & Schwarzenberg, München/Berlin, 1953.

A. Schaeffer, *Die Entwicklung der künstlichen organischen Farbstoffe*, herausgegeben vom Verfasser, 1951, Id., Chemie der Farbstoffe und deren Anwendung. Verlag Th. Steinkopff, Dresden, 1963.

O. Bayer, *Die neuere Entwicklung der Farbenchemie* (paper read at German Universities). Published by author, Leverkusen, 1958.

# VII. Explosives (gun powder and blasting agents)

Explosives are chemical compounds or mixtures of substances which undergo a sudden chemical transformation during which considerable amounts of gas and heat are liberated, producing a propelling effect. They are either endothermic compounds the negative heat of formation of which is liberated during decomposition, or mixtures of non-explosive individual components containing unstable groups, e.g. reactive oxygen, which react violently with the other components generating heat and gases. Thus, an explosion may consist in the extremely rapid combustion of either a single substance which contains sufficient oxygen, e.g. nitroglycerine, trinitrotoluene, or a mixture of materials supplying the oxygen, e.g. nitrate or chlorate and carbon or sulphur, concentrated nitric acid, nitrogen tetroxide ($N_2O_4$) or liquid oxygen, with organic materials.

The decomposition of nitroglycerine takes place according to the following equation, with simultaneous formation of 715 litres of gas per kg:

$$4C_3H_5(ONO_2)_3 = 12CO_2 + 10H_2O + 6N_2 + O_2$$

The decomposition of nitroglycol takes place as follows:

$$O_2NOCH_2—CH_2ONO_2 \rightarrow 2CO_2 + 2H_2O + N_2$$

Not every explosive or explosive mixture is suitable for shooting or blasting purposes. A condition for practical use is an adequate chemical stability and not too great a tendency to detonation, i.e. no great sensitivity to heating, impact, shock, and friction. On the other hand, detonation must be reliably initiated by a priming agent. For practical purposes this restricts the very large number of explosive materials to a few chemical groups and a strictly limited number of compounds.

The oldest blasting agent and gun powder is black powder, consisting of 75 per cent $KNO_3$, 10 per cent sulphur, and 15 per cent charcoal (turn of the 13th to the 14th century, Berthold Schwarz and Roger Bacon). It was used exclusively for 500 years until nitroglycerine was discovered (1865) and Alfred Nobel invented dynamite which consisted at first of 75 per cent nitroglycerine and 25 per cent kieselguhr; this mixture is no longer used.

Since that time nitric acid esters have been among the most important explosives. Other materials used include aromatic and aliphatic polynitro substances and mixtures of nitrates, chlorates, and perchlorates and liquid oxygen with organic combustible materials. According to the intended use, these agents are mixed, shaped, and wrapped in completely different ways.

Unlike black powder and nitroglycerine explosives, industrial and military explosives used in practice cannot be reliably detonated by a flame or a black powder fuse. This is an important safety factor against accidental ignition. For the initiation priming explosives are used of which the most important are mercury fulminate, $HgC_2N_2O_2$ (Hg salt of fulminic acid HO—N=C), lead azide (lead salt of hydrazoic acid $HN_3$) and diazodinitrophenol (4,6-dinitro-2-diazo-1-hydroxybenzene). Priming explosives are very sensitive to impact, friction, and heat. They are compressed only in small quantities in copper or aluminium cases (blasting caps, detonators). These can be ignited either by impact, or with the aid of a fuse, or electrically, and they initiate the detonation of the explosive either directly or via a special priming cartridge.

Since during the manufacture of explosives, particularly the nitration of glycerine, glycol, and diglycol, one must expect spontaneous decomposition and explosions, the quantities mixed at any one time are not very large, or the operations are carried out continuously and the finished explosives are currently removed. In an explosives factory the individual groups of machines are separated from each other by banks of earth. The factory buildings are sunk deep into the ground and are covered with an easily crumbling roof on which earth and turf are placed. In this way an explosion does not project heavy components and the force of the explosion is largely checked. Only one wall has windows and in front of it is built a high bank of earth. Access and connecting passages through the banks of earth are brick-lined tunnels with sharp bends. The earth wall restricts an explosion or a fire as far as possible to its point of origin, and a shock wave cannot propagate laterally.

## 1. LIQUID NITRIC ESTERS (NITROGLYCERINE)

In practice the following are used: nitroglycerine (glycerine trinitrate), nitroglycol (glycoldinitrate), and diglycoldinitrate ($O_2NO—CH_2—CH_2—O—CH_2—CH_2—ONO_2$).

Nitroglycerine is very sensitive to impact, the other esters much less so.

### *Manufacture of nitroglycerine*

Very pure glycerine ('dynamite glycerine') free from oxidizable substances is required for the manufacture of nitroglycerine. Formerly, nitration used to be carried out in cylindrical lead vats, with charges of 100–250 kg glycerine. Now, steel equipment is used and the initial mixes yield 2,000 kg nitroglycerine. Esterification is carried out with a mixture containing $NHO_3$ of 99 per cent purity and 25–30 per cent oleum. Thus, only a minimum of spent acid remains which contains very little nitroglycerine in solution (70 per cent $H_2SO_4$, 15 per cent $HNO_3$, 15 per cent $H_2O$). During esterification a large amount of heat is liberated but the nitration temperature must not rise under any circumstances above 30°C. The glycerine is fed under constant stirring with air into the prepared nitrating acid which is kept at a temperature of 15–20°C, or preferably 0–5°C. Local overheating, marked by the appearance of considerable amounts of nitrous gases, can lead to explosive decomposition. Figure 152 shows equipment for the batch nitration of glycerine (from Winnacker-Weingaertner, *Chemische Technologie*).

Underneath every nitrating system there is a safety vat which contains water five times the volume of the charge and is equipped with air agitation. As soon as there is any sign of decomposition starting the charge can be dumped immediately. It is, however, preferable to fill the safety vat with concentrated sulphuric acid.

When nitration is complete, separation follows which lasts a long time because of the strong emulsification of the oil in the residual acid. Addition of small quantities of paraffin oil, NaF, HF, accelerate separation. The separation vessels used to be made of lead, but are now of V2A steel. The shape of the separation vessels is chosen so that the residual acid separates out at the bottom as far as possible and can be drained from there. The raw product which still contains much residual acid, particularly $HNO_3$, in emulsion and solution, is washed with water under constant stirring with air in lead-lined wooden or V2A steel washing devices and made alkaline with soda. The spent acid yields a large amount of oil when left standing for a long time; afterseparation is carried out in conical vessels. Then the spent acid is denitrated by first separating it into 70–72 per cent sulphuric acid and 57–62 per cent nitric acid. This is accomplished by passing the spent acid in countercurrent flow to superheated steam in acid-resistant columns filled with Raschig rings. These acids can be concentrated or used in any other desired way.

Continuous esterification methods were developed by Schmid-Meissner and Biazzi. They have the advantage that the equipment required is very small, only very little finished nitroglycerine is present in the equipment at any one time, heat dissipation is very good, and few operators are required. The number of persons present in the danger zone is, therefore, always small.

Nitration of glycol and diglycol is carried out in a very similar manner. Nitroglycol is rather volatile which necessitates particularly good ventilation. By adding 20–30 per cent nitroglycol, the freezing point of nitroglycerine is depressed considerably.

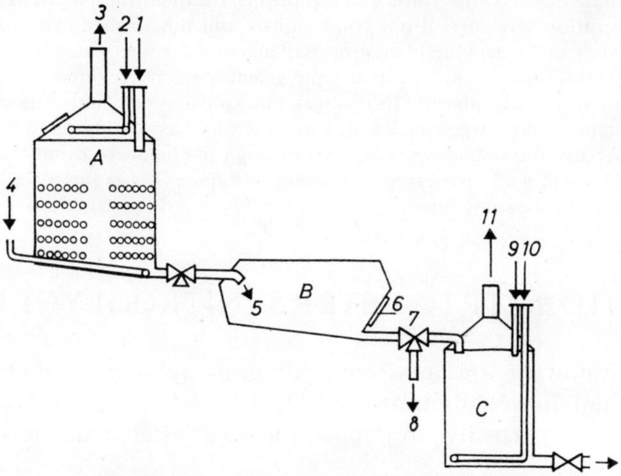


Figure 152. Batch nitration of glycerine
*A* nitration apparatus *B* separator *C* washing installation
*1* acid feed *2* glycerine distributor *3* air exhaust *4* stirring with air *5* inlet *6* inspection glass *7* 3-way tap to separation *8* acid discharge *9* water feed *10* stirring with air *11* air exhaust

Diglycoldinitrate is a component of the 'barrel-preserving' powder with a low combustion temperature. Nitroglycerine and nitroglycol are used mainly for gelatinous explosives (blasting gelatine, gelatine dynamite). Slight heating of nitroglycerine to which 7–8 per cent of suitable collodion cotton has been added yields a gelatinous elastic substance which does not 'exude' oil. This important discovery is also due to Alfred Nobel (1875).

Mixtures of wood meal, ammonium nitrate or soda saltpetre, aromatic nitro substances, potassium chlorate, etc., are produced for a great variety of purposes and are marketed under special names such as Gelatin-Dynamit, Ammon-Gelit, Donarit, Chloratit and Ammon-Gelignite.

For blasting in coal mines 'safety explosives' are used which are to prevent the initation of a firedamp or coal dust explosion.

The principle used is the reduction of the explosion temperature by the addition of salts (e.g. evaporation of water of crystallization, heat absorption by solid particles). In addition to ammonium nitrate, safety explosives contain sodium chloride or sodium bicarbonate (Wetter-Nobelit, Wetter-Detonit, Unigel, Polar Ajax, Polar Viking). Non-freezing gelatinous explosives which consisted of mixtures of nitroglycerine with a rather high setting point and glycerine dinitrate, dinitrochlorohydrin, and diglycerine tetranitrate, are no longer produced. A mixture of 70 per cent nitroglycerine and 30 per cent nitroglycol does not solidify at any winter's temperature.

## 2. NITROCELLULOSE, NITROSTARCH, NITROSUGAR, NITRATED POLYHYDROXY COMPOUNDS

The raw material for nitrocellulose is dry 99 per cent cellulose. Esterification with nitric acid ('nitration') is a rather complicated process. Concurrent with the esterification, decomposition of cellulose dependent on the conditions of the nitration (e.g. temperature, duration) also takes place which affects the solubility of the nitrocellulose. Water content and relative content of sulphuric acid of the terminal acid determine the degree of nitration (N-content).

According to whether the end product is to be gelatinizable collodion cotton (for dynamite) (11·7–12·3 per cent N, soluble in a mixture of ether and ethanol 2:1) or guncotton (13·1–13·3 per cent N, almost insoluble in an ether-ethanol mixture), the final acid is adjusted for a higher or lower water content. The maximum theoretical degree of nitration is 14·1 per cent.

The bulkiness of the cellulose material creates great difficulties in processing. A large excess of nitrating acid is necessary. Only the introduction of the nitration centrifuge (Selwig and Lange, 1891) made nitration economically possible. These centrifuges have since been constantly enlarged and improved. Thus, modern centrifuges frequently have automatic filling and discharging deviceds in addition to automatic travel through the acid-, displacement-, and water centrifuges. The displacement of the acid by as little water as possible is important so as to recover as much as possible of the acid in a high concentration.

Washing with plenty of water can also be carried out in a centrifuge or in large tiled basins with perforated bottoms.

The problem of the stability of nitrocellulose has always caused great difficulties. Insufficiently stable nitrocellulose is subject to continuous uniform decomposition. Pure nitrocellulose, on the other hand, is more stable than nitroglycerine. The fibre retains unstable fractions and traces of acid held very tenaciously which cannot be removed simply by boiling with water or treatment with alkalis.

Stable nitrocellulose is obtained only by first crushing and washing in beaters (see p. 291), further washing in a washing and sieve engine, and finally boiling for many hours with frequent change of water in stabilization boilers (wrought iron vessels with agitators and paddle wheels). To remove most of the water the nitrocellulose is centrifuged or pressed hydraulically. At a water content of 35 per cent it is no longer explosive and is completely harmless and transportable. The same effect is achieved by an alcohol content of 35 per cent.

Formerly, gun cotton played an important role as a military explosive. Now, aromatic nitro substances are used for this purpose. Nitrocellulose is now only used for smokeless powder in compressed form.

As mentioned earlier, collodion cotton is an excellent gelatinizing agent for explosives.

For this purpose the nitrocellulose is treated with solvents producing a colloidal horn-like state, and is pressed into tube, foil, or cube shapes. Another possibility for transforming it into a mouldable product is by first mixing the nitrocellulose under water with nitroglycerine or preferably with diglycoldinitrate, and then to plasticize on rollers heated to 90°C until the water is removed. These powders can also be produced from diglycoldinitrate and nitroguanidine (low calorie content and high gas formation, important for anti-aircraft and anti-tank guns). Smokeless powders are used as a powder train for rockets.

Starch, saccharose, and other sugars can be nitrated similarly to cellulose. Useful explosives are obtained only from starch, or by the nitration of a mixture of sugar and glycerine. Mannitol $HOH_2C—(CHOH)_4—CH_2OH$, and pentaerythrite $C(CH_2—OH)_4$ can also be transformed into nitro esters, mannitol with mixed acid, pentaerythrite by dissolving in 99 per cent nitric acid and precipitating with sulphuric acid. The nitrates are crystalline and can be separated easily. Nitropentaerythrite has the greatest explosive and shattering power of all the homogeneous explosives.

## 3. AROMATIC NITRO SUBSTANCES

Aromatic nitro substances, nitro hydrocarbons, nitro phenols and nitramines with 2–3 nitro groups are explosives. As compared with the nitric acid esters they have the advantage of greater chemical stability, particularly to acids, and insensitivity to heat, impact, and friction. Being nitro substances they are, however, sensitive to alkalis. The more highly nitrated aromatic nitro substances are solid, melt easily, have a high density, and can be pressed and cast safely. All these properties and their high explosive power render them particularly suitable for military purposes. Most modern gun and explosive ammunition contains aromatic nitro substances.

A disadvantage is their negative oxygen balance, causing toxic and combustible gases, like CO, HCN, $CH_4$ and $H_2$, to form during detonation. This limits their use in pure form. By mixing them with compounds containing an excess of oxygen, such as ammonium nitrate, chlorate, and perchlorate, this limitation can be overcome. Aromatic nitro substances can also be used with advantage mixed with nitroglycerine.

Formerly, shells used to be filled mainly with highly explosive picric acid (trinitrophenol). This material does not possess the high resistance to mechanical loaded demanding from modern artillery ammunition. It has, therefore, been largely displaced by the most important military explosive of the present time, trinitrotoluene.

The most important aromatic nitro substances used at present are: picric acid, 2,4,6,-trinitrotoluene, hexanitrodiphenylamine, and tetranitromethylaniline.

*Picric acid*

Picric acid which had long been known as a yellow dye has been used for filling shells only since 1888.

The most economical production process starts from chlorobenzene which is first dinitrated (see p. 262). The material is then saponified to dinitrophenol and nitrated further to picric acid.

Direct nitration of phenol as practised earlier requires very large quantities of nitric acid and sulphuric acid. Picric acid can also be obtained from benzene by oxidizing nitration with nitric acid and mercury nitrate.

The salts of picric acid, e.g. of iron, are very explosive and often cause accidents; only ammonium picrate is not very sensitive. Picric acid can only be stored and shipped safely when moist.

Trinitroanisole, the methyl ether of picric acid, was formerly much used for military purposes. It is obtained by reacting dinitrochlorobenzene with potassium methylate and by subsequent further nitration.

*Trinitrotoluene*

The nitration of toluene was discussed on p. 265. In nitrating to o- and p-nitrotoluene it is important for the m-nitrotoluene content to be as low as possible (low nitration temperature). For pure trinitrotoluenes the m-product is, therefore, removed as far as possible by distillation. Further nitration of o- and p-nitrotoluene yields a mixture of 2,4- and 2,6-dinitrotoluene which can then be nitrated to the desired end product, 2,4,6-trinitrotoluene. Dinitrotoluene itself is not an explosive but is used as an addition to nitroglycerine.

For the nitration to the mono-stage 62 per cent $HNO_3$ and 92 per cent $H_2SO_4$ are sufficient, the di-stage requires 94 per cent $HNO_3$ and 98 per cent $H_2SO_4$.

A suitably adjusted spent acid from the tri-stage can be used for this purpose. In order to obtain the dissolved nitro substances from the spent acid of the di-stage, they are extracted with mononitrotoluene; this operation is often carried out in a similar way in nitration processes. For the tri-stage, 99 per cent excess $HNO_3$, 20 per cent oleum, and a temperature of 80–90°C are required.

Trinitrotoluene used to be separated from the residual acid by crystallization; now separation is carried out from the hot liquid. In order to achieve more complete separation of the product which is slightly soluble in the hot spent acid limited quantities of water are added. The liquid nitro substance is repeatedly hot rinsed with water and then treated with sodium bicarbonate (alkali has to be avoided!) Further purification is carried out by washing with sulphite, destroying selectively by-products such as isomers and tetranitromethane. Pure trinitrotoluene has a melting point of 80·5°C. Cooling by water jet solidifies it to spherical granules, alternatively, a cooling drum rotates in molten dehydrated trinitrotoluene and the solidified layer is scraped off in the form of flakes.

When using trinitrotoluene the aim is to achieve optimum casting density (1·60). Molten trinitrotoluene has a density of 1·467, solidified trinitrotoluene one of 1·663. It will be seen that during solidification strong concentration takes place. During filling of the product into shells, etc., measures have, therefore, to be taken to achieve a casting without cavities and bubbles. A mixture of 60 per cent trinitrotoluene and 40 per cent ammonium nitrate which is frequently used has a pasty consistency at 80°C and a high charge density (1·6). In such mixtures the combustion of trinitrotoluene is more complete.

Dinitro- and trinitronaphthalenes are other nitro substances that are sometimes used. Of the hexanitrodiaryl compounds only hexanitrodiphenylamine, produced from dinitrochlorobenzene and aniline which are nitrated further, has been used on a major scale. Tetranitromethylaniline or tetryl, the most explosive of the aromatic nitro substances, is being produced by nitration of monomethylaniline.

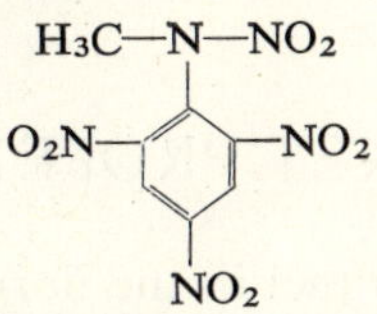


# 4. ALIPHATIC NITRO SUBSTANCES

Of the numerous aliphatic nitro substances tested, such as nitro paraffins, nitro urea, dinitroethylenediamine, only nitroguanidine and cyclotrimethylenetrinitramine (hexogen) have any practical significance.

By melting together dicyandiamide and ammonium nitrate, guanidine nitrate is obtained which is converted into nitroguanidine by removing water with 98 per cent sulphuric acid.

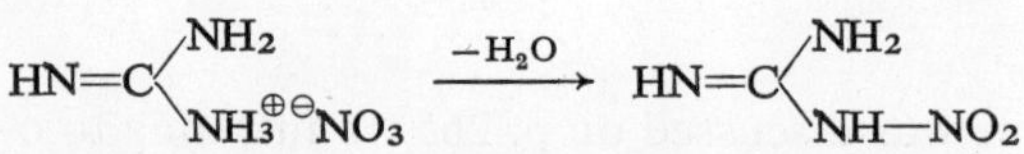

It is a stable explosive of little sensitivity which produces a powerful effect and is suitable for 'cold' powder mixtures because of its low explosion heat.

Hexogen also is a very stable explosive of great force and shattering power. It is one of the most important military explosives, excelling even trinitrotoluene in its effect. It has been known since 1897, but the process for the nitration of hexamethylenetetramine on an industrial scale was not developed until about 1930. Hexogen can also be obtained by reacting ammonium nitrate and acetic anhydride with paraformaldehyde.

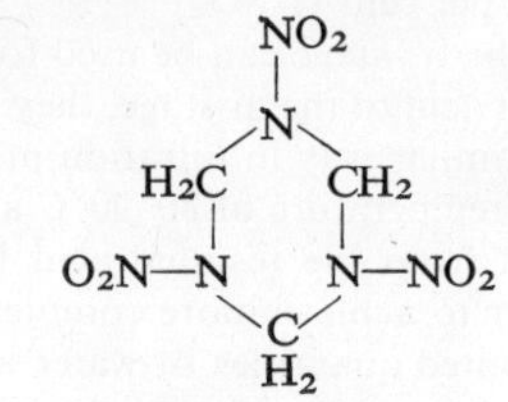

## 5. LIQUID AIR EXPLOSIVES (OXYLIQUITES)

As early as 1900 liquid air with organic substances was used as an explosive. The high nitrogen content strongly reduces the explosive effect. On the other hand, energetic explosives are obtained with liquid oxygen (produced by the Linde method) and absorbent organic materials, such as blotting paper, wood meal, turf, cork meal, and activated carbon. The mixtures are rather dangerous and may only be prepared immediately before use. Owing to the low temperature of liquid oxygen, mercury fulminate is not a reliable primer, but lead azide is.

A number of methods for the chemical analysis, determination of the physical state, and physical measurements of detonation velocity, energy (e.g. in the lead block), and of shattering power by destruction of steel pipes and splinter effect, have been developed for the testing and evaluation of explosives. Testing of sensitivity to mechanical stress, heat, initiating impulse, and of ignitability is also very important.

## 6. ROCKET PROPELLANTS

In contrast to the combustion of a fuel in the normal combustion engine, the energy-supplying reaction in the rocket has to take place in the absence of atmospheric oxygen. A rocket, therefore, carries an oxidant as part of its load. For rocket propulsion use is made of oxidizing reactions, fluorization reactions, and the decomposition of endothermic compounds.

Rocket propellants can be divided into

(a) liquid fuels,

(b) solid fuels.

In liquid fuel rockets the fuel is pumped from storage tanks into combustion chambers, in solid fuel rockets the fuel tank is also the combustion chamber.

Liquid fuels consist of a combination of a fuel (e.g. refined petroleum, alcohol, aniline, hydrazine, liquid ammonia, boranes, dispersions of light metals in mineral oils) and an oxidant (e.g. liquid oxygen, high-strength hydrogen peroxide, fuming nitric acid, liquid fluorine, chlorotrifluoride, mixtures of liquid oxygen and ozone). Both components have to be carried in separate tanks and are brought together only for the reaction.

Fuels which are both fuels and oxidants are also used (e.g. methylnitrate, nitromethane, a mixture of refined petroleum and tetranitromethane, a mixture of methylnitrate and methanol, ethylene oxide, hydrazine, hydrogen peroxide). These fuels are carried in a single tank.

This group of fuels includes the solid rocket fuels which consist mainly of an organic combustible, usually a polymeric substance and an inorganic or organic oxidant (e.g. nitrocellulose, sodium nitrate, ammonium nitrate, potassium perchlorate, ammonium perchlorate, ammonium picrate).

An example of a solid rocket fuel is a mixture consisting of 2 per cent carbon black, 2 per cent plasticizer, 0·4 per cent hardening agent, 0·3 per cent age resister, 0·2 per cent moisture, 10 per cent rubber as a binding agent, to which are added 83 per cent ammonium nitrate as oxidant and 2·3 per cent of a catalyst controlling the development of the oxidation.

## LITERATURE


Winnacker-Weingaertner, *Chem. Technologie*, Organische Technologie, II. Ph. Naoum, and A. Berthmann, Explosivstoffe. Hanser-Verlag, München, 1954.

Ost-Rassow, *Lehrbuch der chemischen Technologie*. 26th ed., J. A. Barth, Leipzig, 1955.

H. Mielke, *Raketentechnik*. VEB Verlag Technik, Berlin, 1959.

J. G. Tschinkel, *Chem. Engng. News*, 1954, 32, 2582.

Rocket Propellants, *Ind. Engng. Chem.*, 1956, 48, 721–777.

T. Urbanski, *Chemie und Technologie der Explosivstoffe*. vols. I and II. Dt. Verl. d. Wiss., Berlin, 1961 and 1963.

# Subject Index



36*